THE
Complete
DINOSAUR

CONTRIBUTORS

R. McNeill Alexander

Reese E. Barrick

Michael J. Benton

José F. Bonaparte

M. K. Brett-Surman

Kenneth Carpenter

Ralph E. Chapman

Karen Chin

Edwin H. Colbert

Philip J. Currie

Peter Dodson

James O. Farlow

Catherine A. Forster

Peter M. Galton

Nicholas Geist

David D. Gillette

Donald F. Glut

Douglas Henderson

Willem Hillenius

Karl F. Hirsch

Thomas R. Holtz, Jr.

Terry Jones

John R. Lavas

Andrew Leitch

Martin G. Lockley

Teresa Maryańska

John S. McIntosh

Ralph E. Molnar

Michael Morales

Frank V. Paladino

J. Michael Parrish

R. E. H. Reid

Bruce M. Rothschild

John Ruben

Dale A. Russell

Scott Sampson

William A. S. Sarjeant

Mary Higby Schweitzer

Paul C. Sereno

William J. Showers

James R. Spotila

Michael K. Stoskopf

Hans-Dieter Sues

Bruce H. Tiffney

Hugh Torrens

Jacques VanHeerden

Darla K. Zelenitsky

INDIANA UNIVERSITY PRESS   BLOOMINGTON AND INDIANAPOLIS

THE
Complete
DINOSAUR

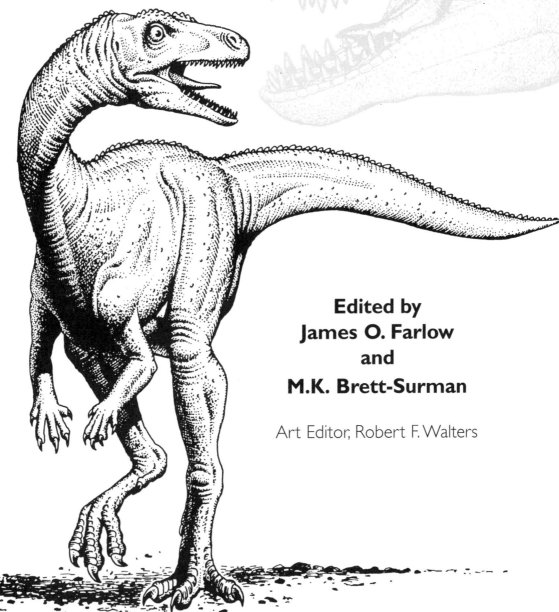

**Edited by
James O. Farlow
and
M.K. Brett-Surman**

Art Editor, Robert F. Walters

The paper used in this publication meets the minimum
requirements of American National Standard for Information
Sciences—Permanence of Paper for Printed Library Materials,
ANSI Z39.48-1984.

Manufactured in the United States of America

Library of Congress Cataloging-in-Publication Data

The complete dinosaur / edited by James O. Farlow and M. K.
    Brett-Surman : art editor, Robert F. Walters.
        p.   cm.
    Includes index.
    ISBN 0-253-33349-0 (alk. paper)
    1. Dinosaurs.   I. Farlow, James Orville.   II. Brett-
Surman, M. K., date
QE862.D5C697   1997
567.9—dc21   97-23698

            4   5   02   01   00   99   98

*Publication of this book is made possible in part with the
assistance of a Challenge Grant from the National Endowment
for the Humanities, a federal agency that supports research,
education, and public programming in the humanities.*

Contents

Preface

James O. Farlow
and
M. K. Brett-Surman

Dinosaurs: The Terrestrial Superlative

In April of 1842, Richard Owen coined the term *Dinosauria* in a footnote on page 103 of his *Report on British Fossil Reptiles*, and defined this new name as meaning "fearfully great, a lizard." Since that time the name has always, incorrectly, appeared as "terrible lizard." How did this etymological and aesthetic error occur? Modern dictionaries always give the meaning of *deinos* as "terrible." This is correct, if one uses the word as an *adjective*—but Owen used the *superlative* form of *deinos,* just as did Homer in the *Iliad.* A check of a Greek-English lexicon from Owen's time will confirm this (Donnegan 1832). Dinosaurs are not lizards, nor are they terrible. They are, instead, the world's most famous "living" superlative!

In most cultures, dinosaurs are the most popular animals of all time. This automatically makes them the most misunderstood animals of all time; most popular books written about dinosaurs are written by non-specialists. (Students are astounded to learn that there are only about eighty professional dinosaur paleontologists in the world.)

More books have been published about dinosaurs since 1990, written by both professionals and amateurs, than were published from 1842 to 1990. Considering the paucity of researchers, dinosaur paleontology, as compared with such fields as mathematics and genetics, is as relatively dynamic as, say, physics. Fortunately, the end of new discoveries about dinosaurs is not in sight.

This book summarizes and celebrates our current knowledge about those superlative, "fearfully great" reptiles. Because the field of dinosaur paleontology has expanded from the emphasis on description and classification prevalent in its infancy to include a broader variety of topics, such as functional morphology and paleobiology, it is no longer possible for any one person to write authoritatively on all aspects of dinosaur studies. Consequently we arranged for the individual chapters to be written by experts on the subjects covered. Our goal throughout this project has been to produce, in one volume, the single most authoritative account of dinosaur paleontology accessible to the general reader.

We have tried to make our book as readable as possible. Authors were instructed to keep technical jargon to a miminum. We know, however, that readers will encounter some unfamiliar and difficult words, so we have provided a glossary of terms at the end of the book.

Another way in which we tried to make the book as easy to use as possible was to provide separate bibliographies for the chapters dealing with technical material. This was done so that the reader would not have to look so hard for the sources of citations. We did, however, depart from this practice in the first section of the book, which deals with the history of dinosaur paleontology. Here the material is less technically formidable, and there is also a fair degree of overlap in references among chapters. We have therefore gathered all of these references and placed them at the end of Part One.

Persons unaccustomed to reading the professional scientific literature may find our way of citing references unfamiliar. Instead of providing sources in footnotes or endnotes, we make such attributions in a different manner. If in 1995 some hypothetical paleontologist named Jones offered the opinion that sauropod dinosaurs discharged unused dietary materials in the form of great big balls of fibrous stuff, and one of our authors wants to refer to the work in which said opinion was expressed, the format for doing so is to cite the author's last name and the year of publication: in this example, it would appear as Jones 1995 (usually in parentheses) or Jones (1995). The reader can then turn to the chapter's bibliography and find the original reference listed alphabetically. Similarly, if Jones had two or more collaborators in what s/he wrote, the work will be cited as Jones et al. 1995, et al. being an abbreviation that means "and others." This way of acknowledging sources may be jarring the first time one encounters it, but it is really a very practical and economical method.

Eagle-eyed readers may notice a seeming inconsistency in the way we identify Chinese and Mongolian authors of articles in chapter bibliographies. For example, some publications by the noted Chinese dinosaur hunter Dong Zhi-Ming are identified as having been written by Dong Z., but other publications by the same author are attributed to Dong, Z.

The problem arises from the fact that many Asian languages reverse the name order from the Western format by giving the family name first and the individual's given name last. In some publications Dong identifies himself as the author by using the Chinese name order, Dong Zhi-Ming. In other publications he identifies himself in Western name format as Zhi-Ming (or Zhiming) Dong. We attempted to determine the name order used in each publication; where Dong used Chinese format, the publication is attributed to Dong Z., but where he used Western format, the publication is credited to Dong, Z.

There is some overlap in content among different chapters, although we have tried to keep this to a minimum. A certain amount of overlap is good, however; different authors may employ the same information in different contexts, or with different emphases—or with different interpretations.

If your goal is to be at the cutting edge of your field, you have to be prepared to risk the occasional loss of professional blood. Such bloodletting takes the form of vigorous disagreements among specialists over topics of contention. Although we have sought to keep our book's discussions of controversial matters civil in tone, it will quickly become apparent to the alert reader that there are sharp divergences of opinion among some chapter authors—and it is worth noting that neither of the editors agrees with all of the opinions expressed in this book! When an author makes an interpretation that other specialists dispute, either elsewhere in this book or in other publications, we often include parenthetical references in the text to draw the reader's attention to those opposing viewpoints.

We could not have put this book together without the help of many other people. First and foremost, of course, we thank our chapter authors, *most* of whom gave us their contributions in a timely fashion (the tardy miscreants out there know who they are . . .). (We sadly note, by the way, that one of our authors, Karl F. Hirsch, died while the book was in press.) We also thank numerous other professional colleagues (particularly John H. Ostrom) who read and critiqued early chapter drafts, thereby helping to ensure the scientific credibility of the book.

We take special pride in the quality of the artwork in this book. Many professional artists graciously permitted the use of their work for far less compensation than they deserved, and we hope that they will feel that the quality of the book justifies their generosity. The persons responsible for the artwork are identified in the individual captions. However, we want to single out for special thanks certain artists whose contributions went light-years beyond what we had any right to expect in exchange for a free copy of the book: Tracy Ford, Brian Franczak, Berislav Kržič, Greg Paul, Mike Skrepnick, and Jim Whitcraft. Special thanks, too, go to Bob Walters, who served as art editor for this project, and suffered through untold miseries (we assume) in the process.

Linda Whitlock had the dubious pleasure of retyping several manuscripts, and in so doing learned more than she cared to know about complicated scientific names. Other persons who assisted us at various stages in the production of this book include Chip Clark, Jenny Clark, Kimberly Brett-Surman, David Steere, Carolyn Hahn, Russell Feather, and Mary Parrish of the Smithsonian Institution; archivist Dean Hannotte; philatelists E. A. Knapp, Fran Adams, Wally Ashby, and Saul Friess; George Stephens and Roy Lindholm of George Washington University; the award-winning author Robert J. Sawyer; and the editor of Asimov's *Analog,* Gardner Dozois.

Thanks also go to Indiana University Press, for agreeing to this project in the first place, and for seeing it through to the not-so-bitter (we hope) end. Special thanks to Robert Sloan for his patience.

Finally, in the interests of domestic bliss, we must thank our wives, Karen and Kimberly, for putting up with our frequent gripes and intermittent bouts of depression during the long months when it seemed that this book project would *never* be finished, and when one of us (who shall go unnamed) claimed to be too busy editing a book to do the dishes.

Our heartfelt gratitude to one and all.

References

Donnegan, M. D. J. 1832. *A New Greek and English Lexicon: Principally on the Plan of the Greek and German Lexicon of Schneider.* First American edition from the second London edition, revised and enlarged by R. B. Patton. Boston: Hilliard, Gray and Co.

Owen, R. 1842. *Report on British Fossil Reptiles.* Part II. Report of the Eleventh Meeting of the British Association for the Advancement of Science for 1841: 60–204.

Calvin and Hobbes

by WATTERSON

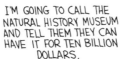

Calvin and Hobbes © 1988 Bill Watterson. Distributed by Universal Press Syndicate. Reprinted with permission. All rights reserved.

The Discovery of Dinosaurs

Content that his reputation had preceded him, Cope continued. "Marsh purchases so many fossils that some have never been unpacked. He does not understand anatomy—an army of myrmidons studies his specimens and writes his papers. They are negligently paid and forbidden to carry on their own researches. . . ." He did not include the fact that he had attempted to stir them to revolt.

"Marsh doesn't read the journals, leading him to duplicate others' work. Yet, with all this, they call him a scientist! In '72 Mudge intended to send me the 'bird with teeth' which has made Marsh's reputation. Marsh heard of the fossils and convinced Mudge to give them to him. At Bridger Basin his men took my bones. And he has instructed his collectors to smash duplicates and other bones—to actually *destroy* fossils to keep them from me!"

> —Sharon N. Farber, "The Last Thunder Horse West of the Mississippi"

Like any other area of human intellectual endeavor, the study of dinosaurs has a rich history, filled with strong, colorful—and often truly eccentric—personalities. This section briefly summarizes the contributions of workers from around the world who brought us to our present understanding of dinosaurs. We begin with an account of how humanity began to get an inkling that there was once a time when our planet was ruled by fearfully great reptiles. We then consider the activities of some of the more notable figures in the history of dinosaur paleontology, beginning in Europe, where the earliest contributions to the field were made. Just as paleontology had its beginnings in Europe, and spread from there to the rest of the world, our account of the history of study of dinosaurs finally turns its focus away from Europe to all parts of the globe.

The chapters in this section are arranged topically, based on geographic area. For a strictly chronological summary of some of the major events in the history of dinosaur studies, see the appendix on p. 707.

The Earliest Discoveries

William A. S. Sarjeant

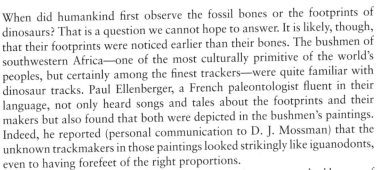

When did humankind first observe the fossil bones or the footprints of dinosaurs? That is a question we cannot hope to answer. It is likely, though, that their footprints were noticed earlier than their bones. The bushmen of southwestern Africa—one of the most culturally primitive of the world's peoples, but certainly among the finest trackers—were quite familiar with dinosaur tracks. Paul Ellenberger, a French paleontologist fluent in their language, not only heard songs and tales about the footprints and their makers but also found that both were depicted in the bushmen's paintings. Indeed, he reported (personal communication to D. J. Mossman) that the unknown trackmakers in those paintings looked strikingly like iguanodonts, even to having forefeet of the right proportions.

A different, but still quite logical, interpretation was attached by one of the primitive peoples of Brazil to footprints of carnivorous dinosaurs exposed on a bedding-plane surface in Paraíba. Symbols graven into the rock alongside show that these were considered to be the footprints of giant running birds, comparable to the living rhea of South America (G. Leonardi in Ligabue 1984).

In Europe and the European settlements in America, fossil footprints of dinosaurs and their progenitors were a source for legends. Footprints exposed in the Rhine Valley, western Germany, may have inspired the story of the slaying of the dragon by the hero Siegfried (Kirchner 1941). When a footprint in Triassic sandstone of *Chirotherium*—the trackmaker was a

1

non-dinosaurian early archosaur—was incorporated into the stonework of Christ Church, Higher Bebington, Cheshire, England, it came to be known locally as the "Devil's Toenail." Tracks of a dinosaur in red sandstones of the Connecticut Valley, noticed at the beginning of the nineteenth century by a Massachusetts farm boy, Pliny Moody, were solemnly considered— despite their size—to be those of the raven that was sent out by Noah from the Ark to seek land but perversely failed to return. Even when restudied by Edward Hitchcock and other U.S. naturalists, the footprints were long believed to be those of birds (Sarjeant 1987).

The fossil shells of invertebrates and the fossil teeth of fishes (especially sharks) are such commonly encountered objects that they have been collected for purposes of decoration, ornamentation, or sorcery since time immemorial (Abel 1939; Oakley 1965; Bassett 1971; Vitaliano 1973; Zammit-Maempel 1982). They have likewise inspired many stories of magic and the supernatural. Fossil bones were also widely a basis for legend, but those were typically the bones of animals more recently extinct—most often the remains of primitive whales (zeuglodonts) or of elephants or their relatives (mastodonts and mammoths). It was these that inspired stories and epics concerning dragons, unicorns, and giant humans. Perhaps certain of the Chinese legends of dragons may have been based on dinosaur remains—for example, a report of dragon bones from Wucheng, in the Qinling Mountains of Sichuan, by one Cheng Qu during the period of the western Jin dynasty, A.D. 265–317 (Dong 1988: 18–19)—but the evidence is equivocal, at best.

In only one instance can a legend be directly associated with dinosaur bones. Jean-Baptiste L'Heureux, a French-Canadian traveler who lived during the early nineteenth century among the Piegan people of Alberta, Canada, recorded that bones shown to him in what is now Dinosaur Provincial Park were revered as those of "the father of buffaloes" (Spalding 1993: 5)—a reasonable concept, since the buffalo was the largest animal known to the Plains Indians.

The problems in arriving at a proper interpretation of chance-discovered dinosaur bones were threefold. First of all, their texture had normally been altered by the processes of petrifaction and their color changed by iron or humic staining; they did not look much like ordinary bones. Second, dissociated bones are infinitely more common—and, to an uninformed observer, infinitely less striking—than complete or near-complete skeletons; nor are the latter usually very fully exposed. Third, the bones which we would now find most impressive—the great vertebrae, the ribs, and the limb bones—were essentially too big to be noticed—or, if noticed, to be taken seriously as bones of animals. When discovered, dinosaur remains might be used as artifacts or for decorative purposes: dinosaur eggshell fragments, carefully arranged into patterns by the early people of the Gobi Desert of Mongolia, were reported by Roy Chapman Andrews (1932: 209), but they were not normally perceived as vestiges of the animal life of past times.

Again and again, it had been recognized that fossils were the remains of once-living creatures—in ancient Greece by Xenophanes of Colophon and Xanthos of Sardis (Adams 1938: 11–12), in the fifteenth century by Leonardo da Vinci, and in the sixteenth by the Dane Niels Stensen, called Steno—but the ideas of these intellectual pioneers were either kept secret, as were Leonardo's till his codified notes were first read four centuries later, or else firmly rejected by subsequent writers, as were Steno's by Martin Lister of London's Royal Society. Indeed, Lister (1671) stated categorically that fossils were "never any part of an animal." The Welsh naturalist Edward Lhuyd temporized (1699), believing them to be a product of minute spawn of animal life, carried inland by vapors arising from the ocean and growing within the rocks. Even Robert Hooke's careful demonstration of the organic nature of fossils, presented to the Royal Society of London sometime after 1668 but published only posthumously (1705), did not convince all his contemporaries. Robert Plot (Fig. 1.1), curator of Oxford's Ashmolean Museum, considered this evidence but concluded instead (in the second [1705] edition of a book initially published in 1677) that the fossil shells of invertebrates in rocks were merely *Lapides sui generis,* stones "formed into an Animal Mould" by "some extraordinary plastic virtues latent in the Earth" to serve as ornaments for the earth's secret places, in the fashion that flowers adorned its surface.

Figure 1.1. Robert Plot (1640–1696), the first illustrator of a dinosaur bone.

The Earliest Reports of Dinosaurs

Yet when Plot discovered and illustrated a dinosaur bone, he recognized it correctly as being "a real *Bone,* now petrified"—more specifically, "the lowermost part of the Thigh-Bone." From its great size, "In Compass near the *capita Femoris,* just two foot, and at the top above the Sinus . . . about 15 inches," he concluded (1677) that it "must have belonged to some greater *animal* than an *Ox* or *Horse;* and if so in all probability it must have been the *Bone* of some *Elephant,* brought hither during the Government of the Romans in Britain." Plot's bone came from the Middle Jurassic strata of Cornwell, near Chipping Norton, Oxfordshire. It is now lost, but his illustration and its dimensions indicate that it was part of the femur of a *Megalosaurus.*

This bone was reillustrated by Richard Brookes in 1763, just after the publication of Carl von Linné's *Systema Naturae* (1758). The illustration was captioned *Scrotum humanum.* From comparison with the labeling of Brookes's other illustrations, it is evident that this was merely a descriptive appellation; he knew quite well that Plot's specimen was part of a bone. However, the name was taken very seriously by a French philosopher, Jean-Baptiste Robinet, who held the eccentric concept that fossils were attempts by nature to reproduce in other fashions the organs of humankind. Robinet (1768) not only accepted that Plot's specimen was a scrotum, but he also believed that it showed the musculature of the testicles and the vestiges of a urethra!

Though Robinet's concept was not taken seriously by other savants, it has been mischievously suggested that *Scrotum humanum* was the earliest scientific name for a dinosaur! (Halstead 1970; Delair and Sarjeant 1975). However, that proposition has now been firmly rejected by the *cognoscenti* of zoological taxonomy (Halstead and Sarjeant 1993).

The eighteenth century saw other dinosaur discoveries in England. In the posthumously published catalogue (1728) of the fossil collection of John Woodward, professor at Gresham College, London, specimen A1 consists of portions of a dinosaur limb bone, perhaps also a megalosaur. These fragments, preserved in the Woodward Collection of the Sedgwick Museum, University of Cambridge, represent the earliest-discovered identifiable dinosaur bone.

The next find was also made in Oxfordshire. Joshua Platt, an English dealer in curiosities, found three large vertebrae, surely of dinosaurs, at Stonesfield in 1755. Unwisely he sent them to Peter Collinson, a Quaker merchant and botanist, for examination; Collinson did nothing with them, and their fate is uncertain. Subsequently, however, Platt found an enormous thighbone at Stonesfield, again probably of *Megalosaurus;* though incomplete, it measured 2 feet 5 inches (approx. 74 cm) in length, its width across the condyle being 8 inches (20 cm) and across the shaft, 4 inches (10 cm). Platt reported this second discovery in a short note (1758) accompanied by a careful illustration. The bone was listed in an unpublished catalogue of Platt's collection (1773), but has since been lost (Delair and Sarjeant 1975: 10).

Part of a scapula of *Megalosaurus*, again from Stonesfield, was presented to what was then the Woodwardian Museum of Cambridge University in 1784; though it survives, it has never been described or illustrated. The centrum of a caudal vertebra of a much larger dinosaur—certainly a sauropod, probably a cetiosaur—was discovered in 1809 at Dorchester-on-Thames, Oxfordshire; this bone also survives in the Cambridge collections, but 166 years were to elapse before it was first reported and illustrated (Delair and Sarjeant 1975: 11 and fig. 3). Before 1816, the Cretaceous strata of the Isle of Wight yielded bones—surely of dinosaurs—to the geologist Thomas Webster. These were mentioned in a general work on the island (Englefield 1816) but were never described and cannot now be identified.

Discoveries outside England

Across the Channel, the Normandy coast may also have yielded dinosaur bones during the eighteenth century. Bones collected from the Vaches Noires cliffs by the Abbé Dicquemare, and reported in 1776, may have included vertebrae and a femur of a dinosaur (Taquet 1984); however, the descriptions were brief and unaccompanied by illustrations. Vertebrae illustrated by Georges Cuvier (1808), from the vicinity of Honfleur, were certainly those of a theropod dinosaur (Lennier 1887) but were misinterpreted as having belonged to crocodiles of an unusual type.

The earliest dinosaur bones discovered in North America likewise merit only brief attention in this history. The first observation must have been that by William Clark. In the course of his exploratory expedition through the recently acquired Louisiana Purchase with Meriwether Lewis in 1806, Clark noted a large rib bone in a cliff on the south bank of the Yellowstone River, below Pompey's Tower (now Pompey's Pillar), close to the site of present-day Billings, Montana. In his journal, Clark noted it as being 3 feet (91 cm) in length, "tho' a part of the end appears to have been broken off," and about 3 inches (7.6 cm) in circumference. He obtained "several pieces of this rib: the bone is neither decayed nor petrified but very rotten" and thought it to be a bone of an immense fish (Clark, quoted in Simpson 1942: 171–172). This find was in the Late Cretaceous Hell Creek Formation and was surely a dinosaur bone.

In contrast, fossil bones discovered by Solomon Ellsworth, Jr., during excavations for a well near East Windsor in the red sandstones of the Connecticut Valley were small enough to be misinterpreted as human bones (Smith 1820). Almost a century passed before Richard M. Lull recognized their dinosaurian character, considering them to be bones of a small coelurosaur. A more recent reexamination by Peter M. Galton (1976) indicates instead that they are bones of a prosauropod.

In the century or so succeeding Plot's discovery, then, dinosaur bones had been repeatedly found and just as often misinterpreted. It was a further find from Stonesfield, Oxfordshire, that at last properly launched the scientific study of dinosaurs.

William Buckland and *Megalosaurus*

When the Reverend William Buckland (Fig. 1.2; see Sues, chap. 2 of this volume, for biographical information), Reader of Mineralogy at the University of Oxford, first made this discovery is uncertain, but a date around 1815 seems likely. His find comprised several huge teeth, recurved and with serrated edges, and a partial lower jaw with a tooth; these were early recognized to be reptilian and compared with the very similar, albeit much smaller, teeth of the living monitor lizard. Cuvier saw these specimens while visiting Oxford in 1818 and later reported (1824) that they had been found several years before his visit. In a letter to Buckland written from Cuvier's laboratory in 1821, the Irish naturalist Joseph B. Pentland was already inquiring plaintively: "Will you send your Stonesfield reptile or will you publish it yourself?" (Sarjeant and Delair 1980: 262).

However, Buckland was a man of diverse concerns that ranged beyond his clerical duties and geological interests; he was a veritable polymath. With so much else to do, he did not make rapid progress toward publishing his discovery. Moreover, Buckland was anticipating the cooperation of his clerical and geological colleague, the Reverend William Daniel Conybeare. In a letter written to Pentland on 11 July 1822 and hitherto unpublished, Buckland stated that Conybeare

> is about to take up immediately the Stonesfield Monitor & to publish a joint paper with me on that Animal, but we have not yet determined through which Channel to give it publication. My great object will be that it will be *in time* for Cuvier's Book. Tell me what time that will be.

As early as 1821, Conybeare had given passing mention to the "Huge Lizard" of Stonesfield, incidentally in an account of a fossil marine reptile; but this joint project did not come to fruition. Yet Buckland's discovery was already becoming quite well known. In a general paleontological text published in 1822, James Parkinson illustrated one of the teeth and wrote:

> Megalosaurus (*Megalos* great, *saurus* a lizard). An animal apparently approaching the *Monitor* in its mode of dentition, and not yet described. It is found in the calcareous slate of Stonesfield. . . . Drawings have been made of the most essential parts of the animal, now in the [Ashmolean] Museum of Oxford; and it is hoped a description may shortly be given to the public. The animal must in some instances, have attained a length of 40 feet [12 m], and stood eight feet [2.4 m] high.

On the strength of this mention, Parkinson is sometimes credited with the authorship of the name *Megalosaurus*. However, that was a period before the rules of taxonomy had been properly formulated, let alone rigorously applied, and such borrowings of information were not yet considered improper. Another unpublished letter, this time from Buckland to Cuvier himself and written on 9 July 1823, makes it quite clear that Buckland was not only the author of that name, but also close to publishing his researches on the "Stonesfield Monitor":

> My Dear Baron, Herewith I send you Proof Plates of the great Animal of Stonesfield, to which I mean to give the name of Megalosaurus & which I shall publish either in 2nd part of V. 5 or the 1ᵗ [first] part of V. 6 of the Geological Transactions.

Cuvier was indeed becoming impatient. Pentland, again acting as his

Figure 1.2. William Buckland (1784–1856), the first scientist to describe and name a dinosaur.

amanuensis, transmitted the impatience in a letter of 28 February 1824:

> Our friend Cuvier has this moment requested me to write to you on the subject of the paper which you proposed publishing on the Stonesfield reptile the Megalosaurus. He is now at that part of his work where he intends speaking of your reptile, and wishes to know if your paper has been yet published—and in what form? And in what work? (Sarjeant and Delair 1980: 304)

However, before this letter was sent, Buckland was at last reading his paper to the Geological Society at its London headquarters on 20 February 1824. Its publication later that year constituted the earliest scientific description of a dinosaur—though, of course, that name for those reptiles had not yet been formulated.

Gideon Mantell and *Iguanodon*

The role played in the discovery of the dinosaurs by another English paleontologist, the Sussex surgeon Gideon Algernon Mantell (Fig. 1.3; see Sues, chap. 2 of this volume, for biographical details), has been widely misrepresented. An oft-repeated story (e.g., Colbert 1983: 13–15) tells how, while her husband was visiting a patient early in 1822, Mantell's wife, Mary Ann, found some fossil teeth in a pile of road metal (stone rubble used for roadmaking); that, excited by her find, Gideon ascertained from which quarry the road metal had come, finding more teeth and bones there; and that these events marked the beginning not merely of his concern with *Iguanodon,* but also of the scientific study of dinosaurs, Buckland being spurred into publishing results amassed later only by the fear that Mantell might anticipate him.

Unfortunately, as Dean (1993: 208–211) demonstrates, this romantic story does not withstand scrutiny. In his book *The Fossils of the South Downs,* published in May 1822, Mantell makes mention that "teeth, vertebrae, bones and other remains of an animal of the lizard tribe, of enormous magnitude . . . [have been] discovered in the county of Sussex." Dean points out also that, though Mary Ann may indeed have found *Iguanodon* teeth for him, Gideon subsequently named himself as their discoverer.

Even with such problems and uncertainties set aside, the story of Mantell's recognition of *Iguanodon* remains a fascinating one. When he displayed some of his finds at the Geological Society on 21 June 1821, he aroused little interest. When Charles Lyell took one of the teeth to Paris for Cuvier (Fig. 1.4) to examine, it was dismissed as being the upper incisor of a rhinoceros, while some metatarsals were considered to belong to some species of hippopotamus (Mantell 1850: 195). However, Mantell was not deterred. While accepting that the teeth were those of a herbivore, he was already beginning to suspect that they were reptilian. The crucial breakthrough came when he was examining bones and teeth in the Hunterian Museum of the Royal College of Surgeons in London. Another visitor on that day, Samuel Stutchbury, drew Mantell's attention to the dentition of the living iguana. Why, the Sussex teeth were merely gigantic equivalents of the iguana's teeth!

When Mantell wrote again to Cuvier, the great French scientist was convinced by the new interpretation, writing: "N'aurons nous pas ici un animal nouveau, un reptile herbivore? [Have we not here a new animal, a herbivorous reptile?]" (quoted in Mantell 1851: 231–232)

Mantell seems originally to have proposed calling his "new" reptile *Iguanosaurus* (Anonymous 1824), but this name was dropped in favor of the more euphonious *Iguanodon,* proposed to him by Conybeare. On 10 February 1825, almost a year after Buckland's paper, Mantell proudly reported his discovery to the Royal Society of London His account was published in its *Philosophical Transactions* later that year.

Megalosaurus had been interesting, but after all, gigantic reptilian carnivores were already well known; were there not very large living crocodiles which attained lengths of nearly 30 feet [9 m]? A giant reptilian herbivore, though; why, that was indeed a wholly novel concept!

Buckland's Other Discoveries

If Buckland had been less preoccupied with other matters, he might have anticipated Mantell in the description of *Iguanodon.* Adam Sedgwick noted (1822) that Buckland had discovered "cetacean" bones at Sandown Bay, Isle of Wight, before Christmas 1822; as Buckland himself later reported (1824: 392), these were in fact bones of *Iguanodon.* He had also already obtained for the Oxford museum a bone of even more gigantic size, found in fragments by the geologist Hugh Strickland in a railway cutting near Enslow Bridge, Oxfordshire. Carefully pieced together by Buckland, these proved to constitute a femur 4 feet 3 inches (1.3 m) long (Strickland 1848). After being carefully cemented and bound with wire, this bone was "long the object of admiration in the Oxford classroom for geology" (Phillips 1871: 247). Buckland also acquired or examined other bones of comparable character, among them vertebrae from Middle Jurassic localities near Chipping Campden and near Thame, Oxfordshire, plus a "blade bone" of enormous size from the latter place, while a whole batch of fossil bones were obtained for him by William Stowe from near Buckingham. All of these, as Buckland wrote in a letter to Stowe (quoted in Phillips 1871: 245), belonged to "some yet undescribed reptile of enormous size, larger than the Iguanodon, and of which I am collecting scattered fragments into our museum, in hope ere long of being able to make of its history."

Yet Buckland never did write further on these gigantic bones. That task was left to Richard Owen, who described and named *Cetiosaurus*—the earliest sauropod dinosaur to be discovered—in an address to the Geological Society of London on 30 June 1841. This served as prelude to Owen's major account of British fossil reptiles, given to the British Association in August 1841 (see Torrens, chap. 14 of this volume).

Further Finds in England

In the meantime, other dinosaur bones were being discovered. Saurian bones from strata exposed at Swanwich (now Swanage) Bay, Dorset, were given passing mention by the geologist William H. Fitton (1824). Vertebrae and an imperfect femur were dug up at Headford Wood Common, Sussex, in 1824 (Murchison 1826). Both discoveries were made in Wealden (Lower Cretaceous) strata. Probably both were *Iguanodon* bones, but the specimens (if they survive) have not been identified and studied.

While Buckland was being again dilatory, Mantell remained energetic. He was accumulating further bones of *Iguanodon* from the Wealden, including bone fragments from the cliffs about Sandown, Isle of Wight, and most notably a slab from Maidstone, Kent. This displayed a partial skeleton

(Fig. 1.5), a discovery so important that it has been facetiously styled his "mantel-piece" (Norman 1993); it allowed him to attempt a reconstruction upon which the early restorations of that creature were to be based. Mantell also discovered the remains of an armored dinosaur—postcranial bones associated with dermal elements and armor plates—which he named *Hylaeosaurus;* this was the first ankylosaur to be described (1833). Before his death in 1852, he was to describe and name two further dinosaur genera.

Samuel Stutchbury reentered the story when, in association with S. H. Riley, he reported the earliest reptilian remains from the English Triassic—specifically from the so-called "Magnesian Conglomerate" of the Bristol district (1836–1840). Three genera were recognized, two of which (*Thecodontosaurus* and *Palaeosaurus*) are accepted nowadays as dinosaurs; the former is now considered a prosauropod, while the latter is of dubious affinity.

Discoveries Elsewhere in Europe

The first recognition of dinosaurs in France came in 1828, when A. de Caumont reported *Megalosaurus* bones from the Middle Jurassic oolite of Caen, Normandy. In the ensuing years the paleontologist Jacques-Amand Eudes-Deslongchamps (1794–1867) painstakingly assembled bones and

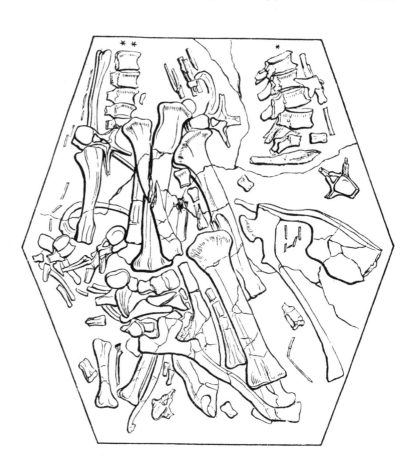

Figure 1.5. Gideon Mantell's "mantel-piece": the slab from Maidstone, Kent, showing the incomplete skeleton of an *Iguanodon* (after Mantell's drawing).

William A. S. Sarjeant

bone fragments from the quarries and construction sites around Caen. When he became confident that he had enough material, he published a description of the partial skeleton thus reconstituted. In tribute to the first scientific discoverer of the giant fossil reptiles, he named it *Poekilopleuron bucklandi* (1838). Eudes-Deslongchamps thought the animal to have been largely marine, though well able "to rest on the shore and bask in the sun" (Buffetaut et al. 1993: 162). This was the last of the misapprehensions about dinosaurs to precede Owen's establishment of the Dinosauria— though, in justice, one must note that there have been many since!

The first finds in Germany were in Triassic red sandstones and followed close upon the discoveries near Bristol. In 1837 Christian Erich Hermann von Meyer (1801–1869) described and named *Plateosaurus,* which was long destined to remain the most fully known of the dinosaurs we now call prosauropods. (More than 100 skeletons have been found in southern Germany and Switzerland.) A tooth recovered from the Jurassic strata of southern Russia by A. Zborzewski was rather unnecessarily given its own generic name, *Macrodontophion* (1837); it was probably that of a carnosaur.

Though the beginnings of human awareness of dinosaurs must be matter for speculation, the footprints of dinosaurs seem to have attracted attention earlier than their osteological remains. The first recorded discovery, and most other early finds, were from the Jurassic strata of England, though there were early observations also in France and North America. During almost 150 years following Robert Plot's first illustration of a dinosaur bone in 1677, these fossils were misinterpreted in a variety of fashions—as the remains of elephants, crocodiles, and fish, even as simulacra of human genitalia. The earliest recognition that they were the bones of long-extinct creatures was by William Buckland, in his study of the carnivorous *Megalosaurus* (1824). Gideon Mantell's researches on *Iguanodon,* the first of the herbivorous dinosaurs to be identified, overlapped Buckland's work and were published only a year later (1825). Though Buckland accumulated sauropod bones also, he did not describe them: this was done eventually by Owen, in his preliminary description of *Cetiosaurus* (1841). In addition to further discoveries in England, giant reptiles recognized and named before Owen's "creation" of the dinosaur (1842) included *Plateosaurus* von Meyer (1837), from Germany, and *Poekilopleuron* Eudes-Deslongchamps (1838), from France, plus a single tooth from Russia.

European Dinosaur Hunters

Hans-Dieter Sues

2

Unlike many other branches of science, paleontology is a highly individualistic enterprise. As Colbert (1968) noted in his classic book on the history of research on dinosaurs, one needs to understand the individual scientists as human beings in order to understand their work. Although Europe was home to the first students of dinosaurs, it was eclipsed by the United States in the latter half of the nineteenth century, when the combined efforts of men such as Edward Drinker Cope, Joseph Leidy, and Othniel Charles Marsh resulted in many remarkable discoveries of dinosaurian skeletons. As a result, the contributions of the early European researchers, especially those from continental Europe, are generally underappreciated by present-day students. This chapter provides a brief introduction to some of the principal European "dinosaur hunters" of the nineteenth and early twentieth centuries and their scientific contributions.

Some of these workers collected dinosaurian remains and left the study of these fossils to others. Most researchers both collected and described dinosaurian specimens. A few students of dinosaurs mentioned in this chapter did little if any collecting of their own, but made very important contributions to the scientific study of these animals.

During the nineteenth century, most of the dinosaurian fossils in England and on the Continent were collected by dedicated clergymen, medical doctors, merchants, and quarrymen. These collectors would subsequently lend, present, or sell their treasures to the great naturalists of the day—often themselves members of the clergy or the medical profession—for scientific study and formal description. (In England, most of the important discoveries, including the recent find of the unusual theropod *Baryonyx walkeri* from the Wealden, are still being made by private collectors.) In addition to drawing on this source, the two most eminent early European researchers, Baron Georges Cuvier (1769–1832) and Sir Richard Owen (1804–1892), enjoyed immense influence and patronage at the highest levels of state and would not hesitate to marshal the resources of their nations to secure fossils of interest to them.

Early English Pioneers

The Reverend William Buckland (1784–1856; see Sarjeant, chap. 1 of this volume, for a picture) is generally credited with publishing the first scientific description of a dinosaur, *Megalosaurus,* based on a dentary fragment with teeth and some postcranial bones from the Middle Jurassic Stonesfield Slate of Oxfordshire (Buckland 1824). However, Delair and Sarjeant (1975) noted that *Megalosaurus* was not Buckland's only discovery of dinosaurian remains. In 1822, the Reverend Adam Sedgwick (1785–1873), another eminent British geologist, reported the discovery by Buckland of "cetacean" bones in Wealden strata at Sandown Bay, Isle of Wight, sometime around Christmas 1821. Later, in his 1824 paper on *Megalosaurus,* Buckland himself mentioned these finds in passing and referred them to *Iguanodon.* Buckland unfortunately did not keep records of the actual dates and detailed circumstances surrounding his various discoveries of dinosaurian fossils. Cuvier (1825: 344) mentioned that he had seen bones of *Megalosaurus* in Buckland's collection during a visit to Oxford in 1818, but Buckland did not publish a full account on *Megalosaurus* until 1824.

Buckland was a brilliant man and a veritable polymath. Born at Axminster as the son of a country clergyman in 1784, he became interested in natural history at a young age. He entered Exeter College at Oxford, later was elected fellow of that college, and was ordained in 1809. Early in life Buckland had shown a strong interest in the earth sciences. He was appointed Reader in Mineralogy at Oxford in 1813. By all contemporary accounts, he was an entertaining and very popular lecturer. In 1818 Buckland successfully petitioned the prince regent through the Tory prime minister, Lord Liverpool, to be given the additional title of Reader in Geology (Edmonds 1979). When he was tempted to accept a better-paid ecclesiastical position elsewhere, Buckland was appointed canon of Christ Church College, again through Liverpool's patronage, to keep him at Oxford. The substantial income of £1,000 per year allowed him to marry. Buckland became president of the Geological Society in 1824. After Buckland was appointed dean of Westminster in 1845, he remained a prominent figure in British public life, but his scientific research essentially ceased.

Buckland was a notorious eccentric whose exploits and home life were legendary even by Victorian standards. According to Charles Darwin's autobiography, the great naturalist disliked Buckland, "who though very good humoured and good-natured seemed to me vulgar and almost coarse man. He was incited more by a craving for notoriety, which sometimes made him act like a buffoon, than by a love of science" (De Beer 1974: 60). Buckland identified the ever-liquefying "martyr's blood" on the pavement of a European cathedral as bat urine—simply by kneeling down and having a lick. Similarly, on a visit to Palermo, Buckland reidentified the bones of Santa Rosalia as those of a goat. He also had a keen interest in classifying animals by taste and would go to great efforts to procure specimens for this "research." Buckland kept numerous animals at his house in Islip, the most famous of which was a young bear named Tiglath Pileser or "Tig." Tig was properly outfitted with cap and gown and for some time resided at Christ Church College, attending wine parties and even the meeting of the British Association at Oxford in 1847. He was subsequently expelled from the college and, having become unruly after developing a "sweet tooth," ended his life in a zoo.

Buckland traveled far and wide in the pursuit of fossils and other objects of natural history. He attempted to use geology and paleontology to support traditional Christian teachings by providing evidence for a universal deluge and the pervasiveness of divine design during the history of life on Earth (Rudwick 1972).

Although Buckland's find of dinosaurian bones on the Isle of Wight clearly predates the discovery of *Iguanodon* by the Sussex doctor Gideon Algernon Mantell (1790–1852), he did not realize the true nature of those specimens until Mantell's work. In his methodical search for dinosaurian remains, Mantell represented the first "dinosaur hunter" (Dean 1993 and in press).

Born and raised in Lewes, a small town some eight miles from the seaside resort of Brighton, Mantell (see Sarjeant, chap. 1 of this volume, for a picture) spent much time as a boy exploring the surrounding countryside, and became intrigued by fossils. He was interested in medicine and, at the age of fifteen, became apprenticed to a respected local doctor. He later studied at St. Bartholomew's Hospital and qualified as a licentiate of Apothecaries' Hall in 1811. After his return to Lewes, Mantell rejoined and later took over his mentor's medical practice upon the latter's retirement. He married a local woman and started a family. His duties as surgeon to three parishes, and as a successful male midwife (Swinton 1975), necessitated many house calls in the country. On these trips, Mantell always had his eyes trained on the ground for fossils.

According to Mantell's often-repeated account, his wife went for a stroll along a country road while her husband was paying a house call in the Cuckfield district of Sussex early in 1822. She supposedly noticed some peculiar teeth in a roadside pile of crushed stone that was being used for road metal. These teeth were unlike anything Mantell had ever seen, and he correctly sensed that his wife had made an important discovery. An excited Mantell searched the quarries of the region, known as the Tilgate Forest, which exposed Early Cretaceous strata of the Wealden. He soon discovered the source of the fossiliferous road metal but never named the quarry because he wanted to keep it secret. It was probably located near Whiteman's Green, Sussex.

Dean (1993 and in press) has shown that Mantell's charming anecdote concerning his wife's discovery is entirely fictional and is not supported by historical evidence. He found no evidence from Mantell's casebooks that Mantell's wife ever accompanied her husband on visits to patients. Mantell himself later (1851) referred to "my first discovery of a tooth in a quarry near Cuckfield" in a discussion of the teeth of *Iguanodon*.

After a lengthy period of comparisons and discussions with the leading experts of the day, Mantell presented a detailed account on the fossils of *Iguanodon* from the Tilgate Forest to the Royal Society of London on February 10, 1825. His paper was read by Davies Gilbert, then treasurer of the society, and subsequently published in May of 1825 in the society's *Philosophical Transactions* (Mantell 1825). Mantell himself was elected a fellow of the Royal Society on November 25 of the same year, in recognition of his momentous discovery.

Now famous, Mantell continued his quest for fossils. In 1833 he relocated his family to Brighton with hopes of a more satisfying life. His hopes were dashed, however, and his medical practice increasingly was overshadowed by his collecting efforts, putting severe strains on the Mantell household. In 1834, workmen discovered the partial skeleton of an *Iguanodon* in a quarry near Maidstone, Kent. The fossil quickly attracted considerable public attention. It was only after much haggling, and raising the requisite £25 with the help of several friends, that Mantell was able to secure this important specimen for his private museum.

In spite of all his accomplishments, including the 1838 publication of his well-received book *Wonders of Geology*, Mantell's fortunes continued to decline. In 1838, he was forced to sell his enormous collection to the trustees of the British Museum for a mere £4,000 (Cleevely and Chapman 1992). He used money from this transaction to establish a medical practice in Clapham Common. Soon thereafter, his wife and children left him. A carriage accident in October 1841 left him with a debilitating back injury. In 1844, Mantell moved once more, this time to London, where he set up practice in the West End. In London he was active in the affairs of the Royal Society, which awarded him its Gold Medal in 1849 (over the objections of Sir Richard Owen, who had become his bitter enemy; see Torrens, chap. 14 of this volume). In addition to *Iguanodon*, Mantell also discovered and described the ankylosaur *Hylaeosaurus*, the sauropod *Pelorosaurus*, and the problematic *Regnosaurus*. A lonely, disappointed man in private life, Mantell died in November 1852.

The Reverend William Fox (1813–1881) was an important but now mostly forgotten collector of Early Cretaceous dinosaurian fossils from the Isle of Wight (Blows 1983). Born and raised in Cumberland, Fox became a clergyman and moved to the Isle of Wight in 1862 as curate to the parish of Brixton (now Brighstone). He befriended the poet Alfred Lord Tennyson, who resided at Farringford near Freshwater, only a few miles west of Brixton. When Fox first became interested in dinosaurs is not known, but he read Sir Richard Owen's monographs and frequently corresponded with the great English anatomist. Owen later described many of the important finds made by Fox during the 1860s. Fox resigned his post in 1867, but he continued to live in Brixton and to collect in the area. In 1875 he became curate of Kingston, near Shorewell, on the Isle of Wight. He died in 1881. In 1882, his collection of more than five hundred specimens was purchased by the trustees of the British Museum (Natural History) in London.

Another important private collector was George Bax Holmes (1803–1887), a Quaker "gentleman" from the Horsham, Sussex. A highly intelligent man, he was independently wealthy as the result of a considerable inheritance and dedicated himself to fossil collecting, hunting, and shooting (Cooper 1992). Holmes started collecting in the early 1830s, and by 1840 his growing collection of vertebrate fossils, including many specimens of *Iguanodon*, had gained the attention of Richard Owen, who first studied Holmes's collection in the summer of that year. Holmes offered his entire collection to Owen for study, with the expectation that many of his fossils would be described and figured (Cooper 1993). Subsequently Holmes became disappointed when Owen did not illustrate more of his specimens, and their relationship turned sour in the 1850s when he discovered that Owen had catalogued some specimens borrowed from Holmes into the collections of the Royal College of Surgeons. His pride was apparently hurt by the fact that Owen increasingly drew on fossils from other private collections (such as that of the Reverend Fox). After Holmes's death in 1887, his collection was sold and is now housed in the Booth Museum of Natural History in Brighton.

Harry Govier Seeley (1839–1909) was a leading Victorian student of dinosaurs (Fig. 2.1), but he also published many important papers on other groups of non-mammalian vertebrates. The son of an impoverished London artisan, Seeley was forced to work in a pianoforte factory at a young age. He became interested in geology through attending popular lectures in London. Seeley entered Cambridge University but apparently never earned

Figure 2.1. Portrait of Harry Govier Seeley. Courtesy of W. E. Swinton.

a degree. Instead he became an assistant to the Reverend Adam Sedgwick at the Woodwardian Museum in Cambridge in 1859. The Woodwardian housed a large collection of Late Cretaceous reptilian fossils from the Cambridge Greensand, which attracted Seeley's attention and formed the subject of his early papers. In 1872, shortly before Sedgwick's death, Seeley gave up his assistantship and settled in London. He subsequently engaged in writing several books and, for a number of years, went on public lecture tours all over the British Isles on behalf of the Gilchrist Trust (Swinton 1962). In 1873, Seeley took the Chair of Physical Geography at Bedford College. Three years later he moved on to King's College, London, where he was appointed to the Chair of Geography and, in 1896, also assumed the Chair of Geology and Mineralogy. Apparently sensitive about his humble background in the rigidly stratified Victorian society, Seeley was highly opinionated, and perennially disagreed with speakers at meetings of the Geological Society of London.

Seeley published extensively on dinosaurian specimens. In 1887, he first proposed the fundamental division of dinosaurs into Ornithischia and Saurischia based on the structure of the pelvic girdle. The later years of his career were mostly devoted to the scientific study of Permian and Triassic therapsids and other tetrapods from the Karoo of South Africa.

Early French Dinosaur Hunters

Although nineteenth-century French vertebrate paleontologists were mainly interested in fossil mammals, there were several notable contributors to the study of dinosaurs (Buffetaut et al. 1993). Vertebrae of a dinosaur from the Jurassic of Honfleur, Normandy, were described and figured in an early study on fossil crocodilians by Cuvier (1808). A professor of natural history at the University of Caen, Jacques-Amand Eudes-Deslongchamps (1794–1867), made the first major discovery of dinosaurian remains in Normandy, a partial skeleton of a large theropod dinosaur from the Middle Jurassic (Bathonian) "Calcaire de Caen." Although much of the specimen, embedded in a large block of limestone, had already been scattered and damaged by the time Eudes-Deslongchamps learned about its existence from a local physician, he managed to salvage a number of caudal vertebrae, many ribs and gastralia, and bones of the fore and hind limbs. Eudes-Deslongchamps (1838) carefully described this material and recognized its affinities to Buckland's *Megalosaurus*. However, he decided to place it in a separate genus, *Poekilopleuron*, which he considered a huge, predominantly marine lizard. Regrettably, Eudes-Deslongchamps's fossils were destroyed during the Allied liberation of Normandy in 1944.

The Upper Cretaceous strata of Provence are now famous for the presence of numerous well-preserved dinosaurian eggs. They have also yielded bones of a variety of dinosaurs, which may have been known as early as 1840 (Buffetaut et al. 1993). These fossils were not clearly identified as dinosaurian until 1869, when a geologist from Marseilles, Philippe Matheron (1807–1899), published an account on well-preserved bones of a dinosaur, which he called *Rhabdodon priscus* and correctly interpreted as related to *Iguanodon*. He also reported on bones of a "monstrous saurian" that he called *Hypselosaurus priscus* (now considered a sauropod) and interpreted as an aquatic reptile similar to crocodilians. Remarkably, Matheron first suggested a possible association of two eggshell fragments with the latter animal.

Friedrich von Huene

Although central Europe has very few occurrences of dinosaur-bearing Mesozoic strata, Germany was home to a group of men who made enormous contributions to the study of dinosaurs. The most famous of these men was Friedrich von Huene (1875–1969) (Fig. 2.2), a scion of an old dynasty of German nobles in the Baltic region. During his remarkably long career, from 1899 to 1966, Huene studied not only dinosaurs but also many other groups of non-mammalian vertebrates. A man of distinguished if ascetic appearance and legendary stamina, Huene was an indefatigable worker who produced hundreds of publications, including several large monographs and books. The son of a Lutheran minister, he was deeply religious and initially contemplated a career as a theologian. His faith shaped his prolific scientific career: Huene repeatedly wrote that through his research he intended to show the intricacies of divine creation to those with eyes to see (e.g., Huene 1944). No hardship and sacrifice would prevent him from carrying out this self-imposed obligation. Huene claimed that he never sought promotion to full professor because the administrative duties associated with this position would have taken away precious time for research.

Figure 2.2. Portrait of Friedrich von Huene. From frontispiece to Huene 1944.

As a young boy in Basel, Switzerland, where his father had moved to teach at a school for Protestant preachers, Huene was an avid collector of fossils. He went on to study geology and biology, first at the University of Basel and subsequently at the famous University of Tübingen in southern Germany. After receiving his doctorate in 1898 for a study of certain Ordovician brachiopods from the Baltic region, Huene joined the staff of the Institute for Geology and Paleontology at Tübingen. The head of the institute suggested to him a revision of the dinosaurs from the Keuper (Upper Triassic) of southern Germany as a suitable topic for his *Habilitation* (a second major thesis required by German universities in order to qualify for a professorial appointment). Huene quickly immersed himself in the study of Triassic reptiles with an emphasis on dinosaurs. His extensive comparative studies of dinosaurian and other reptilian fossils in European and American museums led to a steady stream of monographs and papers, only briefly interrupted by his military service during World War I. Huene was a major proponent of Seeley's (1887) division of the Dinosauria into Saurischia and Ornithischia.

Unlike some of his European contemporaries, Huene not only studied fossils in existing collections, but was also an enthusiastic worker in the field. Just prior to World War I, skeletal remains of the Late Triassic prosauropod dinosaur *Plateosaurus* had been discovered near the small town of Trossingen in southern Germany. Huene was eager to explore this occurrence more thoroughly, but the economic collapse of postwar Germany made it impossible for him to secure funding for paleontological field work. A lucky turn of events came during a visit to Tübingen by the eminent vertebrate paleontologist William Diller Matthew (1857–1930), then at the American Museum of Natural History. Matthew proposed a collaboration between his institution and the University of Tübingen. Under this agreement, the American Museum provided the necessary funds and some collecting expertise, and the resulting collections were to be divided between the two parties, but their scientific study was assigned to Huene. During three consecutive summer seasons, from 1921 to 1923, Huene's teams collected parts of about fourteen skeletons, two of which were virtually complete.

On New Year's Day 1923, Huene received a letter from the museum in La Plata, Argentina, inviting him to study old and new collections of sauropod bones from the Upper Cretaceous of Patagonia. He accepted this offer and set out on a long journey that eventually also took him to South Africa. Not content with merely studying the collections in the La Plata Museum, Huene wanted to examine their geological setting. During a field trip to Patagonia, he actually discovered another promising site in late 1923. Later he published a monograph on the Late Cretaceous dinosaurs of Patagonia (1929), which laid the foundation for all subsequent work on this important material.

After returning from his long trip early in 1924, Huene once again commenced work on numerous papers. Late that year, a German geologist wrote to him about some vertebrate fossils from Santa Maria, southern Brazil, and on Christmas Day, Huene received a crate of Triassic reptile bones from this region collected by a local German doctor. Although this box and subsequent shipments did not contain dinosaurian bones, Huene was excited by this material and raised funds for an expedition, which he undertook in 1928–29. His work uncovered the first major assemblage of tetrapods of Late Triassic age from South America. A few bones collected by Huene and his team were later described as dinosaurian by him. More important in the present context, Huene's success encouraged a team from the Museum of Comparative Zoology at Harvard to undertake further exploration of the Santa Maria region in 1936. This expedition resulted in the discovery of a partial skeleton subsequently identified as one of the oldest known dinosaurs, *Staurikosaurus pricei*.

The culmination of Huene's studies on dinosaurs was the publication in 1932 of a massive, two-volume monograph on the evolutionary history of the Saurischia, in which he reviewed all material referable to this group then known. Aside from a major study on Late Cretaceous dinosaurs from central India (with C. A. Matley 1933), Huene subsequently worked mostly on other groups of reptiles.

The Dinosaurs of Tendaguru

In 1907, Eberhard Fraas (1862–1915), a curator at the Royal Naturaliensammlung (the precursor of today's State Museum of Natural History) in Stuttgart, availed himself of an opportunity to join two German businessmen on a trip to the German Protectorate East Africa (present-day Tanzania). The businessmen were interested in developing the economic potential of the region, and they hoped to benefit from Fraas's geological expertise. Fraas (Fig. 2.3) was already a well-known student of Mesozoic reptiles, and had previously undertaken extensive paleontological field work in Egypt and German Southwest Africa (present-day Namibia) (Wild 1991). On the day of his departure for Africa, Fraas received news from a member of the Commission for the Geographic Exploration of the Protectorates that a Mr. Bernhard Sattler, an engineer with the Lindi-Schürfgesellschaft, a Hanover-based mining and exploration company, had happened upon a gigantic saurian bone weathering out of a bush path near Tendaguru Hill in the interior of the colony. Sattler had dutifully reported his unusual find to his superiors, who in turn had notified the chairman of the commission in Berlin.

On his arrival in Dar es Salaam, Fraas received an official request from the commission to follow up on Sattler's report. Thus he was forced to

Figure 2.3. Portrait drawing of Eberhard Fraas by A. Robbi (1915). Courtesy and copyright of Staatliches Museum für Naturkunde Stuttgart.

organize an expedition on the spot; fortunately, he received logistical support from the German colonial authorities for this venture. After conducting two other geological reconnaissance trips through the territory and into British-ruled Uganda and Kenya, Fraas, weakened by amoebic dysentery, left the town of Lindi for Tendaguru on August 31, 1907, accompanied by a local German official, a military doctor, and sixty native helpers. It took a five-day trip on foot through the coastal plain and across a densely forested high plateau to reach the region around Tendaguru Hill. Sattler met the expedition and took Fraas to the site of his discovery. Fraas was overwhelmed by the large numbers of huge bones and bone fragments of dinosaurs weathering out of the ground at Sattler's site. Despite his rapidly deteriorating health, he initiated excavations at several points in order to obtain unweathered, articulated skeletal remains. Sattler apparently supervised the excavation and conservation of specimens for transport. Fraas's illness rapidly grew worse and finally forced him to return home to Germany in late September of that year. He never recovered his health and died a few years later.

One of the greatest dinosaur graveyards in the world had been discovered. The imperial German government enacted special protection of the Tendaguru area in 1908, the year Fraas published the first technical account on the dinosaurian remains from this region. The director of the Museum of Friedrich-Wilhelm-University in Berlin, Wilhelm Branca, became very interested in following up on Fraas's discovery. With great enthusiasm, Branca set out to raise the large amount of money necessary to organize and undertake a well-staffed expedition for the purpose of excavating skeletons of large dinosaurs and transporting them back to Berlin for preparation, scientific study, and exhibition (see Branca 1914). His persistence paid off: The Prussian Ministry of Culture, the Prussian Academy of Sciences, several learned societies, and, solicited by a committee chaired by the regent of Brunswick, almost one hundred influential private individuals donated money, raising the enormous sum of nearly 200,000 marks (roughly equivalent to about U.S. $600,000 today). Werner Janensch (1878–1969), an expert on fossil reptiles at the Berlin Museum, was placed in charge of the project and led three highly successful expeditions to Tendaguru between 1909 and 1911. At the end of the 1911 field season, the research team decided that more work remained to be done, and Hans Reck (1886–1937) led a fourth and final expedition in 1912.

Janensch and his co-workers identified two horizons rich in dinosaurian remains at Tendaguru, both of which they determined to be Late Jurassic (Kimmeridgian) in age, based on associated marine invertebrate fossils. (During his preliminary geological reconnaissance, Fraas had interpreted the strata at Tendaguru as Late Cretaceous in age.) The rolling, densely vegetated countryside at Tendaguru, along with the tropical climate, posed daunting challenges to field work. The thick cover of vegetation, for the most part composed of scrubs and small trees, did not permit standard quarrying operations. Numerous pits and trenches had to be dug by large crews of workers in the vicinity of Tendaguru Hill. The enormous load of the crated fossils was transported on the backs and heads of hundreds of porters in some 5,400 four-day-long marches across forested terrain to the coastal town of Lindi for shipment by sea to Germany. Although native labor was cheap, the funds were rapidly expended on the huge crews necessary for the formidable task at hand: The Berlin team employed 170 African workers during the 1909 field season, 400 in 1910, and 500 each in 1911 and 1912. The total fossil freight shipped to Berlin amounted to

some 235 metric tons. At the conclusion of the 1912 season, some 136,000 marks had been expended.

Many years of intensive preparation of the rich collections followed. According to Branca's (1914) report, one sauropod vertebra required some 450 hours of cleaning and conservation, and it took 160 hours to piece a 2-meter-long sauropod scapula back together from 80 individual fragments of bone. The outbreak of World War I in 1914 dashed plans by Janensch and his associates to return to East Africa. The German Protectorate East Africa fell under British colonial rule at the end of the war and became the Tanganyika Territory of British East Africa. In 1924, an expedition from the British Museum returned to Tendaguru to secure dinosaurian material for that institution. It was led by the Canadian William E. Cutler, an assistant at the University of Manitoba who already had substantial experience in collecting dinosaurs in the badlands of Alberta. One notable participant in Cutler's expedition was a young Englishman named Louis S. B. Leakey (1904–1982), who went on to become one of the most celebrated explorers of human origins. After eight months of hard work at Tendaguru, Cutler succumbed to malaria in Lindi at the age of forty-two. He was succeeded by an English explorer named F. W. H. Migeod, who had experience living in Africa but no paleontological training, and in 1927 by John Parkinson, who later wrote a book about his experiences (Parkinson 1930). The British efforts yielded little additional dinosaurian material and apparently no new forms. In recent years, attempts by several investigators to secure permits to return to Tendaguru have been unsuccessful.

Meanwhile, Janensch studied the vast collections of dinosaurian remains from Tendaguru in Berlin and supervised the skillful assembling of several skeletons, including that of the gigantic *Brachiosaurus brancai*, at the Berlin Museum in the late 1930s. A series of monographic studies published by Janensch between 1925 and 1961 reflect a lifetime of careful work on this remarkable collection of Late Jurassic dinosaurs. Almost miraculously, the bulk of the collections has survived both the destruction of Berlin during World War II and some forty-five years of subsequent neglect by the Communist rulers of East Germany.

The Man Who Would Be King

One of the most unusual figures ever to grace dinosaurian paleontology—a field replete with colorful personalities—was Franz (Ferenc) Baron Nopcsa von Felsö-Szilvás (1877–1933) (Fig. 2.4). Regrettably, his unorthodox behavior has received far more attention in the English-speaking literature than his scientific work.

The last male representative of one of the most ancient noble families in Transylvania, Nopcsa was very much a product of the turbulent times that formed the heyday of the Austro-Hungarian Empire prior to World War I. A highly cultivated man with a remarkable aptitude for languages, Nopcsa apparently became a paleontologist by mere chance. The discovery of dinosaurian bones on the estate of his sister Ilona in 1895 led him to visit the famous Austrian geologist Eduard Suess (1831–1914) at the University of Vienna. Suess identified the material as dinosaurian, but when Nopcsa pressed him for further details, Suess told the young man to undertake a detailed study himself. Nopcsa obliged and soon published a detailed description of the skull of the hadrosaur *"Limnosaurus"* (now *Telmato-*

Figure 2.4. Portrait sketch of Franz (Ferenc) Nopcsa von Felsö-Szilvás by F. Marton (1926). From A. T. Kubacska, "Franz Baron Nopcsa" (1945).

saurus) transsylvanicus (Nopcsa 1900). He continued his research on the distinctive Late Cretaceous dinosaurs from Transylvania, which culminated in a series of important papers. Nopcsa also published extensively on other dinosaurs, especially from England and France, and, heavily influenced by the ideas of Dollo, on dinosaurian paleobiology. Like Huene, he was an avid early supporter of Seeley's fundamental dichotomy in dinosaurian phylogeny. He also took a keen interest in general biological and geological issues, and recognized and tried to address the difficulties of the neo-Lamarckian evolutionary ideas then prevalent in Europe (Lambrecht 1933; Weishampel and Reif 1984).

Yet his original, restless mind was not content with a quiet life of scholarship. Like many a bored young nobleman in Austria-Hungary, Nopcsa sought adventure and excitement. He became fascinated with the land and people of Albania, then a remote, rebellious outpost of the Ottoman Empire on the Balkan peninsula. Nopcsa devoured the available literature on Albania and its culture, learned the dialects of the region, and traveled extensively through that country. In 1913, a London conference of Europe's leading powers designated most of the territory populated by the Albanian people as an independent country. Austria, worried about a power vacuum developing in its "backyard," decided to install a puppet regime in this new nation.

Nopcsa considered himself eminently qualified to rule Albania, and promptly submitted a takeover plan to the Imperial High Command. He wanted to stage an invasion with five hundred soldiers disguised in civilian clothing, some artillery pieces, and two fast steamboats, which he proposed to purchase with his own funds. Once a beachhead was established, Nopcsa would install himself as *mbret* (ruler) and set up a regime friendly to Austria. He ingeniously proposed to generate the necessary cash flow for the impoverished country by marrying a wealthy American woman, since he was firmly convinced, probably not without cause, that many American millionaires had daughters eager to marry genuine European noblemen. Unfortunately for Nopcsa, his vision was not shared by the imperial government in Vienna, which instead installed Prince Wilhelm zu Wied as ruler of Albania. (Prince Wilhelm was so thoroughly ignorant of Albania that he could not even pronounce his new title. His tenure there proved to be very short-lived—Wilhelm and his family were forced to flee for their lives just a few months later.)

During World War I, Nopcsa served as an officer in the Imperial Austro-Hungarian Army. He carried out dangerous undercover missions, disguised as a peasant, in the heavily contested border region between Hungary and Romania and, of course, in his beloved Albania. A nobleman on the losing side of the war, Nopcsa found his fortunes collapsing in the chaotic years immediately following the war. Some of his estates, now occupied by Romania, had been confiscated without indemnification. During a visit to one of his Hungarian estates, Nopcsa was set upon by a mob of armed, rebellious peasants who took exception to his autocratic manner and nearly killed him. Things seemed to improve when the Hungarian government, aware of his great scientific reputation, put him in charge of the Geological Survey in 1925. Full of innovative plans for reorganizing the survey, Nopcsa quickly ended up at odds with both his superiors and his subordinates, and finally left in a rage in 1929. With his secretary and companion, an Albanian named Bajazid, Nopcsa set out on a 3,500-mile odyssey on a motorcycle through Italy until they ran out of money. He then settled for some time in Vienna to recover his health and resumed scholarly work. Some of this research was innovative and far ahead of its time, especially the application of data from bone histology to dinosaurian classification. Increasingly despondent about the continuing failure of his health and faced with poverty, Nopcsa shot himself after first killing his companion in April 1933.

Louis Dollo: Dinosaurs and Paleobiology

Paleobiology has its intellectual foundations in the work of the Belgian paleontologist Louis Antoine Marie Joseph Dollo (1857–1931) (Fig. 2.5), although this fact is not widely appreciated today (Gould 1970). Dollo is also noteworthy for his role in the first find of numerous well-preserved and articulated dinosaurian skeletons in continental Europe—a series of skeletons of *Iguanodon* discovered in the Saint Barbe coal mine near Bernissart in southern Belgium in 1878.

Figure 2.5. Portrait of Louis Dollo. From frontispiece to "Dollo-Festschrift der Palaeobiologica" (1928).

Of Breton descent, Louis Dollo was born and educated in Lille, France. As an engineering student, he became interested in geology and zoology and took courses in these subjects. Dollo graduated in 1877 as a mining engineer. He went to Brussels in 1882 to study the Bernissart dinosaurs at the Musée Royal d'Histoire Naturelle de Belgique. The Bernissart specimens of *Iguanodon* had been excavated under the able supervision of Louis De Pauw, the chief preparator at that institution, and transported to Brussels for preparation and study (see Casier 1960). Dollo became a Belgian citizen in 1886 and worked at the museum in Brussels for forty-seven years. He was a virtual recluse with almost monastic habits.

Dollo became *conservateur* at the museum in 1891, and was appointed professor of paleontology at the university in 1909 (Van Straelen 1933). He supervised the difficult preparation and mounting for display of several skeletons of *Iguanodon,* both as freestanding mounts and as originally preserved in the rock (*en gisement*). The completed exhibit opened to the public in 1905. Dollo also undertook the first scientific studies of the Bernissart iguanodonts, publishing some nineteen papers on these dinosaurs. His great international reputation provoked the envy of his superiors, who set out to make his life difficult and even temporarily prohibited him from working on fossil reptiles. Dollo's pro-German sympathies during World War I further alienated him from his Belgian colleagues. His situation at the museum improved only toward the end of his life, when one of his former students, Victor Van Straelen, became director of that institution (Abel 1931).

A tireless worker and a highly creative thinker, Dollo published prolifically on dinosaurs and many other fossil reptiles (see bibliography in Van Straelen 1933). He did not undertake any field work of his own (Abel 1931). Dollo eschewed monographic studies in favor of papers ("notes"), renowned for their distinctive telegraphic format of brief numbered statements that reflected his strong mathematical background. Not content with merely describing and naming fossils, he attempted to relate form to function in extinct animals—to develop what he called "ethological paleontology" (Dollo 1910) and what we now refer to as "paleobiology." Curiously, Dollo showed no interest in the geological evidence for reconstructing the ancient environments in which these animals lived and died (Abel 1931). He pondered many issues of dinosaurian paleobiology, such as feeding and locomotion, and his papers laid the foundations for this field of study.

European Dinosaur Hunters in North Africa

Between 1910 and 1914, the professional collector Richard Markgraf (1856–1915) and the German paleontologist Ernst Stromer von Reichenbach (1871–1952) amassed a remarkable collection of bones of Late Cretaceous (Cenomanian) dinosaurs and other vertebrates in strata exposed in the oasis El-Bahariya in the Great Western Desert of Egypt. Despite great difficulties with the English colonial authorities of Egypt, Stromer managed to ship the material to Munich for preparation and scientific study, but much of it only after the end of World War I. Markgraf tried to continue collecting after the outbreak of the war had severed all contacts with Stromer, but he failed, and died in utter poverty in 1915. Between 1914 and 1936, Stromer and his scientific collaborators published a series of monographs on these fossils, especially the dinosaurs, which included the large "fin-backed" theropod *Spinosaurus aegyptiacus*. Unfortunately, all of the dinosaurian specimens from El-Bahariya were destroyed during an Allied bombing raid on Munich in 1944.

French explorers and researchers working in the regions of North and West Africa formerly under French colonial rule discovered numerous important occurrences of dinosaurian fossils as early as 1904. Between 1946 and 1959, the Abbé Albert-Félix de Lapparent (1905–1975) of the Institut Catholique de Paris led nine expeditions to collect Mesozoic vertebrate fossils, especially dinosaurian remains, in various regions of the Sahara Desert. The scion of a family of distinguished French earth scientists, Lapparent became interested in dinosaurs during the course of his doctoral research on the geology of Provence (France). He went on to collect dinosaurian fossils, including eggs and footprints, in France, Spain, Portugal, Spitsbergen, and especially North Africa. In the early 1960s, geologists with the office of the French Atomic Energy Commission discovered a major deposit of Early Cretaceous dinosaurian bones at Gadoufaoua in Niger. Teams from the Muséum National d'Histoire Naturelle (Paris), led by Philippe Taquet, followed up on this discovery and systematically excavated this remarkably fossiliferous site. They collected some 25 tons of material, much of which remains yet to be described (Taquet 1977, 1994).

The early "dinosaur hunters" were dedicated amateurs who labored with much enthusiasm and often at great personal sacrifice. Subsequently, paleontological research, including field work, increasingly became the domain of academically trained professionals. Yet paleontology is one of the few scientific disciplines to which amateurs have continued to make vital contributions.

Since the late nineteenth century, North American paleontologists have tended to dominate research on dinosaurs. Yet important new discoveries of dinosaurian remains are still being made, especially in France (see Buffetaut et al. 1993), but also in England and Spain. In particular, French workers continue to be very active in the field overseas. Scientific work on dinosaurs is a truly international enterprise, which is both its charm and its promise.

North American Dinosaur Hunters

Edwin H. Colbert

The First Discoveries of Dinosaurs in the New World

Perhaps the first North American dinosaur hunters were the unwitting discoverers of unknown animals. In 1802 a New England farm boy named Pliny Moody found some footprints in reddish-brown sandstones near his home at South Hadley, Massachusetts. Because these impressions had the appearance of large bird tracks, they were at the time popularly referred to as trackways made by "Noah's Raven." During the early years of the nineteenth century, other such tracks and trackways came to light, and they soon became objects of study by Professor Edward B. Hitchcock, the president of Amherst College. For several decades, from 1836 to 1865, Hitchcock ranged back and forth and up and down the Connecticut Valley, ferreting out footprints that were widely exposed in the Late Triassic and Early Jurassic sandstones and siltstones of the region. Many of them he collected and removed to Amherst, where in time a museum was built for their reception. In 1858 he published a large monograph entitled *Ichnology of New England,* in which he described most of the footprints as having been made by large birds.

When Hitchcock began his work, the concept of dinosaurs, established by Richard Owen in 1842 (see Torrens, chap. 14 of this volume), was still a matter of future history. But even after Owen's pronouncement, and the pioneer work on dinosaurs in England by Owen, William Buckland, and Gideon Mantell, the possibility of the Connecticut Valley footprints' having been made by dinosaurs did not enter the perception of Hitchcock. So he went to his grave believing that what is now the Connecticut Valley was once inhabited by a varied population of birds, large and small.

3

Needless to say, by midcentury the true nature of these footprints was becoming evident.

In 1855, more than a decade after Owen's recognition and naming of the *Dinosauria,* Dr. Ferdinand V. Hayden, in charge of one of the governmental surveys of the western territories, found some teeth near the confluence of the Judith and Missouri rivers, in what is now Montana, as well as some vertebrae and a toe bone in South Dakota, and these fossils were given to Dr. Joseph Leidy of Philadelphia for identification. The specimens were described by Leidy in 1856. They included several hadrosaur teeth, which he named *Trachodon mirabilis,* as well as some teeth of a carnosaurian dinosaur, designated by Leidy as *Deinodon horridus.* In addition he named two teeth *Troodon* and *Palaeoscincus,* which he indicated as being "lacertilian." These were the first dinosaurs to be named from North America.

Then in 1858 a partial dinosaur skeleton was discovered in a Cretaceous marl pit near Haddonfield, New Jersey, across the Delaware River from Philadelphia. Leidy was instrumental in the excavation of this skeleton, which he described in 1858 and 1859 as *Hadrosaurus foulkii,* the first dinosaur skeleton to be described from North America. Significantly, Leidy recognized the dominant bipedality of *Hadrosaurus,* thereby establishing a basic adaptation in dinosaurian evolution, quite in contrast to Owen's original concept that dinosaurs such as *Iguanodon* were rhinoceros-like quadrupedal tetrapods.

O. C. Marsh and E. D. Cope: High-Stakes Paleontology

Figure 3.1. Edward Drinker Cope.

The discovery of large Mesozoic fossil bones in England and in North America, as well as the realization that such fossils were the remains of a new and previously unrecognized group of extinct reptiles, inspired two particular Americans to become deeply involved with the search for dinosaurs. Edward Drinker Cope and Othniel Charles Marsh (Figs. 3.1, 3.2) entered into this new field of paleontological endeavor with gusto—to such a degree, indeed, that soon their rivalry turned into bitter animosity, resulting in a fossil feud unparalleled in the history of the science.

Leidy, professor of anatomy in the medical school of the University of Pennsylvania, almost immediately found himself in the middle of a vituperative battle between the two warring paleontologists, so he soon retired from this field of research and turned his attention to other matters. He was a quiet, dignified man, who found the loud quarrel between Cope and Marsh too harsh to bear.

Cope was the son of a wealthy Quaker shipping magnate in Philadelphia. Marsh was the nephew of a very wealthy businessman, George Peabody, who spent his adult life in England. Both men had access to ample funds with which to pursue their paleontological quests: Cope as a freelance scholar with ties to the Philadelphia Academy of Science, Marsh as a member of the Yale faculty and director of the Yale Peabody Museum, posts established for him by his munificent uncle. Both were as independent as may be imagined, and both had very strong opinions about fossils, about each other, and about the world in general. The Cope-Marsh feud is so well known that it need not be described here (see Osborn 1931; Schuchert and LeVene 1940; Plate 1964; Colbert 1968; Lanham 1973; Shor 1974; and the appendix of this volume for details). Suffice it to say that as a result of their

Figure 3.2. Othniel Charles Marsh.

rivalry, they made extensive collections of Mesozoic and Cenozoic fossils in western North America, including many skulls and skeletons of dinosaurs. Indeed, their animosity toward each other did have the positive result of opening to the astonished eyes of paleontologists and of the public across the globe the new and exciting world of dinosaurs.

The "Golden Age" of Dinosaur Paleontology

The early period of tentative exploration and research that occupied the first half of the nineteenth century, followed by the fierce Cope-Marsh rivalry of the last three decades of the century, may be regarded as a prologue to the "Golden Age" of dinosaur exploration and research in North America—namely from the so-called "Gay Nineties" until the advent of the "Roaring Twenties." These were the years when paleontologists from the American Museum of Natural History in New York, the United States National Museum (Smithsonian Institution), the Carnegie Museum of Pittsburgh, the National Museum of Canada in Ottawa, and the Royal Ontario Museum in Toronto worked vigorously in the Upper Jurassic and Cretaceous beds of western North America to discover dinosaurian treasures year after year, and to describe them in numerous scientific publications. Of course there were dinosaur hunters from other museums as well, but in North America the institutions listed above were at the forefront of dinosaurian field exploration and laboratory research. These were indeed golden years, filled with the excitement of discovery and the satisfactions of lucid research, frequently followed by the appearance of excellent publications.

One may think of this Golden Age as having been inaugurated by the extensive excavations in the Upper Jurassic Morrison Formation at Bone Cabin, Wyoming—an area near Como Bluff that had been previously worked by Marsh and his associates and assistants, notably Samuel Wendell Williston, Arthur Lakes, William Reed, and others. The force behind the Bone Cabin excavations was Henry Fairfield Osborn, who had come to the American Museum in 1891, where he initiated a very active program in vertebrate paleontology. He was especially interested in Mesozoic mammals that might be found in the Morrison sediments, but at the same time he realized the importance of collecting dinosaur skeletons, both for research and for exhibition. Thus the Bone Cabin program commenced in 1898, with W. D. Matthew, Walter Granger, Barnum Brown, and R. S. Lull, together with Albert Thomson, Peter Kaisen, and others, all working together to unearth gigantic dinosaur skeletons (Fig. 3.3) as well as Jurassic mammals (in a subsidiary quarry known as Quarry Nine). The work at Bone Cabin, lasting through six years, began a program of dinosaur collecting that eventually would enable the American Museum to have the largest collection of these reptiles in the world—in this respect surpassing the collections at Yale and the National Museum that had been assembled by O. C. Marsh and his assistants.

Among the participants in the Bone Cabin project, the names of Barnum Brown and Richard Swann Lull are especially significant. Brown went on from Bone Cabin to devote the remainder of his long life to dinosaurs, especially those from the Cretaceous beds of Alberta, as will be told. He was very much an individualist who preferred to labor in the field by himself, variously assisted by Peter Kaisen and George Olsen of the American Museum fossil laboratory, but usually without other collaborators. And he could be very secretive about what he was up to. What he was

Figure 3.3. American Museum paleontologists at Nine Mile Quarry, Como Bluff, Wyoming, in 1899. *From left to right:* Richard Swann Lull, Peter Kaisen, and William Diller Matthew.

up to generally resulted in amazing collections, properly excavated, commonly accompanied by scanty field notes consisting of locality data in indelible pencil scrawled on scraps of newspaper. He had a habit of locating fossils in one field season, burying them as a dog would bury a precious bone, and then returning to them the next year. Thus he had something "on tap" to ensure the success of a summer to come. This was but one of the facets of a man around whom there is an aura of anecdotes and legends.

Lull devoted his life to dinosaurs, perhaps more in the laboratory than in the field. He was a dignified and imposing person, as befitted his position as a Yale professor. At the university he was noted for his lucid classroom lectures on vertebrate evolution, and for his prodigious efforts as director of the newly erected Peabody Museum, which replaced the old building in which Lull's predecessor, O. C. Marsh, had ruled his paleontological world with cold authority.

In 1909 the Bone Cabin excavations met their match for scientific importance, thanks to work by Earl Douglass for the Carnegie Museum, who discovered articulated dinosaur skeletons in the Morrison Formation, about twenty miles to the east of Vernal, Utah. From this discovery the famous Carnegie Quarry was developed—a dinosaur-collecting project that extended until 1923, resulting in the amassing of prodigious collections of Upper Jurassic dinosaurs. Indeed, Douglass devoted the remainder of his career to the Carnegie Quarry, even to the extent of giving up his life in Pittsburgh and moving to Utah, where he established a homestead near the quarry on which he built a primitive cabin and settled in with his wife and small child. He was truly a dedicated person. In Utah, across the continent from the Carnegie Museum, he was at least spared the dubious pleasure of being under the heavy thumb of W. J. Holland, the dictatorial director of the museum.

The large scope and success of the Carnegie Quarry excavation were due not only to the foresight and the dedicated efforts of Douglass, but also to unlimited financial support from the industrialist Andrew Carnegie, who enthusiastically provided funds so that (among other things) casts of a complete skeleton of *Diplodocus,* excavated from the quarry, could be produced and distributed to various museums throughout the world. And everywhere that *Diplodocus* went, W. J. Holland also went, to supervise the setting up of the plaster bones, and incidentally, to his great satisfaction, to collect honorary degrees. The Carnegie Quarry eventually was enlarged into an *in situ* exhibit that today forms the centerpiece of the Dinosaur National Monument.

Certainly one of the outstanding authorities on dinosaurs during those years when the American Museum and the Carnegie Museum were conducting large quarrying operations in the Morrison Formation was Charles Gilmore of the United States National Museum (Smithsonian Institution). Gilmore, a friendly, amiable person, remarkably modest about his accomplishments, not only spent long hours with his fossils, he also had the admirable habit of writing down his observations and conclusions, thereby publishing a series of exceptionally thorough monographs on Jurassic and Cretaceous dinosaurs. During his long career at the National Museum, he conducted sixteen paleontological expeditions, almost all of them in western North America. Oddly enough, however, only seven of these expeditions were devoted to the collection of dinosaurs—one in the Morrison beds of Utah, the others in the Cretaceous beds in various western states. Almost all of Gilmore's other expeditions were primarily for the purpose of collecting fossil mammals.

During those exciting years of exploration and collecting in the Morrison Formation by the American and Carnegie museums, and in the Cretaceous sediments of various western states by the National Museum, remarkably rich deposits of Cretaceous dinosaurs had also been discovered in the western Canadian provinces, particularly in Alberta. Dinosaur bones of Cretaceous age were first found in western Canada by Dr. George Dawson (Fig. 3.4), the son of Sir William Dawson, one of the giants of

Figure 3.4. George Dawson (the short man standing in the center of the photograph) and his field party at Fort McLeod, British Columbia, in 1879.

Figure 3.5. Charles M. Sternberg (*left*) and an assistant collecting a dinosaur along the Red Deer River.

nineteenth-century geology. George Dawson, working as a Canadian representative of the International Boundary Commission, was a thoroughly trained geologist who had completed his graduate studies in England. He had studied under Thomas Henry Huxley—Darwin's champion—and had the perspicacity to submit his fossils to Cope for verification as to their dinosaurian relationships.

It is an interesting fact that Dawson and two men who followed him as pioneer dinosaur hunters in Canada would all at first glance appear to have been ill suited for their task. Dawson was a hunchback, almost a dwarf, who in spite of this disability was a vigorous field geologist. The other two men, Joseph Burr Tyrrell and Lawrence M. Lambe, both took up field geology and paleontology because they suffered severely from ill health, and decided in each case that only a vigorous outdoor life would ensure for them reasonably long lives. The regimen was truly successful for Lambe; he lived to the ripe old age of ninety-nine.

Tyrrell, one of Dawson's assistants, discovered dinosaur bones in the valley of the Red Deer River in Alberta in 1884. Lambe, a member of the Canadian Geological Survey, made a boat trip down the Red Deer River in 1897, thereby traversing extensive exposures of Upper Cretaceous sediments, from which he collected dinosaur fossils. He studied these specimens in collaboration with Professor Osborn, and they published a joint monograph on the fossils.

Extensive explorations for Cretaceous dinosaurs in Alberta were carried on during the decade of 1910 to 1920 by two teams of paleontologists, who engaged in friendly rivalry as they discovered and collected dinosaurian treasures from the Red Deer River region. One team consisted of Barnum Brown and his assistants, Peter Kaisen and George Olsen, of the American Museum in New York; the other group was made up of Charles H. Sternberg and his three sons, George, Charles M. (Fig. 3.5), and Levi, who were collectors working for the National Museum of Canada and for the Royal Ontario Museum.

North American Dinosaur Hunters   29

Figure 3.6. Barnum Brown's flatboat on the Red Deer River in Alberta.

The saga of the Sternberg family constitutes a remarkable chapter in the history of North American dinosaur collecting. The elder Sternberg and his twin brother migrated from their home in Iowa to Kansas when they were in their late teens, and there Charles became fascinated by various fossils that he discovered in the Cretaceous Dakota sandstone. In short order he was established as a field assistant to Cope, and this was the beginning of a long life devoted to paleontological collecting, a lifestyle that was continued with distinction by his sons.

In the years of the First World War, Brown and the Sternbergs explored the Red Deer River fossil beds by floating down the river on barges (Fig. 3.6), which served as their headquarters. With auxiliary motorboats they ranged up and down the river, climbing up from landing places to explore the exposures. It was a successful technique, and they collected numerous Late Cretaceous dinosaur specimens (Fig. 3.7), including many articulated skeletons, which today can be seen in New York, Ottawa, and Toronto. Today Red Deer River dinosaurs, as represented by original specimens and casts, can be seen in a magnificent display at the Royal Tyrrell Museum in Drumheller, Alberta—a large museum located at the very edge of the Red Deer River badlands.

The "Dark Age" and "Renaissance" of Dinosaur Studies

After the "Golden Age" of dinosaur exploration and research in North America, a time which may more properly be designated as the "First Golden Age," there was something of a lull in the pursuit of dinosaur studies—a lull that extended for perhaps a couple of decades, encompassing the time of the Second World War. Of course the exploration for dinosaurs continued through this period, as did laboratory research and publications, but during those years there was a great emphasis on the

Figure 3.7. Barnum Brown collecting a skeleton of the hadrosaur *Corythosaurus* in Alberta in 1912.

therapsids—the mammal-like reptiles. These tetrapods were of significance because, it was said, they were in the mainstream of evolution—from primitive vertebrates to the advanced mammals. The dinosaurs in the eyes of many paleontologists were "gee-whiz" fossils, nice to have in the exhibition halls where they could impress the public, but generally speaking fossils off on an evolutionary sideline, and therefore of lesser consequence to the evolutionist than were the therapsid reptiles. (I can vouch for this, because I lived through those days.)

The older dinosaurian studies had been largely descriptive, as would be expected with such a wealth of specimens discovered and waiting to be examined. They were in essence osteological studies, aimed at the interpretation of morphological anatomy and taxonomic relationships. But after the Second World War there appeared a new generation of paleontologists, exemplified by John Ostrom, who realized that the skeletons of dinosaurs revealed much more than previously had been seen. Consequently there were ever more sophisticated studies of the bones themselves, such as bone structure, histology, and the implications concerning dinosaurian physiology. There were new interpretations of various bony structures that are so prevalent among the dinosaurs, and the light that such adaptations may throw upon behavior. There were studies of ontogenetic growth patterns and their significance. And there were the new and exciting discoveries of and research on dinosaurian trace fossils and productions, such as tracks and trackways, eggs, nests and other objects in the fossil record.

Furthermore, there was a renaissance of dinosaur collecting, all over the world, not merely in North America, with the resultant discoveries of new dinosaurs of previously unsuspected form and variety. Indeed, our knowledge of dinosaurs has expanded at a truly remarkable rate during the past three or four decades. Therefore perhaps it is appropriate to speak of a "Second Golden Age" of dinosaur collecting and research—the age in which we are now living.

One of the first "big events," if so it may be called, of postwar dinosaur field work was the discovery by the author and his American Museum field crew of a deposit of Late Triassic dinosaurs at Ghost Ranch, New Mexico (Colbert 1995). Here, within a quarry of limited extent, were found literally hundreds of articulated skeletons and partial skeletons of the Triassic theropod *Coelophysis*. Very large blocks containing the intricately interlaced skeletons of this dinosaur were initially collected by the party from the American Museum of Natural History, and subsequently by representatives of the Carnegie Museum, the Natural History Museum of New Mexico, the Ruth Hall Paleontological Museum at Ghost Ranch, the Museum of Northern Arizona, and Yale Peabody Museum. The fossils were and are still being prepared and studied at the several involved institutions, as well as at a considerable number of other paleontological laboratories to which blocks have been distributed.

A discovery of unusual significance was that of *Deinonychus,* collected from the Lower Cretaceous Cloverly Formation in Montana by the Yale Peabody Museum, under the direction of John Ostrom. This discovery, and the research that has resulted from it, as well as from other discoveries of fossils related to *Deinonychus,* has within the past few years opened new horizons of knowledge concerning the dromaeosaurids, which have proved to be among the most aggressive and perhaps the most intelligent of predatory theropods.

Dale Russell (recently retired from the Natural Museum of Canada) has been particularly interested in the subject of dinosaurian intelligence, as a result of his discovery of *Troodon* in the Upper Cretaceous sediments of western Canada. Indeed, Russell and his Canadian confreres, among them Philip Currie and William Sarjeant, have initiated vigorous new programs of dinosaur collecting in the western Canadian provinces.

At the same time, John Horner of the Museum of the Rockies in Bozeman, Montana, has carried on a spectacular program of collecting in the Upper Cretaceous beds of that state, where he has found prodigious deposits of the hadrosaurid *Maiasaura,* consisting not only of individuals spanning an ontogenetic series from hatchling to adult, but also of almost countless numbers of eggs and nests of this dinosaur. As a result of his work, especially histological studies carried on with Armand de Ricqlès of Paris, much new information has been gained not only as to ontogeny, but also as to possible behavior patterns in this dinosaur. This work continues unabated.

Going back in geologic time, mention should be made of the spectacular discoveries made in the Morrison Formation in Utah and Colorado, notably between 1927 and 1967 at the Cleveland-Lloyd Quarry in the San Rafael Swell by Ferdinand F. Hintze, Golden York, William Lee Stokes, and James Madsen and associates of the University of Utah, and from 1972 to 1982 in the Dry Mesa Quarry along the Utah-Colorado border by James Jensen and associates at Brigham Young University. At this latter site, various skeletal elements of prodigious size have been found, indicative of sauropod dinosaurs that surpass even *Brachiosaurus* in size.

Perhaps the ultimate in size among the sauropods, though, was *Seismosaurus,* a skeleton of which was discovered by two hikers in New Mexico in 1979 and subsequently collected by David Gillette and his associates. Work on this enormous dinosaur continues at the present time (Gillette 1994).

Finally, it should be said that dinosaur hunters, not only in North America but also on the other continents, are today giving unprecedented attention to dinosaur tracks and trackways, with a new appreciation of such fossils as keys to dinosaurian locomotion and other behavior traits. Thus the study of footprints has come full circle since the days of Hitchcock and the Connecticut Valley tracks.

For decades the study of dinosaur tracks was a rather static, descriptive branch of paleontological research. But it was given new significance by R. T. Bird's discovery and excavation in 1940 of huge sauropod and theropod trackways of Cretaceous age along the Paluxy River in Texas (Bird 1985). Today dinosaur trackways are being vigorously studied in the field and preserved *in situ* on a worldwide scale. Particular mention might be made of David Gillette, James Farlow, Grace Irby, Martin Lockley, and William Sarjeant in North America, Tony Thulborn in Australia, and Giuseppe Leonardi in Europe and South America.

This survey of the work being carried on by dinosaur hunters of the modern "Golden Age" of dinosaurian research is admittedly incomplete. A comprehensive account of the activities of the new generation of dinosaur collectors would be much too long for inclusion in the space available here. But some of the high spots of dinosaur hunting during the middle and late years of the present century have been described, and perhaps they are sufficient to show that the search for dinosaurs and the resulting research are being carried on today with an intensity beyond anything achieved in the past. Dinosaurs are today a very lively subject, not only in paleontological circles, but also among the general public. Indeed, one is today confronted by dinosaurs everywhere—in museums, in stores, in moving pictures, in books, and always in the field.

Asian Dinosaur Hunters

John R. Lavas

4

Many sites in Asia, particularly within Mongolia and China, are rich in fossil-bearing Mesozoic deposits, and in recent decades more types of dinosaurs have been excavated from here than from any other part of the world. These discoveries are considered to be of prime importance to many aspects of research on dinosaurs, including their Cretaceous evolution and social behavior, the relationships between theropod dinosaurs and birds (Did birds evolve from theropods, or theropods from birds?), dinosaur ontogeny (growth), and paleoembryology. Because vertebrate paleontology first developed in Europe and the United States, much of the initial research in Asia was undertaken by European and American paleontologists, but recent decades have seen local scientists exploiting Asian dinosaur deposits, often with foreign collaboration (Lavas 1993; Spalding 1993). Dinosaurs of most Mesozoic ages are known from Asia, and although the area (which then formed the eastern section of the supercontinent Laurasia) was probably separated from North America by the Bering Strait even well into the Cretaceous, some Asian dinosaurs are very similar to those of western North America. Nonetheless, several major groups of Asian origin apparently remained endemic and are found nowhere else. Others may have evolved in Asia and later dispersed to North America, such as the ankylosaurids, the protoceratopians (also spelled protoceratopsians), and possibly the ornithomimosaurs (though the recent discovery of the Spanish genus *Pelecanimimus* throws the geographic origins of this last group into question).

Because of the area's geographic remoteness, Westerners were slow to infiltrate Asian deserts, and apart from explorers such as Marco Polo, Pavlinoff, and Przhevalski, the Gobi, for example, remained largely unknown as late as the 1920s. For centuries Chinese apothecary owners had crushed "dragon bones" for medicinal use (records of possible dinosaur bones date from the Jin Dynasty), but the first direct evidence of fossils came in 1892, when the geologist Vladimir Obruchev (1863–1956) found some rhino teeth on the caravan trail to the Mongol capital of Urga (now Ulan Bator). Then in 1902, Colonel Manakin of the Russian Army collected remains of the hadrosaur *"Mandschurosaurus"* from the Amur River between Russia and Manchuria, with further material collected during 1915–17 being displayed at the Central Geological Museum in Leningrad (St. Petersburg). In 1920 another Russian, A. A. Borissyak, located rich deposits of Tertiary mammals in Kazakhstan; the geology of this area suggested that major fossil deposits might be found in corresponding Gobi strata.

American Museum Expeditions to the Gobi

In the 1920s, the American Museum of Natural History in New York (AMNH) embarked on a series of pioneering expeditions to the Gobi, conducting work in Mongolia (then called Outer Mongolia), China, and the Chinese-administered territory of Inner Mongolia (Granger 1923; Andrews 1926; Colbert 1968; Lavas 1993). The ventures were promoted to the general public by AMNH zoologist Roy Chapman Andrews (1884–1960), their aim being to find early placental mammals as well as evidence of the earliest humans. Being multidisciplinary (the geology, archaeology, botany, zoology, and topography would all be studied), they were the most expensive land-based ventures organized from the United States at that time. The Central Asiatic Expeditions (CAEs), as they were known, had the blessing of AMNH paleontologist Henry Fairfield Osborn (1857–1935),

Figure 4.1. A camel train bearing supplies for the American Museum's expedition at Bayn Dzak, the "Flaming Cliffs."

Asian Dinosaur Hunters   35

who had previously proposed that Central Asia was the origin of the placental mammals, including humans.

Head paleontologist and scientific coordinator for the CAEs was leading AMNH member Walter Granger (1872–1941), who had collected in the American West (including the Como Bluff and Bone Cabin quarries) and in the Tertiary strata of Egypt (El Fayum). Other AMNH participants of note were George Olsen, Peter Kaisen, and Albert Johnson. The CAEs were supplied by camel caravans (Fig. 4.1) and used specially built Dodge cars for observation work. With their headquarters at Peking (Beijing), they made five summer forays into the Gobi between 1922 and 1930 (winter collecting in Mongolia is not possible because of heavy snows and temperatures as low as –40°C). In addition, Granger organized four winter expeditions to southern and western China.

The highlight of the expeditions came at the Flaming Cliffs of Bayn Dzak in southern Mongolia in July 1923. The Flaming Cliffs are eroded buttresses that border a topographic basin 18 kilometers (11.25 mi.) in diameter, where, in the Cretaceous sandstone, Olsen found the first dinosaur eggs to be positively identified as such (by Granger). Many more eggs were later uncovered, some preserved as large communal nests, while nearby were found over 100 skeletons and skulls of *Protoceratops* from all age groups; *Protoceratops* was the small dinosaur that was then believed to have laid the eggs (Granger and Gregory 1923). Also discovered here were the first placental mammal skulls from Cretaceous deposits, later placed in four new genera. Other dinosaur finds were the theropods *Oviraptor,*

Figure 4.2. The camp of the 1948 Russian expedition to the Nemegt Valley.

Saurornithoides, and *Velociraptor.* In addition to the discoveries at Bayn Dzak, Cretaceous dinosaurs, including *Psittacosaurus,* were found at the Oshih Basin, Iren Dabasu, and On Gong. This material, plus large quantities of Gobi Tertiary mammals also collected, is housed at the AMNH.

Postwar Russian and Polish Expeditions

Politics made further American work in the area unfeasible after 1930, and in 1941 the Russian Paleontological Institute (PIN) was asked by the Mongolians to continue fossil exploration. World War II intervened, however, but directly after the war, in 1946, the PIN sent a reconnaissance party into Mongolia with its director, Yuri Orlov, as chief advisor. Leading vertebrate paleontologist and science fiction writer Ivan A. Efremov (1907–1972) led the expeditions (Efremov 1956), along with one of Russia's best reptile taxonomists, Anatole K. Rozhdestvensky (Rozhdestvensky 1960). Staying in the field for two months, they discovered the Nemegt Valley, a huge oblong-shaped basin 180 kilometers (112 mi.) east to west and 40–70 kilometers (25–43 mi.) north to south. Many dinosaur cemeteries were located across 100 kilometers (62 mi.) of the valley, and this area subsequently became a major site for Cretaceous dinosaur excavations.

The years 1948 and 1949 saw two major PIN expeditions into Mongolia (Figs. 4.2–4.5), with fifteen scientific members in 1948 (including the paleontologist E. A. Maleev and chief geologist N. Novojilov), and a total of thirty-three members in 1949. While ZIL trucks hauled supplies and fossils, four-wheel-drive GAZ field cars were used for observation work. The expeditions covered greater distances than were possible previously, exploring in an arc more than 1,300 kilometers (812 mi.) across Mongolia (Lavas 1993). Cretaceous ankylosaurs (*Pinacosaurus, Talarurus* [Maleev 1956; Tumanova 1987]) were found at Bayn Shireh (eastern Mongolia), while Cretaceous Nemegt deposits produced more

Figure 4.3. Members of the 1948 Russian expedition at Altan Ula. *From left to right:* J. Eaglon, A. K. Rozhdestvensky, and M. Lookijnova.

Asian Dinosaur Hunters   37

Figure 4.4. Russian technician J. Eaglon excavating a skeleton of *Saurolophus* at the Dragon's Tomb in 1948.

ankylosaurs, hadrosaurs (including the giant 14-meter-long, 7.7-meter-tall [45 ft long and 25 ft tall] *Saurolophus* [Rozhdestvensky 1970]), and the equally large, fleet-footed carnivore *Tarbosaurus*, represented by different age groups and several species (some authorities suggest three genera for this grouping of larger theropods: *Maleevosaurus*, *Tarbosaurus*, and *Tyrannosaurus*).

At the "Dragon's Tomb" site (Fig. 4.4) were seven complete *Saurolophus* skeletons, as well as hadrosaur skin impressions, while other Nemegt sites produced many incomplete dinosaurs, such as sauropods,

Figure 4.5. Russian paleontologists in the field. *From left to right:* A. K. Rozhdestvensky, I. A. Efremov, and E. A. Maleev.

John R. Lavas

ornithomimosaurs, and various other small theropods. Gigantic claws, some more than 70 centimeters (28 in.) in length, were the most mysterious discoveries from this site; recent evidence suggests that they belonged to sloth-like theropods (*Therizinosaurus*), which may have used them for opening termite mounds or bending branches to reach fruits (Rozhdestvensky 1970). Many Tertiary mammals were also collected, particularly in 1949 at Altan Teli near Kobdo, western Mongolia, and at Orok Noor in central Mongolia. More than 120 tons (460 crates) of fossil specimens, mostly Nemegt dinosaurs, were recovered on the PIN expeditions; part of this treasure was given to the Mongolians to be housed at the Ulan Bator Municipal Museum. The remainder went to the PIN in Moscow, where it has been joined by many additional fossils collected in subsequent field work (including the joint Sino-Soviet Expeditions of 1959–60), necessitating the construction of another paleontological museum in southwestern Moscow in the 1970s, a complex that today houses the largest paleontological collection in the world.

This Russian work was followed by similarly successful joint Polish-Mongolian expeditions (PMEs) during 1963–65, 1967–69, and 1970–71, involving up to twenty-three members at one time (Kielan-Jaworowska 1969; Kielan-Jaworowska and Dovchin 1969; Kielan-Jaworowska and Barsbold 1972; Lavas 1993). Led by the eminent Polish paleontologist Zofia Kielan-Jaworowska from the Polish Institute of Paleobiology (later of the Paleontological Museum, University of Oslo), the PMEs had the backing of noted Polish scientist Roman Kozlowski. Other Polish paleontologists involved in this work included Magdalena Borsuk-Bialynicka, Aleksander Nowinski, Teresa Maryańska, and Halszka Osmólska (Fig. 4.6). Barsbold Rinchen (the director of the Mongolian Institute of Geological Sciences in Ulan Bator), Dashzeveg Demberlyin, and Perle Altangerel (from the same institute) participated on the Mongolian side.

The PME's highly energetic teams worked a range of localities, includ-

Figure 4.6. Members of the 1965 Polish expedition at Altan Ula. *From left to right:* M. Kuczynski, A. Nowinski, W. Skarzynski, E. Rachtan, D. Walknowski, H. Kubiak, T. Maryańska, J. Malecki, R. Gradzinski, J. Lefeld, H. Osmólska, M. Lepkowski, Z. Kielan-Jaworowska, J. Kazmierczak. and W. Sicinski.

ing Bayn Dzak (where many placental mammals were found), Toogreeg, Tertiary sites in central and western Mongolia (including Altan Teli), and many areas within the Nemegt Valley, particularly the Altan Ula excavations. These produced new sauropods (*Nemegtosaurus, Opisthocoelicaudia* [Borsuk-Bialynicka 1977]), ornithomimosaurs such as *Gallimimus* (Osmólska et al. 1972), a totally new giant type with huge three-clawed forelimbs (*Deinocheirus* [Osmólska and Roniewicz 1969]), small theropods (including coelurosaurs), hadrosaurs (*Barsboldia*), more ankylosaurs (*Saichania, Tarchia* [Maryańska 1977]), and four new genera of pachycephalosaurs (*Goyocephale, Homalocephale, Prenocephale,* and *Tylocephale* [Maryańska and Osmólska 1974]). Sedimentary deposits outside the Nemegt Valley produced new protoceratopians (*Bagaceratops, Microceratops*) from a range of individual age groups, as well as a host of Tertiary mammal remains. The PME's material was divided between the Mongolian Academy of Sciences and the Polish Paleobiological Institute in Warsaw.

Recent Gobi Expeditions

Every season since 1969, the Russian Paleontological Institute has conducted Gobi field work with Mongolian assistance, the Russian participants including paleontologists Sergei Kurzanov, K. E. Mikhailov, T. A. Tumanova, A. V. Sochava, and A. F. Bannikov (Lavas 1993). Many new Cretaceous dinosaurs have been described, including very bird-like theropods (*Avimimus* [Kurzanov 1987], *Mononykus*), ornithomimosaurs (*Anserimimus, Garudimimus, Harpymimus*), tyrannosaurs (*Alioramus*), segnosaurs (*Erlikosaurus, Segnosaurus* [Barsbold and Perle 1980]), iguanodontids (*"Iguanodon"*), hadrosaurs (*"Arstanosaurus"*), protoceratopians (*Breviceratops, Udanoceratops*), and many types of dinosaur nests. During 1993, original specimens from this field work, as well as from the Russian expeditions of the 1940s, were sent to Australia and (later) the United States as part of a comprehensive exhibition of dinosaurs and mammal-like reptiles called "The Great Russian Dinosaurs," a smaller version of which had previously toured Taiwan and Japan.

In 1990, after an absence of some sixty years, the American Museum returned to Mongolia to conduct field work, with participation by such AMNH paleontologists as Mark Norell, Malcolm McKenna, and AMNH dean of science Michael Novacek (Novacek 1996). Dashzeveg Demberlyin (who by then had some thirty years' experience collecting in the Gobi) was the primary Mongolian participant. Work was conducted at several sites, including Bayn Dzak, Toogreeg, and Khulsan, with the emphasis being on placental mammals and small theropods. Significant finds included remains of placentals, dromaeosaurs, oviraptorosaurs (including an embryonic *Oviraptor* showing that some of the Bayn Dzak eggs previously attributed to *Protoceratops* had in fact been laid by this theropod species), ankylosaurs, and more specimens of *Mononykus,* a bird-like, flightless theropod with unusual single-clawed stunted forelimbs (Perle et al. 1993).

The 1990s also saw other countries organize paleontological field work in Mongolia, such as joint French-Italian-Mongolian expeditions beginning in 1991, led by Paris Natural History Museum paleontologists Philippe Taquet and Donald Russell (Taquet 1992), and joint Japanese-Mongolian field work beginning two years later, carried out by Barsbold Rinchen and Japanese field director Mahito Watabe.

Field Work in China and Inner Mongolia

Paleontological field work in China has in most cases been undertaken with less emphasis on labor-intensive expeditions than it has in Mongolia (Spalding 1993). After several dinosaur fossils were found by foreign workers in China in 1913, sites were investigated by Swedish geologist Johann Andersson and Austrian paleontologist Otto Zdansky. After consulting with CAE member Walter Granger (who identified dinosaur remains among Zdansky's collections), Zdansky and Chinese geologist H. C. T'an excavated the Jurassic sauropod *Euhelopus* in 1922, with T'an later finding stegosaur bones as well as remains of the hadrosaur *Tanius*.

In 1926, the remains of Peking Man were found at Choukoutien (southwest of Beijing), where Chinese scientists gained field experience by working with leading anthropologists and paleontologists such as Canadian Davidson Black and the French priest Pierre Teilhard de Chardin. Notable among the Chinese workers was Yang Zhong-jian (1897–1979; known in the west as C. C. Young), who had studied in the U.S., Canada, England, and Germany, and had participated on the last (1930) CAE with Granger and Andrews after returning to his native China in 1928.

From 1927 to 1931, Swedish and Chinese scientists conducted joint field work in northwest China, led by Sven Hedin and F. Yuan. Many fragmentary dinosaurs, including pachycephalosaurs, protoceratopians, and ankylosaurs, were found, and later studied by Bohlin in the 1950s. In 1936 Yang Zhong-jian led a joint Sino-American team to Sichuan, where a skeleton of the sauropod *Omeisaurus* was excavated. By 1959, Yang was director of the Institute of Vertebrate Paleontology and Paleoanthropology (IVPP) in Beijing, which had been the leading Chinese research institute since 1949. Subsequent years saw Chinese dinosaurs of all geological ages excavated, including plateosaurs (*Lufengosaurus*), hadrosaurs (*Tsintaosaurus*, found in 1950 in Shandong Province), and *Shantungosaurus* (the longest hadrosaur at 15.5 meters in length [51.5 ft]), sauropods (*Mamenchisaurus*—the largest Asian dinosaur, possibly reaching lengths of 24 meters [78 ft]), and the stegosaur *Tuojiangosaurus*. After Yang Zhong-jian's death, Dong Zhiming, who had joined the IVPP after graduating from Fudan University in 1962, became China's leading dinosaur paleontologist.

Between 1986 and 1990, extensive multidisciplinary field work was carried out in northwest China and Inner Mongolia by the joint Canadian-Chinese Dinosaur Project (Currie et al. 1993; Currie 1996b), also known as the China–Canada–Alberta–Ex Terra expeditions (sites in Alberta and the Canadian Arctic were also worked). The Gobi field work involved up to forty participants at a time from the IVPP, the Canadian Museum of Nature (Ottawa), and the Royal Tyrrell Museum of Palaeontology (Drumheller) (Grady 1993; Spalding 1993). Canadian members included Philip Currie and Brian Noble (Drumheller), Dale Russell (Ottawa), and Tom Jerzykiewicz (Geological Survey of Canada; he had also previously served on the Polish-Mongolian expeditions), while Chinese participants included Dong Zhiming, Zheng Jiajiang, and Zhao Xi-jin. The main fossil sites were at Junggar Basin (Middle and Upper Jurassic strata), in northwestern China, and at Bayan Mandahu (Upper Cretaceous strata) and Erenhot in Inner Mongolia. More than 60 tonnes of fossils were collected, the dinosaur portion of the fauna (11 new taxa) including psittacosaurs, protoceratopians, ankylosaurs (*Pinacosaurus*), troodontids (*Sinornithoides*), megalosaurs (*Monolophosaurus*), allosaurs (*Sinraptor*), therizinosaurs (*Alxasaurus*), and many small theropods. The remains of pterosaurs, mammal-like therapsid reptiles, mammals, and turtles were also excavated. All material was divided between the IVPP and the Canadian museums. In May 1993 an exhibition containing many of the original specimens, called "Dinosaur World Tour—the Greatest Show Unearthed," began a tour of Canada and (later) Japan.

Field Work Elsewhere in Asia

Apart from the very significant discoveries in Mongolia and China, dinosaur finds have been made in other areas of Asia, particularly in the southern Asian republics of the Commonwealth of Independent States (the CIS—previously Soviet Central Asia). Russian paleontologists such as A. K. Rozhdestvensky, S. M. Kurzanov (Moscow), and the late L. A. Nessov (St. Petersburg) have worked Upper Cretaceous sites in the CIS republics of Kazakhstan, Tajikistan, and Uzbekistan, the dinosaurs found there including ceratopians (also spelled ceratopsians) (*Asiaceratops, Turanoceratops*), hadrosaurs (*Aralosaurus, Jaxartosaurus*), oviraptorosaurs (*Caenagnathasia*), tyrannosaurs (*Itemirus*), and fragmentary remains of other large theropods.

Dinosaurs are also now known from the Southeast Asian countries of Laos and Thailand. The first Laotian finds were isolated bones found in Lower Laos sediments by the French geologist J. H. Hoffet in the 1930s. In 1990 Philippe Taquet located Hoffet's localities (Taquet 1994), and the following year, a Franco-Laotian expedition collected the remains of Lower Cretaceous sauropods, iguanodontids, and theropods. The Thai dinosaurs are all from the Khorat Plateau in northeastern Thailand, and were initially discovered during mineral prospecting in the 1970s. Joint Thai-French expeditions subsequently unearthed the remains of several sauropods, a new species of *Psittacosaurus,* and a variety of theropod bones and footprints, including the earliest tyrannosaurid, *Siamotyrannus.* The finds date from the mid-Jurassic to the Early Cretaceous, and are thus by far the best dinosaur remains known from Southeast Asia (Buffetaut and Suteethorn 1993).

In Japan, dinosaur research also dates from the 1930s, with the discovery of the lambeosaur *Nipponosaurus* on Sakhalin Island, a find described by T. Nagao (Hokkaido University) in 1936. The Japanese lost these islands to Russia during World War II, and little work was conducted on Japanese dinosaurs until the 1980s, when a small number of fragmentary ornithomimosaur, sauropod, and hadrosaur finds were made, some seemingly bearing close similarity to those of China and Mongolia. As mentioned previously, however, Japanese scientists have recently been more involved with organizing joint paleontological ventures with the Mongolians in order to further exploit the extensive Gobi deposits.

Dinosaur Hunters of the Southern Continents

Thomas R. Holtz, Jr.

5

Most North Americans, Europeans, and Asians are understandably more familiar with the dinosaurs that inhabited these regions than they are with the forms that once lived in other parts of the world. With some rare exceptions (in particular, the Humboldt Museum in Berlin), southern dinosaurs are not exhibited in northern museums or featured in popular books written by northern authors. While many dinosaur enthusiasts in North America, for example, are familiar with forms such as *Ankylosaurus, Velociraptor,* and *Tyrannosaurus,* they are mostly unaware of their southern counterparts, such as *Minmi, Noasaurus,* and *Abelisaurus.* (One rare exception to this pattern is the giant sauropod *Brachiosaurus,* which is better known from African material than from the more fragmentary North American and European fossils.)

However, dinosaurs of the southern continents are very important to vertebrate paleontologists. On a simple level, they help to document the overall number of dinosaur species from across the globe at any given time period. Additionally, some fossils from the southern continents demonstrate unusual forms and adaptations unknown from northern dinosaurs, such as the brow-horns of the South American carnivore *Carnotaurus,* the bizarre vertebrae of the Australian armored dinosaur *Minmi,* and the greatly exaggerated dorsal spines of the sauropods *Rebbachisaurus* and *Amargasaurus* (of Africa and South America, respectively). Some of the largest dinosaurs known come from the south: *Brachiosaurus brancai* of Tanzania, Africa, the largest dinosaur known from relatively complete skeletons, and the gigantic *Argentinosaurus huinculensis* (known only from vertebrae and limb bones from South America) indicate the existence of sauropods which surpassed in size any known northern form. Fossils from the southern continents, South America in particular, have been critical in tracing the origin and early evolution of the Dinosauria. Finally, as discussed in later chapters, comparison of fossils from the various continents north and south has been used to trace the history and changes in diversity of Mesozoic terrestrial ecosystems.

By the term *southern continents*, scientists mean the continents and surrounding islands of South America, Africa, Australia, Antarctica, and the Indian subcontinent. The latter may seem an unusual inclusion, as it is currently part of Asia, and so might appear to be part of the northern continents (North America, Europe, and the rest of Asia). However, the great triangular landmass south of the Himalayas has been associated with the truly "southern" lands in both geologic and human history. As is well known, all the continents of Earth formed a single landmass (Pangaea) during the beginning of the Age of Dinosaurs (see also Molnar, chap. 38 of this volume). When Pangaea split up, it divided into two supercontinents, Laurasia and Gondwana (sometimes called Gondwanaland). Laurasia contained North America, Europe, and most of modern Asia, while Gondwana contained the rest of the landmasses. Thus the southern continents discussed in this section once formed a single southern supercontinent. Laurasia and Gondwana continued to break up during the Mesozoic and later Cenozoic eras, forming the modern continental configuration. Sometime during the Cenozoic, the landmass that is now India collided with Asia. The northern coast of India and the southern coast of Asia were crushed, crumpled, and raised up by the collision, forming the Himalaya plateau and the tallest mountains in the world.

The Gondwanan lands share a similar human history. Most were colonies of the European imperial powers during the past five hundred years (and some until the middle decades of this century). Vertebrate paleontology and dinosaur science historically has been predominantly a northern science, with the major research centers in the eastern part of North America and the western and central countries of Europe. During much of the past 150 years, dinosaur research in the remnants of Gondwana has been an exercise in "imperial paleontology" (Buffetaut 1987), with European explorers and scientists traveling to the south to obtain fossils for northern institutions (see Sues, chap. 2 of this volume). Consequently the present section could probably be retitled "Dinosaur Hunters *in* the Southern Continents." However, since the 1960s there has been a growth in dinosaur research by scientists native to Gondwanan continents.

Many of the southern countries are very poor relative to their northern counterparts. Many southern governments must spend most of their time and resources trying to stop the spread of disease, trying to provide enough food, water, and clothing for their people, and fighting in various civil and international conflicts, leaving little or no money or resources for scientific research. Illiteracy is a great problem among southern peoples. Under such conditions, dinosaur science is a luxury which many nations are unwilling or incapable of supporting. It is probably not coincidental that some of the best southern vertebrate paleontology comes from nations (Argentina, South Africa, Australia) which do not suffer from these problems.

Dinosaur Research In India

In 1871, a Mr. Medlicott found a broken bone measuring 117 centimeters (46 in.) in length near Jabalpur. This proved to be the femur (thighbone) of a dinosaur larger than any currently known at that time. Vertebrae of the tail of this animal were found in the same vicinity. On the basis of this material, the noted British paleontologist Richard Lydekker named the creature *Titanosaurus indicus* (1877). This fossil is notable as being the first significant dinosaur discovery of the southern continents. It comes from a family of long-necked sauropod dinosaurs (the Titanosauridae) which once dominated the Gondwanan faunas. Huene (see Sues, chap. 2 of this volume) and C. A. Matley described various Indian dinosaurs during the 1920s and 1930s (Matley 1923; Huene and Matley 1933).

Despite this auspicious beginning, dinosaur research never succeeded as well in India as on the other southern continents. Although dozens of papers by Indian geologists and paleontologists describe individual bones or teeth from across the subcontinent (Loyal et al. 1966), nothing approaching a complete skeleton has been reported. Most of these fossils are from the Cretaceous Period, and are from titanosaurids, large and small predatory dinosaurs, and one or more varieties of armored dinosaur. Notably, a great number of dinosaur eggs are known from the central Indian Lameta Formation (Vianey-Liaud et al. 1987), presumably from the titanosaurids. Indian dinosaurs from the end of the Cretaceous are associated with the Deccan Traps, a vast complex of igneous rocks that indicates that a huge region of the subcontinent was involved in volcanic activity at the end of the Age of Dinosaurs. It is fervently hoped that in upcoming years relatively complete dinosaur fossils will be found on this subcontinent to reveal the exact nature of the animals that once lived here.

An interesting side note: One paleontologist born and educated in India is now a very active dinosaur researcher in the northern continents. Sankar Chatterjee (Fig. 5.1), born in Calcutta and trained at the prestigious Indian Statistical Institute, described various Mesozoic fossils of his homeland,

Figure 5.1. Sankar Chatterjee (*on the ground at left*) in the field. Photograph courtesy of Texas Tech University News and Publications Photo Service.

including the primitive sauropod dinosaur *Barapasaurus* (with Jain and others). However, perhaps his greatest discoveries come from Upper Triassic sediments from central Texas. Since he became a professor of geology at Texas Tech University in 1979, Chatterjee has described many species of dinosaurs and other fossil reptiles from the Dockum Group.

Dinosaur Hunters in and of Africa

Much of the paleontological research in Africa has been conducted by outsiders (mostly Europeans). Some of the major discoveries and expeditions were made by European paleontologists (see Sues, chap. 2 of this volume), and so will be touched on only briefly here. However, renewed interest in African dinosaurs by African and American paleontologists has revealed new forms and localities in this continent.

Dinosaur fossils have been found over most of Africa. Indeed, some of the first fossils from the African region were actually found on the island of Madagascar. Discovered by a doctor named Félix Salètes in the northerly region of Mevarana (sometimes spelled Maevarano), these fossils were described by the French scientist Charles Depéret in 1896. They included a new species of *Titanosaurus,* as well as fragments of a large predatory dinosaur (*Majungasaurus*). Although various French scientists have reported new fragments over the past century, it was only in 1993 that a new paleontological expedition went to Madagascar with the express purpose of finding new Mesozoic fossils. Organized by Madagascan authorities and scientists from the State University of New York, these expeditions have reported the discovery of new, more complete skeletons, including at least three kinds of Mesozoic birds (Forster et al. 1996). It is hoped that continued work will help elucidate this long- but poorly known fauna.

North African countries have revealed many interesting dinosaur species, particularly from the middle of the Cretaceous Period. German expeditions under Ernst Stromer von Reichenbach to Egypt and various French expeditions under R. Lavocat (1954), Albert-Félix de Lapparent (1960), and Philippe Taquet (1976) to western North Africa (Niger) recovered tons of fossil material. It is interesting that each of these men discovered a different dinosaur with a tall dorsal (back) fin: the predatory *Spinosaurus aegyptiacus* by Stromer, the long-necked sauropods *Rebbachisaurus garasbae* by Lavocat and *R. tamesnensis* by Lapparent, and the duckbill relative *Ouranosaurus nigeriensis* by Taquet.

In addition to ongoing studies by French expeditions, North American teams have been investigating North African fossils. Recently, a multinational team led by University of Chicago dinosaur expert Paul C. Sereno recovered the fossils of several sauropod individuals, as well as two new theropods, *Afrovenator* and *Deltadromeus,* and more complete material of a third theropod, *Carcharodontosaurus,* that rivaled the North American *Tyrannosaurus* in size (Sereno et al. 1994, 1996; Currie 1996a). Preliminary studies indicate that some of these are the oldest Cretaceous dinosaurs known from Africa. New discoveries by Canadian paleontologist Dale A. Russell in Morocco may be among the youngest dinosaur fossils of Africa.

Few dinosaurs have been found in west Africa. However, several distinct footprint morphologies and scrappy bones and teeth of ornithopods, theropods, and sauropods are reported from the Koum Basin of Cameroon (Jacobs et al. 1996).

East African paleontology is well known for both early humans and dinosaurs. Separated by tens of millions of years but only some hundreds of kilometers are the early hominid fossils of Olduvai Gorge and Lake Turkana, on one hand, and the spectacular dinosaurs of Tendaguru Hill in Tanzania, on the other. Tendaguru, as described by Sues in chapter 2 of this volume, is the classic imperial paleontological dig, with hundreds of trained local workers under the supervision of a few German scientists. However, Tendaguru is not the only dinosaur locality in eastern Africa. In 1924 a farmer named E. C. Holt in what was then the Nyasaland Protectorate of the British Empire (now independent Malawi) discovered fossil bones. He reported these to F. Dixey, the director of the Geological Survey of the protectorate. First described in 1928 by S. H. Haughton (a South African paleontologist), this titanosaurid dinosaur is now known as *Malawisaurus dixeyi*. Long ignored by scientists, this and other fossils from the Middle Cretaceous of Malawi have received new attention as the result of a series of expeditions in the 1980s (Jacobs et al. 1990, 1993, 1996; Jacobs 1993). Under the direction of Louis L. Jacobs, paleontologist at Southern Methodist University and director of the Shuler Museum of Paleontology in Dallas, Texas, these expeditions have included other Americans such as Dale A. Winkler and William R. Downs as well as native Malawian scientists, in particular Zefe M. Kaufulu of the University of Malawi and Elizabeth M. Gomani of the Department of Antiquities of Malawi. The Malawi expeditions (and the recent Madagascan studies) may prove models for future paleontological expeditions in the poorer countries of the world, where the resources of richer northern institutions and the expertise of both northern and local scientists can work together to uncover new information about the past.

Southern Africa is best known for its fossils older and younger than the dinosaurs. The Karoo Beds contain some of the best Permian and Triassic therapsid ("mammal-like reptiles") fossils in the world, while sedimentary rocks from the last few million years document some of the ape-like ancestors and relatives of humans (*Australopithecus* and *Paranthropus*). However, southern African rocks also contain fossils from the beginning and end of the Jurassic period. The earliest Jurassic rocks of southern Africa have provided excellent fossils of various primitive bird-hipped dinosaurs (especially *Lesothosaurus* and *Heterodontosaurus*), sauropodomorphs (*Massospondylus* and *Vulcanodon*), and the primitive carnivore *Syntarsus rhodesiensis*. The latter was described and studied by Michael Raath of the Port Elizabeth Museum in Humewood, South Africa (Raath 1969, 1990). Like its North American relative *Coelophysis*, *Syntarsus rhodesiensis* is known from dozens of specimens. Late Jurassic fossils of southern Africa are less well known. The South African armored dinosaur *Paranthodon* and the long-necked sauropod *Algoasaurus* are preserved only as fragments, but in Zimbabwe better material of the sauropods *Brachiosaurus*, *Camarasaurus*, *Dicraeosaurus,* and *Janenschia* has been found (Raath and MacIntosh 1987).

South American Discoveries: Keys to Dinosaur Evolution

Of all the remains of Gondwana, South America has arguably proven the most productive in terms of important discoveries. What is unusual is that active programs of dinosaur research in the South American nations are young even by southern standards, even more unusual in light of numerous paleontological discoveries in that continent over the past century and a half. Darwin discovered many unusual mammal and bird fossils in his voyage in the H.M.S. *Beagle,* which were described by his colleague (and eventual foe) Sir Richard Owen. South America, in particular Argentina, produced its own internationally respected vertebrate paleontologists during the latter part of the nineteenth century. Chief among these were the brothers Florentino and Carlos Ameghino and Francisco P. Moreno. These scientists concentrated on the strange and diverse mammals that evolved in South America during its isolation as an island continent for most of the Cenozoic. Moreno reported the presence of dinosaur fossils in 1891, but little scientific work was done with these. As already discussed by Sues (see chap. 2 of this volume), Huene worked at the Museum of La Plata, Argentina, describing many new dinosaur forms during the 1920s.

Argentine paleontologists began to turn to dinosaur science during the middle part of this century. Between 1959 and 1962, various Argentine institutions (in particular the Instituto Miguel Lillo of Tucamán) recovered fossils from the Ischigualasto Basin. Among the various Triassic fossils they discovered were the remains of a primitive meat-eating dinosaur which Osvaldo A. Reig (1963) named *Herrerasaurus ischigualastensis.* An earlier (1936) expedition from the Museum of Comparative Anatomy of Harvard University to Brazil had uncovered a similar dinosaur, but it was not until 1970 that it was described and named *Staurikosaurus pricei* by the American Museum of Natural History's dinosaur expert Edwin H. Colbert. More recently, scientists from Argentina and the University of Chicago have discovered more complete skeletons of *Herrerasaurus* (Sereno and Novas 1992, 1993; Novas 1993; Sereno 1993) as well as the most primitive dinosaur known, which they have named *Eoraptor lunensis* (Sereno et al. 1993).

During the middle part of the century, expeditions to Argentina by Alfred Sherwood Romer (1894–1973) of Harvard University discovered a wealth of fossils from the dawn of the Age of Dinosaurs. Among the most unusual discoveries were the remains of the primitive relatives of dinosaurs, in particular *Lagerpeton* and *Marasuchus.* These discoveries are significant in and of themselves, but one of the greatest benefits to dinosaur research from these expeditions concerns a man rather than a fossil. A protégé of Romer during the American's southern expeditions, José F. Bonaparte, has become one of the most prolific paleontological discoverers and researchers of all time. With dozens of papers on the subject of Mesozoic vertebrates, Bonaparte ranks third among all dinosaur specialists, living or dead, in the number of valid generic names (10) he has assigned to newly discovered dinosaurs; he ranks second among living dinosaur specialists in this regard (data from Dodson and Dawson 1991; Bonaparte has added at least five more names since the time of that compilation).

Nicknamed "Master of the Mesozoic," Bonaparte has greatly expanded our knowledge of South American dinosaurs since 1970. Furthermore, he has been joined by many colleagues and students in his research, including Jamie Powell, Fernando Novas, Rudolpho Coria, Leonardo Salgado, Jorge Calvo, and Rubén Martinez. In addition to the early dinosaurs, Argentine scientists have discovered some of the best fossils from the Middle Jurassic (only poorly known in Europe, unknown in North America, and otherwise well known only from China), unusual forms from the Early Cretaceous (the horned predator *Carnotaurus* and the high-spined sauropod *Amargasaurus*), armored sauropods, and diverse species from the Late Cretaceous (including yet another size rival to *Tyrannosaurus, Giganotosaurus* [Coria and Salgado 1995], and a primitive iguanodontian [Coria and Salgado

1996]). With his student Fernando Novas, Bonaparte (Bonaparte and Novas 1985) first recognized a group of giant flesh-eaters (the Abelisauridae), to which many of the predators from other southern continents appear to belong. New and unusual fossils from South America (particularly Argentina) will no doubt continue to enlighten scientists as to the origin, history, and diversification of the Dinosauria (Kellner 1996; Novas 1996a, 1996b).

Dinosaurs Down Under: Discoveries in Australia and New Zealand

Dinosaurs were first reported in Australia in 1844, based on fragmentary fossils, but very little material has been recovered over the intervening decades. The Land Down Under has a very low population density and lacks the extensive badlands of Mesozoic sedimentary rocks with which the Americas and Asia are blessed. During the 1970s and 1980s, however, Australian dinosaur paleontologists discovered some significant material.

One of the most productive dinosaur workers of modern times is Richard Anthony Thulborn (Fig. 5.2). Tony Thulborn, a vertebrate paleontologist at the Department of Zoology at the University of Queensland, has conducted research in a variety of fields of dinosaur studies. He is one of a small group of experts on dinosaur trackways, and described (with Wade and others) the Lark Quarry tracksite. Located in western Queensland, this locality includes many types of bipedal dinosaur tracks, allowing Thulborn to study not only the diversity of organisms there but also the variety of gaits and other aspects of locomotion that the trackmakers employed (Thulborn and Wade 1984). Additionally, Thulborn has made important discoveries in the early evolution of the bird-hipped ornithischian dinosaurs, including "fabrosaurids" of southern Africa and the primitive armored *Scelidosaurus*. Furthermore, he has written important papers on the origin of birds and on the thermal biology of dinosaurs.

Other paleontologists engaged in active dinosaur research in Australia include Ralph Molnar of the Queensland Museum, Patricia Vickers-Rich of Monash University, and Patricia's husband, Thomas Rich of the Victoria Museum in Melbourne. Of particular interest is Molnar's discovery of an unusual armored dinosaur named *Minmi paravertebrata* (Molnar 1980, 1996). This animal, known from very well preserved fossils, exhibits features from most of the major groups of armored dinosaurs, as well as some unusual unique structures. Further work on it may help elucidate the relationships among the various lineages of armored dinosaurs.

Figure 5.2. R. A. (Tony) Thulborn. Photograph courtesy of Dr. Sue Turner (Mrs. R. A. Thulborn).

The main focus of the dinosaurian research of Rich and Vickers-Rich lies in the discovery of an unusual fauna of dinosaurs from the Middle Cretaceous. It is not the individual dinosaurs themselves that are so unusual, as they are types (small bipedal herbivores called hypsilophodontids, predatory allosaurids, etc.) which are known from other regions at that same time. However, the fossils from Dinosaur Cove in southern Victoria were from animals that lived within the Cretaceous Antarctic Circle (Rich et al. 1988; Rich and Rich 1989; Rich 1996; Vickers-Rich 1996). Even though dynamite was needed to get the fossils out of the hard rock, the bones found showed a diversity of organisms and a high quality of preservation, well worth the effort. These fossils, and others from the

Dinosaur Hunters of the Southern Continents

southern coast of Australia, will help scientists understand the adaptations which dinosaurs used to survive the long (but perhaps not frigid) Cretaceous polar winters.

Although most of the work on dinosaurs in the Australian region has been based on that continent, new discoveries are coming to light in neighboring New Zealand. In a recent review (Molnar and Wiffen 1994), at least five species of dinosaur were recognized from fossils collected on North Island. These included the fragmentary remains of sauropods, theropods, an ankylosaur, and a *Dryosaurus*-like ornithopod. Almost all of these were found by Joan Wiffen, called by some "the Dragon Lady of New Zealand." These fossils hint at the potential of even more dinosaur skeletons waiting to be discovered in this island nation (Wiffen 1996).

The Frozen Lands: Dinosaur Hunting in Antarctica

Antarctica is unlike any other continent on Earth. It is the only major landmass that does not have an indigenous people, and indeed lacks a native terrestrial vertebrate fauna. Much of the continent is covered under hundreds to thousands of meters of glacial ice.

However, these icy conditions began only relatively recently in geologic time. Throughout the Mesozoic, terrestrial animals and plants flourished in Antarctica, which was then united with South America, Africa, India, and especially Australia. Indeed, Mesozoic vertebrate fossils have been known from this southernmost land since Peter Barrett, a New Zealand geologist, discovered a fragment of an Early Triassic amphibian in 1967 (Barrett et al. 1968). It would be two decades, however, before the first Antarctic dinosaurs were found.

Fossil collecting in Antarctica is very difficult, because of both the extreme cold and the lack of extensive outcrop. Dozens of previously unknown dinosaur species may be preserved as fossils in inland Antarctica, but because of the ice cap we may never be able to reach them. Only in those few places where Mesozoic rocks break through the ice on the coastline and islands around Antarctica proper can dinosaur fossils be found.

The first Antarctic dinosaur, a Late Cretaceous armored dinosaur, was discovered in the Santa Maria Formation in 1986 by scientists of the Instituto Antartico Argentino (Gasparini et al. 1987, 1996; Olivero et al. 1991). A medium-sized bipedal herbivore was found in rocks of about the same age shortly afterward by British scientists (Milner et al. 1992). Most recently, unusual Early Jurassic species were found deeper within the Antarctic interior by American scientist William Hammer (Hammer and Hickerson 1993, 1994). Among these Jurassic discoveries is a strange predatory dinosaur with a crest that is perpendicular, rather than parallel, to the long axis of the skull. This bizarre animal, named *Cryolophosaurus ellioti*, appears to be related to the younger allosaurids and sinraptorids of North America, Europe, and Asia. It is hoped that future expeditions to these frozen wastes might reveal what relationships these dinosaurs of the deepest south have with those of other parts of the world.

For most of history, dinosaur research in the southern continents has been conducted by Europeans and North Americans who were after fossils for museums in their native lands. The past thirty years or so have seen a rise in local, southern paleontologists interested in the "fearfully great reptiles." In cooperation with scientists from the north, southern paleontologists have added greatly to our understanding of the origin and history of the dinosaurs and their distributions and diversity through time, and have given us a better understanding of the features, adaptations, and habitats of these ancient creatures. In the next several decades, great new discoveries are expected by dinosaur hunters of the southern continents.

References to Part One

Abel, O. 1931. Louis Dollo. 7. Dezember 1857–19. April 1931. *Palaeobiologica* 4: 321–344.

Abel, O. 1939. *Vorzeitliche Tierreste im deutschen Mythus, Brauchtum und Volksglauben.* Jena, Germany: Fischer.

Adams, F. D. 1938. *The Birth and Development of the Geological Sciences.* Baltimore: Williams and Wilkins. Reprint, New York: Dover Books, 1954.

Andrews, R. C. 1926. *On the Trail of Ancient Man.* New York: G. P. Putnam's Sons.

Andrews, R. C. 1932. *The New Conquest of Central Asia: A Narrative of the Explorations of the Central Asiatic Expeditions in Mongolia and China, 1921/ 30.* New York: American Museum of Natural History.

Anonymous. 1824. Organic remains. *New Monthly Magazine* 12 (December): 575.

Barrett, P. J.; R. J. Braillie; and E. C. Colbert. 1968. Triassic amphibian from Antarctica. *Science* 161: 460–462.

Barsbold, R., and A. Perle. 1980. Segnosauria, a new infraorder of carnivorous dinosaurs. *Acta Palaeontologica Polonica* 25: 187–195.

Bassett, M. G. 1971. "Formed stones," folklore and fossils. *Amgueddfa: Bulletin of the National Museum of Wales* 7: 1–17.

Bird, R. T. 1985. *Bones for Barnum Brown: Adventures of a Dinosaur Hunter.* Fort Worth: Texas Christian University Press.

Blows, W. T. 1983. William Fox (1813–1881), a neglected dinosaur collector of the Isle of Wight. *Archives of Natural History* 11: 299–313.

Bonaparte, J. F., and F. E. Novas. 1985. *Abelisaurus comahuensis,* n. g., n. sp., Carnosauria del Cretácico Tardio de Patagonia. *Ameghiniana* 21: 256–265.

Borsuk-Bialynicka, M. 1977. A new camarasaurid sauropod *Opisthocoelicaudia skarzynskii* gen. n., sp. n. from the Upper Cretaceous of Mongolia. *Palaeontologica Polonica* 37: 5–64.

Branca, W. 1914. Allgemeines über die Tendaguru-Expedition. *Archiv für Biontologie* 3: 1–13.

Brookes, R. 1763. *The Natural History of Waters, Earths, Stones, Fossils, and Minerals, with their Virtues, Properties, and Medicinal Uses: To Which is added, The Method in which Linnaeus has treated these Subjects.* Vol. 5. London: Newberry. 5th ed., 1772.

Buckland, W. 1824. Notice on the Megalosaurus, or Great Fossil Lizard of Stonesfield. *Transactions of the Geological Society of London* 1 (series 2): 390–396.

Buckland, W. 1829. On the Discovery of the Bones of the Iguanodon, and Other Large Reptiles, in the Isle of Wight and Isle of Purbeck. *Proceedings of the Geological Society of London* 1: 159–160.

Buffetaut, E. 1979. A propos du reste de dinosaurien le plus anciennement décrit: L'interprétation de J. B. Robinet (1768). *Histoire et Nature* 14: 79–84.

Buffetaut, E. 1987. *A Short History of Vertebrate Paleontology.* London: Croom Helm.

Buffetaut, E.; G. Cuny; and J. Le Loeuff. 1993. The discovery of French dinosaurs. *Modern Geology* 18: 161–182. (Republished in W. A. S. Sarjeant [ed.], *Vertebrate Fossils and the Evolution of Scientific Concepts* [Reading, England: Gordon and Breach, 1995], pp. 159–180.)

Buffetaut, E., and V. Suteethorn. 1993. The dinosaurs of Thailand. *Journal of Southeast Asian Earth Sciences* 8: 77–82.

Casier, E. 1960. *Les Iguanodons de Bernissart.* Brussels: Editions du Patrimoine de l'Institut Royal des Sciences Naturelles de Belgique.

Caumont, A. de. 1828. *Essai sur la topographie géognostique du département du Calvados.* Caen, France: Chalopin.

Cleevely, R. J., and S. D. Chapman. 1992. The accumulation and disposal of Gideon Mantell's collections and their role in the history of British palaeontology. *Archives of Natural History* 19: 307–364.

Colbert, E. H. 1968. *Men and Dinosaurs: The Search in Field and Laboratory.* New

York: E. P. Dutton. (Reprinted as *The Great Dinosaur Hunters and Their Discoveries* [New York: Dover Publications, 1984], with a new preface by the author.)

Colbert, E. H. 1970. A saurischian dinosaur from the Triassic of Brazil. *American Museum Novitates* 2181: 1–24.

Colbert, E. H. 1983. *Dinosaurs: An Illustrated History.* Maplewood, N.J.: Hammond.

Colbert, E. H. 1995. *The Little Dinosaurs of Ghost Ranch.* New York: Columbia University Press.

Conybeare, W. D. 1821. Notice of the Discovery of a New Fossil Animal, Forming a Link between the Ichthyosaurus and Crocodile, together with General Remarks on the Osteology of the Ichthyosaurus. *Transactions of the Geological Society of London* 5 (series 1): 559–594.

Cooper, J. A. 1992. The life and work of George Bax Holmes (1803–1887) of Horsham, Sussex: A Quaker vertebrate fossil collector. *Archives of Natural History* 19: 379–400.

Cooper, J. A. 1993. George Bax Holmes (1803–1887) and his relationship with Gideon Mantell and Richard Owen. *Modern Geology* 18: 183–208.

Coria, R. A., and L. Salgado. 1995. A new giant carnivorous dinosaur from the Cretaceous of Patagonia. *Nature* 377: 224–226.

Coria, R. A., and L. Salgado. 1996. A basal iguanodontian (Ornithischia: Ornithopoda) from the Late Cretaceous of South America. *Journal of Vertebrate Paleontology* 16: 445–457.

Currie, P. J. 1996a. Out of Africa: Meat-eating dinosaurs that challenge *Tyrannosaurus rex. Science* 272: 971–972.

Currie, P. J. (ed.). 1996b. Results from the Sino-Canadian Dinosaur Project, Part 2. *Canadian Journal of Earth Sciences* 33 (4): 511–648.

Currie, P. J.; Z.-M. Dong; and D. A. Russell (eds.). 1993. Results from the Sino-Canadian Dinosaur Project. *Canadian Journal of Earth Sciences* 30 (10–11): 1997–2272.

Cuvier, G. 1808. Sur des ossemens fossiles de crocodiles, et particulièrement sur ceux des environs du Havre et de Honfleur, avec des remarques sur les squelettes des sauriens de la Thuringe. *Annales Muséum National d'Histoire Naturelle Paris* 12: 73–110.

Cuvier, G. 1824. *Recherches sur les ossemens fossiles du quadrupèdes.* Revised ed. 6 vols. Paris: Dulour et d'Ocagne.

Cuvier, G. 1825. *Recherches sur les ossemens fossiles.* Vol. 5, pt. 2. 3rd ed. Paris: G. Dulour et E. d'Ocagne, Libraires.

Dean, D. R. 1993. Gideon Mantell and the discovery of *Iguanodon. Modern Geology* 18: 209–219. (Republished in W. A. S. Sarjeant [ed.], *Vertebrate Fossils and the Evolution of Scientific Concepts* [Reading, England: Gordon and Breach, 1995], pp. 207–218.)

Dean, D. R. In press. *Gideon Mantell and the Discovery of Dinosaurs.* Cambridge: Cambridge University Press.

De Beer, G. (ed.). 1974. *Charles Darwin–Thomas Henry Huxley Autobiographies.* Oxford: Oxford University Press.

Delair, J. B., and W. A. S. Sarjeant. 1975. The earliest discoveries of dinosaurs. *Isis* 66: 5–25.

Depéret, C. 1896. Note sur les dinosauriens sauropodes et théropodes du Crétacé Supérieur de Madagascar. *Bulletin de Societe Géologique du France, série 3* 24: 176–196.

Desmond, A. 1982. *Archetypes and Ancestors: Palaeontology in Victorian London 1850–1875.* Chicago: University of Chicago Press.

Dodson, P., and S. D. Dawson. 1991. Making the fossil record of dinosaurs. *Modern Geology* 16: 3–15.

Dollo, L. 1910. La paléontologie éthologique. *Bulletin de la Société belge de Géologie, de Paléontologie et d'Hydrologie* 23: 377–421.

Dong Z.-M. 1988. *Dinosaurs from China.* London: British Museum (Natural History), and Beijing: China Ocean Press.

Edmonds, J. M. 1979. The founding of the Oxford Readership in Geology, 1818. *Notes and Records of the Royal Society of London* 30: 141–167.

Efremov, I. A. 1956. *Road of the Wind*. Moscow: All Union Scientific Publishers. (In Russian.)

Englefield, H. C. 1816. *A description of the principal picturesque beauties, antiquities, and geological phoenomena, of the Isle of Wight. . . . With additional observations on the strata of the island, and their continuation in the adjacent parts of Dorsetshire*. London: Payne and Foss.

Eudes-Deslongchamps, J. A. 1838. Mémoire sur le *Poekilopleuron Bucklandii*, grand saurien fossile intermédiaire entre les crocodiles et les lézards. *Mémoires de la Société Linnéenne de Normandie (Calvados)* 6: 37–146.

Fitton, W. H. 1824. Enquiries Respecting the Geological Relations of the Beds Between the Chalk and the Purbeck Limestone in the South-east of England. *Annals of Philosophy* 8: 365–383.

Forster, C. A.; L. M. Chiappe; D. W. Krause; and S. D. Sampson. 1996. The first Mesozoic avifauna from eastern Gondwana. *Journal of Vertebrate Paleontology* 16 (Supplement to no. 3): 34A.

Galton, P. 1976. Prosauropod dinosaurs [Reptilia: Saurischia] of North America. *Postilla* 169: 1–98.

Gasparini, Z.; E. Olivero; R. Scasso; and C. Rinaldi. 1987. Un ankylosaurio (Reptilia, Ornithischia) campaniano en el continente antartico. *Anais do X Congreso Brasileiro de Paleontologia* 1: 131–141.

Gasparini, Z.; X. Pereda-Superbiola; and R. E. Molnar. 1996. New data on the ankylosaurian dinosaur from the Late Cretaceous of the Antarctic Peninsula. *Memoirs of the Queensland Museum* 39: 583–594.

Gillette, D. D. 1994. *Seismosaurus, the Earth Shaker*. New York: Columbia University Press.

Gould, S. J. 1970. Dollo on Dollo's Law: Irreversibility and the status of evolutionary laws. *Journal of the History of Biology* 3: 189–212.

Grady, W. 1993. *The Dinosaur Project: The Story of the Greatest Dinosaur Expedition Ever Mounted*. Edmonton: The Ex Terra Foundation; Toronto: MacFarlane and Ross.

Granger, W. 1923. Paleontological discoveries of the Third Asiatic Expedition. *Bulletin of the Geological Society of China* 2: 105–108.

Granger, W., and W. K. Gregory. 1923. *Protoceratops andrewsi*, a pre-ceratopsian dinosaur from Mongolia. American Museum *Novitates* 72: 1–9.

Halstead, L. B. 1970. *Scrotum humanum* Brookes 1763: The first named dinosaur. *Journal of Insignificant Research* 5: 14–15.

Halstead, L. B., and W. A. S. Sarjeant. 1993. *Scrotum humanum* Brookes: The earliest name for a dinosaur? *Modern Geology* 18: 221–224. (Republished in W. A. S. Sarjeant [ed.], *Vertebrate Fossils and the Evolution of Scientific Concepts* [Reading, England: Gordon and Breach, 1995], pp. 219–222.)

Hammer, W. R., and W. J. Hickerson. 1993. A new Jurassic dinosaur fauna from Antarctica. *Journal of Vertebrate Paleontology* 13 (Supplement to no. 3): 40A.

Hammer, W. R., and W. J. Hickerson. 1994. A crested theropod dinosaur from Antarctica. *Science* 264: 828–830.

Haughton, S. H. 1928. On some reptilian remains from the dinosaur beds of Nyassaland. *Transactions of the Royal Society of South Africa* 16: 67–75.

Hooke, R. 1705. *The posthumous works of Robert Hooke . . . containing his Cutlerian lectures, and other discourses, read at the meetings of the illustrious Royal Society*. London: Waller.

Horner, J. R., and J. Gorman. 1988. *Digging Dinosaurs*. New York: Workman Publishing Co.

Huene, F. von. 1929. Los saurisquios y ornitisquios del Cretáceo Argentino. *Anales del Museo de La Plata* 3 (2): 1–196 + atlas of 44 plates.

Huene, F. von. 1932. Die fossile Reptil-Ordnung Saurischia, ihre Entwicklung und Geschichte. *Monographien für Geologie und Paläontologie* (1) 4: 1–361 + atlas of 56 plates.

Huene, F. von. 1944. *Arbeitserinnerungen*. Halle: Kaiserlich Leopoldinisch-Carolinisch Deutsche Akademie der Naturforscher.

Huene, F. von, and C. A. Matley. 1933. The Cretaceous Saurischia and Ornithischia of the central provinces of India. *Palaeontologia Indica*, n.s. 21 (1): 1–74.

Jacobs, L. L. 1993. *Quest for the African Dinosaurs: Ancient Roots of the Modern World*. New York: Villard Books.

Jacobs, L. L.; D. A. Winkler; W. R. Downs; and E. M. Gomani. 1993. New material of an Early Cretaceous titanosaurid sauropod from Malawi. *Palaeontology* 36: 523–534.

Jacobs, L. L.; D. A. Winkler; and E. M. Gomani. 1996. Cretaceous dinosaurs of Africa: Examples from Cameroon and Malawi. *Memoirs of the Queensland Museum* 39: 595–610.

Jacobs, L. L.; D. A. Winkler; Z. M. Kaufulu; and W. R. Downs. 1990. The dinosaur beds of northern Malawi, Africa. *National Geographic Research* 6: 196–204.

Jain, S. L.; T. S. Kutty; T. Roy-Chowdhury; and S. Chatterjee. 1975. The sauropod dinosaur from the Lower Jurassic of Deccan, India. *Proceedings of the IV International Gondwana Symposium, Calcutta* 1: 221–228.

Kellner, A. W. A. 1996. Remarks on Brazilian dinosaurs. *Memoirs of the Queensland Museum* 39: 611–626.

Kielan-Jaworowska, Z. 1969. *Hunting for Dinosaurs*. Cambridge, Mass.: MIT Press.

Kielan-Jaworowska, Z. 1975. Late Cretaceous mammals and dinosaurs from the Gobi Desert. *American Scientist* 63: 150–159.

Kielan-Jaworowska, Z., and Barsbold R. 1972. Narrative of the Polish-Mongolian Palaeontological Expeditions, 1967–1971. *Palaeontologica Polonica* 27: 5–13.

Kielan-Jaworowska, Z., and N. Dovchin. 1969. Narrative of the Polish-Mongolian Palaeontological Expeditions, 1963–1965. *Palaeontologica Polonica* 19: 7–32.

Kirchner, H. 1941. Versteinerte Reptilfährten als Grundlage für ein Drachenkampf in einem Heldenlied. *Zeitschrift der Deutschen Geologischen Gesellschaft* 93: 309.

Kurzanov, S. M. 1987. Avimimidae and the problem of the origin of birds. *Sovmestnaia Sovetsko-mongol'skaia paleontologicheskaia ekspeditsiia. Trudy* 31: 5–95. (In Russian.)

Lambrecht, K. 1933. Franz Baron Nopcsa +, der Begründer der Paläophysiologie, 3. Mai 1877 bis 25. April 1933. *Paläontologische Zeitschrift* 15: 201–222.

Lanham, U. 1973. *The Bone Hunters*. New York: Columbia University Press.

Lapparent, A. F. de. 1960. Les Dinosauriens du «Continental intercalaire» du Sahara central. *Mémoires de la Société Géologique de France, Nouvelle Série* 39: 3–57.

Lavas, J. R. 1985. Bibliographical review of the Subclass Archosauria (Class: Reptilia), 1960 to 1984. Unpublished manuscript.

Lavas, J. R. 1993. *Dragons from the Dunes: The Search for Dinosaurs in the Gobi Desert*. P.O. Box 14–421, Panmure, Auckland 6, New Zealand.

Lavocat, R. 1954. Sur les Dinosauriens du continental intercalaire des Kem-Kem de la Daoura. *Comptes Rendus de Dix-Neuvième International Géologique 1952* 195–206.

Lennier, G. 1887. Etudes paléontologiques. Description des fossiles du Cap de la Hève. *Bulletin de la Société Géologique de Normandie* 12: 17–98.

Lessen, D. 1992. *Kings of Creation*. New York: Simon and Schuster.

Lhuyd, E. 1699. *Lithophylacii Britannici Ichnographia, sive, lapidum aliorumque Fossilium Britannicorum singulari figura insignium; quotquot hactenus vel ipse invenit vel ab amicis accepit, Distributio classica, Scrinii sui lapidarii repertorium cum locis singulorum natalibus exhibens. Additis raritorum aliquot figuris aere incisis; cum Epistolis ad Clarissimos Viros de quibusdam circa marina Fossilia & stirpes minerales praesertim notandis*. London: Lipsiae, Gleditsch and Weidmann.

Ligabue, G. (ed.). 1984. *Sulle Orme dei Dinosauri*. Esplorazioni e Richerche, vol. 9. Rome and Venice: Erizzo for Le Società del Gruppo ENI.

Lister, M. 1671. Fossil shells in several places of England. *Philosophical Transactions of the Royal Society of London* 6: 2282.

Loyal, R. S.; A. Khosla; and A. Sahni. 1996. Gondwanan dinosaurs of India: Affinities and palaeobiogeography. *Memoirs of the Queensland Museum* 39: 627–638.

References to Part One   55

Lydekker, R. 1877. Notices of new and other Vertebrata from Indian Tertiary and
Secondary Rocks. *Records of the Geological Society of India* 10: 30–43.

Maleev, E. A. 1956. Armoured dinosaurs from the Upper Cretaceous of Mongolia.
Trudy Paleontologicheskogo instituta Akademii Nauk S.S.S.R. 62: 51–92. (In
Russian.)

Mantell, G. A. 1822. *The Fossils of the South Downs; or, Illustrations of the
Geology of Sussex.* London: Relfe.

Mantell, G. A. 1825. Notice on the Iguanodon, a Newly Discovered Fossil Reptile,
from the Sandstone of Tilgate Forest, in Sussex. *Philosophical Transactions of
the Royal Society of London* 115: 179–186.

Mantell, G. A. 1833. *The Geology of the South-East of England.* London: Longman.

Mantell, G. A. 1850. *A Pictorial Atlas of Fossil Remains consisting of coloured
illustrations selected from Parkinson's "Organic Remains of a Former World"
and Artis's "Antediluvian Phytology."* London: Bohn.

Mantell, G. A. 1851. *Petrifactions and their teachings; or, A hand-book to the
gallery of organic remains of the British Museum.* London: Henry G. Bohn.

Maryańska, T. 1977. Ankylosauridae (Dinosauria) from Mongolia. *Palaeontologica
Polonica* 37: 85–151.

Maryańska, T., and H. Osmólska. 1974. Pachycephalosauria, a new suborder of
ornithischian dinosaurs. *Palaeontologica Polonica* 30: 45–102.

Matheron, P. 1869. Notice sur les reptiles fossiles des dépôts fluvio-lacustres
crétacés du bassin à lignite de Fuveau. *Mémoires de l'Académie des Sciences,
(Belles-) Lettres et (Beaux-) Arts de Marseille* 1868–69: 345–379.

Matley, C. A. 1923. The Cretaceous dinosaurs of the Trichinopoly district and the
rocks associated with them. *Record of the Geological Society of India* 61: 337–
349.

Meyer, H. von. 1837. Mitteilung an Prof. Bronn (*Plateosaurus engelhardti*). *Neues
Jahrbuch für Mineralogie, Geologie und Paläontologie* 1837: 317.

Milner, A. C.; J. J. Hooker; and S. E. K. Sequeira. 1992. An ornithopod dinosaur
from the Upper Cretaceous of the Antarctic Peninsula. *Journal of Vertebrate
Paleontology* 12 (Supplement to no. 3): 44A.

Molnar, R. E. 1980. An ankylosaur (Ornithischia: Reptilia) from the Lower
Cretaceous of southern Queensland. *Memoirs of the Queensland Museum* 20:
77–87.

Molnar, R. E. 1996. Preliminary report on a new ankylosaur from the Early
Cretaceous of Queensland, Australia. *Memoirs of the Queensland Museum* 39:
653–668.

Molnar, R. E., and J. Wiffen. 1994. A Late Cretaceous polar dinosaur fauna from
New Zealand. *Cretaceous Research* 15: 689–706.

Moreno, F. P. 1891. Reseña general de las adquisiciones y trabajos bechos en 1889
en el Museo de La Plata. *Revista de Museo de La Plata* 58–70.

Murchison, R. I. 1826. Geological Sketch of the North-Western Extremity of
Sussex, and the Adjoining Parts of Hants. and Surrey. *Transactions of the
Geological Society of London* (Series 2) 2: 103–106.

Nopcsa, F. von. 1900. Dinosaurierreste aus Siebenbürgen. Schädel von *Limnosaurus
transsylvanicus* nov. gen. et spec. *Denkschriften der kaiserlichen Akademie der
Wissenschaften Wien, mathematisch-naturwissenschaftliche Classe* 68: 555–
591.

Norell, M. A.; J. M. Clark; Dashzeveg D.; Barsbold R.; L. M. Chiappe; A. R.
Davidson; M. C. McKenna; Perle A.; and M. J. Novacek. 1994. A theropod
dinosaur embryo and the affinities of the Flaming Cliffs dinosaur eggs. *Science*
266: 779–782.

Norman, D. B. 1993. Gideon Mantell's "Mantel-Piece": The earliest well-preserved
ornithischian dinosaur. *Modern Geology* 18: 225–246. (Republished in W. A.
S. Sarjeant [ed.], *Vertebrate Fossils and the Evolution of Scientific Concepts*
[Reading, England: Gordon and Breach, 1995], pp. 223–243.)

Novacek, M. 1996. *Dinosaurs from the Flaming Cliffs.* New York: Anchor Books/
Doubleday.

Novas, F. E. 1993. New information on the systematics and postcranial skeleton of
Herrerasaurus ischigualastensis (Theropoda: Herrerasauridae) from the Ischi-

gualasto Formation (Upper Triassic) of Argentina. *Journal of Vertebrate Paleontology* 13: 425–450.

Novas, F. E. 1996a. Alvarezsauridae, Cretaceous basal birds from Patagonia and Mongolia. *Memoirs of the Queensland Museum* 39: 675–702.

Novas, F. E. 1996b. Dinosaur monophyly. *Journal of Vertebrate Paleontology* 16: 723–741.

Oakley, K. P. 1965. Folklore of fossils. *Antiquity* 39: 9–16, 117–125.

Olivero, E. B.; Z. Gasparini; C. A. Rinaldi; and R. Scasso. 1991. First record of dinosaurs in Antarctica (Upper Cretaceous, James Ross Island): Paleogeographical implications. In M. R. A. Thomason, J. A. Crane, and J. W. Thomson (eds.), *Geological Evolution of Antarctica: Proceedings of the Fifth International Symposium on Antarctic Earth Sciences* (Cambridge: British Antarctic Survey), pp. 617–622.

Osborn, H. F. 1931. *Cope: Master Naturalist.* Princeton, N.J.: Princeton University Press.

Osmólska, H., and E. Roniewicz. 1969. Deinocheiridae, a new family of theropod dinosaurs. *Palaeontologica Polonica* 21: 5–22.

Osmólska, H.; E. Roniewicz; and R. Barsbold. 1972. A new dinosaur (*Gallimimus bullatus* n. gen., n. sp. (Ornithomimidae)) from the Upper Cretaceous of Mongolia. *Palaeontologica Polonica* 27: 103–143.

Owen, R. 1841. A Description of a Portion of the Skeleton of the Cetiosaurus, a Gigantic Extinct Saurian Occurring in the Oolitic Formations of Different Parts of England. *Proceedings of the Geological Society of London* 3: 457–462.

Owen, R. 1842. Report on British fossil reptiles. *Reports of the British Association for the Advancement of Science* 11: 60–204.

Parkinson, J. 1822. *Outlines of Oryctology: An Introduction to the Study of Fossil Organic Remains, especially of Those Found in the British Strata; intended to aid the student in his enquiries respecting the nature of fossils, and their connection with the formation of the earth.* London: Published by the author.

Parkinson, J. 1930. *The Dinosaur in East Africa.* London: H. F. and G. Witherby.

Perle A.; M. A. Norell; L. M. Chiappe; and J. M. Clark. 1993. Flightless bird from the Cretaceous of Mongolia. *Nature* 362: 623–626.

Phillips, J. 1871. *Geology of Oxford and the Valley of the Thames.* Oxford, England: Clarendon Press.

Plate, R. 1964. *The Dinosaur Hunters: Othniel C. Marsh and Edward D. Cope.* New York: David McKay.

Platt, J. 1758. An Account of the Fossile Thigh-Bone of a Large Animal, dug up at Stonesfield near Woodstock, In Oxfordshire. *Philosophical Transactions of the Royal Society of London* 50: 524–527.

Plot, R. 1677. *The Natural History of Oxfordshire, Being an Essay toward the Natural History of England.* Oxford, England: Published by the author. 2nd ed., London: Brome, 1705.

Raath, M. A. 1969. A new coelurosaurian dinosaur from the Forest Sandstone of Rhodesia. *Arnoldia* 4: 1–25.

Raath, M. A. 1990. Morphological variation in small theropods and its meaning in systematics: Evidence from *Syntarsus rhodesiensis.* In P. Currie and K. Carpenter (eds.), *Dinosaur Systematics: Approaches and Perspectives* (Cambridge: Cambridge University Press), pp. 91–105.

Raath, M. A., and J. S. McIntosh. 1987. Sauropod dinosaurs from the Central Zambezi Valley, Zimbabwe, and the age of the Kadzi Formation. *South African Journal of Geology* 90: 107–119.

Reig, O. A. 1963. La presencia de dinosaurios saurisquios en los "Estratos de Ischigualasto" (Mesotriássico superior) de las Provincias de San Juan y La Rioja (Republica Argentina). *Ameghiniana* 3: 3–20.

Rich, P. V.; T. H. Rich; B. E. Wagstaff; J. McEwen Mason; C. B. Douthitt; R. T. Gregory; and E. A. Felton. 1988. Evidence for low temperatures and biologic diversity in Cretaceous high latitudes of Australia. *Science* 242: 1403–1406.

Rich, T. 1996. Significance of polar dinosaurs in Gondwana. *Memoirs of the Queensland Museum* 39: 711–717.

Rich, T. H. V., and P. V. Rich. 1989. Polar dinosaurs and biotas of the Early Cretaceous of southeastern Australia. *National Geographic Research* 5: 15–53.

Riley, S. H., and S. Stutchbury. 1836. A description of various fossil remains of three distinct saurian animals discovered in the autumn of 1834, in the Magnesian Conglomerate on Durdham Down, near Bristol. *Proceedings of the Geological Society of London* 2: 397–399.

Riley, S. H., and S. Stutchbury. 1840. A description of various fossil remains of three distinct saurian animals discovered in the Magnesian Conglomerate near Bristol. *Transactions of the Geological Society of London* 5 (series 2): 349–357.

Robinet, J.-B. 1768. *Considérations philosophiques de la gradation naturelle des formes de l'être, ou les essais de la Nature qui apprend à faire l'Homme.* Paris: Saillant.

Rozhdestvensky, A. K. 1957. Duckbilled dinosaurs: *Saurolophus*—from the Upper Cretaceous of Mongolia. *Vertebrata PalAsiatica* 1 (2): 129–149. (In Russian.)

Rozhdestvensky, A. K. 1960. *Chasse aux Dinosaures dans le Desert de Gobi.* Paris: Fayard.

Rozhdestvensky, A. K. 1970. Giant claws of enigmatic Mesozoic reptiles. *Paleontological Journal* 4: 117–125.

Rudwick, M. J. S. 1972. *The Meaning of Fossils: Episodes in the History of Palaeontology.* New York: American Elsevier.

Sarjeant, W. A. S. 1987. The study of fossil vertebrate footprints: A short history and selective bibliography. In G. Leonardi (ed.), *Glossary and Manual of Tetrapod Footprint Palaeoichnology* (Brasília: Departamento Nacional da Produção Mineral, República Federation do Brasil), pp. 1–19.

Sarjeant, W. A. S., and J. B. Delair. 1980. An Irish naturalist in Cuvier's laboratory: The letters of Joseph Pentland, 1820–1832. *Bulletin British Museum (Natural History),* Historical series 6: 245–319.

Schuchert, C., and C. M. LeVene. 1940. *O. C. Marsh, Pioneer in Paleontology.* New Haven, Conn.: Yale University Press.

Sedgwick, A. 1822. On the Geology of the Isle of Wight. *Annals of Philosophy,* n.s. 3: 329–335.

Seeley, H. G. 1887. On the classification of the fossil animals commonly named Dinosauria. *Proceedings of the Royal Society of London* 43: 165–171.

Sereno, P. C. 1993. The pectoral girdle and forelimb of the basal theropod *Herrerasaurus ischigualastensis. Journal of Vertebrate Paleontology* 13: 425–450.

Sereno, P. C.; D. B. Dutheil; M. Iarochene; H. C. E. Larsson; G. H. Lyon; P. M. Magwene; C. A. Sidor; D. J. Varricchio; and J. A. Wilson. 1996. Predatory dinosaurs from the Sahara and Late Cretaceous faunal differentiation. *Science* 272: 986–991.

Sereno, P. C.; C. A. Forster; R. R. Rogers; and A. M. Monetta. 1993. Primitive dinosaur skeleton from Argentina and the early evolution of Dinosauria. *Nature* 361: 64–66.

Sereno, P. C., and F. E. Novas. 1992. The complete skull and skeleton of an early dinosaur. *Science* 258: 1137–1140.

Sereno, P. C., and F. E. Novas. 1993. The skull and neck of the basal theropod *Herrerasaurus ischigualastensis. Journal of Vertebrate Paleontology* 13: 451–476.

Sereno, P. C.; J. A. Wilson; H. C. E. Larsson; D. B. Dutheil; and H.-D. Sues. 1994. Early Cretaceous dinosaurs from the Sahara. *Science* 266: 267–271.

Shor, E. N. 1974. *The Fossil Feud between E. D. Cope and O. C. Marsh.* Hicksville, N.Y.: Exposition Press.

Simpson, G. G. 1942. The beginnings of vertebrate palaeontology in North America. *Proceedings of the American Philosophical Society* 85: 130–188.

Smith, N. 1820. Fossil Bones Found in Red Sandstone. *American Journal of Science* 2: 146–147.

Spalding, D. A. E. 1993. *Dinosaur Hunters: 150 Years of Extraordinary Discoveries.* Toronto: Key Porter Books. Rocklin, Calif.: Prima Books, 1995.

Sternberg, C. H. 1909. *The Life of a Fossil Hunter.* New York: Henry Holt and Co. Reprint, Bloomington: Indiana University Press, 1990.

58   **References to Part One**

Sternberg, C. H. 1932. *Hunting Dinosaurs in the Badlands of the Red Deer River, Alberta, Canada.* San Diego: Published by the author.

Strickland, H. E. 1848. On the Geology of the Oxford and Rugby Railway. *Proceedings of the Ashmolean Society* 1848: 192.

Swinton, W. E. 1962. Harry Govier Seeley and the Karroo reptiles. *Bulletin British Museum (Natural History)* 3: 1–39.

Swinton, W. E. 1975. Gideon Algernon Mantell. *British Medical Journal* 1: 505–507.

Taquet, P. 1976. Géologie et paléontologie du gisement de Gadoufaoua (Aptien du Niger). *Cahiers de Paléontologie* 17: 1–191.

Taquet, P. 1977. Dinosaurs of Niger. *Nigerian Field* 42: 1–10.

Taquet, P. 1984. Cuvier-Buckland-Mantell et les dinosaures. In E. Buffetaut, M. Mazin, and E. Salmon (eds.), *Actes du Symposium paléontologique Georges Cuvier (Montbéliard)*, pp. 475–491.

Taquet, P. 1992. *Dinosaures et Mammiferes du Desert du Gobi.* Paris: Muséum National d'Histoire Naturelle.

Taquet, P. 1994. *L'Empreinte des Dinosaures.* Paris: Éditions Odile Jacob.

Thulborn, R. A., and M. Wade. 1984. Dinosaur trackways in the Winton Formation (mid-Cretaceous) of Queensland. *Memoirs of the Queensland Museum* 21: 213–517.

Tumanova, T. A. 1987. The armoured dinosaurs of Mongolia. *Sovmestnaia Sovetsko-mongol'skaia paleontologicheskaia ekspeditsiia* 32: 1–80.

Van Straelen, V. 1933. Louis Dollo (1857–1931). Notice biographique avec liste bibliographique. *Bulletin du Musée Royal d'Histoire Naturelle de Belgique* 9 (1): 1–29.

Vianey-Liaud, M.; S. L. Jain; and A. Sahni. 1987. Dinosaur eggshells (Saurischia) from the Late Cretaceous Intertrappen and Lameta formations (Deccan, India). *Journal of Vertebrate Paleontology* 7: 408–424.

Vickers-Rich, P. 1996. Early Cretaceous polar tetrapods from the Great Southern Rift Valley, southeastern Australia. *Memoirs of the Queensland Museum* 39: 719–723.

Vitaliano, D. B. 1973. *Legends of the Earth: Their Geologic Origins.* Bloomington: Indiana University Press.

Weishampel, D. B. 1990. Dinosaurian distribution. In D. B. Weishampel, P. Dodson, and H. Osmólska (eds.), *The Dinosauria* (Berkeley: University of California Press), pp. 63–139.

Weishampel, D. B., and W.-E. Reif. 1984. The work of Franz Baron Nopcsa (1877–1933): Dinosaurs, evolution and theoretical tectonics. *Jahrbuch der Geologischen Bundes-Anstalt Wien* 127: 187–203.

Wiffen, J. 1996. Dinosaurian paleobiology: A New Zealand perspective. *Memoirs of the Queensland Museum* 39: 725–731.

Wild, R. 1991. Die Ostafrika-Reise von Eberhard Fraas und die Erforschung der Dinosaurier-Fundstelle Tendaguru. *Stuttgarter Beiträge zur Naturkunde*, Serie C 30: 71–76.

Woodward, J. 1728. *Fossils of all kinds, digested into a method suitable to their mutual relation and affinity; with the names by which they were known to the ancients, and those by which they are this day known; and notes conducing to the setting forth the natural history and the main uses, of some of the most considerable of them, as also several papers tending to the further advancement of the knowledge of minerals, of the ores of metalls, and of subterranean productions.* London: Published by the author.

Zammit-Maempel, G. 1982. The folklore of Maltese fossils. *Papers in Mediterranean Social Studies* no. 1.

Zborzewski, A. 1834. Aperçu des recherches physiques rationelles, sur les nouvelles curiosités Podolie—Volhyniennes, et sur les rapports géologiques avec les autres localités. *Bulletin de la Société des Naturalistes de Moscou* 7: 224–254.

References to Part One   59

The Study of Dinosaurs

Part Two

Any one of the larger carnivorous dinosaurs would meet the case. Among them are to be found all the most terrible types of animal life that have ever cursed the earth or blessed a museum.

—Sir Arthur Conan Doyle, *The Lost World*

As summarized in the first part of this book, our knowledge of dinosaurs has accumulated through the combined efforts of many people, professionals and amateurs alike, over the last century and a half. We now know a great deal about these "fearfully great" reptiles, and we are learning more all the time.

So *how* do we know what we know? What are the bases for the statements about dinosaur biology and evolution that will be made by one contributor after another in the remaining sections of this book? Those questions are the subject of Part Two, which describes how paleontologists, and the other professionals who assist them, find, study, and interpret dinosaur fossils.

This section begins by explaining how a paleontologist decides where to look for dinosaur bones, and what is done with them once they are found; both traditional and state-of-the-art methods of collecting and preparing dinosaur fossils are summarized.

It would be very nice to have a living *Anatotitan* or *Triceratops* to study in the field or laboratory, but nature hasn't been that kind to us. Most of our information about dinosaurian evolutionary relationships, or about how the great reptiles functioned as living animals, comes from study of their skeletons. This means that in order to understand how paleontologists interpret dinosaurs, one must have a basic knowledge of the bones of the dinosaur skeleton, and so we devote chapter 7 to a tour of the different bones of a dinosaur's body.

One of the major goals of paleontology is to reconstruct, to the extent that this is possible, the course of evolution. In dinosaur paleontology this involves determining the phylogenetic relationships of the various dinosaur groups to each other, and also to other kinds of animals. How is this done? Chapter 8 explores different approaches to the naming and classifying of organisms, including dinosaurs.

One of the key developments in evolutionary biology over the last generation has been the general acceptance of the principles of phylogenetic systematics (cladistics) in interpreting the evolutionary relationships among different groups of organisms, including dinosaurs. Although a cladistic approach to organizing information about evolutionary patterns is an eminently logical way of doing things, it comes as a shock the first time one encounters it. (Birds *are* dinosaurs? Get out of here. . . .) So our chapter on classification explains how phylogenetic systematics works, and compares its approach to dinosaur classification with a more traditional approach.

To say that dinosaur classification is contentious is like saying that the Atlantic Ocean is a bit damp. The number of different dinosaur classifications operational at any one time can be described by the formula

$$C = (N + A) - 1$$

where C is the number of classifications, A is the number of amateur paleontologists, and N is the number of dinosaur paleontologists. The "−1" represents the true classification, which we shall never know (part of Durham's Law). The stability of any classification can be a double-edged sword. A classification can be stable because we have obtained a close approximation to the actual relationships of the organisms under study. Unfortunately, stability can also reflect consensus due to the lack of an adequate fossil record—or a stagnation of research.

Geologists have constructed a formidable set of terms to describe the intervals of earth history during which dinosaurs and other ancient organisms lived. Readers will not be able to understand how dinosaurs evolved unless they understand the names applied to the various intervals of Mesozoic time. Consequently we include a short chapter to orient the novice in a timely manner.

Although some field and laboratory methods in paleontology have not changed in the last century, new technologies have revolutionized much of the way in which dinosaurs are studied. Chapter 10 describes these new technologies, and the way they are affecting the research methods of paleontologists.

Jurassic Park gripped the imagination of the moviegoing public with the possibility that dinosaurs might be re-created from genetic material in dinosaur blood once imbibed by Mesozoic mosquitoes. Paleontologists are indeed interested in the possibility of recovering dinosaur biomolecules, but there is very little chance that these can be used to populate our zoos with living examples of the fearfully great reptiles. On the other hand, dinosaur biomolecules may well provide us with valuable insights into the relationships of dinosaurs to other animals. It is, in consequence, necessary to include a chapter about the problems and potential of finding and studying such biochemical traces of dinosaurs.

The results of scientific research on dinosaurs are generally published in learned technical journals written by and for scientists. However, the general public has an insatiable interest in dinosaurs, and part of the mission of major natural history museums is to satisfy that curiosity by putting dinosaur fossils, and explanatory material about them, on display. This is not an easy task. Chapter 12 describes all the planning and labor that goes into putting together a successful dinosaur exhibit.

Our most vivid impressions of dinosaurs as living creatures are based on the work of scientific artists. The final chapter of Part Two outlines the thinking and the steps that a paleontological artist goes through in preparing a scientifically accurate drawing or painting of a dinosaur as a living animal.

Hunting for Dinosaur Bones

David D. Gillette

The successful hunt for dinosaur bones requires knowledge of local geology and stratigraphy, time to conduct the search, an ability to distinguish rocks from fossils, and a little bit of luck. There are two cardinal rules of field paleontology: (1) the people engaged in the work of prospecting, mapping, and excavation must have permission from the landowner or land manager; and (2) those involved in field activities must respect the landscape. These practices apply equally to public lands (property owned by the federal, state, or local government), tribal lands, and private property.

Where and How Do You Look?

For effective results, field paleontologists rely on geologic maps (Fig. 6.1) that plot the areal distribution of geologic formations (formally defined mappable units). The maps can be used to narrow the search area so that time is not wasted where fossil bones should not be expected. Although fossils are rarely found in volcanic ash, only sedimentary rocks can be expected to have fossils. Searching for dinosaur bones in areas where there are only igneous rocks is futile. On the other hand, not all sedimentary rocks have dinosaur bones. In general, dinosaurs occur where the sediments were deposited on land by lakes or streams rather than in the sea. Moreover, they are found only in rocks ranging in age from the Late Triassic to the Late Cretaceous.

For example, the Morrison Formation in the American West is an Upper Jurassic sedimentary unit consisting of mudstones, siltstones, and sandstones that can be distinguished from underlying and overlying formations. To a paleontologist who wants to search for Upper Jurassic dinosaurs, geologic maps that show the locations of Morrison Formation exposures are indispensable. Recognizing the target formation in the field requires a basic understanding of the local stratigraphy, or succession of formations. With experience, the field paleontologist can read the rock record to interpret environments of deposition and further narrow the search area to those places where fossil bone may be found.

Once the paleontologist recognizes the formation in the field, the actual search begins. A good approach is to search where other fossils have been recovered. The abundance of fossils is usually patchy: In some places dinosaur bones are numerous, and in other places the rocks may be barren.

Figure 6.1. Simplified geologic map in the vicinity of the Cleveland-Lloyd Dinosaur Quarry, eastern Utah. More than 10,000 bones of the Late Jurassic carnivore *Allosaurus* have been excavated from this site in the Brushy Basin Member of the Morrison Formation (Jmb). The legend is arranged in stratigraphic order, with the oldest formation (Curtis Formation of Jurassic age, Jc) in the map area at the bottom and the youngest (Cedar Mountain Formation of

(continued)

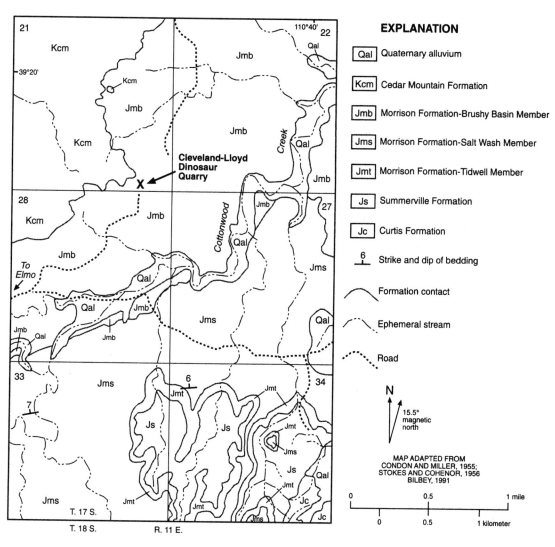

EXPLANATION

Qal — Quaternary alluvium

Kcm — Cedar Mountain Formation

Jmb — Morrison Formation-Brushy Basin Member

Jms — Morrison Formation-Salt Wash Member

Jmt — Morrison Formation-Tidwell Member

Js — Summerville Formation

Jc — Curtis Formation

6 — Strike and dip of bedding

Formation contact

Ephemeral stream

Road

N

15.5° magnetic north

MAP ADAPTED FROM
CONDON AND MILLER, 1955;
STOKES AND COHENOR, 1956
BILBEY, 1991

0   0.5   1 mile

0   0.5   1 kilometer

Hunting for Dinosaur Bones 65

Cretaceous age, Kcm, and alluvium of Quaternary age, Qal, respectively) at the top. The strike and dip symbol indicates the regional tilt of these formations as 6 degrees downward to the north.

Paleontologists often visit old sites in the quest for more bones from a particular individual, or in hopes of finding bones of new dinosaurs exposed by erosion subsequent to the last excavation.

In some formations, any dinosaur bone is important and should be excavated and deposited in a museum with proper field records. In others, dinosaur bones may be so abundant that discretion is called for. Priority may be established for rare or unusual species, for certain bones that are diagnostic, or for articulated skeletons. In all cases, bones discovered in the course of a search should be left undisturbed, including fragments that have eroded out of the rock and been carried downslope by gravity. These are clues to the position of bones hidden in the rocks. There is nothing to be gained by piling the fragments into small mounds, and much to be lost by disturbing these important clues.

A careful search around exposed bones may reveal other bones. Occasionally, dinosaur skeletons are articulated, and the exposure of one or a few bones may indicate a partial or complete skeleton. If the bones seem important, they should be photographed and documented. Photographs ideally should be taken from several angles and distances to show the bones, the setting of the site, and the overall landscape. This information helps the scientist locate the bones and make decisions concerning excavation and further study. If made by an amateur or a scientist whose specialization is not dinosaurs, the discovery should be brought to the attention of a dinosaur paleontologist. Field notes, even if crude, are indispensable for future decision-making; these should include a sketch of the bones, a sketch of the landscape with landmarks for reference, the date, notes on how to get there, the names of the discoverers, and any other information that might be useful.

Whether conducted by amateurs or professionals, the documentation process following discovery is critical for further evaluation. Identification of bones to the species level is often difficult or impossible prior to excavation, but a dinosaur specialist can usually make an educated guess. Researchers seldom recognize immediately that a discovery belongs to a new species, but such a possibility should always be considered in the preliminary evaluation.

Decisions to excavate or not to excavate rest on an informal set of criteria that may change over the course of years or decades. What is recognized as an unimportant discovery by one paleontologist may prove to be critical to others. For this reason, indiscriminate collecting of dinosaur bones for personal collections and commercial gain should be discouraged.

Selecting Which Bones to Excavate

Evaluation of a discovery requires deliberation. The commitment to excavate dinosaur bones requires not only the time and expense of the field activity, but also commitment to laboratory preparation (generally five to ten times more expensive and time-consuming than the field work), study, and storage in a museum. Because dinosaur bones are large and often extremely heavy, storage is expensive. For large bones especially, the space requirements are staggering.

Provided that budget and time are available, the paleontologist responsible for the field work must decide whether to excavate according to the following criteria: (1) Are the bones articulated? (2) Are there unusual or important faunal and floral associations that may have scientific value?

(3) Do the bones belong to a baby, juvenile, or subadult? (4) Are cranial bones likely to be found? (5) Are there paleoenvironmental or stratigraphic relations that make the dinosaur important? (6) Are there unusual features or preservation that might offer important new information? (7) Is there a real possibility that the newly discovered bones represent a new genus or species? (8) Are there any indications of cause of death, circumstances of burial, or other information that relates to the paleobiology of the dinosaurs? (9) Is there a possibility that the skeleton may serve as a display specimen in a museum exhibit?

Other questions may apply, too, according to the research interests of the paleontologist and the needs of the landowner or land manager. Positive answers to any of these questions may be sufficient to justify initiating a test excavation. Conversely, if these or similar questions cannot be answered positively, the bones should not be excavated.

Prior to the launching of a full-scale excavation that might require months or even years of field work and laboratory preparation, a test excavation may be appropriate. Because dinosaur bones are usually encased in hard rock and may require considerable disturbance of the vicinity to remove, the landowner or land manager must participate in the planning of such a test excavation. The goal should be to determine the real or potential extent of bones, the depth of bones beneath the surface, and better answers to the questions posed above. Even a relatively minor disturbance should be evaluated for its potential damage to the landscape: Are rare or endangered species of plants or animals likely to be affected? Will the activities alter drainage or require modification of roads or trails? Is it safe? Will there be an effect on livestock or local wildlife? Are there any indications of archaeological materials that might be disturbed by excavation? In areas where these questions are likely to be important, especially in western North America, specialists may assist in the evaluation. If conducted on federal lands, such an evaluation may be formally prepared as an Environmental Impact Statement as required by law. Although delays and adjustments may be frustrating, this deliberate planning is important for the wise management of an area where other concerns are equally important.

If deemed important after a preliminary test excavation, a full-scale field operation is warranted.

Mapping the Excavation

Dinosaurs are no longer trophies. Instead they are scientific specimens whose context is as important as the bones themselves. As bones are exposed, their occurrence must be mapped with care (Fig. 6.2). A grid system laid out on a regular pattern with stakes and strings is the most commonly adopted procedure. Grid intervals are typically 1 meter in length. For mapping in greater detail, some paleontologists use a portable meter-square frame divided by strings into decimeters (10-cm intervals). The frame can be laid over a meter square in the mapping grid to assist in the drawing of the map. This map becomes a permanent field record that must be deposited with the bones at the museum that receives the fossils. The hand-drawn map, like all other documents produced in the field work, becomes archival information that gives the bones scientific context. On all maps, north-east-south-west orientations are standard and should be laid out with an accurate compass.

Careful photographic documentation often enhances the mapping, but

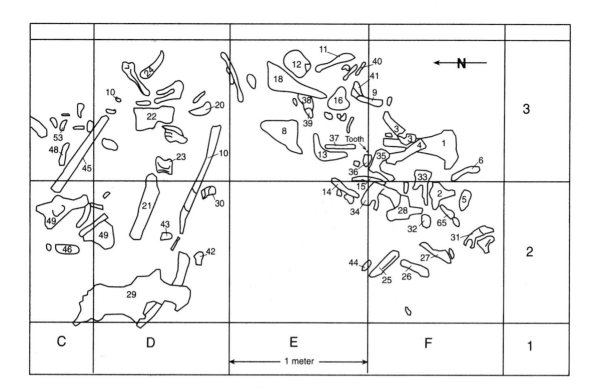

Figure 6.2. Portion of a dinosaur quarry map, from a site near Green River, Utah, in the Upper Jurassic Morrison Formation. Most of the bones are from the theropod dinosaur *Allosaurus*. Numbers correspond to specimen records with notes for reference during laboratory preparation and subsequent research. For example, bone number 29, an ilium of *Allosaurus*, is mainly in the western half of quadrant D–2.

should not be used alone. Photographs should include a scale and, where possible, an exact location in the grid. All photographs should be recorded in a field notebook with the location, date, and subject as a cross-reference for labeling them later. High-quality Polaroid photographs provide an immediate record that can be labeled on site. Like the field maps, the photographic record is an essential part of the archival information concerning an excavation and should be deposited with the collection.

Paleontologists increasingly use technologically advanced survey equipment that records exact position on the earth's surface and exact elevation measured from orbiting satellites with remarkable accuracy, often with a confidence interval of a meter or less. Computer technology that transfers such map information to digital information promises to revolutionize the mapping aspect of field paleontology. Computer-assisted drafting (CAD) applications may be used to store data that can be manipulated, edited, and altered for various forms of presentation. Digitized cameras and associated technology are likely to replace standard photography, but these new applications are expensive and presently impractical for most paleontologists. Such enhancements are useful, but cannot replace on-site mapping and record-keeping of field data.

Field Records

Maps and photographs are a critical part of every excavation. In many respects they are as important as the fossils themselves, for these records are the only permanent documentation of the orientation of the skeleton, the nature of the sediments in which they occur, and their exact location for future reference. These visually oriented field records must also be supplemented by written accounts, best produced on a day-to-day basis. Some paleontologists require all personnel on site to keep a field journal, and insist that a copy of their journal be deposited with the specimens at the conclusion of the field work.

The field notes should accurately record all aspects of the field work, including date, personnel, operations conducted, discoveries and preliminary evaluations, sketches of field observations that can be keyed to the maps, and notes on stratigraphy and geology. The journal may also include a wide variety of other subject material, such as weather, local flora and fauna, and memorable events of the day. Some paleontologists formalize this aspect of field work by requiring the completion of specially prepared forms, while others prefer an unstructured, open format for their journals. While these records are important as a kind of diary for each individual, they are invaluable documents for museum archives.

Excavation Techniques

Excavation of fossil bones is tricky. Although television documentaries may give the false impression that techniques are easy and the results are immediate, in actual practice the work of excavation requires considerable planning, time, and labor. While techniques vary from site to site and from one paleontologist to another, certain aspects are universal. These are described below to provide an overview of the procedures and materials, but the techniques cannot be learned without in-the-field training by an experienced paleontologist.

Moreover, safety is an important aspect of an excavation. Where bones are particularly large, sometimes weighing hundreds of kilograms and occasionally a ton or more, or where diggings are deep, there may be considerable danger. Blocks of rock and bones that weigh from several to as much as ten tons are occasionally removed from quarry sites; such operations are extremely hazardous and should be attempted only by seasoned experts. Similarly, sites themselves can be hazardous, particularly if they are situated in steep terrain or near unstable topography where falling debris or landslides are likely. There are no good excuses for taking shortcuts that sacrifice the safety of the participants. Occasionally, excavations draw a crowd of spectators; their safety, too, is important and cannot be ignored.

Tools and equipment should be chosen to suit the job. Sometimes bulldozers and jackhammers are required for removal of layers of rock above the skeleton. Such extreme procedures should be adopted only with the full consent of the landowner or land manager. The principle of minimum disturbance should be adopted especially when using heavy equipment: Move only enough rock and disturb only enough of the landscape to accomplish the goals of the excavation. More often, small power tools driven by a generator and hand tools are appropriate. For working close to bones, hand tools as small as dental picks may be required.

Initial exposure of bones ideally uncovers only their upper surfaces. The act of lifting or chiseling the rock away may have a profound impact on the bone. Fossil dinosaur bones have been confined by rocks for at least 65 million years, in total darkness, and protected from atmospheric conditions. At the time of exposure, the confining rocks are partially removed, and the bones may expand and crack. They may also undergo disturbingly rapid chemical alteration when exposed to air, especially desiccation, which causes the bones to shrink, further opening any natural cracks in them. Newly exposed bones may also change color, usually as the result of oxidation of minerals contained in the pores of the bones (reduced iron compounds altering to oxidized iron). These changes in the bones can be controlled by application of chemical hardeners that penetrate the open spaces and cracks, slowing or halting the undesirable effect.

Repairs should be made immediately, if practicable, but may have to wait for the stability and security of a laboratory. Chemical hardeners, adhesives, and glues vary in their quality and suitability. The guidance of technical personnel acquainted with excavation materials is essential. Use of an inappropriate glue or hardener may irreparably damage the bones. No one glue or adhesive is universally satisfactory, because their suitability varies with weather conditions, rock type, and moisture in the bones and surrounding rock.

After the upper surfaces of the bones are exposed, stabilized, and mapped, a plan for excavation must be developed, much like the children's game of "pick-up sticks." If bones are close to each other, or even one atop another, the order of their removal becomes a critical concern. Sometimes several bones can be removed together, or individual bones can be isolated and removed separately. Occasionally when a skeleton is articulated, the problem is reversed: for ease of excavation and for safety, the bones must be separated along natural cracks.

With a well-reasoned plan prepared in advance, the next step is to isolate each bone or group of bones by digging vertical trenches around their perimeter. The trenches produce a pedestal capped by the bones and supported by the underlying rock. This activity must be conducted with care, because collapse of the rocks and the bones can destroy hours of work and require days of repair work in the laboratory. For best results, the bones should not be exposed on their sides or undersurface but instead should be left with adhering rock to keep natural cracks in them from expanding, causing the bones to fall apart. Any new bone surface exposed during trenching may be stabilized with chemical hardeners, if necessary. Each bone must be numbered with permanent ink, and the numbers recorded with the map records and field notes.

At some point in the trenching activity, the bone must be further stabilized by application of a jacket that makes the block rigid. This is usually burlap soaked in plaster, although fiberglass and other materials are sometimes used instead. For economy and versatility, however, the burlap-and-plaster techniques are by far the most frequently used. The plaster must not be set directly on the bone. Instead, the bones should first be covered with successive layers of damp tissue, paper towels, or strips of newspaper that separate the plaster from the bone and provide a tight cushion for transport.

This first application of burlap and plaster "bandages" locks the bones and rock into place; it is similar to the medical practice of setting a broken bone with plaster and gauze. After the jacket sets, the pedestal must be carefully undercut (Fig. 6.3) and new bandages applied to the side and undersurface as the pedestal is narrowed. If the block is large or heavy, lumber or steel might be required. These can be fixed to the top and sides and should be completely encased in additional layers of plaster bandages. For huge blocks, steel banding and timbers may be required. During the undercutting stage, the block should be stabilized with props to prevent its shifting and consequent injury to workers or damage to the bone.

Eventually the undercutting activity and successive layers of bandages

Figure 6.3. The pedestal of rock that these paleontologists are removing from beneath the block is replaced by props to hold the block in place. As the undercutting proceeds, bandages of plaster and burlap are carefully applied to the sides and undersurface to lock the rock and bones of a partial skeleton of *Allosaurus* into place. Once the undercutting is completed, the block will be turned upside down to complete the application of the plaster case. The quarry floor was exposed by removing nearly a meter of overburden from above the bone level. The site is near Green River, Utah, in the Morrison Formation.

produce a plaster-encased block standing on a narrow pedestal that can be turned over without damage to the contents. Turning small blocks that weigh 10 or 20 kilograms is simple, but the larger the block, the more difficult the process, because there is always a danger that the contents of the block will shift and break loose from their confinement. For very large blocks, weighing hundreds of kilograms, there is also an obvious safety hazard akin to that of felling trees in the forest. The exact direction and movement of the block as it is rotated can be predicted and to some extent controlled, but often a large block pivots differently than expected. This procedure should be conducted only by an expert.

Prior to the block's being turned over, the outer surface must be labeled with permanent ink, indicating the field number for each bone contained inside, a north arrow for orientation, the site name and number, the date, the collector's name, and other information that will be pertinent to museum storage. This information is critical because the blocks may not be opened for laboratory preparation for weeks, months, or even years.

After the block is upside down, the small opening must be covered with plaster bandages and the block entirely closed. The block can now be transported without damage to the bones, although each block should be padded and locked into place to minimize jostling and bouncing. All blocks should be handled with care during transport.

The transfer of the bones to the museum involves proper transfer of documentation at the same time. The fossils become the responsibility of the museum's collection manager, who is concerned with the care of the fossils. The blocks may be placed in permanent or temporary storage, where they will be secure until readied for laboratory preparation, or they may be sent directly to a technical laboratory, called the preparation laboratory, for in-house removal of the bones from the remaining rock, repairs, and stabilization. Ultimately, the bones are to be made ready ("prepared") for study, storage, or exhibit. Field notes should include on-site treatment of bones, listing which glues, hardeners, or adhesives were applied. The field records of materials used in the excavation become vital to the museum preparators, who must continue the process that was begun in the field.

Technological Applications in Excavations

On the ground, the prediction of where bones are located in the subsurface is difficult or impossible. At best, the experience of the seasoned field worker is the most reliable source of information. The goal is to predict how much of the skeleton remains buried, its orientation, its depth, and its extent. If the paleontologists know this information in advance of the commencement of an excavation, they can plan a complete excavation that will discover all the bones, disturb the minimum amount of rock, and complete the work in minimum time at minimum expense.

While most of the techniques discussed above have changed but little over the past century, certain new applications of modern technology hold considerable promise for improving efficiency in the field. Although every field paleontologist dreams of being able to locate bones from aerial photos or satellite images, the best these techniques can do is decipher certain aspects of the surface geology. In North America and in many parts of the world, such applications of airborne remote sensing are superfluous because geologic maps are cheaper and the interpretations have already been conducted.

On the other hand, ground-based remote sensing may be used profitably at many sites. Presently, no single technique has been developed for paleontology, but borrowed or hired equipment that has been developed for other disciplines such as hydrology or archaeology is often available. Paleontological applications have special problems, however, and the service of an experienced engineer familiar with the equipment and the theory behind its design is essential. Ideally, paleontologists need a device that can "take an x-ray" of the ground to see the bones beneath the surface. To date, this idea remains a fantasy, despite movie depictions that show entire skeletons in exquisite detail.

The most promising technique is ground-penetrating radar. From a mobile unit about the size and shape of a power lawn mower and mounted on wheels or carried on poles, radio waves are transmitted into the ground along a preestablished grid pattern. Reflections along boundary layers are recorded on the receiving device and show a continuous subsurface profile along each grid line (Fig. 6.4). The layers may be changes in rock type or composition, changes in saturation of water, natural cracks in the ground, or the boundary between rock and bone. Rather than the image of a bone being produced in the record, the indication of possible bone is a disturbance pattern in the profile. Deciphering the patterns requires experience and considerable intuition. In one excavation where the techniques were tried as experiments, such disturbance patterns represented bones with about 50 percent accuracy. Such applications have proven more successful in archaeological sites, where large disturbances such as buildings or caverns can be detected.

A different technique that uses seismic data also holds considerable promise for the future. Acoustic diffraction tomography, or seismic tomography, requires strategically placed vertical holes that penetrate far beneath the projected level of the bones, preferably at least twice the depth. Specially designed hydrophones in a fixed array are suspended in these holes, which are sealed with pipe and filled with water. On the surface, a high-powered 8-gauge Magnum shotgun mounted on wheels, affection-

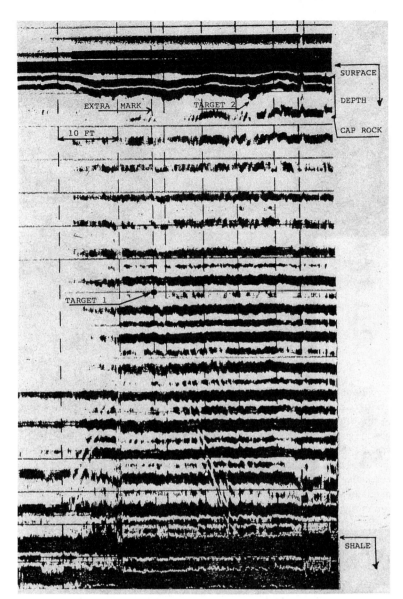

Figure 6.4. This record of a single traverse along a grid line for ground-penetrating radar at the *Seismosaurus* site (Gillette 1994) in the Morrison Formation in New Mexico shows two possible "targets" as the peaks of the parabolic disturbance patterns. Ground surface is at the top of the recording, target 1 is at a depth of 8 feet. Target 1 marked the position of the vertebral column of the skeleton of *Seismosaurus hallorum* from the top of a sandstone mesa that capped the bone level 2 to 3 meters below. Note that the horizontal scale is compressed. Courtesy Sandia National Laboratory, Albuquerque, New Mexico.

ately called a "betsy" (Fig. 6.5), sends a lead slug into the ground at predetermined points. The shock wave that passes from that point is transmitted into the ground in all directions. The hydrophones receive the shock waves and record the exact time of arrival. Reception times along the array of hydrophones that are anomalous, either too slow or too fast, indicate the existence of something buried in the subsurface and in a direct line between the shotgun blast and the vertical array of hydrophones. Geometric calculations can locate the buried object, which the paleontologists hope is a bone or a skeleton, and through repeated tests from different points, the depth and extent of the object can be resolved. Like ground-penetrating radar, acoustic diffraction tomography requires the services of trained engineers and is so expensive that its potential lies only in large excavations.

Hunting for Dinosaur Bones   73

Figure 6.5. Photograph of a "betsy" in action at the *Seismosaurus* site in New Mexico. This modified 8-gauge Magnum shotgun on wheels blasts a lead shot into the ground to generate a shock wave that passes through the ground. With sensors placed in bore holes, the positions of bones may be determined by the arrival times of the shock waves. Courtesy Southwest Paleontology Foundation, Inc.

Magnetic properties of buried bones are sometimes different from those of surrounding rocks because of their chemical composition. Measurement of this difference is the objective of proton free precession magnetometry, which records the intensity of the earth's magnetic field on a predetermined grid at the surface of the earth. The remarkably precise instrument is mounted on a pole and has the decided advantage of mobility so that it can be carried into remote areas with ease. Variations in the intensity of the magnetic field as plotted on a grid map may indicate the existence of something other than rock beneath the surface. Fossil bones are one possibility, but other subsurface materials may also produce an anomaly.

In some places, fossil bones and fossil wood contain radioactive isotopes of uranium. These isotopes decay, and the emission of energy, or ionizing radiation, may be recorded by hand-held counters. The traditional Geiger counter measures radiation of this sort, but is not sufficiently sensitive to detect radiation emitted from buried objects. Instead, radiation can be detected from subsurface sources by scintillation counters that take measurements of radiation on a grid pattern. Unusually high counts that differ from the normal background radiation may indicate bone beneath the surface. This technique seems to be useful, if at all, only for bones that are quite shallow, no more than a half-meter from the surface.

Laboratory Preparation of Bones

At the museum, the tedious process of preparation of the bones is the final step leading to their exhibit, study, or storage in the permanent cataloged collections. The purposes of preparation are to fully expose the bones, complete all necessary repairs, and stabilize them for curation. For dinosaur bones this procedure is time-consuming and expensive because of their size. Like other stages in discovery and excavation, laboratory

David D. Gillette

preparation techniques are best learned by hands-on experience under the supervision of an experienced preparator. In general, hand tools and dental picks are appropriate for most preparation, but often microscopes and needles are required. For removal of rock far from bone, the gentle impact of pneumatic tools that operate like miniature jackhammers is quite effective.

Repairs should be made with glues and adhesives approved by the chief preparator or the collections manager. As with the chemicals applied during excavation, no single hardener or glue is universally used, and suitability varies with location and the nature of the fossils. In recent years, collection managers have begun to recognize that many adhesives and preservatives, though useful during preparation, actually lose their effectiveness with age and accelerate the degradation of the fossil bone. One widely used chemical, commonly called glyptal, which was popular for more than four decades, disintegrates with age and is no longer recommended. Loss of effectiveness of glues and chemical hardeners can be devastating to museum collections. In general, the guiding principle in the use of chemicals and adhesives is "least is best."

If the bones are weak, structural supports may be necessary. These are custom-designed to fit the bone, and may consist of plaster-and-burlap bandages that form a cradle, fiberglass, steel bands, lumber, or any other structural material that adds support and strength.

Many paleontologists have adopted the practice of retaining samples of untreated bone and rock for future reference. In some cases, these materials have yielded new fossils of microscopic size, such as pollen, spores, or teeth of small animals, thus greatly enhancing the knowledge of the paleoenvironmental setting of an excavation site. These samples should be curated with the bones produced by the excavation.

In all cases, laboratory preparation records are important. Many preparation laboratories use forms to record which adhesives are used for each bone along with a detailed record of preparation activities on each bone. If in the future the bone needs repair or additional preparation, these treatment reports are critical if the original hardeners or adhesives are to be dissolved and replaced.

Curation and Conservation

Dinosaur bones present special curation problems because of their size. Their weight should be spread evenly through the use of pads or specially designed cradles made of plaster and burlap bandages, or foam carved to fit the bone. They should be stored in a clean, dark place where they are protected from changes in humidity, temperature, and decomposer organisms. Such conditions are best achieved in specially designed storage areas where temperature and humidity are constant. All bones must be labeled, and their locations and associated archives recorded in a collections catalog. They should be monitored periodically for damage to the bones and labels, and for deterioration of hardeners and glues. No one wants to pick up a dinosaur bone from a museum cabinet and have it fall apart because of decay or alteration during curation.

Curation and conservation of dinosaur bones actually begins in the field with proper technique. Museum professionals assigned the responsibility of curation are increasingly concerned with long-term effects of every action taken on a bone. Many of these personnel belong to the professional society called the Society for the Preservation of Natural History Collections. This group meets annually and publishes a wealth of technical information on conservation and curation in the journal *Collection Forum*. Membership is open to anyone interested in the subject of conservation of museum collections. Similarly, conservation research and evaluation of techniques are often reported in the journal *Curator*, published by the American Museum of Natural History. Curation standards for dinosaur bones and fossil vertebrates in general are evolving at a rapid pace. For new museums or personnel newly assigned curation responsibilities, membership in the SPNHC is a step in the right direction.

Collection Management

Besides direct responsibility for the care of the bones, collection management entails a myriad of other duties. Records associated with the collection are essential, for they are the only primary source of information concerning the excavation. Included in collection management duties are the maintenance of at least three filing systems, preferably on a computerized data management system: archives files, locality files, and specimen catalog files. These documents are among the most critical in the museum, for they give meaning to the collections. They should be managed with the utmost of care, and they should always be kept up to date. In addition, archival records themselves often require special treatment and conservation; photographs may be unstable, paper may become brittle with age or the ink may fade, and decay by fungus or bacteria can destroy paper.

The ultimate purposes of the museum collection are research and education. Thus the fossils should be accessible to qualified personnel for study. This aspect of collections is often overlooked in the architectural design for the collections area of the museum. Layout space, desk space, tables, good lighting, and access to electrical outlets are essential for suitable study of fossils. Microscopes and measuring devices are also important and should be made available to curators and visiting researchers. If specimens are to be displayed, their whereabouts must be recorded in the catalog and a note placed on the shelves where they are ordinarily stored. When visiting researchers need to examine an exhibit specimen, they should be allowed access to the bones, or the bones should be temporarily removed from the display. Museum visitors are seldom disturbed by a note in an exhibit stating that a specimen is removed for study; such a note indicates active use of a museum's collections.

This long road leading to the addition of another dinosaur bone, or a skeleton, to a museum collection began with the discovery in the field. The road never ends, because the bones recovered during an excavation and deposited in a museum collection are available for study for as long as museums exist.

References

Converse, H. H. Jr. 1984. *Handbook of Paleo-Preparation Techniques*. Gainesville: Florida State Museum.

Crowther, P. R., and W. A. Wimbledon (eds.). 1988. *The Use and Conservation of Paleontological Sites*. Special Papers in Paleontology no. 40. London: The Paleontological Association.

Feldmann, R. M.; R. E. Chapman; and J. T. Hannibal (eds.). 1989. *Paleotechniques*. The Paleontological Society Special Publication no. 4. (Chapters 18–29 are dedicated to large fossil vertebrates, including laboratory techniques, chemicals and adhesives, and field techniques for large specimens.)

Fitzgerald, G. R. 1988. Documentation guidelines for the preparation and conservation of paleontological and geological specimens. *Collection Forum* 4 (2): 38–45.

Gillette, D. D. 1994. *Seismosaurus the Earth Shaker.* New York: Columbia University Press.

Kummel, B., and D. Raup. 1965. *Handbook of Paleontological Techniques.* San Francisco: W. H. Freeman.

Leiggi, P., and P. May. 1994. *Vertebrate Paleontological Techniques.* Vol. 1. New York: Cambridge University Press.

Rixon, A. E. 1976. *Fossil Animal Remains: Their Preparation and Conservation.* London: Athlone Press.

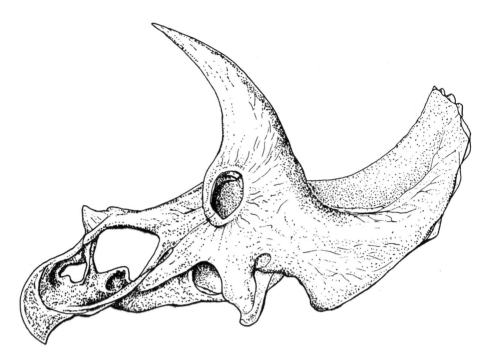

The Osteology of the Dinosaurs

Thomas R. Holtz, Jr.,
and
M. K. Brett-Surman

7

Dinosaurs are known only by their bones and teeth in most cases. The soft parts of the body, the skin, muscles, and other organs, were destroyed by decay processes fairly quickly after death. Only bones and teeth, the hard mineralized parts of a dinosaur, are durable enough to be preserved over tens of millions of years. Except for footprints, and much rarer traces such as eggs and skin impressions, fossilized skeletal material represents the only physical remains of the ancient dinosaurs. Thus the osteology (the study of bones) of dinosaurs is our main source of knowledge about these extinct animals.

Dinosaurs are tetrapod vertebrates—in other words, animals with bony skeletons and four limbs. All tetrapods, including amphibians, mammals, turtles, lizards, and birds, are built along the same general body plan. For example, the forelimb, or arm, of all tetrapods has one upper arm bone closest to the body, two bones below the elbow, several small bones in the wrists, and then a series of longer finger bones. (In some animals, such as snakes, the forelimbs have disappeared, but the ancestors of these animals had arms of the basic structure.) The reason all these animals share this common body plan is that all are descended from the same ancestral stock with that plan: the differences between the particular shapes of the bones arise from the same body plan having been modified, or adapted, to different uses (for example, the wings in birds or bats, the digging claws of moles, the grasping hand of a *Velociraptor,* or the pillar-like forefoot of a *Brachiosaurus*). Because of this common descent, we can recognize bones that are homologous; that is, they represent bones descended from the same original structure. Thus the upper arm bone of any tetrapod is homologous to the upper arm bone of any other tetrapod.

For an example of homology, compare the right hand of a human and the right "forepaw" of the plant-eating dinosaur *Iguanodon* (Fig. 7.1). These hands are oriented in the same direction, with the back of the hand facing us and the palm facing away. In the dinosaur, the homologue to the thumb has been fused into a large pike. The last digit, homologous to the "pinky" of a human, has evolved into an opposable finger. Opposability (the ability to place the digit on the palm) is characteristic of the thumbs of humans. Two different anatomical features in two different animals which have the same function but which are evolved from different parts of the body are called analogous. Thus the opposable digit of the *Iguanodon* is analogous to the human thumb, but homologous to the human pinky.

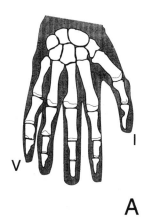 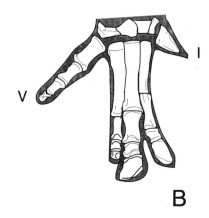

A   B

Figure 7.1. The right hands (manus) of (A) a human (*Homo sapiens*) and (B) the herbivorous dinosaur *Iguanodon mantelli* in anterior view. The spike in the hand of *Iguanodon* is homologous to (occupies the same anatomical position as) the human thumb. The opposable digit in the hand of *Iguanodon* is homologous to the human "pinky" but is analogous to (has the same function as) the human thumb.

It must be noted that in comparative anatomy the term *homologous* was not originally used in an evolutionary context. Sir Richard Owen, who coined the term *Dinosauria* in 1842 (see Torrens, chap. 14 of this volume), is also responsible for the use of homology in an anatomical context. Owen (1846, 1849) believed that for each major group of organisms there was a single basic body plan or "blueprint," of which all species in that group are variations. This body plan, the archetype, was not considered to have ever existed in the physical universe, but was a mental construct representing the simplified anatomical organization of each major group of organisms, such as vertebrates, mollusks, or insects. In this context the pectoral fin of a trout, the wings of birds, the forelimbs of horses, and the arms of humans were considered homologous since each was a variation of the same structure in the vertebrate archetype.

Sir Charles Darwin and his primary advocate, Thomas Henry Huxley, co-opted the concept of homology into the new theory of evolution by means of natural selection. In their view, where all animals of the same body plan have a common origin, a homologous structure in two or more organisms represents variations of the same structure that was present in a real common ancestor. (See Desmond 1982 for a detailed discussion of the social and political as well as the scientific conflict between Owen and Huxley over the concepts of "archetypes" and "ancestors.")

The Osteology of the Dinosaurs   79

Anatomical Names, Directions, and Views

Because the basic bony anatomy, or skeleton, of all tetrapods is based on an ancestral body plan, all homologous bones can be given the same name. Since the anatomists who coined the various names used the classical languages Latin and Greek for scientific discourse, as did naturalists, astronomers, and other early scientists, most of these bones are named in Latin. Similarly, other structures in the skeleton (such as the socket for the eye or for the nostril) are also given Latin names.

In the past, many bones were given up to three sets of names. The bones of the body were first formally named in the human skeleton. Another, more general, set of names was applied to the mammals because they were historically the next group to be scientifically studied. Finally the "lower" vertebrates, such as lizards, crocodiles, birds, and amphibians, were given another, often simpler, set of names. For example, the main portion of the human cheekbone is a single unit called the os zygoma (os is Latin for "bone"; *zygoma* is Greek for "cheek"). In mammals, this bone is also called the os zygoma, more commonly just the zygoma. In all other vertebrates it is known as the os jugale (or more commonly just the jugal). However, with some rare exceptions (such as the zygoma/jugal), modern anatomists now use the same names for the homologous structures in all vertebrate species. For some details on the anatomical names in various tetrapod groups, see the *Nomina Anatomica* (1983) for human anatomy, the *Nomina Anatomica Veterinaria* (1983) for general mammalian skeletons, and Baumel and Witmer 1993 for bird osteology.

Before some of the important bones and other structures in the skeleton of dinosaurs are described, the principle of anatomical direction should be discussed. In order to describe the positional relationships of bones to one another in the skeleton of an animal, a series of pairs of directions has been invented. Like "north" and "south," or "up" and "down," these directions always have an opposite, pointing the other way. Unlike "north" and "south" and "up" and "down," however, these directions are not based on the external environment. Instead, they are internal to each organism, regardless of how it may move about in the outside world. The names are based on the standard posture of most tetrapods (that is, all forelimbs on the ground, head pointing forward, belly toward the floor and back toward the ceiling), so that the homologous directions in a human being (with only our feet on the floor, our face pointing the same direction as our belly, our belly pointed forward, and our backs pointed behind us) are oriented in somewhat different external directions. However, a crawling baby is oriented in essentially the same position as most other tetrapods.

The first pair of these are anterior (sometimes called cranial) and posterior (sometimes called caudal). *Anterior* means "toward the tip of the snout," and *posterior* means "toward the tip of the tail." For example, the shoulders are anterior to the hips, the skull is anterior to the neck, and the nostrils are anterior to the eye sockets. Conversely, the hips are posterior to the shoulders, the neck posterior to the skull, and the eye sockets posterior to the nostrils. Because these terms are independent of the external environment, they remain the same regardless of the position an animal is in (i.e., if a cat curls up, its tail is still posterior to the skull).

A second pair of anatomical directions is dorsal and ventral. *Dorsal* means "toward, and beyond, the spine" (or more simply "up"), while *ventral* means "toward, and beyond, the belly" (or, generally, "down"). In the skull, the teeth are ventral to the eyes, and the upper jaw is dorsal to the lower jaw.

The next pair of directions, medial and lateral, refer to directions relative to an imaginary plane through the center of the body, which runs from the tip of the snout through the tip of the tail, bisecting the body into a right and a left half. These two directions refer to the relative positions of bones to each other and to this imaginary midline. *Medial* refers to bones or structures which are closer to the midline (i.e., closer to the center), and *lateral* means farther away from the midline (i.e., farther out, or more right or more left). The shoulder blades are lateral to the ribs, and the spine is medial to the ribs.

A last pair is used primarily for directions within the limbs (the arms and legs) and is sometimes applied to the tail. *Proximal* means "closer to the trunk," while *distal* means "farther out from the trunk." For example, the hip is proximal to the knee, and the wrist is distal to the elbow.

Although these four pairs, anterior/posterior, dorsal/ventral, medial/lateral, and proximal/distal, are generally used to describe the relationships of bones to one another, they can also be used adverbially to describe how an anatomical structure is constructed. For example, the teeth of the upper jaw point ventrally, the snout of most animals projects anteriorly from the eyes, and the ischium (a bone of the hip) points posteroventrally (back and down) in all dinosaurs.

The names of the anatomical directions can be used to describe the particular surface of the bone illustrated in a photograph or drawing. To see the dorsal anatomical view of the skull, for example, means to see the top surface. The ventral view would be a picture of the bottom of a bone or skeleton, the anterior view the front, the posterior view the back. There is both a right lateral and a left lateral view of the skeleton, depending on whether you are viewing the right or left, respectively, of the animal. A medial view would show the surface of a bone which normally faces the midline.

In the following section, we will examine the major bones in the skeletons of dinosaurs. Drawings of various dinosaurs are used to illustrate the position and general shape of these bones. However, the Dinosauria were a very diverse group of animals, so there is considerable variation in the details of their skeletons. Elsewhere in this book you will find drawings of the osteology of the different dinosaur groups.

Sections of the Skeleton

The Skull

The skeleton of a dinosaur, or other tetrapod, can be divided into two main divisions: the skull, which refers to all the bones and teeth of the head, and everything else. The skull is mostly composed of many different bones which, like most of the bones of the body, are paired (i.e., there is one of those bones on the left side of the skull and one on the right side of the skull). There are also bones, though, which are single, and these usually lie on the midline. For example, the supraoccipital bone is right above the hole where the spinal cord enters the skull, and it is not paired. The outlines of the individual bones of the skull are recognized as sutures, where the different bones meet.

The skull itself is divided into two major sections. The upper part of the skull, including the eyes, nostrils, upper jaw, and braincase, is called the cranium (plural *crania*). The lower jaw is composed of the left and right mandibles.

It is sometimes easiest to recognize the different individual bones of the skull by starting out with non-bony landmarks. Landmarks are particular homologous features which are recognizable from animal to animal. Two of the best landmarks are the eye sockets and the nostril openings. The technical term for the eye socket is the orbit, while each individual nostril is called a naris (plural *nares*).

Among the other landmarks of the skull are the teeth. Made of materials (dentine and enamel) even tougher and more durable than true bone, teeth occur only on particular bones of the skull. In the dinosaurian upper jaw are two tooth-bearing bones. The most anterior is the premaxilla, and the posterior one (which is almost always much larger) is the maxilla. The premaxilla is ventral to the naris, and the maxilla is ventral to another opening. This opening is called the antorbital fenestra (plural *fenestrae*), literally "the window anterior to the orbit." This opening is found in many dinosaurs and their closest relatives. In the beaked, ornithischian ("bird-hipped") dinosaurs, the antorbital fenestrae are very reduced or entirely closed over, while in the saurischian ("lizard-hipped") dinosaurs these fenestrae are often very large. The antorbital fenestra sits in a depression, the antorbital fossa (plural *fossae*). In some advanced meat-eating dinosaurs, there are additional openings anterior to the antorbital fenestra. These are called the maxillary fenestra (or sometimes the accessory antorbital fenestra) and the promaxillary fenestra (which is more anterior on the inside of the antorbital fenestra and cannot be seen in this figure). Figure 7.2 illustrates most of these landmarks on the skull of the tyrant dinosaur *Daspletosaurus*.

Figure 7.2. The skull of the tyrant dinosaur (tyrannosaurid) *Daspletosaurus torosus* in left lateral view, illustrating the important bones (in CAPITALS) and landmarks of the skull. The subnarial foramen is a character found only in the Saurischia, and the maxillary fenestra and jugal foramen are unique to certain members of the Theropoda. The nasal rugosities, frontal notch, jugal process, and surangular buttress are specializations of tyrant dinosaurs, and are not found in most dinosaurs. Illustration by Tracy Ford.

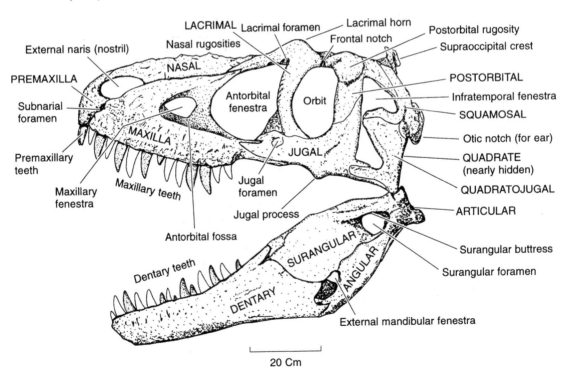

20 Cm

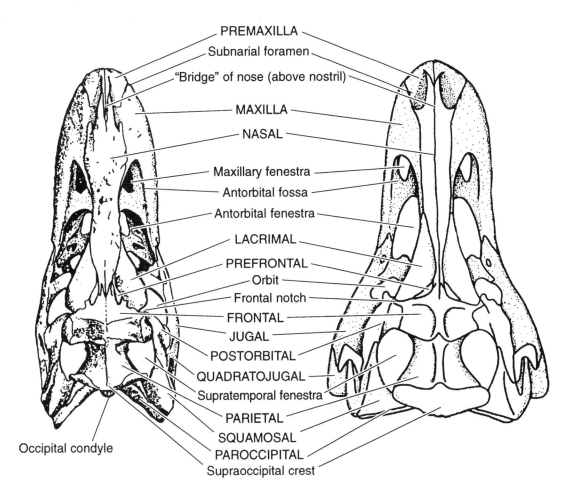

PREMAXILLA
Subnarial foramen
"Bridge" of nose (above nostril)
MAXILLA
NASAL
Maxillary fenestra
Antorbital fossa
Antorbital fenestra
LACRIMAL
PREFRONTAL
Orbit
Frontal notch
FRONTAL
JUGAL
POSTORBITAL
QUADRATOJUGAL
Supratemporal fenestra
PARIETAL
SQUAMOSAL
PAROCCIPITAL
Supraoccipital crest
Occipital condyle

Figure 7.3. The skulls of the tyrant dinosaurs *Daspletosaurus torosus* (left) and *Tyrannosaurus rex* (right) in dorsal view, illustrating the important paired bones (in CAPITALS) and landmarks on the dorsal surface of the skull. Illustration by Tracy Ford.

In some dinosaurs with a horny beak, teeth in the premaxilla, and sometimes even in the maxilla, are absent. When a dinosaur, or a jawbone, has no teeth, it is said to be edentulous. In the ceratopsian (frilled or horned) dinosaurs, there is an additional bone anterior to the edentulous premaxilla. This bone is called the rostral (or "snout") bone. The rostral is a single bone, joining the two premaxillae.

In the rear of the skull, posterior to the orbit, lie additional openings in the skull. These are called the lateral temporal (or infratemporal) fenestra and the supratemporal fenestra. The lateral temporal fenestra is a large opening on the side of the skull, while the supratemporal fenestra is on the dorsal surface of the skull. Both are associated with the attachment of jaw muscles.

From these various landmarks, the positions of some of the other important skull bones can be determined (Figs. 7.2, 7.3). The jugal, or "cheekbone," is posterior to the maxilla and ventral to the orbit. The lacrimal is a small bone between the antorbital fenestra and the orbit. The quadrate is a major bone in the rear of the skull, where the cranium articulates with the mandible (lower jaw). All dinosaurs and birds have a quadrate/articular jaw joint; in other words, a bone in the back of the lower jaw bone, called the articular, articulates with the quadrate bone in the

The Osteology of the Dinosaurs   83

skull. (Mammals have a dentary/squamosal jaw joint, meaning that the bones forming the joint in mammalian jaws are not homologous to the bones forming the joint in dinosaurian jaws.)

A series of paired bones lie along the dorsal and posterior surface of the skull (Fig. 7.3). These bones meet along the midline, and so form "mirror images," right and left, of each other. The most anterior are the nasals, long paired bones on the dorsal surface of the skull, posterior to the premaxilla. Posterior to the nasals are the frontals. The parietals are paired bones above the braincase on the posterior surface of the skull, posterior to the frontal. The squamosals are on the posterior surface of the skull.

There are many bones which are joined together around the brain cavity. These tightly sutured bones are collectively called the braincase and lie inside the outer skull bones listed above. The spinal cord exits from the brain through the foramen magnum (or "great opening") on the posterior of the skull. Beneath the foramen magnum is a structure called the occipital condyle, a rounded knob joint (or condyle), where the cranium articulates with the vertebral column (Fig. 7.4). In humans, other mammals, and our extinct relatives, as well as in amphibians, there are two occipital condyles (one right, one left), but in dinosaurs and other reptiles there is only a single rounded knob directly ventral to the foramen magnum.

Teeth in dinosaurs share a similar overall structure in that they arise out of sockets in the dentary bone (lower jaw) and in the premaxilla and

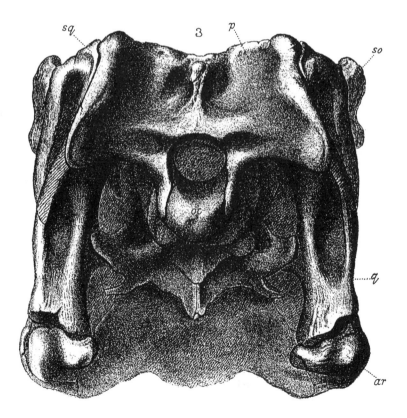

Figure 7.4. The skull of the armored dinosaur *Stegosaurus stenops* in posterior view, illustrating some important bones and landmarks of the rear of the skull. Illustration from Ostrom and McIntosh 1966. Original is a lithograph from a never-completed monograph on the Stegosauria to have been written by O. C. Marsh. Abbreviations: ar = articular; oc = occipital condyle; p = parietal; q = quadrate; sq = squamosal; so = supraorbial.

maxilla in the upper jaw. The teeth were ever-growing and ever-replacing, so that as one tooth was worn out (or ripped out), another grew out of the same socket to replace it. Dinosaurs had a continuing supply of teeth, unlike mammals, which get only two sets. In most dinosaurs (and in most other tetrapods), teeth were formed from a core of dentine and an outer surface of enamel. However, in two groups of herbivorous ornithischian dinosaurs (ornithopods and ceratopsians), the enameled surface was confined to one side of the tooth, and dentine formed only the surface of the other side. Because dentine is softer than enamel, the tooth became self-sharpening as the dentine wore away more quickly than the enamel when the teeth ground against each other. Dinosaur teeth did not really occlude against each other as they do in humans and most other mammals. Mostly they slid past each other in a slicing action. In ceratopsians the cutting surfaces of the teeth are oriented in a vertical plane that produces a scissors-like action. Because of their combination of slicing teeth and huge jaw muscles (ceratopsians may have had the strongest jaw muscles of any herbivorous animal), they have been called the "first Cuisinarts." Hadrosaurs are the only dinosaurs that "chewed" food, in that the teeth came together in a grinding action. Most saurischian dinosaurs simply grabbed food and swallowed it without the benefit of mastication.

In dinosaurs, as in most non-mammalian vertebrates, the mandible is composed of several different bones. The tooth-bearing bone of the mandible is called the dentary, and there are several bones posterior to it, which form the connection with the cranium. In mammals, the mandible is formed exclusively by the dentary. In the Ornithischia (bird-hipped dinosaurs), there is an extra bone in front of the dentaries. Called the predentary, this bone joins the two dentaries and forms a strong beak.

The Axial Skeleton

All the bones in the skeleton except for the skull are collectively referred to as the postcranium ("posterior to the cranium") (Fig. 7.5). The postcranium can be divided into two sections, often called the axial and appendicular skeletons. The axial skeleton is the "core" of an animal, its spine, trunk, and tail (the vertebrae). The appendicular skeleton refers to the fore- and hind limbs, and the girdles that attach the limbs to the trunk.

The most important part of the axial skeleton is the vertebral column. This column, the "backbone," is composed of many individual elements. Each of these bones is called a vertebra (plural *vertebrae*). A vertebra is composed of a large spool- or cylindrical-shaped structure, the centrum ("body," plural *centra*), ventrally, and a neural arch dorsally. On each neural arch are two sets of finger-like projections called the zygapophyses. Directed forward (angled upward and inward) are the prezygapophyses. These articulate with the postzygapophyses (facing backward and angled down and out) on the vertebra in front. These zygapophyses control the amount of movement between two vertebrae. Between the centrum and the neural arch runs the spinal cord. Projecting dorsally from the neural arch is a neural spine, to which are attached the back muscles (and which form the bumps down your spine).

The vertebral column of a dinosaur, like those of most non-mammalian tetrapods, is divided into four major segments (a mammalian column can be divided into five). These sections are the cervical (neck), dorsal (back), sacral (hip), and caudal (tail) vertebrae (Fig. 7.5). (In mammals, the dorsals can be divided into two separate segments: a thoracic, or chest, series, which has large ribs, and a lumbar, or lower back, series, in which there are no ribs.) The sacral vertebrae are often fused into a single structure in dinosaurs, called a sacrum (plural *sacra*). Dinosaurs are recognized as having three or more sacral vertebrae fused together (unlike most other reptiles, such as lizards and crocodiles, which have only two). The vertebrae of each of the different sections are shaped differently, reflecting the different requirements of the various sections of the body (i.e., flexibility in the neck, strength in the hips, etc.). Figure 7.6 illustrates a vertebra of the giant herbivorous dinosaur *Apatosaurus*.

The vertebrae of some dinosaurs are quite complex, with additional ridges, prongs, and other structures. One such structure is the hypantrum (a small anterior projection at the base of the neural spine), which articulates with the hyposphene (a small posterior projection at the base of the neural spine) of the preceding vertebra. Pleurocoels are openings along the lateral surfaces of vertebrae which open to a chamber inside the bone of the centrum and/or the neural arch. A pleurocoel can be a simple cavity, or it can be a very complex system of chambers and channels.

Lateral to the cervical and dorsal vertebrae are the ribs. Ribs (for some reason the only major part of the skeletal anatomy more commonly referred to by their English name than by their Latin name, costae) are long, narrow, paired bones forming a "cage" around the vital organs. The ribs of dinosaurs attach to the vertebrae at the bottom of the neural arch and the top of the centra by two separated projections. The ventralmost of these projections is the capitulum (plural *capitula*; "little head"), while the dorsalmost is the tuberculum (plural *tubercula*; "little lump"). Along the belly of some dinosaurs are gastralia (singular *gastralium*), or "belly ribs," which strengthened the ventral side of the animal and acted as a girdle to

Figure 7.5. The skeleton of the tyrant dinosaur *Daspletosaurus torosus* in left lateral view, illustrating the postcranial skeleton. Illustration by Tracy Ford.

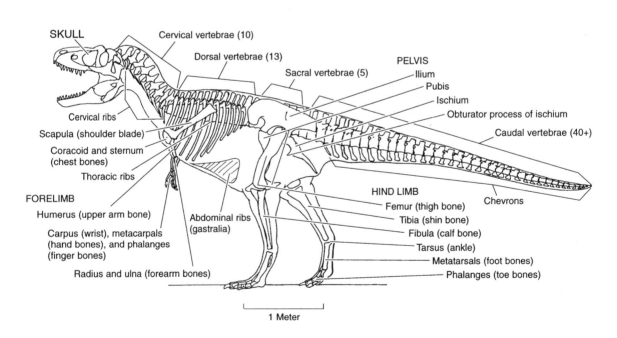

hold in the viscera ("guts"). Ventral to the caudal vertebrae are the chevrons, structures which are something like "tail ribs" or upside-down neural arches.

The Appendicular Skeleton

The appendicular skeleton refers to the fore- and hind limbs, and the girdles that attach these limbs to the body. Although there is a great similarity between the structures of the forelimb and the hind limb, the pattern of the girdles is very different.

The pectoral girdle attaches the forelimb to the trunk (Fig. 7.8). The largest of the bones in the pectoral girdle is the scapula (plural *scapulae*), or "shoulder blade." Ventral to the scapula is the coracoid. On the posterior surface of the girdle, the region where the scapula and coracoid meet forms a circular shoulder joint. In some dinosaurs, clavicles ("collarbones") are present, which attach the shoulder girdle to a series of fused bones along the ventral region of the chest. These fused bones formed the sternum (plural *sterna*), or "breastbone." In other, advanced, bird-like carnivorous dinosaurs there is a furcula ("wishbone," plural *furculae*) instead of clavicles. It is uncertain at present if furculae are formed by the fusion of the clavicles, or if they represent new structures (Bryant and Russell 1993).

Most of the forelimb, or arm, is made up of three bones. There is a single upper arm bone, the humerus (plural *humeri*), which joins the two forearm bones at the elbow. Of the two forearm bones, the ulna (plural *ulnae*) is the larger and more posterior, while the radius (plural *radii*) is generally smaller and more anterior. The many small bones of the wrist are known as the carpals. Distal to the carpals are the long bones of the palm of the hand, the metacarpals. The metacarpals are numbered in Roman numerals, from I to V, with I the most medial (inside, near the thumb) and V the most lateral (outside). The fingers are called digits, and they are numbered with the same scheme as the metacarpals (with digit I the "thumb" and digit V the "pinky"). The individual bones of the finger are called phalanges (singular *phalanx*). The distalmost of the phalanges are sometimes called the unguals, and supported the horny claws or hooves. Collectively, the digits, metacarpals, and carpals form the hand, or manus (plural *manus*).

The hind limbs attach to the axial skeleton at the pelvic girdle (Fig. 7.9). Also known as the pelvis ("hip," plural *pelves*), the pelvic girdle is

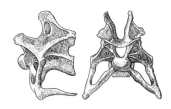

Figure 7.6. The 8th cervical (neck) vertebra of the gigantic sauropod dinosaur *Apatosaurus louisae,* in left lateral (*left*) and anterior (*right*) views. Illustration by Tracy Ford, modified from Gilmore 1936.

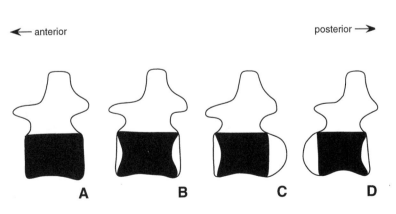

← anterior   posterior →

A   **B**   **C**   **D**

Figure 7.7. Schematic representation of the four main types of vertebral articulations. In each case the anterior end of the vertebra is to the left. (A) *amphiplatyan,* flat on both anterior and posterior ends; (B) *amphicoelous,* concave on both anterior and posterior ends; (C) *procoelous,* concave anterior end, strongly convex posterior end; (D) *opisthocoelous,* strongly convex anterior end, concave posterior end.

The Osteology of the Dinosaurs   87

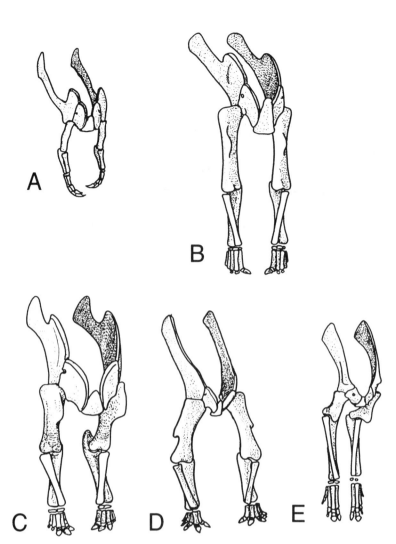

Figure 7.8. Forelimbs of the Dinosauria, in oblique right anterolateral view. (A) *Tyrannosaurus rex;* (B) *Apatosaurus louisae;* (C) *Stegosaurus stenops;* (D) *Chasmosaurus mariscalensis;* (E) *Corythosaurus casuarius.* Not to scale. Illustrations by Tracy Ford; modified from Osborn 1916, Gilmore 1936, Galton 1990, Lehman 1989, and Weishampel and Horner 1990.

A

B

C   D   E

composed of three bones per side. The largest of these is the ilium (plural *ilia*), which is the dorsalmost and which connects to the sacrum. Attaching beneath the ilium are the other two bones. The pubis (plural *pubes*) attaches anteriorly, and the ischium (plural *ischia*) attaches posteriorly. In most of the "lizard-hipped" or saurischian dinosaurs, the pubes point anteroventrally and the ischia posteroventrally (Figs. 7.9A–B). However, in the "bird-hipped" or ornithischian dinosaurs and certain "lizard-hipped" groups, the pubes point posteroventrally as well (Figs. 7.9C–E). Nevertheless, a pubis can always be distinguished from an ischium because the pubis attaches to the ilium anterior to the ischium. The ilium, pubis, and ischium form an open, round hole in the pelvis. Called the acetabulum (plural *acetabula*), this opening is the hip socket. In most tetrapods, including most mammals, turtles, lizards, and crocodiles, the acetabulum has a solid sheet of bone forming the medial wall of the socket. This condition is called a "closed" acetabulum. The Dinosauria, however, are specialized in having

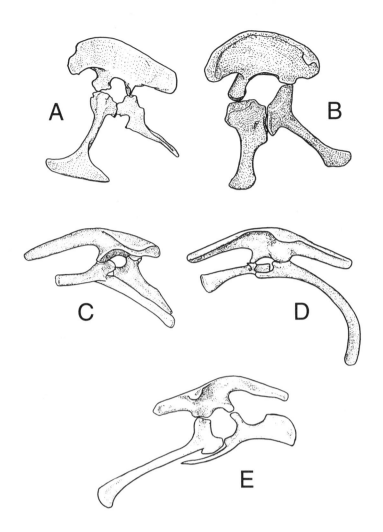

Figure 7.9. Pelvic girdles of the Dinosauria; A–D in left lateral view, E in right lateral view. Saurischian pelves: (A) *Tyrannosaurus rex;* (B) *Apatosaurus excelsus.* Ornithischian pelves: (C) *Stegosaurus stenops;* (D) *Chasmosaurus mariscalensis;* (E) *Corythosaurus casuarius.* Not to scale. Illustration by Tracy Ford.

an "open" acetabulum, one in which there was a hole all the way through the socket, and thus no medial wall of bone.

The pattern of bones in the hind limb, or leg, closely matches that of the forelimb (Fig. 7.10). There is a single upper leg, or thigh, bone, the femur (plural *femora*). The femur joins the lower leg at the knee, but there is no well-formed kneecap in the leg of a dinosaur. There are two bones in the lower leg: the tibia ("shin bone," plural *tibiae*), the larger and more medial of the two, and the fibula (plural *fibulae*), the thinner and more lateral one. Distal to the tibia and fibula are the tarsals, the small bones of the ankle. Unlike the complex ankle region of most tetrapods, the tarsals of dinosaurs are fairly simple. Two proximal elements, the larger, more medial astragalus (plural *astagali*) and the smaller, more lateral calcaneum (plural *calcanea*), adhere to the distal ends of the tibia and fibula, respectively. The other tarsals form a row of bones adhering to the long bones of the foot. There is no pronounced heel (posterior projection) in the ankle of dinosaurs, only a roller-joint between the astragalus/calcaneum and the distal tarsals. The long bones of the foot are called the metatarsals, numbered I to V in a medial-to-lateral fashion. Like the fingers, the toes are called digits,

The Osteology of the Dinosaurs 89

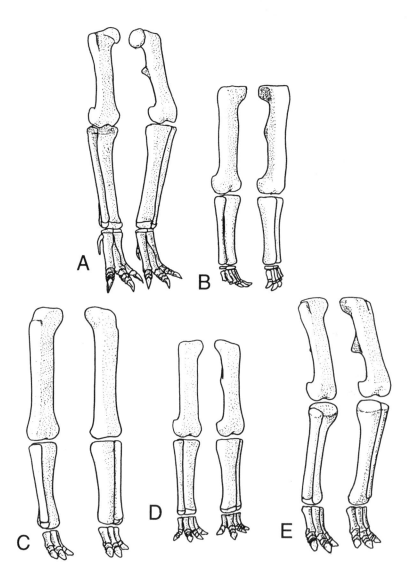

Figure 7.10. Hind limbs of the Dinosauria, in oblique right anterolateral view. (A) *Tyrannosaurus rex;* (B) *Apatosaurus louisae;* (C) *Stegosaurus stenops;* (D) *Chasmosaurus mariscalensis;* (E) *Corythosaurus casuarius.* Not to scale. Illustration by Tracy Ford.

numbered from the medialmost (I, the "big toe" in humans) to the lateralmost (V, the "little toe" in humans). Again as in the manus, each toe bone is a phalanx (plural *phalanges*), and the distalmost phalanges are unguals. The digits, metatarsals, and tarsals are collectively called the foot, or pes (plural *pedes*).

Unlike crocodiles, bears, and humans, which are plantigrade (flat-footed), dinosaurs are digitigrade. This means that dinosaurs walked on their toes, as chickens, cats, and dogs do. In order to distribute the weight of the animal, and act as a shock absorber, there was a pad of cartilage and connective tissues behind the foot. When you see the footprint of a large dinosaur, the front edges are marked by the bony claws, while the main depression is made by the non-bony pad.

Some dinosaurs also have a second set of bones in the body that arise out of the epidermis (outside skin). These bony growths, or osteoderms, form the many and varied patterns of armor seen in many dinosaurs. Most notable of these are the plates and spikes in stegosaurs and ankylosaurs. These were all anchored in the skin by connective tissue.

In order to strengthen the vertebrae in ornithischians, many of the tissues that connected the vertebrae together became filled with calcium and literally "turned to bone." These are the famous "ossified tendons" of the bird-hipped dinosaurs, and look somewhat like parallel strands of spaghetti. Those that occur below the tail and run across the chevrons between the caudal vertebrae are called hypaxial tendons. Those that occur above the vertebral centra and run across the neural spines are called epaxial tendons. In these dinosaurs, the base of the tail is very stiff, and not very mobile relative to the hips, while the tail becomes more mobile further posteriorly. In some saurischians (particularly some of the more advanced theropods), another path is followed. Instead of ossifying the tendons, the prezygapophyses of the vertebrae started to elongate and grow over several vertebrae at one time. In *Deinonychus*, these zygapophyses can cover as many as twelve vertebrae at one time. Similarly, the chevrons of *Deinonychus* were elongated to stiffen the tail. In dinosaurs such as this, the tail was most mobile anteriorly, and immobile distally, the opposite condition from what is found in ornithischians.

References

Baumel, J. J., and Witmer, L. M. 1993. Osteologia. In J. J. Baumel, J. E. Breazile, H. E. Evans, and J. C. Vanden Berge, (eds.), *Handbook of Avian Anatomy: Nomina Anatomica Avium,* 2nd ed., pp. 45–132. Publications of the Nuttall Ornithological Club, no. 23.

Bryant, H. N., and Russell, A. P. 1993. The occurrence of clavicles within Dinosauria: Implications for the homology of the avian furcula and the utility of negative evidence. *Journal of Vertebrate Paleontology* 13: 171–184.

Desmond, A. 1982. *Archetypes and Ancestors: Paleontology in Victorian London, 1850–1875.* Chicago: University of Chicago Press.

Galton, P. 1990. Stegosauria. In D. B. Weishampel, P. Dodson, and H. Osmólska (eds.), *The Dinosauria,* pp. 435–455. Berkeley: University of California Press.

Gilmore, C. W. 1936. Osteology of *Apatosaurus,* with special reference to specimens in the Carnegie Museum. *Memoirs of the Carnegie Museum* 11: 175–300.

Lehman, T. 1989. *Chasmosaurus mariscalensis,* sp. nov., a new ceratopsian dinosaur from Texas. *Journal of Vertebrate Paleontology* 9: 137–162.

Nomina Anatomica. 1983. Baltimore: Williams and Wilkins.

Nomina Anatomica Veterinaria. 1983. Ithaca, N.Y.: World Association of Veterinary Anatomists, Cornell University Press.

Osborn, H. F. 1916. Skeletal adaptations of *Ornitholestes, Struthiomimus, Tyrannosaurus. Bulletin of the American Museum of Natural History* 35: 733–771.

Ostrom, J. H., and McIntosh, J. S. 1966. *Marsh's Dinosaurs: The Collections from Como Bluff.* New Haven, Conn.: Yale University Press.

Owen, R. 1846. Report on the archetype and homologies of the vertebrate skeleton. *Report of the British Association for the Advancement of Science, Southampton Meeting* pp. 169–340.

Owen, R. 1849. *On the Nature of Limbs.* London: Van Voorst.

Weishampel, D. B., and J. R. Horner. 1990. Hadrosauridae. In D. B. Weishampel, P. Dodson, and H. Osmólska (eds.), *The Dinosauria,* pp. 534–561. Berkeley: University of California Press.

The Taxonomy and Systematics of the Dinosaurs

Thomas R. Holtz, Jr., and

M. K. Brett-Surman

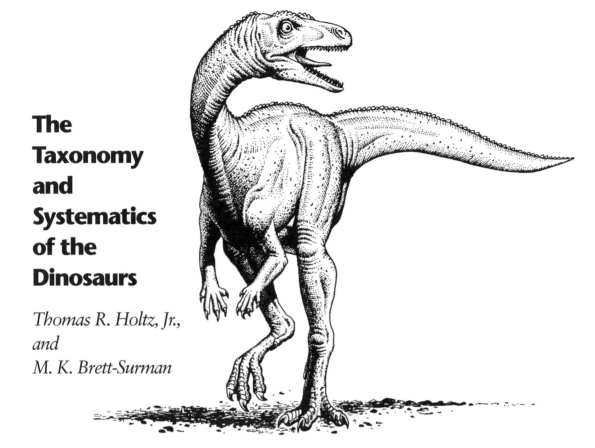

What's in a Name? Taxonomy

Taxonomy, the "naming of names," is the scientific practice and study of labeling and ordering like groups of organisms. It should not be confused with systematics, the scientific study of the diversity of organisms within and among clades (genetically related groups of organisms). Both help us to understand the world of organisms, but each practice helps us in a different way: systematics, to understand relationships among organisms, and taxonomy, to give internationally standardized names to organisms and groups of organisms in order to increase the efficiency of communication among researchers.

All languages have common names for different plants and animals. The main problem is that all languages have *different* names for the same plants and animals! This was not a problem until natural historians started to catalogue and study the floras and faunas from around the world. It was realized by western Europeans that the animals and plants of India, eastern Asia, the Pacific Islands, and particularly the "New World" of North and South America had great economic value as sources of medicine, spices, food, and furs. Whoever was the first to find and identify new plants and animals of economic importance in these regions would have the best access to these resources. Thus it became important to identify and classify these organisms.

There was initially a great deal of confusion as the great exploring and colonizing European countries each used their own names for the plants and animals they were discussing in the scientific literature. The only compromise that pleased all concerned was that every organism would be given a formal, official name based upon Latin or Greek, the language of the most highly educated Europeans and the Catholic church. In the seventeenth century, Caspar Bauhin (1623) and John Ray (1686–1704) invented the precursors of the later binomial (two-name) system. They introduced the concept of genus and species. It was not until the eighteenth century that these names were organized into a hierarchy of divisions (kingdom, class, order, family, genus, species) by Carl Linné (formally known as Carolus Linnaeus, a natural historian and botanist in mid–1700s Sweden) and his successors (Linné 1758).

The basic principle of Linnaean taxonomy is the nested hierarchy, in which each group is nested in a series of larger and larger, and thus more inclusive, groups. Each group is a taxon (plural *taxa*), a named group of organisms. Living taxa are recognized by their unique combinations of anatomical characters—bones, skin, hair/feathers/scales, physiology, DNA sequences, reproductive features, and so on. Extinct vertebrate taxa can be defined only by their bones and teeth.

Taxa are named in Latin or Latinized forms of other languages. Among other languages, Greek is the most common in taxonomy, but any other language (including English, Mongolian, Sanskrit, and the invented Elvish languages of J. R. R. Tolkien!) will do, as long as it has Latinized endings. All taxa, of whatever level, must be in a latinized form. Taxonomic names can be named after many things, including

- features of the anatomy: Mammalia, for the mammary glands;
- general appearance: *Anatotitan,* the "Titanic Duck," for a duckbilled dinosaur;
- behavior (alleged or otherwise): *Tyrannosaurus rex*, the "King Tyrant Lizard";
- name of their discoverer or other significant individual: *Lambeosaurus,* discovered by L. Lambe;
- the location from which it was found: *Edmontonia,* an armored dinosaur first discovered in Edmonton, Canada;
- or anything else the namer decides.

The most basic taxonomic levels, to which every living organism can be assigned, are the genus (plural *genera*) and species. The rules of nomenclature (the official naming of taxa) are based on a species. Species names are listed only along with the genus name, never by themselves. Species are, literally, more specific words than genera: species refer to a smaller total number of organisms than do genera. The international rules for naming a new species are governed by the International Code of Zoological Nomenclature (ICZN).

The exact biological and philosophical boundary where one species ends and another begins (or, more practically, whether a given specimen is assignable to a particular species) is the subject of much debate among biologists and paleontologists. Different criteria are used to define species by different scientists. Some, for example, use the degree of similarity or difference in the genetic code of a newly found organism in comparison with a catalogued specimen to include or exclude the new individual in the catalogued species. Others define the boundary of a species based on evolutionary divergence or "lineage splitting": All individuals who had a more recent common ancestor with a catalogued specimen than with another catalogued specimen are included in the first specimen's species.

The Taxonomy and Systematics of the Dinosaurs   93

One way many biologists determine whether two or more individuals are members of the same species is to observe the results of matings between individuals. If under natural conditions they mate and produce offspring that themselves can produce offspring, then the original two individuals are members of the same species. If the two individuals cannot mate, or do not produce living offspring, or the offspring they produce are sterile, then the two original specimens are not members of the same species. Of course, this is an impossible test with fossil individuals! For dinosaur studies, an individual is assigned to a species only if it shows a high degree of physical similarity in many parts of the skeletal anatomy to others thought to be in that species. The determination of which characteristics of the skeleton to use in classification is somewhat subjective.

A genus is defined as a group of one or more closely related species. If two members of the same genus but different species mate, they may have living offspring, but those will be sterile in almost every case (e.g., mules, the sterile offspring of the horse, *Equus caballus,* and the donkey, *Equus asinus*). Genus names are often listed by themselves. (Most people know dinosaurs only by their generic, not their specific, names; for example, people say *"Triceratops,"* not *"Triceratops horridus."*) Genera are, literally, more generic than species: Genera refer to a larger number of individuals than do species. The Linnaean binomials are thus the reverse of the European style of names of individuals, in which the personal ("Christian") name comes first and the surname last. If the authors of this chapter were to write their names in analogy with a Linnaean binomial, they would be called *Holtz thomas* and *Brett-Surman michael.* The following are some examples of the names of modern animal species: humans, *Homo sapiens* (thinking person [Linnaeus was an optimist]), abbreviated *H. sapiens;* cats, *Felis cattus* (cat cat); dogs, *Canis familiaris* (familiar dog); moose, *Alces alces* (elk elk); and the American alligator, *Alligator mississippiensis* (Mississippi alligator). Some dinosaur species include *Tyrannosaurus rex* (king tyrant lizard), *Apatosaurus excelsus* (surpassing deceptive lizard), *Triceratops horridus* (roughened three-horned face), and *Iguanodon mantelli* ([British naturalist Gideon] Mantell's iguana tooth).

Linnaean species names are written *Genus species,* always in italics (or underlined in the case of handwriting or typescript). The species name is abbreviated *G. species.* Although it once was common practice, the trivial name is never capitalized: for example, *Tyrannosaurus Rex* (a usage seen in many popular books) is incorrect; the proper form is *Tyrannosaurus rex.* It is never proper taxonomic grammar to use a trivial nomen by itself (for the example above, *rex* or *Rex* by itself is never correct; only *Tyrannosaurus rex* or the abbreviation *T. rex* is proper).

Type Specimens, Priority, Synonymy, and Validity

Linnaean taxonomy is based on the idea of type specimens. A type is the actual individual specimen which is first given the name. Types are the "name holders" only—they are not sacred objects that represent what a species should look like. They are simply the first reference specimen to carry a new name. This specimen must be

- deposited in an accredited institution where it is available for study;
- catalogued (for example, the type specimen of the duckbilled dinosaur

Edmontosaurus annectens, in the collections of the National Museum of Natural History [Smithsonian Institution], is catalogued under number USNM 2414); and

• described in the scientific literature where its name is presented.

Additional specimens are assigned to a species (or genus or other taxon) based on how closely a taxonomist believes the new specimen is related to a type. If the specimen is very similar, showing all the features, it is probably the same taxon. However, if the specimen does not show any features that disqualify it, but it shows no features that are definitely distinctive of the taxon, it can be questionably assigned to the taxon. If the specimen shows new features, it may be a new taxon.

For example, William Buckland's first dinosaur was not like any other known reptile, so he considered it a new genus and named it *Megalosaurus.* Gideon Mantell's first dinosaur was likewise not identical to any other known reptile, so he considered it a new genus, and his specimens were made the type of *Iguanodon.*

Each species has a type specimen. Each genus, in turn, has a type species (the first species to be given that generic name).

Not all types are complete specimens. Most fossil vertebrate types are incomplete skeletal material. So it is not uncommon for two (or more) names to be proposed which later to turn out to be the same genus or even the same species! When this happens, the oldest valid name (by date of publication) has priority, and is the official name of that taxon. The younger names are considered junior synonyms and not used. For example, the name *Troodon formosus* was given to a dinosaur tooth named by Joseph Leidy in 1856. Much later, in 1932, Charles M. Sternberg named a very fragmentary skeleton of a small dinosaur *Stenonychosaurus inequalis.* More complete skeletons found after the 1960s showed that the tooth called *Troodon* and the fragments called *Stenonychosaurus* belonged to the same species. Because the former was named seventy-six years earlier, it had priority, and the small bird-like dinosaur is properly called *Troodon formosus.* For a more famous (some would say infamous) example of the use of priority, see chapter 20 of this volume for the history of the names *"Brontosaurus"* (note that invalid names are enclosed in quotation marks) and *Apatosaurus.*

Although some specimens may be named new species on valid grounds at the time, later discoveries can show that these fossils are not distinct. The names based upon these types are then considered invalid. Only those specimens with distinct features can be valid types.

For example, S. H. Haughton in 1928 referred some dinosaur bones from Africa to a new species, *Gigantosaurus dixeyi.* The type material of *Gigantosaurus* (*G. megalonyx* from England, named in 1869 by H. Seeley) turned out to have no characters distinct from other dinosaur species or genera. *"Gigantosaurus" dixeyi* was left without a proper generic name until Jacobs and colleagues transferred the species to a new genus, *Malawisaurus,* in 1992.

Family and Family Group Names

Biologists have long recognized that animals can be grouped together hierarchically not only into species and genera, but also into larger and larger groups. For example, lions (*Panthera leo*) and tigers (*Panthera tigris*) can be grouped with domestic cats (*Felis cattus*) because of many similarities, including their retractable claws. Cats can be grouped with bears and dogs because of their specialized cheek teeth. Cats, dogs, and bears can be grouped with horses, humans, and whales because they all give milk. And so on, throughout the living world.

Each of these groupings can be considered a taxon. For most taxa larger than the genus, there are no specific rules for names, other than the requirement that the names be in Latin or a Latinized form of other languages. For example, Richard Owen used three known genera (*Megalosaurus, Iguanodon,* and *Hylaeosaurus*) that together were so distinct from all others that he gave them their own name, "Dinosauria" (Latinized Greek for "fearfully great lizards"). Unlike species names (which are always of the form *Genus species*), taxa from the genus and higher taxonomic levels have one-word names only: Felidae, Carnivora, Mammalia, etc. Unlike the species or genus names, taxa higher than the genus are never italicized. Other than that, there are few rules for most higher taxonomic names. One special type of taxon which does have special rules of nomenclature is the family (and the related subfamily and superfamily). A family is an assemblage of closely related genera, such as the cats (great and small), the dogs (from foxes to timber wolves), or the ostrich dinosaurs. Each family has a type genus (just as genera have type species and species have type specimens) from which that family gets its name. The family name comes from the name of the type genus (above, *Felis, Canis,* and *Ornithomimus,* respectively), modifying the ending according to Latin rules (generally dropping the *-is* or *-us*), and adding the suffix *-idae* (Latin for "of the family of"). Thus, the cat family is Felidae, the dog family Canidae, and the ostrich dinosaur family Ornithomimidae (remember that since a family is larger and more inclusive than a genus or species, the name is not italicized). When families are spoken of informally, their names are used in the lowercase, and the *-idae* ending becomes *-id*: above, felid, canid, and ornithomimid (see Table 8.1).

Table 8.1
Family Group Suffixes

Rank	Formal Suffix	Vernacular Suffix	Examples (Formal, Vernacular)
Superfamily	-oidea	-oid	Hadrosauroidea, hadrosauroid
Family	-idea	-id	Hadrosauridae, hadrosaurid
Subfamily	-inae	-ine	Hadrosaurinae, hadrosaurine

Traditionally, each genus belonged to a family, even if that genus was the only member of the family. However, some scientists now use families only when two or more genera are grouped together (see below).

Sometimes a family has so many genera that it becomes important to recognize groups within the family that contain more than one genus. A new taxon, the subfamily, is used for these smaller divisions. Subfamily names are formed by taking a type genus (just as in family names) and adding *-inae* instead of *-idae* to the shortened genus name. For example, Hadrosauridae, the family of duckbilled dinosaurs, contains more than thirty distinct genera. Those closer to *Hadrosaurus,* with solid or no crests, broad bills, and very large nostrils, are grouped into the subfamily Hadrosaurinae, while those closer to *Lambeosaurus,* with hollow crests, narrow bills, and smaller nostrils, are grouped into the subfamily Lambeosaurinae. Some taxonomists split subfamilies into even smaller divisions (such as tribes, subtribes, supergenera, etc.), but this practice is not yet common in dinosaur taxonomy.

On the other hand, sometimes taxonomists want to recognize a taxon which includes a family and other closely related families or genera. The most common way to do this is to name a superfamily. Superfamilies are formed by taking a type family and changing the *-idae* ending into *-oidea.* For example, paleontologists recognize that the Allosauridae and Sinraptoridae, two families of carnivorous dinosaurs, are closely related. They are grouped together into the superfamily Allosauroidea.

Systematics

Systematics is the scientific study of the diversity of organisms within and among clades (genetically related groups of organisms). Systematics is related to taxonomy in that the former is the practice of identifying evolutionarily significant groups of organisms, while the latter is the practice of naming those evolutionarily significant groups. Traditionally, there have been two different methods of systematics employed by vertebrate paleontologists, evolutionary systematics (sometimes called evolutionary taxonomy or gradistics) and phylogenetic systematics (often called cladistics). Since the methods vary between these types of systematics, the taxonomy associated with each also varies.

Evolutionary Systematics: Grades

Evolutionary systematics ("gradistics") is an eclectic system of classification based upon morphological similarity and the Linnaean taxonomic hierarchy. Groups of organisms are recognized by their physical resemblances. In order to be considered valid, all members of a gradistic taxon must have a common ancestor which was also considered a member of that taxon (for example, the common ancestor of all lizards must be a lizard, or the common ancestor of all dinosaurs a dinosaur). However, unlike cladistics, a gradistic taxon can exclude descendant groups if the descendants share a great number of anatomical advances not shared by any other member of the larger group. In other words, only those organisms of the same grade of development are included together in a taxon, while descendants of a higher grade of development are excluded from this taxon. For example, snakes all lack legs and eyelids and have many specialized characters not shared by their ancestors (which were lizards), so the snakes (Ophidia) are excluded from the lizards (Lacertilia) under gradistic systematics. Similarly, under gradistics, birds (Aves) are excluded from their reptilian ancestors because birds possess many specialized features (toothless beaks, wishbones, feathers, warm-bloodedness, and many more) not found in turtles, lizards, snakes, crocodiles, and the like. Some taxonomists under gradistics even allow groups of animals of the same grade of organization not descended from the same common ancestor to be placed in their own taxon. Although this practice was never common, some still use this extreme version of gradistics.

The Taxonomy and Systematics of the Dinosaurs   97

Under evolutionary taxonomy, all taxa are assigned a Linnaean rank. The standard Linnaean ranks are phylum, class, order, family, genus, and species. Linnaean taxonomy is a system of nested hierarchies: in practical terms, this means that each phylum contains one or more classes, each class one or more orders, and so on. As commonly used, each species in evolutionary taxonomy must belong to a genus, family, order, class, and phylum, even if that species is the only known representative of each of those higher taxa (a case of redundant taxonomic names). For example, the ancestral bird species *Archaeopteryx lithographica* is the only known species of the Genus *Archaeopteryx*, the Family Archaeopterygidae, the Order Archaeopterygiformes, and the Subclass Sauriurae. "Archaeopterygidae," "Archaeopterygiformes," and "Saururae" are thus redundant taxa.

It has long been recognized that there are subgroups which are intermediate between Linnaean ranks, so various prefixes (*super-*, *sub-*, *infra-*, and many others) have been used for these intermediate ranks (superclass, subfamily, infraorder, for example). Also, some taxonomists have added additional ranks (division, cohort, etc.) which are intercalated between previous ranks. However, this ultimately resulted in a bewildering number of ranks and subranks, as shown in Table 8.2. Because the number of intercalated subranks increased to the point of being unmanageable, there has been a trend to abandon the rank concept (above the family group level) by both gradistics and cladistics.

The gradistic system of taxonomy has been very useful, in various incarnations, since the 1700s. Much of our understanding of major living groups comes from research under the evolutionary system of systematics. A taxon in gradistic systematics must have a common ancestor, but may exclude one or more groups of descendants. For example, the Superclass Tetrapoda ("four-footed ones") has long been considered to be composed of four classes: Amphibia (cold-blooded tetrapods which have no scales and must reproduce in water), Reptilia (cold-blooded tetrapods which have scales and lay their eggs on land), Aves (warm-blooded tetrapods which have scales and feathers and lay their eggs on land), and Mammalia (warm-blooded tetrapods which have hair, give milk, and either lay eggs on land or have internal eggs). Almost all cultures recognized these classes (especially birds and mammals). However, evolutionary biologists soon recognized that reptiles are descendants of extinct amphibians (as "amphibians" were traditionally conceived by Linnaean taxonomists), and that birds and mammals are descendants of different groups of "reptiles" (although mammalian ancestors are now not considered members of the Reptilia; see below).

Until the 1970s, this was the system most widely used. It is a system designed to provide a taxonomy in which like groups are placed into the same hierarchical level (such as families), and to provide "evolutionary statements" that are built into the scheme. One can look at an evolutionary classification and see which groups are most closely related, their level of organization (body plans, or grades) as reflected in their Linnaean ranking (e.g., orders within a class were assumed to be "evolutionary equals"), and sometimes a degree of anatomical complexity. Evolutionary taxonomy recognizes grades of evolution which do not reflect total genetic relationships. A good example of a grade taxon is the reptilian order "Thecodontia." This is a group of archosaurian reptiles that were all placed in the same order because they had thecodont (socket-tooth) dentition. Because they were more "advanced" than earlier reptiles, and more "primitive" than the dinosaurs, birds, pterosaurs, and crocodiles, they were put in their own group, which reflected their level (grade) of evolution rather than their particular relationships to other archosaurs. However, thecodonts were not characterized by any unique features. Instead, they had features shared with all other archosaurs, but at the same time they lacked the specializations of the more advanced forms (dinosaurs, birds, pterosaurs, and crocodiles).

Table 8.2

Linnaean Taxonomic Ranks and the Systematics of the Duckbilled Dinosaur Species *Anatotitan copei*

See text for discussion. Note that most of these ranks are no longer used by either gradistics or cladistics.

Traditional Ranks	Additional Subordinal Ranks
Phylum Chordata	"
Class Reptilia	"
Order Ornithischia	Order Ornithischia
	Parvorder Genasauria
	Nanorder Cerapoda
	Hyporder Euornithopoda
	Suborder Ornithopoda
	Infraorder Iguanodontia
	Gigafamily Dryomorpha
	Megafamily Anklopollexia
	Grandfamily Styracosterna
	Hyperfamily Iguanodontia
	Superfamily Hadrosauroidea
Family Hadrosauridae	Family Hadrosauridae
	Subfamily Hadrosaurinae
	Tribe Edmontosaurini
Genus *Anatotitan*	Genus *Anatotitan*
Species *Anatotitan copei*	Species *Anatotitan copei*

Note: Traditional ranks are those required by standard evolutionary systematics; additional subordinal ranks are ranks below order used to more precisely describe the systematic position of the species. Parvorder Genasauria to Superfamily Hadrosauroidea from Sereno 1986; Family Hadrosauridae to Species *Anatotitan copei* from Brett-Surman 1988.

Phylogenetic Systematics: Clades

Evolutionary taxonomy (gradistics) has been correctly criticized for being too subjective and trying to put too much information into a system that tries to pigeonhole everything into a classification scheme that is a rigid artificial abstraction. To meet the need for a more objective system, one closer to the reality of evolution, with or without all the Linnaean ranks, in the 1950s entomologist Willi Hennig invented what is known as phylogenetic systematics—also known as cladistics (Hennig 1950, 1966). Cladistics has now replaced evolutionary systematics as the method used by most vertebrate paleontologists.

A method was needed to show the closeness of ancestry between two groups, because independent evidence shows that most species arise as splitting events (when two or more parts of a population of organisms, separated by some sort of barrier, follow different evolutionary pathways as a result of natural selection and genetic drift). For any three taxa, two must share a more recent common ancestor than the third. For example,

The Taxonomy and Systematics of the Dinosaurs 99

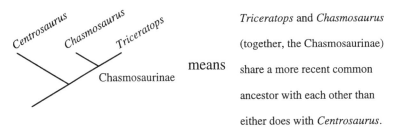

Triceratops and *Chasmosaurus* (together, the Chasmosaurinae) **means** share a more recent common ancestor with each other than either does with *Centrosaurus*.

Figure 8.1. A cladogram representing the phylogenetic relationship among three horned dinosaurs.

the horned dinosaurs *Triceratops* and *Chasmosaurus* are both more closely related to each other than either is to a *Centrosaurus*. This can be shown graphically (Fig. 8.1) by a cladogram, which shows that *Triceratops* and *Chasmosaurus* are joined with each other above the level at which *Triceratops, Chasmosaurus,* and *Centrosaurus* are all joined.

The point where two (or more) lines in a cladogram join is called a node. A node is recognized as a taxon itself—specifically the taxon containing all taxa which join at that node. In the cladogram in Figure 8.1, a node could be the Chasmosaurinae (the subfamily of horned dinosaurs with generally longer frills, longer snouts, and longer brow horns than nasal horns).

The taxon which shares a splitting event with another taxon is called the sister taxon or sister group. In the cladogram in Figure 8.1, *Chasmosaurus* is the sister taxon to *Triceratops, Centrosaurus* is the sister taxon to *Chasmosaurus* plus *Triceratops,* and, conversely, the group *Chasmosaurus* plus *Triceratops* is the sister taxon to *Centrosaurus*. In general practice, when scientists refer to the sister group to a particular taxon, they mean the closest that is known to science, and not just the closest on a very simplified cladogram.

From a cladogram, we can recognize three types of groups (Fig. 8.2). Monophyletic ("single branch") groups are composed of a single ancestor and all of its descendants. Mammals have long been recognized as a monophyletic taxon. Paraphyletic ("nearly a branch") groups are the grades of evolutionary taxonomy: a single ancestor, but not all descendants. "Lizards" are paraphyletic if snakes are excluded from lizards. Similarly, "reptiles" are paraphyletic if birds are excluded from the Reptilia. Polyphyletic ("multiple branches") groups have multiple ancestors, which have long been regarded as invalid by taxonomists. Grouping mammals and birds together without including "reptiles" (as a grade; see below) is a polyphyletic grouping, because mammals had a separate origin within the "reptiles" from that of birds.

Monophyletic groups are called clades ("branches"). Phylogenetic systematics seeks to find the relationships among taxa to form clades. Because our interest is in monophyletic groups, it is very important to use only smaller monophyletic groups while conducting a phylogenetic analysis.

(A parenthetical note: The terms *evolutionary systematics* and *phylogenetic systematics* can be confusing at times. Scientists using the "evolutionary systematics" methodology are interested in determining phylogenies (evolutionary trees which depict ancestor-descendant relationships). Workers employing "phylogenetic systematics" are interested in the recency of common ancestry and the interrelationships of clades, without making any statements about which taxa were ancestral to other taxa. This approach accepts biological evolution as the sole reason for the existence of the branching pattern of life. The terms *gradistics* and *cladistics* more accu-

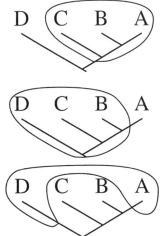

The group (A+B)+C is **monophyletic**, because it includes all descendants of the most recent common ancestor of A, B, and C.

The group (B+C)+D is **paraphyletic**, because it excludes a descendant of the most recent common ancestor of B, C, and D (namely, A).

The group A+D is **polyphyletic**, because A and D do not have a common ancestor which is not also the common ancestor of B and C.

Figure 8.2. A hypothetical cladogram portraying possible relationships among four groups of organisms. *From top to bottom:* monophyletic (an ancestor and all its descendants), paraphyletic (an ancestor but not all of its descendants), polyphyletic (no immediate common ancestor). Traditional gradistic taxonomy accepts the use of both paraphyletic and monophyletic groups, but cladistics requires that all taxa be monophyletic.

rately reflect the procedures used by these different schools of thought: Gradists sometimes accept the use of paraphyletic grades of organisms in their taxonomies, while cladists accept only the use of monophyletic clades.)

Phylogenetic Analyses

Phylogenetic analyses are the various methods used to determine the interrelationships of a group of organisms. For example, we might wish to examine the relationships of the long-necked herbivorous dinosaur group Sauropodomorpha. In particular, we wish to resolve whether the somewhat more primitive sauropodomorphs of the Late Triassic and Early Jurassic form a paraphyletic series to the giant Jurassic and Cretaceous Sauropoda, or whether these genera form their own monophyletic group, the Prosauropoda (Fig. 8.3). In the first case, some basal sauropodomorphs (for example, *Melanorosaurus* or *Riojasaurus*) are more closely related to true sauropods than are other basal sauropodomorphs (for example, *Thecodontosaurus* or *Anchisaurus*). In the second case, the Prosauropoda as a whole forms the sister group to Sauropoda.

Phylogenetic analyses are, thus, searches for clades. How is this done?

Many biologists use genetic or other biomolecular similarities between organisms to search for clades. For fossil groups (because the genes have all decayed over millions of years), the tool for finding clades is the search for shared derived characters (synampomorphies). First, scientists examine the characters of organisms, that is, their physical features (shape of the bones and their relationships to one another, presence or absence of rare structures, etc.). Then they look at how these characters are distributed among various taxa, both within and outside the groups they wish to study. They determine which characters are found in taxa both inside and outside the group of interest, which characters are found only within the group of interest but among all members of that group, and which characters occur only within subsets of the group of interest (see Chapman, chap. 10 of this volume, for more information about how organism characters are coded and analyzed in such a cladistic analysis).

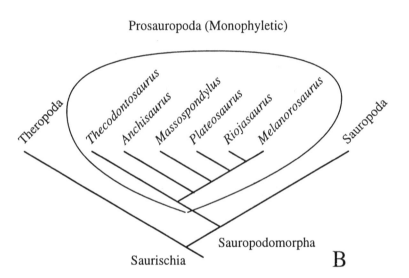

Figure 8.3. Two possible cladograms for the sauropodomorph dinosaurs of the Late Triassic and Early Jurassic (see also VanHeerden, chap. 19 of this volume). (A) Some of the basal sauropodomorphs (*Riojasaurus, Melanorosaurus*) share a more recent common ancestor with the sauropods than they do with more primitive prosauropods. In this scheme, Sauropoda are the direct descendants of the basal sauropodomorphs, and the similarities between *Melanorosaurus* and sauropods represent shared derived characters. (B) The basal sauropodomorphs are all more closely related to each other than any is to the sauropods, making a monophyletic group (clade) Prosauropoda. In this scheme, Sauropoda is the sister taxon to Prosauropoda, and the similarities between *Melanorosaurus* and sauropods represent convergence.

Characters that are found in all the members of the group of interest (and possibly outside the group) are called primitive, and presumably were present in some common ancestor of all of the creatures in which they are now found. For example, five fingers are primitive for mammals, and hair is primitive for a primate. Therefore, the presence of five fingers cannot help us determine which mammals are most closely related to each other, nor can the presence of hair help us understand the cladistic relationships within primates. Primitive characters are considered primitive homologies.

In contrast, characters which are found in only a few groups are probably shared derived characters which evolved in some relatively recent common ancestor. For example, having five fingers is a shared derived character of the tetrapods when compared to all vertebrates (including fishes), and hair is a shared derived character of mammals among tetrapods. Consequently the tetrapods share a common ancestor that lived more recently than the common ancestor of all vertebrates, and mammals share a common ancestor that lived more recently than the common ancestor of all tetrapods. We can therefore use the presence of a five-fingered hand to distinguish the tetrapods from other vertebrates, and hair to distinguish mammals from other tetrapods. Derived characters are advanced homologies. Shared primitive characters do not help resolve a cladogram, while shared derived characters do.

Unique derived characters (those found only within a single group), while important for understanding the biology of animals, do not help to resolve the cladistic relationships among taxa. For example, because feathers are unique to birds among modern tetrapods, they do not help us to recognize which other group of tetrapods is the sister group to birds.

Convergences are a special sort of character. Because of similar functions or behaviors, two or more groups of organisms can independently acquire very similar features. Although at first these resemblances might seem to be shared derived characters, additional evidence shows that they are convergent. For example, it might at first appear that the upright posture of some mammals and dinosaurs is a shared derived character of a mammal-dinosaur group. However, the dinosaurian skull, vertebrae, limbs, tail, and indeed most of the rest of the skeleton share more derived characters with other reptiles than with mammals. Thus the upright posture in mammals and dinosaurs is convergent.

Definition and Diagnosis

As you might imagine, there is a difference between the definition and diagnosis of different groups in the gradistic and cladistic systems of taxonomy. In the former, definition and diagnosis are for all intents and purposes identical, and are character-based. In cladistics, definitions are based on taxa and diagnoses are recognized by characters.

Under gradistics, the definition (the meaning of the taxon name) and the diagnosis (the way in which that taxon is recognized) are essentially the same. Taxa are defined by their characters (derived or primitive), so that a gradistic "Reptilia" could be defined as all amniotes (animals that reproduce by means of a specialized shelled egg, or derivatives of that style of reproduction) that have scales but lack feathers, fur, or warm-bloodedness. The diagnosis of the gradistic "Reptilia" would then be the presence of an amniotic egg and scales, and the lack of feathers, fur, and warm-bloodedness.

Similarly, under gradistics, dinosaurs would be defined as the Superorder (or Class, or Subclass, etc.) Dinosauria, that group of archosaurian reptiles with upright limbs, three or more sacral vertebrae, and perforate acetabula. The Superorder Dinosauria consists of the two orders Saurischia and Ornithischia. The Ornithischia would be considered to have had an ancestor among the Saurischia because the saurischian skeleton is the less "advanced" of the two. This information could be presented as an evolutionary tree showing when each group originated and from which clade each group arose.

Under cladistics, the definition of a taxon is based on the relationships of two or more taxa. De Queiroz and Gauthier (1990, 1992, 1994) recognized two main kinds of phylogenetic definitions, stem-based and node-based. A third form, derived character–based, is unstable (Padian

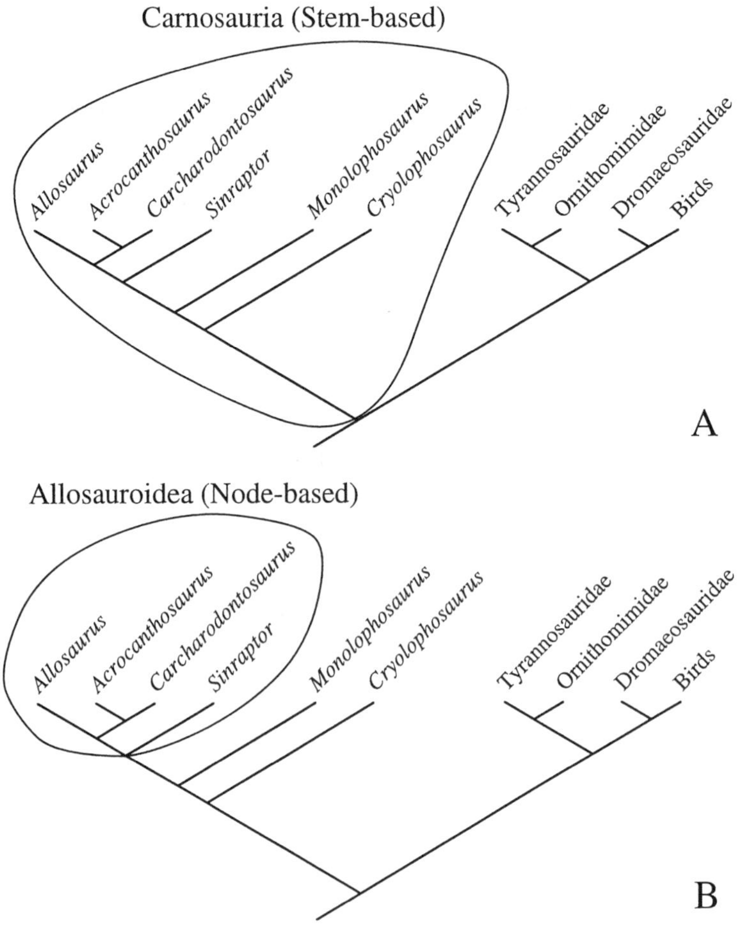

Carnosauria (Stem-based)

A

Allosauroidea (Node-based)

B

Figure 8.4. A cladogram of some carnivorous dinosaurs, showing the two main types of phylogenetic taxon definitions. (A) Carnosauria is a stem-based taxon (*Allosaurus* and all theropods closer to *Allosaurus* than to birds). (B) Allosauroidea is a node-based taxon (all descendants of the most recent common ancestor of *Allosaurus* and *Sinraptor*). Thus *Cryolophosaurus* and *Monolophosaurus* are both carnosaurs, but not allosauroids. *Carcharodontosaurus, Acrocanthosaurus,* and (by definition) *Allosaurus* and *Sinraptor* are carnosaurs and allosauroids.

and May 1993; Bryant 1994; Holtz 1996), in that the character used to diagnose the clade may be found to have evolved independently more than once. On the other hand, stem-based and node-based taxon definitions will always represent natural clades, because all organisms share common ancestry to one degree or another.

Stem-based taxon definitions are of the form "Taxon X and all organisms sharing a more recent common ancestor with Taxon X than with Taxon Y" (see Fig. 8.4A). For example, the carnivorous dinosaurian taxon Carnosauria is defined as *Allosaurus* and all taxa sharing a more recent common ancestor with *Allosaurus* than with birds (Holtz and Padian 1995). Node-based definitions are of the form "the most recent common ancestor of Taxon X and Taxon Y, and all descendants of that common ancestor." For example, Allosauroidea could be defined as the most recent common ancestor of the carnosaur genera *Allosaurus* and *Sinraptor,* and all of that ancestor's descendants.

Looking at a more inclusive group, under cladistics Dinosauria would be defined as the most recent common ancestor of Saurischia and Ornithischia, and all of that ancestor's descendants. Note particularly that Linnaean ranks are not used, and that there is no implication as to which group is the ancestor of the other group (in fact, they are considered sister taxa, and not ancestor and descendant).

Considering still broader, more inclusive groups, the category Reptilia is now considered a node-based taxon: the most recent common ancestor of turtles, lepidosaurs (lizards [including snakes] and the tuatara), and archosaurs (crocodiles and birds and their extinct relatives). Thus Aves (the

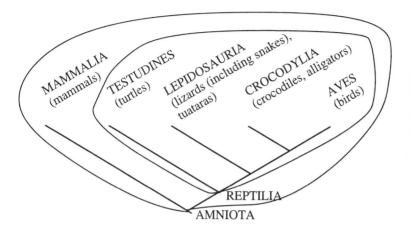

Figure 8.5. Cladogram of the living Amniota. Birds (Aves) are part of the monophyletic taxon Reptilia, but mammals (Mammalia) are not.

birds) is part of the larger monophyletic Reptilia. Mammals, however, are not part of this clade, because our ancestors diverged from the common ancestor of all reptiles (as now defined) before the turtle-lepidosaur-archosaur divergence (see Fig. 8.5). Thus under cladistics, the ancestors of mammals by definition were not reptiles (i.e., were not part of the clade Reptilia), while birds' ancestors and birds themselves are true reptiles (i.e., members of the clade Reptilia). (Some ornithologists take extreme exception to this "demotion"!)

Diagnosis of taxa under the phylogenetic taxonomic system follows definition. After the distribution of shared derived characters within a cladogram is determined, those shared derived characters which unite taxa into a stem-based or node-based taxon are used as the diagnosis of that taxon.

In the chapters that follow, the various contributing authors discuss dinosaurian relationships from both the gradistic and cladistic schools of systematics. It is informative to compare and contrast the conclusions the authors draw about the ancestors and ancestral characteristics of the dinosaur groups they study, based on the different methods they use to understand the evolutionary relationships of these wonderful ancient animals.

References

Bauhin, C. 1623. *Pinax Theatri Botanici*. Basel.
Brett-Surman, M. K. 1988. Revision of the Hadrosauridae (Reptilia: Ornithischia)

and their evolution during the Campanian and Maastrichtian. Ph.D. dissertation, George Washington University.

Bryant, H. N. 1994. Comments on the phylogenetic definition of taxon names and conventions regarding the naming of crown clades. *Systematic Biology* 43: 124–130.

De Queiroz, K., and J. Gauthier. 1990. Phylogeny as a central principle in taxonomy: Phylogenetic definitions of taxon names. *Systematic Zoology* 39: 307–322.

De Queiroz, K., and J. Gauthier. 1992. Phylogenetic taxonomy. *Annual Review of Ecology and Systematics* 23: 449–480.

De Queiroz, K., and J. Gauthier. 1994. Toward a phylogenetic system of biological nomenclature. *Trends in Ecology and Evolution* 9: 27–31.

Haughton, S. H. 1928. On some reptilian remains from the dinosaur beds of Nyassaland. *Transactions of the Royal Society of South Africa* 16: 67–75.

Hennig, W. 1950. *Grundzüge einer Theorie der phylogenetischen Systematik.* Berlin: Deutscher Zentralverlag.

Hennig, W. 1966. *Phylogenetic Systematics.* Urbana: University of Illinois Press.

Holtz, T. R. Jr. 1996. Phylogenetic taxonomy of the Coelurosauria (Dinosauria: Theropoda). *Journal of Paleontology* 70: 536–538.

Holtz, T. R. Jr., and K. Padian. 1995. Definition and diagnosis of Theropoda and related taxa. *Journal of Vertebrate Paleontology* 15 (Supplement to no. 3): 35A.

Jacobs, L. L.; D. A. Winkler; W. R. Downs; and E. M. Gomani. 1993. New material of an Early Cretaceous titanosaurid sauropod from Malawi. *Palaeontology* 36: 523–534.

Leidy, J. 1856. Notices of remains of extinct reptiles and fishes, discovered by Dr. F. V. Hayden in the Bad Lands of the Judith River, Nebraska Territories. *Proceedings of the Academy of Natural Science, Philadelphia* 8: 72–73.

Linné, C. 1758. *Systema Natura per Regina Tria Naturae, Secundum Classes, Ordines, Genera, Species cum Characterisbus, Differentiis, Synonymis, Locis. Editio decima, reformata, Tomus I: Regnum Animalia.* Laurentii Salvii, Holmiae.

Padian, K., and C. May. 1993. The earliest dinosaurs. In S. G. Lucas and M. Morales (eds.), *The Nonmarine Triassic,* pp. 379–380. Albuquerque: New Mexico Museum of Natural History and Science Bulletin 3.

Ray, J. 1686–1704. *Historia plantarum.* 3 vols. London: S. Smith and B. Waldorf.

Sereno, P. C. 1986. Phylogeny of the bird-hipped dinosaurs (Order Ornithischia). *National Geographic Research* 2: 234–256.

Sternberg, C. M. 1932. Two new theropod dinosaurs from the Belly River Formation of Alberta. *Canadian Field-Naturalist* 46: 99–105.

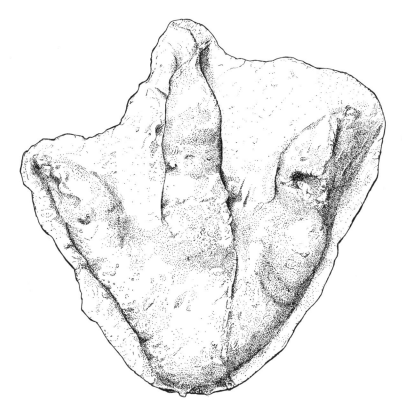

Dinosaurs and Geologic Time

James O. Farlow

When Did Dinosaurs Live?

Geologists subdivide earth history into extremely long intervals of time known as eons, and eons are in turn subdivided into eras, eras into periods, periods into epochs, and epochs into ages. The oldest rocks that have survived to the present day formed about 4 billion (4,000 million) years ago, at the beginning of the Archean Eon. The earliest fossils, representing bacteria or similar organisms, are found in rocks of about this age.

The Archean Eon ended about 2.5 billion (2,500 million) years ago, and the following Proterozoic Eon began. The Proterozoic Eon ended roughly 570 million years ago. During this long interval of time, living things became increasingly diverse; the first true animals appeared toward the end of the Proterozoic Eon.

Animals with sophisticated internal or external skeletons became abundant and diverse with the beginning of the Phanerozoic Eon, which began with the close of Proterozoic times and continues to the present. The Phanerozoic is divided into three eras, the Paleozoic (the oldest), the Mesozoic, and the Cenozoic (in which we still live).

Dinosaurs in the usual sense of the word (that is, excluding birds) lived during the Mesozoic Era. The Mesozoic is subdivided into three periods (Table 9.1), the Triassic, Jurassic, and Cretaceous. Each period is subdivided into epochs, such as the Late Jurassic Epoch of the Jurassic Period, and each of these epochs is subdivided into ages, such as the Campanian Age of the Upper Cretaceous Epoch.

9

The various intervals of geologic time were originally based on rock sequences (cf. Rudwick 1976; Albritton 1986; Dott and Prothero 1994). For example, the Cretaceous Period was defined as that interval of time during which rocks of the Cretaceous System were deposited. Consequently paleontologists frequently use terms such as *Triassic, Jurassic,* and *Cretaceous* in two senses, one referring to the rocks themselves, and the other to the time intervals represented by those rocks. If we talk about the Lower Triassic, for example, we are using the term to refer to the rocks: Lower Triassic rocks occur stratigraphically below Middle and Upper Triassic rocks. On the other hand, if we refer to the Early Triassic, we are designating a time span that occurred before the Middle and Late Triassic. Thus "Lower" in the stratigraphic sense corresponds to "Early" in the temporal sense, and "Upper" to "Late." "Middle" (conveniently) can be used in either the stratigraphic or the time sense.

Adjectives such as *early, middle,* and *late,* in addition to their use as parts of the names of formally defined intervals of geologic time, are also used to describe informal stretches of earth history. For example, one can talk informally about the "later Mesozoic," "the middle Cretaceous," or the "early Cenozoic." These are not defined intervals of the geologic past, but can be useful terms when talking about geologic or evolutionary events. Note that the initial letters of the words *later, middle,* and *early* are not capitalized when one uses them informally. Table 9.1 may seem to contain a daunting number of names, but using this terminology enables paleontologists to pinpoint when particular kinds of dinosaurs lived as finely as the resolution of the stratigraphic record permits. The reader will encounter many of these terms again and again throughout the book; you are encouraged to return to Table 9.1 to help you keep all those chronological names straight. Be aware, though, that some of the contributors to this book use slightly different subdivisions of Mesozoic time (cf. Fig. 16.5, Table 19.3) than those employed here in Table 9.1.

Ghost Lineages

If the remains of a dinosaur are found in rocks that formed during the Kimmeridgian Age of the Jurassic Period, it is obvious that kind of dinosaur lived during that time interval. A more difficult question, and one that confronts paleontologists of whatever specialty, is how literally one should interpret the absence of fossils of a particular kind in rocks of a given age.

Nobody has found non-avian dinosaur remains in late Cenozoic rocks, although it is certainly possible (if improbable, numerous science fiction books and movies to the contrary notwithstanding) that such fossils may one day be found. Neither have dinosaur bones turned up in sedimentary rocks that formed during the early Paleozoic Era. It is extremely unlikely that any ever will, either—unless our understanding of the development of life over time is completely erroneous! So far, so good.

Where it gets trickier is with respect to rather narrower intervals of geologic time. A good case in point has to do with the evolutionary origin of birds. Cladistic analyses of the characters of early birds such as *Archaeopteryx* strongly suggest that the closest relatives of birds are theropod dinosaurs (cf. Chiappe 1995; Chatterjee 1996; Elzanowski 1996; Forster et al. 1996; Paul 1996; and Wellnhofer 1996 for various recent interpretations of this phylogenetic hypothesis). However, the most bird-like theropods occur in rocks that are much younger than those in which *Archaeop-*

108   *James O. Farlow*

Table 9.1
A Mesozoic Time Scale

Adapted from Gradstein et al. 1994

MESOZOIC ERA

Period	Epoch	Million Years Ago	Age
CRETACEOUS	LATE	65 to 71	Maastrichtian
		71 to 83	Campanian
		83 to 86	Santonian
		86 to 89	Coniacian
		89 to 93	Turonian
		93 to 99	Cenomanian
	EARLY	99 to 112	Albian
		112 to 121	Aptian
		121 to 127	Barremian
		127 to 132	Hauterivian
		132 to 137	Valanginian
		137 to 144	Berriasian
JURASSIC	LATE (MALM)	144 to 151	Tithonian/Portlandian
		151 to 154	Kimmeridgian
		154 to 159	Oxfordian
	MIDDLE (DOGGER)	159 to 164	Callovian
		164 to 168	Bathonian
		169 to 177	Bajocian
		177 to 180	Aalenian
	EARLY (LIAS)	180 to 190	Toarcian
		190 to 195	Pliensbachian
		195 to 202	Sinemurian
		202 to 206	Hettangian
TRIASSIC	LATE	206 to 210	Rhaetian
		210 to 221	Norian
		221 to 227	Carnian
	MIDDLE	227 to 234	Ladinian
		234 to 242	Anisian
	EARLY	242 to 245	Olenekian
		245 to 248	Induan

teryx remains are found. Some paleontologists take this (along with other lines of evidence) to mean that theropods are unlikely to have been the ancestors of birds. Some workers suggest that primitive birds could have been ancestral to some theropods (Paul 1988, 1996; Olshevsky 1991). Others go even further, arguing that similarities between the two groups represent evolutionary convergence rather than close relationship (Hou et al. 1996; Feduccia 1996).

Such a conclusion obviously assumes that our knowledge of when the various kinds of dinosaurs lived is fairly complete; it takes the geologic occurrence of dinosaur faunas at close to face value. On the other hand, if dromaeosaurs or other theropods really are the closest relatives (the sister taxon) of birds, as suggested by cladistic analyses, this means that the theropods in question must have originated earlier in time than is suggested by their presently known fossil record. Such hypothetical geological range extensions of groups of organisms prior to their earliest known fossil occurrences are known as ghost lineages (cf. Norell and Novacek 1992a, 1992b; Weishampel and Heinrich 1992; Norell 1993; Benton 1994; Benton and Storrs 1994, 1996, Storrs 1994). If a particular taxon did originate earlier than its presently known first occurrence in the fossil record, then fossil hunting in the older rocks where it is predicted to occur may eventually reveal that the group did in fact exist then. Different contributors to this book differ in how much weight they are willing to give to the earliest known geologic occurrences of particular kinds of dinosaurs, as opposed to cladistic predictions of when those groups should have originated. The reader should keep this in mind in interpreting the authors' hypotheses of relationships among different kinds of dinosaurs.

References

Albritton, C. C. Jr. 1986. *The Abyss of Time: Unraveling the Mystery of the Earth's Age.* Los Angeles: Jeremy P. Tarcher.

Benton, M. J. 1994. Palaeontological data, and identifying mass extinctions. *Trends in Ecology and Evolution* 9: 181–185.

Benton, M. J., and G. W. Storrs. 1994. Testing the quality of the fossil record: Palaeontological knowledge is improving. *Geology* 22: 111–114.

Benton, M. J., and G. W. Storrs. 1996. Diversity in the past: Comparing cladistic phylogenies and stratigraphy. In M. E. Hochberg, J. Clobert, and R. Barbault (eds.), *Aspects of the Genesis and Maintenance of Biological Diversity*, pp. 19–40. Oxford: Oxford University Press.

Chatterjee, S. 1996. Origin and early evolution of birds and their flight. In Society of Avian Paleontology and Evolution, 4th International Meeting, Program and Abstracts, pp. 2–3. Washington, D.C.

Chiappe, L. M. 1995. The first 85 million years of avian evolution. *Nature* 378: 349–355.

Dott, R. H. Jr., and D. Prothero. 1994. *Evolution of the Earth.* 5th ed. New York: McGraw-Hill.

Elzanowski, A. 1996. A comparison of jaws and palate in the theropods and birds. In Society of Paleontology and Evolution, 4th International Meeting, Program and Abstracts, p. 4. Washington, D.C.

Feduccia, A. 1996. *The Origin and Evolution of Birds.* New Haven, Conn.: Yale University Press.

Forster, C. A.; L. M. Chiappe; D. W. Krause; and S. D. Sampson. 1996. The first Mesozoic avifauna from eastern Gondwana. *Journal of Vertebrate Paleontology* 16 (Supplement to no. 3): 34A.

Gradstein, F. M.; F. P. Agterberg; J. G. Ogg; J. Hardenbol; P. van Veen; J. Thierry; and Z. Huang. 1994. A Mesozoic time scale. *Journal of Geophysical Research* 99 (B12): 24,051–24,074.

Hou, H.; L. D. Martin; Z. Zhou; and A. Feduccia. 1996. Early adaptive radiation of birds: Evidence from fossils from northeastern China. *Science* 274: 1164–1167.

Norell, M. A. 1993. Tree-based approaches to understanding history: Comments on ranks, rules, and the quality of the fossil record. *American Journal of Science* 293A: 407–417.

Norell, M. A., and M. J. Novacek. 1992a. Congruence between superpositional and phylogenetic patterns: Comparing cladistic patterns with fossil records. *Cladistics* 8: 319–337.

Norell, M. A., and M. J. Novacek. 1992b. The fossil record and evolution: Comparing cladistic and paleontologic evidence for vertebrate history. *Science* 255: 1690–1693.

Olshevsky, G. 1991. *A Revision of the Parainfraclass Archosauria Cope, 1869, Excluding the Advanced Crocodylia.* Mesozoic Meanderings no. 2.

Paul, G. S. 1988. *Predatory Dinosaurs of the World: A Complete Illustrated Guide.* New York: Simon and Schuster.

Paul, G. S. 1996. Complexities in the evolution of birds from predatory dinosaurs: *Archaeopteryx* was a flying dromaeosaur, and some Cretaceous dinosaurs may have been secondarily flightless. In Society of Avian Paleontology and Evolution, 4th International Meeting, Program and Abstracts, p. 5. Washington, D.C.

Rudwick, M. J. S. 1976. *The Meaning of Fossils: Episodes in the History of Palaeontology.* 2nd ed. New York: Neale Watson Academic Publications.

Storrs, G. W. 1994. The quality of the Triassic sauropterygian fossil record. *Révue de Paléontologie* 7: 217–228.

Weishampel, D. B., and R. E. Heinrich. 1992. Systematics of Hypsilophodontidae and basal Iguanodontia (Dinosauria: Ornithopoda). *Historical Biology* 6: 159–184.

Wellnhofer, P. 1996. The meaning of *Archaeopteryx*, a critical review in the light of new discoveries and recently published literature. In Society of Avian Paleontology and Evolution, 4th International Meeting, Program and Abstracts, p. 20. Washington, D.C.

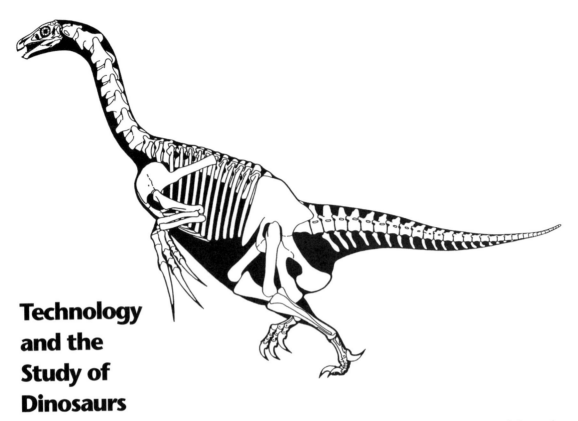

Technology and the Study of Dinosaurs

Ralph E. Chapman

10

The types of research done by dinosaur paleontologists and the tools available to them to carry out this research have changed dramatically over the past 150 years. In other chapters in this book, authors discuss various aspects of the biology of dinosaurs, including behavior, physiology, relationships among dinosaur groups, growth, sexual dimorphism, and feeding. How can scientists make such inferences about organisms that have been dead for at least 65 million years? It isn't easy, but we have a lot of help from technology and data analysis.

In the nineteenth century, the typical procedure was to go to the field and find fossils, prepare, identify, and describe them, and then move on to the next set of samples or go back into the field. To some degree, this basic approach has continued until the present day. But today's dinosaur paleobiologist can use methods far superior to those used by his or her predecessors to look for and collect dinosaurs in the field and then describe and illustrate them. Nowadays such efforts are just the beginning of the work; the paleontologist applies those traditional, descriptive studies within more theoretical contexts. To do this, he or she also must have knowledge of areas such as computers, mathematics, statistics, phylogenetics (the study of relationships among organisms), ecology, taphonomy (the process of fossilization), and a host of others.

Happily, new theoretical approaches and advanced technology are making it easier to study dinosaurs and the world they inhabited. In this chapter I will discuss how advanced technology has affected and will affect the ways we study dinosaurs, and what theoretical approaches now are available to the dinosaur paleobiologist. A big part of technology's impact is seen in the way we now collect, manage, and use data, and how this will change in the future.

Gathering Data

Technology now provides an array of equipment and approaches that can allow the dinosaur paleobiologist to collect and analyze data better, more easily, and more creatively than previously was possible. Some of these methods are well established, others are just now being applied, and others are soon to be realized (e.g., virtual reality).

Describing the Location of Fossils and Fossil Sites

The changes begin in the field. In chapter 6 of this volume, David Gillette discusses new ways of figuring out where the dinosaurs are in the ground, including techniques such as geophysical diffraction tomography (Witten et al. 1992). This is just a beginning, however. Various other types of information must be recorded to allow further analysis, beginning with determining exactly where the fossils have been found. In the old days paleontologists used their trusty compasses to plot locations on maps of varying quality using whatever landmarks they could find. Nowadays, additional help comes from Global Positioning Systems (GPS), which allow users to establish their position on the earth within tens of meters or less. The recent Mongolian expeditions by the American Museum of Natural History used GPS devices to keep track of their positions in a badly mapped and isolated area (McKenna 1992). These location data can then be plotted in the laboratory using Geographic Information Systems (GIS), which automate the coordination of geographic with other types of data.

Once a global position is determined, the distribution and orientation of the fossils should be carefully mapped. This can help the researcher make inferences about how they were deposited, whether different remains are articulated or at least associated, or what the vertical (time) relationships of different specimens are. Previously, compass readings and distances obtained with tape measures were the best tools, and they are still useful. However, Electronic Distance Measurement (EDM) devices and other surveying equipment can allow the positions of fossils to be determined with millimeter resolution in three dimensions over large areas, and to have this information fed directly into the computer. This method has yet to be used extensively with dinosaur sites, but it is well established for paleo-hominid sites, such as Olorgesailie in Kenya (see Jorstad and Clark 1995), where the whole outcrop area has been mapped, as has the position and orientation of each fossil bone. Sonic digitizers have been used in this way to determine and store locations for the Rancho La Brea fossil vertebrate material (Jefferson 1989). Some high-resolution GPS systems also supply three-dimensional data that can be similarly used to produce detailed three-dimensional maps of quarries, which can then be explored in numerous ways using GIS, Computer-Aided Design (CAD), or three-dimensional modeling programs. With this information, the paleontologist then has the data needed to ask important questions, such as whether herd behavior is indicated by footprint evidence, or whether the bones in a quarry have all been oriented by a flooding river to form a bonebed.

Technology and the Study of Dinosaurs   113

Technology in the Laboratory

In the laboratory the potential of technology is just as great, starting with specimen preparation. Computed tomography (CT) and other imaging technologies (see also Sochurek 1987) should progressively revolutionize how we prepare and study fossils. Preparation can be done more intelligently and quickly if we know what material, if any, is available in the matrix before we start. Standard x-rays have been used for many years to study fossils (see Zangerl and Schultze 1989), but three-dimensional techniques are changing things rapidly. Paleontologists have used CT scans for years to see whether fetal dinosaur material is contained in fossil eggs before going through the destructive processes necessary to look for it in more conventional ways (cf. Hirsch et al. 1989). CT scans also have been used to view skulls that were prepared many years ago to see what parts of the dinosaurs are genuine fossil and what parts have been reconstructed by past preparators (Gore 1993). In the relatively near future, preparators should have a three-dimensional representation of their specimens available to view as they work.

Three-dimensional imaging will have the bonus effect of allowing the automated casting of fossils (also known as prototyping or stereolithography). That is, once three-dimensional data are available for specimens, that information can be used non-destructively by computer-driven processes to produce plastic, metal, paper, or other 3-D reconstructions (see Burns 1993). This has been done for human skeletal material, especially teeth but also whole skull regions, and easily can be adapted to dinosaurs. In my laboratory, we have managed to digitize a hadrosaur toe bone and generate a computer model and a non-destructive cast using stereolithography. Currently, the costs can be prohibitive, but they are declining at a fast rate and should be reasonable within a few years.

A further bonus from prototyping will be the ability to produce casts at whatever scale we wish while maintaining the original shapes, allowing a paleontologist studying ceratopsian heads, for example, to have casts of all important material in a single room by scaling them down to a manageable size. Anyone who has tried to take a measurement off a large *Triceratops* skull knows just how helpful that will be! This also should make casts more readily available to researchers, giving them better access to crucial material. Because preparation and casting can be destructive processes, in special cases specimens may not have to be prepared completely or at all. Instead, the three-dimensional scan will be used to reconstruct and produce casts of specimens that are still in matrix.

A very important application will be the publication of high-resolution scans or other three-dimensional representations of important specimens in a form that will allow many paleontologists to explore their anatomy. This is just starting to happen and was first done on a CD-ROM for the skull of the mammal-like reptile *Thrinaxodon* by Tim Rowe, William Carlson, and William Bottorff (1993) of the University of Texas. Using this technique, paleontologists can view successive slices from a high-resolution CT scan of the skull going from front to back, side to side, and top to bottom. Additionally, the major descriptive literature on the specimen was reprinted on the CD-ROM. Clearly, this approach will allow more researchers to have access to the small number of widely dispersed yet crucial dinosaur specimens.

The next step will be to use three-dimensional modeling, virtual reality, and three-dimensional imaging to reduce the need to use the original specimen in many types of research, thus subjecting it to less wear. Many measurements can be made off three-dimensional images taken with sufficient resolution. Furthermore, researchers will have the option of viewing specimens easily from any angle, including from the inside. Such images and methods will have obvious uses in education, illustration, and exhibition, as well as less obvious applications for the analysis of animal function. For example, it should be possible to combine three-dimensional images with assumptions on eye morphology to get a theropod's-eye view of the world.

Thus dinosaur paleontologists will not have to travel as much to examine the specimens they need for their research. Instead, researchers, universities, and museums will trade casts and/or three-dimensional images for most research needs. Many currently costly, long-term transactions will take place almost instantaneously and far less expensively through electronic communication.

After preparation, dinosaur paleobiologists have many options available to them for obtaining data relevant to the study of dinosaur biology. Measurements which once were made by tape measure and ruler, and later calipers, can now be taken using electronic calipers which input their data directly into computers, or two-dimensional and three-dimensional digitizers which allow the coordinates of points to be input from photographs (two-dimensional) or specimens (three-dimensional) at a much faster rate (sometimes more than 100 times as fast), with no loss of resolution, or at even higher resolution than previously available. In my laboratory at the National Museum of Natural History, I have a three-dimensional digitizer that allows me to place a specimen on a dais-like structure and, by putting a stylus on a point, input three-dimensional data with a resolution of about 0.5 millimeters. Other technologies for automatically scanning in three-dimensional surfaces are available at even higher resolutions. Within a few years, many measurements will be made using three-dimensional images within CAD or visualization packages.

Image analysis systems also are being used and will become more and more popular, especially to those studying the microstructure of dinosaur bones, teeth, or eggs. Attaching a video camera to a microscope and viewing slices of bone allows the paleontologist to automate the process of measuring the vascularity of bones, taking it from a very difficult task to one accomplished automatically in less than a few seconds (see Chinsamy 1993a, 1993b).

This is only a small sample of how advanced technology is revolutionizing the way we gather data on dinosaurs. The next steps are to store those data and then do something with them using analytical methods. These also have been areas of incredible change over the past few years because of technological advances.

Data Management

Once you have gathered information, how do you store it? The optimal situation should make it available quickly to the researcher during his or her work, and later have it and other general data available to all researchers in an easily retrievable form.

In the past, specimen data have been kept, sometimes quite haphazardly, by individuals in notebooks, which frequently disappeared when they died. Occasionally field notebooks and laboratory notes were kept and made available to others, but this was far too seldom, and many specimens today are either worthless or less scientifically valuable than they should be because adequate collecting and preparation data are no longer available for them. Some organizations (e.g., the Smithsonian Institution and the United States Geological Survey) have long had procedures in place to retain field and laboratory notes for later access by other workers, but this has tended to be the exception and not the rule.

Clearly this situation must change, and technology is helping considerably. The primary tool for change is the computer database, and many paleontologists have been developing such databases for bibliographic and specimen data for more than two decades. Many museums have computerized their collections and are continuing to improve the quality and quantity of the data stored. Many researchers have their own computerized databases for their bibliographic references, as well as other databases for whatever data they can gather on the specimens they are studying.

Computer databases are collections of records—one for each specimen, bone, bibliographic reference, or whatever unit is being studied. For each record, data are stored in various fields—the individual categories for each type of information being stored. For bibliographic data there will be fields for author(s), date of publication, journal or book references, page numbers, key words, and so on. For specimen data there might be fields for taxonomic data, locality data, important measurements, descriptions of the bones found and what conditions they are in, etc. Databases of information for thousands of references or specimens can then be stored and queried for subsets of data. If a researcher wants to know what papers have discussions on ankylosaurs, or dinosaurs from Thailand, for example, it will be at his or her fingertips in less than a few seconds, even if there are more than 10,000 records in the database. International standards have yet to be developed for such databases, however, and such standards will be necessary before these data are fully accessible to everyone. It is also important to remember that collections databases tend to be very different from research databases. Researchers generally will have to subsample and augment collections databases in developing their own for research.

The most difficult problem is assembling the data and entering them into the database in the first place, which can require a very big time investment. I have already discussed some of the technological advances that will help reduce the time involved in data collection. Scanning devices with optical character recognition (OCR) capabilities also will help to input and interpret data already on paper but currently unavailable in computer-usable form. To reduce the overall time and work involved and combine the efforts of many dinosaur workers, we must develop comprehensive bibliographic and specimen databases for use by the whole research community, and establish international standards for them. The Society of Vertebrate Paleontology is already working on transferring its *Bibliography of Fossil Vertebrates* fully to electronic media. *A Bibliography of the Dinosauria* by Dan Chure and Jack McIntosh (1989) was not released in digital form, but subsequent editions and other bibliographies will be.

The importance of the development, upkeep, and general availability of major databases on dinosaur specimens should not be underestimated. One of the most difficult and costly aspects of a dinosaur paleontologist's research is finding out what important specimens are out there, many of which have lurked in drawers for fifty or more years without being described, and getting to view, photograph, and measure them. Database development should greatly reduce these costs and allow paleontologists to make better use of the limited travel funds they have.

116   *Ralph E. Chapman*

Furthermore, for some important specimens, these data can reduce unnecessary handling and measurement, which can be destructive. In the zoological collections at the National Museum of Natural History (Smithsonian Institution), crucial primate specimens have been measured so often that the bones have been worn down and the measurements can no longer be made accurately. This is a problem that will only get worse, and efforts must be made to provide the needed data and yet protect the specimens. Developing comprehensive databases and making them widely available should go a long way toward reducing these problems.

The development of the Internet also should facilitate this process considerably, although the extremely rapid evolution of the system makes it difficult to keep up with it without support (cf. Levine and Baroudi 1993; Lambert and Howe 1993). Electronic mail communication will continue to greatly improve communication among paleontologists. Computer bulletin boards and discussion groups on the Internet are used moderately to heavily by professionals, and their numbers are increasing rapidly. The Society of Vertebrate Paleontology has a discussion group for vertebrate paleontologists, and others exist for dinosaur enthusiasts that generate mail daily on topics such as bird origins, what new discoveries were announced recently, and general theories on dinosaur biology.

The most exciting part of the Internet for professionals, however, is the ability to explore data files from great distances and transfer data from one system to another. FTP (File Transfer Protocol) and RCP (Remote Copy Protocol) procedures allow researchers to move data files, images, manuscripts, and programs onto their computers from other systems worldwide. Telnet allows users to log on to other computers, or their home systems, from distant or remote sites, including the field if cellular technology is available.

Many systems have been developed to facilitate this global exchange of information using the Internet. The Archie system developed by McGill University allows the available files for many sites to be cataloged, indexed, and searched. WAIS (Wide Area Information Servers) allow the contents of documents to be searched as well as their titles. These systems are then used and searched using Gopher systems and, in some cases, the World Wide Web (WWW).

Gopher systems—the name derived from the mascot of the University of Minnesota (the home of the first Gopher)—have been developed to allow Internet users to explore large sets of data stored in various computers worldwide, through sets of organized menus. For example, the Gopher system at the National Museum of Natural History contains information on various natural history themes, from common names for crayfish, to hadrosaur bibliographic data, to data for various National Museum of Natural History collections.

The World Wide Web is a hypertext/hypermedia-based system for searching and using this system within a mouse-driven Windows context while incorporating text, images, and, in some cases, animation and video. Using Archie, WAIS, Gophers, and the WWW, you can find out what computers have information stored on various topics (such as dinosaurs), search the data, get the information and/or download the files for further use, and view images and animation.

Most museums are proceeding toward making their collections data available to users remotely in some way or another. Combining this with the WWW also makes it possible to tour exhibits graphically from remote distances. The only stumbling blocks appear to be the progressive expansion of the Internet, getting it to more and more people, slow access speeds due to this exponential expansion, especially in viewing images and animation within the WWW, and a host of legal and non-trivial issues concerning data control, copyrights, and related issues. As I write this, discussions are raging on discussion groups regarding how detailed locality data should be on Internet databases. Too much information on localities may leave them vulnerable to being over-collected or even destroyed. It should make for an interesting few years.

Data Analysis

Once the data are collected and stored in a retrievable form, we need to use this information to make inferences about dinosaur biology and earth history. To do this, we apply various data analysis techniques that use statistics, geometry, mathematics, and other procedures. Through the years, paleontological studies have become progressively more rigorous as these methods have been used more and more.

What kind of research are we talking about? Just about anything theoretical. When researchers suggest that some theropods could run in excess of 20 miles per hour, they do so on the basis of analyses of footprint data or biomechanical capabilities inferred from bones and muscle attachments. Other paleontologists can and do dispute these results by offering data of their own. Analysis can show which parts of the Mesozoic world had more diverse dinosaur faunas, suggest whether a group of dinosaur footprints shows herding behavior, or suggest what dinosaurs are most closely related to each other. I will discuss some general categories of analysis and give a few examples. The list will be far from exhaustive because of space constraints, but it should give a general feel for the types of studies being done now.

Modeling, Simulation, and Functional Morphology

Modeling, simulation, and functional morphology all attempt to emulate, describe, and help explain the natural world with mathematical, statistical, algorithmic, logical, and/or visual tools. The techniques used by each vary from study to study, and it consequently is difficult to typify the subjects with a few examples. However, I will give some examples that illustrate the types of approaches possible.

Modeling

Modeling and scientific visualization (see Nielson and Shriver 1990) is an attempt to examine the complex natural world with simplified models of various types. Many forms are applicable to dinosaurs, and I will discuss morphological modeling as an example. Here, complex structures (e.g., skulls, footprints) are used to generate simpler shapes, either by reconstructing them using mathematical equations, or by digitizing the original in two or three dimensions. The results then are viewed graphically using powerful computers and are compared with the original. In some cases, the reason for doing this is simply to be able to view morphology more easily and from all angles. In other cases, the researcher is trying to test the capability of the model to describe the original fossil, and will use this as a way of deducing how the structure evolved or functions. To give a non-dinosaur example, most organisms that coil during growth (e.g., snails, *Nautilus*) can have their morphology modeled using just a few mathematical coefficients. Their evolution and variation can then be studied by varying these few coefficients and viewing the results (e.g., Raup 1966).

One example of this kind of approach is the construction of three-dimensional morphological models of Early Cretaceous dinosaur footprints from Texas developed by Farlow (1991, 1993; see also Farlow and Chapman, chap. 36 of this volume). The data were generated using a three-dimensional digitizer and a software package that could generate topographic and surficial maps (the wire-net diagram). More advanced work in development by Farlow and me will include solid modeling and, we hope, the automated casting in plastic or plaster of the footprint at a much smaller scale as a test for similar work on fossil bone material.

Simulation

Simulations are attempts to model biological or geological processes using algorithmic approaches. This entails starting with a set of instructions and following a series of steps to produce a final outcome, generally with a strong random component added. They have seldom been applied directly to dinosaur data, but there is great potential here. Simulations have illustrated expected lithospheric plate movements during continental drift, and have been used to model climate at various times during the Mesozoic (cf. Molnar, chap. 38 of this volume). Behrensmeyer and Chapman (1993) used simulations to explore the interaction of various factors during the development of vertebrate bonebeds. Alan Cutler, A. K. Behrensmeyer, and I are using these simulations further to examine the Cretaceous-Tertiary boundary event. Results suggest that, even with a catastrophic event, we should not expect to see massive beds of dead animals in the fossil record (Cutler and Behrensmeyer in press). There are many other possibilities for learning about dinosaurs with simulations, and we are just starting to explore them.

Functional Morphology

Functional morphology is the study of how organisms operate as living systems. For dinosaurs, this can include locomotion (Heinrich et al. 1993), architectural analysis of specific bones (Weishampel 1993), and a wide range of other studies. It would include, for example, studies calculating the running or walking speeds of dinosaurs by noting their footprint size and the distance and angles of the footprints in a series made by a single animal (Alexander 1989; Farlow 1991, 1993; Lockley 1991). Another example would be studies determining how large sauropods could structurally support their exceptionally long necks. One of the more interesting series of functional studies was conducted by Weishampel and Norman (1989; many references therein) on the very complex and kinetic chewing mechanism present in advanced ornithopods. Functional studies can be very difficult because they frequently utilize complex mathematics or very detailed anatomical examinations of bone. However, they provide much of the best information we have on the biology of dinosaurs for use in restoring their life appearance (Paul 1987). This can lead to incidental developments such as the development of dinosaur-based robots (Poor 1991: 61–63) and the application of computer animation like that used in

Technology and the Study of Dinosaurs 119

the movie *Jurassic Park* (Shay and Duncan 1993). More accurate morphologic models can be used to explore the capabilities of dinosaurs in real life.

Morphometrics

Morphometrics is the quantitative analysis of shape. Any time a researcher states that one specimen looks like another, or suggests that there are male and female forms of *Tyrannosaurus rex*, or tries to discriminate among species based on morphology, some form of morphometrics is being used. Related to morphometrics is allometry, the study of size and its consequences (cf. Alexander 1989), which, understandably, is of great interest to dinosaur researchers.

In the grand old days of paleontology, a researcher's intuition was the main basis for inferences made about how shape varies in dinosaurs. Later, single measurements and ratios of pairs of standard measurements were used to identify different taxa (singular *taxon*) (e.g., Brown and Schlaikjer's [1943] classic study of pachycephalosaurids, then called troodontid dinosaurs). These univariate (one-variable) and simple bivariate (two-variable) approaches were problematic, however, because the great size changes that typify dinosaur growth series made single measurements generally useless for this purpose. Ratios were just as problematic, because a single species will generally exhibit different ratios during growth, and because more than one species can have the same ratios with very different growth curves.

More sophisticated bivariate allometric studies began with S. W. Gray's (1946; Lull and Gray 1949) analyses of ceratopsian dinosaurs and continue to the present day. It is through these types of allometric studies that functional morphologists can infer the capabilities and behavior of dinosaurs (see Alexander 1989 and chap. 30 of this volume for an excellent discussion of size and its consequences).

Morphometric methods typically become more powerful as more variables are added. Multivariate (many variable) analyses were the next step,

Figure 10.1. Allometry of *Stegoceras*. Measurements used for multivariate morphometric analysis of the cranium of *Stegoceras* (Chapman et al. 1981). (A) Lateral view of a skull with shading indicating the dome elements typically found for that form and used in the study. (B–D) Dorsal, lateral, and ventral views of an element with the 15 variables measured on each indicated. The braincase is the figure 8–shaped element in the ventral view (D).

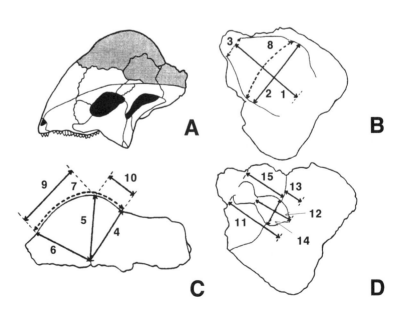

typified by pioneering studies by Peter Dodson (1975, 1976) on lambeo-saurine hadrosaurs and ceratopsians, and later work by various research-ers, such as Chapman et al. (1981) on *Stegoceras* and Weishampel and Chapman (1990) on prosauropods. These were the first powerful morpho-metric studies that argued convincingly for sexual dimorphism (the pres-ence of differently shaped males and females within a single species), and suggested that many species were invalid and could be subsumed (synony-mized) into others because they merely represented different sexes or different growth stages.

Finally, a new series of morphometric procedures have been developed that allow the original geometry of the specimens to be used in place of isolated measurements. These follow pioneering work by D'Arcy Thomp-son in his book *On Growth and Form* (1942). They include techniques developed by Fred Bookstein (e.g., 1991), Jim Rohlf, Richard Benson, me, and a host of other researchers (see references in Rohlf and Bookstein 1990) and fall into two major categories: outline methods and landmark techniques.

Outline methods fit mathematical functions to the outlines of objects, and have been used to study sand grains, foraminiferan tests, dinosaur footprints, and many other things. Landmark approaches use landmark points, those that can be found on all specimens of interest and represent the same position on all of them (e.g., the intersection of cranial bones at the sutures), to look at shape changes between specimens. Outline methods are useful when the outline is all or most of the data that are available. Landmark approaches are more powerful and provide the best way of extracting a lot of information. Outline techniques are being used by me and various others to study the shapes of footprints, stegosaur plates, and other similar objects. Landmark methods have been used extensively by me, Peter Dodson, and Catherine Forster on various dinosaur groups. I will describe two morphometric examples, one using conventional multivariate analysis and the other landmarks.

Chapman et al. (1981) used conventional multivariate analysis in a study of the pachycephalosaurid *Stegoceras*. Domes from a series of specimens were measured for various skull, braincase, and dome variables (Fig. 10.1). These measurements were subjected to a mathematical procedure called Principal Components Analysis (PCA), which allows the fifteen-dimensional data (there were fifteen original measurements) to be reduced mostly to two major axes, or dimensions, of variation. The specimens are replotted for these two axes in Figure 10.2. We can interpret the output from the PCA in relation to the original variables and in doing so come up with a first axis (x-axis in Fig. 10.2) that is a size axis; those specimens with low values are small domes, and those with high values are larger domes. The second axis (y-axis) is one that reflects braincase size and dome size, and reflects sexual dimor-phism in *S. validum*. Those specimens with high values have relatively large braincases and small domes, and those with low values have relatively larger domes and smaller braincases. Assuming that braincase size is consistent in the species and is an indicator of age, axis 2 shows that we have specimens with big domes and small domes for any given age. This is exactly as one would expect for a structure such as the dome that is inferred to have been used in intraspecific combat (combat between members of the same species), typically by males competing for females (cf. Sampson, chap. 27 of this volume). Other taxa also were included in the study, including other species of *Stegoceras* which clustered with the males of *S. validum*. The one specimen that appeared very different was renamed a new genus, *Gravi-tholus*, by Wall and Galton (1979).

Technology and the Study of Dinosaurs 121

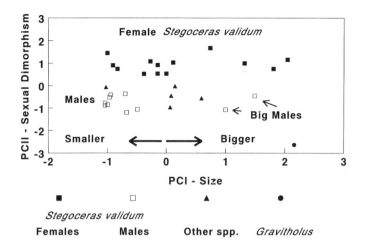

Figure 10.2. Results of multivariate morphometric analysis of *Stegoceras* (Chapman et al. 1981). Plotted are data for 29 specimens for first two axes of Principal Components (PC) Analysis. The x-axis, PC1, is a size axis; specimens plotting to the left are small, and those to the right are big. Those in the middle (0) are about average size. The y-axis (PC2) is the sexual dimorphism axis for *Stegoceras validum*. Specimens with high values for PC2 have relatively large brains and smaller domes. Those with high negative values have large domes and relatively small brains for their dome size. Note that presumed males of *S. validum* have high negative values for PC2, while supposed females high positive values. The other species of *Stegoceras* plot with the males. Note the very large dome for the specimen of *Gravitholus*.

For the landmark analysis I selected excellent illustrations by Greg Paul of two hadrosaurines, *Edmontosaurus* and *Anatotitan*, shown in Figure 10.3. For both dorsal and lateral views of the crania, I selected series of landmark points, indicated in Figure 10.4. I then applied a technique known as Resistant-Fit Theta-Rho Analysis (RFTRA; Chapman 1990a, 1990b; Chapman and Brett-Surman 1990) to each view separately. RFTRA starts with a base specimen (here *Edmontosaurus*) and scales, rotates, and shifts the other skull to optimize its fit onto that specimen without its being distorted. The result is a series of arrows indicating how each landmark point has changed from *Edmontosaurus* to *Anatotitan*, and interpretations can be made from the direction and magnitude of those arrows (Fig. 10.5).

The results of this analysis are rather straightforward. In the top view, the major difference appears to be that the snout of *Anatotitan* is elongated

Figure 10.3. Illustrations used for landmark shape analysis of hadrosaurines. Drawings modified from restorations done by Greg Paul for dorsal (*top*) and lateral (*bottom*) views of *Edmontosaurus* and *Anatotitan*.

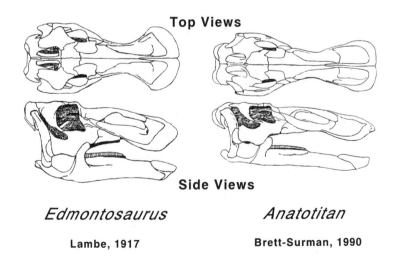

Top Views

Side Views

Edmontosaurus

Lambe, 1917

Anatotitan

Brett-Surman, 1990

122   *Ralph E. Chapman*

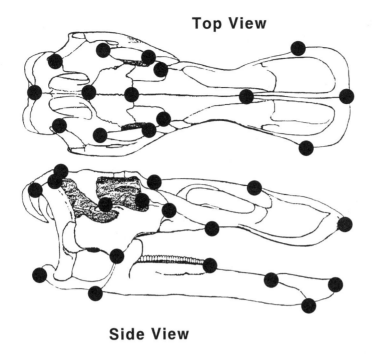

Top View

Side View

Figure 10.4. Landmarks used for shape analysis of hadrosaurines. Each landmark, indicated by the circles, is at intersections of cranial sutures, fenestrae, the orbit, and extreme tips of the snout or jaw. For the top view, 15 landmarks were used, and for the lateral view 17 landmarks.

Top View Results

Side View Results

Figure 10.5. Results from Resistant-Fit Theta-Rho-Analysis (RFTRA) of hadrosaurine crania. Two separate analyses were run, for top and lateral views. Arrows indicate direction and magnitude of change from their position in *Edmontosaurus* to that in *Anatotitan*. Note major changes in the snout area.

relative to that of *Edmontosaurus*. Other differences are small, especially considering that these are two different genera. In side view, the differences become more interesting: not only has the snout been stretched out, but key areas in the jaw area also have shifted. This is to be expected. If a jaw is elongated, this is bound to have a significant effect on the muscles and bones that support chewing by those jaws.

The potential utility of both conventional multivariate morphometric approaches and landmark approaches should be apparent from these two examples. There are other methods that are just as useful, especially within slightly different contexts. Morphometric applications are still relatively rare for dinosaurian material, but this is changing rapidly.

Distributional Studies

Distributional studies attempt to make sense of the distribution of organisms across time and space. When they include paleontological data, they involve the following disciplines: paleoecology, the study of distributions related to ecological parameters; biostratigraphy, the examination of distributions in the rock record with particular reference to time; and paleobiogeography, the analysis of organismal distributions related to geography and tectonics. Not surprisingly, these three disciplines overlap, and major studies must discuss distributions related to all three of these factors.

Dinosaur paleobiologists are only now starting to make headway in these areas because their data have always been so fragmented and limited. Certainly there have been many relevant observations (e.g., noting the similarity of Asian and western North American dinosaurs during specific time periods) and discussions relating dinosaur distributions to continental drift. However, the ongoing collection of new specimens, along with a major compilation of dinosaur localities worldwide (Weishampel 1990), has made it possible to start looking at dinosaur distributions in a more quantitative and synthetic way.

There are many possible approaches to analyzing distributional data. I will give a simple example using one of the more common methods.

Data summarizing the taxa found in various localities, regions, or samples are placed into a data matrix. These data can include the numbers of individuals found for each taxon, the percentage of the fauna they make up, or simply their presence (coded as 1) or absence (coded as 0). For this example I will use presence-absence data for sixteen regions and thirty-one taxa yielding the 16×31 matrix of 1's and 0's shown in Table 10.1 (which also includes data on the age and geographic region for the localities, and the taxa used, mostly families of dinosaurs). The data are derived directly from Weishampel's (1990) study and are a part of a much larger, ongoing study by Weishampel and Chapman.

Data from Weishampel (1990)

Table 10.1.

Data Used for Distributional Analysis of Presence-Absence Data for 31 Dinosaur Taxa for 16 Localities from the Upper Jurassic to the Upper Cretaceous

Data from Weishampel (1990)

Age	Locale	Ref	Taxon Number 1 2 3 1234567890123456789012345678901
JU L	WYO/USA	136	1101111111100000000000000000000
JU L	UTH/USA	137	1111111111100000000000000000000
JU L	COL/USA	138	1111101111110000000000000000000
JU L	CHINA	174	1100100100000000000000000000000
CR E	MON/USA	192	0011000000011011000000000000000
CR E	TEX/USA	202	0010100000011011000000000000000
CR E	ENGL	210	0011001010001101010000000000000
CR E	ENGL	211	1110001010001001000000000000000
CR E	MONG	248	0000000000000011001000000000000
CR E	MONG	252	0000100000010001001000000000000
CR L	ALB/CAN	294	0001000000011010110111111110000
CR L	SAS/CAN	295	0001000000010010100111101100000
CR L	MON/USA	300	0001000000011010110100101110000
CR L	WYO/USA	301	0001000000011110110100101110000
CR L	MONG	363	0000000000010010100100100101000
CR L	MONG	364	0000001100010010110110110111111

0 = Taxon absent from that locality; 1 = Taxon present at that locality.

Abbreviations: JU = Jurassic; CR = Cretaceous; E = Early; L = Late; MONG = Mongolia; ENGL = England; USA = United States of America; WYO = Wyoming; ALB = Alberta; SAS = Saskatchewan; CAN = Canada; TEX = Texas; MON = Montana; COL = Colorado; UTH = Utah; CHINA = People's Republic of China
Reference numbers are locality numbers from Weishampel 1990.

Taxon used for coding; numbers refer to column numbers:
(1) Cetiosauridae; (2) Stegosauridae; (3) Brachiosauridae; (4) Hypsilophodontidae; (5) Allosauridae; (6) Maniraptora I (*Coelurus* and *Ornitholestes*); (7) Camarasauridae; (8) Diplodocidae; (9) Dryosauridae; (10) Camptosauridae; (11) Ceratosauria; (12) Ornithomimosauria; (13) Nodosauridae; (14) Titanosauridae; (15) Dromaeosauridae; (16) Iguanodontia; (17) Hadrosauridae; (18) Pachycephalosauridae; (19) Psittacosauridae; (20) Tyrannosauridae; (21) Elmisauridae; (22) Caenagnathidae; (23) Troodontidae; (24) Segnosauria; (25) Ceratopsidae; (26) Ankylosauridae; (27) Protoceratopsidae; (28) Oviraptoridae; (29) Garudimimidae; (30) Avimimidae; (31) Homalocephalidae.

Technology and the Study of Dinosaurs 125

The next step of the analysis is to decide which units you wish to compare. You have the option of looking at the relationships among the taxa based on their occurrence in the localities, or looking at the relationships of the localities based on the taxa that have been found there. Either way, you have to produce a matrix of inferred differences between the units being studied. This distance matrix is calculated using one of many different mathematical coefficients (see Cheetham and Hazel 1969 for an early review), which have differing properties and are appropriate at different times.

For the present analysis I have used the Dice coefficient because it accommodates data of varying quality pretty well. The result is a distance matrix that is 16 × 16 units for the samples and 31 × 31 units for the taxa. Each entry in the matrix is an estimate of how different each unit is from another unit.

Once you have a distance matrix, the next step is to do something with it, typically by using methods that reduce many-dimensional data to a smaller, more interpretable, number (1–3) of dimensions. These methods include ordinations such as Principal Components Analysis (already discussed under the topic of morphometrics) and cluster analysis. A cluster analysis shows relationships through a dendrogram, a tree-like diagram that clusters or groups together the most similar units being studied, and then progressively adds new units or clusters (groups of units) at larger and larger distances until all are joined. You interpret the resulting patterns by looking for major clusters and noting the relationships of the different units

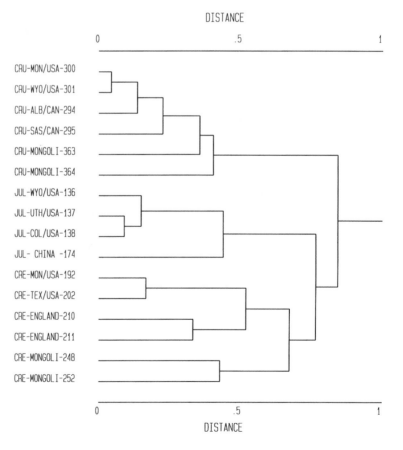

Figure 10.6. Distributional analysis of dinosaur localities; data from Weishampel 1990. Cluster analysis (UPGMA) for 16 × 16 distance matrix (1-Dice Coefficient) calculated for locality data using presence-absence of 31 taxa. Note three major clusters for each time period, and geographic clustering within each major cluster.

126   *Ralph E. Chapman*

or groups of units within these clusters. It is important to note that dendrograms act like mobiles hanging from a ceiling, in that they can rotate at each connection point, so two units that are next to each other may or may not be any more similar to each other than they are to other units. To find the relationship, or inferred difference, between two units, you must follow them to their closest shared connection point.

The results of the cluster analysis run for the localities (Fig. 10.6) show that time is the major control of the dinosaur taxa, with members of each time unit clustering together (note that the three major clusters include those localities with the same age). Within time units, geography contributes next, with North American localities joining together first, later to be joined by European and Asian localities. With more localities and greater control on the environments of deposition (which, unfortunately, may never be possible with dinosaur data), we might expect ecological factors and latitude also to show some effect on the results.

For the analysis of the taxa (Fig. 10.7), the results show clusters of taxa that are found together, again with a strong time imprint. The two main clusters are for the Jurassic and Cretaceous, evident by reexamining the data in Table 10.1. The Psittacosauridae show as an odd group because they are found only in the Early Cretaceous of Mongolia, along with an odd mixture of Cretaceous and more typical Jurassic forms. Within these larger groups, one can also detect clusters, such as those with more typical North American groups, and others with a strong Asian influence. With

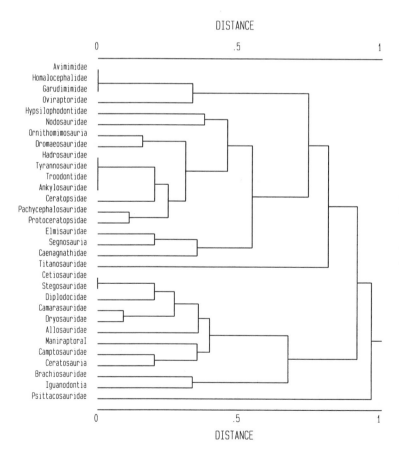

Figure 10.7. Distributional analysis of dinosaur taxa; data taken from Weishampel 1990. Cluster analysis (UPGMA) for 31 × 31 distance matrix (1-Dice Coefficient) calculated for taxon data using their presence-absence at 16 localities. Note major clusters for typical Cretaceous and Jurassic taxa, with some geographic tendencies within each major cluster.

more detailed data, we might be able to detect which taxa tend to inhabit the same ecological areas or might have direct relationships (e.g., predator and prey), but it will be some time, if ever, before we have dinosaur data sufficient for that level of analysis.

Phylogenetic Analysis

Phylogenetic studies attempt to reconstruct the relationships of groups of organisms. Each of these groups typically represents a taxon such as a species, genus, or family. Even under optimal conditions for biological data, developing a single, stable phylogenetic reconstruction is very, very difficult. For paleontological data, this gets progressively more difficult because of missing data resulting from incomplete preservation. Dinosaur data can be among the most limited of all paleontological data, so phylogenetic analyses of dinosaurs can be very difficult to perform. They can still be quite rewarding, however, and provide us with considerable insight to guide further research.

I will demonstrate how paleontologists do phylogenetic analyses with an example analysis on sauropod dinosaurs taken from data published by Russell and Zheng (1993). I have taken their original analysis of nine genera and reduced them to four (*Plateosaurus, Brachiosaurus, Shunosaurus,* and *Apatosaurus*) to simplify the results. The approach I will use is called cladistics, the predominant method used by evolutionary taxonomists at the present time. The example analysis is very simple, but should provide some indication of how such analyses are done. (For good introductions to cladistics, see Wiley et al. 1991; Maddison and Maddison 1992; Swofford and Begle 1993; Holtz and Brett-Surman, chap. 8 of this volume).

The goal of any phylogenetic analysis is to produce a reconstruction or tree for the forms being studied, in this case the four genera. The tree graphically represents the relationships among the taxa by grouping them together. Trees are made up of nodes and branches. Nodes are the taxa which are connected by the branches. The nodes include the taxa being studied (terminal nodes), which are on the tips of the branches, and internal nodes, which are inferred taxa that include all nodes beyond them (closer to the top of the diagram) on the tree. Internal nodes with all their descendants are called a clade. If the taxa being studied are genera, internal nodes represent groupings of the genera and are roughly equivalent to subfamilies, families, and the higher taxa of more classical approaches to classification.

In our example, the genera are the terminal nodes, and an internal node connecting *Brachiosaurus, Shunosaurus,* and *Apatosaurus,* for example, would represent the conventional group Sauropoda, which is a clade. It is important to note that cladistic analyses do not generate internal taxa following conventional taxonomic hierarchies. Instead, internal nodes contain all taxa subsequent to them on the tree. For example, if birds evolved from theropod dinosaurs, then Aves becomes a subgrouping within the clade Theropoda, which itself is a subgroup of the clade Dinosauria. Two related and neighboring taxa on the tree are called sister taxa.

How do we generate trees? Researchers try to obtain the simplest tree possible for the taxa being studied, following the philosophy of parsimony. They do this by encoding the data for the taxa in various ways into characters, which vary within the tree. This coding most often is done as 1's and 0's, with the latter most often representing the primitive state of the character. There are other ways of coding characters as well. Characters that define a group are called apomorphies, or derived characters, and most taxa are characterized by a number of different apomorphies. Benton (1990), for example, lists more than twenty apomorphies defining the group Theropoda.

In our example, Russell and Zheng (1993) were able to encode twenty-one characters for the four genera (Table 10.2). These characters describe the taxa in a variety of ways and include the presence or absence of a structure (e.g., a tail club as part of character #19), counts of the number of other structures (e.g., dorsal vertebrae in character #14), different shapes of structures (e.g., the cross-section of the dentary teeth in character #7), or some other general condition (e.g., whether the axis spine is low or high in character #8).

Table 10.2.

Characters Used for Cladistic Analysis of Four Sauropodomorphs

Data taken from Russell and Zheng 1993. Character 13 was invariant for the four genera selected for this example so it was not included.

Taxon	Character Number																			
	1	2	3	4	5	6	7	8	9	10	11	12	14	15	16	17	18	19	20	21
Plateosaurus	0	0	0	0	0	0	0	0	0	0	0	0	0	0	0	0	0	0	0	0
Brachiosaurus	0	1	1	0	0	1	0	1	0	1	1	0	1	1	0	0	0	0	1	1
Shunosaurus	0	1	0	0	0	0	0	1	0	0	0	0	0	1	1	0	1	2	0	0
Apatosaurus	1	0	1	1	1	1	1	1	1	1	1	1	0	1	1	1	1	1	1	1

Characters (for more details see Russell and Zheng 1993):
 1. Pterygoid flange axis angle orthogonal (0) or acute (1).
 2. Posterior face of quadrate excavated shallow (0) or deep (1).
 3. External mandibular fenestra present (0) or absent (1).
 4. Symphyseal ramus of dentary curved (0) or sharply angled (1).
 5. Alveolar margin extends beyond (0) or is restricted to (1) anterior half of dentary.
 6. Number of dentary teeth more (0) or less (1) than 15.
 7. Dentary teeth laterally compressed (0) or cylindrical (1) in cross-section.
 8. Axis spine low (0) or high (1).
 9. Middle cervical vertebrae spines unsplit (0) or deeply split (1).
 10. Lateral surfaces of cervical centra shallowly excavated (0) or bear deep excavations (1).
 11. Cervical centrum length/dorsal centrum length ratio <2.0 (0) or >2.0 (1).
 12. Cervical ribs greatly exceed centrum (0) or are subequal (1).
 14. Number of dorsals is >12 (0) or ≤12 (1).
 15. Height of dorsal neural arches roughly equal to (0) or twice (1) centrum height.
 16. Dorsal spine surface primitive (0) or with reduced suprazygapophyseal laminae and vertical crest on lateral surface (1 = Russell and Zheng state 2).
 17. Basal caudal centra amphiplatyan (0) or procoelous (1).
 18. Median caudal haemal spines single (0) or divided (1).
 19. End of tail unspecialized (0), whiplash (1), or club (2).
 20. Distal ends of ischia compressed (0) or broadly everted and flat (1).
 21. Astragalus outline in dorsal aspect quadrilateral (0) or triangular (1).

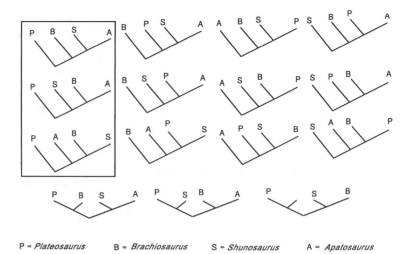

Figure 10.8. Possible tree configurations (dichotomous branching only) for cladistic analysis of four prosauropod/sauropod genera. The basal node for all phylogenies is roughly equivalent to the Sauropodomorpha—the group that contains prosauropods and sauropods. The three trees enclosed in a box are the three phylogenies that result if *Plateosaurus* is used as the outgroup in the analysis. In that case, the internal node just up from the basal node in these three trees would be equivalent to the Sauropoda.

P = *Plateosaurus*   B = *Brachiosaurus*   S = *Shunosaurus*   A = *Apatosaurus*

Characters change within trees in steps. The coding for a specific character may change along a branch from character state 0 to state 1, or from 1 to 0, or from 1 to 2 depending on how the characters vary within the analysis and how they are coded. Each change is a step, and a shift back to the original state is termed a reversal. Characters that change to the same state at two or more different places within the tree show homoplasy, representing either miscoding of the character or the evolutionary phenomenon of convergence or parallelism. The best trees exhibit the smallest number of steps and minimize the number of reversals and the amount of homoplasy. This becomes difficult when the number of characters becomes large, as is shown in Table 10.3. Assuming just dichotomous branching (only two branches allowed from each internal node), the number of potential trees exceeds 1 million with only nine characters. This is why computer programs with efficient testing algorithms are necessary for nearly all analyses. The two most commonly used now are *MacClade* (Maddison and Maddison 1992) and *PAUP* (Swofford and Begle 1993).

Table 10.3.

Number of Topologically Different Phylogenetic Trees Possible with Increase in the Number of Taxa Used

Data from Wiley et al. 1991. Given are the number of taxa, the number of all trees, including those with polytomies (nodes with more than two branches), and the number of dichotomous trees (those restricted to nodes with only two branches).

Number of Taxa	Number of All Trees	Number of Dichotomous Trees
3	4	3
4	26	15
5	236	105
6	2,752	945
7	39,208	10,395
8	660,032	135,135
9	12,818,912	2,027,025
10	282,137,824	34,459,425

130   *Ralph E. Chapman*

The analysis can be improved by using an outgroup, a related taxon that provides the starting value for the characters and serves to root the tree. In our example, *Plateosaurus* serves as the outgroup to the sauropods. The data matrix that was generated by Russell and Zheng (summarized here as Table 10.2) then provides the basic input to the computer programs. Following Table 10.3, and as illustrated in Figure 10.8, there are fifteen possible tree configurations for the four genera. The possibilities are limited to three, however, by the designation of *Plateosaurus* as the outgroup (see the three trees enclosed by a box in Fig. 10.8). These data were run through the two programs.

The output from *PAUP* is given in Figure 10.9 and illustrated more conventionally from *MacClade* in Figure 10.10. Statistics describing the three different trees also are given in Figure 10.9. Note that tree length is smallest for the middle tree in both Figure 10.9 and Figure 10.10, making *Brachiosaurus* and *Apatosaurus* sister taxa, with *Shunosaurus* joining further back to define the Sauropoda. The length was twenty-four steps compared with twenty-eight for the bottom tree in Figures 10.9 and 10.10,

```
Tree description:

    Unrooted tree(s) rooted using outgroup method
    Character-state optimization: Accelerated transformation (ACCTRAN)

Tree number 1 (rooted using user-specified outgroup):

Tree length = 29
Consistency index (CI) = 0.724
Homoplasy index (HI) = 0.276
CI excluding uninformative characters = 0.529
HI excluding uninformative characters = 0.471
Retention index (RI) = 0.111
Rescaled consistency index (RC) = 0.080

                                /------------------- Brachiosaurus
                 /---------------------5------------------- Shunosaurus
/--------------------6------------------------------------- Apatosaurus
\--------------------------------------------------------- Plateosaurus

Tree number 2 (rooted using user-specified outgroup):

Tree length = 24
Consistency index (CI) = 0.875
Homoplasy index (HI) = 0.125
CI excluding uninformative characters = 0.750
HI excluding uninformative characters = 0.250
Retention index (RI) = 0.667
Rescaled consistency index (RC) = 0.583

                                /------------------- Brachiosaurus
                 /---------------------5------------------- Apatosaurus
/--------------------6------------------------------------- Shunosaurus
\--------------------------------------------------------- Plateosaurus

Tree number 3 (rooted using user-specified outgroup):

Tree length = 28
Consistency index (CI) = 0.750
Homoplasy index (HI) = 0.250
CI excluding uninformative characters = 0.562
HI excluding uninformative characters = 0.438
Retention index (RI) = 0.222
Rescaled consistency index (RC) = 0.167

                                /------------------- Shunosaurus
                 /---------------------5------------------- Apatosaurus
/--------------------6------------------------------------- Brachiosaurus
\--------------------------------------------------------- Plateosaurus
```

Figure 10.9. Output for phylogenetic analysis of sauropods from program *PAUP*. Characters taken from Russell and Zheng 1993.

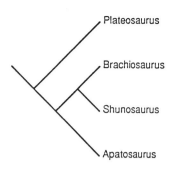

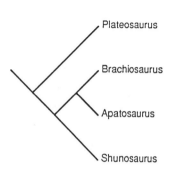

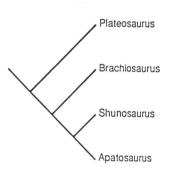

Figure 10.10. Output for phylogenetic analysis of sauropods from program *MacCLADE*. Characters taken from Russell and Zheng 1993.

which connects *Apatosaurus* and *Shunosaurus,* and twenty-nine for the other combination shown in the top of the illustrations.

Other statistics (Fig. 10.9) likewise suggest that the middle tree is best. These include the consistency index, which is an indication of the efficiency of the tree at representing the character changes. This index is the ratio of the shortest possible treelength to the treelength obtained in the analysis, and higher is better. The retention index is similar, but is calculated using a different algorithm. Finally, the *homoplasy index* is an estimate of the effect of homoplasy on the tree, and lower values are better. In all cases, the middle tree in both figures is the best, based on the characters and taxa used.

The next step is to examine the best trees generated by the programs, evaluating the results by noting changes in the characters, and determining which act as apomorphies at each node. In our example, characters 8 and 15 are apomorphies for the Sauropoda, and characters 2, 6, 10, 11, 20, and 21 are apomorphies for the *Brachiosaurus-Apatosaurus* clade. Problem characters, those showing reversals, include 16, 18, 19, and 2; the first three would define an *Apatosaurus-Shunosaurus* clade and the last a *Brachiosaurus-Shunosaurus* clade. Apomorphies defining the terminal nodes include #14 for *Brachiosaurus,* character state 2 for character #19 for *Shunosaurus,* and 6 characters (#1, 4, 5, 7, 12, 17) for *Apatosaurus.*

This example is typical for cladistic analyses, in that there are many reversals and the trees are less than perfect. It does not contain any repeated characters, which is uncommon. The analysis by Russell and Zheng (1993) added many more taxa and arrived at three shortest trees, but favored one that would include *Shunosaurus* and *Apatosaurus* as more closely related. The last step in any phylogenetic analysis is to evaluate the results, add more characters and taxa, and generate new trees. It is an ongoing process, the hope being that each iteration brings us closer to the true tree.

Producing phylogenetic reconstructions is difficult. However, it is an important exercise which forces a paleontologist to get to know his or her specimens and taxa very well, and helps define major areas that need to be studied.

These are exciting times for those of us who study dinosaurs. With better methods for finding, collecting, and identifying fossils, the number of useful specimens is increasing at an amazing rate, especially considering the relatively small number of working paleontologists. We are able to use advanced technology and better methods for extracting and using information, and this is making it easier for us to make strong and testable inferences, not only about dinosaur biology, but also concerning the world that dinosaurs lived in. The dinosaur paleobiologist has a very difficult task because of the relatively small amount of fossil material available, but the methods discussed herein, and others to be developed in the future, should allow us to know more and more about dinosaurs nonetheless.

References

Alexander, R. M. 1989. *Dynamics of Dinosaurs and Other Extinct Giants.* New York: Columbia University Press.
Behrensmeyer, A. K., and R. E. Chapman. 1993. Models and simulations of time-averaging in terrestrial vertebrate accumulations. In S. M. Kidwell and A. K. Behrensmeyer (eds.), *Taphonomic Approaches to Time Resolution in Fossil*

Assemblages, pp. 125–149. Paleontological Society Short Courses in Paleontology no. 6. Knoxville: University of Tennessee.

Benton, M. J. 1990. Origin and interrelationships of dinosaurs. In D. B. Weishampel, P. Dodson, and H. Osmólska (eds.), *The Dinosauria,* pp. 11–30. Berkeley: University of California Press.

Bookstein, F. L. 1991. *Morphometric Tools for Landmark Data: Geometry and Biology.* Cambridge: Cambridge University Press.

Brown, B., and E. M. Schlaikjer. 1943. A study of the troodont dinosaurs with the description of a new genus and four new species. *Bulletin American Museum of Natural History* 82: 115–150.

Burns, M. 1993. *Automated Fabrication: Improving Productivity in Manufacturing.* Englewood Cliffs, N.J.: Prentice-Hall.

Chapman, R. E. 1990a. Conventional Procrustes approaches. In F. J. Rohlf and F. L. Bookstein (eds.), *Proceedings of the Michigan Morphometrics Workshop,* pp. 251–267. Special Publication no. 2. Ann Arbor: University of Michigan Museum of Zoology.

Chapman, R. E. 1990b. Shape analysis in the study of dinosaur morphology. In K. Carpenter and P. J. Currie (eds.), *Dinosaur Systematics: Approaches and Perspectives,* pp. 21–42. Cambridge: Cambridge University Press.

Chapman, R. E.; P. M. Galton; J. J. Sepkoski, Jr.; and W. P. Wall. 1981. A morphometric study of the cranium of the pachycephalosaurid dinosaur *Stegoceras. Journal of Paleontology* 55: 608–618.

Chapman, R. E., and M. K. Brett-Surman. 1990. Morphometric observations on hadrosaurid ornithopods. In K. Carpenter and P. J. Currie (eds.), *Dinosaur Systematics: Approaches and Perspectives,* pp. 163–177. Cambridge: Cambridge University Press.

Cheetham, A. H., and J. E. Hazel. 1969. Binary (presence-absence) similarity coefficients. *Journal of Paleontology* 43: 1130–1136.

Chinsamy, A. 1993a. Bone histology and growth trajectory of the prosauropod dinosaur *Massospondylus carinatus* Owen. *Modern Geology* 18: 319–329.

Chinsamy, A. 1993b. Image analysis and the physiological implications of the vascularisation of femora in archosaurs. *Modern Geology* 19: 101–108.

Chure, D. J., and J. S. McIntosh. 1989. *A Bibliography of the Dinosauria (Exclusive of the Aves), 1677–1986.* Paleontology Series no. 1. Grand Junction: Museum of Western Colorado.

Cutler, A. H., and A. K. Behrensmeyer. In press. Models of vertebrate mass mortality events at the KT boundary. Proceedings of the Conference on New Developments Regarding the KT Event and Other Catastrophes in Earth History, 1994. Geological Society of America, Special Paper.

Dodson, P. 1975. Taxonomic implications of relative growth in lambeosaurid dinosaurs. *Systematic Zoology* 24: 37–54.

Dodson, P. 1976. Quantitative aspects of relative growth and sexual dimorphism in *Protoceratops. Journal of Paleontology* 50: 929–940.

Farlow, J. O. 1991. *On the Tracks of Dinosaurs: A Study of Dinosaur Footprints.* New York: Franklin Watts.

Farlow, J. O. 1993. *The Dinosaurs of Dinosaur Valley State Park.* Austin: Texas Parks and Wildlife Department.

Gore, R. 1993. Dinosaurs. *National Geographic* 183 (1): 2–53.

Gray, S. W. 1946. Relative growth in a phylogenetic series and in an ontogenetic series of one of its members. *American Journal of Science* 244: 792–807.

Heinrich, R. E.; C. B. Ruff; and D. B. Weishampel. 1993. Femoral ontogeny and locomotor biomechanics of *Dryosaurus lettowvorbecki* (Dinosauria, Iguanodontia). *Zoological Journal of the Linnean Society* 108: 179–196.

Hirsch, K. F.; K. L. Stadtman; W. F. Miller; and J. M. Madsen, Jr. 1989. Upper Jurassic dinosaur egg from Utah. *Science* 243: 1711–1713.

Jefferson, G. T. 1989. Digitized sonic location and computer imaging of Rancho La Brea specimens from the Page Museum salvage. *Current Research in the Pleistocene* 6: 45–47.

Jorstad, T., and J. Clark. 1995. Mapping human origins on an ancient African landscape. *Professional Surveyor* 15 (4): 10–12.

Technology and the Study of Dinosaurs 133

Lambert, S., and W. Howe. 1993. *Internet Basics: Your Online Access to the Global Electronic Superhighway.* New York: Random House.

Levine, J. R., and C. Baroudi. 1993. *The Internet for Dummies.* San Mateo, Calif.: IDG Books.

Lockley, M. 1991. *Tracking Dinosaurs: A New Look at an Ancient World.* Cambridge: Cambridge University Press.

Lull, R. S., and S. W. Gray. 1949. Growth patterns in the Ceratopsia. *American Journal of Science* 247: 492–503.

Maddison, W. P., and D. R. Maddison. 1992. *MacClade: Analysis of Phylogeny and Character Evolution, Version 3.* Sunderland, Mass.: Sinauer Associates.

McKenna, P. C. 1992. GPS in the Gobi: Dinosaurs among the dunes. *GPS World* 3 (6): 20–26.

Nielson, G. M., and B. Shriver (eds.). 1990. *Visualization in Scientific Computing.* Los Alamitos, Calif.: IEEE Computer Society Press.

Paul, G. S. 1987. The science and art of restoring the life appearance of dinosaurs and their relatives: A rigorous how-to guide. In S. J. Czerkas and E. C. Olson (eds.), *Dinosaurs Past and Present,* vol. 2, pp. 4–49. Seattle: Los Angeles County Musuem of Natural History and University of Washington Press.

Poor, G. W. 1991. *The Illusion of Life: Lifelike Robotics.* San Diego: Educational Learning Systems.

Raup, D. M. 1966. Geometric analysis of shell coiling: General problems. *Journal of Paleontology* 40 (5): 1178–1190.

Rohlf, F. J., and F. L. Bookstein (eds). 1990. *Proceedings of the Michigan Morphometrics Workshop.* Special Publication no. 2. Ann Arbor: University of Michigan Museum of Zoology.

Rowe, T.; W. Carlson; and W. Bottorf. 1993. *Thrinaxodon: Digital Atlas of the Skull.* Austin: University of Texas Press.

Russell, D. A., and Z. Zheng. 1993. A large mamenchisaurid from the Junggar Basin, Xinjiang, People's Republic of China. *Canadian Journal of Earth Sciences* 30: 2082–2095.

Shay, D., and J. Duncan. 1993. *The Making of Jurassic Park: An Adventure 65 Million Years in the Making.* New York: Ballantine Books.

Sochurek, H. 1987. Medicine's new vision. *National Geographic* 171 (1): 2–41.

Swofford, D. L., and D. P. Begle. 1993. *User's Manual for PAUP: Phylogenetic Analysis Using Parsimony—Version 3.1, March 1993.* Washington, D.C.: Laboratory of Molecular Systematics, Smithsonian Institution.

Thompson, D'A. W. 1942. *On Growth and Form: The Complete Revised Edition.* 1992 ed. New York: Dover Books.

Wall, W. P., and P. M. Galton. 1979. Notes on the pachycephalosaurid dinosaurs (Reptilia; Ornithischia) from North America, with comments on their status as ornithopods. *Canadian Journal of Earth Sciences* 16: 1176–1186.

Weishampel, D. B. 1990. Dinosaurian distribution. In D. B. Weishampel, P. Dodson, and H. Osmólska (eds.), *The Dinosauria,* pp. 63–139. Berkeley: University of California Press.

Weishampel, D. B. 1993. Beams and machines: Modeling approaches to the analysis of skull form and function. In J. Hanken and B. K. Hall (eds.), *The Skull,* vol. 3, pp. 303–343. Chicago: University of Chicago Press.

Weishampel, D. B., and R. E. Chapman. 1990. Morphometric study of *Plateosaurus* from Trossingen (Baden-Württemberg, Federal Republic of Germany). In K. Carpenter and P. J. Currie (eds.), *Dinosaur Systematics: Approaches and Perspectives,* pp. 43–51. Cambridge: Cambridge University Press.

Weishampel, D. B.; P. Dodson; and H. Osmólska (eds.). 1990. *The Dinosauria.* Berkeley: University of California Press.

Weishampel, D. B., and D. B. Norman. 1989. Vertebrate herbivory in the Mesozoic: Jaws, plants, and evolutionary metrics. In J. O. Farlow (ed.), *Paleobiology of the Dinosaurs,* pp. 87–100. Special Paper no. 238. Boulder: Geological Society of America.

Wiley, E. O.; D. Siegel-Causey; D. R. Brooks; and V. A. Funk. 1991. *The Compleat Cladist: A Primer of Phylogenetic Procedures.* Special Publication no. 19. Lawrence: University of Kansas Museum of Natural History.

Witten, A.; D. D. Gillette; J. Sypniewski; and W. C. King. 1992. Geophysical diffraction tomography at a dinosaur site. *Geophysics* 57: 187–195.

Zangerl, R., and H.-P. Schultze. 1989. X-radiographic techniques and applications. In R. M. Feldmann, R. E. Chapman, and J. T. Hannibal (eds.), *Paleotechniques,* pp. 165–178. Special Publication no. 4. Knoxville, Tenn.: The Paleontological Society.

Molecular Paleontology: Rationale and Techniques for the Study of Ancient Biomolecules

Mary Higby Schweitzer

11

Biomolecules as Fossils

Extinction—the word has a finality to it that makes us time-bound humans uncomfortable. That is why, throughout history, there have been legends of prehistoric beasts that survive—the "plesiosaur" that supposedly haunts the depths of Loch Ness, for example, or the rumors of sauropod dinosaurs alive in the impenetrable jungles of Africa or South America. The most recent embodiment of this theme is the blockbuster movie *Jurassic Park*, in which genetic engineering imparts a high-tech angle to the reversibility of extinction and the return of the dinosaurs.

Is this "bring 'em back alive" concept that formed the basis of the movie as fantastic as it seems? Is it the stuff of Hollywood only, or is there a possibility that someday *Jurassic Park* will be a reality? Can we find enough molecules surviving in fossil tissues to reconstruct an organism? To answer this, we must first clarify what is meant by the term *fossil*.

To many, the term *fossilization* implies a process whereby the hard parts of an organism are completely replaced with inorganic minerals, thus preserving form at the expense of organic components. But in reality, a fossil is any naturally occurring evidence of past life (Schopf 1975), regardless of its condition. Therefore, a frozen mammoth, a footprint, and a petrified dinosaur bone are all types of "fossils." The fossil record consists of specimens that exhibit various states of preservation, ranging from complete replacement to very little alteration at all.

However, there is fundamental difference between using a mineralized dinosaur bone to interpret function, movement, body size, and evolutionary relationships, and studying the actual protein or nucleic acid compounds that were part of the bone, produced by the once-living cells of the animal. The former is based upon inference. It compares extinct taxa with living animals that show like characteristics. These inferences are used to form hypotheses which can never be completely verified. The latter approach, if recovery of such molecular compounds proves possible, opens the way to obtaining more objective data about extinct taxa.

The Nature of Biomolecules

Biomolecules are those molecules whose specific arrangements of atoms are observable only as part of living or once-living systems. Nucleic acids, proteins, lipids, and sugars are some examples of molecules that are formed in living organisms. The most elusive, yet also most informative, of the biomolecules targeted by molecular paleontology is deoxyribonucleic acid (DNA), which carries all of the information needed to specify any organism, from bacterium to human. The components of the molecule itself are highly conserved across all taxa. The DNA from a bacterium is virtually indistinguishable in its chemistry from that of a human.

The fundamental unit of DNA, a nucleotide, consists of a five-carbon sugar, deoxyribose, a phosphate, and a nitrogenous base—one of four possible types: adenine, cytosine, thymine, and guanine. The sugar and phosphate molecules form the "backbone" of a double helix—like the sides of a ladder twisted around its length. The specificity, or information content, lies in the order of these bases along the chain of sugars and phosphates, like the rungs of the twisted ladder. The sequence of all the bases on each of the chromosomes in the nucleus of every cell of an organism is the genome, which directs the production of every characteristic of an organism, from individual cell membranes and organelles, to proteins involved in digestion and metabolic maintenance, to height, weight, gender, and eye color. It is this order of bases throughout the entire genome of an organism that makes a tree, a human, a dog, or a dinosaur. By comparing the order of these bases and how they differ among organisms, we can gain insight into how closely groups may be related. In theory, it should be possible to take the genetic information from any cell of any organism and make a new, genetically identical being, which was the premise behind *Jurassic Park*. Reality, however, is far different.

In order to produce a viable being, not only does every base in the sequence have to be present (roughly one billion or more bases per genome), but each base has to be in the proper order. Furthermore, these must be arranged on the correct number of chromosomes, and the correct DNA-associated proteins must be present. Next, all of this chemical information has to be placed within an environment where it can be stimulated by the proper hormones, and triggered by the proper signals accurately to turn each gene "on" or "off" at exactly the right time in development.

For dinosaurs, or any other extinct organism, we don't know the number of chromosomes that specified the taxon, and we haven't yet learned the genetic triggers, timing, or hormonal controls that would have operated for that species. Furthermore, work with ancient DNA just thousands of years old, far younger than that from dinosaurs, has resulted in sequences only a few hundred base pairs long—several orders of magnitude less than necessary to specify any vertebrate species. DNA recovered from ancient tissues under the best of circumstances has been subjected to forces that cause it to degrade into small pieces (Lindahl 1993), and in most cases modify some of the bases that remain (DeSalle et al. 1993), so that much information is lost.

So why do so many scientists devote time, energy, and money to the search for ancient DNA? Despite the ravages of time and chemical degradation, it is thought that much can be learned from these remnant biomolecules, even though bringing extinct species back to life will probably never be possible.

The first question that molecular paleontologists hope to answer has to do with relationships. We know that dinosaurs, crocodiles, and birds share a common ancestor, and that birds are the closest living relatives to the Dinosauria—or, in cladistic terms, *are* dinosaurs (see Feduccia 1996 for a dissenting view). This is based upon analysis of gross features, or morphology, preserved in dinosaur bones (Gauthier 1986). If informative DNA could be recovered from dinosaur bones, it might be possible not only to determine which lineage among dinosaurs is closest to birds, but also to clarify some relationships among the different kinds of dinosaurs. Additionally, it might allow us better to estimate the time of divergence of birds and dinosaurs.

DNA or protein sequences obtained from dinosaur bone specimens, carefully aligned and subjected to analyses, could be used to build phylogenetic trees. As with other forms of phylogenetic analysis, the robustness of the phylogenetic proposals depends upon the number of characters in the analysis. Because each base or amino acid residue of a sequence is a character, if the sequences are of reasonable length and contain a sufficient number of phylogenetically significant sites, the phylogenies derived by this method can be more robust than those derived from more typical morphological characters. Additionally, molecular data may provide a way to test phylogenies based upon morphological studies (cf. Cooper et al. 1992 and Vickers-Rich et al. 1995 for molecular studies of the evolutionary relationships of moa, a group of prehistoric flightless birds from New Zealand), thus avoiding the circular reasoning of testing trees with the same characters upon which they are built.

Recovery of DNA from extinct organisms that show close relationships to modern taxa may also give insight into how DNA changes over time. By comparing sequences of certain genes, we may be able to get a picture of which bases are altered between, for example, a quagga (a recently extinct member of the horse family) and a modern horse (Higuchi et al. 1984). We may also gain information about the biochemical alteration or degradation of DNA and other biomolecules.

But there are other questions besides those of relationships that DNA analysis might be able to answer. For example, there has been much debate about the physiology of dinosaurs. Both morphological and histological (microscopic) analysis of dinosaur bones suggests that their metabolism was at the least very different from that of modern reptiles, and may have been equivalent to that of modern mammals and birds (de Ricqlès 1980; Bakker 1986; Varricchio 1993). Other scientists, however, see "growth rings" in histological sections of dinosaur bone, which is evidence that the metabolism of these beasts may have been slow and growth intermittent, much as in modern crocodiles (Reid 1984, 1985, and chap. 29 of this volume; Chinsamy et al. 1994). Molecular evidence might provide the tools with which to settle this debate.

Methods of Recovering and Studying Ancient DNA

The possibility of using small fragments of DNA recovered from ancient specimens for such studies has only recently become realistic because of the development of a procedure called the polymerase chain reaction (PCR). This technique involves the repeated replication of selected gene regions to quantities that are sufficient for current methods of sequencing and analysis (Fig. 11.1). Theoretically, only one molecule of original template is required for this process.

Amplification of DNA, or the building of several exact copies of the original starting molecules, requires a specific enzyme called a polymerase, as well as other components needed to build DNA—the four nitrogenous bases, buffers, and primer molecules. Amplification of specific genes or parts of genes is determined by the choice of short, specific primer molecules. These primers find the complementary regions on the template molecule, bind to that site, and form a starting point for the enzyme to begin building new molecules. Careful design of primer molecules allows the researcher to select against contaminant molecules and to increase chances of success by amplifying only a certain type of molecule. For example, through analysis of sequences in the data bases, one can find gene regions that are unique to birds and crocodilians. Because of the shared common ancestry of these groups with dinosaurs, it is assumed that sequences conserved between crocodiles and birds must have been present in dinosaurs as well. Using these shared sequences to build primer molecules would optimize chances of amplifying "dinosaur" molecules over other contaminant molecules that may be present. Because the primers will bind only to their corresponding sequences on template molecules, amplification of human and microbial contaminant molecules, which don't contain the exact order of sequences for primer binding, will be eliminated.

In attempting to isolate rare "dinosaur" DNA molecules from bone extracts, one has to assume that the chances of contamination are high, because of fungi and bacteria which are always present in the matrix surrounding such fossils. Also, despite extreme precautions, contamination can be introduced from handling, and from instruments used in the lab for analysis. Therefore, sterile conditions and primer design are two critical conditions for ancient DNA analysis.

In modern taxa, molecular phylogenies, for the most part, support phylogenies derived from morphology, particularly at the higher levels. Because the study of ancient DNA is a new area, and is plagued with enzyme artifact and contamination problems, phylogenetic trees derived from ancient DNA sequences must closely approximate those obtained from traditional morphological analysis, such as those done for the dinosaurs by Gauthier (1986; see also Holtz and Brett-Surman, chap. 8 of this volume). If they don't, or if dinosaurs cluster with other taxa (mammals, for example, or fungi), we must assume either that the alleged dinosaur DNA derived from contaminant, or that there were errors in the amplification process.

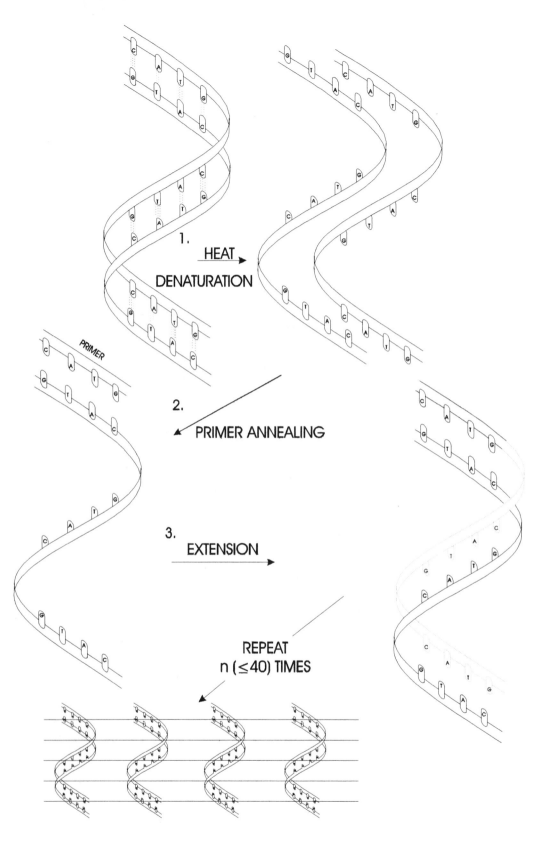

1.

HEAT
DENATURATION

PRIMER

2.

PRIMER ANNEALING

3.

EXTENSION

REPEAT
n (≤40) TIMES

140   *Mary Higby Schweitzer*

Even if the amplified DNA were really dinosaurian in origin, but didn't cluster as expected, or if the sequences from dinosaur material were not consistent or repeatable, it is questionable what could truly be gained from such studies. The possibility remains that our phylogenies of the Dinosauria are incorrect, and that molecular phylogenies will eventually reveal this, but at this early stage of development of the field of molecular phylogeny, more evidence than just fragments of PCR-amplified DNA will be required to reorder our beliefs about the relationships of dinosaurs to modern taxa.

Other Biomolecules Potentially Recoverable from Fossils

DNA is not the only biomolecule with potential for preservation in fossil tissues; nor is it the only molecule that contains information concerning the physiology of the organism from which it is derived. There are also proteins that show strong potential for preservation in fossil tissues, and some of these could be as informative as small fragments of DNA.

Working with proteins, or their amino acid constituents, has some advantages over work with DNA. Because the researcher is not directly amplifying proteins, but working only with what is actually present, problems with contamination may be reduced. Additionally, most amino acids exist in one of two forms, and demonstrate a kind of "handedness." It is known that almost all proteins in all living systems are made up of amino acids in the "L" form. However, over time, as the proteins degrade into small fragments, amino acids are converted or "racemized" to the other "D" form, each amino acid undergoing this conversion at a different rate.

Comparing the ratios of D and L forms of each amino acid has been suggested as a way of aging the protein, or the source of the protein (Bada and Protsch 1973; Bada 1985). Given that each amino acid undergoes these changes at a different rate, comparing the D/L ratios of several amino acids has also been proposed as a way of identifying contamination (Bada 1985; Poinar et al. 1996). This idea is not fully accepted, however, as several factors can affect the rate of racemization, and within each bone there are microenvironments in which preservation may differ greatly. It has been recommended that other methods be used in conjunction with racemization data for a more accurate interpretation (Macko and Engel 1991), and more than one sample of each specimen should be tested, and ratios compared, for an accurate interpretation of the data. Furthermore, it has been noted that both aspartic acid and glutamic acid, two amino acids used as markers for these types of studies (Poinar et al. 1996), can undergo spontaneous reversal to the L form, thus altering the D/L ratios and giving false indications of contamination (Kimber and Griffen 1987). Additionally, some of the endogenous protein fragments may be protected through complexes with minerals, and may show different racemization ratios from those retrieved from the whole bone (S. Weiner, personal communication). For example, amino acids and protein fragments in amber-entombed tissues are greatly protected from amino acid racemization (Poinar et al. 1996), and it is assumed that amber resins will protect other biomolecules as well.

However, because bone tissues are the only tissues of dinosaurs that survive, and dinosaurs are not likely to be trapped in amber, it is reasonable

Figure 11.1. Schematic drawing of the sequence of steps in the polymerase chain reaction. In step 1, double-stranded molecules of DNA are separated to single strands by exposure to high heat. In step 2, the heat is lowered to allow primer molecules to bind to the appropriate sites on the separated template strands. Because these molecules are shorter than the complementary strand which has separated, they will bind more quickly, preventing re-annealing of the original strands. In step 3, the new strand is built, using the separated strand as template, from nucleotides present in the reaction mix. This is accomplished through the action of a heat-resistant DNA polymerase enzyme. The cycle is repeated up to 40 times, yielding exponential growth of specific segments of DNA from rare template. In the figure, the letters A, C, G, and T refer to the bases of the DNA molecule, adenine, cytosine, guanine, and thymidine, respectively. Figure drawn by Matt Schweitzer.

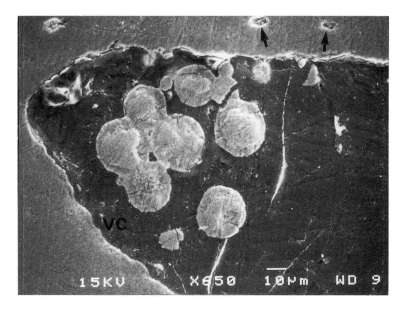

Figure 11.2. Scanning electron micrograph of dinosaur trabecular tissues. Arrows indicate osteocyte lacunae, which in life housed the individual bone-forming cells, or osteocytes. V.C. indicates a blood vessel channel. Tissues were embedded in a synthetic polymer, sectioned and ground to the appropriate thickness, then carbon-coated for use in the electron microscope. Magnifications are as indicated.

to direct protein studies to those bone-associated proteins which would most likely be preserved. Collagen is the protein that predominates in bone, and although it is also produced by some invertebrates, it is not produced by microbes, a common source of contamination. Collagen is a structural protein (i.e., one of the proteins involved as a structural component of the organism, rather than a protein—like enzymes—that functions in the regulatory processes of an organism), and so is more stable than some other proteins. Osteocalcin is another bone protein, produced by osteocytes (bone cells), and therefore occurs only in vertebrates. Albumin is a serum protein, and is frequently associated with blood and bone tissues. The same is true for hemoglobin, the protein that allows red blood cells to transfer oxygen to tissues.

Collagen (Jope and Jope 1989; Baird and Rowley 1990; Tuross and Stathoplos 1993), osteocalcin (Muyzer et al. 1993), albumin (Tuross 1989), hemoglobin (Ascenzi et al. 1985; Smith and Wilson 1990), and other blood- and bone-related proteins (Cattaneo et al. 1992) have already been identified from fossil bone, including dinosaur bone (Muyzer et al. 1993). Hemoglobin protein (Dickerson and Geis 1983; Perutz 1983) and collagen (Har-el and Tanzer 1993) show consistent differences in modern taxa between warm- and cold-blooded species. If these proteins could be identified in dinosaur bone, it might be possible to compare their amino acid sequences with those of modern taxa, and perhaps end the debate on whether or not dinosaurs were warm-blooded, at least for those dinosaurs for whom these molecules could be recovered.

The explosion of technology applicable to the field of molecular biology has made examination of such molecules much easier, with much greater resolution than ever before. With such developments comes the opportunity to apply these techniques to the fields of paleontology and archaeology. I will now consider some of the techniques which have been applied to the study of the preservation of biological compounds in dinosaur bones, with emphasis on a well-preserved specimen of *Tyrannosaurus rex* whose bones are virtually unaltered since the animal's death.

Analyses of the Bony Tissues of a *Tyrannosaurus rex* and Other Dinosaurs

Microscopy

Microscopic analysis has been done on thin sections of dinosaur bone since before the turn of the century (see de Ricqlès 1980). The traditional picture of dinosaurs as sluggish reptiles was challenged by these analyses, as it was noted that microscopically, dinosaur bone tissues exhibited more characteristics in common with the bones of warm-blooded birds and mammals than they did with those of modern ectotherms. However, advances in electron microscopy have provided much greater resolution for microscopic studies than ever before. Scanning electron microscopy (SEM) allows the definition of surface topography of tissues, such as scratches on teeth, or the three-dimensional shape of microscopic inclusions (Schweitzer and Cano 1994). Figure 11.2 is a scanning electron micrograph of dinosaur cancellous (bone marrow) tissues, showing a blood vessel channel and microstructures therein. These have yet to be positively identified, but they may be organic rather than geological in origin (Schweitzer et al. in press). Coupled with Energy Dispersive X-ray (EDX) capabilities, an SEM also allows the identification of elements found within fossil matrices, and helps to determine the amount of diagenetic addition or subtraction of elements (i.e., amount of replacement or permineralization the tissues have undergone). Figure 11.3 shows the results of an elemental analysis of tissues of the *Tyrannosaurus rex*, and

Figure 11.3. Elemental analysis of modern bone and *Tyrannosaurus rex* tissues, prepared as in Figure 11.2. Calcium and phosphorus predominate in these tissues in the same ratios as seen in modern bone. Oxygen and carbon are not figured into the numerical analyses, both because these elements are ubiquitous and because this technique is not sensitive enough with lighter elements to obtain adequate values. Very few trace elements are present, and those that are are also seen in modern bone tissues that have been similarly prepared and analyzed. This indicates few diagenetic additions to or subtractions from the original bone content.

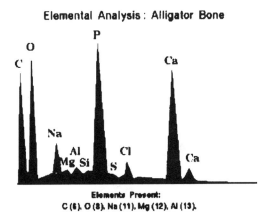

Elemental Analysis : Alligator Bone

Elements Present:
C (6), O (8), Na (11), Mg (12), Al (13),
Si (14), P (15), S (16), Cl (17), Ca (20)

Element	Atom %	Wt %
Na	8.39	5.48
Mg	0.86	0.59
Al	0.60	0.46
Si	0.51	0.41
P	31.56	27.95
Cl	5.12	5.16
Ca	52.30	59.54
S	0.56	0.51

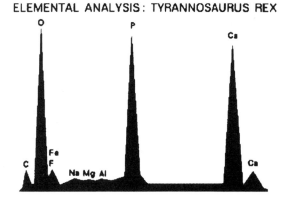

ELEMENTAL ANALYSIS : TYRANNOSAURUS REX

Elements Present:
C (6), O (8), F (9), Na (11), Mg (12), Al (13), P (15), Ca (20), Fe (26)

Element	Atom %	Wt %
Na	0.99	0.65
P	27.91	24.94
Ca	51.34	59.35
Fe	3.88	6.25
Al	.27	.21
F	15.40	8.44
Mg	.22	.15

illustrates the combination of elemental and structural information obtained by these methods.

Transmission microscopy (TEM) allows the analysis of tissues to a much greater degree of resolution. Samples are sectioned to a thickness of from 3 to 5 micrometers for this type of study, as the electrons must be able to pass through the sample to generate an image. This technique allows a more detailed study of the microstructures of tissues than the above methods. Using TEM, we have identified organic fibers (Fig. 11.4) remaining in dinosaur bone tissues after removal of the mineral phase of bone (Schweitzer et al. 1997a).

Using TEM at very high electrical energies (200 KV or more) also allows us to examine the character of the crystals that make up the bone tissues (Zocco and Schwartz 1994). Because in bone formation the collagen fibers are laid down first, and then mineral is secondarily laid down upon them, the mineral crystals are aligned in specific orientations determined by the collagen fibrils. Crystals that were geologically deposited would be more randomly distributed without the collagen fibers to provide orientation.

Immunocytochemistry

Immunocytochemistry includes the binding of a generated antibody to components present in sample tissues, and has been demonstrated to be a useful tool in studies of ancient organisms. This technique is based upon the premise that cell membranes, proteins, and other components in living organisms carry "markers" of small proteins or carbohydrates which aid the body's defense system to recognize "self" and "foreign." These marker molecules have parts which are folded into very specific structures called epitopes. Antibodies are proteins produced by the body to bind to foreign epitopes and target them for destruction. In the case of fossil tissues, if small pieces of protein that contain epitopes remain, then antibodies generated against, for example, the protein collagen may recognize these epitopes and bind to them. A second antibody recognizes the first, and will bind to it. The second antibody carries a chemical or radioactive "tag"

Figure 11.4. Transmission electron micrograph of dinosaur tissues. Tissues were first acid-decalcified to remove the mineral phase, then embedded in a plastic, microtomed, and stained with lead citrate and uranyl acetate to reveal the organic fibers. Magnification × 42,500.

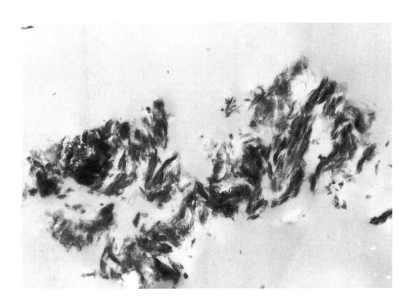

144   *Mary Higby Schweitzer*

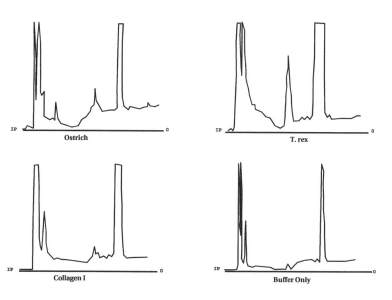

Figure 11.5. High-Performance Liquid Chromatography (HPLC) separations of extractions of dinosaur tissues and controls. Bone was extracted chemically as described in the text, and molecules contained in the resulting supernatants were separated on a reverse-phase HPLC column. Monitoring for absorbance was done at 214 nm, a value indicative of the peptide bonds present in proteins. IP indicates the point of injection of the solution onto the column, and the x-axis is time, in minutes. The line marked "0" indicates no absorbance at this wavelength by the solution components. The extraction from *Tyrannosaurus* bone, like that from ostrich bone and a collagen sample, gives results consistent with the presence of proteins.

which can be detected by fluorescence or exposure to films. Antibody binding can be used to recognize protein remnants in extraction of bone, as well as to localize particular regions of whole tissues that may contain the proteins of interest. Work with antibodies has been successful with very ancient specimens, including dinosaurs (deJong et al. 1974; Lowenstein 1981; Collins et al. 1991; Muyzer et al. 1993; Schweitzer et al. 1997b).

Chromatography/Spectrometry

Chromatography provides a means to separate certain types of molecules from a mixture, so that the resulting purified samples can then be further analyzed and identified by one of several types of spectrometry. One example of chromatography available to the paleontologist is high-performance liquid chromatography (HPLC). This procedure uses a thin column packed with special materials of known properties. The sample in question (in this case, crushed bone treated to extract any organic components, and filtered to remove particles) is injected into this column. A solvent liquid, which changes over a gradient from water, a polar molecule (a molecule with positively and negatively charged ends), to some more hydrophobic (lacking a chemical affinity for water), nonpolar solvent, is moved across the column at high pressure. Organic molecules from a heterogeneous mix can then be separated by their characteristic interactions with these solvents.

By collecting the separated fractions and subjecting them to mass spectrometry, one can identify each fraction by obtaining molecular weights and comparing them with standards. The fractions can also be analyzed by ultraviolet spectrometry and then tentatively identified by their absorbance of certain wavelengths of light. For example, nucleic acids characteristically absorb light at 260 to 265 nanometers. Absorbance by a sample at this wavelength is an indication that nucleic acids may be present. The peptide bond, which ties together the amino acids that make up all proteins, absorbs light at 214 nanometers. This wavelength is one that is routinely used to identify proteins in a solution. Figure 11.5

Molecular Paleontology

illustrates the results of subjecting an extraction of *T. rex* bone to this type of analysis. This shows that there are extractable compounds in these dinosaur tissues consistent with protein fragments.

HPLC technology is also used to identify the amino acid constituents of protein molecules. Extracting any organic molecules from ancient bone tissues and treating the extracts with special chemicals can cause each amino acid present to separate from a mixture in a very specific order. Comparing this order with known, prepared standards allows the investigator to identify the amino acids present in a compound. The pattern of some specific amino acids may give clues as to their protein source. For example, the conformation, or "shape," of the protein collagen is highly constrained. It is twisted like a three-stranded rope into a triple helix, and where the three "strands" of collagen protein come together, there is a size restriction. Therefore, in all collagens every third amino acid of the chain is a glycine, because this amino acid has no side chains, and is the smallest of all amino acids. Two other amino acids, proline and hydroxyproline, are also very common in all collagens, and hydroxyproline is unique to collagen and not found in any other protein. The presence of high proline and glycine ratios in an amino acid analysis is strongly indicative of the collagen protein, and when hydroxyproline is identified, collagen is definitively present. Figure 11.6 shows an amino acid analysis of an extract of the *Tyrannosaurus rex* bone. Both glycine and proline are definitely identified in significant quantities. However, hydroxyproline cannot be positively identified in these runs.

Other Methods

The fields of biophysics and biochemistry have additional techniques useful in analysis of fossil specimens. Some of these hold potential for gaining information about particular compounds which may be present in such specimens.

Nuclear magnetic resonance imaging, once a rare and expensive technology, is now routinely used in medicine as well as analytical labs. This method operates on the principle that atoms in a compound consist of

Figure 11.6. Amino acid analysis of extractions of dinosaur bone. This technique reveals the presence of amino acid constituents of proteins in the extracted tissues of *Tyrannosaurus rex*. Relatively high levels of both proline and glycine are consistent with the presence of the protein collagen in these tissues, inidcating that remnants of the proteins present in the dinosaur bone when the dinosaur was alive can still be detected. The labels are standard abbreviations for amino acids, which can be found in any basic biology text.

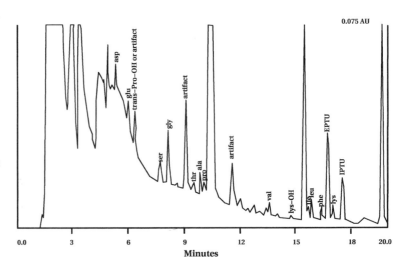

electrons spinning about a nucleus, resulting in a tiny magnetic field. The field can be manipulated by pulsing the compound with a very strong magnetic force, which causes the magnetic fields of the compounds to align with the outside field. This release of energy not only helps with the identification of compounds, but also gives an idea of their structure. For example, the association of proteins with metal atoms, such as occurs in hemoglobin or other metalloproteins, yields very specific patterns by this method, which can be considered "fingerprints" for the presence of these compounds.

Infrared spectroscopy and resonance raman spectroscopy are two methods of analysis that rely on vibration about the chemical bonds to identify a compound. These techniques are sensitive, and can identify very small amounts of a particular molecule within a heterogeneous mix. Resonance raman can specifically describe the environment of the chemical bonds of metalloproteins, and help to identify any other atoms, such as oxygen, which may be bound to the protein remnants. The above methods have been applied to the *Tyrannosaurus rex* bone as well, and preliminary results indicate the presence of a compound which possesses some of the characteristics unique to heme, an important constituent of the hemoglobin molecule.

While the scenes portrayed in *Jurassic Park* are certainly beyond our current capabilities, and most likely will always remain so, much information about the biology and physiology of fossil species may still be contained within their bones. As detection techniques become more sensitive and extraction methods become more refined, the potential for recovering information about the biochemistry of extinct organisms increases.

For example, if protein fragments containing epitopes can be recovered from dinosaur tissues, these can be tested against antibodies from modern taxa. How strongly the antibodies bind to the dinosaur proteins would be an indicator of the closeness of relationships between these taxa and the dinosaur.

Obtaining sequences from hemoglobin proteins and comparing them with those derived from modern animals may indicate the oxygen-carrying capacity of this protein, or give insight into the metabolic rates of dinosaurs. Knowledge such as this may answer questions about how these animals could have grown so large and how they handled oxygen debt, and even, indirectly, allow speculations about atmospheric conditions, such as oxygen partial pressures.

DNA fragments, if they are large enough to be informative, would most certainly shed light on relationships between dinosaurs and modern taxa. In addition, homologies between dinosaur DNA fragments and their modern relatives can be tested by "hybridizing" strands of modern DNA with strands of dinosaur DNA, and testing for bond strengths.

This information would be useful in elucidating and clarifying relationships. In addition, techniques such as these hold potential for determining interactions between biological compounds and secondary geological components. This knowledge may yield information about how these biomolecules degrade, as well as how they are preserved under certain geological conditions. Combining these techniques may allow us to predict what sedimentary environments best preserve fossil specimens for biochemical analysis, and such predictions may optimize our future searches for biomolecules within the tissues of long-extinct animals.

References

Ascenzi, A.; M. Brunori; G. Citro; and R. Zito. 1985. Immunological detection of hemoglobin in bones of ancient Roman times and of Iron and Eneolithic ages. *Proceedings of the National Academy of Sciences USA* 82: 7170–7172.

Bada, J. 1985. Amino acid racemization dating of fossil bones. *Annual Review of Earth and Planetary Sciences* 13: 241–268.

Bada, J. L., and R. Protsch. 1973. Racemization reaction of aspartic acid and its use in dating fossil bones. *Proceedings of the National Academy of Sciences USA* 70: 1331–1334.

Baird, R. F., and M. J. Rowley. 1990. Preservation of avian collagen in Australian Quaternary cave deposits. *Palaeontology* 33: 447–451.

Bakker, R. T. 1986. *The Dinosaur Heresies: New Theories Unlocking the Mystery of Dinosaurs and Their Extinction*. New York: William Morrow.

Cattaneo, C.; K. Gelsthorpe; P. Phillips; and R. J. Sokol. 1992. Detection of blood proteins in ancient human bone using ELISA: A comparative study of the survival of IgG and albumin. *International Journal of Osteoarchaeology* 2: 103–107.

Chinsamy, A.; L. M. Chiappe; and P. Dodson. 1994. Growth rings in Mesozoic birds. *Nature* 368: 196–197.

Collins, M. J.; G. Muyzer; P. Westbroek; G. B. Curry; P. A. Sandberg; S. J. Xu; R. Quinn; and D. MacKinnon. 1991. Preservation of fossil biopolymeric structures: Conclusive immunological evidence. *Geochimica et Cosmochimica Acta* 55: 2253–2257.

Cooper, A.; C. Mourer-Chauviré; G. K. Chambers; A. von Haeseler; A. C. Wilson; and S. Pääbo. 1992. Independent origins of New Zealand moas and kiwis. *Proceedings of the National Academy of Sciences USA* 89: 8741–8744.

deJong, E. W.; P. Westbroek; J. F. Westbroek; and J. W. Bruning. 1974. Preservation of antigenic properties of macromolecules over 70 myr. *Nature* 252: 63–64.

de Ricqlès, A. 1980. Tissue structures of dinosaur bone. In R. D. K. Thomas and E. C. Olson (eds.), *A Cold Look at Warm Blooded Dinosaurs*, pp. 103–139. Selected Symposium 28, American Association for the Advancement of Science. Boulder, Colo.: Westview Press.

DeSalle, R.; M. Barcia; and C. Wray. 1993. PCR jumping in clones of 30-million-year-old DNA fragments from amber preserved termites (*Mastotermes electrodominicus*). *Experientia* 49: 906–909.

Dickerson, R. E., and I. Geis. 1983. *Hemoglobin: Structure, Function, Evolution and Pathology*. Menlo Park, Calif.: Benjamin/Cummings Publishing.

Feduccia, A. 1996. *The Origin and Evolution of Birds*. New Haven, Conn.: Yale University Press.

Gauthier, J. 1986. Saurischian monophyly and the origin of birds. In K. Padian (ed.), *The Origin of Birds and the Evolution of Flight*, pp. 1–55. California Academy of Science Memoir no. 8.

Har-el, R., and M. Tanzer. 1993. Extracellular Matrix 3: Evolution of the extracellular matrix in invertebrates. *FASEB Journal* 7: 1115–1123.

Hedges, S. B. 1994. Molecular evidence for the origin of birds. *Proceedings of the National Academy of Sciences USA* 91: 2621–2624.

Higuchi, R.; B. Bowman; M. Freiberger; O. A. Ryder; and A. C. Wilson. 1984. DNA sequences from the quagga, an extinct member of the horse family. *Nature* 312: 282–284.

Jope, E. M., and M. Jope. 1989. Note on collagen molecular preservation in an 11 ka old *Megaceros* (giant deer) antler: Solubilization in a non-aqueous medium (anhydrous formic acid). *Applied Geochemistry* 4: 301–302.

Kimber, R. W. L., and C. V. Griffen. 1987. Further evidence of the complexity of the racemization process in fossil shells with implications for amino acid racemization dating. *Geochimica et Cosmochimica Acta* 51: 839–846.

Lindahl, T. 1993. Instability and decay of the primary structure of DNA. *Nature* 362: 709–715.

Lowenstein, J. M. 1981. Immunological reactions from fossil material. *Philosophical Transactions of the Royal Society of London* B 292: 143–149.

Macko, S. A., and M. H. Engel. 1991. Assessment of indigeneity in fossil organic matter: Amino acids and stable isotopes. *Philosophical Transactions of the Royal Society of London* B 333: 367–374.

Muyzer, G.; P. Sandberg; M. H. J. Knapen; C. Vermeer; M. Collins; and P. Westbroek. 1993. Preservation of the bone protein osteocalcin in dinosaurs. *Geology* 20: 871–874.

Perutz, M. F. 1983. Species adaptation in a protein molecule. *Molecular Biology and Evolution* 1: 1–28.

Poinar, H. N.; M. Hoss; J. L. Bada; and S. Pääbo. 1996. Amino acid racemization and the preservation of ancient DNA. *Science* 272: 864–866.

Reid, R. E. H. 1984. The histology of dinosaur bone, and its possible bearing on dinosaur physiology. *Symposium of the Zoological Society of London* 52: 629–663.

Reid, R. E. H. 1985. On supposed Haversian bone from the hadrosaur *Anatosaurus*, and the nature of compact bone in dinosaurs. *Journal of Paleontology* 59: 140–148.

Schopf, J. M. 1975. Modes of fossil preservation. *Review of Palaeobotany and Palynology* 20: 27–53.

Schweitzer, M. H., and R. J. Cano. 1994. Will the dinosaurs rise again? In G. D. Rosenberg and D. L. Wolberg (eds.), *Dino Fest*, pp. 309–326. Paleontological Society Special Publication 7. Knoxville: University of Tennessee.

Schweitzer, M. H.; C. Johnson; T. G. Zocco; J. R. Horner; and J. R. Starkey. 1997a. Preservation of biomolecules in cancellous bone of *Tyrannosaurus rex*. *Journal of Vertebrate Paleontology*: 349–359.

Schweitzer, M. H.; M. Marshall; K. Carron; D. S. Bohle; S. C. Busse; E. V. Arnold; D. Barnard; J. R. Horner; and J. R. Starkey. 1997b. Heme compounds in dinosaur trabecular bone. *Proceedings of the National Academy of Sciences, USA* 94: 6291–6296.

Smith, P., and M. T. Wilson. 1990. Detection of haemoglobin in human skeletal remains by ELISA. *Journal of Archaeological Science* 17: 255–268.

Tuross, N. 1989. Albumin preservation in the Taima-taima mastodon skeleton. *Applied Geochemistry* 4: 255–259.

Tuross, N., and L. Stathoplos. 1993. Ancient proteins in fossil bones. *Methods in Enzymology* 224: 121–128.

Varricchio, D. 1993. Bone microstructure of the Upper Cretaceous theropod dinosaur *Troodon formosus*. *Journal of Vertebrate Paleontology* 113: 99–104.

Vickers-Rich, P.; P. Trusler; M. J. Rowley; A. Cooper; G. K. Chambers; W. J. Bock; P. R. Millener; T. H. Worthy; and J. C. Yaldwyn. 1995. Morphology, myology, collagen and DNA of a mummified upland moa, *Megalapteryx didinus* (Aves: Dinornithiformes) from New Zealand. *Tuhinga: Records of the Museum New Zealand Te Papa Tongarewa* 4: 1–26.

Zocco, T. G., and H. L. Schwartz. 1994. Microstructural analysis of bone of the sauropod dinosaur *Seismosaurus* by transmission electron microscopy. *Journal of Paleontology* 37: 493–503.

Dinosaurs as Museum Exhibits

Kenneth Carpenter

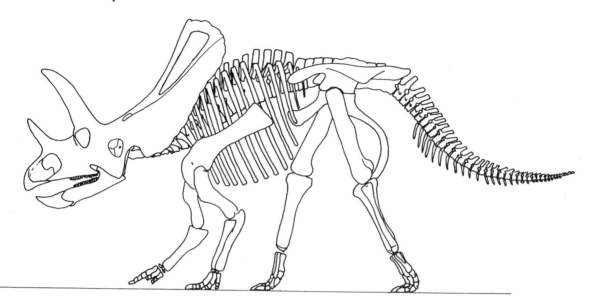

12

Ever since the first dinosaur skeleton was mounted for exhibition in 1868, the public has had a fascination with these extinct animals. Many natural history museums were quick to capitalize upon this interest, with the result that today almost every country has at least one dinosaur skeleton on display. In fact, the steel magnate and philanthropist Andrew Carnegie donated casts or replicas of the sauropod skeleton *Diplodocus carnegii* to the principal museums of South America and Europe (of course, having a dinosaur species named after him may have had something to do with his generosity).

Today we are in the midst of a new Golden Age of Dinosaurology, due in large part to the post–World War II baby boom generation coming of age. With advances in communication, there has been a steady stream of popular dinosaur books and television specials, and at least one sitcom in which dinosaurs were the main stars! Many museums strapped financially have turned to dinosaurs to draw the visitors (and their money). Robotic dinosaurs, in temporary exhibit halls, turn their pneumatic heads, swing their tails, and growl. Meanwhile, the museum gift shop offers stuffed dinosaurs, dinosaur erasers, wooden dinosaur skeleton kits, dinosaur cookie cutters, and anything else remotely dinosaurian to the public as souvenirs.

A more serious approach by some museums has been to present to the public what dinosaur paleontologists have recently learned. Gone is the view that *Tyrannosaurus* was a lumbering giant that stood up on its hind legs, tail dragging on the ground, and head high in the clouds (as another bipedal dinosaur is depicted in Fig. 12.1A). Instead, *Tyrannosaurus* is seen as a moderately swift predator, using its tail to counterbalance a horizontal body (Fig. 12.1B). The public has responded favorably to these new exhibits (meaning that attendance is up), and as a result, increasing numbers of museums have mounted or remounted dinosaur skeletons.

Dinosaur Skeletons in the Public Eye

Almost as soon as fossilized vertebrate skeletons were found, attempts were made to assemble them for the public. One of the first skeletons mounted was that of a mastodon displayed in 1806 at Charles Peale's museum in Philadelphia, Pennsylvania. The legs of the skeleton may be seen in Peale's 1822 painting *The Artist in His Museum* (Alexander 1983: Fig. 3). The techniques used in the mount were unfortunately not recorded.

This exhibit of a long-extinct animal skeleton had a tremendous impact on the public, and it became clear to entrepreneurs that there was money to be made with similar displays. Soon traveling exhibitions featured extinct animals in many cities, both in the United States and in Europe. For a price, the public could see these fossilized skeletons. Competition for the public's money resulted in some shady one-upmanship. One entrepreneur claimed to have the largest mastodon skeleton in the world. Actually, its exaggerated length of 10 meters (32 feet) and height of 4.5 meters (15 feet) were due to several partial skeletons' being used to make a composite skeleton (Simpson 1942).

Yet another entrepreneur boasted of having the skeleton of a "sea monster," which was actually a composite archaeocete whale with an exaggerated length of 35 meters (114 feet) (Kellogg 1936). This mount was rather crude, using boards and metal bars to hold the skeleton together (Fig. 12.2). Nevertheless, this innovation made it easier to disassemble the skeleton and move it from city to city for repeated exhibition. Several of these prehistoric animal skeletons were displayed around Europe as well.

It was only natural once fossil skeletons began to be displayed that someone would mount a dinosaur skeleton. In 1868 Waterhouse Hawkins, a well-known sculptor from England, was asked to re-create various prehistoric animals for a museum in New York City. Hawkins spent time at the Academy of Natural Sciences of Philadelphia, where some of the more important fossil specimens were curated. Among these were the skeletons of the dinosaurs *Hadrosaurus foulkii* and *Dryptosaurus aquilunguis*.

In order to get various body proportions correct for his sculptures, Hawkins approached Joseph Leidy about molding and casting these skeletons and mounting the real bone skeletons for exhibit at the academy. Hawkins convinced Leidy that mounting the skeletons would increase visitor attendance. Neither skeleton was complete (Fig. 12.3B), and Hawkins had to rely a great deal upon his own imagination to reconstruct the missing bones. Hawkins utilized many of the techniques still used today: vertebrae were drilled and strung onto a steel rod, limb bones were held to the supporting armature with steel bands, missing bones were modeled as mirror images of their preserved counterparts, or modeled from the best analogous living animal (Fig. 12.3A).

The tripodal pose used by Hawkins for the *Hadrosaurus* mount was based upon Leidy's observation that "the enormous disproportion between

the fore and hind parts of the skeleton of *Hadrosaurus* has led me to suspect that this great herbivorous Lizard sustained itself in a semierect position on the huge hinder extremities and tail" (Leidy 1865: 97). The reconstructed mammalian-like characters used in the mount, including the seven cervicals (neck vertebrae) and the shape of the scapula (shoulder blade), reflect the influence of the kangaroo skeleton. The public saw the final result of Hawkins's hard work in 1868, and, as Hawkins had predicted, visitor attendance at the academy soared.

While preparing the *Hadrosaurus* skeleton for display, Hawkins also made plaster of paris cast skeletons. One set of casts was to be used in the museum in New York City, while others were distributed to the Smithsonian Institution in Washington, D.C., to Princeton University, and to the Field Museum of Natural History in Chicago. Once assembled, these were some of the first casts of a dinosaur skeleton ever displayed.

Figure 12.1. (A) *Edmontosaurus annectens*, an example of a traditional dinosaur mount in which the tail is like a prop and the head is high in the air. (B) A new, dynamic dinosaur mount. This plastic cast or replica of *Tyrannosaurus rex* is mounted horizontally, with the tail counterbalancing the body over the legs. Although most of the skeleton is supported internally by a steel armature through the legs, steel cables from the ceiling support the neck and tail. Compare the completed base with the base in Figure 12.11. Signage presenting information about *Tyrannosaurus* is displayed on the railing (*arrow*).

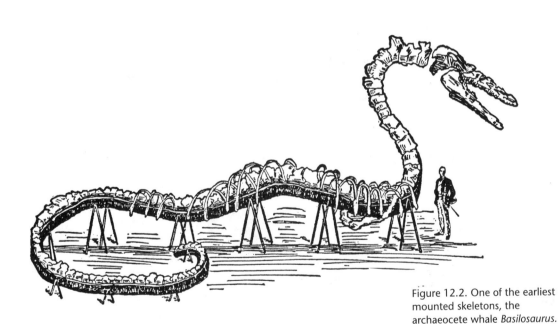

Figure 12.2. One of the earliest mounted skeletons, the archaeocete whale *Basilosaurus*. The mounting technique used wood and steel, permitting the skeleton to be dismantled and moved to a new location. From Lucas 1902.

Dinosaurs as Museum Exhibits   153

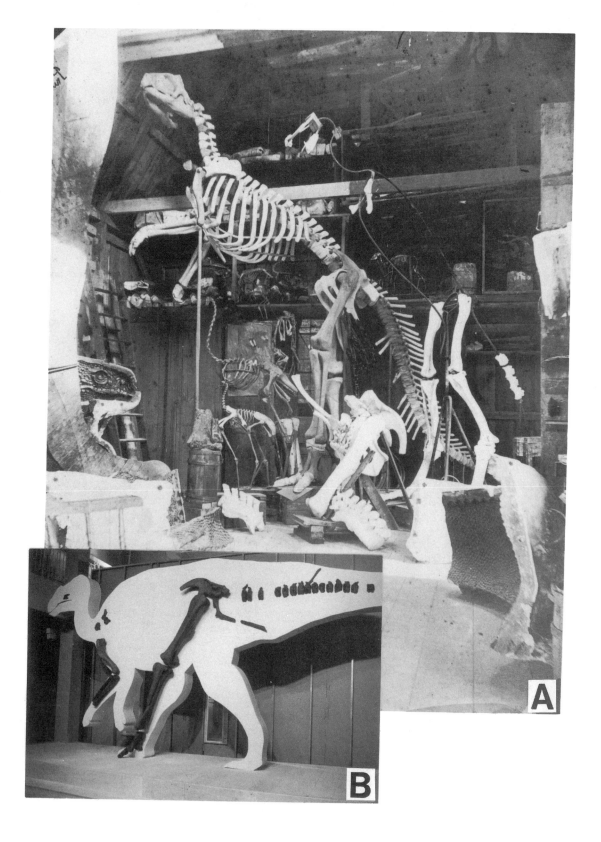

Many of the mounting techniques used by Hawkins were improved during the late 1870s and early 1880s by Louis Dollo at the Museé Royal d'Histoire Naturelle in Brussels. Dollo and his assistants mounted a group of *Iguanodon* skeletons collected from a coal mine at Bernissart in Belgium. Finding so many of the skeletons with all of their bones articulated was a big bonus for Dollo (Fig. 12.4). There could be no question about where each bone belonged. As an aid for mounting, life-sized drawings of the skeletons were prepared. Individual bones or sections of bone were suspended from a wooden scaffolding in the position they would assume in the drawings, attached by metal bands to an iron armature (Fig. 12.5). Dollo used the skeletons of an emu and a kangaroo as references in mounting the dinosaur skeletons. This ensured that the bones were articulated as accurately as possible.

The use of comparative anatomy in the mounting of dinosaur skeletons became even more important when the first sauropod skeleton was mounted in 1905. In preparation for mounting this skeleton, William Matthews and Walter Granger of the American Museum of Natural History dissected several modern reptiles in order to better understand how muscles and joints worked. The muscle scars on the reptile limb bones were matched with those on sauropod limbs. Matthew and Granger then used strips of paper to connect the origin and insertion points of the sauropod muscles. The bones were adjusted so as not to violate probable muscle movement. Once the bones were properly positioned, they were fastened permanently to a steel armature.

Figure 12.3. (A) Workshop of Waterhouse Hawkins showing the mounted skeleton of *Hadrosaurus foulkii* (fossil bone dark) and modern skeletons used for comparative purposes. Partially mounted skeleton to the right of *Hadrosaurus foulkii* is that of *Dryptosaurus aquilunguis*; note the use of steel rods to support the bones. (B) Cast of *Hadrosaurus* bones as currently mounted at the Academy of Natural Sciences. No attempt was made to reconstruct the missing bones as Hawkins had done previously. 12.3A courtesy of the Academy of Natural Sciences of Philadelphia Library.

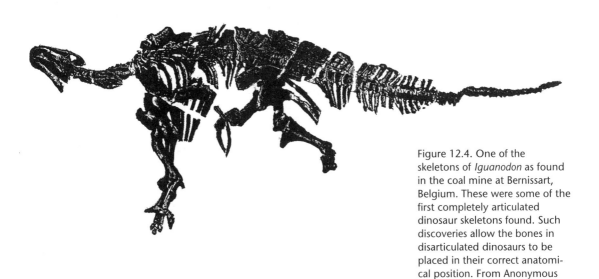

Figure 12.4. One of the skeletons of *Iguanodon* as found in the coal mine at Bernissart, Belgium. These were some of the first completely articulated dinosaur skeletons found. Such discoveries allow the bones in disarticulated dinosaurs to be placed in their correct anatomical position. From Anonymous 1897.

Dinosaurs as Museum Exhibits   155

Figure 12.5. To mount an *Iguanodon* skeleton, Louis Dollo and his crew suspended bones in front of a life-sized drawing of the skeleton. Once the bones were positioned, they were attached to a metal armature. The skeletons of an ostrich and kangaroo stand just in front of the dinosaur's knees. Note the influence that Hawkins and Leidy had with the tripodal stance. From Anonymous 1897.

The use of dissections and comparative and functional anatomy remains important in the mounting of dinosaur skeletons today. This is especially true now that newer dinosaur mounts attempt to "breathe life" into the fossil bones. Skeletons are postured so as to be dynamic, inviting the viewer to imagine the living animal walking, running, or fighting.

Skeletons and Muscles at Work

Let us look briefly at why a knowledge of comparative and functional anatomy is so important in the mounting of dinosaur skeletons. The vertebrate skeleton is the framework for the body. It has many functions, among them providing attachment sites for muscles. Muscles work by pulling, not by pushing. They do this through the shortening and thickening of individual muscle fibers. Thus, a muscle connecting opposite sides of a joint will pull the two bones toward each other (Fig. 12.6). How much motion is possible depends on the type of joint. For example, the elbow is a simple hinge that allows the forearm to move up and down in a single plane.

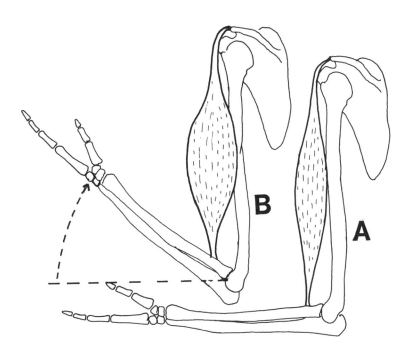

Figure 12.6. To understand how limbs in dinosaur skeletons moved, it is important to understand how muscles and joints work in living animals. To move the lower arm from its position in (A), the deltoid muscle contracts, pulling the arm toward it, in (B). The elbow is a simple hinge joint allowing movement only in one plane.

From knowledge of joints in living animals, it is possible to interpret joint movement in dinosaurs. Often the amount of motion possible, though, is less than the bones alone would indicate (Fig. 12.7A). That is because joints are capped by cartilage and encased in connective tissue to keep the bones together (Fig. 12.7B). This cartilage and connective tissue often leaves a scar around the rim of the joint surface (Fig. 12.7C). The scar delineates the maximum amount of motion possible. To move the limbs beyond the limits imposed by the scars would, in life, indicate a damaged joint. In a dinosaur skeleton, a damaged joint would be implied if, for example, the cartilage scars of the head of the humerus (upper arm bone) were incorporated into the shoulder socket (compare Figs. 12.7D and 12.7E).

The stance of mounted bipedal dinosaur skeletons has changed recently because the living animals are no longer thought to have used their tails as a prop, kangaroo fashion. Instead, the tail was carried in the air as a counterbalance to the body over the hind legs (Fig. 12.1B). The tail moved slightly from side to side to maintain balance alternately over the leg during each step. This change in what we think the tail was for is based upon the straight position of the back and tail in dinosaurs found as articulated skeletons (Fig. 12.4). Furthermore, dinosaur trackways typically lack tail drag marks. This can result in some rather surprising mounts, such as the *Diplodocus* skeleton with the tail in the air at the Denver Museum of Natural History. Nevertheless, the trackways of sauropods lack tail drag marks, indicating that even they carried their long tails in the air!

Dinosaurs as Museum Exhibits   157

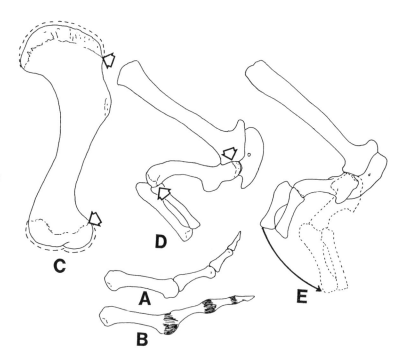

Figure 12.7. The amount of motion at a joint is constrained by cartilage and connective tissue. (A) Hypothetical maximum upward movement of the human finger bones based on joint structure. (B) Actual possible movement as restricted by connective tissue. This example shows that the actual range of movement based on joint structure is usually less than the joints might seem to indicate. (C) The upper arm bone, or humerus, of *Allosaurus,* showing the scar made by the edge of the cartilage cap (*arrows*). Dashed lines show approximate extent of the cartilage caps. (D) The right arm of *Allosaurus,* mounted without consideration of the cartilage that once covered the joints. Arrows point to where the shaft of the humerus has been inproperly incorporated into the shoulder socket and crook of the elbow. (E) Another specimen of *Allosaurus* in which the actual range of movement of the arm bones was determined from the cartilage scars. Note that the bones could not have been positioned as in D. 12.7E is based on Carpenter and Smith, in preparation.

Mounting Skeletons

The work to extract dinosaur bones from the ground, clean them of the encasing rock, restore the missing parts, and mount the skeleton for display can take years. At least seven years were needed for the *Apatosaurus* skeleton on display at the American Museum of Natural History in New York City.

It is because of the time and expense involved in preparing skeletons for exhibition that some museums have begun mounting plastic casts of skeletons. This has generated a controversy that does not have a satisfactory solution. On one hand, the paleontologist wants access to the real bones for study, and on the other, the museum visitor wants to see real fossil bones. In the end the decision is made on the basis of several factors, including what the visitor expects to see, and whether the bones can be exhibited so as to be accessible to the paleontologist. Other considerations include whether the desired specimen for the exhibit is in the museum's collection, or failing that, the availability and costs of obtaining and preparing the required specimen versus the cost of obtaining a cast, and finally the exhibit schedule. Generally, museums that have had a history of collecting dinosaurs use real bone skeletons for the exhibit, while newer museums often have to rely upon casts.

Many older museums have also begun a program of dismantling and remounting their dinosaur skeletons in order to reflect the new information paleontologists have about how the animals stood or moved (compare postures of Figs. 12.1A and 12.3B). Remounting skeletons is often part of a larger program to renovate the dinosaur exhibits. Computer stations with touch screens now dot the exhibit hall, and with a touch anyone can call up information about a particular dinosaur. Remounting a skeleton is often cheaper than preparing a new specimen or purchasing and mounting a cast. The results often make the skeleton look so different that it seems to the visitor that there is a new skeleton on exhibit.

Assembling skeletons involves more than fastening bones to a metal frame. A considerable amount of planning is necessary to ensure that the mount is a good one. Sketches or scale models are prepared showing the finished mount in various views. Such views allow for changes to be made in the posture before assembly makes this impossible or too costly to alter. Sketches and models also show whether the allotted exhibit space is adequate, not cramping the skeleton. A sketch of the skeleton can also become a "blueprint," allowing the steps and materials used in the mounting process to be thought out in detail (Fig. 12.8; compare with Fig. 12.9).

In planning for the mount, it is important to know if the animal was habitually bipedal, walking only on the hind legs, or quadrupedal, walking on all four limbs. Mounting a bipedal dinosaur can be difficult because the forelimbs cannot be used to help support the skeleton. Instead, steel cables from the ceiling can be used to hold the front of the skeleton (Fig. 12.1B), or the skeleton can rest atop a vertical pipe (Figs. 12.8, 12.9A).

It is rare for a dinosaur skeleton chosen to be mounted to have all of its bones preserved. That is because erosion, which uncovers most fossils, destroys what it uncovers. Missing bones can be replaced with casts or actual bones from another individual. Sometimes the missing parts can be sculpted from wood, plaster of paris, papier-mâché, epoxy putty, or

Figure 12.8. A sketch of the dinosaur *Corythosaurus* showing the proposed position of the limbs and tail. Notes were also made on the sketch for how the armature was to be made. Much of the armature was internal or on the back side of bones to minimize their visibility (*dashed lines*). Compare this with Figure 12.9 of the completed skeleton.

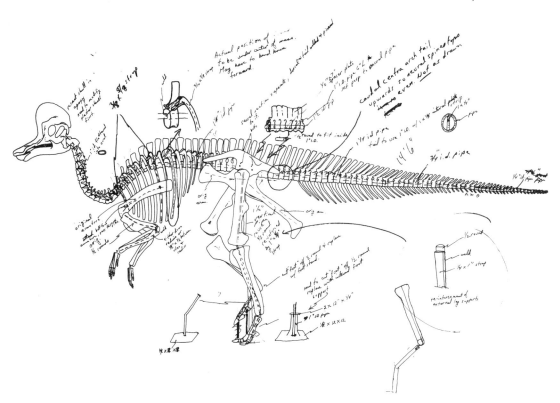

Figure 12.9. The skeleton of *Corythosaurus* as mounted based on the sketch shown in Figure 12.8. During mounting it became necessary to add a vertical pipe to support the front of the body rather than cable the skeleton from the ceiling. Other changes from the original plan in Figure 12.8 include turning the neck and head, making the arms straighter, and adding acrylic rods on the back and tail to replicate ossified tendons. Displayed at the Philadelphia Academy of Natural Sciences.

ceramic clay fired in an oven. Sometimes two or more partial or fragmentary skeletons can be combined to make a single skeleton. This has been done, for example, with the *Triceratops* skeleton at the Smithsonian Institution.

The tools and equipment needed for mounting a skeleton can be obtained from a hardware store; rarely are specialized tools used. These common tools include hammers, saws, pliers, wrenches, nuts and bolts, electric drills, a bench vise, paints or wood stain, and artist brushes of various sizes.

Bones are attached in some way to a support armature, usually made of steel (Fig. 12.10). Because fossilized bone is brittle, the weight of each bone (except perhaps the smallest, lightest ones) is borne by the armature (Fig. 12.10B). This armature may be custom-formed to the surface of the bones, or placed inside bones that have been drilled out. If the armature is outside, it is placed on the underside or back side of bones so that it is least noticeable to the visitor. The steel selected must be of a diameter large enough to support the bone or skeleton without being too obvious. The armature can be welded (Fig. 12.11) or bolted together. The bones are then attached with steel straps, pins, or bolts (Fig. 12.10B).

For vertebrae, a hole is sometimes drilled through the center and the

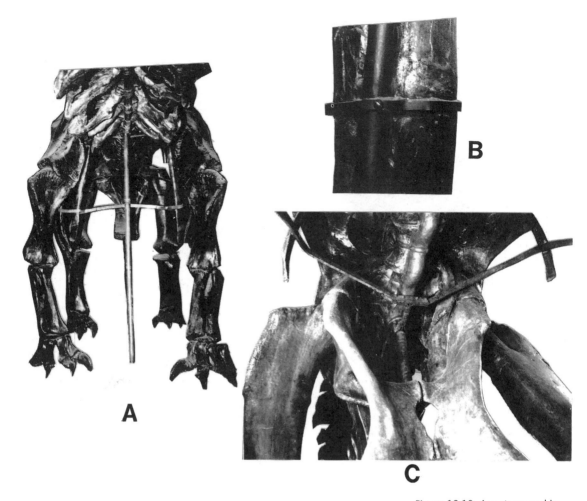

A

B

C

Figure 12.10. Armature used in the mounting of dinosaur skeletons. (A) Vertical pipes used to support a sauropod skeleton at the Peabody Museum (Yale University). (B) Detail showing a metal strap to hold a leg bone to the armature of the *Corythosaurus* shown in Figure 12.9. (C) Underside of a sauropod skeleton at the Denver Museum of Natural History. The vertebrae rest atop a custom-cast steel armature. A steel support for the ribs is also visible.

bones strung like beads on a steel pipe (Fig. 12.12). A custom-cast steel armature can also be made to support the vertebrae from the underside (Fig. 12.10C). But this practice is now rare because it requires a nearby steel foundry, and most of the foundries in the United States have closed down.

Another method for displaying dinosaur skeletons is the panel mount. The skeleton may be left embedded in the rock in which it was found (Fig. 12.13) or mounted so that only one side of the skeleton is visible (Fig. 12.3B). Such mounts have been used to display skeletons as they were originally found, or if one side of the skeleton has been damaged through erosion. The damaged side is embedded in plaster of paris simulated to resemble rock, while the bones of the uneroded side are cleaned for display.

Once a freestanding mount has been completed, the base must be finished in some way to make it visually appealing (compare Figs. 12.1B and 12.12). Sometimes a barrier needs to be erected to protect the fossil from souvenir collectors. Finally, with all mounts, labels must be prepared giving the name of the dinosaur and some information about the animal. Once all of this is completed, the skeleton is ready for its public debut.

Figure 12.11. Welding a section of the armature for the dinosaur *Maiasaura*. Because the skeleton is on all four limbs, the support armature extends down all four legs. The base in this mount is a steel frame.

Other Dinosaur Exhibits

There are several types of dinosaur exhibits other than permanent exhibits of skeletons in museums. Temporary or traveling exhibitions have gained in popularity in recent years. Such exhibits tend to remain at one museum for a few months before being shipped to another museum. One advantage of such exhibits is that they expose the museum visitor to a greater variety of dinosaur skeletons than would otherwise be possible. For example, several exhibitions of Chinese dinosaurs have traveled around North America, Japan, and Europe. These exhibits make it affordable for the museum visitor to see and appreciate these Chinese dinosaurs without the expense of traveling to China.

Most traveling dinosaur exhibits are designed so as to make it easy for the skeletons to be assembled for display. Skeletons are often modular, with pins or bolts holding segments of armature with bones already attached to them. Both cast and real bone skeletons are used in these traveling exhibitions.

Another type of dinosaur exhibit uses restorations of the animals as they might have appeared in life. The first attempts were the fanciful imagination of Waterhouse Hawkins and Richard Owen (who gave Dinosauria their name). Hawkins created life-sized restorations of several extinct animals, including the dinosaurs *Iguanodon*, *Megalosaurus*, and *Hylaeosaurus*. These

Figure 12.12. Mounting of the *Tyrannosaurus* skeleton shown in Figure 12.1B. A steel pipe for the vertebrae extends from the pelvis at the top of the picture. A wooden base was built for this skeleton.

sculptures were displayed in 1853 at Crystal Palace, one of the first World's Fairs. These restorations can still be seen in Hyde Park, London.

Since the exhibition at Crystal Palace, life restorations of dinosaurs have appeared at several World's Fairs, including those in St. Louis (1904) and New York City (1964). Richard Lull attempted another version of the life restoration of a dinosaur at the Peabody Museum at Yale University, using a technique pioneered by the Denver Museum of Natural History in 1918. Lull mounted a skeleton of the dinosaur *Centrosaurus* inside a partial life restoration of the animal. For accuracy, he used actual skin impressions of ceratopsians to replicate the skin surface. The result can still be seen at the Peabody Museum. On the right side, the fleshed-out animal is seen, while on the left side, a mounted skeleton inside a partial body shell is visible.

More recently, robotic dinosaurs have become popular. Pioneered by the Disney Company in the 1960s, robotic dinosaurs attempt to make

Dinosaurs as Museum Exhibits   163

dinosaurs come alive. Many of the early attempts contained anatomical mistakes so that the dinosaurs did not look very lifelike. Today, under the guidance of dinosaur paleontologists, the robotic dinosaur industry pays more attention to details. The results are thought to more closely resemble what we think dinosaurs actually looked like. Robotic dinosaurs seem to be popular, and many museums use them in temporary exhibits to increase museum attendance.

The Future

It is difficult to know what changes will occur with museum exhibits of dinosaurs. If the past is any guide, dinosaur skeletons will always be popular with the museum visitor. Armatures will probably be less noticeable with the use of carbon fiber and epoxy resins in place of steel. New casting materials may be stronger and have a more fossil-like appearance. Finally, walking robotic dinosaurs may appear, stalking the halls of the museum (but probably not dinosaurs cloned from DNA).

References

Alexander, E. 1983. *Museum Masters; Their Museums and Their Influence.* Nashville: American Association for State and Local History.
Anonymous. 1897. *Guide dans Les Collections: Bernissart et Les Iguanodons.* Brussels: Museé Royal D'Histoire Naturelle.
Kellogg, R. 1936. A review of the Archaeoceti. *Carnegie Institute of Washington Publication* 482: 1–366.
Leidy, J. 1865. Memoir on the extinct reptiles of the Cretaceous formations of the United States. *Smithsonian Contributions to Knowledge* 14: 1–135.
Lucas, F. 1902. *Animals of the Past.* New York: McClure, Philips and Co.
Simpson, G. 1942. The beginnings of vertebrate paleontology in North America *Proceedings of the American Philosophical Society* 86: 130–188.

Figure 12.13. A panel mount of the armored dinosaur *Sauropelta* at the American Museum of Natural History.

Restoring Dinosaurs as Living Animals

Douglas Henderson

The illustration of dinosaurs as living animals is an interpretive work of imagination based on scientific inquiry. Such "paleoillustration" depicts, in views and scenes, the living appearance of ancient life, presented in a form borrowed from our direct experience and familiarity with the natural world as we see it today. The interpretive role of paleontological art has put the work of both paleontologists and artists before a wide audience.

Paleontological artists may be scientists or amateurs—anyone who shares an interest in the arts and the earth sciences. Paleoillustration incorporates a traditional approach to art, a development and use of style, medium, and approach to subject unique to each artist. It requires some introduction to the sciences and an appreciation for the limitations in producing complete images from the fossil record. The work that results from a collaborative effort between artist and scientist turns paleontology and related earth sciences into a rich and modern form of storytelling.

Depicting dinosaurs as living animals has little part to play in the actual science of paleontology. Scientists are mostly concerned with seeing their observations and interpretations expressed and disseminated in objectively written publications intended primarily for the scientific community. While technical drawings of fossil material or simple diagrams and reconstructions may serve this purpose, fully rendered images of great beasts passing in silhouette against western sunsets and the like generally do not. These

13

more romantic works, which tell truths nonetheless, are akin to the arts of theater and literature. However, as Emerson wrote in "Woodnotes" of the poet's expression of a personal, subjective experience with the natural world: "What he knows nobody wants." When the concern of science is to be empirical and documental, imaginative imagery is regarded as the wrong language.

On the other hand, for many people the vision of dinosaurs as living animals has a validity as art for its own sake. In addition, paleontological art has considerable value for writers and editors of books and magazines, museum curators, paleontologists, and others concerned with interpreting and presenting the science of paleontology to the general public.

The Reasoning, Research, and Procedures of Paleoillustration

The illustration of dinosaurs, or any other aspect of earth history, has a speculative component. This is due to several factors, including the incomplete nature of the fossil record, varying interpretations based on fossil material, and our inability to observe the specific behavior and natural history of dinosaurs in life. The dinosaur artist must often contend with many unknowns, many possible scenarios, and only a few sure inferences beyond the fossil material itself.

Factors other than science can influence illustrations, such as the expectations and interpretations of the editors, curators, or others who commission illustrations. In addition, the artist may be influenced by long-standing assumptions about the nature of dinosaurs that retain unquestioned acceptance through long-term repetition in the work of other artists and scientists alike.

The scientific artist consequently cannot be expected to render a scene that is literally true. Paleoillustrations, like the science they reflect, are just ideas. They can honor both the reasoned view of science and the unexplored realms that science suggests to the mind's eye.

The imaginative contribution of artists is very much determined by the knowledge, experience, observation, and slant that they bring to their work. An important basis of paleoillustration is effective drawing skills, including a basic understanding of perspective, a sense of composition, some command of a medium, and practice at life drawing. Drawing in the field, a disciplined form of observation, is a learning process. Our subjective experience of the natural world can be captured in drawings, field sketches, and nature studies. In the process, some knowledge can be gained of nature and the natural composition of landforms, river courses, stands of trees, a certain wonderful dishevelment of natural systems (Fig. 13.1). This familiarity with the modern natural world becomes a guide and a fine gauge of normalcy when representing the prehistoric past.

Paleoillustration allows for a very flexible approach in representing the science of paleontology. Artists have been called upon to summarize the broad evolutionary and anatomical diversity of dinosaurs over the entire Mesozoic Era, but have also depicted the diversity of plants and animals that lived in specific times and places. Illustration has taken the form of murals, scenes of condensed knowledge, showing a whole fauna and flora in a single image. Alternatively, restorations have been cast as a series of natural scenes patterned according to some pace by which nature is experienced, one thing or one herd at a time. Images may be dominated by dinosaurs, or they may show dinosaurs as only part of a larger physical

Figure 13.1. *Starvation Creek Grove,* a nature study based on field sketches made by the author in the Sequoia National Forest. The site is situated along a small tributary of Deer Creek in the southern Sierra Nevada north of Lake Isabella, southeast of the Tule River Indian Reservation and near the headwaters of the Kern River.

landscape or ecological community. Science finds any of these approaches valid. The artist, for reasons of esthetics or some appreciation of the story told by scientific data, may find one approach more desirable than others.

Paleoillustration requires some familiarity with the science of paleontology. To begin with, one needs a broad understanding of Earth's ancient geography, climate, flora, and fauna, and the changes in these over time, apart from information about the specific animals to be featured in a scene. Such general background information is available in textbooks and books written for the general public (such as Russell 1977, 1989).

The scientific literature, though technical and not written for the layperson, is a valuable source of information about particular organisms (Fig. 13.2). However, its availability beyond larger university libraries may be limited. Correspondence and discussions with willing paleontologists provide the best opportunities for instruction and guidance. In addition, paleontologists generally collect an extensive library of data related to their studies, which can become a resource available to the artist.

Visiting museums and private collections provides an opportunity to see, sketch, and photograph all manner of fossil material. This can include full-sized, articulated fossil skeletons of dinosaurs, permitting their features to be studied from many points of view. Exhibits in many museums now include life-sized, lifelike sculptures of dinosaurs as well.

Visiting zoos allows the observation of living animals, which may share

Restoring Dinosaurs as Living Animals 167

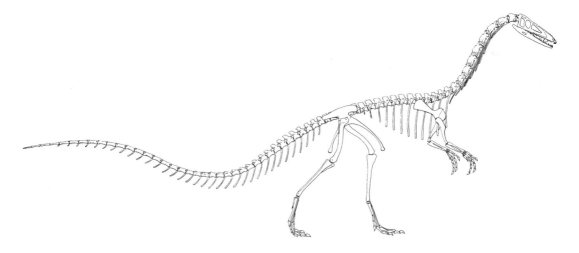

Figure 13.2. Skeletal reconstruction of the Late Triassic theropod dinosaur *Coelophysis* as published by Colbert (1989: Figure 88); figure courtesy of E. H. Colbert.

with dinosaurs some elements of overall body size or anatomy, or which may be close or distant relatives of dinosaurs. Such animals might include large mammals, ground birds, and some reptiles.

Paleoillustration can also require an overview of paleobotany. Textbooks and fossil plant field guides discuss the evolution and classification of plants, and indicate those forms that composed Mesozoic floras. The scientific literature is descriptive and technical, but often unsatisfactory to the artist interested in reconstructing the living appearance of ancient plants. A living plant consists of many anatomical "modules" (roots, stems, branches, leaves, reproductive organs) that commonly separate prior to fossilization. Consequently it is often hard to determine what an entire ancient plant looked like.

However, the fossil record suggests many phylogenetic relationships between Mesozoic plants and living ones. Plants related to those characteristic of Mesozoic floras, such as ferns, tree ferns, cycads, various conifers, and primitive flowering plants (such as magnolia and dogwood), can be found in greenhouses and botanical gardens. Many hardwood trees found in the eastern and southeastern United States have leaves remarkably similar to fossil leaves of Cretaceous age. For artists, this similarity logically suggests the possibility that the living appearance of some modern conifers and hardwood trees presents forms and patterns common to ancient floras. Unfortunately, paleobotanists caution that some of these similarities may be more apparent than real.

Another resource is the work of other scientific artists, especially those who have undertaken a study of dinosaur anatomy, presented reasonable theories of reconstruction, and produced illustrations that depict the living appearance of dinosaurs. The work of Robert Bakker, Mark Hallet, and especially Greg Paul offers convincing restorations, revealing accurate skeletal dimensions, ranges of limb motion, sizes and attachments of muscles, and interpretations of postures and gaits. A familiarity with this work aids in restoring dinosaurs from technical drawings and new or unfamiliar fossil material.

A finished illustration (Fig. 13.3) can be used to describe the steps and considerations involved in producing an image, and the nature of the fossil information on which it is based. The intent of the scene is to show a large group of *Coelophysis,* a 10-foot carnivorous dinosaur, passing through a Triassic forest in an upland environment. The scene is based on fossils collected from the Chinle Formation, a unit of sedimentary deposits left by rivers that crossed the southwestern United States some 225 million years ago.

In 1947, paleontologist Edwin Colbert discovered a fossil bone bed at Ghost Ranch, New Mexico (Colbert 1995). It was composed almost entirely of many articulated and disarticulated *Coelophysis* skeletons, including both adults and juveniles. The large concentration of animals suggested that *Coelophysis* lived or associated in large groups. (It should be noted, however, that some paleontologists think that the Ghost Ranch theropods constitute two different kinds of dinosaur; Sullivan 1994; Sullivan et al. 1996.)

Another goal of the scene is to represent some of the plants of the Chinle flora, including two kinds of ferns, a tree fern, a horsetail, a cycad, a streamside shrub called *Sanmiguelia,* and a large conifer tree, *Araucarioxylon.* Fossil remains of these plants are often found in what during the Triassic were lowland floodplains. *Araucarioxylon* is known from an abundance of fossil logs, the best examples of which are found in Petrified Forest National Park in Arizona. During the Late Triassic, the logs were the remains of trees with their bark and limbs stripped away, the result of stream transport during periodic flood events (Ash 1986; Long and Houk 1988; Vince Santucci, personal communication). The illustration that was prepared represents a forested upland setting where the fossil logs might have originated.

Figure 13.3. Restoration of a large group of the small carnivorous dinosaur *Coelophysis* in the environment of Late Triassic New Mexico.

Restoring Dinosaurs as Living Animals   169

Many details about *Coelophysis,* such as its habits, its preferred environments, its prey, skin patterns, and color, are unknown. What can be inferred from the fossil record, however, affords a wealth of imagery with which to work. This can include some of the following general statements.

The fine preservation of many *Coelophysis* specimens allows the dinosaur to be restored to something close to its living appearance. *Coelophysis* was a bipedal, slightly built, fast-running hunter with small, grasping arms and small, sharp teeth suited for handling prey only smaller than itself. *Coelophysis* shared the Late Triassic American Southwest with a great many other reptiles. It lived amid a flora comprising a rich assemblage of plants. *Coelophysis* may have been common among the variety of upland, forest, river, and lowland floodplain landscapes and habitats available to it. The illustration touches on only a few of these ideas.

The illustration's preliminary outline sketches of *Coelophysis* were based on Colbert's (1989) skeletal reconstruction (Fig. 13. 2). The restoration of its living appearance follows the interpretations of Gregory Paul. The wedge-shaped skull, lacking the complex of facial muscles of mammals, needs only a cover of skin to give it a living appearance. The neck was slender and held in an S curve at rest. There was a slight arch to the spine. The chest narrowed from side to side at the hip region, which was no wider than the connective tissue that spanned the pubis and ischium, the ends of which defined the lower outline of the body. The tail was moderately flexible, like a lizard's. The musculature and general appearance of the neck, legs, and feet were bird-like.

The reconstruction of the Chinle flora is based on publications by paleobotanists and discussions with paleontologists, as well as my own observations. The Chinle flora is known from fossil plant material found in localities across the Southwest. In addition to fossil logs, stems, roots, and a few cones, there is a diversity of leaf impressions. All of this indicates that the Chinle flora included many kinds of plants, many more than are included in the illustration. Some of these fossil plants, especially the ferns and some unique shrubs, can be accurately reconstructed. Others are represented only by isolated leaves or other structures, and the whole plant is unknown.

In illustration, partially known material can be represented by showing the plant structure that is known, or adopting some living appearance based on a relationship or similarity to living plants. In the *Coelophysis* scene, the tree fern *Itopsidema,* known only by its fossil trunk, is reconstructed both as a trunk only and with vague foliage appropriate for a living tree fern.

The conifer tree *Araucarioxylon* is one of three species of trees known only from large and often very well preserved fossil logs. Other conifer fossils include a half-dozen or more types of fossil foliage of small size, superficially resembling sprigs of juniper and cryptomeria, and one larger type resembling fir twigs. None of the foliage can be associated with the logs.

The reconstruction of *Araucarioxylon* can be based on observations of the fossil logs, which show many prominent structures; these include traces of roots, swollen bases, wood surface patterns, long, tapering trunks, and a variety of limb scars. The limb scars are spaced in varying patterns, often sparse, but some along nearly the length of the trunk. They show limbs both large and small, which invariably exited the trunk at an upward and often steep angle. These features suggest that *Araucarioxylon* may have resembled a giant, long-trunked Utah juniper or the tall Monterey cypress that forests a good portion of Golden Gate Park in San Francisco.

Once the purpose of an illustration is decided on and the design of its various elements is considered, the next step is to bring the separate components together into a single scene. Rough sketches of an overall scene are drawn, perhaps many times, to establish some satisfactory composition. Once a composition or design is determined, it becomes a guide to the further preliminary work of refining characters, and also a reference while working on the illustration itself. If the illustration is one requested by a publisher or curator, then the preliminary work may be subject to review and approval by editors, writers, and consultants.

Refining characters can be the most time-consuming portion of illustration, requiring that each individual animal be properly outlined in different poses, angles of view, and size relative to distance across the scene. Artists may refer to skeletal drawings, photographs, or their familiarity with dinosaur proportions. Some artists construct, or refer to, small models and sculptures. In the case of the *Coelophysis* illustration, an army of individual dinosaurs was drawn (Fig. 13.4), and the figures were cut out, shuffled

Figure 13.4. Refined preliminary outlines of the small carnivorous dinosaur *Coelophysis,* showing the appearance of the animal from different views. Most of the work of planning the restoration is done in this step of the project.

Restoring Dinosaurs as Living Animals   171

around, and taped into place around outlined tree trunks sketched on a sheet of paper. This refined the composition into a carefully choreographed outline.

Such an outline can then be transferred to a new surface of paper (or canvas, or whatever surface is needed for the particular project), on which the illustration will be drawn or painted. Over several days, weeks, or even months, the work is slowly built up with line and shading to a finished image (Fig. 13.3). It is a problem-solving process to which each artist will bring her or his own solutions.

The illustration of dinosaurs is an interpretive tool in service of both science and art. It presents views of ancient life that broaden the appeal and understanding of paleontology beyond the circle of scientists. In addition, paleoillustration allows the individual artist to explore the realm of ideas and legitimate fantasy suggested in the objective interpretation of the fossil record.

The work of paleoillustration is unfinished. A search through the sources of paleontological information, from discoveries in the field to the scientific literature, both old and new, reveals a wealth of data that have not been represented in illustration. A collaborative effort of art and science has much left to tell.

References

Ash, S. 1986. *Petrified Forest: The Story behind the Scenery.* Revised ed. Petrified Forest, Ariz.: Petrified Forest Museum Association.

Colbert, E. H. 1989. The Triassic dinosaur *Coelophysis.* Flagstaff: Museum of Northern Arizona Bulletin 57.

Colbert, E. H. 1995. *The Little Dinosaurs of Ghost Ranch.* New York: Columbia University Press.

Long, R. A., and R. Houk. 1988. *Dawn of the Dinosaurs: The Triassic in Petrified Forest.* Petrified Forest, Ariz.: Petrified Forest Museum Association.

Russell, D. A. 1977. *A Vanished World: The Dinosaurs of Western Canada.* Natural History Series, no. 4. Ottawa, Ontario: National Museum of Natural Sciences, National Museums of Canada.

Russell, D. A. 1989. *An Odyssey in Time: The Dinosaurs of North America.* Toronto: University of Toronto Press.

Sullivan, R. M. 1994. Topotypic material of *Coelophysis bauri* Cope and the *Coelophysis-Rioarribasaurus-Syntarsus* problem. *Journal of Vertebrate Paleontology* 14 (Supplement to no. 3): 48A.

Sullivan, R. M.; S. G. Lucas; A. Heckert; and A. P. Hunt. 1996. The type locality of *Coelophysis,* a Late Triassic dinosaur from north-central New Mexico (USA). *Paläontologische Zeitschrift* 70: 245–255.

The Groups of Dinosaurs

Bradley was in the lead when he came suddenly upon a grotesque creature of Titanic proportions. Crouching among the trees, which here commenced to thin out slightly, Bradley saw what appeared to be an enormous dragon. . . . From frightful jaws to the tip of its long tail it was fully forty feet in length. Its body was covered with plates of thick skin which bore a striking resemblance to armor-plate. The creature saw Bradley almost at the same instant that he saw it and reared up on its enormous hind legs until its head towered a full twenty-five feet above the ground. From the cavernous jaws issued a hissing sound of a volume equal to the escaping steam from the safety-valves of half a dozen locomotives, and then the creature came for the man.

—Edgar Rice Burroughs, *Out of Time's Abyss*

The senior editor of this book teaches at a university whose team mascot is the mastodon—an extinct Ice Age relative of modern elephants. Our athletic teams are called the Mastodons (or Dons—and even Tuskers) because a nearly complete skeleton of such a beast was found in the area about the time the campus was being built. This skeleton now lies in state in a big display case on campus, and is viewed by many school groups and other interested parties, the vast majority of whom identify it as a dinosaur, accompanying signs providing information about its true nature notwithstanding.

The senior editor once attended a debate between a creationist and an evolutionist held in a small church in a nearby town. In his peroration, the creationist cited the Komodo dragon (*Varanus komodoensis,* the biggest living lizard) in making a rhetorical point; he observed that, after all, what is the Komodo dragon other than a surviving kind of dinosaur?

These two stories illustrate two common beliefs about what dinosaurs are, or were. The general public regards the word *dinosaur* as a suitable tag for any number of swimming, flying, or lumbering animals, provided that they are big, prehistoric, and extinct, or big, ugly, and reptilian.

As once remarked by that eminent philosopher Sportin' Life, in an entirely different context: It ain't necessarily so. There were many kinds of big, extinct animals (including reptiles) that were not dinosaurs at all, and there are all kinds of big living reptiles that aren't dinosaurs, either.

Part Three addresses the question of just what a dinosaur is. What features define these reptiles, and distinguish them from other kinds of animals? The section begins by placing that question in an historical context, by looking at how and why dinosaurs came to be recognized as a distinct category of reptiles. The story is not one merely of scientific discovery, but also includes conflicting personal ambitions and more than a touch of pure power politics; the concept of the Dinosauria thus emerged in a setting that will be instantly recognizable to anyone who has ever worked in an academic environment.

Several chapters then describe the various groups of dinosaurs, summarizing their key anatomical features. This anatomical information is the foundation of all paleontological research about dinosaurs; it forms the framework on which hypotheses about dinosaur paleobiology and evolution are built. Each chapter therefore considers what we presently know or infer about the biology and evolution of the dinosaur group under consideration.

Politics and Paleontology: Richard Owen and the Invention of Dinosaurs

Hugh Torrens

14

A recent claim is that "dinosaurs are the most American animals that have ever lived" (Kirby et al. 1992: 28). While this is clearly true now, the *invention* of dinosaurs, by the English anatomist Richard Owen (1804–1892), who coined his name for these "fearfully great lizards" in April 1842, was entirely English. All our early knowledge of this new group of reptiles came from England, and all the paleontological material which supported their invention had been found in English rocks.

The work of the French comparative anatomist Georges Cuvier (1769–1832), however, was also crucial. He had declared in 1825 that *Plesiosaurus* (a kind of Mesozoic marine reptile) was "the most heteroclite . . . and the most monstrous [animal], that had yet been found amid the ruins of the former [i.e., fossil] world" (Buckland 1837, vol. 1: 202). This was just after another English contender for "most monstrous" animal had been described: *Megalosaurus,* revealed by the Oxford academic Rev. William Buckland (1784–1856) to the Geological Society of London in 1824. Buckland showed that *Megalosaurus* had possessed five vertebrae in its sacral, or hip, region (Fig. 14.1), which were all ankylosed (also spelled anchylosed), or fused, together (Buckland 1824).

A third "monstrous" contender, *Iguanodon,* followed in 1825, with quite different teeth, teeth that indicated a herbivorous diet. These were revealed by the provincial surgeon and scientist Dr. Gideon Mantell (1790–1852), who was based in Lewes, Sussex, but whose paper was read for him to the rival Royal Society in London (Mantell 1825).

The rivalry between these two societies had started in 1808 when the

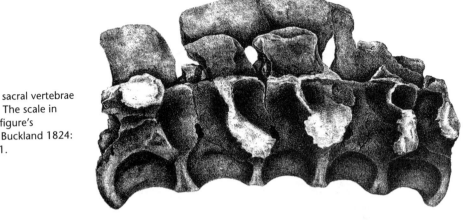

Figure 14.1. The sacral vertebrae of *Megalosaurus*. The scale in inches gives the figure's reduction. From Buckland 1824: Plate 42, Figure 1.

SACRAL. VERTEBRÆ OF MEGALOSAURUS.

Drawn by M. Morland. & 99 Stone by Henry Parry.

Scale of one Inch to one Inch.

long-standing president of the Royal Society, Sir Joseph Banks (1743–1820), with Humphrey Davy (1778–1829) and others had unsuccessfully tried to control the future of the new (1807) Geological Society (Rudwick 1963). Having failed, in 1809 they resigned from it.

The work of William Smith (1769–1839), who had shown that particular rock strata could be identified on the basis of their fossil contents, meant that the relative antiquity of the Middle Jurassic *Megalosaurus* specimens from Stonesfield (from underground mines) could be demonstrated. The stratigraphic horizon of *Iguanodon* proved much more uncertain.

However, if much stratigraphic expertise was available in England by 1820, there was little good comparative anatomy available in Britain. Much of the critical expertise in deciphering such new, and always still fragmentary, fossil vertebrate material still had to come from France, from Cuvier. Cuvier had pioneered the new science of comparative anatomy and had used it to demonstrate both that extinction was a frequent fact in the fossil record and that fossil animals could often be "reconstructed" using surprisingly limited evidence.

Cuvier had again visited England in 1830, and Richard Owen was chosen to show him round. Cuvier's return invitation for Owen to visit Paris, from July to September 1831, greatly influenced Owen in "Cuvierian" methods (Owen 1894, vol. 1: 48–58). But Cuvier died in May 1832, and a scientific vacuum was created. A "power struggle" now developed throughout Europe, to determine who would take over the scientific mantle of Cuvier.

An early contender was the founder of vertebrate paleontology in Germany, Christian Erich Hermann von Meyer (1801–1869). In 1832 he published his first attempt at a taxonomic division of the group of fossil reptiles then classed within the Order *Sauria*. This was based on their organs of locomotion, and in it *Megalosaurus* and *Iguanodon* were grouped together as "Saurians with Limbs similar to those of the heavy Land Mammals" (Meyer 1832: 201). This paper was translated into English in 1837 by Mantell's linguist-curator George F. Richardson (1796–

1848) (Richardson 1837). New footnotes, provided by Mantell, noted that Mantell's new 1833 genus *"Hyleosaurus . . .* probably [also] belongs to this division." Owen was certainly aware of Meyer's work, since he later noted that Meyer had assigned "no other grounds [than their heavy-footedness] for their separation from other Saurians" (Owen 1842: 103).

The decade 1822–1832 was one of political frenzy in England, with the Reform Bill of 1831 and the Reform Act of 1832. These political initiatives brought the first thorough attempts to redraw the political map of Britain and to define who should, and who should not, have the right to vote. The 1832 Act was "one of the most momentous pieces of legislation in the history of modern Britain" (Evans 1983). It changed a system that distributed parliamentary representation through limited patronage to one that relied on more democratically based voting. It also brought an end to fears of political revolution in England. These fears had made many suspicious of all things French, particularly in science, where pre-Darwinian theories of Evolution or Transformism (Laurent 1987) were seen as both highly French and equally revolutionary.

The decade 1822–1832 was also one of major significance for British science. The deaths of two Royal Society presidents, Banks in 1820 and Davy in 1829, had brought debates about how English science, where "the pursuit of science [still] does not constitute a distinct profession" (Babbage 1830: 10), was controlled and how it might be advanced. "The Decline of Science in England" and the too-dominant role of the metropolitan Royal Society in "leading" English science were widely discussed. The British Association for the Advancement of Science (BAAS) came into existence in 1831 as another rival to the Royal Society. BAAS was to be an itinerant and provincial organization, hoping to encourage British scientists *and* to advance its science by annual meetings at provincial centers. Such debates also helped science to gain a better place in popular culture and led to the invention of the word *scientist*. Elitism also suffered, and "upward social mobility based on meritorious accomplishment" would become significantly more common thereafter (Dean 1986). This then encouraged those from lower backgrounds to aspire upward; Gideon Mantell and John Phillips (1800–1874) provide two good examples from geology.

In a scientific England rocked by both the passage of the Reform Act in June 1832 and Cuvier's death in May, reforms in science were soon being promoted by BAAS. Battle lines were also drawn up in the small world of English vertebrate paleontology. The protagonists attempting to usurp Cuvier's reputation for English science were Robert Grant (1793–1874), Gideon Mantell, and Richard Owen.

Grant was the badly paid professor of comparative anatomy and zoology at London's new and "godless" University College. It was called godless because, unlike Oxford and Cambridge, it was open to dissenters. It required no religious tests and erected "no barrier to the education of any sect" (Bellot 1929: 56). It was also based in metropolitan, and thus "central," London, but Grant could be progressively marginalized by Owen, both because of Grant's espousal of evolutionary, and thus revolutionary, French science, and by the low prestige of his new London University chair against those at the older Oxbridge universities, who supported Owen. It did not help that Grant was badly paid and, above all, that his research was badly supported. His work has recently been brought to life by Adrian Desmond, who has revealed how the epithet "the English Cuvier" was first bestowed on Grant (in 1831 [Desmond 1989: 98], and again in 1835–36 [Desmond 1989: 122, 755]). Grant also enters the story of dinosaurs as an early target of Owen, who believed in creation, not evolution (Desmond 1979). But in view of the details now revealed, it is clear that Grant was by 1842 merely a minor target for Owen, since Owen had largely "disposed" of him, before the "invention" of dinosaurs.

It is also clear that Gideon Mantell was the real target of Owen's dino-inventing researches. Mantell was an amateur, the son of a shoemaker in Lewes, Sussex, who was both a political radical (Whig) and a dissenter (Methodist) (Dean 1990: 434). Such a background only encouraged confrontation with Owen's political beliefs. Mantell was also a provincial. As Robert Bakewell (1767–1843), author of the first adequate textbook on geology in England, wrote in 1830 after a visit to Mantell's already wonderful Museum of Fossils in Lewes, "There is a certain prejudice . . . prevalent . . . in large cities, which makes [people] unwilling to believe that persons residing in provincial towns or in the country can do anything important for science" (Bakewell 1830: 10). But Mantell's reputation as a vertebrate paleontologist extended back much further than Owen's, to 1825.

In contrast, Owen was a professional comparative anatomist employed at the Royal College of Surgeons and in metropolitan London. He was nearly a generation younger than Mantell. He was more energetic and by the 1840s in much better health; he was also a centrally placed Christian (Anglican) believer, he was better paid, and his scientific activities were properly patronized by the liberal establishment circles of Oxbridge and the BAAS (Fig. 14.2).

Dinosaurs-to-be became weapons in the ideological wars that developed among these three men. The fossils that collectors such as Mantell had gathered so assiduously had inspired new debates, over whether species had evolved by evolutionary transmutation, from other species, against a belief that they had resulted from separate acts of divine creation. These competing theories caused tensions to build among these three scientists. In addition, there were other tensions between "mere" fossil collectors, such as Mantell, who gathered his own material with real assiduity, and research workers, such as Owen, who did not.

When the BAAS entered the field of vertebrate paleontology, it first supported a young but foreign (Swiss) naturalist, Louis Agassiz (1807–1873), who later made a huge impact in America. Agassiz was granted £210 for his work on British fossil fishes in 1835 and 1836 (1836, *Report of BAAS*, 1835 meeting, 5: xxvii). In 1837 the BAAS asked Owen to undertake a similar commission on the "Fossil Reptiles of Great Britain" and granted him £200 (perhaps near $160,000 in modern money! See Rudwick 1985: 461) in 1838 toward this (1838, *Report of BAAS*, 1837 meeting, 7: xvi, xix, and xxiii; and 1839, ibid., 1838 meeting, 8: xxviii). The three-man committee which granted Owen's funds helpfully included his father-in-law! In the same year the Geological Society's premier Wollaston Medal was awarded to Owen. It had earlier, in 1835, gone to Mantell.

Leading lights in the BAAS now saw it as a slight to British science that British fossil treasures were being revealed by foreigners. Owen was a rising star at the Royal College of Surgeons and St. Bartholomew's Hospital in London. The BAAS chose to support the metropolitan, conservative Owen against the provincial, radical Mantell, who had earlier been encouraged by the BAAS's rival, the London Royal Society. When the BAAS president discussed Owen's first reptilian results in 1840, he duly evoked the name of Cuvier; "could that eminent man view the progress which our young countryman is making towards the completion of the temple of which the French naturalist was the great architect" (1841, *Report of*

Figure 14.2. An engraving of Richard Owen in the 1840s, holding the complete moa femur from New Zealand. From [Timbs] 1852.

BAAS, 1840 meeting, 10: xl). In BAAS eyes there was now a new occupant of Cuvier's mantle—Owen. The previous occupant, Grant, had been marginalized by BAAS and would henceforth have to make do with what little he could earn from teaching and student fees.

Owen's first report on marine British fossil reptiles was read to BAAS in August 1839. It was hailed as the work "of the greatest comparative anatomist living" by William Lucas of Hitchin (Bryant and Baker 1934, vol. 1: 179). It was quickly seen by many others (and the BAAS president) as the work of the new "English Cuvier" (Desmond 1989: 333). By autumn 1840, Owen was gathering data for his second report to the BAAS on the remaining fossil reptiles (Owen 1894, vol. 1: 169–172). At the BAAS Glasgow meeting in 1840, the association's president announced that Owen had already "collected . . . equally numerous materials" toward this second part, and reported the "abundance of new information" which Owen had collected (1841, *Report of BAAS,* 1840 meeting, 10: xli–ii and 443–444). Owen had been greatly helped in this work by the sale of the Mantell collection to the British Museum in 1838 (Cleevely and Chapman 1992). Also helpful to Owen was the offer made to him in 1840 in London by another provincial collector, George Bax Holmes (1803–1887), of his Sussex fossil collections—rather than to the local expert, Mantell (Cooper 1993).

On 2 August 1841, Owen gave his now-famous lecture on the remaining British fossil reptiles to Section C of BAAS. The meeting was held that year in highly provincial Plymouth, in Devon. Neither Mantell nor Grant was present. The newspapers were divided about any interest such provincial science might have. The London *Times,* long unenthusiastic about the BAAS and science, simply complained that Owen's lecture was of "very long detail" (3 August 1841). Most "Devonian" newspapers were more proud of the events taking place in their midst. Three Plymouth newspapers did report Owen's paper, giving an essentially similar notice to prove that syndicated reporting is not new (the *Devonport Independent, Devonport Telegraph,* and *Plymouth, Devonport and Stonehouse Herald,* 7 August 1841). All show how Owen ran systematically through the different groups of fossil reptiles he then recognized in Britain.

After describing the crocodiles, Owen discussed the "extinct species which in their organisation bear a relation to some existing at the present time such as the Iguana." Of these, only *Iguanodon* was separately noted, since there had been a recent discovery of

> the best specimen in . . . a quarry near the town of Horsham . . . ; so enormous was this animal that its claw was six times as large as the claw of an Elephant, all the rest of the body being on the same scale. From the position of the remains, it appeared as if the whole of the skeleton were there, and it was probable that the head of the animal was now under the church!

So Owen still thought that *Iguanodon* was truly gigantic in size when he spoke at Plymouth in 1841.

This was because Owen was still following the lead of Mantell, whose largest estimate for the length of this Horsham, Sussex, specimen was 200 feet, according to its collector (Hurst 1868: 225). But the specimen Owen had used at Plymouth to demonstrate this still-gigantic size had not been "collected" by his new rival Mantell. It had been recovered by Holmes, whose allegiance was to Owen, not Mantell.

A second vital point which these "Devonian" reports reveal is that Owen still thought that these fossil "animal remains . . . never exhibit any indications that their forms graduated, or by any process passed into another species. They appeared to have sprung from one creative act." Owen had thus used *Iguanodon* and allies at Plymouth to loudly and clearly proclaim his anti-evolutionary stance, in continued opposition to Grant.

Other reports of Owen's Plymouth lecture appeared in the London *Athenaeum* (21 August 1841: 649–650) and *Literary Gazette* (7 August 1841: 509–511 and 14 August 1841: 513–519). The *Literary Gazette* even noted that Owen had named two new species of *Cetiosaurus* in his Plymouth address: *C. hypoolithicus* and *C. epioolithicus*. Both are hitherto unnoticed, but since Owen never properly described them, they must remain *nomina nuda!* Other notices of the lecture appeared in French (*L'Institut* 10, no. 420 [13 Janvier 1842]: 11–13—derived from the *Literary Gazette*), in German (*Neues Jahrbuch für Mineralogie, Geographie, Geologie, Petrographie,* Jahrgang 1842, pt. 2: 491–494—from the previous French source), and in American English (*The American Eclectic* 2 [1842]: 587–588). The *Athenaeum* noted that Owen had included all three genera, *Iguanodon, Megalosaurus,* and *Hylaeosaurus,* merely among the "gigantic forms of terrestrial Saurians," not in any new order. Owen had not named dinosaurs at Plymouth. They remained to be "invented."

The fullest report was that given by the *Literary Gazette*. This gives the clearest indication of what Owen actually said, and believed, at Plymouth. It confirms the following:

1. Owen thought that Holmes's new Horsham *Iguanodon* was still "six times greater [in size] than the largest elephant." Clearly Owen still regarded any dinosaurs-to-be as gigantic in size *and* still agreed with Gideon Mantell's existing size estimates.

2. Owen said that "there was no gradation or passage of one form into another, but that [each] were distinct instances of Creative Power, living proofs of a divine will and the work of a divine hand, ever superintending and ruling the existence of our world" among these fossil reptiles. This view was clearly in opposition to that of Grant, as we have seen.

3. Owen claimed that Mantell's name *Iguanodon* was unsuitable and, worse, that it implied a false relationship. The "Devonian" newspaper sources also noted that Owen had observed that "no existing lizard differed more from the *Iguana* than did *Iguanodon* and that the name *Iguanodon* created an erroneous idea of its affinities." This was clearly a new and pointed dig at Mantell, who had named *Iguanodon* in 1825.

4. Crucially, Owen had recognized only "four great families of reptiles," *Sauria, Chelonia, Ophidia,* and *Bactrachia,* and divided the relevant one of these (*Sauria*) into only four groups: enaliosauria—the extinct marine reptiles; loricate or crocodilian sauria; lacertians or squamate sauria; and pterodactyls. Note again the absence of dinosaurs.

Mantell's response to the report of Owen's speech at Plymouth was rapid (1841a). Two of his three points concerned the *Sauria:* (1) Mantell had named *Iguanodon* only because of "the general resemblance in *external form* of [its] fossil teeth with iguana," and (2) Owen's "new" identification of "the supposed teeth of *Hylaeosaurus*" had already been made four years before, but by Mantell!

Mantell's printed response had been moderate, but his true feelings about Owen now were revealed to his friend Benjamin Silliman (1779–1864) in New Haven, Connecticut: "it is too bad [for Owen] to censure those [like Mantell] who have cleared the way for him. I am resolved no longer tamely to submit to such injustice: and if any of my friends again serve me [thus] I will retaliate" (Spokes 1927: 133).

All reports of Owen's Plymouth speech made no mention of any new order (or of dinosaurs). All sources show that Owen had simply grouped *Iguanodon, Megalosaurus,* and *Hylaeosaurus* with lizards, and *within* his "Lacertian or Squamate division of the Saurian Order" in the Class Reptilia. Taxonomy is an inclusive science. The statement that *Iguanodon* and related genera were then placed in a particular division of a named order excludes any possibility that they could have then been included in any other division of this, or any other, order. If one is a postgraduate, one cannot also be an undergraduate at the same institution. These are equally mutually exclusive categories. Owen simply never mentioned dinosaurs at Plymouth in 1841 because he had not yet invented them. We must pay tribute to the Victorian journalists who so accurately reported on Owen's lecture. We may wonder perhaps why today's journalists are in general so different in their attitudes toward the importance of science and in reporting it. In the same way we might ask why historians today are so averse to using such vital sources.

After Plymouth, Owen stayed some time in the West Country, returning to London on 11 September 1841 after forty-seven days away (William Clift's MSS diary, Royal College of Surgeons Library, London). He then got back to work on the third part of his book *Odontography* (Owen 1894, vol. 1: 187), and on gathering data for the further report for the BAAS which he had now been commissioned to undertake, on British fossil mammals.

A letter Owen wrote on 23 December 1841 to paleontologist John Phillips shows that he was then in Cambridge working on vertebrate material in the university's Geological Museum. It ends, "I shall return to London next week to finish the revision of the old Report [that he delivered at Plymouth] before again starting in quest of materials for the new [on mammals]" (Oxford University Museum, Phillips MSS, 1841/65.1).

A major spur for Owen to revise this Plymouth report had come in December 1841, when Mantell's diary noted (between 9th and 31st) that he sent out "copies of my papers on *Iguanodon* and Turtle from the *Philos. Trans.* . . . just published to many friends in England and to many savants in France." Owen would have received one of the first of these. In October 1841 Mantell had suffered a nearly fatal carriage accident, and then paralysis from spinal disease. These crises, with the departure of his wife, who left him in 1839 (Curwen 1940: 140), had brought him to his lowest ebb. Mantell, by now Owen's chief rival in reptilian studies, sent out a hundred offprinted copies of his papers published in the *Philosophical Transactions of the Royal Society of London* in 1841 (Mantell 1841b).

These also helped fan the resentment between the Geological and Royal societies in London. When Mantell published these Wealden reptile studies in the Royal Society's journal, the then-president of the Geological Society, Roderick Murchison (1792–1871), commented, "Whilst I understand . . . the motive which led [Mantell] to communicate his last memoir on the *Iguanodon* to the same Society to which he had addressed his first account of that Saurian, I regret that he should not have communicated to ourselves other paleontological memoirs." Murchison sourly concluded, "So long as the Royal Society produces volumes adorned by the . . . first mathematicians, physiologists and chemists, so long will it maintain its high place, little heeding our humbler pursuits" (Murchison 1842: 653).

Mantell's offprints had been grandly titled "A Memoir on the Fossil Reptiles of the South-East of England by Dr. Mantell." A perhaps unique copy, in its original binding, survives in the library of the Yorkshire Museum (*Yorkshire Evening Press,* 30 June 1993). This was sent out on 1

February 1842, and Mantell had distributed all copies by 3 February 1842 (MSS Diary at Alexander Turnbull Library, Wellington, New Zealand).

Owen's post-Plymouth revisions proved highly significant. On 4 January 1842, Owen's best friend, the barrister William John Broderip (1789–1859), asked Owen's opinion on a piece Broderip had written on spondylid bivalves, "wherein genera are cut down more mercilessly than you have docked the Saurians. I believe with as much justice" (Broderip to Owen, 4 January 1842, Owen MSS, Natural History Museum, London, 5/111). On 14 January 1842, Broderip revealed to William Buckland what Owen's "docking of the Saurians" had been. It had been to enormously reduce their size. Broderip reported, "I am happy to say that I have now before me the beginnings of the end of Owen's *Report* on fossil reptiles—an *opus magnum* in itself though he has reduced the sesquipedality of some of your old friends [*Iguanodon* etc] 'with tails as long at St Martins Steeple'" (British Library Add. Mss 40500, ff. 247–248). The famous steeple of St. Martin in the Fields Church in London was 192 feet high. Owen had now revised his opinion of the size of *Iguanodon*, from that claimed in his Plymouth lecture—where they had been up to six times bigger than the biggest elephant (and thus up to 200 feet)!

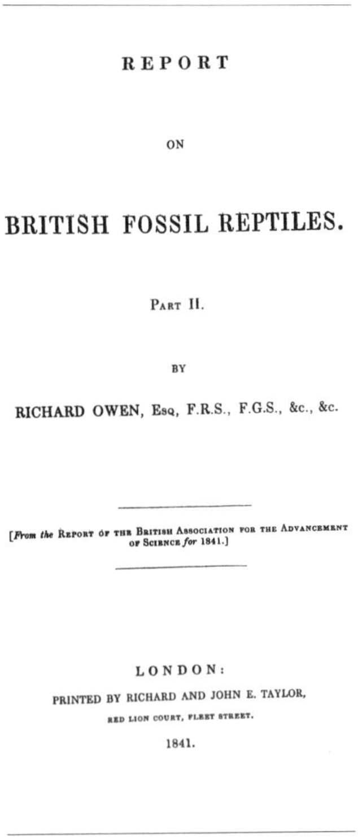

Murchison's presidential address to the Geological Society of London on 18 February 1842 had noted that "Owen . . . will shortly lay before the world the results of his researches into the extinct Saurians of our island" (1842: 652). Owen's long-awaited second report on British fossil reptiles was finally published in London, for the BAAS, in the first fortnight of April 1842, eight months after Plymouth, in an edition of 1,500 copies, priced at 13s/6d (*Publishers Circular*, 15 April 1842, 114, but see also Torrens 1993: 274).

In this publication, Owen at last "invented" dinosaurs. The printers' records survive to confirm the complex history of the volume's gestation and its exact chronology. They demonstrate how many changes Owen made to this particular report. It received numerous corrections in proof as Owen continually revised it (see R. and J. E. Taylor, MSS Check Book 1836–1842, f. 145, and MSS Day Book 1839–1845, f. 187, St. Bride Printing Library, London).

Some of the changes are clear from the printed text. For example, on 23 January 1842, George Bax Holmes had sent Owen details of some new *Goniopholis* (crocodile) scutes. When their discovery was published in April, Owen noted that they had "been discovered since the first sheets of this *Report* went to press" (Owen 1842: 194). The publisher's records for the BAAS reports also survive, and show that sixteen copies of the first fifty copies of the Plymouth *Report* remained unsold by 12 May 1842 (John Murray archives, London). Owen had twenty-five offprints of this paper printed for his own use (see R. and J. E. Taylor, MSS Day Book 1839–1845, f. 187), but to add to the confusion, they carried the incorrect date of 1841—whether to deliberately mislead is not known. The copy illustrated here (Fig. 14.3) was originally Broderip's. The 1841 date has since caused some to believe that such offprints were issued as "preprints," and so to claim that Owen's complete printed report had been ready, and was issued, before his Plymouth lecture (Gardiner 1990). Internal evidence, the printer's *and* publisher's records, with the fact that all the supposed preprints carry the correct pagination from the final, completed volume of 1842, all show how impossible this view is.

Some of Owen's revisions had been inspired by the recent publication of his rival Mantell's two papers (Mantell 1841b). The most important revision was to recognize the new order or suborder of fossil reptiles: the

REPORT

ON

BRITISH FOSSIL REPTILES.

PART II.

BY

RICHARD OWEN, Esq, F.R.S., F.G.S., &c., &c.

[*From the* REPORT OF THE BRITISH ASSOCIATION FOR THE ADVANCEMENT OF SCIENCE *for* 1841.]

LONDON:
PRINTED BY RICHARD AND JOHN E. TAYLOR,
RED LION COURT, FLEET STREET.

1841.

Figure 14.3. Title page of the offprint of Owen's dino-inventing paper, published in April 1842—despite its 1841 date. Author's collection.

Dinosauria. Owen must have had a sudden realization which allowed him to recognize this, in time for the report's publication. The "moment" happened whenever it was that Owen discovered that the sacral vertebrae in *Iguanodon* were fused together, as they were in *Megalosaurus,* in which this feature had long been known. Such fusion, to strengthen the sacral vertebrae, Owen took to be "*the* adaptation of the Dinosaurs to terrestrial life." Owen had found the till-then-unique specimen, showing that this new "character . . . altogether peculiar among Reptiles" (Owen 1842: 103) occurred also in *Iguanodon,* preserved in the museum (Fig. 14.4) of the London wine merchant William Devonshire Saull (1784–1855).

This specimen allowed Owen to infer the new relationship on which he based his new order of Dinosauria. To quote Wittgenstein, "A relationship of similarity or . . . difference does not rest out there waiting for us to find

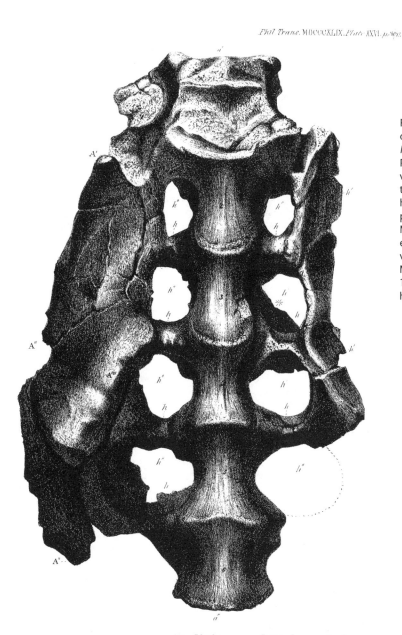

Phil. Trans. MDCCCXLIX. *Plate* XXVI. *p.*300.

Figure 14.4. The sacral vertebrae of the Saull specimen of *Iguanodon.* From Mantell 1849: Plate 26. The fusion of the vertebrae is clearly evident in this, the first illustration of the historic Saull specimen to be published. But the feud between Mantell and Owen extended even to the number of sacral vertebrae which were so fused. Mantell here saw 6—numbered 1–6 downward—while Owen had earlier recognized only 5.

it, but depends upon [our] *classifying* two things as similar or different. In order to say two things are similar . . . we select from the myriad of possibilities the relevant properties whereby we judge two things to be similar" (Cooper 1991: 967).

The shared fusion of the sacral vertebrae was that property. This dino-inventing assessment by Owen was strikingly original. It allowed him to place the two genera *Megalosaurus* and *Iguanodon* in his new order or suborder, Dinosauria, and to infer that the third genus, *Hylaeosaurus,* on the basis of the very limited and fragmentary evidence then known, must also belong there.

Saull had opened his museum in the City of London in 1833. He was a radical socialist who believed in education, and his museum was freely accessible every Thursday to all, even the working classes. His museum was clearly another of the benefits which the 1832 Reform Act had helped to bring about. Owen must have been rather chagrined, as a member of the Anglican Tory establishment, to find that this vital single specimen, on which "the characters of the Order Dinosauria were mainly founded" (Owen 1855: 11), and which had justified his creation of a completely new order of reptiles, was preserved in the museum of a socialist radical! Furthermore, to help in Owen's "war" with Mantell, the specimen had come neither from Mantell's collection nor from Sussex, but from the Isle of Wight in Hampshire. It survives as a *truly* historic object, in the Natural History Museum in London (Fossil Reptilia Reg. no. 37685).

Exactly when Owen first came across the Saull specimen is unknown, but he clearly realized its significance only after the Plymouth meeting. In the reports of his lecture there, such fused sacral vertebrae are noted only in *Megalosaurus* (*Literary Gazette,* 7 August 1841: 517). Later, when Owen claimed that he had found the specimen in 1840, it was in a source which also wrongly claimed that his original paper had been published in 1841! (Owen 1855: 9)

Mantell's reaction to Owen's 1842 publication is recorded in letters to Silliman. On 30 April 1842, Mantell noted that Owen's was "an elaborate and very masterly paper" but that he

> had *again* [emphasis added] to regret a want of honour, and I may say justice, towards those but for whose labor and zeal, [Owen] could never have obtained the materials for his own reputation. He has in several instances behaved very disingenuously towards me; altered names which I had imposed . . . and stated many inferences as if originating with himself, when I had long since published the same. . . . I believe he would have altered *Iguanodon* and *Hylaeosaurus* had I not sent the letter of remonstrance to the *Literary Gazette.*

Silliman agreed, although he had not yet had time to read Owen's memoir: "his treatment of you and that of Agassiz too, is unjust and dishonourable." He then perceptively asked, "Have you reduced the length of the *Iguanodon*? Something in your remarks impressed me as if you had topped his tail a little." Mantell's reply, dated 4 August 1842, confirmed, "You were quite right as to your conjecture that subsequent discoveries have led me to believe that the tail of the *Iguanodon* was short and flat in a vertical direction . . . yet you will see that [in Owen's BAAS Report] he has stated the probable size of the *Iguanodon* to be much shorter, and yet has taken no notice whatever that I had already announced it [in Mantell 1841b: 140]. There are numerous other lamentable offences of the same kind" (Spokes 1927: 135–136).

Owen's bold "invention" of dinosaurs as a separate group of creatures was within highly political science, as has been made clear by Desmond (1989). The members of Owen's new order were still large in size, but *no longer* gigantic. The size of both *Megalosaurus* and *Iguanodon* had been cut down, to put Mantell in his place, to about 30 feet long rather than the previous "up to 200 feet" (Owen 1842: 142–143). Owen saw dinosaurs as four-footed and—crucially—as highly advanced reptiles, like present pachydermal mammals. Such reptiles, vested with advanced mammalian characters, could become a "nail" to be sunk into the progressionists' or transmutationists' "coffin," since such characters clearly could have been given to these extinct reptiles only at their creation, by God. Owen was still in 1842 a creationist. Dinosaurs were a final weapon of support for his anti-evolutionary battle with Grant. His major scaling down of the size of dinosaurs and his deadly realization of what Mantell had failed to realize, that they were not just large lizards (Fig. 14.5), were similar strikes against his now more important rival Mantell.

In 1845 von Meyer formally named the Pachypodes, the group he had earlier separated as his "heavy-footed Sauria." But he had been as heavy-footed as his animals, and was too late to achieve any priority over the term *Dinosauria* (Meyer 1845), although after Mantell's death, Thomas Huxley (1825–1895), Owen's new rival, did try to rehabilitate Meyer's name (Huxley 1870: 32–33).

Late in 1849 Mantell was awarded the Royal Society's Royal Medal for his work on dinosaurs. The history of this again reveals the deep rivalry between Owen and Mantell. Mantell's papers on fossil reptiles (Mantell 1841b) had been refused consideration in 1846, when Owen instead was awarded the medal in November 1846 for his paper on belemnites (a group of extinct mollusks related to squids). This Mantell had, with some justice, noted as "a tissue of blunders from beginning to end. So much for Medalism" (Curwen 1940: 212). Donovan and Crane (1992) have since shown that there is justice in Mantell's claim. In this highly charged atmosphere, there was now a second attempt in 1849 to honor Mantell, with Owen highly active in seeking to deny Mantell any such award! Mantell noted Owen to have said, "All I [Mantell] had done was to collect

Figure 14.5. *Megalosaurus,* as shortened by Owen, showing internally how few bones were then available to help his reconstruction. From Owen 1854: 20.

fossils and get others to work them out"! (Curwen 1940: 243, 245–247). But after skillful lobbying by Charles Lyell (1797–1875), Mantell was awarded the Royal Medal on 4 January 1850 (Curwen 1940: 249) for his work on *Iguanodon*.

This was why Mantell could write to Lyell later in 1850 that "in [Owen's] Report on British Reptiles the only new fact in the osteology of the *Iguanodon*, stated by Owen is the construction of the sacrum which [he] supposes is peculiar to *Iguanodon, Megalosaurus* and *Hylaeosaurus*" (American Philosophical Society, MSS Darwin/Lyell Correspondence, B/D 25, 7 October 1850). Mantell was more generous in print, with a frankness which is both remarkable and praiseworthy in view of the vendettas that had flared up between him and Owen (Benton 1982: 124). He confirmed that "the most important and novel feature in relation to the osteology of the Wealden reptiles enumerated in Professor Owen's *Reports,* was the remarkable structure of the *Sacrum* in the three extinct genera of Dinosaurians. . . . No one had previously suspected that in these reptiles the pelvic arch was composed of more than two anchylosed vertebrae" (Mantell 1851: 268).

It was this newly recognized character which enabled Mantell to recognize at least another two genera of dinosaurs before his death in 1852 (Mantell 1851: 224). One of these was *Pelorosaurus* of 1850 (a sauropod)—his "unusually gigantic & stupendous" saurian of "enormous magnitude" (Mantell 1850)—which Owen (1842: 94–100) had earlier assigned to *Cetiosaurus* (Mantell 1851: 330–332). This Mantell had earlier been intending to call *Colossosaurus!* (See his MSS Diary entry for 4 November 1849, Alexander Turnbull Library, Wellington, New Zealand.) This was Mantell's final attempt to show Owen that some dinosaurs *had* indeed been "stupendous" in size. The fifth genus was *Cetiosaurus* itself, which Mantell finally also recognized as dinosaurian (Mantell 1854: 332). The sixth genus which Mantell had in mind in 1851 was probably his *Regnosaurus* (another sauropod) of 1848 (Mantell 1851: 333), a genus in which the vertebrae were still unknown.

Mantell had noted in 1851 "the spirit of self-aggrandisement and jealousy [by which Owen] had exerted [his] baneful influence over this department of [vertebrate] palaeontology" (Mantell 1851: 226, and see 257, 286, etc.). Huxley also noted in November 1851 that "it is astonishing with what an intense feeling of hatred Owen is regarded by the majority of his contemporaries, with Mantell as arch-hater" (Huxley 1908, vol. 1: 136). A vendetta between the two continued unabated until Mantell's self-induced death late in 1852. This vendetta cost Owen the presidency of the Geological Society of London in 1852, when Leonard Horner and William Hopkins decided that Edward Forbes would be a better candidate in view of Owen's bitter "antagonism to Mantell" (Dawson 1946: 85–86). Owen's obituary of Mantell provides the best evidence of the depth of his antagonism. Mantell's "want of exact scientific . . . knowledge," Owen wrote, had simply "compelled him to have recourse to those possessing it," because of course it was Owen who had "first perceived [Mantell's] error respecting [*Iguanodon*'s] bulk, which had arisen out of an undue enthusiasm touching its marvellous nature" (*Literary Gazette,* 13 November 1852: 842).

In final irony, the first recorded outbreak of Dinomania (Torrens 1993: 279), late in 1853 and 1854 when the Crystal Palace restorations were put on display, used Owen's interpretations of the animals he had "invented" (Fig. 14.6). If only Mantell had been able to react differently to the invitation he received from the Crystal Palace Company. Its board of directors' minutes of 10 August 1852 had resolved "that a Geological

THE SECONDARY ISLAND.

1. Mososaurus.	6. Hylæosaurus.	10, 11. Teleosaurus.	18—20. Labyrinthodo.
2, 3. Pterodactyles.	7. Megalosaurus	12—14. Ichthyosaurus.	21—22. Dikynodon.
4, 5. Iguanodon.	8, 9. Pterodactyles of Oolite.	15—17. Plesiosaurus.	

Figure 14.6. Panorama of the Extinct Animals on the Secondary Island at Crystal Park. From Anonymous 1877: 27. The dinosaurs, numbered 4, 5, 6, and 7, are in the reconstructions envisaged by Owen, not Mantell.

Court be constructed containing a collection of full sized models of the Animals & plants of certain geological periods, and that Dr Mantell be requested to superintend the formation of that collection" (Alexander Turnbull Library, Wellington, New Zealand, MS Papers 0083–032). Mantell's diary noted only, on 20 August 1852, that "Mr [N] Thompson [Thomson], Secretary of the Nat. Hist. Dept. of the New Crystal Palace called, & I found that the plan intended to be carried out as to Geology, was merely to have models of extinct animals. . . . I therefore declined the superintendence of such a scheme" (Dell 1983: 90). Mantell was then less than three months away from his early death and, as his diary for 16 September 1852 had already noted, had already realized that *"in truth I am used up"* (Dell 1983: 91).

If Mantell had undertaken these restorations, we might now see his quite different visions of those wonderful fossil animals that he had largely discovered. These would first have been much larger than the Owen versions which so spectacularly survive in the Crystal Palace Park near London today (Doyle and Robinson 1993). We know that in 1851, Mantell was still urging that *Iguanodon* had been much bigger than Owen's "reduction" of 1842 allowed (Mantell 1851: 312). Mantell had also since discovered, in the autumn of 1849, "the most stupendous humerus of a terrestrial reptile ever discovered; it is 4 and a half feet in length." This became the original for his new genus *Pelorosaurus* (Mantell 1850).

More significant, it is also clear that Mantell had long realized, from study of the well-preserved "Maidstone *Iguanodon*," that in this animal "the hinder extremities, in all probability, resembled the unwieldy contour

Politics and Paleontology 187

of those of the Hippopotamus or Rhinoceros, and were supported by strong, short feet, protected by broad ungual phalanges: the fore feet [however] appear to have been less bulky, and adapted for seizing and pulling down the foliage and branches of trees" (Mantell 1851: 311–313). Mantell had been clearly aware that his *Iguanodon* had had a more kangaroo-like posture since at least 1841 (Mantell 1841b: 140) as opposed to the quadrupedal and more rhinocerine version proposed by Owen, which was that finally re-created for him by Benjamin Waterhouse Hawkins at the Crystal Palace.

There are some lessons for both historians and scientists in the complex story of dino-invention. First, Martin Rudwick's (1985: 465) statement that "the official *Report* of each [BAAS] meeting was not published until several months later [than the meeting which it purported to report] and its summaries cannot be relied on as an accurate record of what was actually read at the time" is abundantly confirmed. Second, it also reveals that Owen, in much modifying his script after it had been delivered at Plymouth, but before it was printed, was simply doing what he had accused another rival, Alexander Nasmyth (died 1848), of having unfairly done just before him in a similar presentation to the BAAS (Nasmyth 1842). Furthermore, it is now clear that Owen was in the habit of secretly altering his BAAS papers after their delivery (Charlesworth 1846: 25–27). Owen might thus have had a vested interest in recording false dates on his offprinted publications (see Fig. 14.3).

A final message from these first dinosaurs is on the political front today. We now can only observe with amazement the "wealth" being created out of Owen's invention of 1842. Of this, Owen could never have had any expectation. But it provides another example of the difficulty, if not stupidity, of politicians' directing scientific research "merely" at "wealth creation." The most unexpected science can create wealth.

One final point is that before April 1842, when dinosaurs were "invented" (but the word only slowly entered the world's languages), there were no such things, and they can technically have no prehistory. Attempts to identify "the first dinosaur bone to come to learned attention" or to be "found in North America, or probably the world" (Simpson 1942: 153, 178) become rather meaningless in historical terms (Desmond 1979: 224). The first dinosaur bone to be found can only be the *Iguanodon* sacrum which Owen found in Saull's Museum. All other dinosaur discoveries were *recognized* to be so, only after this. Thus any other "first" dinosaur bones can be such only with hindsight. The first AIDS case could equally come only after AIDS had been first recognized.

There is, of course, as great and real an interest, and need, to discover when AIDS was first among us as there is to discover when and how dinosaurs evolved, and to uncover the phylogeny of dinosaurs. But we do not have to distort history in either case to pursue such vital inquiries.

References

Anonymous. 1877. *Crystal Palace: A Guide to the Palace and Park*. London: Dickens and Evans.

Babbage, C. 1830. *Reflections on the Decline of Science in England*. London: B. Fellowes and J. Booth.

Bakewell, R. 1830. A visit to the Mantellian Museum at Lewes. *Magazine of Natural History* 3: 9–17.

Bellot, H. H. 1929. *University College London, 1826–1926*. London: University of London Press.

Benton, M. J. 1982. Progressionism in the 1850s. *Archives of Natural History* 11: 123–136.

Bryant, G. E., and G. P. Baker. 1934. *A Quaker Journal*. 2 vols. London: Hutchinson and Co.

Buckland, W. 1824. Notice on the *Megalosaurus* or great Fossil Lizard of Stonesfield. *Transactions of the Geological Society of London* 2 (1): 390–396.

Buckland, W. 1837. *Geology and Mineralogy considered with reference to Natural Theology*. 2 vols. London: W. Pickering.

Charlesworth, E. 1846. On the occurrence of a species of *Mosasaurus*. . . . *London Geological Journal* 1: 23–32.

Cleevely, R. J., and S. D. Chapman. 1992. The accumulation and disposal of Gideon Mantell's fossil collections. *Archives of Natural History* 19: 307–364.

Cooper, C. C. 1991. Social construction of invention through patent management. *Technology and Culture* 32: 960–998.

Cooper, J. A. 1993. George Bax Holmes (1803–1887) and his relationship with Gideon Mantell and Richard Owen. *Modern Geology* 18: 183–208.

Curwen, E. C. 1940. *The Journal of Gideon Mantell*. London: Oxford University Press.

Dawson, W. R. 1946. *The Huxley Papers*. London: Macmillan and Co.

Dean, D. R. 1986. Review [of Rudwick 1985]. *Annals of Science* 43: 504–507.

Dean, D. R. 1990. A bicentenary retrospective on Gideon Algernon Mantell (1790–1852). *Journal of Geological Education* 38: 434–443.

Dell, S. 1983. Gideon Algernon Mantell's unpublished journal, June–November 1852. *Turnbull Library Record* 16: 77–94.

Desmond, A. 1979. Designing the dinosaur: Richard Owen's response to Robert Edward Grant. *Isis* 70: 224–234.

Desmond, A. 1989. *The Politics of Evolution*. Chicago and London: University of Chicago Press.

Donovan, D. T., and M. D. Crane. 1992. The type material of the Jurassic cephalopod *Belemnotheutis*. *Palaeontology* 35: 273–296.

Doyle, P., and E. Robinson. 1993. The Victorian 'Geological Illustrations' of Crystal Palace Park. *Proceedings of the Geologists' Association* 104: 181–194.

Evans, E. J. 1983. *The Great Reform Act of 1832*. London: Routledge.

Gardiner, B. G. 1990. Clift, Darwin, Owen and the Dinosauria. *The Linnean* 6: 19–27.

Hurst, D. 1868. *Horsham, its History and Antiquities*. London: William Macintosh.

Huxley, L. 1908. *Life and Letters of Thomas Henry Huxley*. 3 vols. London: Macmillan and Co.

Huxley, T. H. 1870. On the Classification of the Dinosauria. . . . *Quarterly Journal of the Geological Society of London* 26: 32–51.

Kirby, D.; K. Smith; and M. Wilkin. 1992. *The New Roadside America*. New York: Fireside.

Laurent, G. 1987. *Paléontologie et Évolution en France 1800–1860*. Paris: Éditions du C. T. H. S.

Mantell, G. A. 1825. On the teeth of the *Iguanodon*. *Philosophical Transactions of the Royal Society* 115: 179–186.

Mantell, G. A. 1841a. Fossil Reptiles. *Literary Gazette* (28 August 1841): 556–557.

Mantell, G. A. 1841b. *A Memoir on the Fossil Reptiles of the South-East of England*. An offprinted combination of two papers published in the *Philosophical Transactions of the Royal Society of London*. London: Published privately.

Mantell, G. A. 1849. Additional Observations on the Osteology of the *Iguanodon* and *Hylaeosaurus*. *Philosophical Transactions of the Royal Society* for 1849: 271–305.

Mantell, G. A. 1850. On the *Pelorosaurus;* an undescribed gigantic terrestrial reptile. *Philosophical Transactions of the Royal Society* for 1850: 379–390.

Mantell, G. A. 1851. *Petrifactions and their Teachings*. London: H. G. Bohn.

Mantell, G. A. 1854. *Geological Excursions round the Isle of Wight*. 3rd ed. London: H.G. Bohn.

Meyer, H. von. 1832. *Palaeologica zur Geschichte der Erde und ihrer Geschöpfe*. Frankfurt: Schmerber.

Meyer, H. von. 1845. System der fossilien Saurier. *Neues Jarhbuch für Mineralogie, Geologie und Paläontologie*: 278–285.

Murchison, R. I. 1842. Presidential address to the Geological Society of London. *Proceedings of the Geological Society of London* 3: 637–687.

Nasmyth, A. 1842. *A Letter to the Right Hon Lord Francis Egerton*. . . . London: Churchill.

Owen, R. 1842. Report on British Fossil Reptiles: Part II. *Report of the British Association for the Advancement of Science* for 1841: 60–204.

Owen, R. 1854. *Geology and the Inhabitants of the Ancient World*. London: Crystal Palace Library.

Owen, R. 1855. *Fossil Reptilia of the Wealden Formations, part 2*. London: Palaeontographical Society Monograph.

Owen, R. S. 1894. *The Life of Richard Owen*. 2 vols. London: J. Murray.

Richardson, G. F. 1837. Translation of Hermann von Meyer's *On the structure of the Fossil Saurians*. *Magazine of Natural History*, n.s. 1: 281–293, 341–353.

Rudwick, M. J. S. 1963. The foundation of the Geological Society of London. *British Journal for the History of Science* 1: 325–355.

Rudwick, M. J. S. 1985. *The Great Devonian Controversy*. Chicago and London: University of Chicago Press.

Simpson, G. G. 1942. The beginnings of vertebrate paleontology in North America. *Proceedings of the American Philosophical Society* 86: 130–188.

Spokes, S. 1927. *Gideon Algernon Mantell, Surgeon and Geologist*. London: J. Bale and Co.

[Timbs, J.] 1852. *The Year Book of Facts in Science and Art*. London: Simpkin, Marshall and Co.

Torrens, H. S. 1993. The dinosaurs and dinomania over 150 years. *Modern Geology* 18: 257–286.

Evolution of the Archosaurs

J. Michael Parrish

Although everyone is familiar with groups such as dinosaurs, pterosaurs, and crocodiles, the group to which all of these organisms belong, the Archosauria, is considerably more obscure. The Archosauria was initially erected by Cope (1869) to include dinosaurs, crocodilians, and all their presumed common ancestors. It has been slightly redefined by modern systematists to include the last common ancestor of the two extant groups of archosaurs, the crocodilians and the birds, and all of the descendants of that common ancestor. This is the sense in which I will use the name here.

The amniotes (the evolutionary group containing reptiles, mammals, and birds) have historically been differentiated on the basis of the arrangement of openings in the cheek region of the skull behind the orbit (Fig. 15.1). The pattern that is seen in fishes and amphibians, and is primitive for the amniotes, is for the cheek region to be solid, without any openings. This pattern, termed anapsid, is also seen in early amniotes such as captorhinids and pareiasaurs, and is retained today in turtles, the single group of anapsids that persisted beyond the early Mesozoic. Very early in the evolution of the amniotes, certainly by the middle of the Carboniferous Period (300 million years ago), a group diverged that had a single opening located low on the cheek region (Figs. 15.1, 15.2). This group, the synapsids, represents an important evolutionary line that includes two groups of "mammal-like reptiles," the pelycosaurs and the therapsids, as

15

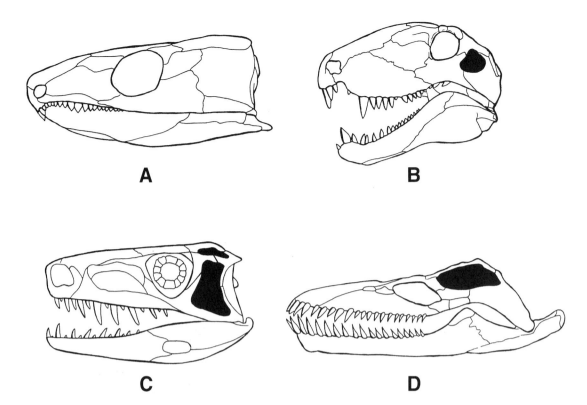

A   B

C   D

Figure 15.1. The four major types of amniote skulls, based on openings in the temporal region. (A) Primitive pattern, retained in modern turtles (anapsid); there are no temporal openings in the cheek region. (B) Synapsid pattern (mammals, mammal-like reptiles), with a single temporal opening (*black*) below the suture between the postorbital and squamosal bones. (C) Diapsid pattern (lizards, snakes, archosaurs, etc.), with two temporal openings separated by the postorbital/squamosal suture. (D) Euryapsid pattern (many marine reptiles). This is probably a modified version of the diapsid pattern, in which the lower temporal opening becomes open ventrally and is lost entirely in many groups. A similar configuration is seen in many lizards. Modified from Colbert and Morales 1991.

well as the mammals themselves, which appeared about the same time that the dinosaurs did, in the later part of the Triassic Period.

A bit later in the Carboniferous, another major group of amniotes appeared, and these animals had two openings in the cheek region. The diapsids, as members of this group are called, include lizards, snakes, crocodiles, birds, and a number of extinct groups. Several groups of extinct amniotes, principally the marine reptile groups Ichthyosauria and Sauropterygia (plesiosaurs and relatives), were formerly thought to form a fourth group, the euryapsids, based on possession of an opening high in the cheek region. The extent to which these euryapsids are closely related to one another is still being debated (e.g., Carroll 1988; Rieppel 1993), but the present consensus is that the euryapsid pattern is simply a modification of the diapsid arrangement, in which the lower opening became open ventrally and was subsequently lost.

Even though these jaw openings represent important evolutionary features, their functional significance remains obscure. However, they appear to correlate with sites of attachment for the major jaw muscles and with sites of possible high stress concentration within the skull.

The earliest diapsid is an animal called *Petrolacosaurus*, known from the Late Carboniferous of Kansas (Reisz 1981). By the beginning of the Permian Period (285 million years ago), the diapsids had split into two divergent lines, one leading to lizards and snakes, the other leading to the archosaurs (Fig. 15.3). The *Lepidosauromorpha* includes a number of groups that are superficially lizard-like, but which lack key features shared by later members of the group (Gauthier et al. 1988). The *Lepidosauria* are a lepidosauromorph subgroup that appeared by the Late Permian, and include all of the modern lepidosauromorphs: lizards, snakes, and the tuatara *Sphenodon*. These groups share distinctive patterns of bone ossification, limb structure, and ear morphology.

The other major group of diapsids is the *Archosauromorpha* (Gauthier 1984). Primitive members of this group are a diverse group of tetrapods, including the beaked, herbivorous rhynchosaurs, the protorosaurs (mostly strange, long-necked aquatic forms such as *Tanyostropheus*), and the trilophosaurs (a group of terrestrial plant eaters with sturdy skulls and distinctive three-cusped teeth). These reptiles share a distinctive tarsal (ankle) pattern that will be discussed below.

Near the end of the Permian Period, another line of archosauromorphs appeared, the *Archosauriformes*. This group was historically identified as the Archosauria, and includes the archosaurs as now recognized, as well as a number of earlier groups (Gauthier 1984). The archosauriforms are distinguished by a number of features, notably the presence of an antorbital fenestra (an opening on the side of the snout between the orbit and the external naris) and of the laterosphenoid bone in the braincase (Clark et al. 1993).

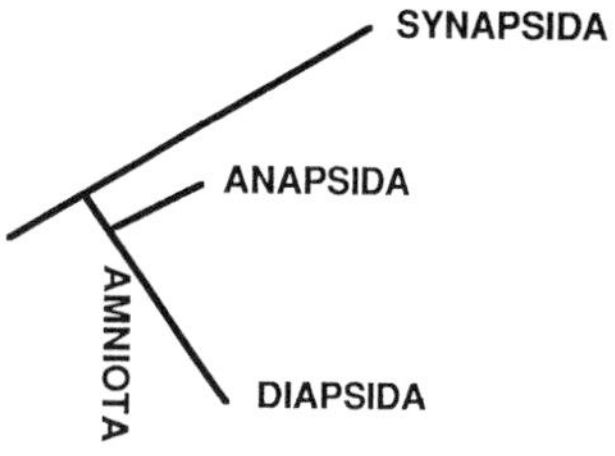

Figure 15.2. Cladogram showing relationships of the major amniote groups.

Evolutionary Patterns

At the end of the Carboniferous and in the Early Permian, most of the large terrestrial vertebrates were basal synapsids that are often grouped together as a paraphyletic group known as pelycosaurs. These forms diminished in abundance near the middle of the Permian, to be succeeded by a variety of therapsids, a group that arose from one group of carnivorous pelycosaurs, the Sphenacodontidae. Archosauriforms appeared near

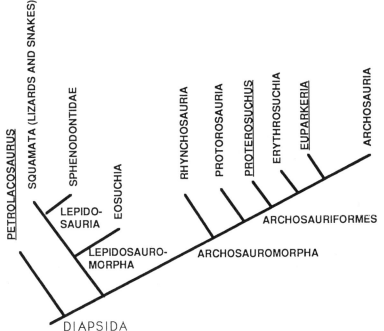

Figure 15.3. Phylogeny of the Diapsida. Modified from Gauthier 1984.

Evolution of the Archosaurs 193

the end of the Permian, but did not become dominant land vertebrates until about the Middle Triassic, as the abundance of therapsids steadily decreased. For the last half of the Triassic, the dominant terrestrial vertebrates were various types of early archosaurs, with the most familiar archosaur groups, the dinosaurs, pterosaurs, and crocodiles, making their appearance in the last third of the period. It was only near the end of the Triassic that the dinosaurs became relatively abundant and assumed the position as the dominant land vertebrates that they held until the end of the Mesozoic Era.

Much has been written about the evolutionary dynamics that drove this succession of dominant vertebrates. A strictly progressive interpretation (e.g., Colbert 1973) was favored until sometime in the 1970s, the idea being that each group that succeeded another was competitively superior to its predecessors. Thus archosaurs were considered to have outcompeted the therapsids, dinosaurs to have outcompeted the other archosaurs, and so on. Benton (1983, and chap. 16 of this volume) offered a more opportunistic interpretation, pointing out that the extinctions of major groups generally preceded the radiation of the forms that later succeeded them. This left open the question of what caused the extinctions, which have generally been linked to environmental changes, or sometimes to catastrophic events such as extraterrestrial impacts.

One way in which the dinosaurs were always considered superior to their predecessors was in their mode of locomotion. The earliest tetrapods, such as *Ichthyostega* from the Devonian Period, used their limbs as little more than points of connection with the ground; most of the muscular force for forward movement was still supplied by wavelike, side-to-side movement of the trunk, a pattern inherited from their fish ancestors. The pattern in early amniotes, and retained in forms such as lizards and turtles, is somewhat similar. The proximal parts of the limbs project horizontally from the body, forming right angles with the distal parts of the limbs, which project downward at the elbow or knee to terminate in the forefeet and hind feet. Such animals exhibited a sprawling gait, with the limbs still less important for forward movement of the body than movements of the trunk, although some forward momentum was imparted by rotation of the proximal parts of the limbs around their long axes (Charig 1972; Brinkman 1980).

Crocodilians and other non-dinosaurian archosauriforms were long considered to be inferior to dinosaurs in their locomotor capabilities. In an influential paper, Charig (1972) depicted crocodilians as exhibiting a "semi-erect" gait, with the capability of bringing the limbs close to the body in a nearly erect "high walk" in addition to walking with a typical sprawling gait. Several recent studies of limb movement in early archosauriforms (Parrish 1986), pterosaurs (Padian 1984), and crocodylomporphs (Walker 1970; Crush 1984; Parrish 1987; but see Gatesey 1991 for a slightly different interpretation of modern crocodilian locomotion) suggest that all but the earliest archosauriforms had what appear to be fully erect gaits, with the limbs aligned within a vertical plane parallel to the body's vertical axis of symmetry through the body's midline. With this arrangement, which humans share with other mammals, birds, and dinosaurs, the limbs move within a single plane, with flexion and extension occurring at the elbow/knee and wrist/ankle joints (Brinkman 1980; Parrish 1986). With such an erect stance, the trunk is usually stiffened to prevent side-to-side movement, and almost all of the muscular effort involved in limb flexion/extension translates into forward movement.

If dinosaurs and most of their archosaurian predecessors and contemporaries had erect stances, then posture alone cannot be cited as a competitive advantage that dinosaurs held with respect to their relatives. Dinosaurs and other erect archosaurs do differ markedly in their limb proportions, as well as in the posture of their feet (Chatterjee 1985; Parrish 1986). In dinosaurs, the tibia and fibula are generally longer than the femur, while the opposite condition pertains in most non-dinosaurian archosaurs. Dinosaurs also held the the metatarsus off the ground in a so-called digitigrade stance, while the other archosaurs mostly retained the primitive arrangement, a plantigrade stance where the entire foot landed on the ground during normal movement.

Early History of the Archosauriformes

The earliest archosauriform is *Archosaurus,* known from fragmentary material from the Late Permian of central Russia (Tatarinov 1960). Even more fragmentary material from the Late Permian of southern Africa may belong to a similar taxon (Parrington 1956). Although these fossils are incomplete, they seem to represent animals very similar to *Proterosuchus,* the next archosauriform to appear in the fossil record. *Proterosuchus* (Fig. 15.4) is quite well known, represented by complete skeletons from South Africa and China, and by more fragmentary material from Argentina and India. This reptile is superficially similar to modern crocodilians in size, body proportions, and inferred ecological habits. *Proterosuchus* appears to have retained the sprawling gait of its earlier amniote ancestors, and its long, low skull and relatively homogeneous, cone-like teeth suggest that this animal may have favored small vertebrates such as fishes rather than larger items as its predominant diet. Initially, a number of different genera and species of earliest Triassic archosauriforms were named from South Africa, where these animals are relatively abundant. However, subsequent study by several workers (Cruickshank 1972; Clark et al. 1993; Welman and Flemming 1993) has shown that the other genera, *Chasmatosaurus* and *Elaphrosuchus,* were based on specimens of *Proterosuchus* in which the skull was crushed dorsoventrally such that the snout drooped downward and the quadrate bone, at the back of the skull, was angled posteriorly. Two other Early Triassic forms, *Kalisuchus* (Thulborn 1979) and *Tasmaniosaurus* (Camp and Banks 1978; Thulborn 1986), are known from Australia, and do appear distinct from *Proterosuchus,* although neither taxon is represented by very complete material.

In the later part of the Early Triassic, two other groups of archosauriforms made their appearance. The first of these, the Erythrosuchidae, are a group of large terrestrial carnivores that persisted into the Middle Triassic, and are represented by at least seven genera (Parrish 1992). The

Figure 15.4. Skeletal reconstruction of *Proterosuchus.* In this and other skeletal drawings in this chapter (all by G. S. Paul, except where otherwise indicated), the size of the animal is indicated by reporting the length of the femur (thighbone). For *Proterosuchus,* femur length is about 150 mm.

Figure 15.5. Skeletal reconstruction of the erythrosuchid *Vjushkovia*. Femur length about 295 mm.

erythrosuchids are the first group of archosauriforms to exhibit a distinctive type of skull, which is tall and compressed mediolaterally, with large, recurved teeth (Fig. 15.5). Erythrosuchids, some of which had skulls more than a meter in length, appear to have been the dominant terrestrial predators of their day. Although information about their limb morphology, and particularly the structure of their feet, is relatively scanty, well-preserved material of *Vjushkovia triplocostata* from southern Russia suggests an animal with a much more erect stance than was present in *Proterosuchus* or earlier amniotes, even though a great deal of rotation of the proximal limb elements probably still took place during locomotion. The most primitive erythrosuchids, *Garjainia* from Russia and *Fugusuchus* from China, were also the earliest to appear. Both retain a relatively narrow snout and skulls that are lower in profile than those of the slightly later *Erythrosuchus* and *Vjushkovia*.

Another important early archosauriform is *Euparkeria* (Fig. 15.6), known from the late Early Triassic of South Africa (Ewer 1965). *Euparkeria* is known from ten specimens from a single locality in the Aliwal North region of South Africa. This was a much smaller animal than other early archosauriforms; the largest specimen would have been less than a meter in total length. *Euparkeria* is the first archosauriform that can be clearly shown to have had armor, a feature retained in some form by most subsequent archosauriforms. Although *Euparkeria* lacks some features that are characteristic of archosaurs, it probably represents the best fossil example we have of what the common ancestor of dinosaurs and crocodilians might have looked like. Often restored in a very dinosaurian bipedal pose, *Euparkeria* was more likely (based on limb proportions) to have been predominantly quadrupedal.

The *Archosauria* in the strict sense originated by the middle of the Triassic. Several features characterize this group, including a new foramen through which the internal carotid artery took a different path into the

Figure 15.6. Skeletal reconstruction of *Euparkeria*. Femur length about 56 mm.

braincase from that seen in other diapsids. Archosaurs are represented by two different lineages: one, dubbed the *Ornithodira* by Gauthier (1984), led to dinosaurs and birds; the other, named the *Crocodylotarsi* (Benton and Clark 1988), led to crocodilians.

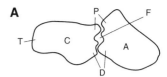

One of the principal features used in elucidating archosaurian phylogeny has been the structure of the ankle (like Fig. 15.7). Archosauromorphs, including early archosauriforms such as *Proterosuchus,* are united by a distinctive ankle pattern in which the two proximal ankle bones, the astragalus and calcaneum, are connected by two pairs of ball and socket joints (Brinkman 1981); the more proximal of these joints consists of a socket on the astragalus and a ball on the calcaneum, while the distal pair has a socket on the calcaneum and a ball on the astragalus. In the crocodylotarsans, the ventral of these joints becomes movable, such that the main joint between the foot and the lower leg is between the astragalus and calcaneum, rather than between those two bones and the distal tarsals. This arrangement was dubbed "crocodile-normal" by Chatterjee (1982), because it is the pattern present in the extant Crocodylia. In the family Ornithosuchidae, the more proximal of the two sets of facets became elaborated, such that a mobile joint developed with the ball on the calcaneum and the socket on the astragalus. This arrangement was termed "crocodile reverse" by Chatterjee (1982). In the Ornithodira, the functional ankle joint is of the primitive type, between the proximal and distal tarsals, although the details of the articulation between the two bones most closely resembles the pattern seen in the Ornithosuchidae (Bonaparte 1982; Chatterjee 1982). This has led to two divergent evolutionary interpretations. The first, held by Thulborn (1980) and Sereno (1991), is that the Ornithodira have retained the primitive, fixed ankle, and that the Ornithosuchidae and Crocodylotarsi constitute a separate evolutionary lineage (Fig. 15.8), the *Crurotarsi* of Sereno and Arcucci (1990), which has developed a moveable joint between the calcaneum and astragalus. The second view, espoused by Gauthier (1984) and Parrish (1986), calls for uniting the ornithosuchids with the Ornithodira in a lineage, the *Ornithosuchia,* which has the crocodile-reverse ankle as a derived character, and whose last common ancestor with the Crocodylotarsi had the primitive, two-faceted type of ankle joint. A more recent study (Parrish 1993) found the two phylogenetic arrangements almost equally likely.

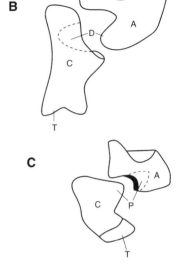

Figure 15.7. Schematic of the major types of archosauriform ankles (after Parrish 1986). (A) Primitive archosauriform pattern with two pairs of articular facets. (B) Crocodile-normal pattern, where the distal of the two facet pairs is modified into a rotary ankle joint. (C) Crocodile-reverse pattern, where the proximal of the two facet pairs is modified into a rotary ankle joint. Abbreviations: A = astragalus; C = calcaneum; P = proximal facet pair; D = distal facet pair; F = perforating foramen. Redrawn from Parrish 1986.

The earliest members of the Ornithodira are the lagosuchids, small, long-legged archosaurs from the Middle Triassic of Argentina (Romer 1971, 1972; Bonaparte 1975a, 1975b; Sereno and Arcucci 1993, 1994). Although the two presently known lagosuchids, *Marasuchus* and *Lagerpeton,* are poorly known, they have pelvic structures, limb proportions, and reduced ankles that clearly link them with the dinosaurs. *Lagerpeton* has a very peculiar hind limb, with short pelvic elements and a foot with elongated lateral digits that may have been utilized for perching.

The other group that appears to belong within the Ornithodira comprises the pterosaurs, or flying reptiles, the earliest well-known forms of which are from the Late Triassic of Italy. A link between the pterosaurs and the other ornithodirans may be provided by a peculiar animal, *Scleromochlus* from the Late Triassic of Scotland, which has a skull structure and limb proportions very similar to those of pterosaurs and an ankle structure very similar to that of lagosuchids (Gauthier 1984; Padian 1984), although not all recent workers endorse a close relationship between *Scleromochlus* and pterosaurs (Sereno 1991). An alternative view, proposed by Wild (1984), holds that the pterosaurs are not even archosaurs, but have closer evolutionary relationships with early lepidosauromorphs, although this is

Evolution of the Archosaurs   197

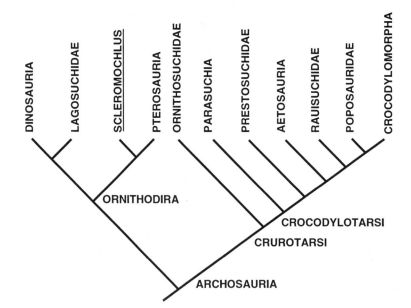

Figure 15.8. Phylogeny of the Archosauria. Simplified from Parrish 1993.

refuted by the considerable character evidence amassed by Gauthier (1984), Padian (1984), and Sereno (1991) in support of a dinosaur/pterosaur clade. The early pterosaurs are clearly flying reptiles, but are considerably different from their later relatives, as they are relatively small and retain long tails. *Eudimorphodon,* one of the earliest well-known pterosaurs, also has complex, multicusped teeth.

Dinosaurs appear in the fossil record by the beginning of the Late Triassic. Three-toed dinosaur-like footprints are known as early as the end of the Early Triassic, but these could also have been made by close relatives of the dinosaurs such as *Marasuchus.*

In the Middle and Late Triassic, most described archosaurs belonged to the Crocodylotarsi, a confusing and until recently little-known group (Benton and Clark 1988; Parrish 1993). The most primitive crocodylotarsans are represented by a group that is not abundant until the Late Triassic, during which they are among the most common of archosaur fossils. The Parasuchia, or phytosaurs (Fig. 15.9), are crocodile-like animals with long, narrow snouts, dorsally placed nostrils, and flattened skulls (Camp 1930; Chatterjee 1978). Their limb structure and the sedimentary settings in which their remains are found suggest that these

Figure 15.9. Skeletal reconstruction of *Parasuchus.* Femur length about 240 mm.

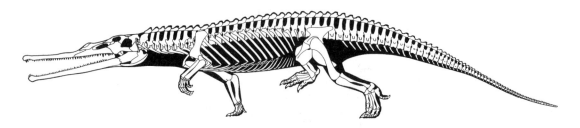

J. Michael Parrish

Figure 15.10. Skeletal reconstruction of the prestosuchid *Saurosuchus*. Femur length about 680 mm.

animals, like modern crocodilians, were amphibious to aquatic. They represent a wide variety of ecological types, from small, presumably mostly piscivorous forms such as *Mystriosuchus,* with rod-like snouts and conical teeth, to gigantic predatory forms such as *Nicrosaurus* with 1.5-meter-long skulls, large teeth, and relatively broad snouts.

One confusing aspect of crocodylotarsan evolution lies in the fact that three distinct groups had the high, narrow skull pattern seen earlier in erythrosuchids and later shared with most meat-eating dinosaurs (Gauthier 1984; Parrish 1993). The earliest of these groups, the Prestosuchidae (Fig. 15.10), are known from South America and Europe in the Middle Triassic, and include gigantic forms such as the Brazilian genus *Prestosuchus* (Huene 1942; Barberena 1978). These seem to have been mostly erect, but do not have any limb adaptations for rapid movement. The Rauisuchidae, which appear in Brazil at the same time as the Prestosuchidae, have a series of derived modifications of the skull, as well as a mediolaterally compressed foot with a reduced lateral digit and specialized ankle joint that indicates the capability for more rapid movement (Bonaparte 1984). Another group, the Poposauridae, appears in the Middle Triassic (Chatterjee 1985). Poposaurs (Fig. 15.11) superficially resemble the others, but have cranial specializations, including a eustachian tube system, that place them as the closest relatives of the crocodylomorphs (Parrish 1987, 1993; Benton and Clark 1988). A relative reduction in size of the forelimbs suggests that poposaurs may have walked bipedally at least some of the time (Chatterjee 1985). Members of a fourth group of terrestrial predators, the Ornithosuchidae, also have skull profiles and body sizes similar to those of the other three groups (Bonaparte 1975b). They also appear to have been capable of relatively rapid movement, but are the only one of these large predator groups that exhibits the crocodile-reverse tarsal pattern. Because all of these groups overlap with one another stratigraphically, their classification has been confusing. Their ecological roles

Figure 15.11. The poposaurid *Postosuchus.* Femur length about 505 mm.

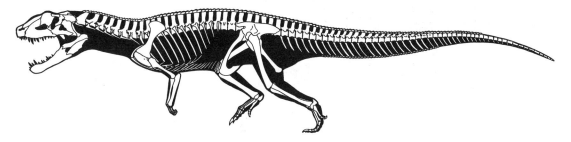

Evolution of the Archosaurs   199

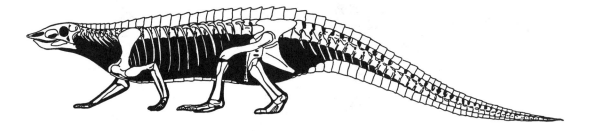

Figure 15.12. Skeletal reconstruction of *Stagonolepis.* Drawing by J. M. Parrish. Femur length about 315 mm.

were probably relatively similar. It seems likely that they differed mainly in the type of prey they fed upon, with groups such as the poposaurs probably specializing in more agile prey than, for example, the prestosuchids.

Before the Late Triassic, virtually all archosauriforms were carnivores. One possible exception is *Lotosaurus,* an enigmatic rauisuchian from southern China that lacked teeth entirely, but instead appears to have had a curving, turtle-like beak (Zhang 1975; Parrish 1993). *Lotosaurus* also had elongated neural spines along the trunk, which suggests that it may have had a fleshy sail like those of some pelycosaurs and dinosaurs. In the Late Triassic, the first group of clearly herbivorous archosaurs appeared, the Aetosauria (Walker 1961; Parrish 1994). Aetosaurs (Fig. 15.12), like *Lotosaurus,* had turtle-like beaks anteriorly, and most had simple, conical teeth in the cheek region. The most primitive aetosaurs, including *Aetosaurus* from Germany, had pointed snouts, but more derived forms, such as *Stagonolepis* and *Desmatosuchus,* had flattened, upturned snouts that appear to have been an adaptation for rooting up vegetation. Other apparent digging adaptations are found in their powerful limbs and broad, elongate digits. The aetosaurs possessed a complete coat of protective dermal armor, which in some derived forms, such as *Typothorax,* forms a broad oval carapace similar to a turtle's shell.

The Crocodylomorpha, including crocodilians and a number of their fossil relatives, appeared in the Late Triassic (Walker 1970; Benton and Clark 1988). Crocodylomorphs can be distinguished by a number of specializations of the skull, including the development of extensive pneumatic cavities and a quadrate bone that tilts anteriorly and develops extensive sutures with other bones at the back of the skull. Other special-

Figure 15.13. The early crocodylomorph *Pseudhesperosuchus.* Femur length about 155 mm.

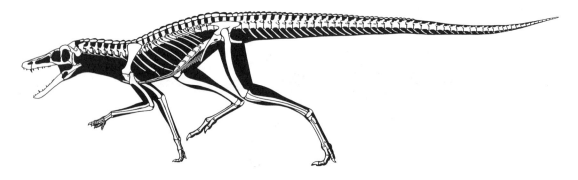

J. Michael Parrish

izations include an elongate wrist and specialized shoulder and pelvic girdles. The earliest crocodylomorphs are a group of forms often united as the Sphenosuchia, a group that originated in the early part of the Late Triassic and persisted until the Late Jurassic. Sphenosuchians (Fig. 15.13) had most of the distinctive crocodylomorph cranial features, but also had narrow, elongate limbs and mediolaterally compressed feet. The sphenosuchians and other early crocodylomorphs such as the Protosuchidae appear to have been erect, fully terrestrial animals, and some, including the appropriately named *Terrestrisuchus*, were probably very agile, rapid runners.

Thus the image we typically have of crocodilians as sluggish, mostly aquatic predators was not true of most of their Triassic and Jurassic ancestors. It was in the Early Jurassic that the first aquatic crocodylomorphs appeared, but these belonged to lines such as the Metriorhynchidae and Teleosauridae, which developed paddle-like appendages from their limbs. The main line of crocodilians did not assume their modern amphibious habits until well into the Mesozoic. The "semi-erect" crocodilian locomotor pattern, instead of being an intermediate stage between sprawling and erect forms, appears to be a modification of the erect stance seen in early crocodylomorphs; it allows modern crocodilians to use their limbs as laterally projecting swimming appendages in addition to the erect "high walk" that they employ when traveling overland (Brinkman 1980; Parrish 1987).

By the end of the Triassic Period, all archosauriforms other than dinosaurs, pterosaurs, and crocodylomorphs had disappeared. Groups underwent steady replacement during that period, and the final decline in diversity near the Triassic-Jurassic boundary does roughly correlate with a dramatic increase in the abundance of dinosaur fossils, notably those of small theropods such as *Coelophysis* and prosauropods such as *Plateosaurus* (Sander 1992).

References

Barberena, M. C. 1978. A huge thecodont skull from the Triassic of Brazil. *Pesquisas* 7: 111–129.

Benton, M. J. 1983. Dinosaur success in the Triassic: A noncompetitive ecological model. *Quarterly Review of Biology* 58: 29–55.

Benton, M. J., and J. M. Clark. 1988. Archosaur phylogeny and the relationships of the Crocodylia. In M. J. Benton (ed.), *Phylogeny and Classification of Amniotes,* pp. 295–338. Systematics Association Special Volume 35A. Oxford: Clarendon Press.

Bonaparte, J. F. 1975a. Nuevos materiales de *Lagosuchus tamalpayensis* Romer (Thecodontia, Pseudosuchia) y su significado en el origen de los Saurischia. *Acta Geologica Lilloana* 13: 5–90.

Bonaparte, J. F. 1975b. The family Ornithosuchidae (Archosauria: Thecodontia). *Colloque International Centre National de la Recherche Scientifique* 218: 485–501.

Bonaparte, J. F. 1982. Classification of the Thecodontia. *Géobios Mémoir Spéciale* 6: 99–112.

Bonaparte, J. F. 1984. Locomotion in rauisuchid thecodonts. *Journal of Vertebrate Paleontology* 3: 210–218.

Brinkman, D. 1980. The hindlimb step cycle of *Caiman sclerops* and the mechanics of the crocodilian tarsus and metatarsus. *Canadian Journal of Zoology* 58: 2187–2200.

Brinkman, D. 1981. The origin of the crocodiloid tarsi and the interrelationships of thecodontian archosaurs. *Breviora* 464: 1–22.

Camp, C. L. 1930. A study of the phytosaurs. *Memoirs of the University of California* 10: 1–161.

Camp, C. L., and M. R. Banks. 1978. A proterosuchian reptile from the Early Triassic of Tasmania. *Alcheringa* 2: 143–158.

Carroll, R. L. 1988. *Vertebrate Paleontology and Evolution*. New York: W. H. Freeman.

Charig, A. J. 1972. The evolution of the archosaur pelvis and hindlimb: An explanation in functional terms. In K. A. Joysey and T. S. Kemp (eds.), *Studies in Vertebrate Evolution*, pp. 121–155. Edinburgh: Oliver and Boyd.

Chatterjee, S. 1978. A primitive parasuchid (phytosaur) from the Upper Triassic Maleri Formation of India. *Paleontology* 21: 83–127.

Chatterjee, S. 1982. Phylogeny and classification of the thecodontian reptiles. *Nature* 295: 317–320.

Chatterjee, S. 1985. *Postosuchus*, a new thecodontian reptile from the Triassic of Texas and the origin of tyrannosaurs. *Philosophical Transactions of the Royal Society of London*, Series B, 309: 395–460.

Clark, J. M.; J. A. Gauthier; J. Welman; and J. M. Parrish. 1993. The laterosphenoid bone of early archosaurs. *Journal of Vertebrate Paleontology* 13: 48–57.

Colbert, E. H. 1973. *Wandering Lands and Animals*. New York: E. P. Dutton.

Colbert, E. H., and M. Morales. 1991. *Evolution of the Vertebrates: A History of the Backboned Animals through Time*. New York: John Wiley and Sons.

Cope, E. D. 1869. Synopsis of the extinct Batrachia, Reptilia, and Aves of North America. *Transactions of the American Philosophical Society* 14: 1–252.

Cruickshank, A. R. I. 1972. The proterosuchian thecodonts. In K. A. Joysey and T. S. Kemp (eds.), *Studies in Vertebrate Evolution*, pp. 89–119. Edinburgh: Oliver and Boyd.

Crush, P. 1984. A late Upper Triassic sphenosuchid crocodile from Wales. *Palaeontology* 27: 133–157.

Ewer, R. F. 1965. The anatomy of the thecodont reptile *Euparkeria capensis* Broom. *Philosophical Transactions of the Royal Society of London,* Series B, 248: 379–435.

Gatesey, S. M. 1991. Hind limb movements of the American alligator (*Alligator mississippiensis*) and postural grades. *Journal of Zoology* 224: 577–588.

Gauthier, J. A. 1984. A cladistic analysis of the higher systematic categories of the Diapsida. Ph.D. dissertation, University of California, Berkeley.

Gauthier, J. A.; R. Estes; and K. de Queiroz. 1988. A phylogentic analysis of Lepidosauromorpha. In R. Estes and G. Pregill (eds.), *Phylogentic Analysis of the Lizard Families*, pp. 15–98. Stanford, Calif.: Stanford University Press.

Huene, F. von. 1942. *Die fossilen Reptilien des Südamerikanischen Gondwana-landes*. Munich: C. H. Beck.

Padian, K. 1983. A functional analysis of flying and walking in pterosaurs. *Paleobiology* 9: 218–239.

Padian, K. 1984. The origin of pterosaurs. In W. E. Reif and F. Westphal (eds.), *Third Symposium on Terrestrial Mesozoic Ecosystems, Short Papers*, pp. 163–168. Tübingen: Attempto Verlag.

Parrington, F. R. 1956. A problematic reptile from the Upper Permian. *Annals and Magazine of Natural History* 12: 333–336.

Parrish, J. M. 1986. Locomotor evolution in the hindlimb and pelvis of the Thecodontia (Reptilia: Archosauria). *Hunteria* 1 (2): 1–35.

Parrish, J. M. 1987. The origin of crocodilian locomotion. *Paleobiology* 13: 396–414.

Parrish, J. M. 1992. Phylogeny of the Erythrosuchidae. *Journal of Vertebrate Paleontology* 12: 93–102.

Parrish, J. M. 1993. Phylogeny of the Crocodylotarsi and a consideration of archosaurian and crurotarsan monophyly. *Journal of Vertebrate Paleontology* 13: 287–308.

Parrish, J. M. 1994. Cranial osteology of *Longosuchus meadei* and a consideration of the phylogeny of the Aetosauria. *Journal of Vertebrate Paleontology* 14: 196–209.

Parrish, J. M.; J. T. Parrish; and A. M. Ziegler. 1986. Permo-Triassic paleogeography and paleoclimatology and implications for therapsid distributions. In N. Hotton III, P. MacLean, J. J. Roth, and E. C. Roth (eds.), *The Ecology and Biology of the Mammal-Like Reptiles,* pp. 109–132. Washington, D.C.: Smithsonian Institution Press.

Reisz, R. R. 1981. A diapsid reptile from the Pennsylvanian of Kansas. *Special Publications of the Museum of Natural History, University of Kansas* 7: 1–174.

Rieppel, O. 1993. Euryapsid relationships: A preliminary analysis. *Neues Jahrbuch für Geologie und Paläontologie, Abhandlungen* 188: 241–264.

Romer, A. S. 1971. The Chañares (Argentina) Triassic reptile fauna. Part X: Two new but incompletely known long-limbed pseudosuchians. *Breviora* 378: 1–10.

Romer, A. S. 1972. The Chañares (Argentina) Triassic reptile fauna. Part XV: Further remains of the thecodonts *Lagosuchus* and *Lagerpeton. Breviora* 394: 1–7.

Sander, M. 1992. The Norian *Plateosaurus* bonebeds of central Europe and their taphonomy. *Palaeogeography, Palaeoclimatology, Palaeoecology* 93: 255–299.

Sereno, P. C. 1991. Basal archosaurs: Phylogenetic relationships and functional implications. Memoir 2. *Journal of Vertebrate Paleontology* 11 (Supplement 4): 1–53.

Sereno, P. C., and A. B. Arcucci. 1990. The monophyly of crurotarsal archosaurs and the origin of bird and crocodile ankle joints. *Neues Jahrbuch für Geologie und Paläontologie, Abhandlungen* 180: 21–52.

Sereno, P. C., and A. B. Arcucci. 1993. Dinosaur precursors from the Middle Triassic of Argentina: *Lagerpeton chanarensis. Journal of Vertebrate Paleontology* 13: 385–399.

Sereno, P. C., and A. B. Arcucci. 1994. Dinosaur precursors from the Middle Triassic of Argentina: *Marasuchus lilloensis,* gen. nov. *Journal of Vertebrate Paleontology* 14: 53–73.

Tatarinov, L. P. 1960. Otkrytie psevdozukhii v verkhnie Permi SSSR. *Paleontologicheskii zhurnal* 1960: 74–80.

Thulborn, R. A. 1979. A proterosuchian thecodont from the Rewan Formation of Australia. *Memoirs of the Queensland Museum* 19: 331–355.

Thulborn, R. A. 1980. The ankle joints of archosaurs. *Alcheringa* 4: 141–161.

Thulborn, R. A. 1986. The Australian Triassic reptile *Tasmaniosaurus triassicus* (Thecodontia, Proterosuchia). *Journal of Vertebrate Palentology* 6: 123–142.

Walker, A. D. 1961. Triassic reptiles from the Elgin area: *Stagonolepis, Dasygnathus,* and their allies. *Philosophical Transactions of the Royal Society of London,* Series B, 244: 103–204.

Walker, A. D. 1970. A revision of the Jurassic crocodile *Hallopus,* with remarks on the classification of crocodiles. *Philosophical Transactions of the Royal Society of London,* Series B, 257: 323–372.

Walker, A. D. 1990. A revision of *Sphenosuchus acutus* Haughton, a crocodylomorph reptile from the Eliot Formation (Late Triassic or Early Jurassic) of South Africa. *Philosophical Transactions of the Royal Society of London,* Series B, 330: 1–120.

Welman, J., and A. Flemming. 1993. Statistical analysis of skulls of Triassic proterosuchids (Reptilia, Archosauromorpha) from South Africa. *Paleontologia Africana* 30: 113–123.

Wild, R. 1984. Flugsaurier aus den Obertrias von Italien. *Naturwissenschaften* 71: 1–11.

Zhang F. 1975. A new thecodont *Lotosaurus,* from the Middle Triassic of Hunan. *Vertebrata Palasiatica* 13: 144–148. (In Chinese, English summary.)

Origin and Early Evolution of Dinosaurs

Michael J. Benton

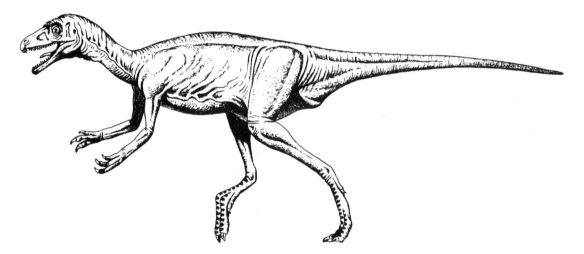

The dinosaurs arose in the Triassic, probably during the Late Triassic. They entered a world that was very different from the typical "age of dinosaurs" scenes, a world in which the dominant herbivores were therapsids ("mammal-like reptiles") and rhynchosaurs, and the carnivores were cynodonts (predatory therapsids) and basal archosaurs of various kinds, commonly called "thecodontians" (see Sues, chap. 2 of this volume). Into this world came the dinosaurs, initially small bipedal carnivores, and they rose to dominance at some point during the Triassic. Certainly by the end of the Triassic Period, dinosaurs were abundant and reasonably diverse, and all the major lineages had diversified.

The origin of the dinosaurs has for a long time been somewhat shrouded in mystery, for a number of reasons: problems in dating the rocks; problems in defining what a dinosaur is; problems of spurious early records; limited knowledge of the first genuine dinosaurs; and disagreements over models of replacement and radiation. Each of these topics will be considered in turn.

16

Stratigraphy

In early accounts of the history of vertebrate life on land during the Triassic (e.g., Colbert 1958; Romer 1970), the stratigraphic schemes used to identify segments of the Triassic rock record were little more refined than references to "Lower," "Middle," and "Upper." Even in later accounts (e.g., Bonaparte 1982; Benton 1983; Charig 1984), there was little improvement, and most of the stratigraphic assignments of tetrapod-bearing rock units were based on comparisons of the vertebrates themselves. In attempts to study patterns of evolution, it is little use to sequence faunas in terms of the nature of the faunas themselves!

The standard stratigraphic scheme for the Triassic rocks is based on ammonoids (Tozer 1974, 1979), and hence can be applied only to marine rocks. There is an independent palynological scheme, based on pollen and spores, for dating continental Triassic rocks (e.g., Visscher et al. 1980; Visscher and Brugman 1981; Weiss 1989), but this scheme is not always reliable. Tie points have been found here and there, where continental Triassic rocks have been correlated with the marine scheme, and this has led to a more independent dating scheme for Triassic terrestrial tetrapod faunas (Ochev and Shishkin 1989; Benton 1991, 1994a, 1994b; Hunt and Lucas 1991a, 1991b; Lucas 1991). The correlations of major tetrapod faunas of the Late Triassic are shown in Figure 16.1.

Definition of the Dinosauria

Richard Owen assumed in 1842 that his new group, the Dinosauria, was a real group, a monophyletic group, or a clade, in modern parlance (see Torrens, chap. 14 of this volume). In other words, he assumed that the Dinosauria had a single ancestor, and that the group included all the descendants of that ancestor. This view was commonly held during most of the nineteenth century, but was shaken by Harry Seeley's demonstration in 1887 that there were two major dinosaur groups, the Saurischia and Ornithischia, distinguished by the nature of their pelvic arrangements (see Holtz and Brett-Surman, chap. 7 of this volume).

Perhaps, thought Seeley, the Saurischia and Ornithischia were distinct evolutionary branches that had arisen much earlier from quite separate ancestors. Seeley was merely espousing a commonly held view at the time called "the persistence of types." The idea was that major changes in form could not happen readily in evolution, and the fossil record showed how major groups retained their main characteristics for long periods of time. Hence, paleontologists had to look for very long periods of initial evolution that led up to each major group, and often these initial spans of evolution were missing from the fossil record.

Seeley's view dominated during most of the twentieth century, and many dinosaur paleontologists made it even more complex. Not only had the Saurischia and Ornithischia evolved from separate ancestors, but so too had some of the subdivisions within those two groups, probably the two saurischian groups, the Theropoda and Sauropodomorpha, and possibly even some of the main ornithischian groups known in 1900, the Ornithopoda, Ceratopsia, Stegosauria, and Ankylosauria. In the end, the dinosaurs became merely an assemblage of large extinct reptiles of the Mesozoic which shared very little in common. Hence, dinosaurs were seen as a polyphyletic group, deriving from two, three, or more sources among the basal archosaurs (reviewed in Benton 1990).

The collapse of the polyphyletic view came quickly and dramatically about 1984. This had been presaged in the 1970s in short papers by Bakker and Galton (1974) and Bonaparte (1976), who saw many unique characters shared by both saurischian and ornithischian dinosaurs. Some brief papers published in 1984 were followed by more substantial accounts (Gauthier 1986; Sereno 1986; Novas 1989, 1994, 1996; Benton 1990); all applied strictly cladistic approaches to the data, and they independently

Origin and Early Evolution of Dinosaurs   205

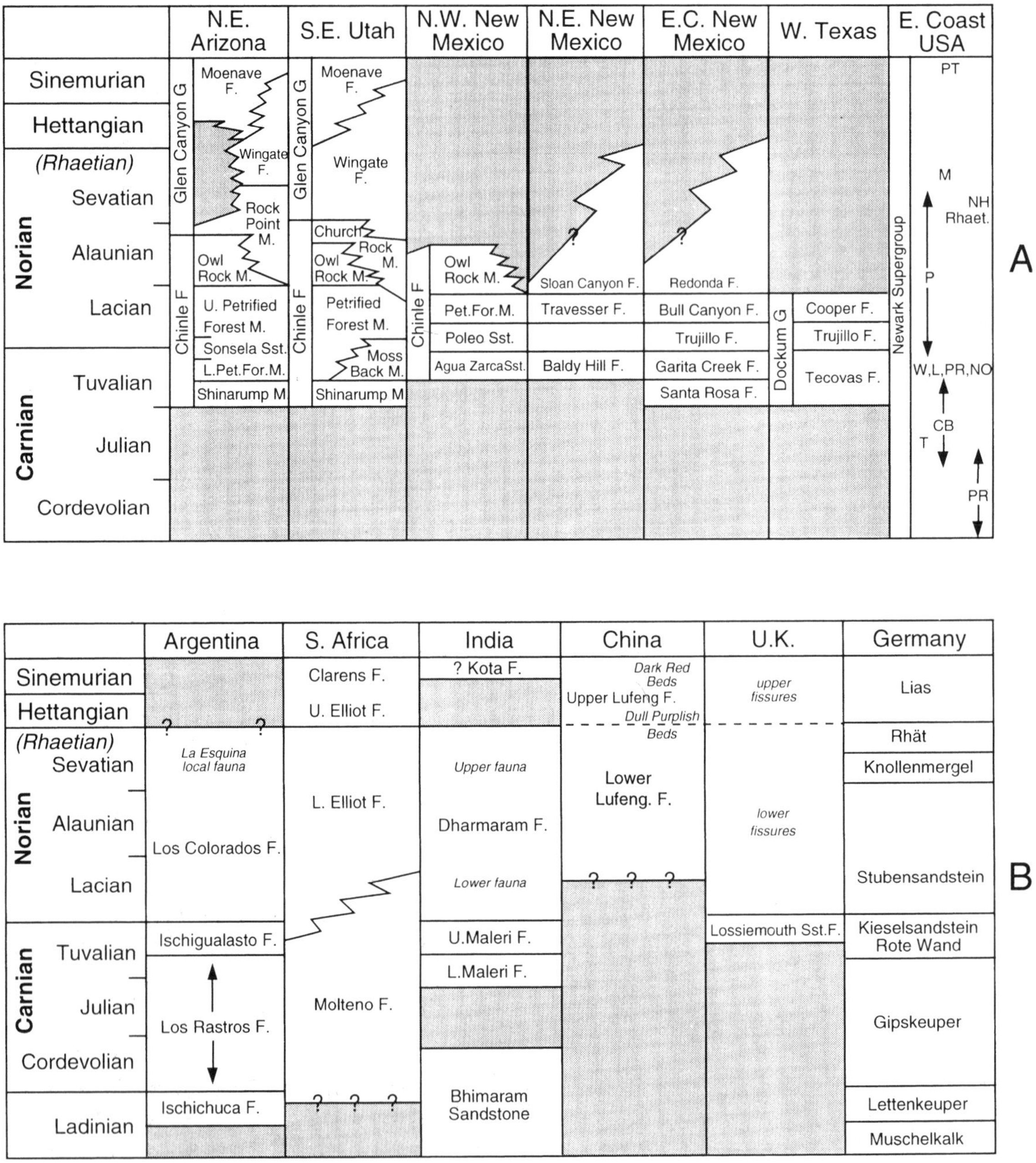

Figure 16.1. Stratigraphy of vertebrate-bearing sequences in the Late Triassic and earliest Jurassic of North America (A) and various parts of Gondwanaland (B). The dates are based largely on comparisons of tetrapods with the German sequence by Olsen et al. (1987, 1990), Hunt and Lucas (1991a, *(continues)*

agreed strongly that the Dinosauria of Owen (1842) is a monophyletic group, defined by a number of synapomorphies, including (cf. Holtz and Brett-Surman, chap. 7 of this volume)

- absence of a postfrontal bone (Fig. 16.2B);
- an elongate deltopectoral crest on the humerus (Fig. 16.2A);
- three or fewer phalanges in the fourth finger of the hand (Fig. 16.2C);

A

		N.E. Arizona	S.E. Utah	N.W. New Mexico	N.E. New Mexico	E.C. New Mexico	W. Texas	E. Coast USA
	Sinemurian	Moenave F. (Glen Canyon G)	Moenave F. (Glen Canyon G)					PT
	Hettangian	Wingate F.	Wingate F.					
	(Rhaetian)							M / NH Rhaet.
Norian	Sevatian	Rock Point M.	Church Rock M.					
Norian	Alaunian	Owl Rock M. (Chinle F)	Owl Rock M. (Chinle F)	Owl Rock M. (Chinle F)	Sloan Canyon F.	Redonda F.		P
Norian	Lacian	U. Petrified Forest M.	Petrified Forest M.	Pet.For.M.	Travesser F.	Bull Canyon F.	Cooper F. (Dockum G)	
Norian	Lacian	Sonsela Sst.		Poleo Sst.		Trujillo F.	Trujillo F.	
Carnian	Tuvalian	L.Pet.For.M.	Moss Back M.	Agua Zarca Sst.	Baldy Hill F.	Garita Creek F.	Tecovas F.	W,L,PR,NO
Carnian	Tuvalian	Shinarump M	Shinarump M			Santa Rosa F.		
Carnian	Julian							CB / T
Carnian	Cordevolian							PR

(Newark Supergroup — E. Coast USA)

B

		Argentina	S. Africa	India	China	U.K.	Germany
	Sinemurian		Clarens F.	? Kota F.	Dark Red Beds / Upper Lufeng F.	upper fissures	Lias
	Hettangian		U. Elliot F.		Dull Purplish Beds	upper fissures	Rhät
Norian	(Rhaetian) Sevatian	La Esquina local fauna	L. Elliot F.	Upper fauna	Lower Lufeng. F.	lower fissures	Knollenmergel
Norian	Alaunian	Los Colorados F.	L. Elliot F.	Dharmaram F.	Lower Lufeng. F.	lower fissures	Stubensandstein
Norian	Lacian	Los Colorados F.		Lower fauna	? ? ?		Stubensandstein
Carnian	Tuvalian	Ischigualasto F.	Molteno F.	U.Maleri F.		Lossiemouth Sst. F.	Kieselsandstein / Rote Wand
Carnian	Julian	Los Rastros F.	Molteno F.	L.Maleri F.			Gipskeuper
Carnian	Cordevolian	Los Rastros F.					Gipskeuper
	Ladinian	Ischichuca F.	? ? ?	Bhimaram Sandstone			Lettenkeuper / Muschelkalk

- three or more sacral vertebrae (Fig. 16.2A);
- a fully open acetabulum;
- a ball-like head on the femur;
- a cnemial crest on the tibia; and
- a well-developed ascending process on the astragalus, fitting on the anterior face of the tibia.

Many other dinosaur-like features had arisen lower down the cladogram in close outgroups of Dinosauria, such as *Lagosuchus*, *Marasuchus*, and Pterosauria (see Parrish, chap. 15 of this volume).

What Is the Oldest Dinosaur?

Over the years, many claims have been made about the date of the first dinosaurs. Earliest supposed records have been based on isolated bones, groups of bones, and footprints from Middle Triassic, Lower Triassic, and even Permian rocks. With a clear cladistic definition of the Dinosauria, it should now be possible to weed these out.

Many supposed early dinosaurs were described on the basis of isolated vertebrae, skull elements, and limb bones from the Triassic of Germany.

1991b), Benton (1994a, 1994b), and others. Abbreviations:
CB = Cow Branch Formation;
L = Lockatong Formation;
M = McCoy Brook Formation;
NH = New Haven Arkose;
NO = New Oxford Formation;
P = Passaic Formation;
PK = Pekin Formation;
PT = Portland Formation;
T = Turkey Branch Formation;
W = Wolfville Formation.
Modified from Benton 1994a.

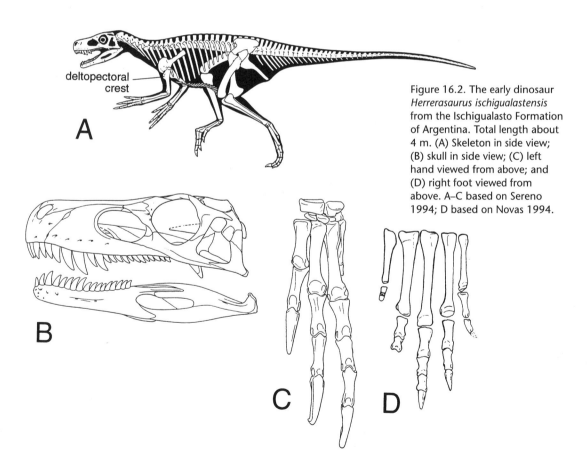

deltopectoral crest

A

B

C

D

Figure 16.2. The early dinosaur *Herrerasaurus ischigualastensis* from the Ischigualasto Formation of Argentina. Total length about 4 m. (A) Skeleton in side view; (B) skull in side view; (C) left hand viewed from above; and (D) right foot viewed from above. A–C based on Sereno 1994; D based on Novas 1994.

Origin and Early Evolution of Dinosaurs   207

Many of these elements have turned out to belong to prolacertiforms or rauisuchid archosaurs, or to be indeterminate (Benton 1986a).

Unusually early records of dinosaurs based on footprints extend back into the Middle and Lower Triassic, and even the Permian. These early finds have all been based on three-toed footprints, a good indication that a print was made by a dinosaur, since most other Permo-Triassic tetrapods left plantigrade four- and five-toed prints. However, the unusually early records are all isolated specimens, and these have proved to be either broken fragments of five-toed prints, invertebrate tracks (king crabs leave tiny "three-toed" impressions), or inorganic sedimentary structures (Thulborn 1990; King and Benton 1996).

The First Dinosaurs

The first unquestionable dinosaurs are all Carnian in age, and late Carnian at that (Fig. 16.1). Although rare elements in their faunas, late Carnian dinosaurs are now known from many parts of the world, and these are catalogued here. Most of the specimens are incomplete, and they will be summarized briefly first, followed by a fuller account of the superb South American early dinosaurs.

The only possible Carnian dinosaur from Europe is *Saltopus* from the Lossiemouth Sandstone Formation of Elgin, Scotland (Huene 1910), but the single specimen is equivocal, and it could be a non-dinosaurian archosaur (Benton and Walker 1985). From Africa comes *Azendohsaurus* from the Argana Formation of Morocco, based on a tooth (Gauffre 1993). *Alwalkeria* from the Maleri Formation of India, based on a partial skull and skeleton, appears to be a small theropod (Chatterjee 1987). Unnamed dinosaur fragments have also been reported from the late Carnian Petrified Forest Formation of Arizona (Lucas et al. 1992).

The late Carnian Santa Maria Formation of Brazil and the Ischigualasto Formation of Argentina have been much more productive than all the other units of the same age elsewhere in the world. This, and the fact that the lagosuchids, close outgroups of the Dinosauria, are exclusively South American, suggests that the dinosaurs perhaps arose in that continent.

The Santa Maria Formation has been the source of *Staurikosaurus pricei,* and the Ischigualasto Formation has yielded *Herrerasaurus ischigualastensis, Eoraptor lunensis,* and *Pisanosaurus mertii.* These Carnian dinosaurs were moderately sized animals, all lightweight bipeds less than 6 meters long. The first three taxa were carnivorous, and the last was a herbivore.

Staurikosaurus and *Herrerasaurus* are members of the Family Herrerasauridae. *Herrerasaurus* (Fig. 16.2) is known in some detail (Novas 1989, 1992, 1994; Sereno and Novas 1992, 1994; Sereno 1994) from eleven specimens, including some partial skeletons. These show a slender, lightweight biped (Fig. 16.2A) ranging in length from 3 to 6 meters. The skull (Fig. 16.2B) is narrow and low. There is a sliding joint in each lower jaw, which allowed the jaws to flex and grasp struggling prey. The neck is slender. The forelimbs are less than half the length of the hind limbs, and the hand is elongated. Digits IV and V of the hand (Fig. 16.2C) are reduced, and the long penultimate phalanges of the hand indicate that it was adapted for grasping. There are two sacral vertebrae, a loss of one from the normal dinosaurian condition, and the acetabulum is perforate. The femur has an inturned subrectangular head to fit into the pelvic bowl, and the tibia bears a cnemial crest. The foot (Fig. 16.2D) is digitigrade (the animal stands high on its toes), the calcaneum is reduced in size, and the astragalus bears an ascending process on the front of the tibia. Two other species from the Ischigualasto Formation, *Ischisaurus cattoi* and *Frenguellisaurus ischigualastensis*, are probably synonyms of *Herrerasaurus ischigualastensis*.

Eoraptor is another small carnivorous dinosaur (Fig. 16.3A, B), only 1 meter long (Sereno et al. 1993). It has a lower snout than *Herrerasaurus*, there is no intramandibular joint, and the hand is shorter. *Pisanosaurus* is a tiny dinosaur (Bonaparte 1976) known from incomplete remains (Fig. 16.3C). The lower jaw (Fig. 16.3D) indicates that this specimen is an ornithischian dinosaur: the teeth are broad and diamond-shaped, with a spatulate outer face, and there is a prominent shelf along the outer face of the lower jaw which marks the bottom of a soft cheek. The hind limb and foot appear dinosaurian, with a reduced calcaneum, an ascending process on the astragalus, and a functionally three-toed digitigrade foot.

Relationships of the First Dinosaurs: Saurischia and Ornithischia

The systematics of the early dinosaurs has been disputed. Until recently, *Herrerasaurus* and *Staurikosaurus* were generally regarded as primitive

Figure 16.3. Early dinosaurs. (A), (B) Skeleton and skull of *Eoraptor lunensis* from the Ischigualasto Formation of Argentina, both in side view. The dinosaur's total length was about 1 m. (C), (D) Partial skeleton and lower jaw of *Pisanosaurus mertii* from the Ischigualasto Formation of Argentina. A and B based on Sereno et al. 1993; C and D based on Bonaparte 1976.

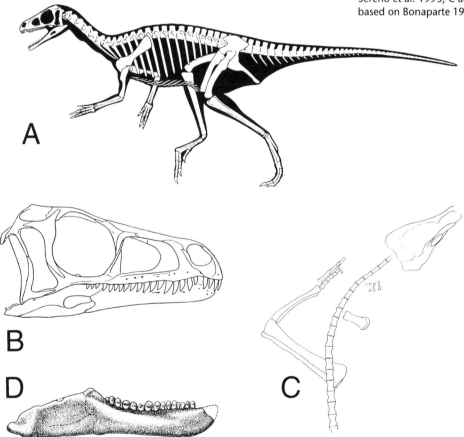

A

B

D

C

Origin and Early Evolution of Dinosaurs   209

forms that were neither saurischians nor ornithischians. However, Sereno and Novas (1992, 1994), Sereno et al. (1993), Novas (1994), and Sereno (1994) have argued that the Herrerasauridae and *Eoraptor* are basal theropods (Fig. 16.4). This means that the split into Saurischia and Ornithischia must have happened at the latest during the Carnian, before the time of the first known dinosaur fossils.

The Saurischia may be diagnosed by a number of characters of the skull, vertebrae, and limbs (Sereno et al. 1993), including the presence of a subnarial foramen, a small opening below the nostril, and a wedge-shaped ascending process of the astragalus. The shape of the saurischian pelvis is not useful diagnostically, because this pattern is primitive and is shared with the ancestors of the dinosaurs, and indeed with most other reptiles.

The Ornithischia have long been recognized as a monophyletic group, and they are diagnosed by the presence of triangular teeth, with the largest tooth in the middle of the tooth row. In addition, there is a coronoid process behind the tooth row in the lower jaw (Fig. 16.3D) and a reduced external/mandibular fenestra (cf. Fig. 16.3B, a saurischian, and 16.3D, an ornithischian). The classic ornithischian trademark, a predentary bone at the front of the lower jaw, is not seen in *Pisanosaurus*, because of incomplete preservation (Fig. 16.3D).

Both *Eoraptor* and *Herrerasaurus* share many typically theropod features: prong-shaped cervical epipophyses (additional processes at the back of the vertebrae), extreme hollowing of the centra and long bones, vestigial nature of digits IV and V in the hand, extensor depressions on metacarpals I through III, and a hand that is more than 50 percent of the combined length of the humerus and radius. *Herrerasaurus* shares with other theropods some additional characters: an intramandibular jaw joint (Fig. 16.2A), elongate prezygapophyses (processes on the front) on distal caudal vertebrae, a strap-shaped scapular blade, digits I through III of the hand with long penultimate phalanges and recurved unguals, and a pubic foot.

Figure 16.4. Cladogram showing suggested relationships of the basal dinosaurs. See text for characters supporting each branching point. Based on information in Sereno et al. 1993. See also Sereno 1997; Novas 1996.

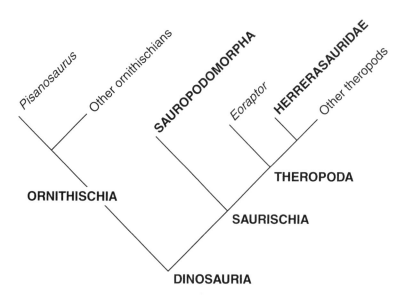

Ecological Models for the Origin of the Dinosaurs

A major faunal turnover took place on land during the Late Triassic. Various long-established groups, sometimes termed paleotetrapods (mammal-like reptiles, "thecodontians," temnospondyl amphibians, rhynchosaurs, prolacertiforms, procolophonids), were replaced by new reptilian types, sometimes termed neotetrapods (turtles, crocodilians, dinosaurs, pterosaurs, lepidosaurs, mammals). It was formerly assumed that this replacement was a long-drawn-out affair, involving competition (Bakker 1977; Bonaparte 1982; Charig 1984), with the dinosaurs leading the way in driving out the mammal-like reptiles, rhynchosaurs, and thecodontians. The success of the dinosaurs was explained by their superior adaptations, such as their upright posture, their initial bipedalism, their speed and intelligence, or their posited endothermy.

I have opposed this assumption of long-term competitive replacement (Benton 1983, 1986a, 1991, 1994a, 1994b). The first contrary evidence came from quantitative studies of tetrapod faunas through the Triassic (Benton 1983), which did not show a long-term decline of paleotetrapod groups and a matching rise of neotetrapod groups. Indeed, new groups generally did not supplant previously existing groups. The study revealed that there was a dramatic changeover from late Carnian faunas containing rare dinosaurs, to those in the early and middle Norian (Fig. 16.1), where dinosaurs dominate.

Later studies (Benton 1986a, 1991, 1994b) clarified this changeover. There had been a mass extinction of previously dominant reptile groups at about the Carnian-Norian boundary, 225 million years ago, the Carnian-Norian extinction event. The dicynodonts, chiniquodontids, and rhynchosaurs all died out. These three families had made up 40 to 80 percent of all the late Carnian faunas, representing the dominant medium-sized and large herbivores worldwide. Other groups that disappeared at this time were temnospondyl amphibians (Mastodonsauridae, Trematosauridae), archosauromorphs (Prolacertidae), basal archosaurs (Proterochampsidae, Scleromochlidae), and some dinosaurs (Herrerasauridae, Pisanosauridae). Hence, ten of the twenty-four late Carnian families died out (a loss of 42 percent), and continental tetrapod faunas were dramatically depleted in terms of diversity and abundance. I suggest (Benton 1991, 1994b) that the apparent gap in the fossil record of terrestrial tetrapods in the early Norian may be a true indication of reduced abundance following a mass extinction event.

A detailed study of the Ischigualasto Formation by Rogers et al. (1993) has confirmed the view that there was no long-term ecological replacement of paleotetrapods by neotetrapods, because members of both assemblages co-occur without evidence of any decline in diversity or abundance of the former and a rise of the latter. Dinosaurs appear early in the Ischigualasto sequence, but they never increase above a diversity of three species or a percentage representation of 6 percent of all specimens collected.

There were other extinctions at, or close to, the Carnian-Norian boundary among marine organisms (foraminifera, ammonoids, bivalves, bryozoans, conodonts, coral reefs, echinoids, and crinoids; Benton 1986b; Sepkoski 1990; Simms and Ruffell 1990). At this time, there was a worldwide series of climatic changes from humid to arid (Simms and Ruffell 1990) that may have been triggered by events associated with the beginning of rifting of the supercontinent Pangaea. Major floral changes occurred, too, with the disappearance of the *Dicroidium* floras of southern continents, and the spread worldwide of northern conifer-dominated floras. Perhaps the drying climates favored conifers over seed-ferns such as *Dicroidium,* and perhaps the dominant Carnian herbivores were unable to adapt to new kinds of vegetation, and died out.

First Radiation of the Dinosaurs

Dinosaurs diversified to a limited extent during the Carnian, and increasingly after the Carnian-Norian extinction event. In the Norian, for the first time, mass accumulations of dinosaur skeletons are found, such as the famous death assemblage of several hundred individuals of the theropod *Coelophysis* at Ghost Ranch, New Mexico, in the Early Norian Upper Petrified Forest Member of the Chinle Formation (Schwartz and Gillette 1994). For the first time, too, dinosaurs became relatively diverse, and they began to exhibit that feature for which the group is famous: large size. Specimens of *Plateosaurus* from the Norian Stubensandstein and Knollenmergel of Germany reached lengths of 6 to 8 meters. Overall, dinosaurs had switched from being minor players in the Carnian, at faunal abundances of less than 6 percent, to being the dominant land reptiles, with abundances of 25 to 60 percent in the Norian.

This model of Late Triassic dinosaurian radiation has been downplayed in some recent studies of mass extinction events. Olsen et al. (1987, 1990) and Hallam (1990) claimed that the Carnian-Norian extinction event played a minor role, and that a second mass extinction event, at the Triassic-Jurassic boundary 202 million years ago (Fig. 16.5), triggered the radiation of the dinosaurs.

There is no question that there was a Triassic-Jurassic mass extinction event, which had major effects in the oceans (Sepkoski 1990). On land, too, several families of reptiles disappeared, particularly the last of the basal archosaurs ("thecodontians") and some mammal-like reptiles. In addition, a major impact crater site, the Manicouagan structure in Quebec, has been identified as the smoking gun for a catastrophic extraterrestrial impact at the Triassic-Jurassic boundary (Olsen et al. 1987, 1990). Shocked quartz has been found at a Triassic-Jurassic boundary section in Italy (Bice et al. 1992), possible evidence of a major extraterrestrial impact (see Russell and Dodson, chap. 42 of this volume). However, the nature of the lamellae in

Figure 16.5. Time scale of major events in the Late Triassic, showing the time line of the early dinosaurs from the Ischigualasto Formation of Argentina, the current date of the Manicouagan impact, and the two mass extinctions.

Period	Stage	Boundary date (Ma)	Events
Jurassic	Hettangian		
Late Triassic	Norian	202	Triassic-Jurassic mass extinction
			214-Ma Manicouagan impact
		220	Carnian-Norian mass extinction
	Carnian		228-Ma Ischigualasto dinosaurs
		230	

Note: Ma = million years ago.

212   *Michael J. Benton*

the shocked quartz is not adequate to rule out other explanations, such as a volcanic source for the material (Bice et al. 1992). Further, the Manicouagan impact structure has been redated (Hodych and Dunning 1992) away from the Triassic-Jurassic boundary (202 million years ago), with an age of 214 million years ago (Fig. 16.5).

Olsen et al. (1987, 1990) argued that there was a dramatic radiation of dinosaurs and other tetrapods after the Triassic-Jurassic mass extinction. This is debatable (Benton 1994b): The major dinosaur lineages were set out in the late Carnian (Theropoda, Sauropodomorpha, Ornithischia), and the first two in particular radiated during the Norian (Podokesauridae, Thecodontosauridae, Plateosauridae, Melanorosauridae). New families of dinosaurs arose in the Early Jurassic (Ceratosauridae, Anchisauridae, Massospondylidae, Yunnanosauridae, Vulcanodontidae, Fabrosauridae, Heterodontosauridae, Scelidosauridae, Huayangosauridae), but most of these are represented by single species.

References

Bakker, R. T. 1977. Tetrapod mass extinctions: A model of the regulation of speciation rates and immigration by cycles of topographic diversity. In A. Hallam (ed.), *Patterns of Evolution as Illustrated by the Fossil Record*, pp. 439–468. Amsterdam: Elsevier.

Bakker, R. T., and Galton. 1974. Dinosaur monophyly and a new class of vertebrates. *Nature* 248: 168–172.

Benton, M. J. 1983. Dinosaur success in the Triassic: A noncompetitive ecological model. *Quarterly Review of Biology* 58: 29–55.

Benton, M. J. 1986a. The Late Triassic tetrapod extinction events. In K. Padian (ed.), *The Beginning of the Age of Dinosaurs: Faunal Change across the Triassic-Jurassic Boundary*, pp. 303–320. Cambridge: Cambridge University Press.

Benton, M. J. 1986b. More than one event in the Late Triassic mass extinction. *Nature* 321: 857–861.

Benton, M. J. 1990. Origin and interrelationships of dinosaurs. In D. B. Weishampel, P. Dodson, and H. Osmólska (eds.), *The Dinosauria*, pp. 11–30. Berkeley: University of California Press.

Benton, M. J. 1991. What really happened in the Late Triassic? *Historical Biology* 5: 263–278.

Benton, M. J. 1994a. Late Triassic terrestrial vertebrate extinctions: Stratigraphic aspects and the record of the Germanic Basin. *Paleontologia Lombarda*, n.s. 2: 19–38.

Benton, M. J. 1994b. Late Triassic to Middle Jurassic extinctions among tetrapods: Testing the pattern. In N. C. Fraser and H.-D. Sues (eds.), *In the Shadow of the Dinosaurs: Triassic and Jurassic Tetrapod Faunas*, pp. 366–397. Cambridge: Cambridge University Press.

Benton, M. J., and A. D. Walker. 1985. Palaeoecology, taphonomy, and dating of Permo-Triassic reptiles from Elgin, north-east Scotland. *Palaeontology* 28: 207–234.

Bice, D. M.; C. R. Newton; S. McCauley; P. W. Reiners; and C. A. McRoberts. 1992. Shocked quartz at the Triassic-Jurassic boundary in Italy. *Science* 255: 443–446.

Bonaparte, J. F. 1976. *Pisanosaurus mertii* Casamiquela and the origin of the Ornithischia. *Journal of Paleontology* 50: 808–820.

Bonaparte, J. F. 1982. Faunal replacement in the Triassic of South America. *Journal of Vertebrate Paleontology* 21: 362–371.

Charig, A. J. 1984. Competition between therapsids and archosaurs during the Triassic Period: A review and synthesis of current theories. *Symposia of the Zoological Society of London* 52: 597–628.

Chatterjee, S. K. 1987. A new theropod dinosaur from India with remarks on the Gondwana-Laurasia connection in the Late Triassic. *Geophysics Monographs* 41: 183–189.

Colbert, E. H. 1958. Tetrapod extinctions at the end of the Triassic. *Proceedings of the National Academy of Sciences of the U.S.A.* 44: 973–977.

Gauffre, F.-X. 1993. The prosauropod dinosaur *Azendohsaurus laaroussii* from the Upper Triassic of Morocco. *Palaeontology* 36: 897–908.

Gauthier, J. A. 1986. Saurischian monophyly and the origin of birds. *Memoirs of the California Academy of Sciences* 8: 1–55.

Hallam, A. 1990. The end-Triassic mass extinction event. *Geological Society of America Special Paper* 247: 577–583.

Hodych, J. P., and G. R. Dunning. 1992. Did the Manicouagan impact trigger end-of-Triassic mass extinction? *Geology* 20: 51–54.

Huene, F. v. 1910. Ein primitiver Dinosaurier aus der mittleren Trias von Elgin. *Geologische und Paläontologische Abhandlungen, Neue Folge* 8: 315–322.

Hunt, A. P., and S. G. Lucas. 1991a. The *Paleorhinus* Biochron and the correlation of the nonmarine Upper Triassic of Pangaea. *Palaeontology* 34: 487–501.

Hunt, A. P., and S. G. Lucas. 1991b. A new rhynchosaur from the Upper Triassic of west Texas, U.S.A., and the biochronology of Late Triassic rhynchosaurs. *Palaeontology* 34: 927–938.

King, M. J., and M. J. Benton. 1996. Dinosaurs in the Early and Mid-Triassic? The footprint evidence from Britain. *Palaeogeography, Palaeoclimatology, Palaeoecology* 122: 213–225.

Lucas, S. G. 1991. Sequence stratigraphic correlation of nonmarine and marine Late Triassic biochronologies, western United States. *Albertiana* 9: 11–18.

Lucas, S. G.; A. P. Hunt; and R. A. Long. 1992. The oldest dinosaurs. *Naturwissenschaften* 79: 171–172.

Novas, F. E. 1989. The tibia and tarsus in Herrerasauridae (Dinosauria, incertae sedis) and the origin and evolution of the dinosaurian tarsus. *Journal of Paleontology* 63: 677–690.

Novas, F. E. 1992. Phylogenetic relationships of the basal dinosaurs, the Herrerasauridae. *Palaeontology* 35: 51–62.

Novas, F. E. 1994. New information of the systematics and postcranial skeleton of *Herrerasaurus ischigualastensis* (Theropoda: Herrerasauridae) from the Ischigualasto Formation (Upper Triassic) of Argentina. *Journal of Vertebrate Paleontology* 13: 400–423.

Novas, F. E. 1996. Dinosaur monophyly. *Journal of Vertebrate Paleontology* 16: 723–741.

Ochev, V. G., and M. A. Shishkin. 1989. On the principles of global correlation of the continental Triassic on the tetrapods. *Acta Palaeontologica Polonica* 34: 149–173.

Olsen, P. E.; S. J. Fowell; and B. Cornet. 1990. The Triassic/Jurassic boundary in continental rocks of eastern North America: A progress report. *Geological Society of America Special Paper* 247: 585–593.

Olsen, P. E.; N. H. Shubin; and M. H. Anders. 1987. New Early Jurassic tetrapod assemblages constrain Triassic-Jurassic tetrapod extinction event. *Science* 237: 1025–1029.

Owen, R. 1842. Report on British fossil reptiles. Part II. *Report of the British Association for the Advancement of Science* 1842: 60–204.

Rogers, R. R.; C. C. Swisher III; P. C. Sereno; C. A. Forster; and A. M. Monetta. 1993. The Ischigualasto tetrapod assemblage (Late Triassic) and ⁴⁰Ar/³⁹Ar calibration of dinosaur origins. *Science* 260: 794–797.

Romer, A. S. 1970. The Triassic faunal succession and the Gondwanaland problem. In *Gondwana Stratigraphy: IUGS Symposium Buenos Aires 1967*, pp. 375–400. Paris: UNESCO.

Schwartz, H. L., and D. D. Gillette. 1994. Geology and taphonomy of the *Coelophysis* Quarry, Upper Triassic Chinle Formation, Ghost Ranch, New Mexico. *Journal of Paleontology* 68: 1118–1130.

Seeley, H. G. 1887. On the classification of the fossil animals commonly called Dinosauria. *Proceedings of the Royal Society* 43: 165–171.

214   *Michael J. Benton*

Sepkoski, J. J., Jr. 1990. The taxonomic structure of periodic extinction. *Geological Society of America Special Paper* 247: 33–44.

Sereno, P. C. 1986. Phylogeny of the bird-hipped dinosaurs (Order Ornithischia). *National Geographic Research* 2: 234–256.

Sereno, P. C. 1994. The pectoral girdle and forelimb of the basal theropod *Herrerasaurus ischigualastensis. Journal of Vertebrate Paleontology* 13: 425–450.

Sereno, P. C. 1997. The origin and evolution of dinosaurs. *Annual Review of Earth and Planetary Sciences* 25: 435–489.

Sereno, P. C.; C. A. Forster; R. R. Rogers; and A. M. Monetta. 1993. Primitive dinosaur skeleton from Argentina and the early evolution of Dinosauria. *Nature* 361: 64–66.

Sereno, P. C., and F. E. Novas. 1992. The complete skull and skeleton of an early dinosaur. *Science* 258: 1137–1140.

Sereno, P. C., and F. E. Novas. 1994. The skull and neck of the basal theropod *Herrerasaurus ischigualastensis. Journal of Vertebrate Paleontology* 13: 451–476.

Simms, M. J., and A. H. Ruffell. 1990. Climatic and biotic change in the Late Triassic. *Journal of the Geological Society of London* 147: 321–327.

Thulborn, R. A. 1990. *Dinosaur Tracks.* London: Chapman and Hall.

Tozer, E. T. 1974. Definitions and limits of Triassic stages and substages: Suggestions prompted by comparisons between North America and the Alpine-Mediterranean region. *Schriftenreihe der Erdwissenschaftlichen Kommissionen, Osterreichische Akademie der Wissenschaften* 2: 195–206.

Tozer, E. T. 1979. Latest Triassic ammonoid faunas and biochronology, western Canada. *Paper of the Geological Society of Canada* 79: 127–135.

Visscher, H., and W. A. Brugman. 1981. Ranges of selected palynomorphs in the Alpine Triassic of Europe. *Review of Palaeobotany and Palynology* 34: 115–128.

Visscher, H.; W. M. L. Schuurman; and A. W. Van Erve. 1980. Aspects of a palynological characterisation of Late Triassic and Early Jurassic "standard" units of chronostratigraphical classification in Europe. *Proceedings of the IVth International Palynological Conference, Lucknow (1976–77)* 2: 281–287.

Weiss, M. 1989. Die Sporenfloren aus Rät und Jura Südwest-Deutschlands und ihre Beziehung zur Ammoniten-Stratigraphie. *Paläontographica, Abteilung B* 215: 1–168.

Theropods

Philip J. Currie

17

The meat-eating dinosaurs, or Theropoda, survived more than 160 million years. In fact, when birds are taken into consideration, the theropod lineage has been around for 230 million years and is still very prominent today. The Theropoda includes some of the earliest dinosaurs known. The ancestor of all dinosaurs would have been a meat-eating animal that would have had the general appearance of a theropod. But that animal would have lacked a number of characteristics that define Theropoda.

Specimens discovered in 230-million-year-old Upper Triassic rocks in Argentina include the oldest and most primitive dinosaurs presently known. One of these animals, *Herrerasaurus,* was for a long time thought to be too primitive to be classified as either a saurischian or an ornithischian, and therefore it was considered to be representative of the common ancestor of both lineages. However, the recovery of better specimens in recent years has shown that this animal is a primitive theropod (Novas 1993; Sereno 1993; Sereno and Novas 1993). *Herrerasaurus* has a movable intramandibular joint (Fig. 17.1) in the middle of the lower jaw (Sereno and Novas 1993), like all but a few theropods that have secondarily fused this articulation. Another animal found in the same rocks as *Herrerasaurus* is anatomically closer to what would be considered the ideal ancestor of both saurischian and ornithischian dinosaurs. This dinosaur, *Eoraptor,* was only a meter in total length. Because it lacks the extra joint in the lower jaw, some workers do not consider it to be a theropod (Sereno et al. [1993], however, do consider it to be a theropod), but it was evidently trending toward becoming one. One of the characteristics of the Theropoda is the

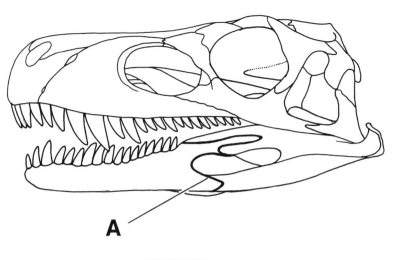

Figure 17.1. Skull of *Herrera-saurus*. (A) Intramandibular joint. Redrawn from Novas 1993; this and other drawings in this chapter by Mike Skrepnick.

reduction in size, or the complete loss, of the outer fingers of the hand. Although *Eoraptor* still retains five fingers, the outer ones (IV and V) are not as stout as would be expected in the ideal dinosaur ancestor.

Characteristic Features of Theropods

Theropods are generally easy to recognize. Because they were flesh-eaters, most have blade-like teeth with serrated ridges along the front and back margins (Fig. 17.2). Under a microscope, each serration usually looks like a miniature tooth with a recurved, sharp tip to hook and separate meat fibers, and its own blade-like ridges of enamel to cut those fibers (Abler 1992). In some large theropods such as the tyrannosaurs, the serrations became wide and chisel-shaped so that they would not break when they came in contact with bones in prey. The claws, especially on the hand, are often recurved and taper to sharp points. Usually theropods are slender, rather long-legged, bipedal animals that were built to move quickly (Fig. 17.3). None of these characters is unique to theropods, even though they are significant aspects of their appearance. Serrated teeth have evolved independently many times, and can be found in such diverse carnivores as sharks and saber-toothed "tigers." Most carnivorous tetrapods, and many herbivores as well, have sharp claws. Bipedality was characteristic of ancestral dinosaurs, and is widespread in both the Saurischia and the Ornithischia.

Unlike most ornithischian dinosaurs, theropods always have hollow limb bones. The carnivorous dinosaurs also show a tendency toward having air-filled (pneumatic) bones in the front part of the body. Air sacs from the nasal cavity extended laterally into a big pocket on each side of the face called the antorbital fossa (Fig. 17.4). From there, air tubes (pneumatic diverticula) invaded the bones of the roof of the mouth and around the eye. At the back of the skull, an elaborate system of tubes carried air from the throat and through the middle ear to invade the bones of the braincase in advanced theropods and birds. Vertebrae and ribs in the neck and at the front of the body were pneumatized by air sacs and diverticula connected

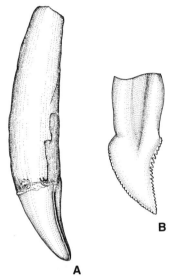

Figure 17.2. Teeth of (A) *Tyrannosaurus* and (B) *Troodon*, not to scale. Both teeth are drawn with the crown, the portion of the tooth that protruded above the gums, pointed down. The crown length, from base to tip, of the *Tyrannosaurus* tooth is several centimeters; the crown length of the *Troodon* tooth is about 1 cm. In both teeth the root of the tooth, the portion embedded in the jaw, is as long as or longer than the tooth crown. Note the cutting edges on both teeth. In the *Tyrannosaurus* tooth, the individual serrations are too small to be readily distinguishable, given the scale of the drawing. The *Troodon* tooth, in contrast, has very coarse serrations for its size.

Theropods 217

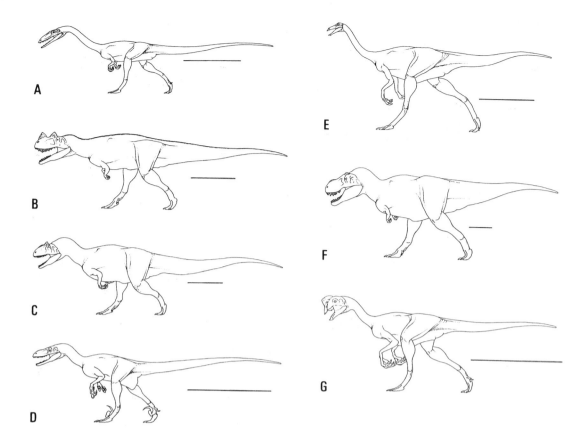

Figure 17.3. Restorations of representative theropods. The scale bar in each drawing is 1 m in length. (A) *Coelophysis;* (B) *Ceratosaurus;* (C) *Allosaurus;* (D) *Deinonychus;* (E) *Ornithomimus;* (F) *Tyrannosaurus;* (G) *Oviraptor.*

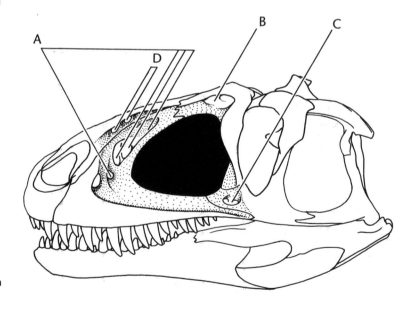

Figure 17.4. Skull of *Sinraptor,* redrawn from Currie and Zhao 1993. The antorbital fossa is shown in the stippled pattern, and the antorbital fenestra that sits within the fossa in black. Lines point to pneumatopores in bones that bound the antorbital fossa: (A) maxilla, (B) lacrimal, (C) jugal, and (D) nasal.

to the lungs (see Reid, chap. 32 of this volume). In some theropods, pneumatic bones are found as far back as the middle of the tail, and in birds, most of the bones of the limbs and limb girdles can also be air-filled. These elaborate air systems have been very good for assessing the interrelationships of theropods and birds, but are not unique to the Theropoda. Antorbital fossae are characteristic of all archosaurs, including crocodilians and pterosaurs. The elaborate systems of invasion of vertebrae by extensions of the lungs are very similar in pterosaurs, sauropods, and theropods. This suggests that vertebral pneumaticity is a primitive characteristic that was secondarily lost in ornithischians and prosauropods.

The presence of certain unique characters in the skeleton allows paleontologists to identify theropods. These include the following features:

- the bone in front of the eye (the lacrimal) extends onto the top of the skull;
- there is an extra joint in the lower jaw;
- there are prominent processes (epipophyses) on the neck vertebrae (Fig. 17.5);
- elongate prezygapophyses are found on tail vertebrae (Fig. 17.6 shows an extreme case);
- the scapula (shoulder blade) is strap-like;
- the upper arm bone (humerus) is less than half the length of the upper leg bone (femur);
- the hand is elongate, but has reduced the size of, or lost, the outer two fingers;
- there are distinctive pits on top (back) of the bones (metacarpals) in the "palm" of the hand for the attachment of ligaments;
- the bones between the last and second-to-last joints of the fingers are elongate;
- there is an expansion on the distal end of the pubis (one of the hip bones); and
- there is a shelf-like ridge of bone near the head of the upper leg bone (femur) for the attachment of muscles.

Within the Theropoda, certain evolutionary trends can be observed over the course of their Mesozoic history. In the skull, pneumatization of the snout becomes more pronounced in advanced theropods, and accessory antorbital openings appear in front of the antorbital fenestra. By the Cretaceous, most theropods seem to have had a degree of stereoscopic vision, with the eyes facing more forward than they had been in earlier forms. The brain had become relatively larger, especially in the smaller theropods. As previously mentioned, pneumatization of the braincase is more elaborate in Cretaceous theropods.

One of the unifying characters of the Theropoda, the presence of the

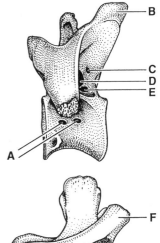
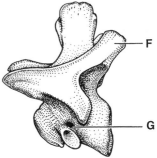

Figure 17.5. Tenth cervical (neck) vertebra of *Carnotaurus* (above) and fifth cervical of *Allosaurus* (below). Scale bar is 10 cm long. In both drawings, the topmost line (B, F) points to the epipophysis. Other lines (A, C–E) indicate pneumatopores. Vertebrate redrawn from Bonaparte et al. 1990 and Madsen 1976.

Figure 17.6. Caudal (tail) vertebra of *Deinonychus*. Scale bar is 10 cm long. Extending to the right of the main part of the bone is an extremely long prezygapophysis (*above*) and a shorter (but still impressive) chevron bone (*below*). Redrawn from Ostrom 1969.

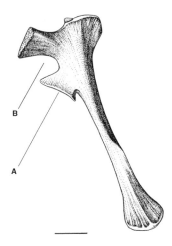

Figure 17.7. Ischium of *Allosaurus*. Scale bar is 10 cm long. (A) Obturator process; (B) obturator notch. Redrawn from Madsen 1976.

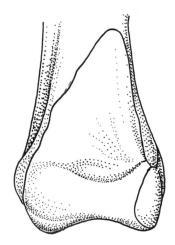

Figure 17.8. Lower end of the lower leg–ankle complex of *Ornithomimus*. The triangular sheet of bone extending up the front of the lower leg is the ascending process of the astragalus, one of the ankle bones.

epiphyses in cervical vertebrae, tended to be lost in more advanced theropod lineages as the neck took on a stronger curvature and the vertebrae became regionally more distinct from each other.

The fingers tended to become more elongate, and there was a strong tendency for further reduction of the number of fingers. Many of the Jurassic theropods had three functional fingers with only a vestige of the fourth, and the fourth was completely lost in allosaurids. *Compsognathus* and tyrannosaurids independently reduced the number of fingers to two, and *Mononykus* retained only the first finger.

In the hips, the distal expansion of the pubis expands into an enlarged "boot" in many of the theropod families, and the obturator foramen of the ischium opens up into a notch above the obturator process (Fig. 17.7). The limbs of theropods became progressively longer over their history, and bones of the lower leg (tibia and fibula) and the long bones at the base of the foot (the metatarsus) increased in length relative to the femur. The lengthening of leg elements and changes in proportions reflect faster running abilities. As theropods became faster, they needed more control and better shock absorption in their feet. The ascending process of the astragalus (Fig. 17.8) became much higher, extending as much as 25 percent of the length of the tibia. The lower end of the fibula and the calcaneum had a tendency to become smaller (Fig. 17.9), and the latter element was lost as a distinct element in some families. The ankle bones closest to the foot were thin and flat. Theropods had three functional toes (the second, third, and fourth), supported by three metatarsals (in humans the metatarsals are in the flat of the foot). In addition to these, the first toe was retained as a "dew claw" on the inside or back of the foot. The upper part of the middle (third) bone of the metatarsus became narrow and splint-like (Fig. 17.10) in most of the Late Cretaceous theropods, and metatarsals II and IV contacted each other at the top (Holtz 1994b). The lower end of the third metatarsal would have contacted the ground first when a theropod was running. Through a system of bony contacts and elastic ligaments, the shock of impact was transferred to, and dissipated through, the adjacent metatarsals, thereby reducing the chance of foot injuries. In elmisaurids, avimimids, and birds, the heads of the three main metatarsals fused with the distal tarsals to form a unified tarsometatarsal bone.

Specializations within the Theropoda

Cranial Ornamentation

Theropods developed some amazingly diverse features that are best interpreted as display characters for attracting potential mates and/or warning off potential rivals.

Horns on the skull are common in large theropods. There is a nasal horn in *Ceratosaurus* (Fig. 17.11), and rugosities on the nose of the tyrannosaurid *Alioramus* suggest that it may have had a series of small protuberances. In *Carnotaurus* big frontal horns extend dorsolaterally above the eyes. Most large theropods have low horns in front of their eyes on the lacrimals, and they are best developed in *Allosaurus* and some of the tyrannosaurids. The skull roof of *Tyrannosaurus* had become very thick, and there are big rugosities above and behind the eye that may have been covered by horny epidermal sheaths.

Some of the early large theropods developed elaborate crests on their skulls. *Dilophosaurus,* as its name ("two-crested reptile") implies, had a pair of hatchet-like crests (Fig. 17.11), one on either side of the skull roof extending from the nasal opening to the orbit. A recently described Early Jurassic theropod from Antarctica, *Cryolophosaurus,* had a thin crest that extended from side to side over the eyes (Hammer and Hickerson 1994). The Middle Jurassic *Monolophosaurus* from China had a single crest (Fig. 17.11) that was formed by folding of the bones from the side of the skull up and over the middle of the skull. This crest is hollow and has pneumatic connections with the nasal cavities, suggesting that in addition to being a visual identifier, it may have functioned as a resonating chamber for the alteration and amplification of sounds produced in the throat.

The skulls of oviraptorids are very unusual, and look like the skulls of highly derived birds like the cassowary (Fig. 17.12). *Oviraptor* has an inflated crest formed of the nasal, frontal, and sometimes parietal bones. The crest is highly pneumatic, and its interior spaces are continuous with the nasal chambers.

Toothless Theropods

Theropod teeth tend to be rather conservative in overall shape, although as we have seen, they can be complex on the microscopic level. Although most non-flying (non-avian) theropods retained teeth, some lineages replaced their teeth with horny bills. In Late Triassic times, one possible theropod from Texas, *Shuvosaurus,* was toothless. Small Jurassic theropods, including the earliest known bird (*Archaeopteryx*), retained their teeth, although the fossil record for that time is too incomplete to state that there were no toothless species. By Cretaceous times, however, oviraptorosaurs (Caenagnathidae, Oviraptoridae), ornithomimids, and many birds appear to have become toothless. This appears to have occurred independently, because the earliest known unquestionable ornithomimid (*Harpymimus*) retained vestiges of teeth at the front of its jaws, a European ornithomimosaur (*Pelecanimimus*) had more than 200 teeth in its jaws, and many Jurassic and Cretaceous birds had full complements of teeth.

The jaws of oviraptorosaurs were highly specialized, although it is still not known what they were specialized for. The name *Oviraptor* means "egg stealer," and it is possible that these animals were egg-eaters. The skull had been pulled back into a more erect position than in most other theropods so that the entrance to the throat was positioned underneath the middle of the lower jaws. The jaws were narrow at the front and back (Fig. 17.13) but were more widely separated at midlength. There were complementary ridges on upper and lower jaws, and a bony process protruded from the roof of the mouth. Complete eggs could have been picked up in the mouth and pushed partway down the throat. They would have been punctured at this time by the bony protrusion from the palate, which looks similar to the "egg teeth" found in the throats of egg-eating snakes. Once punctured, the egg could have been further broken by muscular or elastic contraction in the throat. This system would have allowed the ingestion of eggs without loss of any of their juices.

Other functions have been proposed for oviraptorosaur jaws, such as crushing the shells of mollusks or shearing fibrous plant tissues (Smith 1992). Although the jaws were clearly adapted to produce very powerful bites, the beaks were hollow and air-filled, and were inappropriate for crushing anything with hard shells. Furthermore, oviraptorids are most common in semi-arid to arid paleoenvironments, from which clams have

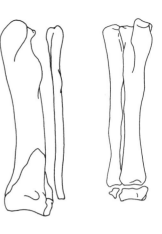

Figure 17.9. Left lower leg and ankle of *Gorgosaurus* (left) and right lower leg and ankle of *Herrerasaurus* (right), viewed from the front. In both drawings, the larger, more massive leg bone is the tibia (the bone on the left in the case of *Gorgosaurus,* and on the right in *Herrerasaurus*), and the less stoutly built leg bone beside it is the fibula. Beneath the tibia is a large ankle bone, the astragalus; in *Gorgosaurus* the astragalus has a large ascending process that extends up the front of the tibia (as in *Ornithomimus,* Fig. 17.8). In both dinosaurs there is a small ankle bone, the calcaneum, beneath the fibula, next to the astragalus. In *Herrerasaurus* this bone, though smaller than the astragalus, is still reasonably good-sized. In *Gorgosaurus,* however, the calcaneum has become the small, New Jersey–shaped piece of bone attached to the much larger astragalus. Figures redrawn from Lambe 1917 and Reig 1963. Not to scale.

Figure 17.10. Left hind foot of *Struthiomimus,* redrawn from Osborn 1916. The middle (third) metatarsal pinches out near the top (proximal end) of the foot.

not been reported. The retention of sharp claws and body proportions appropriate for speed and agility argues against the hypothesis that oviraptorosaurs had become leaf- or nut- and seed-eating herbivores. The sharp-edged beaks and powerful biting forces could have been used to kill smaller animals such as lizards and mammals, but the structure of the jaws suggests that the prey would have then been swallowed whole.

Ornithomimids are in some ways even more difficult to interpret, because their toothless jaws were elongate and slender, and would have been incapable of producing enough power to tear apart carcasses or to kill any but the smallest animals. It has been long assumed that they were omnivores, eating small vertebrates and invertebrates, eggs, and easily digested plant parts such as fruit and seeds. A recent biochemical analysis of stable isotopes in two ornithomimid bones produced ambiguous results, suggesting that one animal was eating meat, and that the other was omnivorous (Ostrom et al. 1993).

Elongate Vertebral Spines

Several unrelated theropods developed elongate spines on their vertebrae during Early Cretaceous times. The neural spines are five times the height of the centra in *Altispinax,* and eleven times in *Spinosaurus* (Fig. 17.14). The spines are also elongate in *Acrocanthosaurus,* and probably compare best in relative height with those of *Altispinax.* The high neural spines would have formed a skin-encased sail-like structure, the function of which is unknown. Because of the large size of all three genera, the height of the spines may have been required for support and rigidity. Clearly, however, the height exceeds this requirement. Because *Spinosaurus* and *Acrocanthosaurus* lived in the tropics close to sea level, the "sails" may

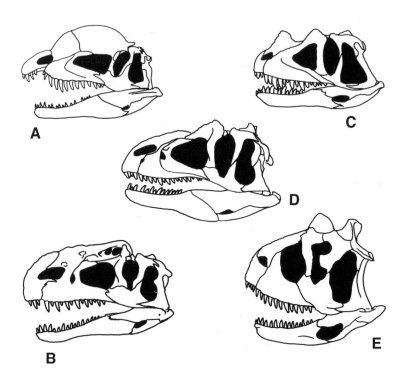

Figure 17.11. Horns, crests, and bumps on theropod skulls (not to scale). (A) *Dilophosaurus;* (B) *Monolophosaurus;* (C) *Ceratosaurus;* (D) *Allosaurus;* (E) *Carnotaurus.* Redrawn from Gilmore 1920; Welles 1984; Paul 1988; Bonaparte et al. 1990; and Zhao and Currie 1993.

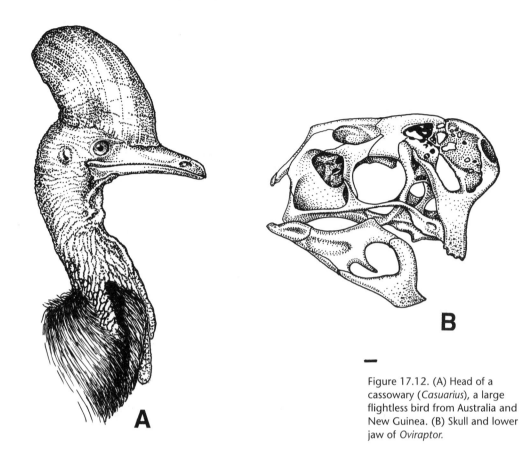

Figure 17.12. (A) Head of a cassowary (*Casuarius*), a large flightless bird from Australia and New Guinea. (B) Skull and lower jaw of *Oviraptor*.

have been required to cool the animals off. A sauropod (*Rebbachisaurus*) and an ornithischian dinosaur (*Ouranosaurus*) that lived in the same region as *Spinosaurus* also had "sails," which gives some credence to this theory. A third possible function is visual identity. Larger individuals would have had relatively higher "sails" for either attracting potential mates or scaring off potential rivals.

Legs, Feet, and Running

It has already been noted that theropods had a tendency to become relatively faster over time, and the legs of Cretaceous species are generally longer. They are also proportioned differently, with the bones of the lower leg and metatarsus being more elongate. Increased speed was also reflected in foot structure, with the development of a more elastic, flexible metatarsus for reducing and distributing the stresses caused by impact of the foot with the ground when the animal was running (Holtz 1994b). The lower leg bones (tibia and fibula) of an ornithomimid are longer than the upper leg bone (femur), and the metatarsus is elongate. These proportions, plus structural changes, suggest that ornithomimids were the fastest dinosaurs, and they are often compared with living ostriches, which can run up to 80 kilometers per hour. Mechanically, there is nothing in ornithomimid skeletons to prevent them from having attained similar speeds. Trackways can be used to give estimates of speeds that animals were traveling when

Theropods   223

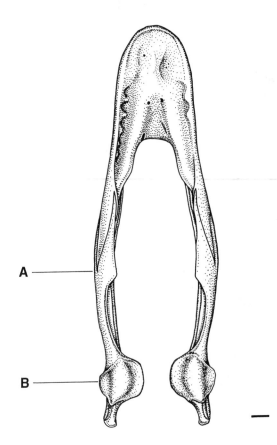

Figure 17.13. Lower jaws of the oviraptorosaur *Caenagnathus;* scale bar is 1 cm. Line A shows the outward bowing along the midpoint of the length of the jaw, and line B indicates the jaw articulation. Redrawn from Currie et al. 1993.

A ——

B ——

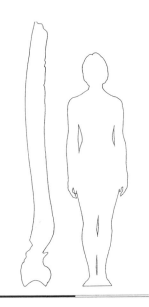

Figure 17.14. Vertebra of *Spinosaurus,* with a human figure for comparison. Scale bar is 1 m. Vertebra redrawn from Stromer 1915.

they left their footprints (Alexander 1976). Although trackways of running ornithomimids are not known, theropods from Texas were calculated to have run across a muddy surface at speeds of up to 40 kilometers per hour (Farlow 1981). This is unlikely to have been the highest speed that theropods could run, but it presently is the highest confirmable speed.

An unusual foot structure was first described for *Deinonychus* (Fig. 17.15) from the Lower Cretaceous rocks of Montana. The first (innermost) toe of a theropod was raised off the ground as a "dew claw," and normally the next three toes carried the weight of the animal. In dromaeosaurids, troodontids, and possibly *Noasaurus* (from Argentina), the second toe was also raised off the ground. The claw on this toe had become larger, more recurved, and more sharply pointed, and the joints gave it a large range of movement reminiscent of the retractable claws of cats. The second toe had apparently become an offensive weapon, and the other two toes carried the weight of the animal. There are some fundamental differences in the skeletons of dromaeosaurids and troodontids that suggest that these theropods were not closely related, and that in fact the specialized raptorial claw had evolved independently in the two lineages. Supporting this idea is the fact that two modern South American bird species, seriemas, have independently developed a strongly raptorial claw on the second toe, which is usually held off the ground.

224   *Philip J. Currie*

Ecology

Theropods had a universal distribution, and specimens have been found from the North Slope of Alaska to Antarctica. Nevertheless, the ecological preferences of theropods are poorly understood, mostly because of the rarity of specimens in the vast majority of the species.

Wherever herbivorous dinosaurs are found, theropods are also recovered. Larger theropods are always found associated with the larger herbivores. The extremes of depositional environments were the lush coastal lowlands of Late Cretaceous western North America and the deserts of central Asia. The Djadokhta Formation (Upper Cretaceous) of Mongolia and equivalent beds in China represent dry, stressful environments suitable for relatively few species of dinosaurs. Although hundreds of skeletons have been collected at these sites, the dinosaurian diversity is low, and none of the animals was longer than 4 meters. With the exception of a few isolated teeth and bones that appear to have been washed in from other environments, tyrannosaurids are not found in the Djadokhtan beds. Small theropods, on the other hand, are well represented in these stressed environments by dromaeosaurids, oviraptorids, and troodontids.

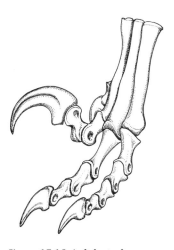

Figure 17.15. Left foot of *Deinonychus* showing raptorial claw on digit II. Redrawn from Ostrom 1969.

Theropod genera were less restrained than ornithischian genera by climate and vegetation, and seem to have had wider geographic ranges, as in modern Carnivora (Farlow 1993). For example, Dinosaur Provincial Park and Devil's Coulee represent two synchronous but distinct habitats separated by about 300 kilometers in present-day Alberta. There appears to have been more rainfall at Dinosaur Park 75 million years ago, which produced a well-watered, well-vegetated ecosystem. Devil's Coulee was farther from the coastline, may have been in a rain shadow, and was undoubtedly drier on at least a seasonal basis. In spite of the close proximity of the two sites, there are significant differences in the composition of the populations of herbivores. One of the most obvious contrasts is the presence of the hadrosaurs *Corythosaurus* and *Lambeosaurus* in Dinosaur Provincial Park, whereas the most common herbivore of Devil's Coulee was *Hypacrosaurus,* a hadrosaur that was once thought to have been found only in younger rocks. The recovery and analysis of isolated teeth also suggests that ankylosaurs were much more common in the drier region. Food and climate controlled the distribution of hadrosaurs, and probably most other herbivores. But evidence suggests that the carnivores of Dinosaur Park and Devil's Coulee are the same taxa—*Gorgosaurus libratus, Saurornitholestes langstoni,* and *Troodon formosus* being the most common forms.

Theropods were greatly outnumbered by herbivorous dinosaurs. Comparing the number of articulated skeletons of hadrosaurs, ceratopsians, and ankylosaurs to tyrannosaurids in Dinosaur Provincial Park produces a ratio of 10:1 (sample size is 148 excavated skeletons). This ratio increases to 11:1 when uncollected articulated skeletons are added into the census (boosting the sample size to 207). Comparing the number of isolated identified bones and teeth collected for the same animals in the same place produces a lower figure (7:1 for a sample size of more than 12,000 catalogued specimens in the collections of the Royal Tyrrell Museum of Palaeontology, although there is an unquestionable collecting bias favoring theropods in this sample). Other ways to estimate the number of theropod individuals range from 5:1 to 511:1 (Farlow 1993).

Small theropods were more common than tyrannosaurids as isolated bones and teeth, but rarer as articulated skeletons. It is more difficult to determine how large their populations may have been in relation to prey species because they would have eaten a wide range of non-dinosaurian animals, small species of herbivorous dinosaurs, and the young of large herbivorous species. Ornithomimids may also have been omnivorous, which further complicates such calculations. Nevertheless, the catalogued collections suggest that theropods overall made up less than 20 percent of the dinosaur fauna at Dinosaur Provincial Park. Even though theropods were greatly outnumbered by herbivorous dinosaurs, they display relatively greater diversity. Sixteen of the thirty-five species of dinosaurs from this site are theropods, representing seven of the thirteen families. This suggests that they were more specific in partitioning their food resources. This is reflected by the interspecific variability of theropod jaws and teeth, and the great range in size.

It is possible that immature large theropods competed with small species of theropods, and this could explain the virtual absence of small theropods at most known dinosaur localities. However, it is also possible that in some large species the juveniles associated with mature animals in packs or family groups. In this scenario, large theropods of all ages would have eaten prey killed by the more mature individuals in the pack, and thereby the younger individuals would not have been competing directly with the small species of theropods. There is some evidence to suggest that tyrannosaurids maintained family groups, so it is not surprising that the greatest diversity of small theropods is found in paleoenvironments where tyrannosaurids were the large predators.

Behavior

Insight into theropod behavior is provided by a few rather exceptional finds. One has to be careful how this information is applied to other theropods, however, because the behavior of carnivorous dinosaurs was unquestionably as diverse as that of modern mammalian carnivores.

At the Ghost Ranch site in New Mexico, larger specimens of *Coelophysis* have been recovered with smaller specimens inside the rib cage. Although it is possible that these represent unborn juveniles, the evidence suggests that the juveniles were eaten by the adults. Cannibalism is widespread in modern reptiles, birds, and mammals, so it is not surprising that it was practiced by at least some theropods.

There has been a long-standing debate concerning whether large theropods were active hunters or scavengers. Much of this discussion has focused on *Tyrannosaurus rex* because it is the largest and most famous theropod known. One tyrannosaurid even bears the name *Albertosaurus sarcophagus* ("corpse-eating reptile from Alberta") because the paleontologist who described it could not imagine its being an active hunter.

The stout teeth of tyrannosaurids gave them the capacity to bite deeply, even allowing them to bite into bone (Erickson and Olson 1996), but this would have been useful regardless of whether they were active predators or scavengers. In any case, the hunter/scavenger debate is somewhat meaningless in that most carnivores behave in both capacities. Hyenas, which are among the most efficient scavengers known, hunt and kill their own prey whenever possible, whereas even the big cats will scavenge carcasses that they find by chance. Theropods would also have been opportunists, but it is unlikely that they would have found enough dead animals to meet their dietary needs.

226 *Philip J. Currie*

There are many other arguments to suggest that large theropods were hunters. The legs of *Tyrannosaurus*, for example, have the length and proportions expected of an animal that was faster than the assumed prey (hadrosaurs and ceratopsians), and these features are even better developed in the juveniles, which were clearly fast, agile animals. The forward-facing eyes may have provided stereoscopic vision, a characteristic that is more meaningful for a hunter than a scavenger. The disproportionately long teeth would have been useful for killing prey, but it is hard to imagine why they would have had a selective advantage in a scavenger. Assuming that the evidence for pack behavior is correct (see below), *Tyrannosaurus* could also have improved its chances of successful food acquisition by cooperative hunting strategies.

In the Upper Jurassic beds of Colorado, sauropod bones have been discovered with bite marks from large theropods. The fact that some of these damaged bones rehealed shows that the intended prey survived the attacks. This in turn demonstrates that large theropods did indeed attack living animals. A Lower Cretaceous trackway site in Texas also seems to document an incident in which a large theropod was following a sauropod (Farlow 1987). If Jurassic and Early Cretaceous large theropods were active hunters, arguments suggesting that the more sophisticated tyrannosaurids were strictly scavengers seem unfounded.

There is at least one example of a small theropod that attacked a herbivore while it was still alive during Late Cretaceous times in Mongolia. Perhaps the most remarkable dinosaur discovery ever made, a *Velociraptor* was found lying on its side next to a *Protoceratops* (see Chin, chap. 26 of this volume). Both specimens are virtually complete, with the protoceratopsian lying upright on its stomach. It is now known that these animals were living in a desert environment with active sand dunes. The hind feet of the theropod are in position to rake the flanks of the *Protoceratops*, and the left hand is clasping the back of the herbivore's skull. Unfortunately for the theropod, its right arm is in the jaws of the intended prey. A possible scenario is that the *Protoceratops* was hiding behind a sand dune in a sandstorm when it was attacked by the *Velociraptor*. It died from its injuries, but not before locking its jaws on the arm of the predator. Under normal circumstances, the theropod probably would have been able to free itself eventually. However, in this case it was not able to do so before being buried by sand carried by wind over the top of the dune, and it suffocated.

Rocks in the same region produced another remarkable specimen in 1923. A skeleton of *Oviraptor* was found next to a nest of what were assumed to be *Protoceratops* eggs. Because of the peculiar jaw apparatus of the theropod and its proximity to the nest, it was concluded that the *Oviraptor* was pillaging the nest when it was overcome and buried during a sandstorm—which is why *Oviraptor* was assigned a name meaning "egg stealer." It is hard to imagine why the theropod did not simply leave the nest when the sand became too deep. Recent expeditions turned up more nests of the same kind of eggs, in some of which an *Oviraptor* skeleton was preserved sitting on top of the nest (Norell et al. 1995; Dong and Currie 1996). It now appears likely that the theropods were protecting their own nests, rather than pillaging the nests of other species. The identification of the eggs as oviraptorid was confirmed when a joint American Museum of Natural History and Mongolian expedition discovered an egg with an oviraptorid embryo in it (Norell et al. 1994). Although this reopens the question of what oviraptorids ate, it does suggest that some small theropods tended their nests, possibly incubating, or at least guarding, the eggs

Theropods 227

(Norell et al. 1995; Dong and Currie 1996; Geist and Jones 1996a, 1996b; Norell and Clark 1996).

At Ghost Ranch in New Mexico, there is a remarkable bonebed that contains full and partial skeletons of at least 1,000 individuals of *Coelophysis* (Schwartz and Gillette 1994). The recovery of so many individuals from a single site strongly suggests that this dinosaur was gregarious. A similar accumulation of skeletons in Zimbabwe of the closely related *Syntarsus* supports this conclusion. A Lower Jurassic trackway site at Holyoke (Massachusetts) shows evidence of twenty theropods moving as a group. It is hard to imagine why so many Late Triassic and Early Jurassic sites suggest that theropods collected into large groups. It may have been that such large packs formed only for short periods of time, possibly for breeding or migration. By Late Jurassic times, this kind of behavior may have changed, because with one possible exception, there is no further evidence of such large groupings.

At the Late Jurassic Cleveland-Lloyd Quarry in Utah, the remains of at least forty-four individuals of *Allosaurus* were found in a bonebed. Evidence at this site suggests that the accumulation of bones occurred over a relatively long period of time, and that the site may have been a predator trap, like Rancho La Brea in Los Angeles. In predator traps, herbivores become mired in quicksand (the possible source of the sediments at Cleveland-Lloyd), tar (Rancho La Brea), or mud. The death cries and struggles of the herbivores attract the attention of carnivores, which in turn become trapped. Their meat rebaits the trap and brings in more predators/scavengers. The net result is that more carnivores are trapped than herbivores. Assuming that Cleveland-Lloyd is a predator trap, it provides no evidence of gregarious behavior in *Allosaurus*.

Lower Cretaceous footprint sites in British Columbia suggest that small theropods moved in packs of a half-dozen or so individuals, while large theropods moved in smaller groups of two or three animals. Sites of similar age in Montana produced the remains of at least three *Deinonychus* associated with a single broken-up skeleton of the herbivorous *Tenontosaurus* (Maxwell and Ostrom 1995). It was suggested that the *Deinonychus* were part of a pack that had attacked the much larger *Tenontosaurus,* and were casualties of the ensuing fight. Evidence favoring pack behavior in tyrannosaurids is not as strong. A large, well-preserved *Tyrannosaurus* skeleton found in 1990 in South Dakota by the Black Hills Institute was associated with the remains of several smaller specimens of *Tyrannosaurus,* suggesting the possibility of a family grouping. Because mature tyrannosaurids hunted animals the same size as or smaller than themselves, there may not have been any need for packs larger than the family units. Allosaurids, on the other hand, may have had to collect into larger packs to hunt and kill the much larger sauropods (Farlow 1976). The size of theropod packs may also have depended on the openness of the environments. One is more likely to find large packs of carnivores today in open environments such as the plains of western North America or the African veldt, than in heavily forested areas.

Migration

A large tyrannosaurid, probably *Albertosaurus,* is known from the North Slope of Alaska and from one of the Canadian Arctic islands. The evidence is based on teeth, but clearly shows that large theropods lived within the Arctic Circle during Late Cretaceous times. Bonebeds document the presence of herds of hadrosaurs in the same region, and it has been suggested that these animals migrated into the Arctic to take advantage of the high plant productivity during the summer months. In the winter these herds may have pushed south again to avoid the twenty-four-hour nights of mid-winter, when all plants would have been dormant. If the hadrosaurs were indeed on a north-south migratory pattern, it is conceivable that tyrannosaurids were following the herds, picking off the young, weak, sick, and old individuals. Unfortunately, the idea of large theropods migrating is only speculation.

Evolution and Systematics

Diversity

As previously discussed, theropods tend to be rare but diverse at most localities. The Ghost Ranch *Coelophysis* bonebed and the *Allosaurus* domination of the Cleveland-Lloyd Quarry are unusual occurrences, because theropods normally constitute less than 20 percent of the fossils recovered at any site. Of the approximately 300 genera of dinosaurs presently recognized as valid, about 40 percent are theropods. Almost 50 percent of the recognized families of dinosaurs are theropods, which compares well with mammals, almost half of the recognized families of which include carnivores, insectivores, piscivores, and/or omnivores. There are preservational, collecting, and research biases that have not been accounted for in the dinosaur figures, but refinement of the calculations is unlikely to significantly alter the conclusion that theropod diversity is high, even though population counts would have been low. Because of the problem of greater complexity with less information, it is not surprising that theropod interrelationships are not as well understood as the evolution of most taxa of herbivorous dinosaurs, and that theropod taxonomy is more volatile and susceptible to change.

Classification

For many years, theropods were divided into two lineages. Carnosauria included all of the large-bodied theropods, whereas Coelurosauria included all of the small ones. This simple classification did not reflect the true relationships of theropod taxa, and is no longer accepted by anyone who works on these animals. Several alternative classifications have been proposed (Gauthier 1986; Holtz 1994a; Sereno et al. 1994, 1996), but none is universally accepted by theropod researchers.

Theropods form a monophyletic group (they share specialized characteristics that were inherited from a common ancestor), within which *Herrerasaurus* is the most primitive theropod presently known. By the end of the Triassic, more characteristic forms were *Coelophysis* and *Syntarsus*. *Dilophosaurus* was a related genus that had attained a relatively large size by Early Jurassic times. These three animals have been united with *Ceratosaurus* of the Late Jurassic into the Ceratosauria, although this is an unfortunate union that many paleontologists disagree with.

The remaining theropods are usually classified as the Tetanurae, whose distinctive characters include the presence of a large opening (maxillary fenestra) in front of the antorbital fenestra, and extensive contact between the first and second bones (metacarpals) in the flat of the hand. Other characters that were used to set up the Tetanurae, such as the loss of the fourth digit and the presence of an obturator process on the ischium, are not as universal as they were once thought to have been.

Carnosaurs appeared in Middle Jurassic times and persisted until the end of the Cretaceous. The large theropods include a diverse assemblage of families (Abelisauridae, Megalosauridae, Sinraptoridae, Allosauridae, Carcharodontosauridae) that are nevertheless unified by a suite of shared unique characters (synapomorphies), including the universal presence of neck vertebrae that have a ball joint on the front and a socket on the back. Tyrannosaurids, which have neck vertebrae that are almost flat on both ends, have traditionally been included in the Carnosauria, with the notable exception of a classification done by Huene (1926). A recent cladistic analysis (Holtz 1994a) confirms that tyrannosaurids are more closely related to small theropods of the Cretaceous, and therefore do not belong in the Carnosauria.

Gauthier (1986) redefined the Coelurosauria to include ornithomimids and a new clade he referred to as the Maniraptora. As already stated, tyrannosaurids should also be classified as coelurosaurs. Maniraptorans, which include birds, are characterized by a pulley-like wrist joint (Fig. 17.16) that allowed the hand to be folded back against the body. It is possible that this characteristic was secondarily reduced or lost in ornitho-mimids and tyrannosaurids, and that Coelurosauria and Maniraptora are synonymous. Maniraptorans are a diverse assemblage made up of oviraptorosaurs (including caenagnathids and oviraptorids), dromaeosaurids, troodontids, therizinosaurs, and birds. Elmisauridae is probably a junior synonym of Caenagnathidae, and the names Therizinosauroidea and Segnosauria probably refer to the same animals.

The Origin of Birds

There has been a lot of speculation and controversy about the origin of birds, but the strongest evidence suggests that birds are the direct descendants of theropod dinosaurs. Alternative theories (Chatterjee 1991; Martin 1991; Feduccia 1996) suggesting that birds were derived from ornithosuchids or crocodylians (some workers prefer the spelling "crocodilians") are not supported by the presence of uniquely ornithosuchid or crocodylian characters in birds, whereas there are more than 120 characters that are found only in dinosaurs and birds.

It should be pointed out that there are relatively few characters that distinguish the earliest birds from theropods. This is why two specimens of *Archaeopteryx* that do not include feather impressions were initially identified as the theropod *Compsognathus*. From the other side of the problem, dinosaurs such as *Caenagnathus* and *Avimimus* were originally thought to have been birds. Pneumatic, hollow bones, endothermy, and even feathers are a few characters that were probably inherited by birds from their theropod ancestors. The only substantial character to define birds is their ability to fly.

Within the Theropoda, there has been considerable debate about which families are most closely tied to the origin of birds. The most bird-like theropods known are mostly from Upper Cretaceous rocks (see, for instance, Novas and Puerta 1997), although this simply reflects preservational biases. Small theropods are very poorly known in older rocks, and most of the Late Cretaceous families have clearly had long independent histories. Avimimids, oviraptorids, and ornithomimids were the most bird-

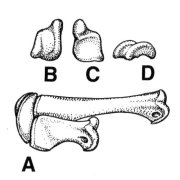

Figure 17.16. Portion of the hand and wrist of *Deinonychus*. Scale bar is 1 cm. (A) The crescent moon–shaped (semilunate) wristbone, the radiale (the leftmost bone in the drawing), as it fits against two hand (metacarpal) bones, the short and stout metacarpal of digit I and the longer metacarpal of digit II. Note the pulley-shaped keel on the side of the radiale away from the hand bones. The top set of drawings shows the radiale in proximal (B), distal (C), and ventral (D) views. Redrawn from Ostrom 1969.

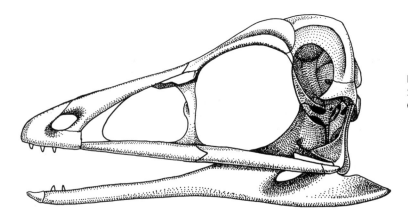

Figure 17.17. Skull of *Protoavis*. Scale bar is 1 cm. Redrawn from Chatterjee 1991.

like in appearance, although this is largely attributable to the convergent evolution of toothless, beaked skulls. Dromaeosaurids and troodontids, sometimes incorrectly included in a single taxon referred to as the Deinonychosauria, are more likely avian ancestral stocks. Both families are represented by well-preserved skeletons from Lower Cretaceous rocks, and Late Jurassic dromaeosaurid and troodontid teeth have also been reported.

In 1986, the discovery of some fossils in Upper Triassic rocks of Texas aroused considerable interest in the paleontological community. *Protoavis* (Fig. 17.17) was described as an early bird (Chatterjee 1991), which supposedly had some avian characters that were more advanced than those of *Archaeopteryx*. If birds were present during the Late Triassic, then theropods are less likely to have been ancestral to birds, and *Archaeopteryx* would have been a relict in the Late Jurassic that was off the main line of bird evolution. With the exception of the braincase, most of the skeleton of *Protoavis* is poorly known and not especially bird-like. Pneumatic characteristics of the *Protoavis* braincase are characteristic of birds, troodontids, some dromaeosaurids, and other theropods, and therefore are not characters found only in birds. The anatomy of newly described birds from the Lower Cretaceous of China (*Sinornis*) and Spain (*Iberomesornis*) confirms that *Archaeopteryx* is close to the main line of bird evolution. In light of present evidence, it is more logical to consider *Protoavis* as a peculiar small theropod.

References

Abler, W. L. 1992. The serrated teeth of tyrannosaurid dinosaurs, and biting structures in other animals. *Paleobiology* 18: 161–183.

Alexander, R. McN. 1976. Estimates of the speeds of dinosaurs. *Nature* 261: 129.

Bonaparte, J. F.; F. E. Novas; and R. A. Coria. 1990. *Carnotaurus sastrei* Bonaparte, the horned, lightly built carnosaur from the Middle Cretaceous of Patagonia. Contributions in Science, Natural History Museum of Los Angeles County, no. 416.

Chatterjee, S. 1991. Cranial anatomy of a new Triassic bird from Texas. *Philosophical Transactions of the Royal Society of London*, Series B, 332: 277–346.

Currie, P. J.; S. J. Godfrey; and L. Nessov. 1993. New caenagnathid (Dinosauria: Theropoda) specimens from the Upper Cretaceous of North America and Asia. *Canadian Journal of Earth Sciences* 30: 2255–2272.

Currie, P. J., and X.-J. Zhao. 1993. A new carnosaur (Dinosauria, Theropoda) from the Jurassic of Xinjiang, People's Republic of China. *Canadian Journal of Earth Sciences* 30: 2037–2081.

Dong Z.-M. and P. J. Currie. 1996. On the discovery of an oviraptorid skeleton on a nest of eggs at Bayan Mandahu, Inner Mongolia, People's Republic of China. *Canadian Journal of Earth Sciences* 33: 631–636.

Erickson, G. M., and K. H. Olson. 1996. Bite marks attributable to *Tyrannosaurus rex:* Preliminary description and implications. *Journal of Vertebrate Paleontology* 16: 175–178.

Farlow, J. O. 1976. Speculations about the diet and foraging behavior of large carnivorous dinosaurs. *American Midland Naturalist* 95: 186–191.

Farlow, J. O. 1981. Estimates of dinosaur speeds from a new trackway site in Texas. *Nature* 294: 747–748.

Farlow, J. O. 1987. *Lower Cretaceous Dinosaur Tracks, Paluxy River Valley, Texas.* Waco, Tex.: South Central Section, Geological Society of America, Baylor University.

Farlow, J. O. 1993. On the rareness of big, fierce animals: Speculations about the body sizes, population densities, and geographic ranges of predatory mammals and large carnivorous dinosaurs. *American Journal of Science* 293-A: 167–199.

Feduccia, A. 1996. *The Origin and Evolution of Birds.* New Haven, Conn.: Yale University Press.

Gauthier, J. 1986. Saurischian monophyly and the origin of birds. In K. Padian (ed.), *The Origin of Birds and the Evolution of Flight,* pp. 1–55. San Francisco: California Academy of Sciences.

Geist, N. R., and T. D. Jones. 1996a. Juvenile skeletal structure and the reproductive habits of dinosaurs. *Science* 272: 712–714.

Geist, N. R., and T. D. Jones. 1996b. Dinosaurs and their youth. *Science* 273: 166–167.

Gilmore, C. W. 1920. Osteology of the carnivorous Dinosauria in the United States National Museum, with special reference to the genera *Antrodemus* (*Allosaurus*) and *Ceratosaurus.* Bulletin 110, United States National Museum (Smithsonian Institution).

Hammer, W. R., and W. J. Hickerson. 1994. A crested theropod dinosaur from Antarctica. *Science* 264: 828–830.

Holtz, T. R. Jr. 1994a. The phylogenetic position of the Tyrannosauridae: Implications for theropod systematics. *Journal of Paleontology* 68: 1100–1117.

Holtz, T. R. Jr. 1994b. The arctometatarsalian pes: An unusual structure of the metatarsus of Cretaceous Theropoda (Dinosauria: Saurischia). *Journal of Vertebrate Paleontology* 14: 480–519.

Huene, F. von. 1926. The carnivorous Saurischia in the Jura and Cretaceous formations principally in Europe. *Museo de La Plata, Revista* 29: 35–167.

Lambe, L. M. 1917. The Cretaceous theropodous dinosaur *Gorgosaurus.* Memoir 100, Canada Department of Mines, Geological Survey.

Madsen, J. H. Jr. 1976. *Allosaurus fragilis:* A revised osteology. Utah Geological and Mineral Survey, Bulletin 109. Salt Lake City: Utah Department of Natural Resources.

Martin, L. D. 1991. Mesozoic birds and the origin of birds. In H. P. Schultze and L. Trueb (eds.), *Origins of the Higher Groups of Tetrapods,* pp. 485–540. Ithaca, N.Y.: Cornell University Press.

Maxwell, W. D., and J. H. Ostrom. 1995. Taphonomy and paleobiological implications of *Tenontosaurus-Deinonychus* associations. *Journal of Vertebrate Paleontology* 15: 707–712.

Norell, M. A., and J. M. Clark. 1996. Dinosaurs and their youth. *Science* 273: 165–166.

Norell, M. A.; J. M. Clark; L. M. Chiappe; and Dashzeveg D. 1995. A nesting dinosaur. *Science* 378: 774–776.

Norell, M. A.; J. M. Clark; Dashzeveg D.; Barsbold R.; L. M. Chiappe; A. R. Davidson; M. C. McKenna; Perle A.; and M. J. Novacek. 1994. A theropod dinosaur embryo and the affinities of the Flaming Cliffs dinosaur eggs. *Science* 266: 779–782.

Novas, F. E. 1993. New information on the systematics and postcranial skeleton of *Herrerasaurus ischigualastensis* (Theropoda: Herrerasauridae) from the Ichigualasto Formation (Upper Triassic) of Argentina. *Journal of Vertebrate Paleontology* 13: 400–423.

Novas, F. E., and P. F. Puerta. 1997. New evidence concerning avian origins from the Late Cretaceous of Patagonia. *Nature* 387: 390–92.

Osborn, H. F. 1916. Skeletal adaptations of *Ornitholestes, Struthiomimus, Tyrannosaurus. Bulletin American Museum of Natural History* 35: 733–771.

Ostrom, J. H. 1969. Osteology of *Deinonychus antirrhopus,* an unusual theropod from the Lower Cretaceous of Montana. Bulletin 30, Peabody Museum of Natural History, Yale University.

Ostrom, P.; S. A. Macko; M. H. Engel; and D. A. Russell. 1993. Assessment of trophic structure of Cretaceous communities based on stable nitrogen isotope analyses. *Geology* 21: 491–494.

Paul, G. S. 1988. *Predatory Dinosaurs of the World: A Complete Illustrated Guide.* New York: Simon and Schuster.

Reig, O. A. 1963. La presencia de dinosaurios saurisquios en los "Estratos de Ischigualasto" (Mesotriásico Superior) de las Provincias de San Juan y La Rioja (República Argentina). *Ameghiniana* 3: 3–20.

Russell, D. A., and Dong Z.-M. 1993. A nearly complete skeleton of a troodontid dinosaur from the Early Cretaceous of the Ordos Basin, Inner Mongolia, China. *Canadian Journal of Earth Sciences* 30: 2163–2173.

Schwartz, H. L., and D. D. Gillette. 1994. Geology and taphonomy of the *Coelophysis* quarry, Upper Triassic Chinle Formation, Ghost Ranch, New Mexico. *Journal of Paleontology* 68: 1118–1130.

Sereno, P. C. 1993. The pectoral girdle and forelimb of the basal theropod *Herrerasaurus ischigualastensis. Journal of Vertebrate Paleontology* 13: 425–450.

Sereno, P. C.; D. B. Dutheil; M. Iarochene; H. C. E. Larsson; G. H. Lyon; P. M. Magwene; C. A. Sidor; D. J. Varricchio; and J. A. Wilson. 1996. Predatory dinosaurs from the Sahara and Late Cretaceous faunal differentiation. *Science* 272: 986–991.

Sereno, P. C.; C. A. Forster; R. R. Rogers; and A. M. Monetta. 1993. Primitive dinosaur skeleton from Argentina and the early evolution of Dinosauria. *Nature* 361: 64–66.

Sereno, P.C., and F. E. Novas. 1993. The skull and neck of the basal theropod *Herrerasaurus ischigualastensis. Journal of Vertebrate Paleontology* 13: 451–476.

Sereno, P. C.; J. A. Wilson; H. C. E. Larsson; D. B. Dutheil; and H.-D. Sues. 1994. Early Cretaceous dinosaurs from the Sahara. *Science* 266: 267–271.

Smith, D. 1992. The type specimen of *Oviraptor philoceratops,* a theropod dinosaur from the Upper Cretaceous of Mongolia. *Neues Jahrbuch für Geologie und Paläontologie Abhandlungen* 186: 365–388.

Stromer, E. 1915. Wirbeltier-Reste der Baharije-Stufe (unterstes Cenoman). Das Original des Theropoden *Spinosaurus aeqyptiacus* nov. gen. nov. spec. *Abhandlungen. Bayerische Akademie der Wissenschaften. Mathematisch-Naturwissenschaftliche Klasse* 28: 1–32.

Welles, S. P. 1984. *Dilophosaurus wetherilli* (Dinosauria, Theropoda) osteology and comparisons. *Palaeontographica A* 185: 85–110.

Witmer, L. M. 1997. The evolution of the antorbital cavity of archosaurs: A study in soft-tissue reconstruction in the fossil record with an analysis of the function of pneumaticity. Society of Vertebrate Paleontology Memoir 3.

Zhao, X.-J., and P. J. Currie. 1993. A large crested theropod from the Jurassic of Xinjiang, People's Republic of China. *Canadian Journal of Earth Sciences* 30: 2027–2036.

Theropods 233

Segnosaurs (Therizinosaurs)

Teresa Maryańska

The segnosaurians are medium-sized to large (3 to 7 meters in length) dinosaurs, characterized by a relatively small skull with straight to flattened, serrated teeth, a massive and long neck, an unusual structure of the opisthopubic pelvis (with the pubic bones directed backward), a hand ending with large claws, and broad and short four-toed feet. As presently known, segnosaurians constitute a rare, exclusively Cretaceous group of Asian saurischians, most probably of theropod affinity. They are represented by a few genera, discovered in Cretaceous sediments in the territory of Mongolia and China. Most segnosaurian specimens are known from disarticulated skeletons. Scientific study of the group, and discussion about its relationship to other dinosaurs, began in the 1970s, with the description of the middle Late Cretaceous form *Segnosaurus galbinensis* from southeastern Mongolia (Perle 1979) and the Late Cretaceous species *Nanshiungosaurus brevispinus* from Guandong, People's Republic of China (Dong 1979). A probable Early Cretaceous segnosaurian species, *Alxasaurus elesitaiensis,* from Alxa (Alashan) Desert, Inner Mongolia, People's Republic of China, was described by Russell and Dong (1993).

Segnosaurian Anatomy

The only known segnosaurian skull is that of *Erlikosaurus* (Fig. 18.1). It is low and long with large, elongated external nasal openings and a toothless

18

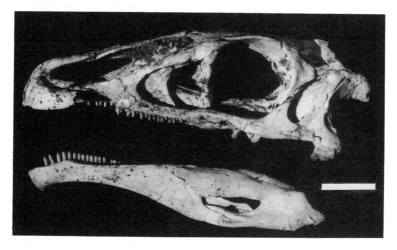

Figure 18.1. Skull and mandible of *Erlikosaurus adrewsi* (holotype: GIN 100/111, Baysheen Tzav, southeastern Mongolia). Scale = 4 cm.

beak. On both sides of the skull, between the nostril cavities and eye sockets, there is a large opening (antorbital fenestra). The hard palate is formed by horizontal processes of the maxillary and premaxillary bones, while more posteriorly, the mouth cavity is highly vaulted. The base of the skull and the ear region are swollen, and the bones there are strongly pneumatized (that is, they contained air chambers). Complete mandibles (lower jaws) are known only in *Erlikosaurus* and *Segnosaurus* (Figs. 18.1, 18.2); the lower jaw of *Alxasaurus* is very fragmentary. The lower jaw of *Erlikosaurus* and *Segnosaurus* is shallow, with the anterior portion curved downward. The rostral-most portion of the lower jaw in *Erlikosaurus* and *Segnosaurus* is toothless. In *Alxasaurus*, however, teeth are still present here.

The numerous teeth in the upper jaw are almost uniform in shape. They are straight, narrow, and slightly flattened laterally, and have serrated edges. The first five mandibular teeth in *Segnosaurus* and *Erlikosaurus* are straight, and larger then the others. The posterior teeth are flattened, with serrated edges, and clearly diminish in size toward the rear.

Figure 18.2. Left mandible of *Segnosaurus galbinensis* (holotype: GIN 100/80, Amtgay, southeastern Mongolia) in lateral view. Scale = 4 cm.

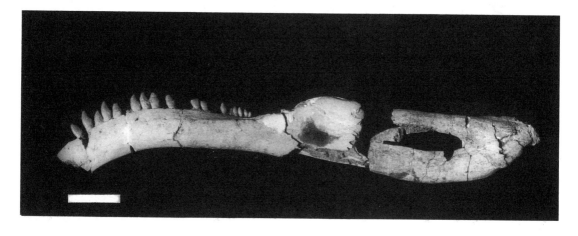

Segnosaurs (Therizinosaurs)   235

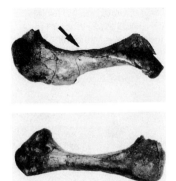

Figure 18.3. Left humerus of *Erlikosaurus andrewsi* in anterior and posterior views. Arrow indicates the position of a sharply pointed boss on the posteromedial surface of the bone. Scale = 4 cm.

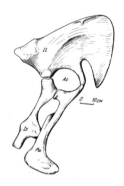

Figure 18.4. Pelvis of *Segnosaurus galbinensis* in right lateral view. After Perle 1979.

In their general shape, the segnosaurian skull, mandible, and teeth are more like those of prosauropods than those of theropods.

An almost complete series of vertebrae is known in *Nanshiungosaurus* (only the tail is missing). In *Erlikosaurus* and *Segnosaurus*, the vertebral column is only fragmentarily preserved. In *Alxasaurus* the column is partly preserved but crushed, but the tail is articulated and nearly complete. The segnosaurian neck and anterior trunk vertebrae are large and strongly pneumatized, and bear low neural spines. The short and broad cervical (neck) ribs are fused to the vertebrae in *Erlikosaurus* and *Nanshiungosaurus,* but not in *Alxasaurus*. The structure of the neck vertebrae is comparable to that of sauropods. The dorsal (trunk) ribs are flat and broad. Six fused sacral (hip) vertebrae are present in *Segnosaurus;* there are five sacrals in *Alxasaurus* and *Nanshiungosaurus*. The tail was moderate in length; judging by those tail vertebrae that are known, its anterior portion had to be massive.

The pectoral girdle is known in *Segnosaurus, Therizinosaurus,* and *Alxasaurus*. The straight and narrow scapula and large coracoid are very similar to those of theropods. The forelimbs are characterized by a massive humerus (arm bone), with strongly expanded ends and a large, bony crest (deltopectoral crest) running along two-thirds of the humeral length in *Therizinosaurus,* more than half the length of the humerus in *Segnosaurus,* but less than half the humeral length in *Erlikosaurus* (Fig. 18.3). The humerus in *Alxasaurus,* as reconstructed by Russell and Dong (1993), is slender, and its deltopectoral crest is not expanded. A characteristic feature of the segnosaurian humerus is the presence of a sharply pointed boss on the posteromedial surface at the midlength of the bone. This boss is present in *Segnosaurus, Erlikosaurus,* and *Therizinosaurus,* but it is not observed in *Alxasaurus*.

An almost complete manus (hand), with metacarpal bones and three digits, is known only in *Therizinosaurus* and *Alxasaurus;* in general appearance it resembles that of theropods. The manual phalanges (finger bones) are short in *Therizinosaurus* but relatively long in *Alxasaurus*. The manual claws in *Therizinosaurus* are very strange: compressed, weakly curved, and very long, exceeding the forearm length; the claw preserved in the holotype of *Therizinosaurus cheloniformis* is about 70 centimeters long! The hand claws of *Segnosaurus* are also compressed but strongly curved, and are relatively shorter and more massively built.

The most striking segnosaurian feature is the structure of the pelvis, which is known in *Segnosaurus* (Fig. 18.4), *Nanshiungosaurus, Enigmosaurus,* and *Alxasaurus*. It is short and firmly fused to the hip vertebrae, and has widely separated ilia (upper hip bones). The ilium is very deep in front of the acetabulum (hip socket) and flares strongly laterally. The postacetabular portion of the ilium is very short and ends in a knob-like protuberance. The two other pelvic bones—the pubis and the ischium—are parallel to each other, and both are directed downward and backward, forming the so-called opisthopubic type of pelvis. In *Alxasaurus,* the deep anterior wings of the ilia flare outward only moderately, and the pelvis is relatively longer and more similar to that in opisthopubic theropods such as *Adasaurus* and *Velociraptor*.

Almost complete hind limbs are known in *Segnosaurus,* and possibly in *Therizinosaurus* (see below). In *Erlikosaurus* and *Alxasaurus,* the hind limbs are fragmentary. The length of the shank is more than 80 percent of the length of the thigh. The tibia is strongly expanded distally. The slender fibula closely presses against the tibia. On both preserved proximal tarsal (ankle) bones, the astragalus bears a tall, laterally curved ascending process, very similar to that in some theropods. The segnosaurian pes (hind foot) is four-toed, short, and broad. The metatarsus (cannon bone) is composed of five elements; metatarsals I–IV are massive, while metatarsal V is vestigial (Fig. 18.5). The toe phalanges are short and robust in the Late Cretaceous forms, but relatively longer in *Alxasaurus*. Toe claws are pointed, recurved, and strongly laterally compressed in *Erlikosaurus,* but large, massive, recurved, and only slightly compressed in *Segnosaurus.* In *Alxasaurus* they are relatively shorter. The short and broad segnosaurian foot is more like that of some prosauropods than that of theropods, but the proximal portion of metatarsal I probably does not articulate with the ankle bones.

Segnosaurian Taxa and Relationships with Other Dinosaurs

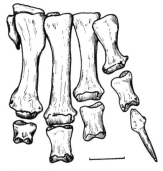

Figure 18.5. Incomplete right pes of *Segnosaurus galbinensis* in dorsal view. After Perle 1979. Scale = 10 cm.

The segnosaurians are represented by only a few genera. Each genus is represented by but a single species. *Segnosaurus galbinensis* (Perle 1979), *Erlikosaurus andrewsi* (Barsbold and Perle 1980), and *Enigmosaurus mongoliensis* (Barsbold 1983) come from the middle Late Cretaceous (?Cenomanian-Turonian [the question mark indicates uncertainty about the age]) Baynshirenskaya svita of southeast Mongolia. *Nanshiungosaurus brevispinus* (Dong 1979) is known from the Late Cretaceous (?Maastrichtian) Nanxiong Formation of Guandong, People's Republic of China. *Therizinosaurus cheloniformis* (Maleev 1954), from the Late Cretaceous (Campanian/Maastrichtian) Nemegt Formation of southern Mongolia, and *Alxasaurus elesitaiensis* (Russell and Dong 1993), from the Early Cretaceous (?Albian) of the Alashan Desert, Inner Mongolia, People's Republic of China, are probably segnosaurians as well.

With two pedal digits and large, recurved claws, described by Dong (1979) as a carnosaurian, *Chilantaisaurus zheziangensis,* from the Late Cretaceous of Zheizjang, People's Republic of China, probably also belongs to the segnosaurians (Barsbold and Maryańska 1990), as does a humerus (American Museum [AMNH] 6368) described by Gilmore (1933) as a tyrannosaurid, *Alectrosaurus olseni,* from the Late Cretaceous of the Iren Dabasu Formation, Inner Mongolia (Mader and Bradley 1989). A frontal bone from the Judith River Formation in Alberta, Canada, attributed by Currie (1987) to *Erlikosaurus,* appears suspect (Barsbold and Maryańska 1990).

Since 1979, when *Segnosaurus* and *Nanshiungosaurus* were described, several authors have discussed the interrelationships of the segnosaurian taxa, and the affinities of this group with other dinosaurs. Perle (1979), who erected a new family, the Segnosauridae, to include *Segnosaurus,* assigned this new family to the Theropoda, while Dong (1979) considered *Nanshiungosaurus* a sauropod. A year later, a new infraorder, the Segnosauria—a separate line of predatory dinosaurs within the suborder Theropoda—was proposed (Barsbold and Perle 1980). In 1983, Barsbold and Perle erected a second family within the Segnosauria, the Enigmosauridae, for *Enigmosaurus.* This family name was synonymized with Segnosauridae by Barsbold and Maryańska (1990). However, the theropod affinity of the Segnosauria, suggested in some papers by Perle and Barsbold, was weakened by the same authors in other papers. For example, Perle (1981) argued that representatives of the Segnosauria do not show characters that unequivocally justify their assignment to the Theropoda or any other suborder of dinosaurs.

Segnosaurs (Therizinosaurs)   237

A different interpretation of segnosaurian relationships was offered by Paul (1984). According to him, segnosaurian dinosaurs do not display any theropod-like characters, and are instead derived prosauropods that show some ornithischian characters and adaptations (especially to a herbivorous diet); they thus constitute late representatives of the prosauropod-ornithischian transition.

According to Gauthier (1986), segnosaurians are remarkably specialized, and most probably are related to broad-footed sauropodomorphs. Barsbold and Maryańska (1990) considered the Segnosauria to be a herbivorous group of the Saurischia, probably more closely related to sauropodomorphs than to theropods. Dong (1992) recognized segnosaurians as a new dinosaur order, Segnosaurischia [*sic*].

The systematic position of the Segnosauria has generated a considerable diversity of opinions because of the unusual suite of characters found in Late Cretaceous representatives of this group. The situation has changed since the description of the Early Cretaceous *Alxasaurus elesitaiensis* (Russell and Dong 1993). This species is represented by an almost complete, but partially crushed and disarticulated, postcranial skeleton, and is regarded by Russell and Dong as the most primitive known member of the segnosaurian dinosaurs. Some of the highly derived and specialized characters of the Late Cretaceous segnosaurians are not yet present in *Alxasaurus,* and others are less well developed. Consequently Russell and Dong concluded that *Alxasaurus* is undoubtedly a theropod, whose morphology puts it close to the ancestry of the Late Cretaceous therizinosaurs (= segnosaurs). They therefore considered the question of theropod affinities of the segnosaurians as resolved.

The assignment of *Therizinosaurus* to the segnosaurian dinosaurs requires some comments, however. *T. cheloniformis* was established and described by Maleev (1954). The first, very fragmentary remains (large claws and rib) of this species described were thought by Maleev to belong to a new family (Therizinosauridae) of turtles. Later, Barsbold (1976) described new material of *T. cheloniformis* (an articulated pectoral girdle and forelimb with claws) from the Late Cretaceous of Khermeen Tzav, southern Mongolia, and transferred the Therizinosauridae to the Theropoda. Thus *Therizinosaurus* is known only from the forelimb with enormously large claws. Perle (1982) described a segnosaurian-like hind limb as probably belonging to a species of *Therizinosaurus,* and included the family Therizinosauridae within the Segnosauria. The material described by Perle was found in the same horizon and locality as—and close to—the site where the forelimb of *T. cheloniformis* described by Barsbold (1976) was found.

Russell and Dong (1993) accepted Perle's interpretation; according to them, the morphology of *Alxasaurus* confirms a very close relationship of *Therizinosaurus* to other segnosaurians. In consequence, Russell and Dong considered the family name Segnosauridae (Perle 1979) to be a junior synonym of Therizinosauridae (Maleev 1954). They proposed a new classification of this group of theropod dinosaurs: Superfamily Therizinosauroidea, with two families, Alxasauridae for *Alxasaurus* and Therizinosauridae (= Segnosauridae) for the remaining genera. Thus, according to Russell and Dong, the discovery and description of *Alxasaurus* not only resolved the problem of affinity of segnosaurians with theropods, but also "transformed" segnosaurians into therizinosauroids, a tetanuran theropod group more closely related, in Russell's and Dong's opinion, to such theropods as oviraptorosaurs and troodontids than to tyrannosaurids and dromeosaurids. Clark et al. (1994) redescribed the skull anatomy of

Erlikosaurus mongoliensis. The skull features of this species not only support the relationship of therizinosauroids with theropods, but also corroborate the tetanuran affinity of therizinosauroids postulated by Russell and Dong. However, detailed phylogenetic relationships of the group still need examination.

Segnosaurian Paleobiology

Information about segnosaurian paleobiology is scant. The remains of segnosaurians were discovered in sediments of rivers or lakes. According to Barsbold and Perle (1980), the segnosaurian mode of life was different (not very active) from that of other theropods, and these authors suggested that segnosaurians may have been amphibious fish-eaters. Paul (1984) was the first to propose that these dinosaurs were herbivores. Several features speak in favor of this interpretation of segnosaurian diets: the structure of the teeth (although the teeth seem poorly constructed for dealing with resistant plant material), the toothless condition of the anterior portion of the snout (which was probably covered by a horny beak), the partial development of lateral dentary shelves (suggesting the presence of incipient cheeks). Paul's thesis was accepted by Barsbold and Maryańska (1990), Russell and Russell (1993), and Russell and Dong (1993).

Judging by their skeletal morphology, segnosaurians were slow-moving animals with a bulky trunk (another feature consistent with a herbivorous diet), a long and far-reaching neck, highly mobile forelimbs with large hands, and massive feet. Dong (1992) reconstructed *Nanshiungosaurus* as a quadrupedal animal. According to Russell and Russell (1993) and

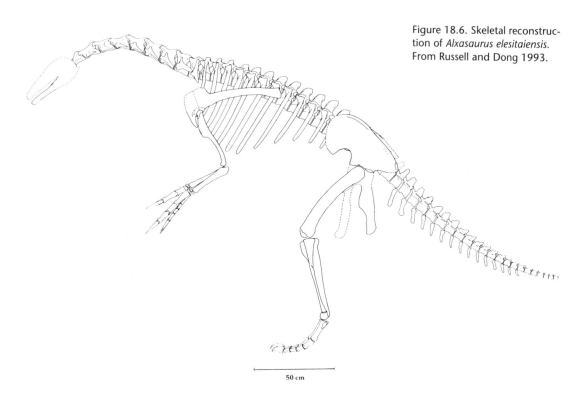

Figure 18.6. Skeletal reconstruction of *Alxasaurus elesitaiensis*. From Russell and Dong 1993.

50 cm

Segnosaurs (Therizinosaurs)   239

Russell and Dong (1993), however, the morphology of the dorsal vertebrae and the natural curvature of the hip and trunk vertebrae (observed in *Nanshiungosaurus* and *Alxasaurus*) suggest that the backbone rose up in front of the hips, and that the animal was able to hold its body in a bipedal stance (Fig. 18.6). The forelimbs and the massive anterior portion of the tail could serve as props when the animal was walking or sitting. The capacity to stand and walk in bipedal position, the long movable neck, and the construction of the forelimbs, which probably acted as an efficient grasping organ, speak in favor of the hypothesis that segnosaurians were high-browsing animals (Russell and Russell 1993).

During approximately fifteen years of study of segnosaurians, some very controversial opinions as to their affinities with other saurischians have been expressed. Perle's (1979) initial statement that segnosaurians represent an aberrant line of theropod dinosaurs is confirmed, according to Russell and Dong (1993) and Clark et al. (1994), by the postcranial anatomy of their most primitive representative, *Alxasaurus,* and skull features of *Erlikosaurus.* However, detecting interrelationships between different segnosaurian genera, and resolving to which theropod group they are most closely related, requires further discoveries and more studies; the presently known, very fragmentary and mostly uncomparable, skeletons of the Late Cretaceous forms are not sufficient for resolving this problem. My field experience in this matter makes me optimistic. At present, *Erlikosaurus* is known from one specimen—the holotype of *E. andrewsi.* However, in 1989, during a two-day stay at some Mongolian localities, I saw in the rocks the remains of two skeletons of this species. This may mean that segnosaurians were not as rare as we think, giving me hope that more complete fossils of these enigmatic dinosaurs will ultimately be found.

References

Barsbold R. 1976. New data on *Therizinosaurus* (Therizinosauridae, Theropoda). *Soviet-Mongolian Paleontological Expedition, Transactions* 3: 76–92. (In Russian.)

Barsbold R. 1983. Carnivorous dinosaurs from the Cretaceous of Mongolia. *Soviet-Mongolian Paleontological Expedition, Transactions* 19: 1–117. (In Russian.)

Barsbold R. and T. Maryańska. 1990. Segnosauria. In D. B. Weishampel, P. Dodson, and H. Osmólska (eds.), *The Dinosauria,* pp. 408–415. Berkeley: University of California Press.

Barsbold R. and Perle A. 1980. Segnosauria, a new infraorder of carnivorous dinosaurs. *Acta Palaeontologica Polonica* 25: 185–195.

Clark, J. M.; Perle A.; and M. A. Norell 1994. The skull of *Erlikosaurus andrewsi,* a Late Cretaceous "Segnosaur" (Theropoda: Therizinosauridae) from Mongolia. *American Museum Novitates* 3115: 1–39.

Currie, P. J. 1987. Theropods of the Judith River Formation of Dinosaur Provincial Park, Alberta. Fourth Symposium on Mesozoic Terrestial Ecosystems, Short Papers. *Occasional Paper of the Tyrrell Museum of Palaeontology* 3: 52–60.

Dong Z. 1979. The Cretaceous dinosaur fossils in southern China. In *Mesozoic and Cenozoic Redbeds in Southern China,* pp. 342–350. Beijing: Science Press. (In Chinese.)

Dong Z. 1992. *Dinosaurian Faunas of China.* Beijing: Ocean Press, and Berlin: Springer Verlag.

Gauthier, J. 1986. Saurischian monophyly and the origin of birds. *Memoirs of the California Academy of Scences* 8: 1–55.

Gilmore, C. W. 1933. On the dinosaurian fauna of the Iren Dabasu Formation. *Bulletin of the American Museum of Natural History* 67: 23–95.

Mader, B. J., and R. L. Bradley. 1989. A redescription and revised diagnosis of the syntypes of the Mongolian tyrannosaur *Alectrosaurus olseni*. *Journal of Vertebrate Paleontology* 9: 41–55.

Maleev, E. A. 1954. New turtle-like reptile in Mongolia. *Priroda* 3: 106–108. (In Russian.)

Paul, G. S. 1984. The segnosaurian dinosaurs: Relics of the prosauropod-ornithischian transition? *Journal of Vertebrate Palaeontology* 4: 507–515.

Perle A. 1979. Segnosauridae—a new family of theropods from the Late Cretaceous of Mongolia. *Soviet-Mongolian Palaeontological Expedition, Transactions* 8: 45–55. (In Russian.)

Perle A. 1981. A new segnosaurid from the Upper Cretaceous of Mongolia. *Soviet-Mongolian palaeontological Expedition, Transactions* 15: 50–59. (In Russian.)

Perle A. 1982. On a new finding of hind limb of *Therizinosaurus* sp. from the Late Cretaceous of Mongolia. *Problems of Mongolian Geology* 5: 94–98. (In Russian.)

Russell, D. A., and Dong Z. 1993. The affinities of a new theropod from the Alxa Desert, Inner Mongolia, People's Republic of China. *Canadian Journal of Earth Sciences* 30: 2107–2127.

Russell, D. A., and D. E. Russell. 1993. Mammal-dinosaur convergence. *National Geographic Research and Exploration* 9: 70–79.

Segnosaurs (Therizinosaurs)   241

Prosauropods

Jacques VanHeerden

19

During the Late Triassic Age, some 230 million years ago, a group of dinosaurs with long necks, small heads, and large bodies made their appearance on Earth. They are thought to have been (mainly) herbivorous, but they never attained the huge size or the specialized adaptations of later herbivorous dinosaurs. They disappeared toward the end of the Early Jurassic Age.

In 1920 the great German paleontologist Friedrich von Huene proposed the name Prosauropoda for this group. Huene (1914, 1920, 1932, 1956) grouped the sauropods, prosauropods, and carnosaurs together, with the coelurosaurs as a sister group to that assemblage. This classification came under attack in the 1960s, with important contributions coming from Colbert (1964), Charig et al. (1965), and Bonaparte (1969). The real "red herring" in this instance (first recognized by Walker [1964]) was the *loose* association of carnosaur-like teeth with prosauropod-like limb bones. With the discovery of new forms and the reassessment of previously described Triassic saurischians, it became clear that Huene's classification was no longer tenable.

The earliest known prosauropods are already too specialized to be the ancestors of the sauropods. It follows that the name "prosauropod," which means "precursor of the sauropods," is a misnomer. But even though alternative names have been proposed (such as Edwin Colbert's Plateosauria [1964]), the name Prosauropoda has remained.

In a sense the prosauropods are the stepchildren of dinosaur research. Perhaps this is because they are an early, and quite possibly sterile, branch of dinosaur evolution. It may also be because there are relatively few specimens, of which only a handful are adequately preserved. But there is no doubt that they have never captured the imagination of either the layperson or the specialist as the Jurassic and Cretaceous dinosaurs have done. It is therefore sobering to consider that they spanned the full first quarter of the 160 million years that dinosaurs existed on Earth!

At the time when prosauropods first appeared (the Late Triassic), all the continents were still united in a single supercontinent known as Pangaea. By the end of the Triassic Period, Pangaea had started to break up into a northern and a southern supercontinent, known as Laurasia and Gondwana, respectively. The oldest known prosauropod genera are from central Gondwana, from what is now called Africa. It would therefore appear that the prosauropods originated in this part of the world and then spread to other continents.

Prosauropods were among the first dinosaurs to be described. *Thecodontosaurus* was described in 1836, not long after the discovery of *Megalosaurus* and *Iguanodon* in Britain. In the next year, 1837, prosauropod remains from the Norian Age of Germany were described as *Plateosaurus* by Hermann von Meyer. Today this is the best-known of all prosauropods, with more than a hundred fragmentary to complete skeletons known, as well as several skulls. When Meyer first described *Plateosaurus*, very little was known about the range of diversity among individuals of a population, and the very anatomy of dinosaurs was still poorly understood.

About this time, Owen (1842) coined the term *dinosaur* (which means "fearfully great lizard") to describe these large reptiles. Later some other plateosaurid remains were described and given the name *"Dinosaurus"* by Ludwig Rütimeyer (1856a). However, this name had previously been used for a lizard species, so it had to be changed. Rütimeyer (1856b) then renamed his fossil *Gresslyosaurus*. It is now generally accepted that this is another name (officially a "junior synonym") of *Plateosaurus* (cf. Galton 1986a, 1990). What all this means is that *Plateosaurus* has a special association with the name "dinosaur"—in a sense it is the "first" dinosaur. And even though there is now no dinosaur with the genus name *Dinosaurus*, the concept so captured the minds of people that this whole group of reptiles became known as "dinosaurs." And if *Plateosaurus* warranted the characterization "terrible lizard," what would we call *Tyrannosaurus* or *Diplodocus*?

In the past 160 years, much has been added to our knowledge of the dinosaurs. Many new species of prosauropods have since been described, and we now have a better conception of their anatomy, and even aspects of their physiology, behavior, and ecology. Because *Plateosaurus* is the best-known form, we will use it as the basis for the description of the prosauropods. The description is based mostly on the extensive work done by Friedrich von Huene (1926, 1932) and Peter Galton (1984, 1985c, 1986a, 1990).

Anatomy of *Plateosaurus*

Plateosaurus (Fig. 19.1) has a small skull, a relatively long neck, and a more or less pear-shaped trunk, with the body weight concentrated around the pelvis. The skull is rather peculiar in being very narrow compared to its height and length. Seen from above, the skull is long and pointed. In side view the portion behind the eye sockets appears to be turned downward.

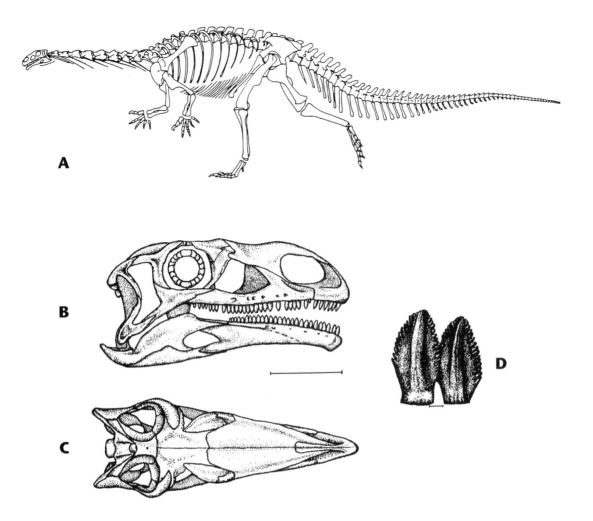

Figure 19.1. *Plateosaurus:*
(A) Reconstruction by Weishampel and Westphal, from Galton 1990. (B) and (C) Skull in lateral and dorsal view. (D) Side view of a couple of teeth; after Galton 1990. Scale for A = 50 cm; for B and C = 10 cm; for D = 1 mm.

The result is that the top margin of the hind part (the temporal region) forms an angle of 20 degrees, and the bottom margin an angle of 30 degrees, with the long axis of the skull. The back of the skull (occiput) is more or less rectangular. The paired nostrils are large, but the eye sockets (orbits) are small compared to those of other prosauropods. Endocranial casts of the nasal region and the braincase indicate that smell played an important role in the lives of these animals. The eyes were directed to the side rather than to the front. Depth perception was therefore reduced, but a wide field of vision is important for spotting predators.

The upper tooth row is long (5 to 6 teeth in the premaxilla, 24 to 30 in the maxilla) and extends to midway under the orbit. The lower jaw has from twenty-one to twenty-eight teeth, and this set is only three-quarters the length of the upper tooth row. The jaw articulation is situated well below the tooth row. The jaws did not close in a scissors-like fashion, but the upper and lower tooth rows cut past each other in a nutcracker action. The tooth-bearing front part of the lower jaw is fairly shallow, while the back third is much deeper, providing room for the insertion of the jaw-closing muscles. The teeth are set in sockets (the thecodont condition) and

244   *Jacques VanHeerden*

were regularly replaced. The crowns of the front teeth of both jaws taper toward the tip and are curved slightly backward. The back teeth have broad, flat tips, and their crowns are set slightly obliquely relative to the long axis of the jaw. The back edge of one tooth, therefore, slightly overlaps the leading edge of the following tooth. The tooth margins are serrated. There are no wear facets on the teeth, indicating that the teeth did not grind against one another during feeding. There is evidence that *Plateosaurus* had a soft secondary palate which separated the nasal passage and the mouth cavity, so that breathing was not interrupted by food-cropping. Another feature which may be related to feeding is the presence of a prominent ridge on the lateral surface of the mandible. It has been suggested that the ridge served as attachment for a muscular cheek. Such a cheek would have been used for temporary food storage, much as in modern chipmunks.

The axial skeleton includes a proatlas, ten neck, fifteen trunk, three sacral (connected to the pelvis), and about fifty tail vertebrae, plus ribs and gastralia (the so-called abdominal ribs). The first neck vertebrae are long and low, but farther away from the head they become higher and axially (lengthwise) shorter. The ribs of the neck vertebrae are thin and delicate and point backward (i.e., in line with the axis of the neck).

The trunk (dorsal) vertebrae increase gradually in size toward the pelvis. The ribs of the first eight are robust structures. In life their distal ends were continued in cartilage, forming a proper rib cage. From the ninth pair onward, the ribs show a progressive decrease in length, and they were not connected to one another. In addition to the ribs there were slender, rod-like gastralia, which supported the ventral abdominal wall.

The sacrum consists of three strongly built vertebrae. The sacral ribs are fused to form a rigid structure. The articulations between the sacral vertebrae are arranged in such a fashion that sideways movement between the sacrals was reduced. This is important in view of the fact that the pelvis is attached to the sacrum, and this was the main support point of the whole body.

There are approximately fifty tail vertebrae. The first few have relatively high and short centra. The short centra made it possible for the animal to bend the tail rather sharply upward near its base. This was important when the animal reared up against trees. Toward the tip of the tail, the caudal vertebrae become smaller. Below the caudal vertebrae, and articulating with them, are Y-shaped elements (the so-called chevron bones), which increased the height of the tail, especially near its root. They served for muscle insertion.

The shoulder girdle and forelimb (Fig. 19.2) consist of the bones typically found in tetrapods. The scapula is a long, blade-like bone with expanded ends. The lower (distal) end is thickened and forms half of the articulation (glenoid) facet for the humerus. The part of the coracoid that forms the other half of the glenoid facet is equally thickened, while the rest of the bone is thin and more or less oval in shape. A slender, rod-like clavicle has been reported in *Plateosaurus* (Huene 1926).

Unlike the other long bones of the body, the humerus appears to be "flat." It looks like a figure 8 that has been twisted so that the two heads are not in the same plane. The proximal expansion (i.e., nearest to the scapula) is drawn out into a ridge laterally, the deltopectoral crest, which points anteriorly. The radius is a little more than half the length of the humerus. The ulna is slightly longer, being almost three-quarters the length of the humerus.

The carpal bones, which form the wrist, are usually found in two rows in tetrapods. In *Plateosaurus,* no remains of the proximal carpals (closer to the radius and ulna) have been found; they were probably cartilaginous

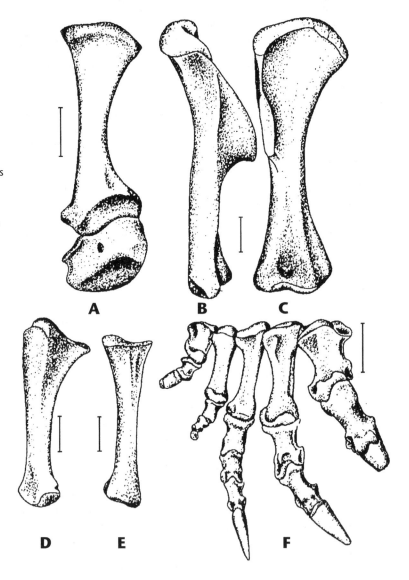

Figure 19.2. Right shoulder girdle and forelimb of *Plateosaurus,* all to the same scale. (A) Scapula and coracoid in lateral view. (B) and (C) Humerus in lateral and anterior views. (D) Ulna in lateral view. (E) Radius in lateral view. (F) Manus in anterior view. After Galton 1990. Scale for A = 10 cm; scale for B–F = 5 cm.

A   B   C

D   E   F

and therefore were not preserved. From this, one would conclude that the forelimbs did not carry much weight during walking, but that manipulation with the hands was more important. Two distal carpals (I and II) are usually present, while a third may sometimes be preserved. Both have irregular edges, indicating that they were "rounded off" in cartilage. The distal carpals capped the proximal ends of the like-numbered metacarpals.

The metacarpals form the palm of the hand. Metacarpal I is a short, stocky bone. Its distal articular surface (ginglymus) is inclined at an angle of about 45 degrees. This means that the first digit pointed obliquely inward (medially) when the hand was placed with the palm downward (i.e., pronated). Metacarpal II is the longest of the metacarpals, with III slightly shorter and IV about 70 percent the length of II; there is also a decrease in the robustness of these elements from II through IV. Metacarpal V is both the shortest and the least massive of these elements. The proximal ends of the five metacarpals form more or less a straight line, thereby ensuring good articulation between palm and wrist.

The phalangeal (finger bone) formula of the hand is probably 2–3–4–3–2 (each number indicates how many phalanges there are in each finger, beginning with digit I and ending in digit V). Phalanx 1 of the first finger is the longest of all phalanges in the hand. The ungual (nail-bearing) phalanx of the first finger is the largest of all unguals. Its size and sharpness indicate that it was probably used for defense. The second and third fingers are shorter, with the phalanges (including the ungual) of II more robust than those of III. The phalanges of the fourth and fifth fingers are rather small, and these digits lack unguals.

The pelvis (Fig. 19.3) consists of the paired ilia, ischia, and pubes. The ilium is approximately one and a half times as long as it is high (deep). In contrast to the condition in carnosaurs and even many sauropods, the anterior process of the iliac blade is pointed. The posterior process is more triangular in shape. The acetabulum is open (perforate), and there is a ridge, the supra-acetabular crest, which borders the dorsal rim of the

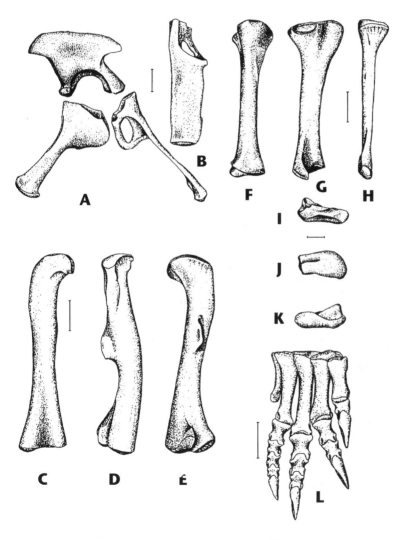

Figure 19.3. Right pelvis and hind limb of *Plateosaurus,* all to the same scale. (A) Pelvis (ilium, ischium, and pubis slightly separated) in lateral view. (B) Pubis in anterior view. (C–E) Femur in anterior, lateral, and posterior views, respectively. (F) and (G) Tibia in anterior and lateral view. (H) Fibula in lateral view. (I–K) Astragalus in anterior, proximal (dorsal), and posterior view. (L) Pes in anterior view. After Galton 1990. Scale = 10 cm (A–H, L) or 5 cm (I–J).

acetabulum, effectively deepening the socket for the head of the femur (thighbone).

The pubis, like the scapula, is a blade-like bone. The part of the bone closest to the ilium is twisted through an angle of about 90 degrees to form part of the rim of the acetabulum. This part of the pubis is pierced by a large opening, the so-called obturator foramen. Below the obturator foramen, the two pubes meet in the midline, thus forming a broad plate (or "apron") that extended downward and forward.

The ischium is about 80 percent of the length of the pubis. The proximal part of the element is blade-like but has a thickened rim where it forms the border of the acetabulum. Below the acetabulum this end meets the pubis in a symphysis. The distal part of the ischium is shaft-like.

Viewed from the side or the back, the femur has the shape of a flattened S. The head of the femur, which fits into the acetabulum, is not rounded (as one would expect), but is rather flat. The fourth trochanter, an important muscle insertion area, is very prominent. The position of this trochanter is a rough indication of the measure of "bipedality"; in bipedal carnosaurs it is above the middle of the length of the femur, and in the heavy sauropods lower down. In *Plateosaurus* its lower end lies in the distal half of the femur, indicating that this dinosaur was probably not as typical a biped as, for instance, the theropods.

The tibia is about 75 percent the length of the femur; this low ratio of tibia to femur length indicates that *Plateosaurus* was not a fast runner. The lower (distal) end of the tibia has a groove into which fits the ascending process of the astragalus. The fibula is a relatively slim bone with a triangular, flattish proximal end and a broad distal end.

The tarsal bones form the ankle, and like the carpals they are found in two rows. The proximal tarsals are the astragalus and the calcaneum. The astragalus is a thick disk with a hollowed lateral surface which gave a tight fit with the calcaneum. The calcaneum is much smaller. Together, the astragalus and calcaneum form a functional unit with the tibia and fibula. The ankle joint is between the proximal tarsals (astragalus and calcaneum) and the distal tarsals. This is known as the "advanced mesotarsal ankle joint" and is found in all dinosaurs, pterosaurs, and birds.

There are two distal tarsals which cap metatarsals III and IV; both are disk-like and roughly triangular in shape (Huene 1926: plate 6).

The metatarsals form the "ball" of the foot (equivalent to the palm of the hand). Metatarsal III is the longest of the metatarsals, and is about half the length of the tibia. Metatarsals II and IV are slightly shorter. V is less than half the length of III, while I is a little longer. All the metatarsals, with the exception of V, are more or less the same thickness.

The phalangeal formula of the foot is 2–3–4–5–1, with the third toe the longest and the fourth just slightly shorter. As a result of the orientation of the distal articulation surfaces on the metatarsals, the foot is slightly splayed between the second and third toes. None of the unguals is particularly large. They decrease gradually in size from the first to the fourth. The single phalanx on the fifth toe is vestigial.

Other Prosauropods

The anchisaurids and thecodontosaurids (Fig. 19.4A) differ from the plateosaurids in being smaller and more lightly built. Their heads and eye sockets are relatively larger, the nostrils are smaller, and their tooth rows shorter (Fig. 19.5A–D). The jaw articulation is slightly below the upper

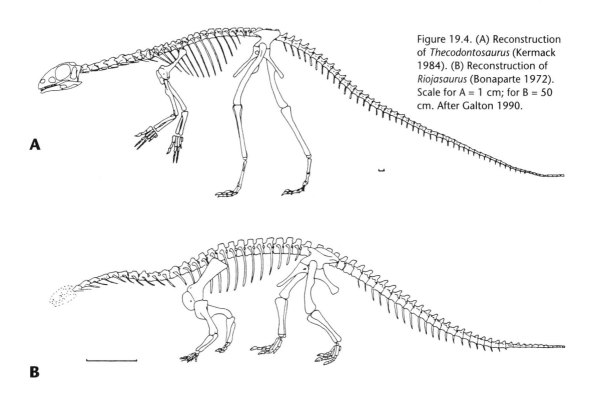

Figure 19.4. (A) Reconstruction of *Thecodontosaurus* (Kermack 1984). (B) Reconstruction of *Riojasaurus* (Bonaparte 1972). Scale for A = 1 cm; for B = 50 cm. After Galton 1990.

A

B

tooth row, so that the jaws closed in a scissors-like fashion. As a result, the back teeth of the upper and lower jaws first passed each other, and then the bite moved progressively forward as the jaws continued to close. The bite force was therefore applied over a short distance at any given time, which means that greater force could be exerted on the food at the bite point. This indicates the processing of more resistant food. This contention is supported by the fact that the jaws were also comparatively short.

The melanorosaurids and blikanasaurids, on the other hand, are fairly heavily built. Unfortunately both families are still inadequately known. The blikanasaurids are known from a single lower hind limb (Galton and

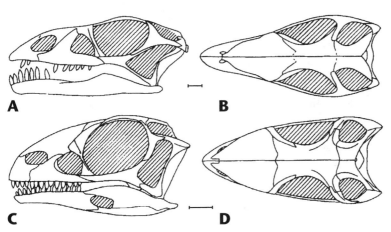

Figure 19.5. Skulls of *Anchisaurus* (A, B) and *Thecodontosaurus* (C, D) in lateral and dorsal view respectively Galton 1990. Scale = 1 cm.

A

B

C

D

VanHeerden 1985). The best-known melanorosaurid is the South American genus *Riojasaurus* (Fig. 19.4B), several partial skeletons of which have been uncovered (Bonaparte 1972). Recently Bonaparte (1994) also described the skull. At one stage the melanorosaurids were thought to be sauropod ancestors; however, even the earliest known melanorosaurids have a number of derived characters that exclude them from direct sauropod ancestry. Nevertheless, they represent an early evolutionary development toward large, heavy-bodied, quadrupedal herbivores, in which they paralleled the later sauropods.

Posture and Locomotion

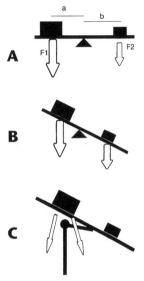

Figure 19.6. (A) A simple lever system with a large weight on one side and a smaller weight on the other side of the pivot point. The anti-clockwise moment is the product of distance *a* and force *F1*; the clockwise moment is *b* times *F2*. By adjusting the distances from the weights to the center, one can balance the lever so that it remains horizontal. Then *F1* × *a* = *F2* × *b*. (B) Simulation of the lever system in a bipedal dinosaur body; here the lever must be inclined at an angle. (C) When the femur is held more or less horizontally, and the femur rotation in the acetabulum is restricted, the main pivot point is at the knee; this effectively shifts part of the body weight to behind the pivot point, indicated by the two arrows.

Extant reptiles are quadrupeds, but I have observed lizards lifting their forelegs off the ground when running fast. There are unconfirmed reports that even the crocodile does that. The South African monitor lizard (a smaller cousin of the Komodo dragon) is able to stand up on its hind legs against a tree trunk, an indication that there is enough flexibility in its backbone for such a posture. However, it cannot raise its forequarters without such support. Where support is lacking, the energy cost for maintaining such a posture is understandably high; lizards run only short distances bipedally and then become exhausted. In contrast, several extinct reptile groups, mostly dinosaur taxa, were habitual bipeds. How did they manage this? And what was the normal posture of the prosauropods?

At first glance, the ratio between forelimb and hind-limb lengths should be a reliable indication of whether an animal is a biped or a quadruped—that is, the shorter the forelimbs are compared to the hind limbs, the better the chances that the animal was a biped. However, the situation is not quite so simple. In many sauropods and stegosaurs, the forelimb was relatively short, but the spine was sharply curved and the animals walked on all fours. On the other hand, several ornithomimid coelurosaurs had relatively long forelimbs (Barsbold and Osmólska 1990), but these were too slender to have supported the animals' body weight.

Another, more reliable, anatomical ratio which is often used to determine the method of locomotion (cf. Galton 1976) is that between the hind limb and the length of the trunk. The hind-limb length is taken as the total of the lengths of the femur, the tibia, and metatarsal III. The trunk length is equal to the total length of the centra of all the dorsal vertebrae. But why is this is a more reliable indicator of bipedality or quadrupedality?

A quadruped has four supports—its four legs. A stationary bipedal dinosaur could have three supports, those being the two hind legs and the tail. Alternatively, it had to counterbalance the weight of the anterior part of its body (in front of the pelvis) with the weight of the tail (see Molnar and Farlow 1990). This is analogous to balancing a ruler on one's finger. The force exerted by any object (such as an eraser) placed on the ruler is calculated by multiplying its weight by the distance from the pivot; this is called the moment (Fig. 19.6A). In large bipeds such as *Allosaurus* and *Tyrannosaurus*, which had a large head and heavy trunk, the total moment on that side is reduced by shortening of the vertebral column; this is usually done by the shortening of individual vertebrae, rather than a reduction in the number of vertebrae. At the same time we may see an elongation of the tail, which increases the total moment in the opposite direction (Fig. 19.6B).

Another way of adjusting the moments is as follows: Normally the pivot point is in the acetabulum (where the head of the femur fits into the pelvis), but when the femur is inclined forward rather than downward, the knee joint becomes the pivot point (Fig. 19.6C). Now the heavy pelvis and the tissues it encloses effectively lie *behind* the pivot point. In bipeds (such as birds) that balance themselves in this way, the femur (upper leg) is shorter than the tibia (shin bone). It follows that where the femur is appreciably longer than the tibia (as in prosauropods), the only way they could have balanced the moments on the two sides was by shortening the trunk (and possibly the neck). In some dinosaurs (but probably not in prosauropods), a rather long neck was held in an S-shaped curve, effectively reducing the moment around the pivot.

When an animal moves forward, the moments change constantly and their analysis becomes more complex. If the forequarters are not too heavy, an animal can rear up on its hind legs (much as chipmunks do). If the hind legs are long enough to place the feet far forward, well in front of the center of gravity, the animal can walk on its hind legs. It is presumed that many prosauropods actually moved this way; they are said to have been facultatively bipedal.

Various ratios for different prosauropods are given in Table 19.1. In bipedal dinosaurs, the ratio of hind limb to trunk can be as high as 1:22 (e.g., in hadrosaurs). In *Triceratops* and *Stegosaurus*, the ratio is 0.9; in some other quadrupedal dinosaurs it is considerably less (Galton 1976). On the basis of this ratio, *Sellosaurus* and *Plateosaurus* may have been facultatively bipedal. On the other hand, on the basis of the structure of its vertebral column, *Plateosaurus* was interpreted as a habitual quadruped by Christian and Preuschott (1996).

Table 19.1.

Body Measurements and Proportions in Prosauropod Dinosaurs

MCII = second metacarpal (hand bone); MTIII = third metatarsal (foot bone). Measurements in millimeters.

	Humerus + ulna + MCII	Femur + tibia + MTIII	Trunk length	Forelimb/ hind limb ratio	Hind limb/ trunk ratio	MTIII/ tibia ratio
Anchisaurus	150 + 105 + 36	211 + 145 + 98	507	0.64	0.90	0.68
Sellosaurus	175 + 100 + 53	226 + 216 + 131	537	0.57	1.07	0.61
Thecodontosaurus	220 + 150? + 50	255 + 180 + 110	640	0.77	0.85	0.61
Plateosaurus	400 + 270 +100	680 + 490 + 240	1415	0.55	1.00	0.49
Riojasaurus	470 + 310 + 150?	600 + 510 + 200	1600?	0.71	0.82	0.39
Massospondylus				0.44	0.95	0.42–0.53
Lufengosaurus	327 + 215 + 56	555 + 357 + 216	1216 (approx.)	0.53	0.92	0.61

Sources of data: Anchisaurus (A. polyzelus), based on YPM 1883; data from Galton 1976. Sellosaurus (S. gracilis) based on Efraasia diagnostica, as described by Galton (1973); the lengths of the ulna and metatarsal III were obtained by extrapolation from a second specimen; the series of dorsal vertebrae is incomplete and the length is estimated. Thecodontosaurus (T. antiquus) based on Huene 1932: 116, plate 54 (1); the trunk length is reconstructed. Plateosaurus based on Huene 1926. Riojasaurus from Bonaparte 1972; length of trunk estimated from reconstruction drawing. Massospondylus based on specimen QG 1559 described by Cooper (1981), who gave only the ratios, not the actual measurements; Cooper also used the length of metacarpal III instead of II for his calculations, which would make a small difference. Lufengosaurus is based on Young 1941; most figures are averages of the measurements of the left and the right side; the length of the centrum of the 15th dorsal was taken as 80 mm.

Another clue to the mode of progression is to be found in the structure of the forelimb and especially the hand (manus). Where the forelimbs carry a substantial part of the body weight, they will be robust, and the metacarpals and phalanges (respectively the palm and the fingers) fairly short. In contrast, Huene (1926) described the manus of *Plateosaurus* as a "grasping hand" and reconstructed the animal in a bipedal pose.

Closely linked to the foregoing is the evidence supplied by tracks. Although such evidence should be treated with caution, there are instances in which tracks can be fairly confidently related to fossil remains (cf. Baird 1980; Olsen and Galton 1984; Farlow 1992; Lockley et al. 1992; Lockley and Hunt 1995). Several ichnogenera (i.e., genus names based on footprints) have been attributed at least in part to prosauropods. Baird described footprints in the Navajo Sandstone (North America) as *Navahopus falcipollex* (Fig. 19.7A). He attributed the tracks to *Ammosaurus*, a plateosaurid prosauropod from the same horizon. Ellenberger (1972) described similar footprints from the Lower Elliot Formation (South Africa) as *Tetrasauropus unguiferus* (Fig. 19.7B). Both sets of prints have a five-toed foot (pes) with a reduced fifth digit, as well as a hand (manus) with an offset first digit and an unusually large ungual.

According to Olsen and Galton (1984), the most abundant recognizable tracks from the Elliot Formation appear to belong to the ichnogenus *Brachychirotherium*. The longest digit in the pes is the third, followed by IV, II, I, and V (Fig. 19.7C). Where manus impressions occur, the five digits are subequal in length. Such prints could have been made by large thecodontians, perhaps rauisuchids. There are many instances, however, where manus impressions are absent, and the pedal prints could have been made by prosauropods. Ellenberger (1972) described such tracks from the

Figure 19.7. (A) Prints of the right manus and pes in the Navajo Sandstone (Early Jurassic), described as *Navahopus falcipollex,* and attributed to the plateosaurid *Ammosaurus* (after Baird 1980: Fig. 12.3). (B) Similar impressions from the Elliot Formation (Late Triassic) of South Africa described as *Tetrasauropus unguiferus* by Ellenberger (1972; figured in Olsen and Galton 1984). (C) Footprints of a bipedal reptile from the Elliot Formation, described as *Pseudotetrasauropus bipedoida* by Ellenberger (1972) and referred to *Brachychirotherium* by Olsen and Galton (1984). Scale = 20 cm.

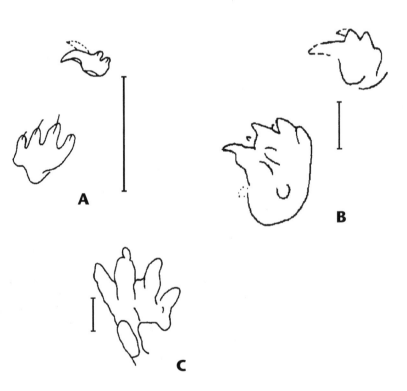

A

B

C

Elliot formation as *Pseudotetrasauropus bipedoida*, which were called "bipedal *Brachychirotherium*" tracks by Olsen and Galton.

Farlow (1992) mentioned two types of footprints from the Pliensbachian Age of Morocco (North Africa); he suggested that one of the two might have been made by very large descendants of melanorosaurid or blikanasaurid prosauropods. Lockley et al. (1992) identified tracks from the Late Triassic and Early Jurassic of North America (*Pseudotetrasauropus, Tetrasauropus,* and *Otozoum*) as belonging to prosauropods. Those authors concluded that the *Pseudotetrasauropus* tracks were made by a gracile, bipedal animal, and the *Tetrasauropus* prints by heavy quadrupedal animals. This also applies to the South African footprints with the same names.

Galton (1990) concluded that the majority of prosauropods were probably facultatively bipedal, although light-bodied forms such as *Anchisaurus* and *Sellosaurus* might have been "more bipedal," and *Thecodontosaurus* fully bipedal. Although it is not possible to link the skeletal remains directly with tracks, the fact that what appear to be bipedal prosauropod tracks have now been described lends weight to Galton's conclusion. The quadrupedal forms, on the other hand, could rear up on their hind legs and maintain this position by using the tail for additional support (cf. Bakker 1978). In such a position they could use the large first ungual phalanx of the hand for defense or even offense (Galton 1976, 1990). Judging from the hand structure and fossil trackways, however, it would appear that most prosauropods *moved* bipedally. The exceptions were when they went up a slope (as in Baird's [1980] reconstruction of the *Navahopus*-maker), or in the case of the more heavily built melanorosaurids and blikanasaurids. In the South American melanorosaurid *Riojasaurus,* the first ungual of the manus is relatively small compared to the same in *Plateosaurus.* The first finger is also not so strongly deflected. Such a hand is better adapted to support the weight of the forequarters than is the case in *Plateosaurus.* The trunk to hind-limb ratio in *Riojasaurus* also indicates that it was quadrupedal rather than bipedal.

As regards cursorial ability (i.e., how fast the animal could move), the ratio between the metatarsal and the tibia is a good guide (Ostrom 1978). In bipedal dinosaurs this ratio can be as low as 0.45 (*Allosaurus;* Madsen 1976). In *Sellosaurus* (= *Efraasia*) it is 0.53, and in *Anchisaurus* 0.68 (Cooper 1981), which indicates that these two could run fairly swiftly.

Sense Organs

In some small prosauropods (especially the thecodontosaurids), the orbits were large and the nostrils small. This probably means that vision was more important in these animals than smell. But even in *Thecodontosaurus* the eyes were directed sideways rather than anteriorly, resulting in a broad field of vision rather than stereoscopic vision. In *Plateosaurus* and related forms, the nostrils and olfactory nerves were relatively large and the eye sockets small, indicating that smell played an important role in their behavior.

Osmólska (1979) suggested that the large nostrils found in most herbivorous dinosaurs (including the prosauropods) could have housed salt glands to help rid the animals of the excess of potassium ions taken in with their plant food. Whybrow (1981), on the other hand, argued that there would probably not have been sufficient room for respiratory structures *and* salt glands in the nostrils. It is possible that the antorbital fossa housed

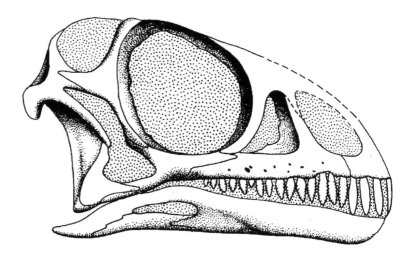

Figure 19.8. Side view of the skull of *Massospondylus* (after Cooper 1981). Scale = 2 cm.

a scent gland similar to that found in some modern antelope species. The scent secreted by the gland could have been used to mark territories or for the release of mating pheromones.

Diet and Feeding

There has been much speculation about what the prosauropods ate. Their masticatory apparatus appears to have been ill suited to deal with resistant plant material. First, the teeth are rather "flat" (spatulate; cf. Fig. 19.1) and lack the grinding surfaces usually found in herbivore teeth. Second, with the exception of *Yunnanosaurus*, there is no indication of tooth-on-tooth wear. Third, there is no possibility that the lower jaw could have moved laterally, as is the case in mammalian herbivores. In fact, the whole skull structure is rather flimsy, and would probably not have been able to withstand the continued chewing of hard material.

To confuse the issue still further, large "carnosaur"-like teeth have been found with prosauropod remains. This led many to suppose that the melanorosaurids (and perhaps other prosauropods) were carnivorous. It was Mick Raath (in Cooper 1980) who pointed out that these teeth frequently lack roots, which indicates that they were probably broken off during the life-and-death struggle between predator and prey. The broken teeth could belong to any of a number of large carnivorous groups, including non-dinosaurian taxa.

Cooper (1981) advanced numerous reasons why he regarded *Massospondylus* (Fig. 19.8) to have been a carnivore, "an opportunistic feeder eating small reptiles, amphibians, mammals, insects and perhaps even fish" (Cooper 1981: 814) as well as carrion (the last was originally suggested by Mick Raath, as quoted by Cooper).

These ideas were discussed by Galton (1984, 1985b, 1986b), who argued that prosauropods were herbivores. He contrasted the typical adaptations of carnivorous and herbivorous reptiles (Galton 1986b: Table 16.1). His table is reproduced here in somewhat simplified form, with my addition of a final comparison (Table 19.2).

254 *Jacques VanHeerden*

Table 19.2.

A Contrast of Features of the Jaws and Teeth of Carnivorous and Herbivorous Reptiles

Carnivorous Reptiles	Herbivorous Reptiles
Jaw articulation in line with tooth rows	Jaw articulation below the tooth rows
Teeth have large gaps between them	Teeth are closely spaced (small or no gaps)
Tooth crowns oriented in line along the middle of the jaw	Crowns are "twisted" with regard to the long axis of the jaw, and the teeth overlap one another in side view (*en echelon* arrangement)
Tooth crowns taper continuously from the root to the tip	Tooth crowns first widen before tapering (i.e., they are leaf-shaped)
Teeth have fine serrations which are at right angles to the edge of the crown	Teeth have coarse serrations which are at about 45 degrees to the edge of the crown, and occur more toward the tip
Teeth are of different sizes; new teeth are continuously formed to replace broken and lost ones (e.g., crocodile; *Allosaurus*, cf. Madsen 1976)	Teeth are more or less the same size, because there is less breakage

Where the jaw articulation is in line with the tooth row, the jaws close in a scissors-like fashion. This means that the bite force is concentrated at one point at any one time. Where the jaw articulation is vertically offset from the tooth row, in contrast, the bite force is distributed over a longer distance of the jaw, and less force is applied at any one point.

Widely spaced teeth make better piercing structures, while closely spaced teeth form a cutting edge. When teeth are set one behind another, struggling prey is less likely to break off any teeth; the teeth also inflict more damage to the prey animal should it try to escape. When the teeth are placed so that the long axes of their crowns are at an angle to the length of the jaw, struggling prey can easily snap off the teeth. For the same reason, leaf-shaped teeth would be more prone to break off just above the root if they were to meet with real resistance. The spike-like teeth of carnivorous types, on the other hand, work like daggers and can easily be extracted from the prey when the jaws reopen. Furthermore, with teeth such as these, set one behind another, the struggling of the prey would cause the tear produced by one tooth to run into the next one, thereby causing more damage.

With respect to tooth serrations, anyone who has used a hacksaw will know that a blade with fine serrations cuts metal, but coarse serrations are suitable only for softer material. Although it is generally true that plant material is more resistant than animal flesh, it is likely that dinosaurian herbivores were rather selective and avoided the woody material in plants.

From the foregoing it is clear that the majority of prosauropods (if not all of them) would have been ill adapted for a carnivorous diet. It is also clear that they would not have been able to handle resistant plant material, but it is likely that they would have had an ample supply of softer plant material, particularly leaves of lower plants (horsetails, psilophytes, ferns, and seedferns). The oblique orientation of the teeth in the jaws indicates that they cropped the leaves with a sideways motion of the head. In *Plateosaurus* the jaws closed in a nutcracker-like fashion; because of the narrow beak, this would have caused the plant material to be cut into short sections. It is possible that there was a shunting of food material between the cheeks and the teeth, because the soft secondary palate points to extended food processing in the mouth.

But what about the anchisaurids and thecodontosaurids, which had more of a scissors-like jaw action, because the jaw articulation was almost on the same level as the (lower) tooth row? These animals could apply the bite force over short distances as the jaws closed, a jaw action like that of carnivores. However, the teeth of these dinosaurs are leaf-shaped (Galton 1990: Fig. 15.4D), although their base is broader than in *Plateosaurus*. Their tooth serrations are set at about 45 degrees to the tooth's edge, and they are somewhat finer than the tooth serrations of *Plateosaurus*, indicating a more resistant diet. However, it seems unlikely that these dinosaurs would have made good predators. It would also appear that we can exclude carrion feeding, because they show none of the adaptations of modern carrion-feeders (strong jaws in hyenas, hooked bills in vultures). And without stereoscopic vision, they would not have been very good at catching insects on the wing, although they could probably have eaten crawling insects. The evidence therefore favors herbivory, or at best omnivory. Perhaps these reptiles fed on tougher plant material than leaves, such as shoots. Their long neck would have suited them well for reaching to the growing tips of tall plants.

Additional mechanical breakdown of food occurred in the stomach through the action of a gastric mill consisting of gastroliths (small stones). Gastroliths have been reported for *Massospondylus* (Bond 1955; Raath 1974) and *Sellosaurus* (Huene 1932; Galton 1973; *Efraasia* is a junior synonym of *Sellosaurus*, cf. Galton and Bakker 1985). Farlow (1987) suggested that they could have opted for high passage rates, selecting the more nutritive components of their food. Galton (1990) concluded that they could have had low turnover rates, with short small intestines and long large intestines, a partitioned colon, and a well-developed symbiotic microflora breaking down the plant material.

Activity Levels

We can only guess what the general activity levels in these early dinosaurs must have been. Activity is based not only on the proportions of different body parts (e.g., legs to trunk length) but also on many physiological factors, such as respiration rate, speed of food digestion, and blood circulation. Where trackways can be reasonably associated with fossil animals, we can obtain an indication of the length of stride, but such evidence is also not totally unambiguous. There is therefore no unassailable evidence, but the hind-limb proportions given in Table 19.1 (see especially the ratio in the last column) can be compared with those of living mammals to give us at least an indication of the speed of movement (Coombs 1978). The smaller prosauropods (anchisaurids, thecodontosaurids, massospondylids, and *Sellosaurus* and *Ammosaurus* among the plateosaurids) were probably slow runners, while the larger forms were heavier and moved much more slowly.

Evolutionary Relationships

Sereno (1989) suggested that the Sauropodomorpha comprises two clades, one of these being the Sauropoda, and the second clade consisting of the Prosauropoda and the Segnosauria. Sereno listed several synapomorphies among the prosauropods which indicate that they constitute a monophyletic group (i.e., they had a common origin). The characters shared by the segnosaurs and prosauropods all refer to the skull, and therefore cannot be used directly in those cases where the skull is still unknown. Sereno also noted features that are shared by all sauropodomorphs. In his view there is a strong case for a monophyletic Sauropodomorpha and Saurischia.

Galton (1990) gave the following diagnosis of the Prosauropoda: skull about half the length of the femur; jaw articulation slightly below the level of the maxillary tooth row; teeth small and similar in structure, or only weakly differentiated, leaf-like and with coarse marginal serrations; first finger with a large, pointed ungual, and fourth and fifth fingers reduced; distal part of pubes forming a broad "apron"; foot with a vestigial fifth toe.

Galton's cladogram of the Prosauropoda is reproduced here (Fig. 19.9). The cladogram reflects the assumption that the ancestral saurischian was a

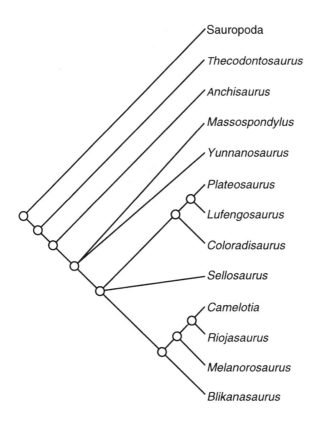

Figure 19.9. Cladogram of the Prosauropoda (Galton 1990).

relatively small, bipedal, erect, and cursorial animal. *Herrerasaurus, Staurikosaurus,* and *Lagosuchus* were used for outgroup comparisons (Gauthier 1986; Galton 1990). Among the known prosauropods, *Thecodontosaurus* is the form closest in structure to such a hypothetical ancestor: this prosauropod was small, bipedal, and at least subcursorial (i.e., slow-running).

The way that prosauropod taxa group together in the cladogram suggests that evolution of the group was characterized by the following trends: progressive enlargement of the nostrils, increase in length of the tooth row, a progressively more ventrally offset jaw articulation, increase in length of the neck and trunk (relative to hind-limb length), elongation of the forelimb, "straightening" of the femur, and downward movement of the muscle attachment on the middle of the femur (the fourth trochanter).

The proposed cladogram is not without its problems. As already stated, it is based on the assumption that the ancestral saurischian was relatively small, bipedal, erect, and cursorial, an idea that was supported, *inter alia,* by Huene (1932) and Romer (1956, 1966). However, when one considers the known fossil record of prosauropods (Table 19.3), it is evident that *Thecodontosaurus* and the other light-bodied prosauropods are relative latecomers, whereas the very early forms, such as *Riojasaurus, Melanorosaurus,* and *Euskelosaurus,* were large, heavily built, and rather ponderous animals. The fact that these forms come from what constituted a fairly large area in the southern part of Gondwana, and are *succeeded* in the fossil

Table 19.3.

Stratigraphic Distribution of Prosauropod Genera

Epoch	Age	Million Years Ago	Genera	
Early Jurassic	Toarcian	194		?*Anchisaurus*
	Pliensbachian	200	*Massospondylus* *Yunannosaurus* *Lufengosaurus*	*Ammosaurus*
	Sinemurian	206		
	Hettangian	213		
Late Triassic	Rhaetian	219	*Camelotia*	*Thecodontosaurus*
	Norian	225	*Coloradisaurus* *Euskelosaurus* *Mussaurus* *Plateosaurus* *Sellosaurus* ?*Melanorosaurus* *Riojasaurus*	
	Carnian	231	*Azendohsaurus* *Euskelosaurus* *Melanorosaurus* *Blikanasaurus*	

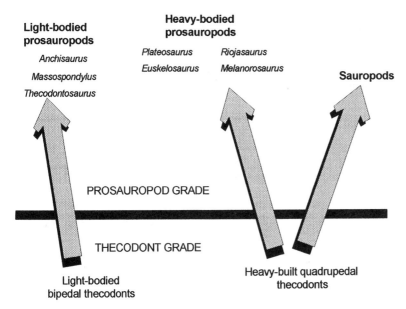

Light-bodied prosauropods

Anchisaurus

Massospondylus

Thecodontosaurus

Heavy-bodied prosauropods

Plateosaurus   *Riojasaurus*

Euskelosaurus   *Melanorosaurus*

Sauropods

PROSAUROPOD GRADE

THECODONT GRADE

Light-bodied bipedal thecodonts

Heavy-built quadrupedal thecodonts

Figure 19.10. A conservative representation of "prosauropod" phylogeny. Two separate lines of evolution are shown, the light-bodied forms on the left and the heavy-bodied forms on the right, both transgressing the thecodont-prosauropod division. The sauropods may have had a separate origin from the thecodontians, rather than sharing their descent with the heavily built prosauropods.

record by forms such as *Anchisaurus, Massospondylus,* and *Thecodontosaurus,* indicates that this is not just a local aberration in the fossil record. To suppose that light-bodied prosauropods were present but not preserved (or have not yet been found) simply to fit the cladistic hypothesis is a rather tenuous line of reasoning.

In view of the foregoing, alternative hypotheses need to be considered. One possibility is that the prosauropods stemmed from a heavily built, quadrupedal, and perhaps herbivorous thecodontian, which first gave rise to forms such as *Riojasaurus, Melanorosaurus,* and *Euskelosaurus.* It is possible that the more lightly built, bipedal prosauropods stemmed from dinosaurs such as these. If this is true, the Saurischia could still be a monophyletic group, although our concept of the course of their evolution would be changed.

A second alternative hypothesis is that the prosauropods are polyphyletic (Fig. 19.10). If it is true that *Thecodontosaurus* is anatomically close to the supposed bipedal, cursorial (and carnivorous) "saurischian" ancestor, this and other light-bodied forms may have a separate origin among the thecodonts, separate from the ancestor of more ponderous prosauropods. The presence of light-bodied forms such as *Staurikosaurus* (and the somewhat more robust *Herrerasaurus*) in Carnian Age sediments in South America, contemporaneous with *Blikanasaurus, Euskelosaurus,* and *Melanorosaurus,* appears to strengthen such a hypothesis. This would mean that the "Prosauropoda" is not a clade (monophyletic) but rather a grade. In other words, there were two or more separate evolutionary lines leading from thecodontians to saurischians. This would also imply that the Saurischia—and the Dinosauria—are polyphyletic (see Benton, chap. 16 of this volume, for an opposing view).

It is suggested that many "prosauropod" characters are, in fact, linked to the method of locomotion in these animals. Thecodontians had an imperforate acetabulum because the femur was directed laterally, and the pressure point in the pelvis was in the center of the acetabulum. In the

Prosauropods   259

rauisuchid thecodontians the gait was "semi-improved"—that is, the femur was oriented more vertically (at an angle of about 45 degrees from the vertical). The ilium was tilted so that pressure exerted on the pelvis during locomotion was still directed toward the center of the acetabulum. The perforate acetabulum and vertical femur found in "prosauropods" and other dinosaur taxa could arise from the rauisuchid condition by further adduction of the femur and reorientation of the ilium to lie more vertically. Alternatively, these reptiles could have arisen from the normal thecodontian condition (a vertical ilium) by adduction of the femur from the horizontal position to a vertical orientation. Under either scenario, the body weight would be transferred through the supra-acetabular ridge on the ilium (rather than through the acetabulum) to the head of the femur. With the loss of pressure in the acetabulum, this would become perforated. Loss of bony tissue due to lack of pressure (and vice versa) is a normal phenomenon in bone development.

All three hypotheses of prosauropod relationships have their merits—and their weaknesses. At the present moment there does not appear to be enough evidence to settle the question of prosauropod origin(s) one way or another. A final answer will be possible only when more material becomes available.

Leaving their origin(s) still unresolved, we must now consider the descendants of the prosauropods. Several authors (e.g., Charig et al. 1965; Cruickshank 1975; VanHeerden 1978; Galton 1986b, 1990; Sereno 1989) have indicated that the known prosauropods could not have given rise to the sauropods. A possible link between very early melanorosaurid prosauropods and the sauropods,however, may be the large and very heavily built saurischian *Vulcanodon* (Raath 1972). This dinosaur comes from the Hettangian (Early Jurassic) of Zimbabwe, southern Africa, and is regarded as a very primitive sauropod (Cooper 1984; McIntosh 1990). Cruickshank (1975) noted that *Vulcanodon* has an unreduced fifth metatarsal, a feature found in *Herrerasaurus* and the sauropods, but not in prosauropods. All known prosauropods have a reduced metatarsal V, and so *Vulcanodon* might be a member of a paraphyletic group close to the ancestry of the Sauropoda.

But did the prosauropods disappear without any trace? Sereno (1989) and Barsbold and Maryańska (1990) indicated that the enigmatic Cretaceous segnosaurs are allied to the prosauropods. Unfortunately the segnosaurs themselves are not too well known so it is not possible to give a reliable answer (see Maryańska, chap. 18 of this volume). If these two dinosaur groups are closely related, it would mean that the prosauropods must have existed in one form or another until the Cretaceous, most probably in environments where the chances of fossilization were (almost) zero.

References

Baird, D. 1980. A prosauropod dinosaur trackway from the Navajo Sandstone (Lower Jurassic) of Arizona. In L. L. Jacobs (ed.), *Aspects of Vertebrate History: Essays in Honor of Edwin Harris Colbert,* pp. 219–230. Flagstaff: Museum of Northern Arizona Press.

Bakker, R. T. 1978. Dinosaur feeding behaviour and the origin of flowering plants. *Nature* 274: 661–663.

Barsbold R. and T. Maryańska. 1990. Segnosauria. In D. B. Weishampel, P. Dodson, and H. Osmólska (eds.), *The Dinosauria,* pp. 408–415. Berkeley: University of California Press.

Barsbold R. and H. Osmólska. 1990. Ornithomimosauria. In D. B. Weishampel, P. Dodson, and H. Osmólska (eds.), *The Dinosauria*, pp. 225–244. Berkeley: University of California Press.

Bonaparte, J. F. 1969. Dos nuevas "faunas" de reptiles Triásicos de Argentina. *I Gondwana Symposium, Mar del Plata Ciencas Tierra* 2: 283–306.

Bonaparte, J. F. 1972. Los tetrápodos del sector superior de la Formación Los Colorados, La Rioja, Argentino (Triásico Superior). I Parte. *Opera lilloana* 22: 1–183.

Bonaparte, J. F. 1994. Dinosaurs of South America. Part I: Triassic dinosaurs of South America. *Gakken Meok* 5: 4–25.

Bonaparte, J. F., and M. Vince. 1979. El hallazgo del primer nido de Dinosaurios Triásicos (Saurischia, Prosauropoda), Triásico superior de Patagonia, Argentina. *Ameghiniana* 16: 173–182.

Bond, G. 1955. A note on dinosaur remains from the Forest Sandstone (Upper Karroo). *Arnoldia* 2: 795–800.

Charig, A. J.; J. Attridge; and A. W. Crompton. 1965. On the origin of the sauropods and the classification of the Saurischia. *Proceedings Linnean Society of London* 176: 197–221.

Christian, A., and H. Preuschott. 1996. Deducing the body posture of extinct large vertebrates from the shape of the vertebral column. *Palaeontology* 39: 801–812.

Colbert, E. H. 1964. Relationships of the saurischian dinosaurs. American Museum *Novitates* 2181: 1–24.

Coombs, W. P. Jr. 1978. Theoretical aspects of cursorial adaptations in dinosaurs. *Quarterly Review of Biology* 53: 393–418.

Cooper, M. R. 1980. The first record of the prosauropod dinosaur *Euskelosaurus* from Zimbabwe. *Arnoldia* 9: 1–17.

Cooper, M. R. 1981. The prosauropod dinosaur *Massospondylus carinatus* Owen from Zimbabwe: Its biology, mode of life and phylogenetic significance. *Occasional Papers, National Museums and Monuments of Rhodesia*, Series B, 6: 689–840.

Cooper, M. R. 1984. A reassessment of *Vulcanodon karibaensis* Raath (Dinosauria: Saurischia) and the origin of the Sauropoda. *Palaeontologica africana* 25: 203–231.

Cruickshank, A. R. I. 1975. The origin of sauropod dinosaurs. *South African Journal of Science* 71: 89–90.

Dutuit, J. M. 1972. Découverte d'un dinosaure ornithischien dans le Trias supérieur de l'Atlas occidental marocain. *Comptes Rendue Academie Science Paris* D275: 2841–2844.

Ellenberger, P. 1972. Contribution à la classification des Pistes de Vertébrés du Trias: Les types du Stormberg d'Afrique du Sud (I). *Palaeovertebrata* Memoire Extraordinaire, Montpelier.

Farlow, J. O. 1987. Speculations about the diet and digestive physiology of herbivorous dinosaurs. *Paleobiology* 13: 60–72.

Farlow, J. O. 1992. Sauropod tracks and trackmakers: Integrating the ichnological and skeletal records. *Zubía* 10: 89–138.

Galton, P. M. 1973. On the anatomy and relationships of *Efraasia diagnostica* (v. Huene) n. gen., a prosauropod dinosaur (Reptilia: Saurischia) from the Upper Triassic of Germany. *Paläontologische Zeitschrift* 47: 229–255.

Galton, P. M. 1976. Prosauropod dinosaurs (Reptilia: Saurischia) of North America. *Postilla* 169: 1–98.

Galton, P. M. 1984. Cranial anatomy of the prosauropod dinosaur *Plateosaurus* from the Knollenmergel (Middle Keuper, Upper Triassic) of Germany. Part I: Two complete skulls from the Trossingen/Württ. with comments on the diet. *Geologica et Palaeontologica* 18: 139–171.

Galton, P. M. 1985a. Notes on the Melanorosauridae, a family of large prosauropod dinosaurs (Saurischia: Sauropodomorpha). *Géobios* 19: 671–6.

Galton, P. M. 1985b. Diet of prosauropod dinosaurs from the Late Triassic and Early Jurassic. *Lethaia* 18: 105–123.

Galton, P. M. 1985c. Cranial anatomy of the prosauropod dinosaur *Plateosaurus* from the Knollenmergel (Middle Keuper, Upper Triassic) of Germany. Part II: All the cranial material and details of soft-part anatomy. *Geologica et Palaeontologica* 19: 119–159.

Galton, P. M. 1986a. Prosauropod dinosaur *Plateosaurus* (= *Gresslyosaurus*) (Saurischia: Sauropodomorpha) from the Upper Triassic of Switzerland. *Geologica et Palaeontologica* 20: 167–183.

Galton, P. M. 1986b. Herbivorous adaptations of Late Triassic and Early Jurassic dinosaurs. In K. Padian (ed.), *The Beginning of the Age of Dinosaurs*, pp. 203–221. Cambridge: Cambridge University Press.

Galton, P. M. 1990. Basal Sauropodomorpha—Prosauropoda. In D. B. Weishampel, P. Dodson, and H. Osmólska (eds.), *The Dinosauria*, pp. 320–344. Berkeley: University of California Press.

Galton, P. M., and R. T. Bakker. 1985. The cranial anatomy of the prosauropod dinosaur *"Efraasia diagnostica,"* a juvenile individual of *Sellosaurus gracilis* from the Upper Triassic of Nordwürttemberg, West Germany. *Stuttgarter Beiträge Naturkunde,* Series B, 117: 1–15.

Galton, P. M., and J. VanHeerden. 1985. Partial hindlimb of *Blikanasaurus cromptoni* n. gen. and n. sp., representing a new family of prosauropod dinosaurs from the Upper Triassic of South Africa. *Géobios* 18: 509–516.

Gauthier, J. 1986. Saurischian monophyly and the origin of birds. In K. Padian (ed.), *The Origin of Birds and the Evolution of Flight*, pp. 1–55. Memoirs California Academy of Science no. 8.

Haughton, S. H. 1924. The fauna and stratigraphy of the Stormberg Series. *Annals South African Museum* 12: 323–497.

Huene, F. von. 1907–08. Die Dinosaurier der europäischen Triasformation mit Berücksichtigung der aussereuropäischen Vorkommnisse. *Geologische und Paläontologische Abhandlungen*, Supplement 1: 1–419.

Huene, F. von. 1914. Saurischia et Ornithischia triadica ("Dinosauria" triadica). *Fossilium Catalogus. I. Animalia* 4: 1–21.

Huene, F. von. 1920. Bemerkungen zur Systematik und Stammesgeschichte einiger Reptilien. *Zeitschrift für Induktive Abstammungs- und Vererblehre* 24: 162–166.

Huene, F. von. 1926. Vollständige Osteologie eines Plateosauridien aus dem schwäbischen Trias. *Geologische und Paläontische Abhandlungen* 15: 129–179.

Huene, F. von. 1929. Los Saurisquios y Ornithisquios de Cretacéo Argentino. *Annales Museo de La Plata*, Series 2, 3: 1–196.

Huene, F. von. 1932. Die fossile Reptil-Ordnung Saurischia, ihre Entwicklung und Geschichte. *Monograph Geologie Paläontologie*, Series 1, 4: 1–361.

Huene, F. von. 1956. *Paläontologie und Phylogenie der Niederen Tetrapoden.* Jena: Gustav Fischer.

Huxley, T. H. 1866. On some remains of large dinosaurian reptiles from the Stormberg Mountains, South Africa. *Geological Magazine* 3: 563.

Kermack, D. 1984. New prosauropod material from South Wales. *Zoological Journal of the Linnean Society* 82: 101–117.

Lambert, D. 1983. *A Field Guide to Dinosaurs.* New York: Avon Books.

Lockley, M.; K. Conrad; M. Paquette; and J. Farlow. 1992. Distribution and significance of Mesozoic vertebrate trace fossils in Dinosaur National Monument. Sixteenth Annual Report of the National Park Service Center, University of Wyoming, pp. 74–85.

Lockley, M., and A. P. Hunt. 1995. *Dinosaur Tracks and Other Fossil Footprints of the Western United States.* New York: Columbia University Press.

Lydekker, R. 1890. Contributions to our knowledge of the dinosaurs of the Wealden, and the Sauropterygia of the Purbeck and Oxford clay. *Quarterly Journal Geological Society of London* 6: 36–53.

Madsen, J. H., Jr. 1976. *Allosaurus fragilis,* a revised osteology. *Utah Geological and Mineral Survey Bulletin* 109: 1–163.

Marsh, O. C. 1885. Names of extinct reptiles. *American Journal of Science*, Series 3, 29: 169.

Marsh, O. C. 1891. Notice of new vertebrate fossils. *American Journal of Science*, Series 3, 42: 265–269.

Marsh, O. C. 1895. On the affinities and the classification of the dinosaurian reptiles. *American Journal of Science,* Series 3, 50: 483–498.

McIntosh, J. S. 1990. Sauropoda. In D. B. Weishampel, P. Dodson, and H. Osmólska (eds.), *The Dinosauria,* pp. 345–401. Berkeley: University of California Press.

Meyer, H. von. 1837. Mitteilung an Prof. Bronn (*Plateosaurus engelhardti*). *Neues Jahrbuch für Mineralogie, Geologie und Paläontologie* 1837: 317.

Molnar, R. E., and J. O. Farlow. 1990. Carnosaur paleobiology. In D. B. Weishampel, P. Dodson, and H. Osmólska (eds.), *The Dinosauria,* pp. 210–224. Berkeley: University of California Press.

Olsen, P. E., and P. M. Galton. 1984. A review of the reptile and amphibian assemblages from the Stormberg of southern Africa, with special emphasis on the footprints and the age of the Stormberg. *Palaeontologia africana* 25: 87–110.

Osmólska, H. 1979. Nasal salt glands in dinosaurs. *Acta Palaeontologica Polonica* 24: 205–215.

Ostrom, J. H. 1978. The osteology of *Compsognathus longipes* Wagner. *Zitteliana* 4: 73–118.

Owen, R. 1842. Report on British Fossil Reptiles. Part II. *Report of the British Association for the Advancement of Science* for 1841: 60–204.

Owen, R. 1854. *Descriptive Catalogue of the Fossil Organic Remains of Reptilia Contained in the Museum of the Royal College of Surgeons of England.* London: British Museum of Natural History.

Raath, M. A. 1972. Fossil vertebrate studies in Rhodesia: a new dinosaur (Reptilia: Saurischia) from near the Triassic-Jurassic boundary. *Arnoldia* 5: 1–37.

Raath, M. A. 1974. Fossil vertebrate studies in Rhodesia: Further evidence of gastroliths in prosauropod dinosaurs. *Arnoldia* 7: 1–7.

Riley, H., and S. Stutchbury. 1836. A description of various fossil remains of three distinct saurian animals, recently discovered in the Magnesian Conglomerate near Bristol. *Transactions Geological Society of London,* Series 2, 5: 349–357.

Romer, A. S. 1956. *Osteology of the Reptiles.* Chicago: University of Chicago Press.

Romer, A. S. 1966. *Vertebrate Paleontology.* 3rd ed. Chicago: University of Chicago Press.

Rütimeyer, L. 1856a. (*Dinosaurus gresslyi*). *Bibliothèque Universelle des Sciences Belles-Lettres et Arts,* Genève Sept. 1856: 53.

Rütimeyer, L. 1856b. Reptilienknochen aus dem Keuper. *Allgemeine Schweizerische Gesellschaft für de Gesammten Naturwissenschaften* 41: 62–64.

Sereno, P. C. 1989. Prosauropod monophyly and basal sauropodomorph phylogeny. *Journal of Vertebrate Paleontology* 9 (Supplement no. 3): 38A.

VanHeerden, J. 1978. *Herrerasaurus* and the origin of the sauropod dinosaurs. *South African Journal of Science* 74: 187–189.

Walker, A. D. 1964. Triassic reptiles from the Elgin area: *Ornithosuchus* and the origin of carnosaurs. *Philosophical Transactions of the Royal Society of London* B 248: 53–134.

Whybrow, P. J. 1981. Evidence for nasal salt glands in the Hadrosauridae (Ornithischia). *Journal of Arid Environments* 4: 43–57.

Young, C.-C. 1941. A complete osteology of *Lufengosaurus huenei* Young (gen. et. sp. nov.). *Palaeontologica Sinica,* Series C, 7: 1–53.

Young, C.-C. 1942. *Yunnanosaurus huangi* (gen. et sp. nov.), a new Prosauropoda from the Red Beds at Lufeng, Yunnan. *Bulletin Geological Society of China* 22: 63–104.

Sauropods

John S. McIntosh,
M. K. Brett-Surman,
and James O. Farlow

20

The Discovery of Sauropods

As with many other major dinosaur groups, the first bones of the great sauropods were unearthed in England. Some vertebrae and fragmentary limb bones from rocks of the Middle Jurassic period were reported to the Geological Society of London by John Kingston on June 3, 1825. No record survives of what was said, or of the bones themselves, but it is clear from the fragmentary nature of the fossils that there can have been no idea of the nature of the animal to which they belonged—except that it was very large. More scattered bones were found in Jurassic rocks during the next sixteen years, and when Richard Owen conducted his major study of all the British fossil reptiles in 1841 (published as Owen 1842), he found there was a group of bones of a very large animal which could not be assigned to any known form, but which resembled one another by the spongy texture of the bones, somewhat like that of whales. Consequently he named this animal *Cetiosaurus,* the "whale-lizard." He concluded that *Cetiosaurus* was "strictly aquatic and most probably of marine habits" and was related to the crocodiles.

Strictly speaking, *Cetiosaurus* was not the first named sauropod genus, for Professor Owen, in a book published a few months earlier, had named two heart-shaped teeth *Cardiodon,* without knowing to what type of animal they belonged. Various authors have speculated that *Cardiodon* teeth may have belonged to *Cetiosaurus,* but this has never been proved.

During the next two decades after Owen's study, a few more reports of *Cetiosaurus* bones were made, including one of a 129-centimeter-long femur found eight miles from Oxford. Dr. Gideon Mantell, the author of *Iguanodon*, named a large humerus *Pelorosaurus* in 1850. In 1860, Owen established a new suborder of crocodiles, the Opisthocaudia, for *Cetiosaurus* and *Streptospondylus,* the latter of which later proved to be a theropod. This was based on the fact that some of the dorsal vertebrae of each of these animals are opisthocoelous (having vertebral centra which are concave behind, convex in front). This was in contrast to the vertebrae of true crocodiles, which are procoelous (concave in front and convex behind).

A breakthrough from this very unsatisfactory knowledge of the sauropods finally occurred some thirty years after the naming of *Cetiosaurus,* with the discovery of the major part of a skeleton of that animal from the same quarry at Gibralter, near Oxford, whence came the 129-centimeter femur mentioned above. The skeleton was described in detail by Professor John Phillips of Oxford University in his book *The Geology of Oxford,* published in 1871. The bones recovered included representatives of almost all the limb and girdle bones, as well as vertebrae from different parts of the column, but no skull material except for an imperfect tooth, and only very imperfect foot elements. Except for the incorrect orientation of the ischium and pubis, Phillips's identifications of the bones were quite accurate, and his observations gave the first real picture of what a sauropod dinosaur was like. Without reaching a final conclusion, he compared *Cetiosaurus* not only to the crocodiles but also to the lizards and dinosaurs. He discussed the huge size of the animal and concluded from the tooth that "the animal was nourished by . . . vegetable food." He further stated that "the femur by its head projecting freely from the acetabulum seems to claim a movement of free stepping more parallel to the line of the body and more approaching the vertical than the sprawling gait of the crocodile. The large claws concur in this indication of terrestrial habits. But on the other hand, these characters are not contrary to the belief that the animal may have been amphibious." After more arguments concerning the tail vertebrae, he concluded, "We have therefore a marsh living or river side animal." During the 1870s, various scattered remains of individual vertebrae and isolated limb bones were reported from England, but in the latter years of that decade, discoveries of much more complete animals were made in North America.

As in England, the first report of what later proved to be a sauropod dinosaur in North America was of two teeth named *Astrodon* in 1856, from Lower Cretaceous beds in Maryland. But 1877 was to be the decisive year in which knowledge of the sauropods flowed in from the western United States. Edward Drinker Cope reported a partial forelimb in February from Utah (actually collected in 1859) under the name *Dystrophaeus*—without any clear indication of its systematic relationships. Then during the summer, fall, and winter, and continuing into 1878 and 1879, there was a rapid series of papers by O. C. Marsh and Edward Cope on remains from Garden Park and Morrison, Colorado, and Como Bluff, Wyoming. Their well-documented, bitter rivalry (see Colbert, chap. 3 of this volume) resulted in the establishment of a number of new genera and species, often named on the basis of partial shipments from their collectors while major portions of the skeletons were still in the ground. This resulted in considerable synonymy (in which different names were assigned to what turned out to be the same kind of animal). Those genera based on substantial material were *Camarasaurus, Apatosaurus,* "*Morosaurus*" (later shown to be a synonym of *Camara-*

saurus), *Diplodocus*, and *"Brontosaurus"* (later shown to be a synonym of *Apatosaurus;* see Box 20.1).

Some of the most important developments of this period came in three contributions by Marsh: in 1878, when he established the name Sauropoda for a new suborder of dinosaurs and provided ten distinguishing characters of the group; in the same paper, in which he figured the fore and hind limb and girdle bones of *"Morosaurus" grandis* in proper position for the first time; and in 1879, when he briefly described for the first time a sauropod skull, based on the disarticulated and somewhat incomplete skull of *"Morosaurus" grandis*. This was followed by Marsh's (1883) publishing the first restoration of a complete sauropod skeleton, that of *"Brontosaurus" excelsus*. This was based on a single animal which, though lacking the skull, front part of the neck, end of the tail, ulna, and much detail of the feet, gave a reasonably good idea of what a sauropod dinosaur really looked like. Further contributions were made by Marsh in the 1880s and early 1890s, including the first complete and articulated sauropod skull, that of *Diplodocus*, in 1884, and new genera from Maryland and South Dakota.

The late 1890s and early 1900s were some of the most productive years in understanding the diversity of North American sauropods, with many expeditions to the West, most notably by the American Museum of Natural History in New York and the Carnegie Museum (now the Carnegie Museum of Natural History) in Pittsburgh, but also by many other institutions, including the Field Columbian Museum (now the Field Museum of Natural History) in Chicago and various universities, among them the University of Kansas and the University of Wyoming. Most notable was the American Museum's Bone Cabin Quarry northeast of Medicine Bow, Wyoming, which produced hundreds of specimens, including many articulated fore and hind limbs, together with, for the first time, many articulated fore and hind feet. This work resulted in major monographs by Henry Fairfield Osborn and C. C. Mook (1921), John Bell Hatcher (1901), William J. Holland (1910), and Elmer S. Riggs (1903), in which *Diplodocus*, *Apatosaurus*, and *Haplocanthosaurus* were described and figured in detail. It also led in 1905 to the first mounted sauropod skeleton in New York, that of *"Brontosaurus"* (= *Apatosaurus*) *excelsus*, and shortly thereafter of *Diplodocus carnegii* in Pittsburgh. Casts of the latter were sent to museums all over the world: in London, Paris, Berlin, Vienna, St. Petersburg, Mexico City, La Plata (Argentina), and Munich (never mounted). The climax of the North American sauropod discoveries came with the discovery in 1909 by Earl Douglass, of the Carnegie Museum, of the great dinosaur quarry at what later became Dinosaur National Monument in eastern Utah. From this quarry have come several nearly complete articulated skeletons each of *Apatosaurus*, *Diplodocus*, *Camarasaurus*, and *Barosaurus*, including many fine skulls of the first three of these.

At the same time as all this was going on in the United States, in the mid–1890s sauropods began to be found in Africa. Lydekker described titanosaurids from the Cretaceous of Argentina in 1892. These were closely related to very fragmentary material which he had reported from India in 1877. First came a number of finds of scattered material in Madagascar referred to the brachiosaurid *Bothriospondylus*. This was followed in the first two decades of the twentieth century by well-organized German expeditions to German East Africa (now Tanzania), which produced good partial skeletons of *Brachiosaurus*, *Dicraeosaurus*, an animal referred to the American genus *Barosaurus*, and another now known as *Janenschia*. Skeletons of the first two of these were complete enough to form the bases

Box 20.1.

Why Did
the Name
"Brontosaurus"
Change to
Apatosaurus?

"Brontosaurus" is one of the best-known, most intellectually pleasing, and most familiar names of all the dinosaurs. Why then do all recent books refer to it as *Apatosaurus?* The answer involves the celebrated Marsh-Cope rivalry and the incomplete understanding of dinosaurs and their anatomy in the early days of excavations.

It all started in 1877, when Arthur Lakes sent Marsh a sacrum and a few other vertebrae of a very large dinosaur which he had collected near the little town of Morrison, Colorado. At the time, both Marsh and Cope were trying to beat each other in naming as many dinosaurs as possible. The unwritten (at that time) law of priority states that the first name published with an adequate description of the specimen it is based on is entitled to be accepted as the official name. Any later published name becomes what is known as a "synonym" and cannot be used again. In the cases of both Marsh and Cope, they described new animals based on the first specimens they received from their field collectors, even when the rest of the skeleton was still in the ground!

Such was the case with the sacrum that Marsh received from Lakes. In 1877, Marsh named it *Apatosaurus ajax.* More of the skeleton was sent to Marsh in 1878, and it was briefly and adequately described and figured.

In the summer of 1879, Marsh's famous field collector William H. Reed discovered a new dinosaur quarry at Como Bluff, Wyoming, which contained a single large skeleton. Among the first bones shipped was a sacrum, which Marsh promptly named *Brontosaurus excelsus.* During the next several years, the rest of the skeletons from Morrison and Como Bluff were excavated and shipped to Marsh. Both proved to be represented by most of the skeletal elements, except the skulls. Because the skeleton of *Brontosaurus* was one of the most complete skeletons ever found of a sauropod, it was figured and described more completely than that of *Apatosaurus.* The two animals were similar in size and design, and Marsh referred them both to the same family but kept them as separate genera because *Apatosaurus* had three vertebrae in the sacrum and *Brontosaurus* had five. The famous paleontologist S. W. Williston was at this time one of Marsh's assistants. He recognized the similarity of the two animals and, as we now know from his unpublished notes, suspected that they were, in fact, the same genus. Although the *Apatosaurus* skeleton was larger than the one for *Brontosaurus,* it was from a younger individual. At that time, it was not known that two more vertebrae fused onto the original three sacral vertebrae (one at each end) as the animal got older.

In 1903, Elmer Riggs, of the Field Columbian Museum in Chicago, was describing another *Apatosaurus* skeleton which he had found near Fruita, Colorado. He realized that both *Brontosaurus* and *Apatosaurus* belonged to the same genus. Because of the law of priority, the name *Apatosaurus,* which came first by two years, must be preserved. In his paper, Riggs acknowledged his consultation with Williston, who was by then a professor at the University of Chicago. Although the original *Apatosaurus* skeleton was not as complete as the *Brontosaurus* skeleton, it was more than adequate to show that the two belong in the same genus. That is why the name *"Brontosaurus"* is now always published inside quotation marks—to show that it is no longer valid.

for mounted skeletons in Berlin. These very productive efforts were unfortunately brought to an end by World War I.

This same time period was also to yield the rear half of a skeleton of *Cetiosauriscus* from England, one of only three or four good associated specimens from Europe. Later finds included a partial brachiosaurid skeleton from France and several other sauropod skeletons from Portugal.

Sauropods 267

The skeleton of *"Brontosaurus" excelsus* that was excavated by Marsh's crews at Como Bluff, Wyoming, proved to be the most complete sauropod skeleton found as of the late 1880s. However, as with many dinosaur skeletons, the head was missing. Thus, in order to complete the two skeletal restorations published in 1883 and 1891, Marsh chose two other large skulls from the collections at Yale. These two isolated skulls were deemed by Marsh to be likely candidates for *"Brontosaurus"* because they were from the Upper Jurassic Morrison Formation. The first was found several miles to the east of the headless skeleton, and the second came from Garden Park, Colorado, again from the Morrison Formation. These skulls possessed large, spoon-shaped teeth like those of *Camarasaurus,* so when skeletons of *"Brontosaurus"* were mounted around the country, the skulls of *Camarasaurus* were used as a model. At that time, no one had found a complete skull in articulation with a skeleton of either *"Brontosaurus"* or *Camarasaurus,* so the correct association of skull and skeleton was problematic.

In the early 1900s, a skull attributed to *Diplodocus* was found lying in association with two *Apatosaurus* skeletons at Dinosaur National Monument. (In 1903 the name *"Brontosaurus"* was correctly changed to *Apatosaurus.*) This new skull was lying over the middle of the neck of one of the skeletons and only a few feet from the forefoot of the other. It had peg-like teeth typical of *Diplodocus,* and quite unlike those of *Camarasaurus.* This latter skeleton is now mounted at the Carnegie Museum in Pittsburgh. At that time, the Carnegie's director, W. J. Holland, suspected that this skull was the true head of *Apatosaurus,* but he refrained from mounting it because of opposition from the powerful and influential H. F. Osborn, the director of the American Museum in New York, who had mounted his *Apatosaurus* with a head modeled from *Camarasaurus.* A further complication arose when a mixup occurred in the cataloguing of the skull from Dinosaur National Monument and a second skull from the same quarry. The second skull was bigger than the original skull, and it was assumed that the smaller of the two skulls could not be large enough to come from such a large *Apatosaurus* skeleton.

Further studies of more complete skeletons showed that *Apatosaurus* belonged to the Diplodocidae, and not to the Camarasauridae. Field records and detailed measurements by Earl Douglass (the discoverer of Dinosaur National Monument, and the person who excavated both skulls) revealed conclusively the solution to the mixup. It was the larger skull that belonged to the *Apatosaurus* skeleton. Consequently, *Apatosaurus* had a skull like that of *Diplodocus* (with peg-like teeth), not like that of *Camarasaurus* (with spoon-shaped teeth).

For complete details on this detective story, see McIntosh and Berman 1975 and Berman and McIntosh 1978.

The 1920s and 1930s were significant for massive monographs on *Camarasaurus* by H. F. Osborn and C. C. Mook (1921), on *Apatosaurus* by C. W. Gilmore (1936), and in 1929 an exhaustive study by Huene of a number of genera of the still-enigmatic titanosaurids found in the Late Cretaceous of Argentina.

During the 1930s, a spectacular Upper Jurassic sauropod site, the Howe Quarry near Shell, Wyoming, was excavated by the American Museum of

Natural History, but this material remains undescribed and largely unprepared. More significant, this decade marked the start of a series of papers on Chinese Jurassic sauropods by the legendary Chinese scientist Yang Zhong-jian, known in the West as C. C. Young. Several good partial skeletons of the Chinese *Helopus* (now *Euhelopus*) had been described by Wiman in 1929, but it was Young's papers over the next forty years that began to reveal the rich sauropod fauna of China. His work culminated in a 1972 paper with Chao on the "ridiculously" long-necked *Mamenchisaurus*.

The important sauropod contributions of the past two decades have come mostly from South America and China. Bonaparte and collaborators have described generalized Middle Jurassic genera from Argentina and a number of new animals from the Early and Late Cretaceous, many of them titanosaurids. Among the spectacular discoveries was the dermal armor of *Saltasaurus*. Even more spectacular was the discovery of the large Middle Jurassic dinosaur quarry at Dashanpu near Zigong in southern Sichuan Province of China. From these have come a large number of skeletons of the relatively primitive *Shunosaurus*, some of them virtually complete with skulls. Also present were at least ten skeletons of *Omeisaurus* and less complete material of *Datousaurus*. All three of these animals add greatly to our knowledge of the early evolution of the sauropod. Other discoveries by Dong Zhi-ming and others in various provinces of China continue to add to our knowledge of these animals.

Sauropod Anatomy

The sauropods were quadrupedal animals ranging in adult body length from about 7 meters up to perhaps as much as 40 meters. They were characterized by having the relatively smallest skulls (compared to body mass) of all the dinosaurs, the lowest dinosaurian EQs (encephalization quotient: the ratio of estimated brain mass to estimated body mass; *not* to be confused with "IQ," something that obviously cannot be measured for any dinosaur), long necks, long tails, massive limbs, and five-toed hands and feet. The very weak joint connecting the atlas (first neck vertebra) and the occipital condyle of the skull resulted in sauropod heads' often being separated from their necks and lost after death. Thus a large number of otherwise well-known genera lack any knowledge of the head—unfortunately, in most cases, the most important element in diagnosing the genera. Fortunately this disadvantage has been somewhat mitigated by the enormous diversity in the development of the vertebral columns. In order to reduce the weight of these animals, the vertebrae, particularly those of the neck and trunk, developed all sorts of cavities (called pleurocoels) and struts—a marvelous job of engineering for maximum strength with minimum bone that has not yet been equaled since the Mesozoic!

Broadly speaking, one may divide the sauropod skulls into two functional types (Fig. 20.1): large forms with thick spoon-shaped teeth, typified by *Camarasaurus* and *Brachiosaurus,* and smaller, lighter, longer-snouted forms with slender peg-like teeth, such as *Diplodocus* and its allies.

Important in defining the various genera is the total number and distribution of vertebrae in the neck and trunk (presacral vertebrae). The primitive number is twenty-five in both prosauropods and sauropods—ten cervicals (neck) and fifteen dorsals (trunk) in a typical prosauropod such as *Plateosaurus*. A basal sauropod such as *Shunosaurus* has twelve neck

Figure 20.1. Flesh restoration of the heads of *Diplodocus* (upper left) and *Camarasaurus* (lower right). This and other drawings in this chapter are by Gregory S. Paul.

vertebrae, thirteen dorsals, and no dorso-sacrals, for a total of twenty-five. During sauropod evolution, a process of "cervicalization" took place, in which vertebrae from the trunk were converted into neck vertebrae and, in addition, up to six more new vertebrae were added to the neck and trunk. Furthermore, the individual neck vertebrae became greatly elongated in some animals (more than 1 meter long in some cases).

The primitive sacrum consists of two vertebrae (sacral numbers 2 and 3). To this were added two more from the tail (numbers 4 and 5). Most Late Jurassic forms added another in front, from the trunk (number 1, the dorso-sacral), to produce the standard number of five, but the Late Cretaceous titanosaurs added a second dorso-sacral in front to increase the total number to six.

Another feature, first noted by Cope, was the development in many genera of a mechanism for strengthening the trunk portion of the vertebral column. In addition to the usual method of articulation, in which a pair of smooth processes (postzygapophyses) on the posterior side of the neural arch overlie and articulate with upwardly directed paired processes (prezygapophyses) on the anterior side of the arch of the following vertebra, there is a secondary means of articulation. Below the postzygapophyses there developed a block-like projection (hyposphene) which inserted into a cavity (hypantrum) below the prezygapophyses of the succeeding vertebrae, tending to lock the vertebrae together. This limited the flexibility of this part of the trunk, but gave much added strength. Another common variation in these vertebrae involves the amount and manner of hollowing out of the bodies (centra).

McIntosh, Brett-Surman, and Farlow

Variations in the tail among sauropod taxa involve an increase in the number of tail vertebrae from forty-four in early forms such as *Shunosaurus*, to fifty-three in *Camarasaurus*, to more than eighty in *Apatosaurus* and *Diplodocus*, including as many as forty rod-like elements which form a "whiplash" at the end of the tail. A further feature of the tail is the peculiar development of the chevron bones (the bones pointing downward from beneath the tail vertebrae) in some primitive forms such as *Shunosaurus*, and carried to extremes in *Diplodocus*. Instead of simply pointing downward, the middle chevrons are directed backward, at the same time developing a forwardly directed branch with the result that, when viewed from the side, the chevrons assume the shape of an isosceles triangle with the long side parallel to the ground.

The pectoral (shoulder) girdle consists of a large, elongated scapula expanded at both ends and an oval or quadrilateral-shaped coracoid, which fuses to the scapula in the adult. Small slender bones found with several forms, e.g., *Shunosaurus*, *Omeisaurus*, and *Diplodocus*, have been described as clavicles (collarbones), but this identification remains uncertain in some species. The upper arm bone (humerus) is broadly expanded both proximally and distally and is always longer than the forearm. There are three wrist bones in most Middle Jurassic forms, and two in other sauropods except *Apatosaurus*, which has but one.

There are five sturdy metacarpals in the hand. These are so arranged that when viewed from above, their upper ends form about two-thirds of a circle. They stand nearly vertically so that the animal walked on its fingertips (digitigrade), supported by a large pad in the palm, but this is not obvious in some sauropod tracks such as *Brontopodus* (Farlow et al. 1989). The number of phalanges (finger bones) is greatly reduced, and in most cases only the inner digit bears a claw. Indeed, there is some indication that in some of the latest sauropods, all claws and even all of the phalanges may have vanished.

The three pelvic bones are large and are usually separate, but particularly in early forms the ischia and/or pubes may fuse along the median line, and occasionally in a very old individual the ischium and pubis may fuse. The femur (thighbone) is sturdy and in most sauropods is the longest bone in the skeleton, *Brachiosaurus* with its long forelimb being an exception. The femur is always longer than the bones of the lower leg (tibia and fibula). The ankle joint is reduced to two elements in most sauropods: a large astragalus which locks to the lower end of the tibia, and a much smaller, spherical calcaneum, which disappears in the diplodocids. The rounded undersurface of the astragalus produces a very flexible joint with the underlying five metatarsals (main foot bones).

Unlike the bones of the forefoot, the metatarsals are arranged in semicircular fashion and are directed downward and outward in semiplantigrade fashion. This results in a much larger footprint for the hind foot relative to that of the forefoot. A large fleshy "cushion" formed a sort of heel somewhat like that in elephants. The number of phalanges is considerably greater than in the hand, only the outer two digits being reduced in most forms. Thus the three inner toes bear claws, and occasionally one remains on the fourth toe as well.

Sauropod Evolution

The ancestry of the sauropods, before they burst onto the world scene on almost every continent in the Middle Jurassic, is obscure. The frequent

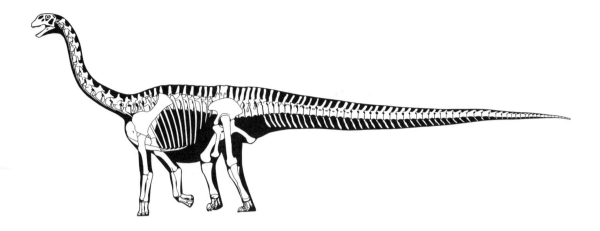

Figure 20.2. Skeletal reconstruction of *Patagosaurus*. Femur length 1360 mm.

assumption that they arose from prosauropods, probably melanorosaurids, has yet to be verified. Furthermore, some characters, including the reduction of the fifth digit of the hind foot in all prosauropods (but not in sauropods), suggest that these animals were already too specialized to have served as sauropod ancestors, and that the sauropods may have arisen as a sister group in the Late Triassic. Indeed, the traditional assumption that the quadrupedal sauropods developed from bipedal prosauropods was questioned as early as 1965 by A. J. Charig, J. Attridge, and A. W. Crompton.

The first animal that can reasonably be recognized as a probable sauropod is *Vulcanodon* from the Triassic-Jurassic boundary in Zimbabwe, and uncertainties about its affinities still remain. The all-important skull and vertebrae (except for those of the sacrum and tail) are unknown. Overall, *Vulcanodon* can be recognized as a quadruped about 10 meters long. Known portions of the animal's skeleton show a mix of prosauropod-like and sauropod-like characters. *Zhigongosaurus,* from the Lower Jurassic of China, and *Ohmdenosaurus,* from rocks of similar age in Germany, may be members of the same family.

The ecological role occupied by the prosauropods in the Late Triassic and Early Jurassic was filled in the Middle Jurassic by the sauropods, with the

Figure 20.3. Skeletal reconstruction of *Haplocanthosaurus*. Femur length 1745 mm.

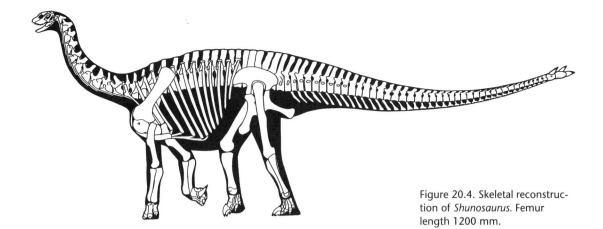

Figure 20.4. Skeletal reconstruction of *Shunosaurus*. Femur length 1200 mm.

disappearance of the former group. A number of different genera with relatively simple vertebrae appeared all around the world and are often lumped together as cetiosaurids (Clark et al. 1995). These include *Cetiosaurus* in Great Britain, *Patagosaurus* (Fig. 20.2) in Argentina, *Barapasaurus* in India, and *Rhoetosaurus* in Australia. A related animal, *Haplocanthosaurus* (Fig. 20.3), survived into the Late Jurassic in the American West. Complete skulls are not known for any of these animals, but an offshoot of this group, consisting of several different genera with well-preserved skulls and forked chevron bones, reminiscent of the later diplodocids, appeared in the Middle Jurassic of China. The first of these, *Shunosaurus* (Fig. 20.4), is

Figure 20.5. Skeletal reconstruction of *Mamenchisaurus*. Femur length 1350 mm.

Sauropods 273

known from a number of articulated skeletons. A second, more advanced form, *Omeisaurus,* was also found in China. Yet another advanced Chinese form, the enormously long-necked *Mamenchisaurus* (Figs. 20.5, 20.6) from the Late Jurassic, was surely one of the strangest sauropods of all, with the ultimate development of the neck, which had nineteen elongated vertebrae. This animal appears to be close to the ancestral line of the diplodocids, but retained a number of primitive characters. It had well-developed forked chevrons and divided vertebral spines in the shoulder region, but still had fairly broad spatulate teeth. How the tail terminated is not known.

The Middle Jurassic of Madagascar has yielded fine sauropod material of the related brachiosaurids, including many bones of juveniles. These remains have been referred to the ill-defined English genus *Bothriospondylus,* a forerunner of the great Late Jurassic *Brachiosaurus* (Fig. 20.7).

Bones of *Brachiosaurus* have been found in such widely separated areas as Colorado, Tanzania, and Portugal. This huge animal, whose head towered 13 meters above the ground, had forelimbs longer than hind, an enormously long neck, and a short tail. Its teeth were broad and spoon-shaped. The spines of the neck and back vertebrae were undivided and the chevrons simple. A striking feature was the diminution in size of the single claw in the hand. Footprints (given the name *Brontopodus;* Farlow et al. 1989) of presumably related animals (*Pleurocoelus*) from the Early Cretaceous of Texas suggest that the claw may have, by that time, vanished altogether.

The camarasaurids appeared in North America and Europe in the Late Jurassic. *Camarasaurus* (Fig. 20.8) was the most common sauropod in the Morrison Formation of the western United States, and along with *Shunosaurus* (Fig. 20.4) is the only sauropod whose osteology is completely known (Madsen et al. 1995; McIntosh et al. 1996). By comparison with other sauropods, the skull was massively built, with a blunt, bulldog-like muzzle and large spoon-shaped teeth. There were twelve vertebrae in the neck, twelve in the trunk, and one dorsal-sacral for a total of twenty-five. Vertebrae of the neck and front half of the trunk had deeply divided neural spines, forming a U-shaped trough, in which lay large ligaments (neural ligaments) connecting the head with the vertebral column back to the sacrum. The tail was only moderately long, with simple chevrons. The hind limb was longer than the forelimb. Related animals have been found in rocks of about the same age in Spain (*Aragasaurus*) and Portugal. *Euhelopus* (Fig. 20.9) from China had a skull resembling that of *Camarasaurus,* but it had seventeen vertebrae in the neck and only slightly divided vertebral spines. A doubtfully associated very long forelimb suggests a relationship with the brachiosaurids, but future discoveries will be needed to determine the true relationships of *Euhelopus.* Yet another sauropod questionably affiliated with the camarasaurids is *Opisthocoelicaudia* (Fig. 20.10) from the Upper Cretaceous of Mongolia.

Another important Late Jurassic family, the diplodocids, was common in North America and East Africa. It includes two of the best-known North American Jurassic dinosaurs, *Diplodocus* (Fig. 20.11) and *Apatosaurus* (Fig. 20.12), animals which attained an overall length of 22 to 24 meters. The family was characterized by an elongated skull with a protruding muzzle and nostrils that opened on the top of the skull near the eyes (Fig. 20.1). However, it was the teeth that were most distinctive. They were small, slender, and pencil-like, and confined to the front of the jaws.

Figure 20.6. Flesh restoration of the sauropod *Mamenchisaurus* being attacked by the theropod *Yangchuanosaurus*.

Figure 20.7. Flesh restoration of a *Brachiosaurus* group.

Sauropods 277

Figure 20.8. Skeletal reconstruction of a juvenile (*above*) and adult (*below*) *Camarasaurus*. Femur length of the juvenile 567 mm, and of the adult 1525 mm.

Vertebrae from the trunk had become incorporated into the neck; there were fifteen neck, ten trunk, and one dorso-sacral vertebrae in both *Diplodocus* and *Apatosaurus*. A deep V-shaped cleft in the spine of these vertebrae housed the ligaments. The tail was enormously long: There were eighty-two caudal vertebrae in *Apatosaurus,* of which the last forty or so were merely long rods of bone forming a "whiplash." Very characteristic were the diplodocid middle chevrons; their skid-like shape reached its greatest development in *Diplodocus,* and has led to much speculation concerning the dinosaur's feeding habits, as described below. Another important characteristic of this family was the significant shortening of the forelimb. A most unexpected recent discovery was a complete set of gastralia (abdominal ribs) in a skeleton of *Apatosaurus*. This suggests that gastralia were probably present in all sauropods.

Diplodocus is distinguished from *Apatosaurus* in having very slender limb bones and elongated neck and tail vertebrae, while *Apatosaurus* has massive limbs and shorter vertebrae. Robert Bakker once fittingly termed *Apatosaurus* a fat *Diplodocus*. Another animal closely related to *Diplodocus* was *Barosaurus* (Fig. 20.13), an American sauropod which differed from *Diplodocus* by having enormously elongated neck vertebrae, rivaling in size those of *Brachiosaurus*. Two gigantic diplodocids have been reported from North America. The first, *Supersaurus,* may have reached 38 meters in length, but a full characterization of this animal must await discovery of more material. (Interestingly, some of the sauropod skeletal material previously named *Ultrasauros* actually belongs to *Supersaurus;* Curtice et al. 1996.) The other, *Seismosaurus,* was equally large and appears to be most closely related to *Diplodocus* itself, but so far its skeleton has been only partially prepared. An interesting feature was the discovery of a large number of gastroliths (stomach stones) in the gut (see also Calvo 1994). Finally the rear two-thirds of a primitive diplodocid, *Cetiosauriscus* from England, remains one of the most complete sauropod skeletons yet found in Europe.

A peculiar offshoot of the diplodocid family occurred in the southern supercontinent Gondwana. *Dicraeosaurus* (Fig. 20.14) from the Late Jurassic of Tanzania had a *Diplodocus*-like skull but a more primitive vertebral distribution—twelve in the neck, twelve in the trunk, and one dorso-sacral. The most striking feature was the enormously high, deeply cleft spines in the neck and trunk vertebrae. Even more extreme was the Early Cretaceous *Amargasaurus* (Fig. 20.15) from Argentina. Here the spines are even much higher, producing the appearance of a sail, but the number of vertebrae was reduced.

The final group of sauropods, the titanosaurids, arose in the Late Jurassic or Early Cretaceous and remained very common in parts of the former southern continent, Gondwana, up to the very end of the Cretaceous. They were less common in Europe, and disappeared altogether for a time in North America (Lucas and Hunt 1989) before being reintroduced from South America in the Late Cretaceous and working their way up as far north as Utah (*Alamosaurus*) before the final extinction of dinosaurs. The titanosaurids present the greatest problem to sauropod scholars, since no complete articulated skeleton, and no complete skull, has ever been found.

Titanosaurids range in size from 7-meter-long *Saltasaurus* to gigantic animals 25 to 30 meters in length. Their teeth are all very small and slender, superficially resembling those of the diplodocids. This feature led a number of paleontologists for many years to refer the diplodocids and titanosaurids to the same family, even though their postcranial skeletons are very different. The all-important vertebral spines of the titanosaurids are all single (unforked), the chevrons are single (not skid-like as in the diplodocids), and a second trunk vertebra has been added to the sacrum to give a total of six. The tail vertebrae are very distinctive; their bodies (centra) are typically procoelous (i.e., they have a ball-and-socket relationship to one another, in which the socket is on the front of the centrum while the ball is on the back). Most interesting is the fact that some or perhaps all of the titanosaurids have bony body armor consisting of embossed plates and small ossicles. A large number of genera have been named, most of them on only a small part of the skeleton. One of the better-articulated skeletons belongs to the large North American genus *Alamosaurus*. Other well-represented forms are the small *Saltasaurus* and *Neuquensaurus* and the large *Argyrosaurus* and *Antarctosaurus,* from Argentina. Titanosaurids have also been reported from the Late Cretaceous of Brazil, Egypt, India, Vietnam, Thailand, England, France, Romania, Spain, and Texas.

Classification

Ideas concerning sauropod classification have changed widely over the years, and continue to do so. Marsh's multifamily scheme with variations prevailed until 1929, when Janensch proposed a two-family scheme (each with four subfamilies) based on the teeth. Those with broad spatulate teeth were placed in the Brachiosauridae, while those with slender peg-shaped teeth were included in the Titanosauridae. The result of this scheme was to bracket in the same family the diplodocids and the titanosaurids, two groups of sauropods as widely divergent from one other as possible. They shared one character in common, the teeth. Most students of the sauropods today recognize four or five well-established families and several less well grounded ones. In this chapter a conservative approach (see Table 20.1) will be taken, with the full knowledge that future discoveries will probably require the establishment of more families (McIntosh 1989, 1990). For an alternative classification, see Upchurch 1995 and Calvo and Salgado 1995.

Table 20.1
Classification of the Sauropods

Sauropoda
      Vulcanodontidae
      Cetiosauridae
            Cetiosaurinae
            Shunosaurinae
      Brachiosauridae
      Camarasauridae
      Diplodocidae
            Diplodocinae
            Dicraeosaurinae
      Titanosauridae
            Titanosaurinae
            Andesaurinae (?)

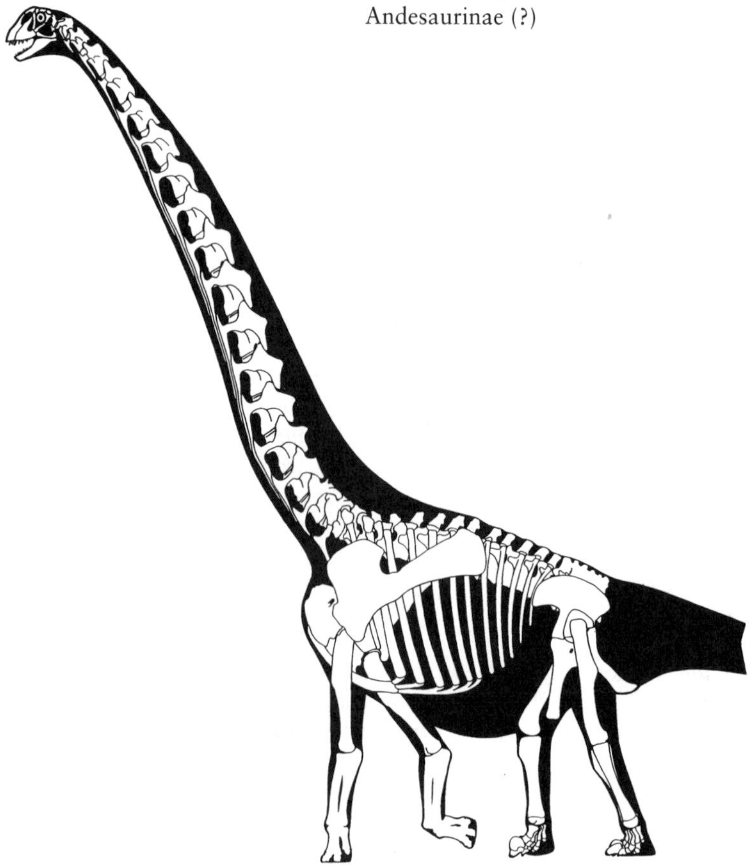

Figure 20.9. Skeletal reconstruction of *Euhelopus*. Femur length 955 mm.

Paleobiology

As already noted, the sauropods were large, quadrupedal herbivores which inhabited all the continents except possibly, but not probably, Antarctica. They ranged in adult length from about 7 to 40 meters. They had the longest necks and tails of all the dinosaurs, and the limbs were held upright in a pillar-like fashion, as opposed to the crocodilian manner. Their overall weight is the subject of considerable controversy, because this depends on how one "dresses up" models with muscles and other sinews, based on the locations and sizes of the muscle scars on the individual bones. In one study, E. H. Colbert (1962) estimated the weight of *Apatosaurus* (*"Brontosaurus"*) at about 30 tons, *Diplodocus* at 12 tons, and *Brachiosaurus* at 80 tons. At the other extreme, Béland and Russell (1980) estimated the body mass of *Brachiosaurus* to be as little as 15,000 kilograms. More recent body mass estimates of the better-known Morrison Formation sauropods are roughly in the middle of these extremes (Anderson et al. 1985; Coe et al. 1987); the very largest sauropods could well have exceeded 50,000 kilograms in body mass, and some might even have reached 100,000 kilograms (Gillette 1991, 1994).

Some sauropods possessed a row of conical epidermal spines down the tail and back (Czerkas 1992, 1994) reminiscent of Dinny in the old *Alley Oop* comic strip. There is not, and can probably never be, any direct evidence about colors and color patterns of the sauropod integument, but

Figure 20.10. Skeletal reconstruction of *Opisthocoelicaudia*. Femur length 1395 mm. The cross-section of the animal is split down the middle, from top to bottom. The left side of the cross-section shows the hind limb as seen from the rear, with the tail removed. The right side of the cross-section shows the forelimb as seen from the front, with the neck removed.

Figure 20.11. Skeletal reconstruction of *Diplodocus*. Femur length 1542 mm.

Sauropods 281

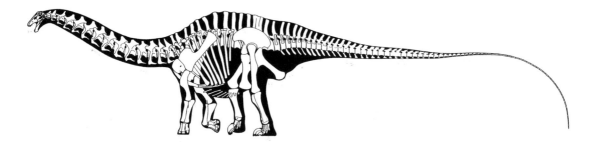

Figure 20.12. Skeletal reconstruction of *Apatosaurus*. Femur length 1785 mm.

it is unlikely that they were all the monochrome drab green or gray creatures pictured in early restorations.

Questions concerning the "lifestyle" of sauropods have spawned many controversies over the years, but conclusions are based largely on indirect evidence. Very useful in this respect have been footprints (Lockley et al. 1994a). Sauropod trackways have been cited as evidence that these dinosaurs could swim, but such "manus-dominated" trails may merely reflect conditions of footprint preservation (Lockley et al. 1994c). Tail drag marks are almost never found (Lockley et al. 1994a), which suggests that the tail may have floated or, an even more sweeping conclusion, that the sauropod tail was always carried free of the ground. This makes good sense in the case of short- or moderately long-tailed animals such as *Brachiosaurus,* but questions remain concerning the diplodocids, where the last forty or eighty tail vertebrae were simply rods of bone without neural arches or any other visible attachments for ligaments or sinews of any kind. Certainly these animals could swing their tails far off the ground, but could they hold them there while at rest? Indeed it has been suggested that this "whiplash" was exactly that—a weapon of defense. The "clubbed" bony tails of the short-tailed *Shunosaurus* may have been used for this purpose.

One question which did lead to much controversy in the early part of this century may be said to have been resolved. O. C. Marsh, in his first (1883) reconstruction of a sauropod skeleton (*"Brontosaurus"*), showed the limbs directed downward in a mammal-like position, rather than sprawling out horizontally as in crocodiles. H. F. Osborn and J. B. Hatcher agreed with this interpretation. However, in 1908 O. P. Hay argued for a more crocodilian-like limb carriage, and he was seconded in this opinion in a number of papers by Gustave Tornier. As a result of these papers, a *Diplodocus* skeleton in Frankfurt, Germany, which had originally been mounted in the traditional pose was dismantled and remounted in the sprawling position. The question was definitely resolved

Figure 20.13. Skeletal reconstruction of *Barosaurus*. Femur length 1520 mm.

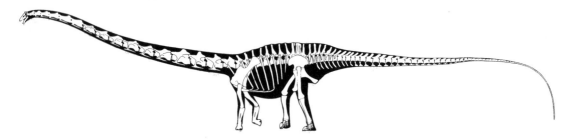

McIntosh, Brett-Surman, and Farlow

Figure 20.14. Skeletal reconstruction of *Dicraeosaurus*. Femur length 1220 mm.

in Marsh's favor by W. J. Holland, who in 1910 took casts of the hip (pelvic) and thigh bones of a *Diplodocus*, and showed that when the head of the latter was placed in the hip socket and oriented horizontally, the thigh was locked into place, preventing all movement. If more than that was needed, the icing on the cake was provided by the discovery of well-preserved sauropod trackways in the Early Cretaceous Glen Rose Limestone of Texas; these showed a relatively narrow width that was inconsistent with a sprawling limb carriage (see Farlow and Chapman, chap. 36 of this volume).

The first complete sauropod skull to be found (and for many years the only one) was that of *Diplodocus*, with nostrils opening upward on the top of the head. This led Marsh to suggest that the sauropods were semiaquatic, and this proposal remained unchallenged for many years. Paleontologists agreed that these animals could, and did, emerge from the lakes and rivers occasionally, perhaps to lay eggs or migrate, but that their normal condition was partial submersion. Subsequently K. A. Kermack (1951), and especially R. T. Bakker (1971) and W. P. Coombs (1975) have strenuously argued a different view, namely that the sauropod's anatomy, in many ways elephant-like, pointed to a land-dwelling creature (summarized in Desmond 1975). The present consensus is that sauropods were indeed terrestrial, but that, as the many sauropod trackways testify, they did enter wet, muddy areas on occasion.

Sauropod bones and footprints occur in a variety of sedimentary contexts (Dodson 1990; Farlow 1992), suggesting that these dinosaurs lived in a variety of habitats. However, Lockley (1991) argued, on the basis of the paleoecological and paleogeographic occurrence of sauropod footprint localities, that sauropods as a group showed a preference for low-latitude situations, as opposed to the high-latitude settings in which footprint assemblages are dominated by ornithopod trackways. Farlow (1992) examined the paleogeographic occurrences of sauropod skeletal

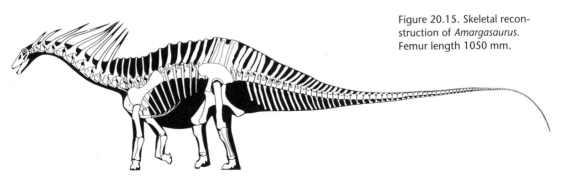

Figure 20.15. Skeletal reconstruction of *Amargasaurus*. Femur length 1050 mm.

sites as well as tracksites, and found the evidence for a paleolatitudinal separation of ornithopods and sauropods less persuasive.

Farlow's conclusions were in turn challenged by Lockley et al. (1994b; also see Lockley, chap. 37 of this volume). These authors added data from new sauropod tracksites and found strong support for the pattern described by Lockley (1991). However, they counted different footprint-bearing layers at particular localities as separate sites, even when these layers were separated by less than a meter of sediment thickness. Furthermore, these authors counted as separate sites footprint localities separated by relatively short geographic distances. Although track occurrences as close together in space and time as these might well be considered different sites, whether the approach employed by Lockley et al. (1994b) is biologically reasonable is arguable. Given the large home range sizes of individual animals that one would expect for sauropod-sized creatures, and the large geographic ranges likely to be occupied by such enormous animals as sauropods (Brown 1995), along with the possibility that multiple closely spaced track-bearing layers at a given site might not be separated by long enough time intervals to record the presence of different sauropod species, there is a real possibility that Lockley et al. (1994b) introduced significant biases into their data set. If a few sauropod species were repeatedly represented in the data, to an extent that other sauropod species were not, this could give a misleading impression about the latitudinal preferences (if any existed) of sauropods as a group.

R. T. Bakker (1986) vigorously argues that sauropods (and other dinosaurs as well) were fully warm-blooded, with high food-consumption rates. This conclusion has been challenged by other workers (e.g., Weaver 1983; Spotila et al. 1991; Daniels and Pratt 1992; Farlow et al. 1995; Paladino et al., chap. 34 of this volume). Perhaps the majority of dinosaur scholars, while admitting that other dinosaurs (particularly theropods—but see Ruben et al., chap. 35 of this volume) may have been warm-blooded, now believe that sauropods did not possess an internal biological "thermostat set on high" as adults, but that they were homeothermic—i.e., heat absorbed by their massive bodies, and produced by their own metabolism, was sufficient to mimic modern "endothermy." Thus a moderate body temperature could be maintained even at rest. When one considers that the Mesozoic was much warmer than today, and that sauropods had the smallest surface area/body volume ratio, the sauropods' main problem may have been overheating. If one compares a sauropod to an elephant skeleton, it is apparent that the main body cavities are similar in size, yet the sauropods have noticeably longer legs, necks, and tails, which are cylindrical in shape and very thin (compared to their lengths). This gives the sauropods essentially six long cylinders with a high proportion of surface area to mass—in other words, six cooling towers to gain or lose heat.

The shapes of sauropod teeth provide clear evidence that these animals were not meat-eaters. On the other hand, wear facets on the teeth of both the camarasaurid and diplodocid types (Fiorillo 1991; Calvo 1994; Barrett and Upchurch 1994) show that the diet did not consist merely of succulent ferns, which formed the major part of the low-lying ground flora. Cycads, high-rising tree ferns, and conifers no doubt played a role. The recent discovery of a large number of gastroliths (stomach stones) in the gut of *Seismosaurus* (Gillette 1991) provided confirming evidence that these small polished stones, found in a number of sauropod-bearing strata, were indeed analogous to those used by some animals, for example certain crocodiles, to help grind up food and aid in the digestion process (Stokes 1942; Farlow, in Bird 1985).

The question of whether sauropods could rear up on their hind legs to feed has been argued for years, and continues to be debated today. Noting the unusual development of the chevrons in the midtail region of *Diplodocus,* which lie horizontally rather than vertically, Hatcher (1901) suggested that the animal reached up to feed on high branches, using the hind limbs and tail as a tripod. This idea was reiterated by Bakker in 1986. In 1993 a skeleton of the close *Diplodocus* relative *Barosaurus* was mounted in this position at the American Museum of Natural History in New York. Critics of this idea have been many, arguing that pumping blood up the vertically directed neck in a rearing sauropod would be impossible, and that the muscular requirements to lift the front end of so heavy a body would be impossible. To answer the first of these objections, some bizarre-sounding suggestions have been put forth—including the notion that sauropods such as *Barosaurus* might have employed several auxiliary hearts to pump blood up their long necks (Choy and Altman 1992)!

Regardless of *how* the blood transfer was accomplished, it is clear that in *Brachiosaurus,* for example, the neck was naturally directed upward and indeed blood did reach the head. (Giraffes have a series of valves in their carotid arteries to prevent the backflow of blood.) As to the biomechanics of lifting the front end of the animal, a definitive study has not yet been concluded, but one point should be noted: In the diplodocids, the trunk and especially the forelimbs have been significantly shortened at the same time as the neck and tail sections have been greatly lengthened. The result is the transfer of the center of gravity rearward, and the reduction of the torque about it if the animal tried to rear up. Thus although it is quite clear that *Brachiosaurus* (*Jurassic Park* notwithstanding) and perhaps most other sauropods did not rear up to feed, there remains the distinct possibility that animals such as *Diplodocus* and *Barosaurus* did so (cf. Alexander 1985, 1989, and chap. 30 of this volume).

A final question concerns whether the sauropods laid eggs or brought forth live young. In the nineteenth century, large numbers of eggs were discovered in southern France in the same strata as bones of the titanosaurid sauropod *Hypselosaurus* (Buffetaut and Le Loeuff 1994). Consequently many writers have accepted their association, and concluded that the sauropods were egg-layers (oviparous). While it still appears likely that the sauropods did indeed lay eggs (cf. Carpenter and McIntosh 1994), one must be careful in accepting this sort of association (see Hirsch and Zelenitsky, chap. 28 of this volume). It should be noted that apparently all other saurischians, as well as birds and crocodilians, laid (or lay) eggs.

The most common question asked about sauropods is, Why did they get so big? (See Fig. 20.16.) This is not an easy question to answer. Body size is linked with a variety of physiological and ecological parameters in a very complex fashion (Peters 1983; Calder 1984; Dunham et al. 1989; Spotila et al. 1991; Brown 1995), and picking out which of the many features of an animal's biology has the greatest influence on its body size is problematical. However, we can identify certain ways in which gigantism may have been advantageous to sauropods.

The first advantage is that the larger a herbivore is, the more food it can get to. Sauropods were not only long, but tall. This allowed them to reach into the foliage of trees where the other dinosaurs could not feed. Thus they had a feeding advantage over other, smaller species. If parental care existed in sauropods, then they could also get to a food source that would not put them in competition with their own juveniles, and it also may have given them the ability to get food for juveniles that they could not get themselves.

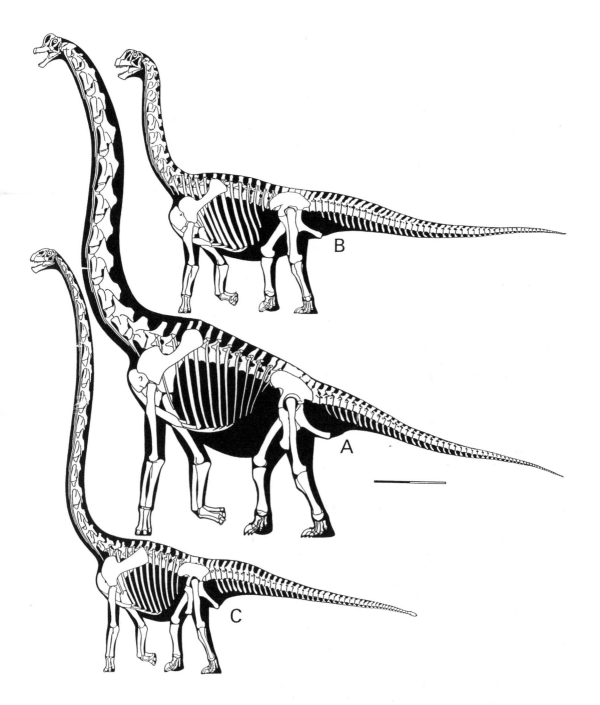

Figure 20.16. A size comparison of (A) *Brachiosaurus,* (B) *Camarasaurus,* and (C) *Omeisaurus.* Scale bar equals 2 m.

A second advantage of large size is defense. The larger an animal is, the better it can defend itself. Size itself can be a major advantage. Sauropods could not outrun any theropod. Their sheer mass gave them enough power that any blow from a limb or tail would have had devastating effects.

It may be instructive to think of sauropod gigantism as an extreme case of the tendency toward very large size that was characteristic of dinosaurs as a group (Hotton 1980; Peckzis 1994). Farlow et al. (1995) contrasted the large body sizes seen in dinosaurian faunas with those of mammalian faunas, and noted that dinosaurs evolved gigantism to a degree never seen in terrestrial mammals. They considered several factors that might have contributed to dinosaurian gigantism, but concluded that the most important factor was probably lower food requirements per animal than would be necessary for equally large mammals. This in turn would make it easier for species of huge animals to maintain large enough population sizes to survive the kinds of environmental calamities that constantly threaten species with extinction.

Whatever the factors that contributed to sauropod gigantism, they resulted in a group of dinosaurs that enjoyed an impressive ecological and evolutionary success—whether one thinks in terms of diversity of taxa or of the stratigraphic longevity of the group as a whole.

References

Alexander, R. McN. 1985. Mechanics of posture and gait of some large dinosaurs. *Zoological Journal of the Linnean Society* 83: 1–25.

Alexander, R. McN. 1989. *Dynamics of Dinosaurs and Other Extinct Giants.* New York: Columbia University Press.

Anderson, J. F.; A. Hall-Martin; and D. A. Russell. 1985. Long-bone circumference and weight in mammals, birds and dinosaurs. *Journal of Zoology, London* A 207: 53–61.

Bakker, R. T. 1971. The ecology of the brontosaurs. *Nature* 229: 172–174.

Bakker, R. T. 1986. *The Dinosaur Heresies: New Theories Unlocking the Mystery of the Dinosaurs and Their Extinction.* New York: William Morrow.

Barrett, P. M., and P. Upchurch. 1994. Feeding mechanisms of *Diplodocus. Gaia* 10: 195–203.

Béland, P., and D. A. Russell. 1980. Dinosaur metabolism and predator/prey ratios in the fossil record. In R. D. K. Thomas and E. C. Olson (eds.), *A Cold Look at the Warm-Blooded Dinosaurs,* pp. 85–102. American Association for the Advancement of Science Selected Symposium 28. Boulder, Colo.: Westview Press.

Berman, D. S., and J. S. McIntosh. 1978. Skull and relationships of the Upper Jurassic sauropod *Apatosaurus* (Reptilia: Saurischia). *Bulletin of the Carnegie Museum of Natural History* 8: 1–35.

Bird, R. T. 1985. *Bones for Barnum Brown: Adventures of a Dinosaur Hunter.* Fort Worth: Texas Christian University Press.

Brown, J. H. 1995. *Macroecology.* Chicago: University of Chicago Press.

Buffetaut, E., and J. Le Loeuff. 1994. The discovery of dinosaur eggshells in nineteenth-century France. In K. Carpenter, K. F. Hirsch, and J. R. Horner (eds.), *Dinosaur Eggs and Babies,* pp. 31–34. Cambridge: Cambridge University Press.

Calder, W. A. 1984. *Size, Function, and Life History.* Cambridge, Mass.: Harvard University Press.

Calvo, J. O. 1994. Jaw mechanics in sauropod dinosaurs. *Gaia* 10: 183–193.

Calvo, J. O., and L. Salgado. 1995. *Rebbachisaurus tessonei* sp. nov., a new sauropod from the Albian-Cenomanian of Argentina: New evidence of the origin of the Diplodocidae. *Gaia* 11: 13–33.

Carpenter, K., and J. McIntosh. 1994. Upper Jurassic sauropod babies from the Morrison Formation. In K. Carpenter, K. F. Hirsch, and J. R. Horner (eds.), *Dinosaur Eggs and Babies,* pp. 265–278. Cambridge: Cambridge University Press.

Charig, A. J.; J. Attridge; and A. W. Crompton. 1965. On the origins of the sauropods and the classification of the Saurischia. *Proceedings of the Linnaean Society of London* 176: 197–221.

Choy, D. S. J., and P. Altman. 1992. The cardiovascular system of barosaurus [*sic*]: An educated guess. *Lancet* 340: 534–536.

Clark, N. D. L.; J. D. Boyd; R. J. Dixon; and D. A. Ross. 1995. The first Middle Jurassic dinosaur from Scotland: A cetiosaurid? (Sauropoda) from the Bathonian of the Isle of Skye. *Scottish Journal of Geology* 31 (2): 171–176.

Coe, M. J.; D. L. Dilcher; J. O. Farlow; D. M. Jarzen; and D. A. Russell. 1987. Dinosaurs and land plants. In E. M. Friis, W. G. Chaloner, and P. R. Crane (eds.), *The Origins of Angiosperms and Their Biological Consequences*, pp. 225–258. Cambridge: Cambridge University Press.

Colbert, E. H. 1962. The weights of dinosaurs. *American Museum Novitates* 2076: 1–16.

Coombs, W. P. Jr. 1975. Sauropod habits and habitats. *Palaeogeography, Palaeoclimatology, Palaeoecology* 17: 1–33.

Curtice, B. D.; K. L. Stadtman; and L. J. Curtice. 1996. A reassessment of *Ultrasauros macintoshi* (Jensen, 1985). In M. Morales (ed.), *The Continental Jurassic*, pp. 87–95. Bulletin 60. Flagstaff: Museum of Northern Arizona.

Czerkas, S. 1992. Discovery of dermal spines reveals a new look for sauropod dinosaurs. *Geology* 20: 1068–1070.

Czerkas, S. 1994. The history and interpretation of sauropod skin impressions. *Gaia* 10: 173–182.

Daniels, C. B., and J. Pratt. 1992. Breathing in long necked dinosaurs: Did the sauropods have bird lungs? *Comparative Biochemistry and Physiology* 101A: 43–46.

Desmond, A. J. 1975. *The Hot-Blooded Dinosaurs: A Revolution in Palaeontology.* London: Blond and Briggs.

Dodson, P. 1990. Sauropod paleoecology. In D. B. Weishampel, P. Dodson, and H. Osmólska (eds.), *The Dinosauria*, pp. 402–407. Berkeley: University of California Press.

Dunham, A. E.; K. L. Overall; W. P. Porter; and C. A. Forster. 1989. Implications of ecological energetics and biophysical and developmental constraints for life-history variation in dinosaurs. In J. O. Farlow (ed.), *Paleobiology of the Dinosaurs*, pp. 1–19. Special Paper 238. Boulder, Colo.: Geological Society of America.

Farlow, J. O. 1992. Sauropod tracks and trackmakers: Integrating the ichnological and skeletal records. *Zubía* 10: 89–138.

Farlow, J. O.; P. Dodson; and A. Chinsamy. 1995. Dinosaur biology. *Annual Review of Ecology and Systematics* 26: 445–471.

Farlow, J. O.; J. G. Pittman; and J. M. Hawthorne. 1989. *Brontopodus birdi*, Lower Cretaceous sauropod footprints from the U. S. Gulf Coastal Plain. In D. D. Gillette and M. G. Lockley (eds.), *Dinosaur Tracks and Traces*, pp. 371–394. Cambridge: Cambridge University Press.

Fiorillo, A. R. 1991. Dental microwear on the teeth of *Camarasaurus* and *Diplodocus*: Implications for sauropod paleoecology. In S. Kielan-Jawowrowska, N. Heintz, and H. A. Nakrem (eds.), *Fifth Symposium Mesozoic Terrestrial Ecosystems and Biota*, pp. 23–24. Contributions of the Paleontology Museum of the University of Oslo 364.

Gillette, D. D. 1991. *Seismosaurus halli* (n. gen., n. sp.) a new sauropod dinosaur from the Morrison Formation (Upper Jurassic–Lower Cretaceous) of New Mexico, U.S.A. *Journal of Vertebrate Paleontology* 11: 417–433.

Gillette, D. D. 1994. *Seismosaurus the Earth Shaker.* New York: Columbia University Press.

Gilmore, C. W. 1936. Osteology of *Apatosaurus* with special reference to specimens in the Carnegie Museum. *Memoirs of the Carnegie Museum* 11: 175–300.

Hatcher, J. B. 1901. *Diplodocus* Marsh: Its osteology, taxonomy, and probable habits, with a restoration of the skeleton. *Memoirs of the Carnegie Museum* 1: 1–63.

Holland, W. J. 1910. A review of some recent criticisms of the restorations of sauropod dinosaurs existing in the museums of the United States, with special reference to that of *Diplodocus carnegiei* in the Carnegie Museum. *American Naturalist* 44: 259–283.

Hotton, N. III. 1980. An alternative to dinosaur endothermy: The happy wanderers. In R. D. K. Thomas and E. C. Olson (eds.), *A Cold Look at the Warm-Blooded*

Dinosaurs, pp. 311–350. American Association for the Advancement Science Selected Symposium 28. Boulder, Colo.: Westview Press.

Huene, F. von. 1929. Los saurisquios y Ornithisquios de Cretacéo Argentino. *Annals Museum La Plata,* Series 2, 3: 1–196.

Janensch, W. 1929. Material und Formengehalt der Sauropoden in der Ausbeute der Tengaguru-Expedition. *Palaeontographica,* Supplement 7, 2: 1–34.

Kermack, K. A. 1951. A note on the habits of sauropods. *Annals and Magazine of Natural History,* Series 12, 4: 830–832.

Lockley, M. 1991. *Tracking Dinosaurs: A New Look at an Ancient World.* Cambridge: Cambridge University Press.

Lockley, M. G.; V. F. dos Santos; C. A. Meyer; and A. Hunt (eds.). 1994a. Aspects of sauropod paleobiology. *Gaia* 10: 1–279.

Lockley, M. G.; C. A. Meyer; A. P. Hunt; and S. G. Lucas. 1994b. The distribution of sauropod tracks and trackmakers. *Gaia* 10: 233–248.

Lockley, M. G.; J. G. Pittman; C. A. Meyer; and V. F. dos Santos. 1994c. On the common occurrence of manus-dominated sauropod trackways in Mesozoic carbonates. *Gaia* 10: 119–124.

Lucas, S. G., and A. P. Hunt. 1989. *Alamosaurus* and the sauropod hiatus in the Cretaceous of the North American Western Interior. In J. O. Farlow (ed.), *Paleobiology of the Dinosaurs,* pp. 75–85. Special Paper 238. Boulder, Colo.: Geological Society of America.

Lydekker, R. 1877. Notices of new and other Vertebrata from Indian Tertiary and Secondary rocks. *Records of the Geological Survey of India* 10: 30–43.

Madsen, J. H. Jr.; J. S. McIntosh; and D. S. Berman. 1995. Skull and atlas-axis complex of the Upper Jurassic sauropod *Camarasaurus* Cope (Reptilia: Saurischia). *Bulletin of the Carnegie Museum* 31: 1–115.

Mantell, G. A. 1850. On the *Pelorosaurus;* an undescribed gigantic terrestrial reptile, whose remains are associated with those of the *Iguanodon* and other saurians in the strata of the Tilgate Forest, in Sussex. *Philosophical Transactions of the Royal Society of London* 140: 391–392.

Marsh, O. C. 1883. Principal characters of American Jurassic dinosaurs. Part VI: Restoration of *Brontosaurus. American Journal of Science,* Series 3, 26: 81–85.

McIntosh, J. S. 1989. The Sauropod dinosaurs: A brief survey. In K. Padian and D. J. Chure (eds.), *The Age of Dinosaurs,* pp. 85–99. Short Courses in Paleontology no. 2. Knoxville: Paleontological Society, University of Tennessee.

McIntosh, J. S. 1990. Sauropoda. In D. B. Weishampel, P. Dodson, and H. Osmólska (eds.), *The Dinosauria,* pp. 345–401. Berkeley: University of California Press.

McIntosh, J. S., and D. S. Berman. 1975. Description of the palate and lower jaw of *Diplodocus* (Reptilia: Saurischia) with remarks on the nature of the skull of *Apatosaurus. Journal of Paleontology* 49: 187–199.

McIntosh, J. S.; W. E. Miller; K. L. Stadtman; and D. D. Gillette. 1996. The osteology of *Camarasaurus lewisi* (Jensen, 1988). *Brigham Young University Geological Studies* 41: 73–116.

Osborn, H. F., and C. C. Mook. 1921. *Camarasaurus, Amphicoelias,* and other sauropods of Cope. *Memoirs of the American Museum of Natural History,* n.s. 3: 247–287.

Ostrom, J. H., and J. S. McIntosh. 1966. *Marsh's Dinosaurs.* New Haven, Conn.: Yale University Press.

Owen, R. 1842. Report on British fossil reptiles. Part II. *Report British Association for the Advancement of Science* for 1841: 60–204.

Peckzis, J. 1994. Implications of body-mass estimates for dinosaurs. *Journal of Vertebrate Paleontology* 14: 520–533.

Peters, R. H. 1983. *The Ecological Implications of Body Size.* Cambridge: Cambridge University Press.

Phillips, J. 1871. *The Geology of Oxford and the Valley of the Thames.* Oxford: Clarendon Press.

Riggs, E. S. 1901. The largest known dinosaur. *Science,* n.s. 13: 549–550.

Riggs, E. S. 1903. Structure and relationships of opisthocoelian dinosaurs. Part 1:

Sauropods 289

Apatosaurus Marsh. *Publications of the Field Columbian Museum, Geology* 2: 165–196.

Spotila, J. R.; M. P. O'Connor; P. Dodson; and F. V. Paladino. 1991. Hot and cold running dinosaurs: Body size, metabolism and migration. *Modern Geology* 16: 203–227.

Stokes, W. L. 1942. Some field observations on the origins of the Morrison "gastroliths." *Science* 95 (2453): 18–19.

Upchurch, P. 1995. The evolutionary history of sauropod dinosaurs. *Philosophical Transactions of the Royal Society of London*, Series B, 349: 365–390.

Weaver, J. C. 1983. The improbable endotherm: The energetics of the sauropod dinosaur *Brachiosaurus*. *Paleobiology* 9: 173–182.

Young, C. C., and H. C. Chao. 1972. *Mamenchisaurus hochuanensis* sp. nov. *Insitute of Vertebrate Paleontology and Palaeoanthropology Mongraph*, Series A, 8: 1–30.

Othniel Charles Marsh and *Stegosaurus*

Stegosaurs

Peter M. Galton

Othniel Charles Marsh of Yale College erected the Stegosauria in 1877 as a new order of large extinct reptiles from the Upper Jurassic of Morrison, Colorado. He considered that the back of what he then thought was a new aquatic reptile, *Stegosaurus armatus* (Greek *stege*, "roof"; *saurus*, "reptile"; Latin *armatus*, "armed"), was covered by large osteoderms or dermal plates (one of which measured more than a meter in length)—large sheets of bone completely embedded in the skin and comparable to those of *Protostega,* a large aquatic turtle from the Late Cretaceous of Kansas. Marsh (1877: 513) noted that the bones were "embedded in so hard a matrix that considerable time and labor will be required for a full description." Fortunately for O. C. Marsh, his workers in the field located quarry sites that yielded better material at Garden Park, Colorado, and Como Bluff, Wyoming (Ostrom and McIntosh 1966), in much softer rocks of what would become known as the Morrison Formation. Unfortunately for science, most of the original material from Morrison at the Yale Peabody Museum of Natural History is still unprepared and undescribed.

Marsh (1880) illustrated the limbs and plates plus dermal spines of *Stegosaurus ungulatus* and reduced the Stegosauria to a suborder within the Dinosauria. He recognized that the smaller plates were actually held erect, but his reconstruction kept the largest plates flat on the back. Marsh suggested that the dermal spine might be associated with the forefoot. In

21

Figure 21.1. Skeletal reconstructions of stegosaurs (A–D) and *Scelidosaurus* (E). (A–B) Stegosaurid *Stegosaurus* from the Upper Jurassic of the western United States: (A) First published reconstruction of a stegosaur, by Othniel Charles Marsh in 1891, composite based on specimens of *Stegosaurus ungulatus* (YPM 1853, 1854, 1858) for vertebrae, girdles, limbs, and four pairs of tail spines plus *S. stenops* (USNM 4934) for skull and dermal armor. (Of the seventeen plates, four anterior plates were paired but not indicated as such on the reconstruction [Marsh 1896], and plate 17 was omitted to make room for the two extra pairs of tail spines in *S. ungulatus* [Czerkas 1987].) (B) *Stegosaurus stenops,* mostly based on USNM 4934, plus USNM 4714 for last half of tail and a tail spine.

(continues)

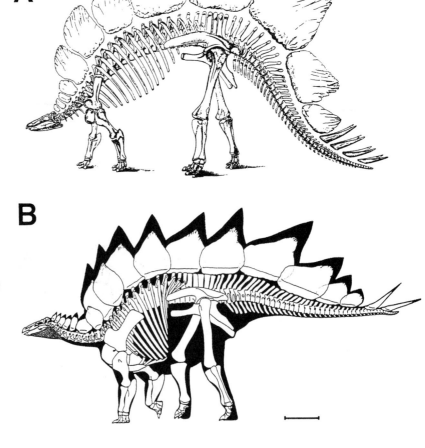

this he followed Richard Owen (1875), who, in his description of the reasonably complete but headless skeleton of *Omosaurus armatus* from the Upper Jurassic of England, described a dermal spine that was compared with the wrist spine of the ornithopod dinosaur *Iguanodon.* Marsh (1880) included *Omosaurus* in his new family, the Stegosauridae. The discovery of an almost complete skeleton in 1887 with the plates and spines preserved in articulation showed that all the plates of *Stegosaurus* were held vertically and that the spines belonged to the tail (Marsh 1887). The first skeletal reconstruction of *Stegosaurus* showed the plates in a single row along the midline plus four pairs of tail spines (Marsh 1891, 1896; Fig. 21.1A; see below and Czerkas 1987 for more details on restorations of plates subsequent to Marsh's).

Changing Concepts of the Armored Dinosaurs

The first subdivision of Richard Owen's (1842) Dinosauria was made by Thomas Henry Huxley (1870), who referred all forms with dermal

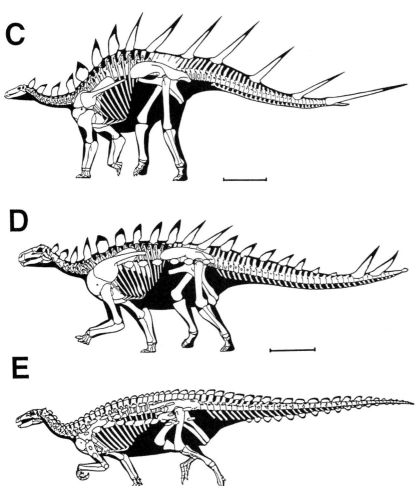

C

D

E

(C) Stegosaurid *Kentrosaurus* from the Upper Jurassic of Tanzania, East Africa, based on mounted skeleton in HMN (see Janensch 1925). (D) Huayangosaurid *Huayangosaurus* from the Middle Jurassic of the People's Republic of China, based on ZDM T7001 (see Zhou 1984), with parasacral spine after Sereno and Dong 1992. (E) The basal thyreophoran *Scelidosaurus* from the Lower Jurassic of England, from Paul 1987. Abbreviations for museums: HMN = Humboldt Museum für Naturkunde, Berlin, Germany; USNM = United States National Museum, Washington, D.C.; YPM = Peabody Museum of Natural History, Yale University, New Haven, Conn.; ZDM = Zigong Dinosaur Museum, Sichuan, People's Republic. Scale lines represent 50 cm. Figures B–E kindly supplied by Gregory S. Paul, who retains the copyright.

armor to the Scelidosauridae. This included *Scelidosaurus,* a reasonably complete skeleton from the Lower Jurassic of England (Fig. 21.1E; Owen 1861, 1863). When Harry Govier Seeley (1887) split the Dinosauria into the Saurischia and Ornithischia, the Scelidosauridae went to the latter. Marsh (1889, 1896) referred all the quadrupedal ornithischians with dermal armor (along with forms with a similar tooth morphology) to the Stegosauria. On the basis of differences in the pelvic girdle, Alfred Sherwood Romer (1927) first made a case for the separation of the suborder Ankylosauria ("fused or joined-together reptiles"), erected without definition by Henry Fairfield Osborn (1923), from the suborder Stegosauria. Romer restricted the Stegosauria to the Stegosauridae plus the Scelidosauridae (for *Scelidosaurus* plus a few poorly known forms). This restricted use of the Stegosauria for just the plated dinosaurs was followed by most paleontologists in the United States, Canada, and England. However, other workers followed Marsh (1889, 1896) with a more inclusive use of Stegosauria (e.g., Hennig [1924] and Lapparent and Lavocat [1955] as Stegosauroidea; Nopcsa [1915, 1928] as Thyreophora, the "shield bearers," in a version that included the Ceratopsia, or horned dinosaurs). The suborder

Figure 21.2. Skulls of stegosaurs and *Scelidosaurus* in left lateral view (A, C, E), ventral view (B, without lower jaw), and dorsal view (D, F). (A, B) Stegosaurid *Stegosaurus stenops* from the Upper Jurassic of Colorado, after Sereno and Dong 1992. (C, D) Huayangosaurid *Huayangosaurus* from the Middle Jurassic of the People's Republic of China, after Sereno and Dong 1992. (E, F) Basal thyreophoran *Scelidosaurus* from the Lower Jurassic of England, after Coombs et al. 1990. Abbreviations: as = anterior supraorbital (or palpebral); d = dentary; m = maxilla; ms = medial supraorbital (or palpebral); n = nasal; o = orbit; pa = palatine; pd = predentary; pm = premaxilla; ps = posterior supraorbital (or palpebral); pt = pterygoid; q = quadrate; s = supraorbital (palpebral); sq = squamosal; v = vomer. Scale lines represent 5 cm.

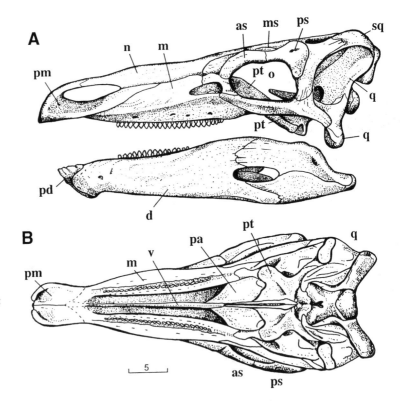

Ankylosauria was not widely accepted until a lengthy diagnosis was provided by Romer (1956).

As a result of several analyses in the 1980s, the Thyreophora of Franz Baron Nopcsa (1915) was reinstated in the sense of Friedrich von Huene (1956; i.e., without the Ceratopsia) to include stegosaurs and ankylosaurs plus their basal relatives, the best-known of which are the Lower Jurassic genera *Scelidosaurus* from England, *Scutellosaurus* from Arizona (Colbert 1981), and the recently described *Emausaurus* from Germany (Haubold 1990). Thyreophorans are united by having a transversely broad bar posterior to the orbit (Fig. 21.2A, C, E) and dermal armor consisting of parasagittal (parallel to midline) rows of low-keeled scutes on the dorsal body surface plus additional rows of lateral low-keeled scutes (Fig. 21.1D, E; Sereno 1986). The presence of a sinuous curve to the lower tooth row (on the dentary), the incorporation of a supraorbital bone (palpebral bone, originally associated with eyelid) into the skull roof to form the middle part of the superior (upper) orbital margin, and other cranial characters identify *Scelidosaurus* (Fig. 21.2E, F) as the sister group (closest relative) of the Eurypoda (= Stegosauria + Ankylosauria; Sereno 1986).

There are many derived (unique) characters linking Stegosauria and Ankylosauria (hence these characters are primitive for Stegosauria). In the skull (Fig. 21.2A–D), these include a superior orbital margin that is formed mostly from three supraorbitals (palpebral bones), the lack of a distinct notch for the anterosuperior part of the ear drum, and a deep median keel that extends the length of the palate. The forefoot (Fig. 21.3A) has relatively short metacarpals (bones between the wrist and "fingers"), and

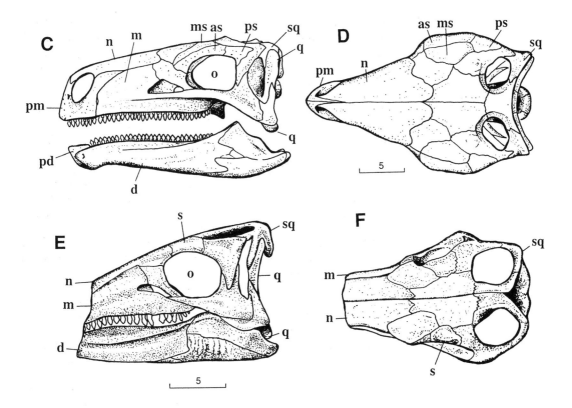

the digits ("fingers") bear hoof-like unguals (the last bone of a digit). The ilium (upper hip bone, Fig. 21.1A–D) is distinctive, with a very short postacetabular process (the part posterior to the hip joint), and a very long preacetabular process that is directed at least 35 degrees lateral to the midline. In adults the tibia and fibula (bones of the shin region) are fused distally with the astragalus and calcaneum (bones of ankle). The hind foot (Fig. 21.3B) has short, spreading metatarsals (bones between the ankle and toes) and hoof-like unguals. The osteoderms or dermal armor include elevated spines (Fig. 21.1A–D).

The term *armored dinosaur* now refers to a member of the Ankylosauria. The Stegosauria, or plated dinosaurs, are now restricted to medium-sized to large (up to 9 m in total body length) quadrupedal ornithischians with an extensive system of erect parasagittal plates and spines along the middle of the back and tail (Fig. 21.1A–D; Dong 1990; Galton 1990; Sereno and Dong 1992; Olshevsky and Ford 1993).

Distribution of Stegosaurs in Time and Space

The earliest record of the group consists of fragmentary remains from the Middle Jurassic (Lower Bathonian) of England (Galton and Powell 1983), but almost complete skeletons with skulls of *Huayangosaurus* (Figs. 21.1D, 21.2C, D) are known from the Bathonian-Callovian of the People's Republic of China (Zhou 1984; Sereno and Dong 1992).

Stegosaurs are best represented in the Upper Jurassic, with articulated

Stegosaurs 295

skeletons (but mostly lacking skulls) known for many of the genera. Stegosaurs from Europe, and England in particular (Galton 1985), were usually discovered in commercial quarries involved in the production of bricks before the widespread use of large machinery. This stegosaur material was originally included in the genus *Omosaurus* Owen (1875), but because this generic name had already been applied to a crocodile by Joseph Leidy (1856), it was preoccupied, and the dinosaur needed to be renamed. *Dacentrurus* Lucas (1902) was applied to *Omosaurus armatus* Owen (1875) for the specimens from the Kimmeridge Clay. *Lexovisaurus* Hoffstetter (1957) was applied to *Omosaurus durobrivensis* Hulke (1887) for the material from the older Lower Oxford Clay that represents a separate genus.

Abundant bones of *Kentrosaurus* (Fig. 21.1C) were excavated by hand labor under the direction of Werner Janensch and Edwin Hennig of Berlin from 1909 to 1912 in the torrid climate at Tendaguru, in what was German East Africa (now Tanzania) (Hennig 1924; Janensch 1925). A partial skeleton of *Chialingosaurus* was described by Chung Chien Young (1959) from the Shangshaximiao Formation of Sichuan, and thanks to the efforts of Zhimin Dong, additional parts of this skeleton, plus skeletons of two additional genera, *Tuojiangosaurus* and *Chungkingosaurus* (Dong et al. 1983), have been excavated from this formation. The People's Republic of China now has the longest and most diverse fossil record for stegosaurs (see Dong 1990; Galton 1990; Olshevsky and Ford 1993).

Stegosaurs are rare in the Cretaceous (Galton 1981), but surprisingly the earliest descriptions of stegosaurs were of *Regnosaurus* and *Craterosaurus* from the Lower Cretaceous of southern England. Gideon Algernon Mantell (1841) described a small piece of a dentary (lower jaw) as the ornithopod *Iguanodon*, but other specimens showed that this was incorrect (Mantell 1848). This jaw, which contains only the roots of a few of the teeth, was referred to the Scelidosauridae, Ankylosauria, and Sauropoda before it was placed in the Stegosauria (Olshevsky and Ford 1993; Barrett and Upchurch 1995). *Craterosaurus* was described by Harry Govier Seeley (1874) as a braincase, presumably dinosaurian, but as Franz Baron Nopcsa (1912) showed, it is actually part of the neural arch of a stegosaurian dorsal vertebra. These fragmentary remains may represent two genera or just one, which would be *Regnosaurus* because it was named first. Stegosaurs are also present in the Lower Cretaceous of South Africa (Galton 1981) and the People's Republic of China (Dong 1990). The most recent record is a partial skeleton of the problematic *Dravidosaurus* from the Upper Cretaceous (Coniacian) of India. This was described by P. Yadagiri and K. Ayyasami (1979), who also reported still-undescribed stegosaur material from the latest Cretaceous (Maastrichtian) of India.

Skull

Compared to the skull of *Huayangosaurus* (from the Middle Jurassic, in the family Huayangosauridae with *Regnosaurus*), that of *Stegosaurus* (and probably the rest of the stegosaurs, all in family Stegosauridae) is shallower, and the skull roof above the orbits is narrower (relative to skull length; Fig. 21.1A–D). As in other ornithischians, the tips of the anterior bones of the skull, the upper premaxillae and the lower predentary, were covered by rhamphothecae (horny sheaths) that were used to crop vegetation. In *Huayangosaurus* each premaxilla has seven teeth, but this bone lacks teeth in the Stegosauridae, and the rhamphothecae were proportionally larger (Fig. 21.2A–C). Compared to teeth, a horny sheath is especially suitable as a cropping structure. It always has a continuous edge that can be self-sharpening, it can have a cutting edge and/or a flat crushing area, and it is rapidly replaced as it is worn. In the lower jaw, the first dentary tooth is next to the predentary in *Huayangosaurus,* while in *Stegosaurus* there is an edentulous gap or diastema comparable to that present in many herbivorous mammals. The cheek tooth crowns are simple in form, with a prominent encircling cingulum (swelling) close to the root, and show a few facets produced by tooth-to-tooth wear.

Ornithischian dinosaurs have been commonly restored with cheeks (1908–1940 and from 1973 on; see below) or without cheeks (1940–1973). Like other ornithischians, stegosaurs have a space external to the maxillary (upper) and dentary (lower) tooth rows that is roofed by a prominent horizontal ridge on the maxilla and floored by the massive dentary (Fig. 21.2A, C). This ridge was one area of attachment (there was a corresponding ridge on the dentary) for a structure that, although it was not close against the teeth, performed some of the functions of the cheeks of mammals. This cheek-like structure bordered a vestibule or space that would have received any chewed food that ended up lateral to (or outside) the lower teeth.

This sort of reconstruction was originally suggested for the ceratopsian *Triceratops* in 1903 by Richard Swann Lull, who restored cheeks with a buccinator muscle for this genus in 1908. Barnum Brown and Erich Maren Schlaikjer (1940) argued against cheeks in ornithischians because the buccinator, a facial muscle innervated by a branch of the seventh cranial nerve, is characteristic of mammals and is not present in any living reptile. However, this argument does not preclude ornithischians (including stegosaurs) from having a cheek-like structure that lacked a buccinator muscle (see Galton 1973 for details). Consequently, chewing in ornithischians was occasionally interrupted so that the long, narrow-based and slender tongue could scoop up the residual food from the vestibule and return it to the mouth to be chewed again or swallowed. Our cheeks are very close to our cheek teeth and usually keep food between them. However, sometimes we have to use the tongue to retrieve food (e.g., peanut butter) from the outside of the teeth and gums because it cannot be kept between the teeth by the muscular efforts of the buccinator muscle.

A deep median keel along the length of the palate (formed by vomers, palatines, and pterygoids; Fig. 21.2B) probably provided support for a soft secondary palate that was supported laterally by the maxillae. This palate separated the more dorsal nasal passages from the oral cavity so that stegosaurs could continue to breathe while chewing food. The lower jaw articulates with the quadrate, the upper end of which is fused to the squamosal in adult individuals of *Stegosaurus* (Fig. 21.2A). The skull formed a solid box, and because of the fixed quadrate, the jaw-closing part of the chewing cycle involved an upward movement of the teeth of the lower jaw against those of the upper jaw without the complexities present in the Ornithopoda.

Vertebrae

The dorsal vertebrae of all stegosaurs are tall. The increase in height is in the region between the centrum (or body) and the diapophyses (transverse processes that carry ribs) (Fig. 21.3C), so the height of the body cavity is increased, which would provide more room for viscera. In other dinosaurs with tall dorsals, the increased height is due to elongation of the neural spines. The neural canal is larger in posterior cervical vertebrae to

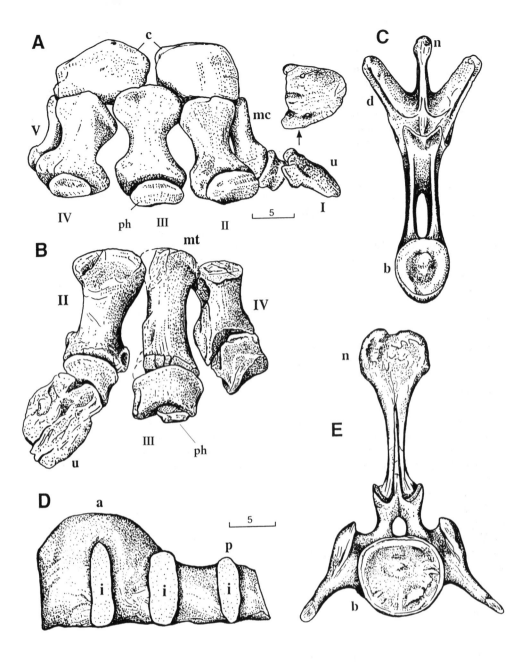

accommodate the brachial enlargement, that part of the spinal cord
involved with the brachial nerve plexus that supplied nerves to the muscles
of the massive forelimbs (Lull 1917). Some mid- and posterior dorsals
appear to have a dilated neural canal in *Stegosaurus* and especially in
Kentrosaurus. However, the size of the canal is considerably reduced by a
thin transverse septum of bone that bridges the upper part of the opening.
Consequently, the increase in height of the pedicel region of the dorsals was
not to provide room for a larger neural canal.

The diapophyses are directed upward by as much as 50 to 60 degrees
from the horizontal in mid-dorsal vertebrae (Fig. 21.3C); this angle decreases to about 25 to 40 degrees in anterior and posterior dorsals. This

increased angle would have provided better support for the overlying osteoderms, the parasagittal orientation of which would have concentrated the weight along the region adjacent to the ends of the diapophyses. This region is also supported by the shafts of the ribs, which have a T-shaped cross-section in the mid-dorsal region with the flat surface on the outside.

In the Stegosauridae, the diapophyses and the T-shaped sacral ribs fuse together to form an almost solid dorsal plate that extends from the base of the neural spines to the flat dorsal surface of the ilium, with which it is continuous. This plate strengthened the sacrum, and appears to be associated with the proportionally long femur that is present in stegosaurids. This elevated the hip joint well above the shoulder joint so that the femur supported much of the body weight.

The neural canal of the sacrum is extremely dilated to form an endosacral enlargement that is extremely large anteriorly, with a smaller enlargement posteriorly (Fig. 21.3D). This region was described by O. C. Marsh (1881: 168) as a "posterior brain case," which may explain the origins of the popular misconception that Stegosaurus had "two sets of brains, one in his head, the usual place, and the other at his spinal base." However, this enlargement did not represent a "sacral brain" (Edinger 1961). Some of this space (anterior enlargement) was occupied by an enlarged spinal cord, which was associated with the increased size of the sacral plexus that sent nerves to the vertically held, weight-supporting hind limbs. In addition, a posterior enlargement housed nerves that ran to the caudifemoral muscle, which functioned to pull the hind limb backward during walking, as well as in lateral movements of the tail with its osteoderms (Wiedersheim 1881; Lull 1910a, 1910b, 1917; see Giffin 1991 for details). The rest of the space was probably occupied by a glycogen body similar to that seen in birds (Krause 1881), the function of which is not clear (Giffin 1991).

The posterior surface of the last centrum of the sacrum is markedly concave transversely, whereas the adjacent surface of the first caudal is markedly convex; both surfaces are vertically straight. The shape of this joint would have facilitated lateral movements of the tail but restricted vertical movements. The ends of the neural spines of the anterior caudal vertebrae are transversely expanded (Fig. 21.3E), so the ends are broader in anterior view than in lateral view. The centra or bodies of the posterior caudal vertebrae are almost square in shape (Fig. 21.1A–D), rather than elongate as in most other dinosaurs. These modifications of the tail aided in supporting the weight of the osteoderms of this region.

There are several very well preserved articulated skeletons of stegosaurs, but none shows any trace of an ossified tendon. Given the ubiquitous occurrence of ossified tendons in all the other groups of ornithischians, it is reasonable to assume that their absence in stegosaurs represents a secondary loss. It has been suggested by Paul Sereno and Zhimin Dong (1992) that the increased height of the pedicels of the dorsal vertebrae may have provided additional attachment surface to maintain trunk rigidity in the absence of ossified tendons.

Girdles and Limbs

The lower part of the scapula (shoulder bone) forms a broad plate. The humerus (upper arm bone) is short but massive, with expanded ends (Fig. 21.1A–D). Consequently there was plenty of room for the attachment of

Figure 21.3. Anatomy of *Stegosaurus* from the Late Jurassic of the western United States. (A) Right forefoot and proximal carpals (wrist bones) in anterior view plus ungual I in dorsal view (ungual II not preserved). (B) Incomplete hind foot (drawn as a left) in anterior view. (C) Mid-dorsal vertebra in anterior view. (D) Sacral endocast in left lateral view. (E) Anterior caudal vertebra in anterior view. A, B after Gilmore 1914; C–E after Marsh 1896. Abbreviations: a = anterior enlargement; b = body or centrum; c = carpals; d = diapophysis; i = cast of intervertebral foramen; mc = metacarpals; mt = metatarsals; n = neural spine; p = posterior enlargement; ph = phalanx; u = ungual phalanx; I, II, IV, V = digits I, II, IV, and V. Scale lines represent 5 cm.

powerful shoulder and pectoral muscles to support the body weight anteriorly. The forefoot is quite elephant-like, being short and relatively inflexible. It has one or two large, block-like proximal carpals (wrist bones) separated by flat intercarpal articulations; there are no distal carpals. Five short and robust metacarpals bear short digits, at least two of which terminate in a hoof-like ungual (Fig. 21.3A). The pubis (anteroventral hip bone, Fig. 21.1A–D) has a relatively long anterior process (at least 40 percent of the length of the posterior process) and an oval-shaped, laterally directed acetabular surface (Sereno and Dong 1992). The femur is slender in side view but broad in anterior view, with a straight shaft of nearly uniform width; in all stegosaurs except *Huayangosaurus*, the femur is long compared to the humerus (and to the tibia; Fig. 21.1A–D), indicating a graviportal (elephantine) mode of locomotion. The stocky proportions of the hind foot (Fig. 21.3B) also indicate a graviportal mode of locomotion. The pes is relatively symmetrical about digit III, a result of the loss of digit I and a great reduction in the size of metatarsal V so that this digit was non-functional. The three central weight-supporting digits (II–IV) are robust and bear hoof-like unguals.

Dermal Armor

A pair of low-angled spines of *Lexovisaurus* were first referred to the outside of the scapula by Franz Baron Nopcsa (1911), but similar isolated plates with a much larger base were placed on the ilium in *Kentrosaurus* (Hennig 1924; Janensch 1925). However, articulated skeletons of *Huayangosaurus* and *Tuojiangosaurus* have these plates adjacent to the pectoral girdle (Sereno and Dong 1992); this pair of parascapular spines (Fig. 21.1C, D) is secondarily lost in *Stegosaurus* (Fig. 21.1B). In *Huayangosaurus* there are low-keeled, lateral osteoderms (similar to those of *Scelidosaurus*), but these are lost in the Stegosauridae (Fig. 21.1). The rest of the osteoderms of stegosaurs are situated above the vertebral column, rather than being spread over the back as a whole, and are angled upward and slightly outward. Viewed from the side, the osteoderms of most stegosaurs except *Stegosaurus* (see next paragraph) form a series that grade from short, erect plates anteriorly to longer, posterodorsally angled spines posteriorly (Fig. 21.1C, D). In all stegosaurs, the series ends with a proportionally narrow pair of spines that continue beyond the last tail vertebra (Fig. 21.1A–D). In all stegosaurs except *Stegosaurus*, all of the dorsal osteoderms are paired, so there is a left and a right representative of each type of plate and spine, and there is a gradual anteroposterior transition passing along the series, from small to larger plates, ending in the spines that cover most of the tail (Fig. 21.1C, D).

In the articulated skeleton of *Stegosaurus stenops* (USNM 4934), the dermal armor consists of a series of seventeen erect and thin plates of varying sizes, and extending along most of the tail, plus only two pairs of spines at the end of the tail (Gilmore 1914). The plates have been reconstructed in different patterns (Czerkas 1987): as a single median row (Fig. 21.1A; Marsh 1891, 1896; Ostrom and McIntosh 1966; Czerkas 1987), or as two parasagittal rows with the plates arranged either in pairs (Lull 1910a, 1910b; Paul 1987) or as a staggered (alternating) series (Fig. 21.1B; Gilmore 1914; Farlow et al. 1976; Bakker 1986; Paul 1992). No two plates have exactly the same shape or size in *Stegosaurus stenops*, in which three of the seventeen preserved plates are in an alternating pattern (Gilmore 1914, USNM 4934). The same plate arrangement also occurs in another specimen of *S. stenops* from Garden City, Colorado (Paul 1992; Carpenter and Small 1993), and in a specimen of an as yet unidentified species of *Stegosaurus* found near Dinosaur National Monument, but there may have been different arrangements of armor in other species. Thus in *Stegosaurus armatus* there is at least one pair of large plates that are the same size and shape (Ostrom and McIntosh 1966: plates 59–1, 60). All of the plates may have been arranged in pairs in this species (Paul 1987), in which there were four pairs of tail spines (Fig. 21.1A; Marsh 1891, 1896; Ostrom and McIntosh 1966: plates 55, 56).

Systematics

The Huayangosauridae (*Huayangosaurus* and *Regnosaurus*) is the sister taxon to the Stegosauridae (all other stegosaurs), which are characterized by the following features: loss of the premaxillary teeth, fusion of the dorsal margins of the sacral ribs to form a nearly solid plate between the ilia (rather than having large dorsal empty spaces between the ribs), increased length of the prepubic process, loss of the row of scutes along either side of the trunk, and elongation of the femur to at least 1.5 times the length of the humerus (vs. 1:1 in *Huayangosaurus*) (Figs. 21.1A–D; Sereno and Dong 1992). However, the systematic relationships among the stegosaurs within this more derived group remain unclear (Sereno and Dong 1992; Sereno 1997).

Taphonomy and Paleoecology

The European record provides no information on the habitat of stegosaurs because it consists of single isolated carcasses that drifted downstream and disintegrated to varying degrees before being deposited in marine sediments. The Tendaguru fauna of Tanzania occurs in near-shore deposits that came from land subjected to a warm climate with periodic droughts. *Kentrosaurus* is a relatively minor element of the fauna, and preservation favored medium-sized individuals (Russell et al. 1980). In the main Tendaguru stegosaur quarry, there was an enormous concentration of partly articulated and partly sorted remains (Hennig 1924).

A detailed study of the taphonomy of Morrison dinosaurs was performed by Peter Dodson et al.(1980), who showed that the common genera are broadly distributed in a variety of different sediments. *Stegosaurus* occurs more frequently in channel sands that represent a concentration of bones from animals that probably spent much of their lives in floodplain areas. However, *Stegosaurus* may have inhabited areas farther from sources of water, so it was probably somewhat separated ecologically from sauropods. Morrison dinosaur carcasses typically decomposed in open, dry areas, or spent a considerable time in channels prior to deposition. Consequently, *Stegosaurus* occurs only occasionally as an articulated skeleton; usually it is part of an accumulation of twenty to sixty skeletons of other dinosaurs with only moderate to low degrees of articulation.

Several skeletons of stegosaurs from Sichuan, China, include skulls, and because the degree of articulation of the skeletons is greater than those from the Morrison Formation, the carcasses were buried more rapidly in Sichuan. *Huayangosaurus* occurs in sandstones of the Xiashaximiao Formation that were deposited in a lakeshore shallow-bank environment under low-energy conditions (Xia et al. 1984).

Paleobiology and Behavior

Stegosaurs were graviportal, or elephantine in their locomotion (Coombs 1978), with an occasional more upright bipedal pose with the body supported by a tripod formed by the hind limbs and tail (Alexander 1985; Bakker 1986). However, as with the bipedal pose in elephants, it is unlikely that the tripodal pose was commonly used for feeding. Consequently, stegosaurs were probably important low-level browsers up to about the 1-meter level, the maximum "comfortable" height that could be reached while on all fours.

The encephalization quotient (ratio of measured brain size to "expected" brain size for an archosaur of the same body size) of stegosaurs is comparable to that of ankylosaurs, and less than that of all other dinosaurs except sauropods. This low value for both groups was attributed by James A. Hopson (1980) to their reliance on defensive armor and tail weapons, rather than speedy flight, to cope with predators.

The tail spines of stegosaurs were originally covered with horn and would have been formidable weapons. Robert T. Bakker (1986) suggested that the loss of ossified tendons in stegosaurs was correlated with increased flexibility of the tail. He visualized *Stegosaurus* using the strong shoulder muscles to pivot its body on the very tall hind limbs, the main weight supporters, as it arched and twisted its tail so the spines were driven into the body of an attacker.

The overall pattern of the plates and spines (Fig. 21.1B–D) is characteristic for each species, so it was probably important for the recognition of other individuals of the same species and for sexual displays; L. S. Davitashvili (1961) suggested that this was probably the original function of the erect osteoderms. The armor of all stegosaurs is ideally arranged for maximum effect during a lateral display (Spassov 1982).

It has been suggested that the plates of *Stegosaurus* were normally held horizontally, to provide a flank defense with the largest plates over the vulnerable hind limb, but that the plates could be suddenly erected by muscles to startle and deter an attacker, or to ward off attack from above (Hotton 1963; Bakker 1986). However, studies of the histology of the plates of *Stegosaurus* by Vivian de Buffrénil et al. (1986) suggest that this kind of plate mobility is unlikely. Surface markings on the basal third of the plate indicate that it was embedded symmetrically in the thick, tough skin. Furthermore, the plates of *Stegosaurus* are unlikely to have functioned as armor because they do not consist of thick, compact bone.

In contrast, apart from any use in sexual display, the plates could have functioned in temperature regulation. In an alternating arrangement, the plates would have worked well as a forced convection fin to dissipate heat, and possibly as heat absorbers from solar radiation (Farlow et al. 1976; Buffrénil et al. 1986). The plates formed a scaffolding for the support of a richly vascularized skin which would have acted as an efficient heat-exchange structure. A heat-absorbing role for the plates makes sense if *Stegosaurus* was an ectotherm ("cold-blooded"), whereas heat loss by radiation or forced convection would be useful if *Stegosaurus* was ectothermic or to any degree endothermic ("warm-blooded"; Buffrénil et al. 1986). These conclusions also apply to the similarly large and thin dorsal plates of *Lexovisaurus*, but in other stegosaurs, display was probably the main function of the plates. In all stegosaurs, the terminal tail spines (Thagomizer) presumably played a role in defense.

References

Alexander, R. McN. 1985. Mechanics of posture and gait of some large dinosaurs. *Zoological Journal of the Linnean Society* 83: 1–25.

Bakker, R. T. 1986. *The Dinosaur Heresies: New Theories Unlocking the Mystery of the Dinosaurs and Their Extinction.* New York: William Morrow.

Barrett, P. M., and P. Upchurch. 1995. *Regnosaurus northamptoni*, a stegosaurian dinosaur from the Lower Cretaceous of southern England. *Geological Magazine* 132: 213–222.

Brown, B., and E. M. Schlaikjer. 1940. The structure and relationships of *Protoceratops*. *New York Academy of Science, Annals* 40: 133–266.

Buffrénil, V. de; J. O. Farlow; and A. de Ricqlès. 1986. Growth and function of *Stegosaurus* plates: Evidence from bone histology. *Paleobiology* 12: 459–473.

Carpenter, K., and B. Small. 1993. New evidence for plate arrangement in *Stegosaurus stenops*. *Journal of Vertebrate Paleontology* 13 (Supplement to no. 3): 28A–29A.

Colbert, E. H. 1981. A primitive ornithischian dinosaur from the Kayenta Formation of Arizona. *Museum of Northern Arizona Bulletin* 53: 1–61.

Coombs, W. P. Jr. 1978. Theoretical aspects of cursorial adaptations in dinosaurs. *Quarterly Review of Biology* 53: 393–418.

Coombs, W. P. Jr.; D. B. Weishampel; and L. M. Witmer. 1990. Basal Thyreophora. In D. B. Weishampel, P. Dodson, and H. Osmólska (eds.), *The Dinosauria*, pp. 427–434. Berkeley: University of California Press.

Czerkas, S. A. 1987. A reevaluation of the plate arrangement on *Stegosaurus stenops*. In S. J. Czerkas and E. C. Olson (eds.), *Dinosaurs Past and Present*, vol. 2, pp. 83–99. Seattle: University of Washington Press.

Davitashvili, L. 1961. *The Theory of Sexual Selection.* Moscow: Izdatel'stvo Akademii Nauk SSSR. (In Russian.)

Dodson, P.; A. K. Behrensmeyer; R. T. Bakker; and J. S. McIntosh. 1980. Taphonomy and paleoecology of the Upper Jurassic Morrison Formation. *Paleobiology* 6: 208–232.

Dong, Z. 1990. Stegosaurs of Asia. In K. Carpenter and P. J. Currie (eds.), *Dinosaur Systematics: Approaches and Perspectives,* pp. 255–268. Cambridge: Cambridge University Press.

Dong, Z. M.; S. W. Zhou; and Y. H. Chang. 1983. The dinosaur remains from Sichuan Basin, China. *Palaeontologica Sinica* 162 (C) 23: 1–166. (In Chinese with English summary.)

Edinger, T. 1961. Anthropocentric misconceptions in paleoneurology. *Rudolf Virchow Medical Society of New York, Proceedings* 19: 56–107.

Farlow, J. O.; C. V. Thompson; and D. E. Rosner. 1976. Plates of the dinosaur *Stegosaurus*: Forced convection heat loss fins? *Science* 192: 1123–1125.

Galton, P. M. 1973. The cheeks of ornithischian dinosaurs. *Lethaia* 6: 67–89.

Galton, P. M. 1981. *Craterosaurus pottonensis* Seeley, a stegosaurian dinosaur from the Lower Cretaceous of England, and a review of Cretaceous stegosaurs. *Neues Jahrbuch für Geologie und Paläontologie, Abhandlungen* 161: 28–46.

Galton, P. M. 1985. British plated dinosaurs (Ornithischia, Stegosauria). *Journal of Vertebrate Paleontology* 5: 211–254.

Galton, P. M. 1990. Stegosauria. In D. B. Weishampel, P. Dodson, and H. Osmólska (eds.), *The Dinosauria*, pp. 435–455. Berkeley: University of California Press.

Galton, P. M., and H. P. Powell. 1983. Stegosaurian dinosaurs from the Bathonian (Middle Jurassic) of England: The earliest record of the Stegosauridae. *Géobios* 16: 219–229.

Giffin, E. B. 1991. Endosacral enlargements in dinosaurs. *Modern Geology* 16: 101–112.

Gilmore, C. W. 1914. Osteology of the armored dinosaurs in the United States National Museum, with special reference to the genus *Stegosaurus*. *United States National Museum, Bulletin* 89: 1–136.

Gilmore, C. W. 1918. A newly mounted skeleton of the armored dinosaur *Stegosau-*

rus stenops in the United States National Museum. *United States National Museum, Proceedings* 54: 383–396.

Haubold, H. 1990. Ein neuer Dinosaurier (Ornithischia, Thyreophora) aus dem unteren Jura des Nördlichen Mitteleuropa. *Revue de Paléobiologie* 9: 149–177.

Hennig, E. 1924. *Kentrurosaurus aethiopicus,* die Stegosaurier-funde von Tendaguru, Deutsch-Ostafrika. *Palaeontographica,* Supplement 7, 1 (1): 103–254.

Hoffstetter, R. 1957. Quelques observations sur les Stégosaurinés. *Muséum National d'Histoire Naturelle de Paris, Bulletin* 2 (29): 537–547.

Hopson, J. A. 1980. Relative brain size in dinosaurs: Implications for dinosaurian endothermy. In R. D. K. Thomas and E. C. Olson (eds.), *A Cold Look at the Warm-Blooded Dinosaurs,* pp. 287–310. Selected Symposium 28, American Association for the Advancement of Science. Boulder, Colo.: Westview Press.

Hotton, N. III. 1963. *Dinosaurs.* New York: Pyramid Publications.

Huene, F. 1956. *Paläontologie und Phylogenie der Niederen Tetrapoden.* Jena: Fischer.

Hulke, J. W. 1887. Note on some dinosaurian remains in the collection of A. Leeds, Esq., of Eyebury, Northamptonshire. *Geological Society of London, Quarterly Journal* 43: 695–702.

Huxley, T. H. 1870. On the classification of the Dinosauria with observations on the Dinosauria of the Trias. *Geological Society of London, Quarterly Journal* 26: 31–50.

Janensch, W. 1925. Ein aufgestelltes Skelett des Stegosauriers *Kentrurosaurus aethiopicus* E. Hennig aus den Tendaguru-Schichten Deutsch-Ostafrikas. *Palaeontographica,* Supplement 7, 1 (1): 257–276.

Krause, W. 1881. Zum Sacralhirn der Stegosaurier. *Biologisches Zentralblatt* 1: 461.

Lapparent, A. F. de, and R. Lavocat. 1955. Dinosauriens. In J. Piveteau (ed.), *Traité de Paléontologie,* vol. 5, pp. 785–962. Paris: Masson et Cie.

Leidy, J. 1856. Notice of extinct animals discovered by Prof. E. Emmons. *Academy of Natural Sciences of Philadelphia, Proceedings* 8: 255–256.

Lucas, F. A. 1902. Paleontological notes: The generic name *Omosaurus. Science* 19: 435.

Lull, R. S. 1903. Skull of *Triceratops serratus. American Museum of Natural History, Bulletin* 19: 685–695.

Lull, R. S. 1908. The cranial musculature and the origin of the frill in the ceratopsian dinosaurs. *American Journal of Science* 4 (25): 387–399.

Lull, R. S. 1910a. The armor of *Stegosaurus. American Journal of Science* 4 (29): 201–210.

Lull, R. S. 1910b. *Stegosaurus ungulatus* Marsh, recently mounted at the Peabody Museum of Yale University. *American Journal of Science* 4 (30): 361–376.

Lull, R. S. 1917. On the functions of the "sacral brain" in dinosaurs. *American Journal of Science* 4 (44): 471–477.

Mantell, G. A. 1841. Memoir on a portion of the lower jaw of the *Iguanodon,* and on the remains of the *Hylaeosaurus* and other saurians, discovered in the strata of Tilgate Forest, in Sussex. *Royal Society of London, Philosophical Transactions* 131: 131–151.

Mantell, G. A. 1848. On the structure of the jaws and teeth of the *Iguanodon. Royal Society of London, Philosophical Transactions* 138: 183–202.

Marsh, O. C. 1877. New order of extinct Reptilia (Stegosauria) from the Jurassic of the Rocky Mountains. *American Journal of Science* 3 (14): 513–514.

Marsh, O. C. 1880. Principal characters of American Jurassic dinosaurs. Part III. *American Journal of Science* 3 (19): 253–259.

Marsh, O. C. 1881. Principal characters of American Jurassic dinosaurs. Part IV: Spinal cord, pelvis and limbs of *Stegosaurus. American Journal of Science* 3 (21): 167–170.

Marsh, O. C. 1887. Principal characters of American Jurassic dinosaurs. Part IX: The skull and dermal armor of *Stegosaurus. American Journal of Science* 3 (34): 413–417.

Marsh, O. C. 1889. Comparison of the principal forms of the Dinosauria of Europe and America. *American Journal of Science* 3 (37): 323–331.

Marsh, O. C. 1891. Restoration of *Stegosaurus*. *American Journal of Science* 3 (42): 179–181.

Marsh, O. C. 1896. Dinosaurs of North America. *United States Geological Survey, 16th Annual Report 1894–95*: 133–244.

Nopcsa, F. 1911. Notes on British dinosaurs. Part IV: *Stegosaurus priscus,* sp. nov. *Geological Magazine* 5 (8): 109–115.

Nopcsa, F. 1912. Notes on British dinosaurs. Part V: *Craterosaurus* (Seeley). *Geological Magazine* 5 (9): 481–484.

Nopcsa, F. 1915. Die Dinosaurier der siebenburgischen Landesteile Ungarns. *Mittheilungen aus dem Jahrbuch der Ungarischen geologischen Reichsanst* 23: 1–26.

Nopcsa, F. 1928. The genera of reptiles. *Palaeobiologica* 1: 163–188.

Olshevsky, G., and T. Ford. 1993. The origin and evolution of the stegosaurs. *Gakken Mook* 4: 65–103. (In Japanese.)

Osborn, H. F. 1923. Two Lower Cretaceous dinosaurs from Mongolia. *American Museum Novitates* 95: 1–10.

Ostrom, J. H., and J. S. McIntosh. 1966. *Marsh's Dinosaurs: The Collections from Como Bluff*. New Haven, Conn.: Yale University Press.

Owen, R. 1842. Report on British fossil reptiles. Part II. *British Association for the Advancement of Science, Annual Report* for 1841: 60–204.

Owen, R. 1861. A monograph of the fossil Reptilia of the Lias formations. Part I: *Scelidosaurus harrisonii. Palaeontographical Society Monographs* 13: 1–14.

Owen, R. 1863. A monograph of the fossil Reptilia of the Lias formations. Part II: *Scelidosaurus harrisonii* Owen of the lower Lias. *Palaeontographical Society Monographs* 14: 1–26.

Owen, R. 1875. Monographs of the fossil Reptilia of the Mesozoic Formations. Parts 2 and 3 (genera *Bothriospondylus, Cetiosaurus, Omosaurus). Palaeontographical Society Monographs* 29: 15–94.

Paul, G. S. 1987. The science and art of restoring the life appearance of dinosaurs and their relatives. In S. J. Czerkas and E. C. Olson (eds.), *Dinosaurs Past and Present,* vol. 2, pp. 4–49. Seattle: University of Washington Press.

Paul, G. S. 1992. The arrangement of plates in the first complete *Stegosaurus,* from Garden Park. *Garden Park Paleontological Society, Tracks in Time* 3 (1): 1–2.

Romer, A. S. 1927. The pelvic musculature of ornithischian dinosaurs. *Acta Zoologica* 8: 225–275.

Romer, A. S. 1956. *Osteology of the Reptiles*. Chicago: University of Chicago Press.

Russell, D.; P. Béland; and J. S. McIntosh. 1980. Paleoecology of the dinosaurs of Tendaguru (Tanzania). *Société Géologiques de France, Mémoires,* n.s. 1980, no. 139: 169–175.

Seeley, H. G. 1874. On the base of a large lacertian cranium from the Potton Sands, presumably dinosaurian. *Geological Society of London, Quarterly Journal* 30: 690–692.

Seeley, H. G. 1887. On the classification of the fossil animals commonly called Dinosauria. *Royal Society of London, Proceedings* 43: 165–171.

Sereno, P. C. 1986. Phylogeny of the bird-hipped dinosaurs (Order Ornithischia). *National Geographic Research* 2: 234–256.

Sereno, P. C. 1997. The origin and evolution of dinosaurs. *Annual Review of the Earth and Planetary Sciences* 25: 435–489.

Sereno, P. C., and Dong Z. 1992. The skull of the basal stegosaur *Huayangosaurus taibaii* and a cladistic analysis of Stegosauria. *Journal of Vertebrate Paleontology* 12: 318–343.

Spassov, N. B. 1982. The bizarre dorsal plates of *Stegosaurus:* Ethological approach. *Comptes Rendus de l'Academie Bulgare des Sciences* 35: 367–370.

Wiedersheim, R. 1881. Zur Paläontologie Nord-Amerikas. *Biologisches Zentralblatt* 1: 359–372.

Xia W.; X. Li; and Z. Yi. 1984. The burial environment of dinosaur fauna in Lower Shaximiao Formation of Middle Jurassic at Dashanpu, Zigong, Sichuan. *Chengdu College of Geology, Journal* 2: 46–59. (In Chinese with English summary.)

Yadagiri, P., and K. Ayyasami. 1979. A new stegosaurian dinosaur from Upper Cretaceous sediments of south India. *Geological Society of India Journal* 20: 251–530.

Young, C. C. 1959. On a new Stegosauria from Szechuan, China. *Vertebrata PalAsiatica* 3: 1–8.

Zhou S. W. 1984. *The Middle Jurassic Dinosaurian Fauna from Dashanpu, Zigong, Sichuan.* Vol. II: *Stegosaurs.* Chongqing: Sichuan Scientific and Technical Publishing House. (In Chinese with English summary.)

Plate 1. An archaic conception of dinosaurs and other prehistoric animals: a scene painted in the mid-nineteenth century by Archibald M. Willard, who later (1876) achieved fame for his patriotic painting *The Spirit of '76.* Used by permission of Don Glut.

Plate 2. On a sunny day early in the Mesozoic Era, a group of prosauropods is disturbed while feeding. Drawing by James Whitcraft.

Plate 3. *Dilophosaurus*, a crested theropod from the Early Jurassic of western North America. Painting by Michael W. Skrepnick.

Plate 4. Mass confusion during Middle Jurassic times in China. Two theropods (*Gasosaurus*) stampede a herd of sauropods (*Shunosaurus*) and a pair of stegosaurs (*Huayangosaurus*). Painting by Gregory S. Paul.

Plate 5. A day at the beach in Germany during the Late Jurassic. Two primitive birds (*Archaeopteryx*) confront a pterosaur (*Pterodactylus*). Painting by Gregory S. Paul.

Plate 6. Two sauropods (*Nemegtosaurus*) stroll along the edge of a lake in eastern Asia during the Late Cretaceous. Painting by Berislav Kržič.

Plate 7. A herd of diplodocid sauropods having a very bad day in the Late Jurassic of Wyoming, at what would become the Howe Quarry site. Dead and dying diplodocids, mired in a drying mudhole, attract the attention of pterosaurs and theropods large (*Allosaurus*) and small (*Ornitholestes* and a compsognathid). Painting by Mark Hallett © 1992.

Plate 8. A South American
scene around the middle of the
Cretaceous Period, depicting
three individuals of the
gigantic sauropod
Argentinosaurus. Painting by
Gregory C. Wenzel.

Plate 9. A middle Cretaceous nodosaurid ankylosaur walks through a gloomy forest in the western United States. Painting by Brian Franczak.

Plate 10. Implements of destruction: the hands and feet of the large middle Cretaceous theropod *Utahraptor* from the western United States. Painting by Donna Braginetz.

Plate 11. Two individuals of the enormous hadrosaur *Shantungosaurus* splash through a Late Cretaceous wetland in China. Painting by Brian Franczak.

Plate 12. *Parasaurolophus*, a crested hadrosaur from the Late Cretaceous of western North America, sports its breeding colors. Painting by Larry Felder.

Plate 13. *Mononykus,* a bizarre theropod (and possibly primitive bird), dashes across a desert in the Late Cretaceous of Mongolia. Painting by Michael W. Skrepnick.

Plate 14. The theropod *Aublysodon* dodges a brush fire in a Late Cretaceous forest in western North America. Painting by Brian Franczak.

Plate 15. *Bactrosaurus,* a hadrosaur from the Late Cretaceous of eastern Asia. Painting by Berislav Kržič.

Overleaf

Plate 16. *Giants in the Earth:* Late Cretaceous dinosaurs from western North America. A tyrannosaurid (*Gorgosaurus,* sometimes called *Albertosaurus*) makes a menacing move toward a group of hadrosaurids (*Gryposaurus*) and a ceratopsid (*Centrosaurus*). Painting by Bob Walters.

Plate 17. Still up to no good: *Gorgosaurus* scrutinizes the hadrosaur *Lambeosaurus* with a gourmand's intent. Painting by Michael W. Skrepnick.

© B. KRŽIČ '95

THE WORLD OF DINOSAURS

A scene in Colorado, 150 million years ago

A scene in Montana, 75 million years ago

Plate 18. The World of Dinosaurs, a series of stamps issued by the United States Postal Service in 1997. The upper scene depicts dinosaurs from the Late Jurassic, and the lower scene shows Late Cretaceous dinosaurs, all from western North America. Painting by James Gurney. Stamp design copyright © U.S. Postal Service. Reproduced with permission.

Plate 19. The *Stegosaurus* model used in the 1933 monster movie *King Kong*. See Fig. 43.5 to see how the model looked in the film. Used by permission of Forrest J Ackerman.

Plate 20. *Deinonychus* (under the alias *Velociraptor*), the chief villain from the 1993 movie *Jurassic Park*. Copyright © 1993 by Universal City Studios, Inc. Courtesy of MCA Publishing Rights, a division of MCA Inc. All Rights Reserved.

Plate 21. Cover art from an issue of the Dell comic book series "Turok, Son of Stone." Each issue had a vividly painted cover featuring dinosaurs or other prehistoric animals.

Plate 22. Bill Watterson's comic strip *Calvin and Hobbes* often had dinosaur-related themes. Watterson's dinosaurs were drawn with considerable sophistication and anatomical accuracy. *Calvin and Hobbes* © 1988 Bill Watterson. Distributed by Universal Press Syndicate. Reprinted with permission. All rights reserved.

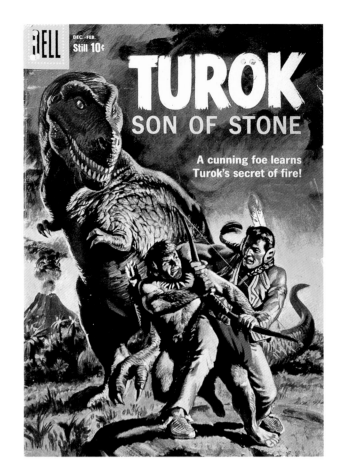

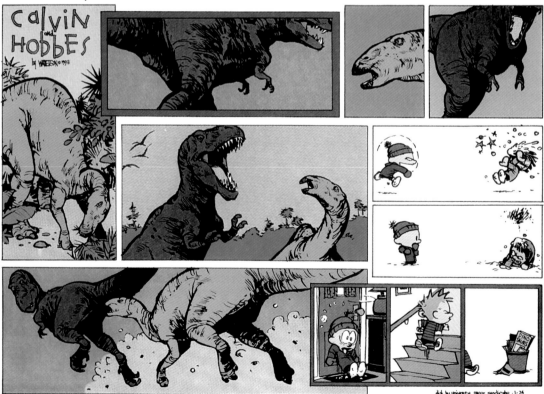

Ankylosaurs

Kenneth Carpenter

Ankylosaurs were short-limbed, four-legged, armor-plated dinosaurs with a long, wide body. The most prominent armor, or scutes, consisted of keeled or unkeeled plates of bone embedded in the skin. These scutes were sometimes supplemented with spines or spikes on the body and tail. In at least one group, there was also a bone club on the end of the tail. The ankylosaurs were so well armored that even the head was sometimes encased in bone scutes. This skull armor could extend even to bony eyelids and scutes covering the cheeks.

For a long time the taxonomy of ankylosaurs was confusing. That is because so many of the specimens consisted of fragmentary material. The most recent classification by W. Coombs and T. Maryańska (1990) recognizes two families, the Nodosauridae (informally called nodosaurids) and the Ankylosauridae (or ankylosaurids). The characters separating the two groups are many, occurring primarily in the skull, shoulder blade, and body armor.

Skeletal Features

Viewed from above, the skull of the nodosaurids is typically long and narrow, with a tapering muzzle (Fig. 22.1A). In contrast, the skull of the ankylosaurids is as wide as it is long, and the muzzle is usually broad (Fig. 22.1B). The long, narrow skull of nodosaurids is more similar to that of other ornithischian dinosaurs, such as *Stegosaurus*, suggesting that it is the

22

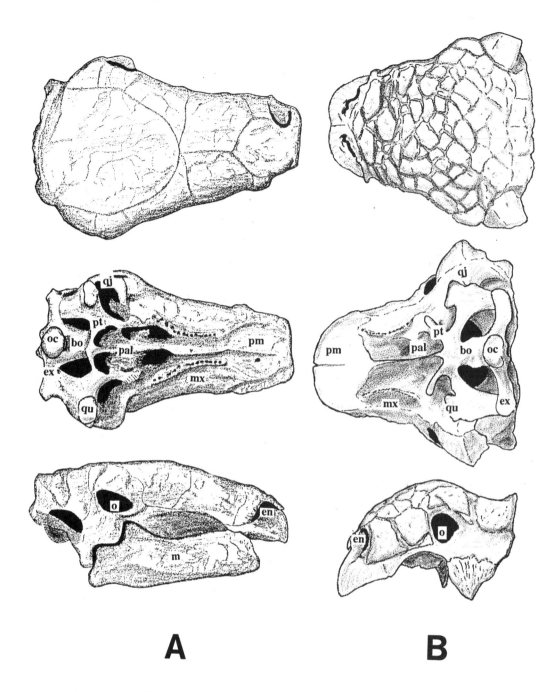

A   **B**

Figure 22.1. Comparison of the skulls of the nodosaurid *Edmontonia* (A) and the ankylosaurid *Euoplocephalus* (B). Abbreviations: bo = basioccipital; en = external nares; ex = exoccipital; m = mandible; mx = maxilla; pal = palatine; pm = premaxilla; pt = pterygoid; o = orbit; oc = occipital condyle; qj = quadratojugal; qu = quadrate.

primitive condition. By primitive we do not mean inferior, but that the feature is more like the ancestral condition.

The scutes on the nodosaurid skull are large and symmetrically arranged on both sides of the skull. In contrast, the scutes are small, numerous, and asymmetrically arranged on the top of the skull in ankylosaurids. In addition, ankylosaurids have horn-like, triangular scutes at the rear upper and lower corners of the skull.

Of the five major skull openings typically seen in the dinosaurs (external nares, antorbital fenestra, supratemporal fenestra, and lateral temporal fenestra), ankylosaurs have only two or three. A skull of *Pinacosaurus* shows that the skull bones expanded to cover the antorbital, supratemporal, and lateral temporal fenestrae in the ankylosaurids. In the nodosaurids, all but the lateral temporal fenestra are covered. On the other hand, the external nares, or nostril openings, are typically large in ankylosaurs, especially *Tarchia*. They usually face forward in ankylosaurids because of the broad muzzle, except in *Shamosaurus* and *Ankylosaurus*, where they face to the sides, as they do in the more narrow-muzzled nodosaurids. In advanced ankylosaurids, such as *Euoplocephalus,* air passed through looped sinuses during respiration. Why such a complex configuration developed is not known, although several hypotheses have been proposed, including that these animals had an especially keen sense of smell, or that this passage was used to warm and moisten the air, or as a resonating chamber (Maryańska 1977). The orbit, or eye socket, of ankylosaurs is usually large and circular or elliptical in shape. The dorsal or upper rim of the orbit is composed of several small bones called supraorbitals. In *Euoplocephalus* there is even a curved disk of bone protecting the eyelid.

The teeth of the ankylosaurs are shaped like tiny hands with the fingers close together. The teeth are small, although those of ankylosaurids are typically smaller than those of the nodosaurids. Exceptions do occur, especially in the primitive *Shamosaurus*, in which the teeth are large and nodosaurid-like. Near the base of the tooth crown in nodosaurids, just above the roots, there may be a shelf, or cingulum; in ankylosaurids the base of the crown is swollen instead. In primitive ankylosaurs, there are conical teeth at the front of the snout in the premaxillaries. These teeth are absent in advanced forms, being replaced by a sharp cutting edge formed by the premaxillaries; in life, there was probably a sharp horny beak. The upper cheek teeth of ankylosaurs are well inset from the sides of the skull, forming a wide shelf that is unique to the Ankylosauria, the function of which will be discussed below.

The vertebral column is divided into seven or eight cervical vertebrae in the neck, about sixteen dorsal vertebrae in the back, and three or four sacral vertebrae in the pelvis (Fig. 22.2). The tail has about twenty caudal vertebrae in ankylosaurids and up to forty in nodosaurids. The cervicals are short, resulting in a short neck to support the large, heavy skull. The dorsals are long, producing a long body. Furthermore, the diapophyses, where the ribs attach to the vertebrae, are angled upward between 30 and 50 degrees. This high angle makes the ribs arc outward, producing a barrel chest. The last three to six dorsals are often fused together and with the sacrals, producing a structure called a synsacrum. This immobile rod of vertebrae may be related to the modification of the pelvis into a wide, flat structure, which we will examine further below.

The caudals are elongated, except near the pelvis. In ankylosaurids the caudals in the last half of the tail are modified to form a "handle" for a large bone club on the end of the tail. This club is formed by several large bone

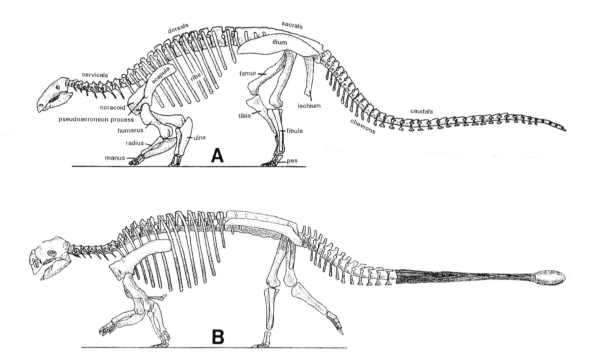

Figure 22.2. Comparison of the skeletons of the nodosaurid *Nodosaurus* (A) and the ankylosaurid *Pinacosaurus* (B). Length of *Nodosaurus* about 5 m, and of *Pinacosaurus* about 3.5 m.

scutes fused together and to the last few tail vertebrae. In nodosaurids, the caudals are not modified to carry a bone club. Instead, the tail was long and slender.

The front and rear limbs of ankylosaurs are short, especially in the ankylosaurids (Coombs 1978, 1979). As a result, the body looked long and low (Fig. 22.2). The scapula (shoulder blade) of the nodosaurids has a spur called the pseudo-acromion process above the shoulder socket for the attachment of the scapulo-humeralis anterior muscle. The position of this spur varies among the different species of nodosaurids, making the shoulder blade important in defining different nodosaurid species.

The upper arm bone, or humerus, is short and stout in ankylosaurs. There is a very prominent large flange of bone on the upper part of the humerus called the deltopectoral crest. Although all dinosaurs have this crest, in ankylosaurs it is especially well developed. The crest was the origination site for several major muscles used in locomotion (Fig. 22.3). The lower arm consists of a large, robust ulna and a smaller radius. At the elbow, the ulna has a very tall, projecting olecranon to which the extensor muscles were attached. The forefeet are short and broad to bear the great weight of the body. There are five digits in primitive ankylosaur species, and four in advanced species.

The pelvis is considerably modified from that seen in other ornithischians. The upper pelvic bone, the ilium, is very large and is oriented almost horizontally. Muscle scars on the underside of the ilium show that the muscles for moving the hind leg were large and powerful, especially those for pulling the leg forward. The front lower pelvic bone, the pubis, is reduced to a very small and rectangular bone. The rear lower pelvic bone, the ischium, projects almost straight down, rather than down and back as in other ornithischians. The acetabulum, or hip socket, differs from that of other dinosaurs in that it is cup-like, rather than just an opening.

The femur (thighbone) is pillar-like, being straight and massive to carry the enormous weight of the ankylosaur body. The femur rests atop two short but massive shin bones, the tibia and fibula. The hind feet are short and stout, and usually (but not always) have four toes in nodosaurids and three in ankylosaurids.

The most distinctive feature about the ankylosaurs as dinosaurs is their armor (Carpenter 1982, 1984, 1990). This armor is composed of large flat or keeled scutes on the neck and shoulders, and smaller keeled scutes on the back and tail. The space between the scutes is filled with a mosaic of smaller scutes or irregular marble- to pea-sized bone. This small armor may cover the belly and limbs as well. Nodosaurids also have spines along the sides or top of the neck and shoulder regions. In ankylo-

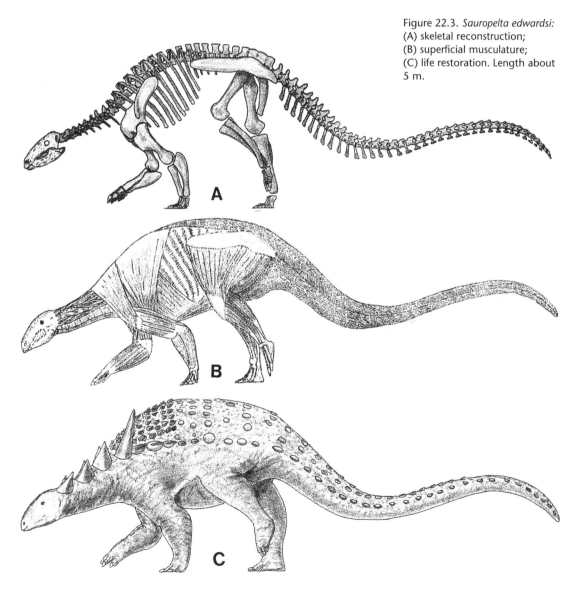

Figure 22.3. *Sauropelta edwardsi:*
(A) skeletal reconstruction;
(B) superficial musculature;
(C) life restoration. Length about 5 m.

saurids, modified scutes fuse together into a terminal club. A fossilized club is very heavy as a result of the minerals that fill the very porous internal structure. In life, organic tissue filled these spaces, making the club strong and resilient.

Origin and Evolution

The origin of the ankylosaurs is not well understood because so few specimens are known from the Early and Middle Jurassic, when ankylosaurs apparently first appeared and diversified. The oldest relatives are the thyreophorans *Scelidosaurus* from the Lower Jurassic of England and *Emausaurus* from the Lower Jurassic of Germany. *Scelidosaurus* is more ankylosaurian because there are thin sheets of dermal armor fused to parts of the skull. These "protoankylosaurs" have a body encased in conical scutes, but the skull still retains the antorbital and supratemporal openings, as well as other primitive features.

True ankylosaurs are known by the Middle Jurassic, and include *Sarcolestes* from England and *Tianchiasaurus* from China. Unfortunately, both genera are represented by fragmentary remains. Early Jurassic ankylosaurs are only slightly better known, including a new skull, as yet unnamed, from the Upper Jurassic of Wyoming. This specimen shows a mixture of both nodosaurid and ankylosaurid characters supporting a monophyletic origin of the Ankylosauria. Nodosaurid characters include tall neck spines, while ankylosaurid features include numerous small scutes fused to the skull. CAT scans of the skull show that the supratemporal and antorbital fenestra are fused closed beneath the armor. Two other ankylosaurs known from this time include the nodosaurids *Mymoorapelta* from the western United States and *Dracopelta* from Portugal.

By the Early Cretaceous, the ankylosaurids are represented by *Shamosaurus* from Asia, and the subfamily Polacanthinae by *Polacanthus* from England and the United States. Nodosaurids are better represented at this time, including *Hylaeosaurus* and *Acanthopholis* from England, and *Sauropelta* and *Pawpawsaurus* from the United States. The enigmatic *Minmi* is known from two nearly complete skeletons with armor found in Queensland, Australia. *Minmi* and an unnamed specimen from Antarctica remain the only known ankylosaurs from the Southern Hemisphere.

The greatest diversification of ankylosaurs occurred during the Late Cretaceous. *Maleevus, Pinacosaurus, Saichania, Talurus,* and *Tarchia* are known from Asia. Most of these taxa are known from skulls, and some by nearly complete skeletons with armor preserved *in situ*. North American ankylosaurids include *Ankylosaurus* and *Euoplocephalus,* both known from skulls and partial skeletons. Nodosaurids are not known from Asia, but are well represented in North America by *Edmontonia, Niobrarasaurus, Nodosaurus, Panoplosaurus,* and *Silvisaurus*. In Europe, *Struthiosaurus* is known.

The distribution of ankylosaurs suggests that they first appeared in Europe in the early Middle Jurassic from some primitive thyreophoran similar to *Scelidosaurus*. The presence of both nodosaurid and ankylosaurid features in some ankylosaurs, such as the unnamed specimen from Wyoming and *Minmi*, indicates that nodosaurids and ankylosaurids split from some common ancestor soon after the first ankylosaur appeared. Nodosaurids soon established themselves in North America and Europe, ankylosaurids established themselves primarily in Asia and North America, and the polacanthines occurred in Europe and North America for a brief time during the Early Cretaceous. *Minmi* appears to retain primitive characters, suggesting that it descended from the basal ankylosaur stock, perhaps derived before the two major families split.

Biology and Behavior

With short legs, ankylosaurs were built low to the ground and were probably restricted to feeding on the lower two meters of plants. They apparently had a corneous covering on the beak, allowing them to crop these plants. The type of plants ankylosaurs ate depended on where the animals lived and on the shape of the muzzle. Most of the ankylosaurids from Asia, such as *Saichania*, lived in arid or semi-arid environments where the plants were adapted to water stress. On the other hand, many ankylosaurids from North America lived in well-watered coastal environments, where vegetation was lush and lacked these protective structures. Interestingly, however, the teeth of the Asian and North American ankylosaurids

Figure 22.4. Spines as an antipredator device: *Edmontonia* keeping an *Albertosaurus* at bay. By keeping the spines facing the predator, *Edmontonia* could easily defend itself.

Figure 22.5. The tail club as an antipredator device: *Dyoplosaurus* defending itself against an *Albertosaurus*.

show no major differences in size or structure. There are, however, noticeable differences between the teeth of nodosaurids and ankylosaurids. Both groups were contemporaneous in the Late Cretaceous of North America, and differences in the teeth show that they probably avoided competition for food. In addition, the shape of the muzzles differs between the two families, showing that how the food was gathered also differed. The broad muzzle of the ankylosaurid *Euoplocephalus* suggests that this animal was a generalized feeder, cropping low plants with its wide beak (Carpenter 1982). On the other hand, its contemporary, the nodosaurid *Edmontonia*, had a narrow muzzle, so it could selectively crop vegetation. This partitioning of food resources enabled the ankylosaurs to cohabit the Late Cretaceous coastal plain.

Ankylosaurs apparently had a very mobile tongue, because of the large size of the throat bones (hyoids) that lay at the base of the tongue. Having a mobile tongue would allow the ankylosaurs to push and roll a wad of vegetation around in the mouth while chewing. Processing the food in this way requires some way to keep the food in the mouth, and that is the function of cheeks. The presence of muscular cheeks is inferred from the inset cheek teeth and the well-developed maxillary shelf. Once the food was swallowed, it was further processed by the grinding action of stones in the gut called gastroliths. Such stones, measuring several centimeters in diameter, have been found in only a few specimens, such as the

314   *Kenneth Carpenter*

nodosaurid *Panoplosaurus*. Their apparent absence in other ankylosaurs, such as the intact ankylosaurid skeletons recovered from sand dune deposits in Mongolia, is puzzling. Perhaps not all ankylosaurs used gastroliths.

Because cellulose, which forms the cell walls of plants, is difficult to digest, nutrients can be extracted only by microbial fermentation. In living mammals, this may occur in special chambers of the stomach, as in the cow, or in the intestines, as in the horse. Ankylosaurs most likely utilized rear gut, or intestinal, fermentation because the very broad hips suggest a very large or long rear gut (Bakker 1986).

The armor of ankylosaurs may have been produced in the skin in a manner similar to that of crocodiles. Traditionally this armor is thought to have provided protection against large predators. Among nodosaurids, the forward-projecting neck spines of *Edmontonia* could have been used as a weapon. The largest spine is braced against the shoulder, and the most obvious purpose of such a spine is as an "antipredator device," used to keep predators at bay (Fig. 22.4). In other ankylosaurs, the tail might have been used for defense, especially if a tail club was present (Coombs 1979), as in the ankylosaurids (Fig. 22.5).

Many other nodosaurids, such as *Sauropelta*, have spines that project upward, suggesting that not all spines were used against predators (Fig. 22.6A). Such spines are certainly noticeable, and may have been similar to the "showy" structures used by living animals in behavioral interactions, such as horns in antelopes. Showy structures are important among many living animals in sexual and agonistic display. Sexual display is used to attract a potential mate, whereas agonistic display is used to drive away a rival. Agonistic behavior can be divided into threat and intimidation. Threat behavior involves the prominent display of a weapon toward the opponent, thereby signaling a willingness to fight. Intimidation, on the other hand, is a form of "psychological warfare" in which the displaying animal tries to make itself appear larger.

While the spines in *Sauropelta* emphasize height, those of *Edmontonia* project outward, making the animal seem more squat and massive than it really is (Fig. 22.6B). Ankylosaurids and the nodosaurid *Panoplosaurus* lack large projecting neck spines, suggesting that frontal intimidation display may not have been important in these animals (Fig. 22.6C). Instead, side intimidation display may have been used, as inferred from the behavior of modern hornless (i.e., primitive) ungulates. In this type of intimidation, the animals probably stood parallel to each other with much hissing, growling, or whatever sounds ankylosaurs made.

It is also possible that ankylosaurs could have "blushed" so that the armor had a pink tint as a result of the infusion of blood under the horny covering of the armor. The vascular grooves covering the surface of most ankylosaur armor indicate that a rich supply of blood was present. Imagine two-ton ankylosaurs with pinkish armor bellowing and growling at each other!

Edmontonia may have had an additional use for the shoulder spine other than to keep predators at bay. The spine is bifurcated, and this structure may be analogous to the tines on a deer antler. In deer, the antlers are engaged by their tines during intraspecific fighting. This keeps the antlers from slipping past one another and possibly injuring the combatants. It is possible that the bifurcated spine of *Edmontonia* may have functioned like an antler tine, allowing the animals to engage in shoving matches without serious injury (Fig. 22.7).

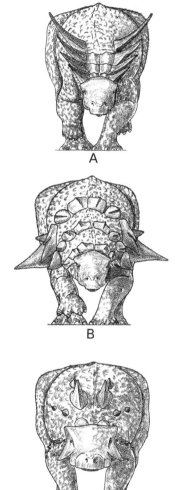

Figure 22.6. Armor in display: (A) *Sauropelta;* (B) *Edmontonia;* (C) *Dyoplosaurus.*

Ankylosaurs 315

Figure 22.7. Use of the bifurcated shoulder spines in intraspecific fights by *Edmontonia*.

References

Bakker, R. T. 1986. *The Dinosaur Heresies: New Theories Unlocking the Mystery of the Dinosaurs and Their Extinction.* New York: William Morrow.

Carpenter, K. 1982. Skeletal and dermal armor reconstruction of *Euoplocephalus tutus* (Ornithischia: Ankylosauridae) from the Late Cretaceous Oldman Formation of Alberta. *Canadian Journal of Earth Sciences* 19: 689–697.

Carpenter, K. 1984. Skeletal reconstruction and life restoration of *Sauropelta* (Ankylosauria: Nodosauridae) from the Cretaceous of North America. *Canadian Journal of Earth Sciences* 21: 1491–1498.

Carpenter, K. 1990. Ankylosaur systematics: Example using *Panoplosaurus* and *Edmontonia*. In K. Carpenter and P. J. Currie (eds.), *Dinosaur Systematics: Approaches and Perspectives*, pp. 281–298. Cambridge: Cambridge University Press.

Coombs, W. P. Jr. 1978. Forelimb muscles of the Ankylosauria (Reptilia, Ornithischia). *Journal of Paleontology* 52: 642–658.

Coombs, W. P. Jr. 1979. Osteology and myology of the hindlimb in the Ankylosauria (Reptilia, Ornithischia). *Journal of Paleontology* 53: 666–684.

Coombs, W. P. Jr., and T. Maryańska. 1990. Ankylosauria. In D. B. Weishampel, P. Dodson, and H. Osmólska (eds.), *The Dinosauria*, pp. 456–883. Berkeley: University of California Press.

Maryańska, T. 1977. Ankylosauridae (Dinosauria) from Mongolia. *Paleontologia Polonica* 37: 85–151.

Margino-
cephalians

*Catherine A. Forster
and
Paul C. Sereno*

The Marginocephalia: A Common Ancestry

The Marginocephalia are a diverse clade of ornithischian dinosaurs composed of two distinct subgroups: the pachycephalosaurs, or "thick-headed" dinosaurs, and the ceratopsians, or horned dinosaurs (Fig. 23.1). Marginocephalians are relative latecomers in dinosaur evolution, appearing first in Lower Cretaceous sediments of Asia. They successfully diversified and spread throughout the Northern Hemisphere, persisting up to the Cretaceous-Tertiary boundary as some of the last of the dinosaurs (Weishampel 1990). In the context of ornithischian evolution, marginocephalians are the sister group to the Ornithopoda, a group arising in the Early Jurassic. The base of the marginocephalian branch must therefore also extend back to the origin of the ornithopod branch. So while known marginocephalians are restricted to the Cretaceous, they have an earlier, as yet undiscovered, history extending back to the early Jurassic.

By the time pachycephalosaurs and ceratopsians first appear in the fossil record, millions of years of evolution have left only traces of their common ancestry. These traces reside in the shared characteristics of their bony skeletons that were inherited from a common ancestor. One important shared trait occurs in the skull of pachycephalosaurs and ceratopsians, but nowhere else among dinosaurs: the posterior edge of the skull roof overhangs the occiput (the area in the back of the skull where the neck attaches) as a narrow shelf. This overhanging shelf, or

23

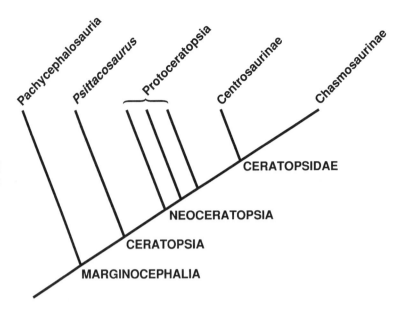

Figure 23.1. This simple cladogram depicts relationships among the Marginocephalia. The Pachycephalosauria constitute the first, or most primitive, branch of the group. The Ceratopsia include the small genus Psittacosaurus (occupying the first branch), the Proto-ceratopsia (the next few twigs), and finally the Ceratopsidae (further subdivided into Centrosaurinae and Chasmosaurinae).

"margin," inspired the collective name for these ornithischians, the Marginocephalia, or "margin-headed" dinosaurs. Other features characterizing marginocephalians involve changes in pelvic structure: loss of a bony process on the ischium, loss of the pubic symphysis (the pubic bones no longer touch each other), and hip sockets that are spaced more widely apart (Sereno 1986).

By the end of the Cretaceous, the top of the skull in each subgroup had become modified and hypertrophied (expanded) in two very distinctive patterns. In pachycephalosaurs, the skull expanded upward, thickening over the braincase into a knobby, sometimes spike-studded dome of solid bone. In ceratopsians, the shelf on the back of the skull was emphasized, extending posteriorly as a frill, or neck shield, and their skulls were ornamented with nasal and supra-orbital horns. Traditionally these bony head accessories have been viewed as weaponry for defense against predators (e.g., Colbert 1948). More recently these structures have been reinterpreted on the basis of living analogs (horned ungulates such as antelopes and deer, or horned lizards such as chameleons) as secondary sexual characteristics employed primarily for intraspecific rivalries (Farlow and Dodson 1975; Forster 1990; Sampson 1993, and chap. 27 of this volume; Dodson 1996). Many, if not the majority, of these striking structural modifications are probably the product of competition and display between individuals within a species during courtship or territorial fights. This type of behavior, inferred from skeletal features, may also be a characteristic of the group.

Ceratopsians have been found in Asia and North America, pachycephalosaurs in Asia, Europe, and North America. Thus marginocephalians have been found only in the Northern Hemisphere, and appear to be absent on all southern continents (see Molnar, chap. 38 of this volume, for a possible exception). Like all ornithischians, marginocephalians were herbivores, and like their ancestors, pachycephalosaurs and the primitive ceratopsians were bipedal, while more derived ceratopsians were secondarily quadrupedal.

Pachycephalosauria: Thick Skulls and Ornamented Domes

Pachycephalosaurs are first recorded in Lower Cretaceous sediments of Europe, where the cranial remains of a very small form, *Yaverlandia,* are found. *Yaverlandia* already shows the vertical thickening of the skull roof that characterizes all pachycephalosaurs. Complete skulls and partial postcrania (body skeleton) of more advanced forms are known from the Upper Cretaceous of North America and Mongolia. Some of the smallest pachycephalosaurs have been found recently in the Upper Cretaceous of Alberta and have mushroom-shaped domes only two inches in diameter. The other size extreme is represented by another North American form, *Pachycephalosaurus,* with a skull approximately two feet in length.

The thickening of the skull that characterizes all pachycephalosaurs occurs principally in two bony elements located over the brain, the frontal and parietal (Maryańska and Osmólska 1974; Maryańska 1990). These bones can achieve a vertical thickness of several inches, as in *Pachycephalosaurus.* Adjacent bones on the rear corners of the skull, the squamosals, are often ornamented with prominent tubercles. The postcranial skeleton is characterized by relatively short forelimbs, emphasizing that pachycephalosaurs were obligate bipeds. One of the most unusual and unique features in the pachycephalosaur skeleton is a woven mesh of ossified tendons that surrounds the distal half of the tail. This bony mesh likely stiffened the tail, although the purpose of this remains obscure. No complete skeleton of a pachycephalosaur has yet been found.

The pachycephalosaurs have often been divided into primitive "flat-headed" and derived "dome-headed" subgroups (e.g., Sues and Galton 1987). However, Sereno (1986) feels that this simple division may not hold; pachycephalosaurs may instead be divided into a series of forms, having both high and low domes, progressing toward a separate subgroup, the Pachycephalosauridae, all with high domes. The most primitive pachycephalosaur yet found is a Chinese form, *Wannanosaurus.* Other primitive pachycephalosaurs, including *Goyocephale, Homalocephale, Ornatotholus,* and *Yaverlandia,* show a progression of characters, including a reduction in the size of the circular openings on the top of the skull above the temporal region (supratemporal openings), increased thickening of the parietal bone, and co-fusion of the frontal and parietal bones on the skull roof.

Many, but not all, of these more primitive pachycephalosaurs are "flat-headed." The skull roof in these "flat-headed" forms is thickened evenly from side to side, with a flat, pitted dorsal surface. A short canine tooth is present in both the upper and lower jaws, as preserved in the Mongolian *Goyocephale,* with the lower canine projecting into a special notch in the upper jaw. The remainder of the dentition is typically ornithischian and consists of small crowns with serrate margins and sharp cutting edges for slicing plant material.

More advanced pachycephalosaurs (Fig. 23.2), the Pachycephalosauridae, occur in both North America and Asia, and include *Stegoceras, Microcephale, Tylocephale, Prenocephale,* and *Pachycephalosaurus.* In these forms, the frontal and parietal are fused into a single massive element and thickened into a high, solid dome of columnar bone. Although this gives the false impression that brain volume has been substantially increased, the brain remains very small relative to the volume of the

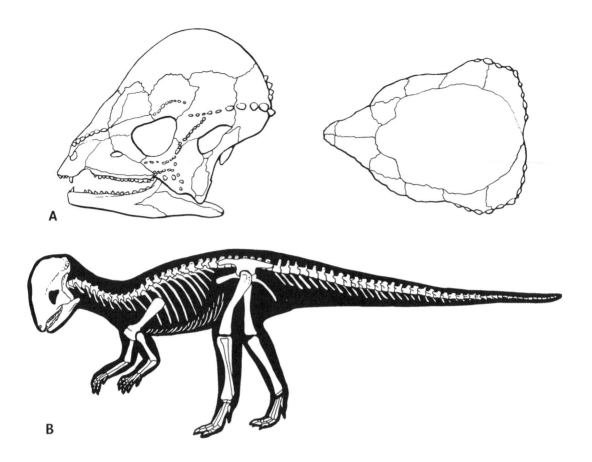

A

B

Figure 23.2. (A) Left lateral (*left*) and dorsal (*right*) view of the skull of the dome-headed pachycephalosaur *Prenocephale*. (B) Skeleton with body silhouette of *Stegoceras*. Total length of *Stegoceras* about 2m.

dome. The external surface of the dome is often very smooth, although some forms ornament the sides of the dome with knobs and short spikes. In more advanced forms, such as the Mongolian *Prenocephale*, the circular temporal openings are completely closed by the surrounding bones, and all that remains is a pit. The vertical thickness of the dome is relatively greater and incorporates all of the surrounding elements of the skull. The most common pachycephalosaurid in North America is *Stegoceras*, which still retains a shelf around the growing dome. The fused, solid dome is very resistant to weathering and is frequently found in isolation in the field.

The basic function of the thickened skull was likely head-butting with fellow pachycephalosaurs, a behavior similar to that in living bighorn sheep (e.g., Galton 1970). Head-butting in living sheep (and other bovids) is a necessary part of social interaction, providing the means for establishing "pecking order" and for fighting over mates and territories. By inference from living animals, it is easy to imagine pachycephalosaurs engaging in similar dominance displays among themselves. In *Pachycephalosaurus* the dome attains a thickness of nine inches over the small brain cavity and presumably withstood tremendous forces upon impacting a rival. The domes are sometimes marked by irregularities and other scars, presumably the result of head-to-head impact. Since all pachycephalosaurs have a thickened skull roof to some degree, head-butting may have constituted an ancestral behavioral pattern that was retained in all members of the group (see Alexander, chap. 30, and also Sampson, chap. 27, of this volume for additional discussion of head-butting in these dinosaurs).

The Ceratopsia: Parrot Beaks, Horns, and Frills

The Ceratopsia, consisting of *Psittacosaurus* and the Neoceratopsia ("protoceratopsians" plus ceratopsids; Fig. 23.1), display a dazzling variety of horns and neck frills, include both bipedal and quadrupedal forms, and range from turkey- to elephant-sized animals. Despite this range in ornamentation, stance, and size, ceratopsians share a number of skeletal innovations that attest to their common ancestry, notably a skull with a very narrow but deep beak-like snout and widely flared cheeks, which give the skull a nearly triangular shape when seen from above. The posterior margin of the skull overhangs the occiput, as in all marginocephalians, but in ceratopsians the margin is very thin and is formed primarily by a single roofing bone, the parietal. The final hallmark of ceratopsians is their development of an additional median bone, the rostral, that forms their upper beak (Sereno 1986; Dodson and Currie 1990). The surface of the rostral is textured and pitted, indicating that it was covered by a horny bill in life. This unique element is found nowhere else in the Dinosauria.

In the early days of dinosaur discovery in North America, ceratopsians originally included only large-bodied Upper Cretaceous quadrupeds, heavily ornamented with horns and large neck frills, of which *Triceratops* is the most familiar. Shortly after the turn of the century, however, smaller-bodied ceratopsians such as *Leptoceratops,* a small form from the Upper Cretaceous of Montana and Alberta, were discovered that lacked cranial horns and large neck frills. Well-preserved remains of small-bodied ceratopsians were also discovered in the Lower Cretaceous rocks of Mongolia in the 1920s during the Central Asiatic Expeditions of the American Museum, and were described as *Protoceratops*. These smaller-bodied forms and others found since then in North America and Asia are often called "protoceratopsids." Contemporaneous with ceratopsids, none of them appears to be a direct ceratopsid ancestor (Forster 1990).

While the discovery of these small, hornless "protoceratopsians" changed our definition of a ceratopsian, another discovery by the Central Asiatic Expeditions was eventually to make a larger impact. This was the earliest and most primitive ceratopsian, *Psittacosaurus,* from Lower Cretaceous sediments in Mongolia. Initially psittacosaurs were allied with more distant ornithischians, rather than with other ceratopsians (e.g., Osborn 1924). This misunderstanding, which persisted for decades, is understandable; the ceratopsian features of *Psittacosaurus* appear only in the detailed structure of the skull. Their postcranial skeleton is similar to that of primitive bipedal ornithischians and, by itself, is nearly as close to that of other ornithischian subgroups as to that of ceratopsians.

Ceratopsians occur exclusively in rocks of Cretaceous age in western North America and Asia. During the Late Cretaceous, intermittent land bridges across the Bering region allowed restricted interchange of faunas. Although no ceratopsian (or pachycephalosaur) species occurs on both sides of the Pacific Ocean, the evolutionary history of each group clearly indicates that there were multiple dispersal events from one side to the other.

Psittacosaurus

Psittacosaurus (Fig. 23.3) is recorded from Lower Cretaceous (Aptian-Albian) rocks in Russia, the People's Republic of Mongolia, and the People's Republic of China. Seven species are currently recognized, including *P. guyangensis, P. meileyingensis, P. mongoliensis, P. sinensis, P. youngi,* and *P. xinjiangensis* (Sereno 1990). Many complete skulls and skeletons of *Psittacosaurus* have been collected, including one tiny juvenile with a skull only 2.8 centimeters long. Adult psittacosaurs probably reached 1.5 to 2 meters in total length.

The name *Psittacosaurus,* meaning "parrot-lizard," derives from the tall, parrot-like outline of the snout in side view. The psittacosaur snout, from the orbit to the front of the skull, is proportionately shorter than in any other dinosaur. The external naris (nasal opening) is very small and is positioned higher on the snout than in other ornithischians. An unusual feature of the psittacosaur postcranial skeleton is the reduction of the two outer digits of the hand, leaving only three functional fingers. The forelimb, however, is not reduced in length or size in psittacosaurs, and may have been employed for walking or browsing on low vegetation.

Two specimens of *Psittacosaurus* have been found with compact concentrations of small, rounded stones within their rib cage (Osborn 1924). These stones, or gastroliths, formed part of the psittacosaur digestive system, grinding up plant material to render it more easily digestible. This system is similar to that seen in some birds and reptiles, which swallow small rocks that are retained in their gizzard. As the ingested food passes

Figure 23.3. (A) Left lateral (*left*) and dorsal (*right*) view of the skull of *Psittacosaurus*. (B) Skeleton with body silhouette of *Psittacosaurus*. Total length about 2m.

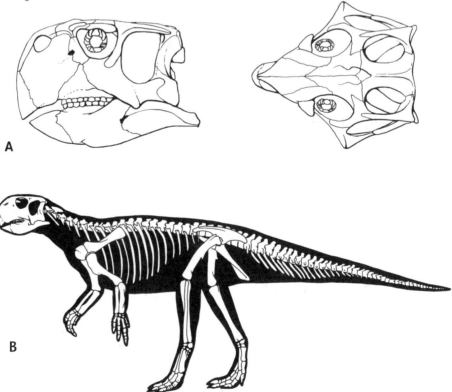

A

B

through the gizzard, it is ground between the stones and broken down for easier processing.

Neoceratopsia

The Neceratopsia contain two general subgroups, the small-bodied protoceratopsids and the large-bodied ceratopsids. Neoceratopsians had developed a fan-shaped bony sheet, or frill, that projected back over the neck as a very elongate version of the primitive marginocephalian shelf. The frill was often rather thin, and did not offer a strong means of defense. Rather, it probably functioned primarily in display behavior to members of the same species. In anterior view, the broad surface of the frill was fully visible, and in ceratopsids its margins were usually ornamented with a variety of ossicles, horns, and fluted protuberances. When viewed head-on, the broad frill must have made an impressive display, making the animal appear larger and possibly more threatening (cf. Sampson, chap. 27 of this volume). Other unique characters of neoceratopsians include a large head relative to body size, fusion of the first three vertebrae to help support their large heads, an upwardly hooked lower beak, and the development of an accessory bony protuberance, called an epijugal, on each of the laterally expanded cheeks (Dodson and Currie 1990; Forster 1990; Lehman 1990).

Protoceratopsians

An expedition from the American Museum of Natural History unearthed the first protoceratopsian, *Leptoceratops* (Fig. 23.4), in 1910 from the Maastrichtian Scollard Formation of Alberta (Brown 1914). Although *Leptoceratops* appears late in time, it is a very primitive short-frilled form, and serves as a reminder that primitive does not necessarily mean early in time. North American protoceratopsians (*Leptoceratops* and *Montanoceratops*) are found only in Maastrichtian rocks, while those from Asia (*Protoceratops, Graciliceratops, Bagaceratops*) occur in Santonian through Campanian sediments.

Protoceratopsids have often been thought to be monophyletic (e.g., Dodson and Currie 1990), but recent studies suggest that they form a progressive series leading to the Ceratopsidae (Sereno 1986; Forster 1990). The most primitive protoceratopsian, *Leptoceratops,* is a small form with a short, solid frill, a short face, and a curved, rocker-shaped lower jaw. In more advanced protoceratopsians, the parietals and squamosals had progressively elongated to extend the frill backward, large bi-lateral openings, or fenestrae, had developed in the frill, the teeth had become packed closer together, the vertebral spines had coalesced over the hips, and a broad-based nasal hump had begun to develop over the nares.

The American Museum of Natural History collected dozens of skulls and skeletons of *Protoceratops*, from small juveniles to large adults, in Mongolia. After studying this series of specimens, Dodson (1976) observed that the juvenile specimens all looked alike until they were "teenagers." They then began to diverge into two increasingly different forms: one large and robust with a pronounced nasal hump, and one smaller, more gracile, with no nasal hump. He interpreted this as evidence of sexual dimorphism—that is, the increasing physical differences between male and female *Protoceratops* as they became mature. Although sexual dimorphism is common in many animals, even among humans, this marked the first time it was observed in a dinosaur.

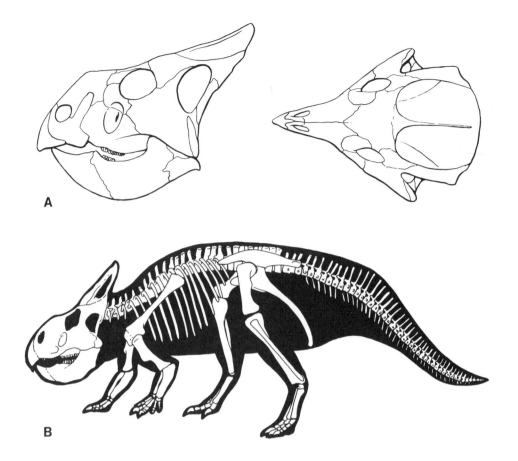

A

B

Figure 23.4. (A) Left lateral (*left*) and dorsal (*right*) view of the skull of *Leptoceratops*. (B) Skeleton with body silhouette of *Protoceratops*. Total length of *Protoceratops* about 2.5 m.

Along with the discovery of *Protoceratops* in Mongolia in the 1920s came the first report of nests of dinosaur eggs. These eggs, which were extremely common in the area, were thought to belong to the ubiquitous *Protoceratops*, although no embryos were found in these eggs to confirm this association. Recently, a *"Protoceratops"* egg was discovered with an embryo inside—of a carnivorous dinosaur called *Oviraptor* (Norell et al. 1994). While this matter awaits further study, it now appears likely that *"Protoceratops"* eggs belong to another dinosaur.

Ceratopsidae

The long horns and expansive, studded neck frills of the Ceratopsidae (Fig. 23.5) make them one of the most easily recognized groups of dinosaurs. Their enormous skulls, reaching well over six feet in length in forms such as *Torosaurus*, are the largest of any land animal that ever lived. Ceratopsids are found only in western North America, from Alaska to New Mexico, in sediments from the latest Cretaceous (Campanian and Maastrichtian).

The first pieces of a ceratopsid to be discovered were a pair of large supraorbital horns found in 1887 in the Denver Formation of Colorado. Having never seen a horned dinosaur before, O. C. Marsh (1887) named this incomplete specimen *Bison alticornis*, thinking such large horns must have

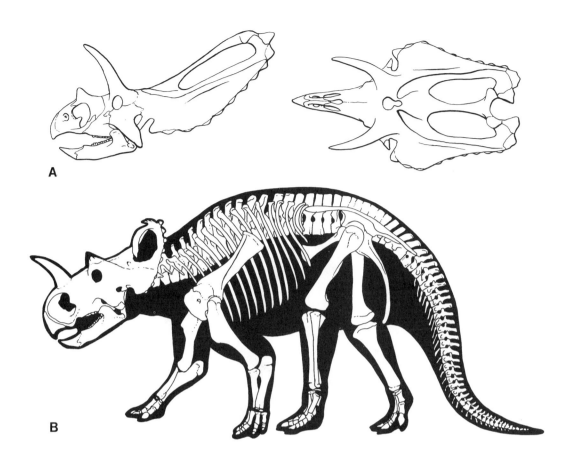

A

B

Figure 23.5. (A) Left lateral (*left*) and dorsal (*right*) view of the skull of chasmosaurine ceratopsid *Pentaceratops*. (B) Skeleton with body silhouette of centrosaurine ceratopsid *Centrosaurus*. Total length of *Centrosaurus* about 5 m.

come from an extinct bison. Soon complete ceratopsid skulls were found in Wyoming, and, realizing his mistake, Marsh renamed this first specimen *Triceratops alticornis*. Since the late 1800s, more than one hundred ceratopsid skulls and skeletons have been collected in western North America.

Ceratopsids share a large array of unique characters that attest to their common ancestry, including extremely large nasal openings, the development of secondary "roofing" over the braincase that opens onto the top of the skull, large body size, and double-rooted teeth (Dodson and Currie 1990; Forster 1990). Another of their innovations is the evolution of a complex dental battery, in which adjacent teeth are locked together in longitudinal rows and vertical columns. As the teeth along the cutting edge of the jaws were worn down, they were shed and replaced from below by new teeth. In ceratopsids, there are at least three teeth in each vertical column: one on the cutting edge, and two waiting as replacements. This process, which continued throughout life, provided a self-sharpening cutting edge without gaps. The cutting edges were oriented nearly vertically, allowing the animal to shear plant material almost like a scissors. The evolution of the ceratopsid dental battery parallels a similar development among the contemporaneous duck-billed hadrosaurs. Both groups have developed efficient dental magazines, incorporating hundreds of teeth in the jaws of a single individual, and constitute an impressive example of parallel evolution among dinosaurs.

Marginocephalians   325

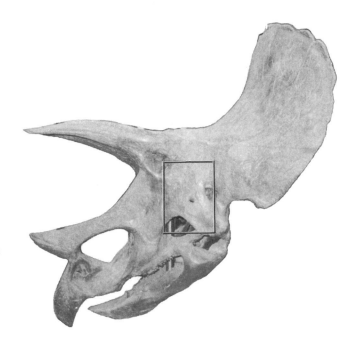

Figure 23.6. *Triceratops* specimen showing a healed wound that may be the result of a fight with another *Triceratops.* The top photo shows the left side of the skull with the location (area in the box) of the close-up photo shown below. The wound passes all the way through the cheek as two open holes. This specimen is at the Science Museum of Minnesota in St. Paul.

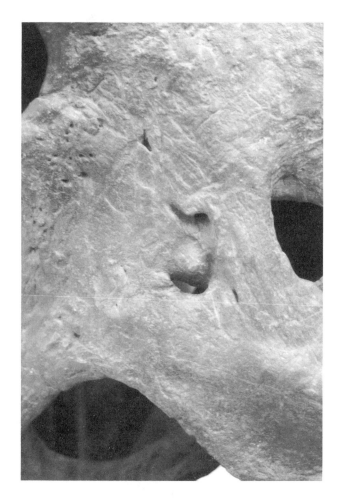

Like their relatives the pachycephalosaurs, ceratopsids also seem to have been well armed for head-to-head sparring (cf. Sampson, chap. 27, and also Alexander, chap. 30, of this volume). While some ceratopsids had long horns over their eyes and nose, others had shorter horns or, in *Pachyrhinosaurus,* a thickened pad of bone over the top of their skull. Like living antelope, deer, and chameleons, ceratopsids may have locked horns in shoving matches to determine territorial rights or dominance over a rival. Although the object was not to wound or kill the opponent, healed scars and punctures in some ceratopsid skulls show that injuries were fairly common (Fig. 23.6; but cf. Rothschild, chap. 31 of this volume). As in living antelopes, fighting likely constituted an important social interaction within a ceratopsid species. Yet when confronted with a predator, an antelope will use its horns for defense. Confronted with *Tyrannosaurus,* it is easy to imagine *Triceratops* defending itself with its horns. In all likelihood, these ornamentations in ceratopsids were multipurpose structures, and evolved for more than a single reason.

While the skulls of different ceratopsids vary greatly in their ornamentation, their postcranial skeletons are all very similar. Their long hind limbs extend vertically to the ground in an upright stance. While some reconstructions also show an upright stance for their front legs (e.g., Bakker 1986), a careful look at the bones reveals that they cannot articulate in this manner, but instead have a semi-erect orientation (Johnson and Ostrom 1995; see Lockley and Hunt 1995 for a different view). All leg bones in ceratopsids were thick and heavily built, with the lower portions of the legs shorter than the upper portions (fast runners, such as horses, have proportionally long lower limbs). This heavy build, coupled with semi-erect front limbs, indicates a powerful animal, but one not built for speed. Strength, horns, and perhaps some social "herding" behavior were their defense against predators, rather than fast escape. Although ceratopsids were probably capable of "galloping" occasionally, their skeletons suggest a slower and more sedentary lifestyle.

Ceratopsids can be divided into two groups, the Centrosaurinae and the Chasmosaurinae. While sharing many ceratopsid features, each group also has its own unique morphologies.

Centrosaurines can be recognized by many characters, including long nasal horns, hooks and processes on the parietals (sometimes developed into long spikes as in *Styracosaurus*), short and square squamosals, and a bony "finger" projecting into the back of the nares. Another unique ornamentation is found in *Pachyrhinosaurus,* in which the orbital and nasal horns had turned into irregular, vertically thickened pads of bone that covered the top of the skull. These pitted and grooved bony pads may have formed the base for horny spikes or protuberances, similar to those found on the face of a rhinoceros.

Large "bonebed" accumulations of centrosaurines have been found in Montana and Alberta. Each bonebed contains hundreds of pieces of skulls and skeletons from a single species, including small juveniles to adults. Scientists studying these "ontogenetic series"—that is, how the animals change in form from birth to adulthood—found that all centrosaurine species, whether *Centrosaurus, Styracosaurus,* or *Pachyrhinosaurus,* looked identical when they were young. It was not until they approached adulthood and became sexually mature that they grew the horns, spikes, and bony pads that identify their genus (Tanke 1988; Ryan 1992; Sampson 1993). Thus taxa based on juvenile animals, such as *Avaceratops, Brachyceratops,* and *Monoclonius,* are difficult to evaluate; for instance, some think that *Brachyceratops* is actually a juvenile *Styracosaurus* (e.g., Sampson 1993).

Chasmosaurines also have their own characters, including a very complex narial opening with numerous fenestrae and bony processes, long orbital horns and short nasal horns, a large conical epijugal bone on the flared cheeks, and an extra "epinasal" bone augmenting the front of the nasal horn. No large chasmosaurine bonebeds have been found, so little is known about juveniles and changes with growth in this group.

Advanced chasmosaurines, such as *Triceratops* and *Torosaurus*, developed large spaces, or frontal sinuses, between the base of the horns and the braincase (Forster 1990, 1996). These frontal sinuses were an expansion of the secondary skull "roofing" seen in all ceratopsids. Similar sinuses are found in living bovids, such as antelope and goats, where they act as shock absorbers to take up stress from blows on their horns (imagine two bighorn sheep ramming together). The sinuses shield the brain from injury during impact, and they likely served the same function in long-horned chasmosaurines. Other chasmosaurines include *Chasmosaurus, Pentaceratops, Arrhinoceratops,* and *Anchiceratops.*

References

Bakker, R. T. 1986. *The Dinosaur Heresies: New Theories Unlocking the Mystery of the Dinosaurs and Their Extinction.* New York: William Morrow.

Brown, B. 1914. *Leptoceratops,* a new genus of Ceratopsia from the Edmonton Cretaceous of Alberta. *Bulletin American Museum of Natural History* 33: 567–580.

Colbert, E. H. 1948. Evolution of the horned dinosaurs. *Evolution* 2: 145–163.

Dodson, P. 1976. Quantitative aspects of relative growth and sexual dimorphism in *Protoceratops. Journal of Paleontology* 50: 929–940.

Dodson, P. 1996. *The Horned Dinosaurs.* Princeton, N.J.: Princeton University Press.

Dodson, P., and P. J. Currie. 1990. Neoceratopsia. In D. B. Weishampel, P. Dodson, and H. Osmólska (eds.), *The Dinosauria,* pp. 593–618. Berkeley: University of California Press.

Farlow, J. O., and P. Dodson. 1975. The behavioral significance of frill and horn morphology in ceratopsian dinosaurs. *Evolution* 29: 353–361.

Forster, C. A. 1990. The cranial morphology and systematics of *Triceratops,* with a preliminary analysis of ceratopsian phylogeny. Ph.D. dissertation, Department of Geology, University of Pennsylvania. 227 pp.

Forster, C. A. 1996. New information on the skull of *Triceratops. Journal of Vertebrate Paleontology* 16: 246–258.

Galton, P. M. 1970. Pachycephalosaurids: Dinosaurian battering rams. *Discovery* 6: 23–32.

Johnson, R. E., and J. H. Ostrom. 1995. The forelimb of *Torosaurus* and an analysis of the posture and gait of ceratopsian dinosaurs. In J. J. Thomason (ed.), *Functional Morphology in Vertebrate Paleontology,* pp. 205–218. Cambridge: Cambridge University Press.

Lehman, T. M. 1990. The ceratopsian subfamily Chasmosaurinae: Sexual dimorphism and systematics. In K. Carpenter and P. J. Currie (eds.), *Dinosaur Systematics: Approaches and Perspectives,* pp. 211–229. Cambridge: Cambridge University Press.

Lockley, M. G., and A. P. Hunt. 1995. Ceratopsid tracks and associated ichnofauna from the Laramie Formation (Upper Cretaceous: Maastrichtian) of Colorado. *Journal of Vertebrate Paleontology* 15: 592–614.

Marsh, O. C. 1887. Notice of new fossil mammals. *American Journal of Science,* series 3, 34: 323–331.

Maryańska, T. 1990. Pachycephalosauria. In D. B. Weishampel, P. Dodson, and H. Osmólska (eds.), *The Dinosauria,* pp. 564–577. Berkeley: University of California Press.

Maryańska, T., and H. Osmólska. 1974. Pachycephalosauria, a new suborder of ornithischian dinosaurs. *Paleontologia Polonica* 30: 45–102.

Norell, M. A.; J. M. Clark; D. Demberelyin; B. Rinchen; L. M. Chiappe; A. R. Davidson; M. C. McKenna; P. Altangerel; and M. J. Novacek. 1994. A theropod dinosaur embryo and the affinities of the Flaming Cliff dinosaur eggs. *Science* 266: 779–782.

Osborn, H. F. 1924. *Psittacosaurus* and *Protiguanodon:* Two Lower Cretaceous iguanodonts from Mongolia. *American Museum Novitates* 127: 1–16.

Ryan, M. 1992. The taphonomy of a *Centrosaurus* (Reptilia: Ornithischia) bone bed (Campanian), Dinosaur Provincial Park, Alberta, Canada. Master's thesis, University of Calgary, Alberta.

Sampson, S. D. 1993. Cranial ornamentation in ceratopsid dinosaurs: Systematic, behavioral, and evolutionary implications. Ph.D. dissertation, Department of Zoology, University of Toronto. 298 pp.

Sereno, P. C. 1986. Phylogeny of bird-hipped dinosaurs (Order Ornithischia). *National Geographic Research* 2: 234–256.

Sereno, P. C. 1990. Psittacosauridae. In D. B. Weishampel, P. Dodson, and H. Osmólska (eds.), *The Dinosauria,* pp. 579–592. Berkeley: University of California Press.

Sues, H.-D., and P. M. Galton. 1987. Anatomy and classification of North American Pachycephalosauria (Dinosauria: Ornithischia). *Palaeontographica* A 178: 183–190.

Tanke, D. H. 1988. Ontogeny and dimorphism in *Pachyrhinosaurus* (Reptilia: Ceratopsia), Pipestone Creek, N. W. Alberta, Canada. *Journal of Vertebrate Paleontology* 8 (Supplement 3): 27A.

Weishampel, D. B. 1990. Dinosaur distributions. In D. B. Weishampel, P. Dodson, and H. Osmólska (eds.), *The Dinosauria,* pp. 63–139. Berkeley: University of California Press.

Ornithopods

M. K. Brett-Surman

24

The Ornithopoda ("bird-feet"), commonly called the ornithopods, were small (less than 1 meter tall and 2 meters long) to large (about 7 meters tall and 20 meters long) bipedal herbivorous dinosaurs that existed from the earliest Jurassic to the end of the Cretaceous. The groups that make up the ornithopods are, more or less in the sequence of their appearance in the fossil record, from the Jurassic: the heterodontosaurids, hypsilophodontids, dryosaurids, camptosaurids, tenontosaurs, iguanodontids, and hadrosaurs. They lived on every continent, including Antarctica. In a world dominated by theropods, the ornithopods had neither armor like the Thyreophora, nor horns like the Ceratopsia. They were the first herbivorous dinosaurs to have multiple tooth rows, cheek pouches, and true mastication ("chewing"). At the time that they were alive, they were the most derived herbivores on Earth. They occupied the niches occupied today by such medium-sized herbivores as antelopes, tapirs, moose, and horses. They were the first herbivorous dinosaurs to engage in "selective feeding" because they had very narrow muzzles that could selectively crop specific parts of plants. They were the first bipedal herbivores to occupy nearly every size range. From the Jurassic to the end of the Cretaceous, they continued to diversify and were the most successful of the herbivorous dinosaurs, both in the numbers of individuals per fauna and in the total number of ornithischian species.

History of Knowledge of the Group

O. C. Marsh of Yale University first named the Ornithopoda in 1881. A revised diagnosis published one year later (Marsh 1882) is paraphrased as follows: ornithopods walked on their toes (not flat-footed), and have five functional fingers on the hand and three on the foot; the pre-pubic bone projects forward and away from the midline of the body (in contrast to the theropods, where the pubes meet and fuse in the midline), and a post-pubic bone is present; the vertebrae are not hollowed out (as in saurischians); the front limbs are small and all the limb bones are hollow; the premaxillary bone (the upper lip bone) has no teeth. In Marsh's scheme, the group included the "camptonotids" (later renamed the camptosaurids), the iguanodonts, and the hadrosaurs.

Because the ornithopods showed none of the elaborate horns, frills, spikes, or body armor found in other ornithischians, they were once considered to be the basic, or stem, group from which other ornithischian lineages arose. Originally, when *Iguanodon* was only the second dinosaur known (1825), it was assumed that this animal was a quadruped. It was Joseph Leidy of Philadelphia who first suggested in the 1860s that *"Trachodon"* (1856), *Hadrosaurus* (1858), and *Iguanodon* might be bipedal (see Torrens, chap. 14 of this volume). This seemed to be confirmed in 1878 when multiple complete skeletons of *Iguanodon* were unearthed in Bernissart, Belgium.

As the years progressed, it became clear that the features used to define ornithopods were also present in other ornithischians. Consequently the ornithopods gradually came to be regarded as "essentially bipedal ornithischians" (Steel 1969). All bipedal ornithischians were therefore assigned to the ornithopods, including pachycephalosaurs, *Stenopelix*, psittacosaurids, and the "fabrosaurs." It was not until the 1970s, and later in Sereno's (1986) cladistic classification, that a reclassification of all the ornithischians resulted in a redefinition of the ornithopods. It became apparent that bipedalism was simply an ancestral character shared with many other dinosaurs. The pachycephalosaurs were pulled out and placed into their own group, united with the Ceratopsia (including the Psittacosauridae) as the Marginocephalia (Maryańska and Osmólska 1974, 1985; Cooper 1985; Sereno 1986; Dodson 1990).

Classification

Today the ornithopods may be partially diagnosed as follows: premaxillary teeth (if present) are on a level lower than the maxillary teeth; the jaw joint is lower than the tooth rows so that the jaws come together like nutcrackers instead of like scissors; the premaxillary bone has a process that extends backward (caudally) toward the orbit (eye); and there is a very large fourth trochanter on the femur for the attachment of the caudifemoralis muscle group. For a more complete classification, with a discussion of the many characters used to define the member clades, see Fastovsky and Weishampel 1996.

Within the Ornithopoda are the Euornithopoda (literally "true ornithopods"), which are distinguished by loss of the fenestrae (windows or holes) in the lower jaw (present in heterodontosaurids; Fig. 24.1); by elongation of the prepubic bone, which extends farther forward than in non-euornithopods; and by the presence of an obturator process on the

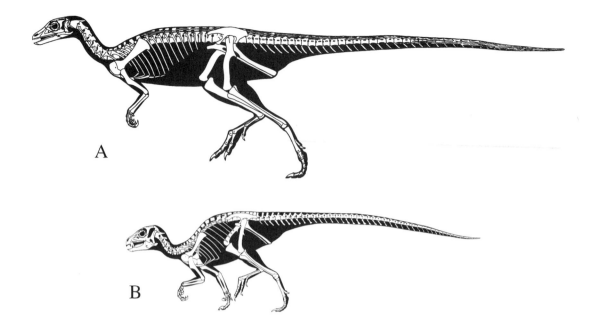

A

B

Figure 24.1. (A) *Lesothosaurus,* a fabrosaurid (?), and (B) *Hetero-dontosaurus.* Length of each animal about 1 m. This and other drawings in this chapter by Gregory S. Paul, who retains the copyright.

ischium. Within the euornithopods are the Hypsilophodontidae and the Iguanodontia. This latter group includes the iguanodontids and hadrosaurs, and is defined by loss of the premaxillary teeth; having a small antorbital fenestra in front of the eye (Fig. 24.2, 24.3A, B), or none at all; an enlarged nasal opening; and a predentary bone with two processes that project backward. All euornithopods have a pleurokinetic skull (discussed below; Norman and Weishampel 1990).

Figure 24.2. *Hypsilophodon* skull. Abbreviations as follows:
AF = antorbital fenestra (fenestra = "window"); AN = angular; CP = coronoid process; D = dentary; EN = external nares; FR = frontal; J = jugal; L = lacrimal; MX = maxilla; N = nasal; P = pal-pebral bone; PD = pre-dentary; PF = prefrontal; PM = premaxilla; PO = post-orbital; Q = quadrate; QJ = quadratojugal; SA = sur-angular; SQ = squamosal; SR = sclerotic rings; TF = temporal fenestra.

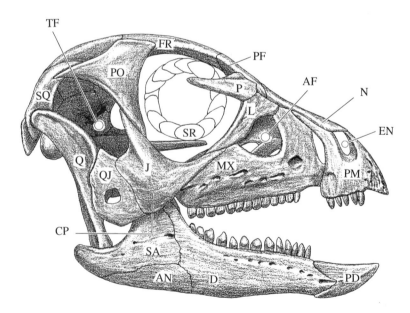

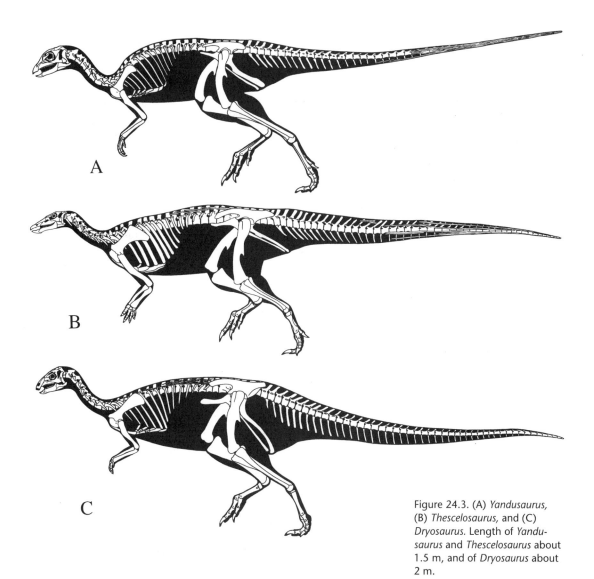

Figure 24.3. (A) *Yandusaurus,* (B) *Thescelosaurus,* and (C) *Dryosaurus.* Length of *Yandusaurus* and *Thescelosaurus* about 1.5 m, and of *Dryosaurus* about 2 m.

The heterodontosaurids (Fig. 24.1B) have several unique features, such as canine-like teeth and relatively long arms with large hands. This was the earliest ornithopod group in which all the cheek teeth are close enough together to form a solid dental battery and the teeth are designed for cutting, instead of "grabbing," vegetation. With a body length of about 1 meter, heterodontosaurs were about half the size of an adult human.

The hypsilophodontids (Figs. 24.2, 24.3A, B) are about 2 meters long and include such famous genera as *Thescelosaurus* and *Hypsilophodon*. This clade was the first ornithopod group to occur worldwide. They retained their premaxillary teeth and had chisel-shaped cheek teeth, and they were lightly built but had relatively heavy hind legs, probably for stability while running. Although about a dozen genera are known, there are only a few complete skeletons known for one genus. New finds in the United States, especially of a new hypsilophodontid from Texas (Winkler et al. 1997), will add new information, especially about growth processes in these dinosaurs. It may turn out that only a subset of this family is monophyletic.

Ornithopods   333

One of the early clades of the Jurassic Iguanodontia, the dryosaurs (Figs. 24.3C, 24.4A), comprises small (just over 2 meters long), lightly built bipedal herbivores—probably the first ornithopod group to exceed 100 kilograms in live body mass. The dryosaurs were the last ornithopod group to have relatively short arms, and this may have prevented them from being functionally quadrupedal. They are the first ornithopods to have a distal expansion on the end of the ischium.

Camptosaurs were the first heavily built ornithopods that were more than 3 meters long, and had relatively longer arms than the dryosaurs. *Camptosaurus* (Figs. 24.4A, 24.5A) is the first ornithopod to have a noticeably elongated muzzle, presumably to increase the amount of food taken and processed per bite. Camptosaurs were also the first group with two functional rows of teeth in each jaw, arranged one above the other in an alternating pattern to form a single chewing unit. Camptosaurs have a very wide pelvis and thick hind limbs. Their front limbs were much shorter than their hind limbs, but could nonetheless reach the ground, permitting four-footed walking. The first metacarpal was reduced to a spur on the

Figure 24.4. Flesh restorations of (A) *Camptosaurus* (left) and *Dryosaurus* (right) and (B) *Muttaburrasaurus*.

A

B

334   *M. K. Brett-Surman*

A

B

Figure 24.5. (A) *Camptosaurus* and (B) *Tenontosaurus*. Length of *Camptosaurus* about 2.5 m, and of *Tenontosaurus* about 4.5 m.

hand, and there was considerable fusion of carpals in the wrist. One rare genus, *Muttaburrasaurus* (Fig. 24.4B) from Australia, is as large as the later iguanodontids, but has not yet been described in detail.

Tenontosaurus (Fig. 24.5B) is an enigmatic genus that has been classified both as a hypsilophodontid and as an iguanodontid (Weishampel and Heinrich 1992). It is about 7 meters long and has a very high skull, an edentulous (toothless) beak, and four digits on the hind foot. One would expect it to be intermediate between the Jurassic camptosaurs and the later Cretaceous hadrosaurs, but it has features that do not place it firmly in either group, such as a relatively more robust pelvis than in other ornithopods. *Tenontosaurus* is famous for being the prey of *Deinonychus* (Maxwell and Ostrom 1995). New finds made in the early 1990s in Montana, Wyoming, Oklahoma, and Texas (Winkler et al. 1997) will finally clarify its phylogenetic position.

The most derived group within the Iguanodontia is the Iguanodontoidea, which includes two groups: the iguanodonts, such as *Iguanodon* and the sail-backed *Ouranosaurus* (Figs. 24.6, 24.7A), and the hadrosaurians, such as *Anatotitan* and *Parasaurolophus* (Figs. 24.7B, 24.8). Hadrosaurs can be characterized as having a wider and more elongated snout than other ornithopods, interlocking teeth in multiple rows or dental batteries, the relatively longest forelimbs of all the ornithopods, and hooflike unguals on the pes. Some genera are noted for expanded nasal crests. Hadrosaurs were the largest of the ornithopods, some forms (*Shantungosaurus,* Fig. 24.8A) approaching sauropods in size. They were mostly Laurasian (that is, occupying the northern continents) in distribution, and have the best fossil record of all the ornithischians.

In previous classifications of the group, the family Hadrosauridae is subdivided into the subfamilies Hadrosaurinae (solid-crested and noncrested genera) and Lambeosaurinae (hollow-crested forms), plus some

Ornithopods   335

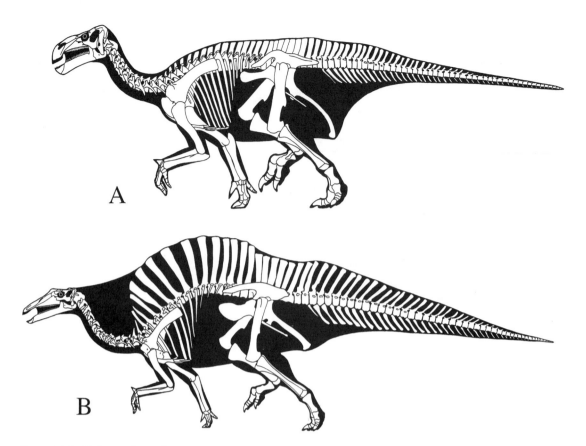

Figure 24.6. (A) *Iguanodon* and (B) *Ouranosaurus.* Length of *Iguanodon* about 9 m, and of *Ouranosaurus* about 6 m.

early forms that do not fit easily into the aforementioned two subfamilies. In recent classifications (Weishampel et al. 1993; Fastovsky and Weishampel 1996), the clades Hadrosaurinae and Lambeosaurinae were placed into the Euhdarosauria, and the earlier forms incorporated into the redefined Hadrosauridae. The clade Hadrosauridae is now used to include the Euhadrosauridae plus *Telmatosaurus, Secernosaurus,* their common ancestor, and all of its descendants. Although there is still no universally accepted classification of the ornithopods as a whole, one classification scheme is as follows (see Fig. 24.9):

Ornithopoda
      Heterodontosauridae
      Euornithopoda
            Hypsilophodontidae
            Iguanodontia
                  Dryomorpha
                  Tenontosaurus
                        Dryosauridae
                        Ankylopollexia
                              Camptosauridae
                              Iguanodontoidea
                                    Iguanodontidae
                                    Hadrosauridae

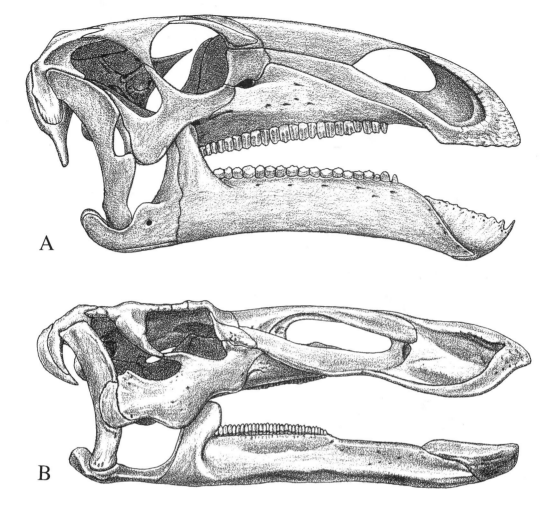

A

B

Figure 24.7. Skulls of (A) *Iguanodon* and (B) *Anatotitan*.

The "Fabrosaur" Problem

"Fabrosaurs" are a Late Triassic and Early Jurassic group of small (less than 2 meters long) bipedal ornithischians that were long placed within the ornithopods (Gow 1981). They were removed from the ornithopods after this group was reclassified in 1986 because they lacked two key features that define ornithopods: an obturator process on the ischium and a tooth row fully recessed from the outer margin of the jaws. This means that fabrosaurids may not have had cheeks. On the other hand, Thulborn (1992) proposed that *Lesothosaurus*, the best-known form to date (Fig. 24.1A), is a fabrosaur, and that this group therefore does have both a recessed tooth row and an obturator process, thereby making these dinosaurs ornithopods. Because there are no complete and associated skulls with skeletons at this time, I have taken the conservative approach and treated the fabrosaurs as outside the ornithopods. A new restoration of *Lesothosaurus* (Fig. 24.1) is presented here to contrast it with the ornithopods.

It is important to note, however, that both the fabrosaurids and the more derived heterodontosaurids and hypsilophodontids establish an early

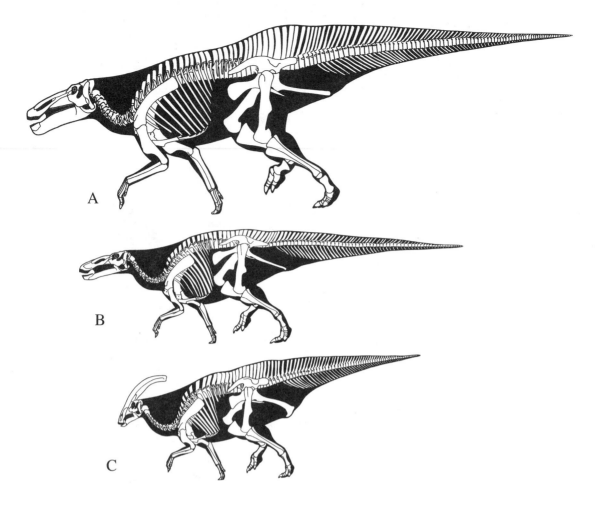

Figure 24.8. (A) *Shantungo-saurus,* (B) *Anatotitan,* and (C) *Parasaurolophus.* Length of *Shantungosaurus* about 17 m, of *Anatotitan* about 12 m, and of *Parasaurolophus* about 9 m.

trend for Jurassic ornithopods—small, lightly built, fast herbivores with narrow snouts for selective feeding on the undergrowth.

Geographic Distribution

Ornithopods occurred in both Gondwana and Laurasia during the Triassic and Jurassic. Only in the Cretaceous do we see evidence of possible provinciality or endemism, with certain groups restricted to particular areas—except for the hypsilophodontids, which continued to occur on every major continent. *Tenontosaurus* seems to have been a Laurasian genus, but this may be an artifact of its limited fossil record; it is known only from certain lower Cretaceous deposits in the United States. Dryosaurs are known from Africa and both South and North America. Iguanodonts are known from both Gondwana and Laurasia, and so are the hadrosaurs, although most forms in the latter group are restricted to Laurasia. During the Late Cretaceous, the continents were breaking apart, making migration routes for land animals increasingly harder to traverse, and this may be responsible for the suggestions of provinciality seen in the distribution of

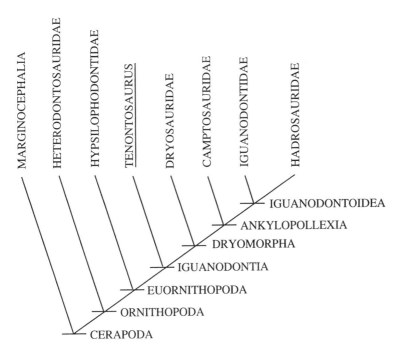

Figure 24.9. Tentative cladogram of the ornithopods. Ornithopod classification changes frequently with new discoveries and better character analysis, and so our understanding of their phylogenetic relationships is in a state of flux. This cladogram draws upon information from several workers on the group. For an alternative classification, see Sereno 1997.

Late Mesozoic ornithopods. On the other hand, the number of presently known fossiliferous sites representing the latest Cretaceous in Gondwana is quite poor; the terrestrial fossil record is heavily biased in favor of Asiamerican (eastern Asia plus western North America) sites. Thus the Late Cretaceous provinciality of ornithopods may be more apparent than real, and ornithopod families may have remained worldwide in their distribution.

Origins and Evolutionary Trends

The evolutionary origins of ornithopods remain obscure. There are fewer than ten species total of ornithischians known from the Triassic. By the Jurassic, only *Heterodontosaurus* (Fig. 24.1B) and *Lesothosaurus* (Fig. 24.1A) are reasonably well known.

When one traces the history of the ornithopods from the Jurassic through the Cretaceous, and from hypsilophodonts through hadrosaurs, several consistent trends appear. The snout becomes progressively longer and broader, and toothless at its front end. This allowed the animal to gather in more food per mouthful and to reach deeper into the vegetation.

There is an increase in the number of cheek teeth for more efficient grinding and slicing. Early ornithopods have just one row of teeth in each jaw, whereas hadrosaurs have three interlocking rows in their jaws, forming a grinding pavement or battery.

A most remarkable elaboration can be seen in the nasal apparatus of some of the iguanodonts and most of the hadrosaurs. In these animals the premaxillary and nasal bones extended backward over the nasal opening and sometimes over the entire length of the skull. This may reflect modifications of the nasal region for vocalization (see below).

Ornithopods   339

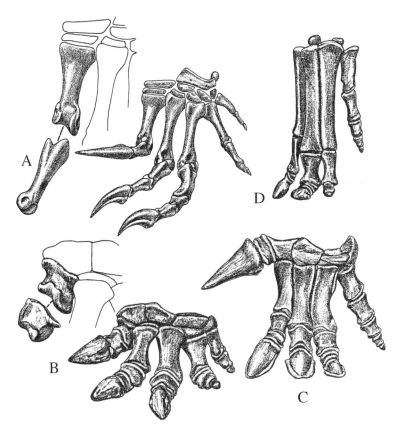

Figure 24.10. The hand (manus) of several ornithopods. (A) The primitive ornithopod *Heterodontosaurus,* showing the details of metacarpal I and the natural medial articulation of the first digit, which might have allowed some grasping ability. (B) *Camptosaurus,* also showing details of metacarpal I, with slightly less medial orientation of the first digit. (C) *Iguanodon.* (D) *Anatotitan.* Note the trend over time (that is, from *Heterodontosaurus* through progressively more derived forms) for a shift from a grasping ability of digit I to a stronger support function for the entire hand. There is also a loss of elements in digit I, loss of carpals, and an elongation of the median metacarpals. Not to scale.

Ornithopods nicely illustrate two of the "laws" of evolution recognized by paleontologists. Cope's Rule states that over geologic time, in a lineage, size will increase. This is obviously true for ornithopods: The Early Jurassic *Heterodontosaurus* is about 1 meter long, while the Late Cretaceous *Shantungosaurus* is into the sauropod size range (more than 20 meters long). Marsh's "Law" states that over geological time, the encephalization quotient (EQ) of a lineage will increase. EQ is defined as the ratio of the actual brain mass to the expected brain mass for a reptile of a certain size, based on samples of modern animals. The brain size is estimated by measuring the volume of the brain cavity. Early ornithopods have EQs well within the range of other ornithischians, but the later ornithopods have the highest EQs of all the ornithischians (Hopson 1977).

The forelimbs became progressively longer with the appearance of each successive ornithopod group. Eventually this allowed the later ornithopods to become what can be called "facultative quadrupeds" or "non-obligatory bipeds." This means that when the arms became long enough relative to the hind limbs, these ornithopods could walk, or in some cases trot, as quadrupeds. At higher speeds, however, these ornithopods probably reverted to bipedalism. Accompanying forelimb elongation, the hands of later ornithopods became more robust, and the claw-like fingers became more hoof-like (Fig. 24.10). Beginning with the iguanodonts, the first and fifth digits of the hands were reduced, and the middle three fingers became elongated to bear the weight of the front limb when touching the ground.

In order to support the weight of the body with increasing size, the number of sacral vertebrae (which connect the pelvis to the backbone) increased. In hadrosaurs, the number of sacral vertebrae can be as many as ten. The pelvic girdle also became somewhat broader, presumably to provide more space for the digestive system, to allow a wider stance for stability, and to provide firmer support for the increased mass of the animal.

Functional Morphology

The ornithopod chewing apparatus was in some ways functionally superior to that of many present-day herbivorous animals. Early ornithopods had slender jaws with small teeth in a single row that could cut vegetation but did not grind or mash the plant fodder. During ornithopod evolution, the cheek teeth became larger, with thicker enamel for increased strength. The edges of the tooth along the crown had denticles to assist in cutting, and ridges were present on the enameled face of the tooth. By the Late Jurassic, when the camptosaurs appeared, there were two staggered rows of teeth positioned one atop the other in both the upper and lower jaw, the two vertical rows of teeth forming a single chewing unit. By the latest Cretaceous, hadrosaurs had three vertical tooth rows with perfectly interlocking diamond-shaped teeth in a massive tooth battery. These teeth were ever-replacing, which means that they emerged from the jaw as though riding on a conveyor belt, to be gradually worn away at the top of the dental battery. The jaws are deep because they were producing hundreds of teeth at any one time. The teeth are enameled on one side only, with softer dentine exposed on the other side, the side that formed the actual chewing surface. As the upper and lower tooth batteries ground against each other, the dentine wore away faster than the enamel, so that the teeth were self-sharpening.

The jaws of derived ornithopods were kinetic, which means that they had the ability to rotate with respect to one another (Norman 1984; Norman and Weishampel 1985; Weishampel and Norman 1989; and Weishampel and Horner 1990). The lower jaw could slide forward and backward slightly on its articulation with the upper jaw. At the same time, the upper jaws were hinged against the skull, enabling them to swing outward independently (like paired trap doors opening in opposite directions), a condition known as pleurokinesis. All of this allowed for more effective grinding of plant food, and gave the jaws some ability to serve as shock absorbers of chewing forces (see Weishampel 1984).

As in other advanced ornithischians, the tooth row of ornithopods was medially recessed from the jawbones, indicating that in life there were fleshy cheeks. The occlusal plane (the surface along which upper and lower teeth met) slanted downward and outward, so that the plant bolus dropped into the cheek pouch during chewing.

The front part of the mouth of later ornithopods had no teeth, and the front end of the cheek teeth was well behind the beak. This gap behind the beak and in front of the teeth, the diastema, allowed the tongue to slide around into the cheeks and manipulate the food bolus back toward the dental battery. The elongation of the jaws in succeeding clades can be partly explained by the addition of the increased number of tooth rows (more than sixty in hadrosaurs).

The beak became wider and was covered by a horny covering. Constant abrading and cutting of resistant plant material acted to sharpen the cutting edge of the beak, enhancing its value as a cropping device.

If iguanodontian ornithopod jaws are compared to those of the other prominent group of Late Cretaceous Asiamerican large herbivores, the ceratopsians, two things are noteworthy: ornithopod jaws are relatively weaker than ceratopsian jaws, and ornithopod jaws are specialized for grinding, as opposed to the slicing action of ceratopsian jaws. Such differences in the way they processed their fodder may be indicative of ways in which ornithopods and ceratopsians avoided direct competition for food. Ornithopods may have preferentially fed on softer vegetation, while ceratopsians may have concentrated on tougher-fiber plants. The interaction of plants and herbivorous dinosaurs undoubtedly affected the evolutionary pathways of both groups (Wing and Tiffney 1987; Tiffney, chap. 25 of this volume).

The snouts and nasal crests of hadrosaurs have long been interpreted in terms of improvement of their owners' vocal abilities (Abel 1924; Weishampel 1981). There are two basic kinds of nasal crests in hadrosaurs—solid and hollow. Solid-crested (saurolophines) and non-crested (edmontosaurines and kritosaurines) hadrosaurs had enlarged external nasal chambers. This enlargement was the result either of simple expansion, as in *Kritosaurus*, or of the formation of expanded and folded pockets of bone at the front of the nose, as in *Edmontosaurus*. Soft tissues in the nasal chambers probably could resonate in a manner analogous to reed instruments such as oboes. Hollow-crested hadrosaurs (lambeosaurines), such as *Corythosaurus* and *Parasaurolophus* (Fig. 24.8), had nasal chambers that functioned like French horns or trombones.

Such elaborate crests probably served many simultaneous functions, such as species recognition, species-specific "hooting" to maintain contact with fellow hadrosaurs when direct visual contact was not possible, possible sexual identification, and age indication. The crests may also have increased the surface area for olfactory tissues, improving the sense of smell.

The vast variety of hadrosaur cranial crests prompted pioneering paleontologists to put these dinosaurs into a plethora of named species, but Dodson (1975), in one of the first major morphometric studies of dinosaurs using multivariate statistical analyses, showed that many of the differences in shape of the crests were due to growth and sexual variation, and were therefore of limited taxonomic use.

Large ornithopods have sometimes been pictured as semi-aquatic animals. Even the name "duckbill" suggests a duck-like proficiency in water. Without armor, or faster-than-theropod speeds, ornithopods are often pictured as running into the water to get away from theropods. Unfortunately, ornithopods were probably poorer swimmers than theropods.

There are three basic means of propulsion in water for a terrestrial vertebrate: paddling with the forearms, sculling with the tail like a crocodile, and using the hind legs as the main propulsive organ. The first two methods would have been very inefficient for the larger dinosaurs because of the speed of resistance from the water. The arms and hands of ornithopods are small compared to the cross-section of the body, so an ornithopod could not generate much forward propulsion. Picture a cross-section of the body of the hadrosaur *Anatotitan* (Fig. 24.8) compared to a cross-section of its hand. For this dinosaur, attempting to paddle with its hands would have been like trying to row a wide canoe with a spoon for a paddle.

The hands of some ornithopods, especially hadrosaurs, are sometimes pictured as webbed, hence an aquatic adaptation. Mummified hands of hadrosaurs, however, show that they were not webbed but padded, a terrestrial adaptation.

Efficient tail sculling is also ruled out for ornithopods because their backbones and the proximal third of their tail had ossified tendons that severely restricted side-to-side movement (but see Coombs 1975 for an alternative hypothesis). These ossified tendons form a tight bundle surrounding the tail in some primitive ornithopods, and a rhomboid latticework of overlapping tendons in advanced ornithopods. There can be as many as nine tendons in two overlapping series per neural spine (in hadrosaurs), but fewer tendons per bundle in less advanced ornithopods. These ossified tendons are well developed on the dorsal part of the tail (epaxial series), but are less well developed on the ventral side of the tail (hypaxial series). Bony processes that project laterally from the tail vertebrae (caudal transverse processes), and that act as the attachment points for muscles that moved the tail from side to side, are small in hadrosaurs, and disappear entirely about vertebra number 16 as one moves backward along the tail. In large theropods, in contrast, these transverse processes, and the tail as a whole, lack ossified tendons, are larger, and extend for more than two-thirds of the length of the tail. Theropod tails were probably much better propulsive devices than were hadrosaur tails.

This leaves leg propulsion as the most likely method for ornithopod swimming, but once again, ornithopods had less powerful (forceful) legs than theropods of a similar size. Even if retreat to water was not an effective way of escaping theropod predators, however, the larger ornithopods probably could swim at least as well as any of the modern terrestrial large herbivores, such as deer or horses.

Most ornithopods were not as well designed for running as theropods. When it came to defense, they had no armor, and they could not depend upon running into the water to escape. So how did they defend themselves? There are two possible answers. The first is safety in numbers. Many living ungulates travel in herds for safety. At least one species of hadrosaur (*Maiasaura*) probably traveled in herds of thousands of individuals. The second possibility is that ornithopods may have been more maneuverable than theropods. This is because ornithopods have a wider pelvis, so they may have been more stable. They also have wider feet for firmer contact with the ground, and a lower center of gravity, which may have given them a smaller turning radius at full speed. Thus the basic method of defense for ornithopods was probably herding behavior on land, where ornithopods could use their advantage of better maneuverability than the theropods.

Large ornithopods were probably quadrupedal walkers and bipedal runners. The forelimb is only two-thirds the length of the hind limb, and the forelimb did not have a wider excursion arc than the hind limb to counteract this difference in size. The scapula could not rotate and thus act as an additional limb element to increase the stride length, as it does in horses, for example. Consequently the ornithopod forelimb could not maintain the same stride length as the hind limb. Whenever an ornithopod went into a full run, it had to retract its forelimbs and run bipedally (see Bennett and Dalzell 1973; Thulborn 1990). (For an alternative hypothesis, see Paul 1987.)

Skin and Eggs

Ornithopods had thick, wrinkled skin with bony knobs of various sizes embedded throughout. This can be seen in "mummified" hadrosaurs at the American Museum in New York and at the Senckenberg Museum in Germany. The hand in later ornithopods was padded much like a snow mitten. One species had a small "dragon frill" down the back that resembled a picket fence in appearance (Horner 1984). There is no evidence for feathers or lizard-like, overlapping scales.

Most of the known ornithopod eggs are represented by the hypsilophodont *Orodromeus* and the hadrosaurs *Maiasaura* and *Hypacrosaurus* (Horner and Currie 1994). Eggs came in many size ranges and were laid in many differing patterns. There are now enough dinosaur nesting sites that a full text has appeared devoted just to the topic of eggs and babies (Carpenter et al. 1994; Hirsch and Zelenitsky, chap. 28 of this volume). On the basis of studies of juvenile *Maiasaura*, it has been suggested that hatchling hadrosaurs were altricial and not precocial (Weishampel and Horner 1994), but this conclusion has been challenged (Geist and Jones 1996).

The Future of Ornithopod Studies

Because ornithopods, especially hadrosaurs, have one of the most complete fossil records known for dinosaurs, representing hatchlings to adults, they are one of the best groups to use in studies of growth series and life histories. The two most fruitful areas for future research will be the often neglected studies of allometry and postcranial functional morphology, and how they relate to ontogeny and phylogeny (Dunham et al. 1989). For example, in some hadrosaurs, characters that appear only in the adult stages of stratigraphically early genera eventually begin to appear in the juvenile stages of stratigraphically later genera (an example of peramorphosis). Paedomorphosis and neoteny are two additional topics that will form the core of future populational studies of dinosaurs. Certain taxonomic characters appear only in the adult stages of development, and this will affect future taxonomies (Brett-Surman 1989).

The largest ornithopods may have lived for so long, and passed through so many size ranges, that they functionally occupied several different niches during their lifespan. This "niche assimilation" would be ecologically equivalent to multiple species of modern mammals. The resulting reduced community structure (as compared to modern mammalian faunas) may have been a factor in dinosaur extinction, because communities with few species are more susceptible to extinction than communities with many species.

References

Abel, O. 1924. Die neuen Dinosaurierfunde in der Oberkreide Canadas. *Naturwissenschaften* (Berlin) 12: 709–716.
Bennett, A. F., and B. Dalzell. 1973. Dinosaur physiology: A critique. *Evolution* 27: 170–174.
Brett-Surman, M. K. 1989. Revision of the Hadrosauridae (Reptilia: Ornithischia) and their evolution during the Campanian and Maastrichtian. Ph.D. thesis, George Washington University.
Carpenter, K., and P. J. Currie (eds.). 1990. *Dinosaur Systematics: Approaches and Perspectives*. Cambridge: Cambridge University Press.
Carpenter, K.; K. F. Hirsch; and J. R. Horner (eds.). 1994. *Dinosaur Eggs and Babies*. Cambridge: Cambridge University Press.
Coombs, W. P. 1975. Sauropod habits and habitats. *Palaeogeography, Palaeoclimatology, Palaeoecology* 17: 1–33.
Cooper, M. R. 1985. A revision of the ornithischian dinosaur *Kangnasaurus coetzeei*

Haughton, with a classification of the Ornithischia. *Annals of the South African Museum* 95 (6): 281–317.

Dodson, P. 1975. Taxonomic implications of relative growth in lambeosaurine dinosaurs. *Systematic Zoology* 24: 37–54.

Dodson, P. 1990. Marginocephalia. In D. B. Weishampel, P. Dodson, and H. Osmólska (eds.), *The Dinosauria*, pp. 562–563. Berkeley: University of California Press.

Dunham, A. E.; K. L. Overall; W. P. Porter; and C. A. Forster. 1989. Implications of ecological energetics and biophysical and developmental constraints for life-history variation in dinosaurs. In J. O. Farlow (ed.), *Paleobiology of the Dinosaurs*, pp. 1–19. Special Paper 238. Boulder, Colo.: Geological Society of America.

Fastovsky, D. E., and D. B. Weishampel. 1996. *The Evolution and Extinction of the Dinosaurs*. New York: Cambridge University Press.

Forster, C. A. 1990. The postcranial skeleton of the ornithopod dinosaur *Tenontosaurus tilletti*. *Journal of Vertebrate Paleontology* 10: 273–294.

Galton, P. M. 1983. The cranial anatomy of *Dryosaurus*, a hypsilophodontid dinosaur from the Upper Jurassic of North America and East Africa, with a review of hypsilophodontids from the Upper Jurassic of North America. *Geologica et Palaeontologica* 17: 207–243.

Geist, N. R., and Jones, T. D. 1996. Juvenile skeletal structure and the reproductive habits of dinosaurs. *Science* 272: 712–714.

Gow, C. E. 1981. Taxonomy of the Fabrosauridae (Reptilia: Ornithischia), and the *Lesothosaurus* myth. *South African Journal of Science* 77: 43.

Hopson, J. A. 1977. Relative brain size and behavior of archosaurian reptiles. *Annual Review of Ecology and Systematics* 8: 429–448.

Horner, J. R. 1984. A "segmented" epidermal tail frill in a species of hadrosaurian dinosaur. *Journal of Paleontology* 58: 270–271.

Horner, J. R., and P. J. Currie. 1994. Embryonic and neonatal morphology and ontogeny of a new species of *Hypacrosaurus* (Ornithischia, Lambeosauridae) from Montana and Alberta. In K. Carpenter, K. F. Hirsch, and J. R. Horner (eds.), *Dinosaur Eggs and Babies*, pp. 312–336. Cambridge: Cambridge University Press.

Marsh, O. C. 1882. Classification of the Dinosauria. *American Journal of Science* (3) 23: 81–86.

Maryańska, T., and H. Osmólska. 1974. Pachycephalosauria, a new suborder of ornithischian dinosaurs. *Paleontologia Polonica* 30: 45–102.

Maryańska, T., and H. Osmólska. 1985. On ornithischian phylogeny. *Acta Paleontologia Polonica* 30: 137–150.

Maxwell, W. D., and J. H. Ostrom. 1995. Taphonomy and paleobiological implications of *Tenontosaurus-Deinonychus* associations. *Journal of Vertebrate Paleontology* 15: 707–712.

Norman, D. B. 1984. On the cranial morphology and evolution of ornithopod dinosaurs. In M. W. J. Ferguson (ed.), *The Structure, Development and Evolution of Reptiles*, pp. 521–547. Symposium 52, Zoological Society of London.

Norman, D. B., and D. B. Weishampel. 1985. Ornithopod feeding mechanisms: Their bearing on the evolution of herbivory. *American Naturalist* 126: 151–164.

Norman, D. B., and D. B. Weishampel. 1990. Iguanodontidae and related ornithopods. In D. B. Weishampel, P. Dodson, and H. Osmólska (eds.), *The Dinosauria*, pp. 510–533. Berkeley: University of California Press.

Padian, K. (ed.). 1986. *The Beginning of the Age of Dinosaurs: Faunal Change across the Triassic-Jurassic Boundary*. Cambridge: Cambridge University Press.

Paul, G. S. 1987. The science and art of restoring the life appearance of dinosaurs and their relatives. In S. J. Czerkas and E. C. Olson (eds.), *Dinosaurs Past and Present*, vol. 2, pp. 5–49. Los Angeles: Natural History Museum, and Seattle: University of Washington Press.

Rosenberg, G. D., and D. L. Wolberg (eds.). 1994. *Dino Fest*. Paleontological Society Special Publication no. 7. Knoxville: Department of Geological Sciences, University of Tennessee.

Sereno, P. 1986. Phylogeny of the bird-hipped dinosaurs (Order Ornithischia). *National Geographic Research* 2: 234–256.

Sereno, P. 1991. *Lesothosaurus*, "fabrosaurids," and the early evolution of the Ornithischia. *Journal of Vertebrate Paleontology* 11: 168–197.

Sereno, P. 1997. The origina and evolution of dinosaurs. *Annual Review of Earth and Planetary Sciences* 25: 435–489.

Steel, R. 1969. *Ornithischia*. Handbuch der Paläoherpetologie 15. Jena: G. F. Verlag.

Sues, H.-D., and D. B. Norman. 1990. Hypsilophodontidae, *Tenontosaurus*, Dryosauridae. In D. B. Weishampel, P. Dodson, and H. Osmólska (eds.), *The Dinosauria*, pp. 498–509. Berkeley: University of California Press.

Thulborn, R. A. (ed.). 1990. *Dinosaur Tracks*. London: Chapman and Hall.

Thulborn, R. A. 1992. Taxonomic characters of *Fabrosaurus australis*, an ornithischian dinosaur from the Lower Jurassic of southern Africa. *Geobios* 25 (2): 283–292.

Weishampel, D. B. 1981. Acoustic analyses of potential vocalization in lambeosaurine dinosaurs (Reptilia: Ornithischia). *Paleobiology* 7: 252–261.

Weishampel, D. B. 1984. Evolution of jaw mechanics in ornithopod dinosaurs. *Advances in Anatomy, Embryology and Cell Biology* 87: 1–110.

Weishampel, D. B.; P. Dodson; and H. Osmólska (eds.). 1990. *The Dinosauria*. Berkeley: University of California Press.

Weishampel, D. B., and R. E. Heinrich. 1992. Systematics of Hypsilophodontidae and basal Iguanodontia (Dinosauria: Ornithopoda). *Historical Biology* 6: 159–184.

Weishampel, D. B., and J. R. Horner. 1990. Hadrosauridae. In D. B. Weishampel, P. Dodson, and H. Osmólska (eds.), *The Dinosauria*, pp. 534–561. Berkeley: University of California Press.

Weishampel, D. B., and J. R. Horner. 1994. Life history syndromes, heterochrony, and the evolution of Dinosauria. In K. Carpenter, K. F. Hirsch, and J. R. Horner (eds.), *Dinosaur Eggs and Babies*, pp. 229–243. Cambridge: Cambridge University Press.

Weishampel, D. B., and D. B. Norman. 1989. Vertebrate herbivory in the Mesozoic: Jaws, plants, and evolutionary metrics. In J. O. Farlow (ed.), *Paleobiology of the Dinosaurs*, pp. 87–100. Special Paper 238. Boulder, Colo.: Geological Society of America.

Weishampel, D. B.; D. B. Norman; and D. Grigorescu. 1993. *Telmatosaurus transsylvanicus* from the Late Cretaceous of Romania: The most basal hadrosaurid. *Palaeontology* 36: 361–385.

Weishampel, D. B., and L. M. Witmer. 1990. Heterodontosauridae. In D. B. Weishampel, P. Dodson, and H. Osmólska (eds.), *The Dinosauria*, pp. 486–497. Berkeley: University of California Press.

Wing, S. L., and B. H. Tiffney. 1987. The reciprocal interaction of angiosperm evolution and tetrapod herbivory. *Review of Paleobotany and Palynology* 50: 179–210.

Winkler, D. A.; P. A. Murry; and L. L. Jacobs. 1997. A new species of *Tenontosaurus* (Dinosauria: Ornithopoda) from the early Cretaceous of Texas. *Journal of Vertebrate Paleontology* 17: 330–348.

Biology of the Dinosaurs

There were, as I say, five of them, two being adults and three young ones. In size they were enormous. Even the babies were as big as elephants, while the two large ones were far beyond all creatures I have ever seen. They had slate-coloured skin, which was scaled like a lizard's and shimmered where the sun shone upon it. All five were sitting up, balancing themselves upon their broad, powerful tails and their huge three-toed hind-feet, while with their small five-fingered front-feet they pulled down the branches upon which they browsed. I do not know that I can bring their appearance home to you better than by saying that they looked like monstrous kangaroos, twenty feet in length, and with skins like black crocodiles.

—Sir Arthur Conan Doyle, *The Lost World*

Many people think that a new classification is the end product of paleontological research. In fact, it is just the beginning. We can think of three levels of intellectual inquiry that result from the discovery of a new dinosaur fossil.

"First level" questions are based on the actual fossils themselves. The primary pieces of evidence come from bones, teeth, tooth wear patterns, skin impressions, muscle attachment sites, joint articulation configurations, bone pathologies, gut contents, coprolites, trackways, and nests, to name the more obvious of these. Such fossils can be used to address questions about the overall size of the dinosaur, changes in its skeletal proportions during growth, the processes by which the animal grew, the kinds of movements of which it was capable, the injuries or illnesses that afflicted it, and something about the likely acuity of its senses, and perhaps even its intelligence.

"Second level" questions try to go beyond these bits and pieces of dinosaur natural history; these questions build upon the answers to "first level" questions, and represent hypotheses about dinosaurs that are not readily testable. What was the dinosaur's likely ecological role? How long did it take to reach sexual maturity, and at what point in the animal's life history did any sexual characters appear? What was its reproductive rate? What was the mortality rate for different age classes of its species?

The final level of inquiry relates to evolutionary theory. How do dinosaurs (and other organisms) fit into the big picture? Can we discern any evolutionary "laws" from major trends in dinosaurian adaptations over time? Are there any patterns in the evolution of dinosaurian clades above the species or genus level? Do dinosaurs fit into any long-term trends in the evolution of terrestrial ecosystems?

Dinosaur fossils do not exist in a vacuum. They are part of an overall fossil assemblage, and a fossil fauna is in turn a component of the sedimentary rocks that accumulated in a particular depositional setting. Consequently we can attempt to fit dinosaurs and other organisms not just into the big picture of biological evolution; we can also consider how the evolution of life has been affected by, and has affected, the development of the earth's physical systems.

In the previous section of this book, each of the chapters describing a dinosaur group presented ideas about what the dinosaurs of that chapter were like as living animals. Part Four returns to that theme, but in a more general manner, by taking a topical approach to various aspects of dinosaur biology. The emphasis in this section, then, is mainly on the second level of inquiry described above, with a few peeks here and there at the big picture as well.

One of the four basic activities of any animal (along with fighting, fleeing, and reproducing) is finding food. The bloodthirsty denizens of Hollywood's version of prehistory notwithstanding, the vast majority of dinosaurs were plant-eaters. To understand dinosaur ecology, therefore, it is necessary to know something about the plant communities in which they lived. This section of the book therefore begins by surveying present knowledge of Mesozoic floras. What kinds of plants were available to herbivorous dinosaurs, what was their quality as fodder, how did plant communities change over the course of the Mesozoic Era, and how did this affect dinosaur communities?

Having considered what plant-eating dinosaurs might have eaten, we then turn to consider the evidence for what herbivorous and carnivorous dinosaurs actually *did* eat. Information from trackways, death assemblages, bite marks, stomach contents, and coprolites is summarized.

If a species is to survive over the long term, reproductive rates must at least balance mortality losses. Reproduction is therefore another essential aspect of animal biology, and so two chapters consider how dinosaurs went about ensuring that there would be a new generation. The first step in making babies is finding a mate, and so we begin by examining how male dinosaurs might have courted their ladies, and at the same time kept rival males out of the picture. This is one area where the analogous behavior of living animals plays a big part in interpretations of dinosaur behavior.

More from lack of fossilizable information than from any excessive prudery on our part, we skip the obvious sequel to success in finding a mate and move right on to how the new generation began life. Most (if not all) dinosaurs were probably egg-layers, and in recent years there has been an explosion of interest in dinosaur nesting sites and eggs. Chapter 28 discusses what we know about dinosaur eggs, and the problems of assigning eggs to the dinosaurs that laid them.

Once it had hatched from its egg, a baby dinosaur was ready to take its place in the Mesozoic world. It, too, would ultimately try to find a mate, but first it would have to grow to adulthood. Chapter 29 considers the processes by which dinosaurs grew, as inferred from the tissues of their bones.

Dinosaurs were not all giants, but a significant trend in dinosaurian evolution was toward gigantism. Dinosaurs included the largest land-living animals in the history of our planet. How did nature engineer such behemoths? Could sauropods rear up on their hind legs to stick their long necks high into treetops? Could big dinosaurs run quickly, or were they restricted to slow, lumbering movements? How did dinosaurs fight? Chapter 30 addresses such questions.

As well-constructed as dinosaurs seem to have been, they were no doubt subject to the same misfortunes that befall all flesh: injuries and diseases. Amazingly, dinosaur bones sometimes record the traces of such maladies, and we devote a chapter to the osteological evidence of dinosaurian injuries and illnesses. This field of study has become an area of paleontology in its own right, paleopathology.

The idea that dinosaurs may have been more like modern birds and mammals than like living reptiles in many features of their biology has its roots in the work of none other than Richard Owen, who gave the Dinosauria their name. As E. D. Cope wrote in 1868: "If he [Cope's "*Laelaps*"] were warm-blooded, as Prof. Owen supposes the Dinosauria to have been, he undoubtedly had more expression than his modern reptilian prototypes possess. He no doubt had the usual activity and vivacity which distinguishes the warm-blooded from the cold-blooded vertebrates." In his first attempt to revise the taxonomy of dinosaur footprints of the Connecticut Valley, R. S. Lull (1904: 475) invoked the idea of warm-blooded dinosaurs as a way to eliminate the possibility that dinosaurs might have had the indeterminate growth typical of reptiles. If dinosaurs, like birds and mammals, grew to a fixed size and then stopped growing, then the ichnologist might not have to face the possibility that footprints of a wide range of sizes could have been made by members of the same species of dinosaur.

Although other paleontologists, such as Richard Owen and L. S. Russell, also considered the possibility of warm-blooded dinosaurs from time to time, it was the work of John H. Ostrom, John R. Horner, and Robert T. Bakker from the 1960s to the 1980s that really put dinosaur physiology on the front burner of paleontological attention. Bakker and Horner in particular were such dynamic and engaging advocates of warm-blooded dinosaurs that the idea received wide attention in the popular media, most notably the movie *Jurassic Park*.

A full examination of the various arguments about dinosaur physiology would require a book in itself. We have chosen to limit coverage in this book to some of the stronger and/or more recently proposed arguments for and against the hypothesis of warm-blooded dinosaurs; the chapters dealing with this topic thus constitute a mini-section within our overall section on dinosaur biology (see the accompanying box for some of our thoughts, though).

We end this section by looking at the dinosaurs that got away: dinosaur footprints. Every dinosaur skeleton represents a tragedy for its erstwhile

Biology of the Dinosaurs   349

Some Irreverent Thoughts about Dinosaur Metabolic Physiology: Jurisphagous Food Consumption Rates of *Tyrannosaurus rex*

M. K. Brett–Surman and
James O. Farlow

It is agreed by all living humans that the highlight of the movie *Jurassic Park* (Universal Studios, 1993) was the consumption of the lawyer by the true hero of the movie, *Tyrannosaurus rex.* This brings up an obvious question: How many lawyers would it take to properly feed a captive *T. rex?* Fortunately science has now progressed to the point where this important question can be answered—and plans made accordingly.

Two pieces of information are needed:

(A) The food requirements of a *T. rex* for one year

(B) The food value of one lawyer

Following the way that it was portrayed in *Jurassic Park,* let us first assume that our *T. rex* is endothermic. Let us also assume that our tyrannosaur weighs 10,000 pounds (4540 kg)—perhaps a bit on the light side (Farlow et al. 1995), but close enough.

Farlow (1990; see Farlow 1976 for details about the data used) published an equation relating the food consumption rate (in watts, or joules/second; that is, the amount of food energy needed per unit time) to body mass (in kilograms) in living endotherms (mammals and birds):

$$\text{consumption rate} = 10.96 \times \text{body mass}^{0.70}$$

For a 4540-kilogram *T. rex,* the equation predicts an average food consumption rate of 3978.8 joules/second. Because we are interested in the time span of one year, we must now multiply this result by 3.1536×10^7, which is the number of seconds in one year (that is, 60 seconds/minute $\times$ 60 minutes/hour $\times$ 24 hours/day $\times$ 365 days/year—unless you are watching golf on TV, in which case this number is much higher), to give us the tyrannosaur's energy needs in joules/year. This results in a big number: 1.2547×10^{11} joules/year.

This gives us the first part of what we need to know in order to begin rounding up enough lawyers to keep our dinosaur content. We must now calculate the energy value of one lawyer.

There are three components of the food value, in joules, of one lawyer: (1) the energy value (in joules) of 1 kilogram of lawyer flesh; (2) the number of kilograms (mass) in our sacrificial lawyer; (3) the digestive percentage, or assimilation efficiency, of a carnivore digesting meat—in the present case, this is the percentage of the lawyer that actually has food value. (We assume that clothing, briefcase, cellular phone, and pocket organizer have no energy value, and so these components of an operational lawyer will be ignored in our calculations.)

We assume that the energy value

occupant—the skeleton would not be there in the rocks had the creature not come to an often untimely end. Footprints, in contrast, were made by living animals going about their business. They may therefore be able to provide us information about dinosaur biology that would be hard to discern from a fossilized carcass.

Two chapters, then, consider the study of dinosaur footprints. Chapter 36 emphasizes the extent to which it is possible to identify footprint makers, the nomenclature applied to dinosaur tracks, and what trackways tell us about dinosaur locomotion. Chapter 37 focuses on the geologic circumstances under which dinosaur footprints are preserved, and on what of lawyer meat, like that of other animals, is 7×10^6 joules/kilogram (Peters 1983). We further assume that our lawyer weighs 150 pounds, or 68.1 kilograms.

The assimilation efficiency of carnivores eating meat is about 90 percent (Golley 1960; this is much higher than for herbivores feeding on high-fiber forage—as presumably was the case for most herbivorous dinosaurs; see Tiffney, chap. 25 of this volume).

The energy value of a single lawyer can now be calculated as

$$68.1 \text{ kg} \times (7 \times 10^6 \text{ joules/kg}) \times 0.9 = 4.2903 \times 10^8 \text{ joules}$$

By dividing the yearly energy requirements of our *T. rex* by the energy value of a single lawyer, we get the yearly lawyer consumption that our dinosaur would need:

$$(1.2547 \times 10^{11} \text{ joules/year})/$$

$$(4.2903 \times 10^8 \text{ joules/lawyer}) = 292 \text{ lawyers/year}$$

The calculations are the same if we assume that our tyrannosaur was an ectotherm, except that we must use an equation relating food consumption rate to body mass in reptiles and amphibians (Farlow 1990; same units as for endotherms):

$$\text{consumption rate} = 0.84 \times \text{mass}^{0.84}$$

For a 4540-kilogram *T. rex*, this equation predicts a feeding rate of 991.3 watts, which works out to 73 lawyers per year.

We can see, then, that genetically resurrected tyrannosaurs would have a far greater predatory impact on the lawyer population if they were endotherms than if they were ectotherms. This is perhaps a good reason for hoping that dinosaurs will turn out to have been endotherms.

References

Farlow, J. O. 1976. A consideration of the trophic dynamics of a Late Cretaceous large-dinosaur community (Oldman Formation). *Ecology* 57: 841–857.

Farlow, J. O. 1990. Dinosaur energetics and thermal biology. In D. B. Weishampel, P. Dodson, and H. Osmólska (eds.), *The Dinosauria,* pp. 43–55. Berkeley: University of California Press.

Farlow, J. O.; M. B. Smith; and J. M. Robinson. 1995. Body mass, bone "strength indicator," and cursorial potential of *Tyrannosaurus rex. Journal of Vertebrate Paleontology* 15: 713–725.

Golley, F. B. 1960. Energy dynamics of a food chain of an old-field community. *Ecological Monographs* 30: 187–206.

Peters, R. H. 1983. *The Ecological Implications of Body Size.* Cambridge: Cambridge University Press.

these trace fossils may be telling us about when and where particular kinds of dinosaurs lived.

References

Cope, E. D. 1868. The fossil reptiles of New Jersey. *American Naturalist* 1: 23–30.

Lull, R. S. 1904. Fossil footprints of the Jura-Trias of North America. *Memoirs Boston Society of Natural History* 5: 461–557.

Land Plants as Food and Habitat in the Age of Dinosaurs

Bruce H. Tiffney

25

Plants in a book on dinosaurs? Not as out of place as you might think. Plants are autotrophs (self-feeders), organisms that are able to capture the sun's energy directly. By contrast, dinosaurs, like all animals, were heterotrophs (other-feeders), animals that have to feed on other organisms in order to live. Because plants lie at the base of the food chain, they have had an immense influence on the evolution of both herbivores and carnivores in Earth's history. The size of the available plants, their rate of growth, their ability to recover from damage, the rate at which they reproduce, their abundance in the environment, and the digestibility of their leaves and reproductive organs all combine to influence the amount of energy that herbivores can draw from them. As these features of plants change through evolutionary time, so also will the nature of the herbivore and carnivore communities dependent upon them.

Additionally, plants are important to animals in that they define the environment within which animals live. By example, a forest forms a barrier to large animals, and is difficult for them to pass through. In contrast, smaller animals perceive the forest as a three-dimensional habitat, and have the option to move vertically as well as horizontally within it. Conversely, the two-dimensional surface of an open "grassland" allows free motion of large animals but limits the options for small animals. Only by burrowing can smaller animals create a three-dimensional environment in open country. This is a common solution for mammals, but of the dinosaurs, only the birds generally were small enough to explore it using preexisting holes.

The interaction of plants and animals is not a one-way street. Herbivores have had an important influence on plant evolution by their choice of food, the volume that they consume, and the frequency and duration of their feeding. Excessive herbivory can destroy environments and place selective pressure on plants, possibly even leading to the extinction of existing lineages or the evolution of new ones.

In this chapter we will survey both the major groups of plants available to dinosaurs as food, and the possible effects of this interaction between plants and dinosaurs on both groups. In looking at this interaction, we will view plants from the perspective of a herbivorous dinosaur, examining those features of plants that influence their quality as food. Similarly, we will look at herbivorous dinosaurs as plant-consuming machines. Readers should be aware of two things: First, this chapter is brief, and presents generalized information for which specific exceptions are often known to exist. Second, the study of the evolutionary interaction of herbivores and the plants they fed upon in the fossil record is in its infancy. Thus, many of the interpretations presented in the second portion of this chapter are really hypotheses.

Groups of Mesozoic Plants

Dinosaurs evolved in a world already populated by two major kinds of land plants, the pteridophytes and the gymnosperms, which were joined by a third kind, the angiosperms, in the middle Cretaceous (Figs. 25.1, 25.5). All three groupings are vascular plants; that is, they possess conducting tissue which transports water and food products within the plant body. This is in contrast to the algae ("seaweeds"), which are almost entirely aquatic, or to the bryophytes (mosses), which are terrestrial but are restricted in their size and significance by the lack of vascular tissue. We will look at the characteristics of the three vascular plant groups in turn, and then at their distribution in space and time during the Mesozoic. We will not discuss terrestrial algae or bryophytes, as they did not form an important part of the diet of herbivorous dinosaurs.

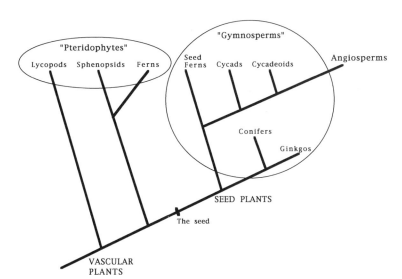

Figure 25.1. General relationships of the major groups of vascular land plants. The branching diagram shows the ancestor-descendant relationships of the groups, while the two circled units indicate the paraphyletic groups "Pteridophytes" and "Gymnosperms." After Doyle and Donoghue 1986; see also Doyle and Donoghue 1992.

Land Plants as Food and Habitat   353

What is the source of the following information? Where fossil plants are similar to living ones, we can infer the biology of the fossil from that of the living counterpart, recognizing that the extant plant cannot be a perfect proxy for the fossil. However, in many cases there are no close living relatives, and inferences must be made from the circumstantial evidence provided by the anatomy and morphology of the fossil, and their parallels in plants of the present day. Further insight may be gained from the depositional situation within which the fossil was found. Such evidence allows us to erect a model of Mesozoic ecosystems, but with the knowledge that this model will surely evolve with new information and understanding. Those wishing further information on Mesozoic plants, or guides to their identification, are referred to Stewart and Rothwell 1993 and Taylor and Taylor 1993.

Pteridophytes

The term *pteridophyte* is used to circumscribe several groups of primitive vascular plants. The pteridophytes are paraphyletic; that is, the group

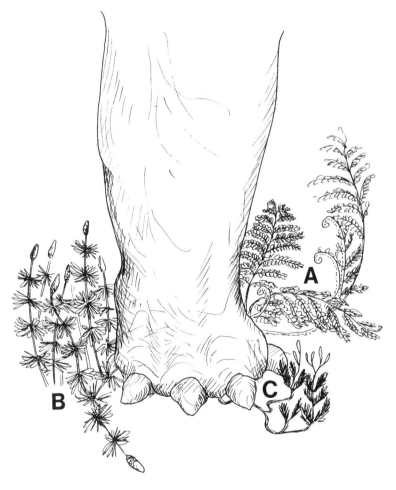

Figure 25.2. Some examples of Mesozoic Pteridophytes. (A) A rhizomatous fern. (B) A sphenopsid (*Equisetum*). (C) A lycopsid. An adult brachiosaur foot is provided for relative scale.

embraces some, but not all, of the descendants of a common ancestor. In this case, the "pteridophytes" all evolved from the first vascular plant, but the group does not include the other descendants of that first land plant, the seed plants. Thus, the pteridophytes are defined by the common character of their aquatic mode of reproduction, rather than by their ancestor-descendant relationships (Fig. 25.1). Of the pteridophyte clades, the most significant to our story are the ferns and their less commonly observed allies, the horsetails (sphenopsids) and the club mosses (lycopsids). All are depicted in Figure 25.2. These three groups share a common history, originating in the Devonian Period. In all three groups, the commonly observed plant is diploid; that is, it possesses two copies of each individual chromosome within each cell. Within specialized structures on this diploid plant, meiosis takes place, resulting in the division of the paired chromosomes into two groups, each containing one copy of each chromosome. This is referred to as the haploid condition. These haploid chromosome groups are encased in a protective structure to form spores, which are released to be blown about in the atmosphere. These spores settle from the air, and if they land in a moist area, they germinate to yield a tiny photosynthetic haploid plant, the gametophyte. The gametophyte bears the haploid egg and sperm. In a process reminiscent of its algal forebears, fertilization occurs when sperm is released from one gametophyte and literally swims in available dew or rainwater to an egg-bearing gametophyte. The fusion of the haploid egg and sperm yields a diploid zygote, which grows into the visible plant (e.g., a fern) familiar to the average viewer. This is an amphibious life cycle, and it restricts pteridophytes in the main to areas where moisture is at least seasonally available. Fertilization can occur only where standing water allows the sperm to swim, and the new diploid plant can grow only where fertilization has taken place. Thus, pteridophytes are uncommon in year-round or strongly seasonally dry environments.

Mesozoic pteridophytes were generally fairly low-growing herbaceous plants without well-developed stems. The most common exceptions to this generalization were tree ferns and, in limited areas during the Triassic Period, a small arborescent (tree-like) lycopod which was the last survivor of the great late Paleozoic tree club mosses. Both possessed an unbranched aerial stem with a single growing point. These tree-like forms excepted, pteridophytes almost universally possess an underground stem or rhizome, which branches to create many growing points, with the subsequent appearance of many new aboveground individuals from one original subterranean rhizome. This offsets the limitations of the sexual reproductive system, because a single successful sexual event can literally result in the growth of a field of plants from subsequent vegetative reproduction. Furthermore, vegetative growth confers the ability to regrow quickly after damage. The loss of existing emergent leaves is countered by the growth of new leaves from the rhizome, which remains safely beneath the ground. Additionally, with sufficient available water, pteridophytes tend to grow fairly rapidly. The foliage of most pteridophytes is relatively succulent, and lacks strong mechanical defenses (tough bark, spines, etc.). However, several living pteridophytes possess chemicals that are carcinogenic in mammals, or are known to inhibit digestion, and past pteridophytes may have had similar effects upon dinosaurs.

Summing the above from the dinosaurian herbivore's point of view, the non-tree-like pteridophytes offered a fast-growing, renewable resource which could be grazed without the death of the whole plant. However, these advantages were offset by the pteridophytic requirement for moisture

Land Plants as Food and Habitat   355

for reproduction. Given the widespread nature of arid and seasonally arid environments in the Mesozoic Era, pteridophytes may not have been generally available.

Gymnosperms

The gymnosperms (naked-seeds) include the primitive groups of seed plants. Seed plants are probably a monophyletic clade (a group that includes an ancestor and all its genetic descendants), because it is generally thought that the seed evolved only once in the Devonian. As the name is commonly applied, the gymnosperms are paraphyletic because they include several clades which have a common ancestor (the first seed plant), but exclude the most advanced of seed plants, the angiosperms (Fig. 25.1). Thus, the gymnosperms are defined by the common character of possessing

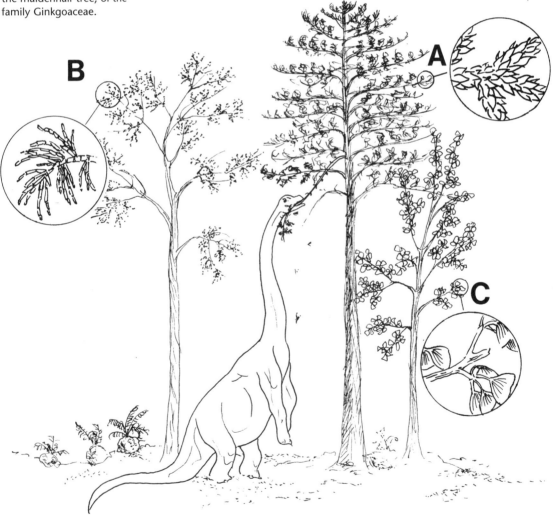

Figure 25.3. Some examples of Mesozoic gymnosperm trees. (A) *Araucaria,* a conifer of the family Araucariaceae. (B) *Pseudofrenelopsis,* a conifer of the family Cheirolepidiaceae (after Alvin 1983). (C) *Ginkgo,* the maidenhair tree, of the family Ginkgoaceae.

a naked seed, as contrasted to the angiosperms, which have a seed borne within a fruit. Several different and distinct branches of seed plants are embraced within the gymnosperms.

The most common gymnosperms in the present day and probably in the Mesozoic are the conifers (Fig. 25.3). These are the "pine trees" and their allies, including the genus *Araucaria* and its relatives, frequently depicted in dinosaur restorations. However, other conifers were also important in the Mesozoic. Some of the more important families include the Cupressaceae (the cypress family), the Taxodiaceae (the bald cypress family), and the Podocarpaceae (the podocarps, today almost entirely limited to the Southern Hemisphere, but probably an important plant type in wetter areas of the Mesozoic). Conifers range in size from small shrubs to large trees, trees being the most common.

A second major group of gymnosperms is the cycadophytes, including the cycads and the cycadeoids (Fig. 25.4). The former group survives in the tropics and subtropics today, while the latter is extinct. Cycads and cycadeoids include both erect forms with narrow, sometimes branching, palm-like trunks, and lower forms with barrel-like or elongate-hemispherical trunks. In both cases, the trunks were capped with a rosette of divided

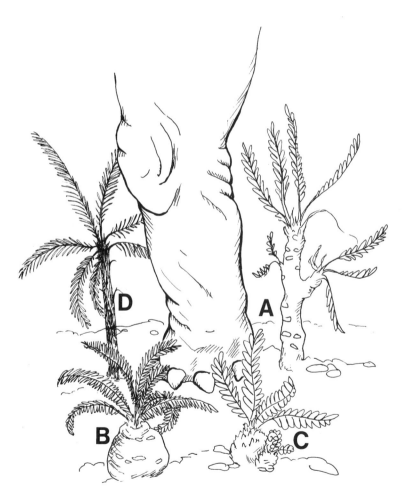

Figure 25.4. Some examples of common smaller Mesozoic gymnosperms. (A) The Cycadeoid *Williamsonia.* (B) The Cycadeoid *Cycadeoidea.* (C) A generalized low-growing cycad. (D) *Leptocycas,* a Late Triassic cycad (after Delevoryas and Hope 1971). An adult brachiosaur leg is provided for relative scale. Note: While all of these taxa lived in the Mesozoic, they did not all live at the same time.

Land Plants as Food and Habitat   357

leaves. Although cycads and cycadeoids looked similar, they had very different modes of reproduction. While the pollen and ovules of cycadeoids were generally borne on a common axis on a single plant, cycads possess pollen and seed-bearing structures borne on separate plants.

Several other gymnosperm clades were clearly important in Mesozoic systems, but are extinct or of very limited importance in the present day and thus difficult to evaluate. These include pteridosperms ("seed ferns"; small to large extinct plants which possessed fern-like foliage but which bore seeds), gnetophytes (survived by a small group of plants including *Ephedra*, commonly called Mormon Tea), Czekanowskiales (an extinct conifer-like group of trees with unusual reproductive characters), and *Ginkgo* (the "maidenhair tree" of many modern city streets; Fig. 25.3), among others.

These plants are united by the common evolutionary novelty of the seed. The seed is to the evolution of terrestrial plants what the amniotic egg (the mode of reproduction in the amniotes, the sauropsids + synapsids) was to the evolution of tetrapods. In both cases this was a major innovation which freed the group from dependence on standing water for sexual reproduction. In the amniotes, fertilization was internal within the egg-bearing parent, and the resulting embryo was surrounded by a protective membrane (the amnion). The embryo was either brought to term within the mother (as in mammals) or encased in a shell containing nutrients and deposited outside the mother to mature and hatch (as in living reptiles and birds).

In plants, this advance involved the retention of the egg-bearing gametophyte on the parent plant, thereby avoiding the stage of the dissemination of spores leading to the growth of a free-living, egg-bearing gametophyte. The sperm is introduced to the egg by the male spore or pollen grain, which is borne to the egg-bearing gametophyte by air currents. The pollen grain germinates in moisture provided by the parent plant that hosts the egg-bearing gametophyte, releasing a sperm which swims to the egg in this moisture. At an early stage, the resulting embryo, plus its store of nutritive tissue and moisture provided by the parent plant, is enclosed in a protective covering to form the seed. The seed may be dispersed from the parent plant by wind, water, or an animal, germinating in a suitable environment at some distance from the parent plant. This set of adaptations results in freedom from the necessity of standing water at the time of fertilization, and to a lesser degree at the time of establishment of the young seedling. While primitive gymnosperms (e.g., pteridosperms) may have maintained some dependence on water for reproduction, the dominant Mesozoic gymnosperms (e.g., cycadophytes, conifers, and others) were able to colonize relatively dry sites. This led to the appearance of a widespread land flora.

How do gymnosperms fare as herbivore food? To present knowledge, all gymnosperms of the Mesozoic and modern day have been and are trees or shrubs, with branching trunks bearing the growing points above ground. There are no "herbaceous" forms in which rhizomes would allow rapid colonization by vegetative growth, and only a very few genera in which other forms of vegetative reproduction are present. Thus, gymnosperms are dependent upon seed for the establishment of new individuals, and are generally not able to respond to external damage by regrowing from underground buds. While some living gymnosperms grow quite quickly (e.g., some species of pine), others (e.g., cycads, some other conifers) are slow to grow and respond to damage. According to both living and fossil evidence, many gymnosperms possess thick bark and tough, resistant, often spinose foliage, generally with abundant thick-walled cells which are hard to break down and digest. Further, extant gymnosperm foliage is often rich in indigestible chemicals and resins. There are exceptions to these generalizations. Cycadophyte foliage from the Jurassic Yorkshire Delta of England is thick, but is not interpreted as tough or resistant (P. Crane, personal communication, 1994). Similarly, *Ginkgo,* a deciduous tree common at higher latitudes in the Mesozoic, possessed soft foliage without great amounts of resin. An extinct group of conifers, the Cheirolepidiaceae, appear to have grown in great numbers in the lower paleolatitudes in the middle and late Mesozoic, and to have borne quantities of quite succulent foliage, although we have no sense of their chemistry or rate of growth.

These general characteristics of substantial height, thick bark, high resin content, and tough foliage are often correlated in the present day with plants that grow in dry environments and/or in environments frequented by fire. Drought and fire may have been common in the early and middle Mesozoic, ensuring the dominance of gymnosperms through this time, and thus helping to determine the nature of the available food for herbivores. Pteridophytes with subterranean rhizomes, particularly ferns, might also be favored by burning in the dry season of a seasonally moist climate.

From a dinosaurian herbivore's point of view, gymnosperms were probably the most significant source of food simply because of their wide dissemination across the land. However, the generally resistant nature of Mesozoic gymnosperm foliage and its rich chemical content probably dictated that large quantities of foliage had to be consumed in order to obtain sufficient nutrition. Furthermore, the presumed generally slow rates of Mesozoic gymnosperm growth and the lack of underground rhizomes suggest that relatively long intervals had to pass between one herbivore visit and the next to a single plant or stand (Bond 1989; Midgley and Bond 1991).

Angiosperms

The angiosperms or flowering plants are the most recent major group of vascular plants to evolve. There are some 220,000 species of angiosperms in the modern world, contrasted to some 750 species of gymnosperms and 10,000 to 12,000 species of pteridophytes. The angiosperms first appeared in the middle Cretaceous, and they diversified to dominate the world flora by the end of the Cretaceous (Niklas et al. 1985; Lidgard and Crane 1990). While the angiosperms are almost certainly a monophyletic group (that is, derived from a single common ancestor), their origins are obscure. It is difficult to define them on the basis of any particular character. In common discussion, they are the "flowering plants," in reference to their possession of flowers. This is often associated with insect pollination, although the latter character is also seen in some gymnosperms. For the present case, it is best to consider them as an "advanced" group of seed plants, as reflected by their position nested within the gymnosperms (Fig. 25.1).

The earliest angiosperms were apparently herbaceous, but they quickly diversified into a wide range of ecological niches and growth forms, generally characterized by a "weedy" ecology of fast growth and disturbance tolerance. Important to our interest in dinosaur fodder, the angiosperms possess modifications of the life cycle and vegetative body that allow them, on the whole, to reproduce and grow more quickly than gymnosperms (Stebbins 1981; Bond 1989; Midgley and Bond 1991). Furthermore, many but not all angiosperms possess either underground

Land Plants as Food and Habitat   359

rhizomes or an ability to sprout from the roots, allowing them to colonize large areas without sexual reproduction and to recover from grazing. While the group is quite variable, on the whole it would be safe to describe them as possessing more succulent foliage with fewer indigestible chemicals than gymnosperms.

In some respects, the angiosperms were an answer to a herbivorous dinosaur's prayer. Because of the seed habit, they grew in a wide range of terrestrial sites, and were thus almost ubiquitously available. Because of their rapid life cycle, rapid growth rate, and ability for vegetative reproduction, they were relatively tolerant of intensive herbivory, and could regenerate rapidly following cropping. All of these features suggest that, compared with gymnosperms, angiosperms could support larger numbers of herbivores. Also, angiosperms possess a wider range of growth forms and a greater diversity of habitat ecologies than gymnosperms. As a result, more of the world probably became vegetated after the appearance of the angiosperms, altering the three-dimensional environment in which dinosaurs lived.

The Succession of Paleofloras through Time

The land flora has not been homogeneous through time. During the late Paleozoic Era, the interplay of continental positions and global climate created many sites for the growth of moisture-loving pteridophytes, and they dominated terrestrial vegetation. After the Late Permian, however, terrestrial climates generally became drier and more seasonal (termed "continental climates"), because of climatic effects of the formation of Pangaea. This placed the reproductive cycle of existing pteridophytes under increasing stress, and the Late Permian and Early Triassic witnessed a major extinction of the dominant late Paleozoic pteridophytes, coupled with a

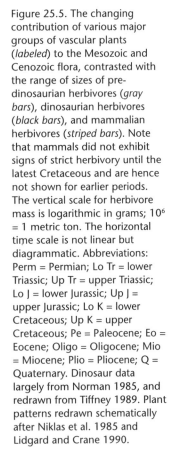

Figure 25.5. The changing contribution of various major groups of vascular plants (*labeled*) to the Mesozoic and Cenozoic flora, contrasted with the range of sizes of pre-dinosaurian herbivores (*gray bars*), dinosaurian herbivores (*black bars*), and mammalian herbivores (*striped bars*). Note that mammals did not exhibit signs of strict herbivory until the latest Cretaceous and are hence not shown for earlier periods. The vertical scale for herbivore mass is logarithmic in grams; 10^6 = 1 metric ton. The horizontal time scale is not linear but diagrammatic. Abbreviations: Perm = Permian; Lo Tr = lower Triassic; Up Tr = upper Triassic; Lo J = lower Jurassic; Up J = upper Jurassic; Lo K = lower Cretaceous; Up K = upper Cretaceous; Pe = Paleocene; Eo = Eocene; Oligo = Oligocene; Mio = Miocene; Plio = Pliocene; Q = Quaternary. Dinosaur data largely from Norman 1985, and redrawn from Tiffney 1989. Plant patterns redrawn schematically after Niklas et al. 1985 and Lidgard and Crane 1990.

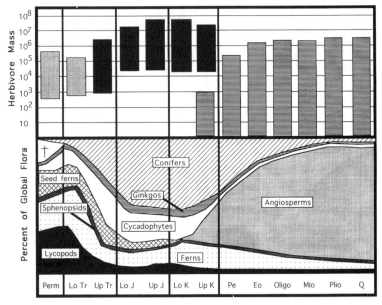

Bruce H. Tiffney

rise in the importance of gymnosperms in the terrestrial flora (Fig. 25.5; Niklas et al. 1985).

The Triassic gymnosperms were dominated by conifers, seed ferns, early cycadophytes, and various less well understood groups. These clades diversified through the Jurassic into the Cretaceous. The seed ferns became less and less important during the Mesozoic, disappearing in the Cretaceous, while the cycadophytes diversified and became increasingly important in the Jurassic and then declined in the Cretaceous. The conifers similarly diversified through the Jurassic into the Early Cretaceous and began to decline in diversity in the later Cretaceous (Niklas et al. 1985; Lidgard and Crane 1990). In the Late Triassic and especially the Jurassic, several modern lineages of ferns evolved, paralleling the breakup of Pangaea and the resulting reduction in the global area dominated by continental climates.

The rise of the angiosperms to dominance in the later Cretaceous depended on both vegetative and reproductive features. The relatively more rapid reproductive cycle and rate of growth of angiosperms may have given them an edge over gymnosperms, particularly in the ability to colonize disturbed sites (Bond 1989). As a result, gymnosperms became increasingly restricted to marginal sites where physical features (low levels of sunlight, low temperatures, poor nutrients) put angiosperms at a greater disadvantage. Further, angiosperms are often pollinated by insects and their seeds frequently dispersed by vertebrates. Both these features could influence rates of gene flow, and thus of speciation. This has led to the hypothesis that angiosperms have a greater ability for speciation and thus for evolutionary flexibility than gymnosperms because of their relationships with animals (Stebbins 1981), although this hypothesis has been questioned (Midgley and Bond 1991).

Paleophytogeography

The various kinds of plants were not distributed homogeneously across the globe in the Mesozoic (see Meyen 1987: 313–323 for a brief summary, and Vakhrameev 1991 for a detailed treatment). Latitudinal climatic gradients and continental positions shifted through the Mesozoic, influencing the distribution of the different kinds of plants and the vegetation that they created. The synthesis of patterns in Mesozoic vegetation on a global level is still in its early stages, and different sources vary in the details. Furthermore, our knowledge of these fossil floras is far better for present-day Northern Hemisphere landmasses than for those of the modern Southern Hemisphere. Thus many of the following generalizations could be demonstrated to be wrong by future research.

The whole of the Mesozoic is marked by the pattern of the poles' being relatively temperate and moist in comparison to the equatorial zone, which was generally warmer and drier, tending to strong aridity at some times. Often the equatorial zone was dominated by drought-adapted conifers and cycadophytes, while communities toward the poles were increasingly dominated by moisture-requiring conifers, other gymnosperms (e.g., ginkgos, Czekanowskiales), and pteridophytes. The arid equatorial belt gradually developed during the Late Triassic through the Middle Jurassic and carried on with little change through the later Cretaceous. While some authors suggest that the arid belt expanded dramatically in the later Jurassic and then contracted in the Early Cretaceous (Hallam 1984, 1993; Vakhrameev 1991), Ziegler et al. (1993) suggest that this is a result of the

Northern Hemisphere continents "drifting" through arid belts, rather than a function of actual global climatic change. The angiosperms first appeared in arid locales in the equatorial belt in the later part of the Early Cretaceous, and quickly spread to the poles by the early Late Cretaceous, invading existing conifer communities and displacing cycadophytes, pteridophytes, and lesser-known groups in the process (Saward 1992; Spicer et al. 1993).

We tend to envision the past in terms of what we are familiar with in the present. However, in doing so, we must be careful not to turn the past *into* the present. Thus, while many reconstructions of the Triassic-Cretaceous Earth depict a planet as densely and continuously vegetated as the present-day Earth, alternative hypotheses should be entertained. The combination of continental configuration, global climate, and available plant types in the Mesozoic suggests that there was no vegetation unit equivalent to the modern "tropical rain forest" (Saward 1992; Ziegler et al. 1993). Rather, most of the continents in the lower and middle latitudes possessed a discontinuous vegetation adapted to at least seasonal drought. Herbivores in equatorial and subequatorial regions would encounter adequate fodder along seashores and riverine lowlands, but might find inland and upland settings to have more patchy vegetation, scattered in an otherwise arid and perhaps vegetation-poor environment. At the higher latitudes, vegetation probably became more continuous and productive with the increasing availability of moisture. This vegetation was probably deciduous at the poles (Saward 1992; Spicer et al. 1993; Ziegler et al. 1993). It has been informally suggested that dinosaur distribution may have paralleled this pattern, because the greatest diversity of dinosaurs has been collected in the Mesozoic mid-latitudes of the Northern and Southern hemispheres (see comments by J. M. Parrish in Hallam 1993: 296). While these patterns are comparable, the perceived distribution of dinosaurs could also reflect where they have been collected, rather than all the places in which they originally lived.

Dinosaurs and Plants: Basic Concepts

Much as we inferred the biology of Mesozoic plants, we have to infer the biology of Mesozoic herbivores from a combination of the morphology and geological context of the fossils and the biology of living vertebrate herbivores, both reptilian (King 1996) and mammalian. From living organisms, we can deduce two broad generalizations about herbivores.

First, the efficiency of food use depends on its digestibility. Plant food may be broadly categorized as resistant to digestion (much foliage, bark) or fairly digestible (fruits, seeds, some starch-rich roots). In order to gain energy from the latter source, herbivores need only crack the protective outer layer and subject the contents to mild chewing. However, to digest leaves, herbivores need either to mechanically break down the walls of the individual leaf cells to expose the cell contents, or to retain the leaves for prolonged times in the gut, thereby exposing them to bacterial fermentation that is capable of breaking down the cell walls. Some dinosaurs apparently ingested stones (gastroliths) which were retained in the digestive tract (Stokes 1987). These may have functioned to aid digestion by one of two means. Classically, the muscular activity of the stomach is supposed to have caused the stones to bounce against each other, thereby crushing the trapped plant material. Alternatively, the stones may have served to mix up the contents of the stomach, ensuring more complete digestion of its contents (Gillette 1994). It is also possible that the tough plant material was broken within a gizzard-like structure, or even within a muscular crop (Farlow 1987; Gillette 1994), although evidence for these soft parts is not found in the fossil record.

Second, food quality and herbivore size are related. Large living herbivores require large amounts of food. Because high-quality food items (fruit, seeds) are widely scattered in the environment, large herbivores cannot afford to seek them out, and thus tend to consume great quantities of easily obtainable but low-quality food (Mellett 1982; Farlow 1987). This generates a self-reinforcing association between large size and low-quality fodder. A greater food intake requires a larger stomach. Further, fermentation requires a long residence time in the stomach, necessitating a larger stomach to hold fodder in various stages of digestion. Both factors dictate a larger whole animal. Such large herbivores become "whole-plant predators," consuming substantial portions of the plant, and exerting a very strong influence on the vegetation they inhabit. For example, elephants in African game reserves can turn forests into grasslands as a result of their intense predation upon trees.

In contrast, small living herbivores do not require as great a volume of food, and can afford to seek out higher-quality food items, such as fruits and seeds, that are scattered in the environment. As a result, small herbivores tend to be more specialized in their food preferences, and to be plant predators at the level of individual plant organs. This creates new possibilities for specific kinds of plant-animal interactions, including pollination or fruit and seed dispersal.

These principles of herbivore-plant interaction are central to what distinguishes the age of dinosaurs from the subsequent age of mammals, as both the size of the herbivores and the nature of the available plant food change from the Mesozoic to the Cenozoic (Fig. 25.5).

Late Triassic

The Triassic spanned a major transition from a seed fern–pteridophyte flora to a conifer-cycadophyte flora, particularly in the equatorial region and southerly higher latitudes. The Early Triassic vegetation included many forms with relatively soft leaves, borne on plants of a variety of heights, from herbs through shrubs and trees. The conifer-cycadophyte vegetation that spread in the Late Triassic was dominated by plants with tougher and often spinose foliage, which was generally rich in chemical defenses. In the case of many conifers, this foliage was borne on trees. The probable reason for this change in vegetation was the aforementioned spread of continental climates accompanying the formation of Pangaea.

This floristic change appears to parallel a transition in herbivorous tetrapods (Benton 1983). The dominant herbivores in the Early and Middle Triassic were therapsids, which tended to be relatively small (up to the size of a pig) and to feed within a few feet of the ground (Zavada and Mentis 1992). By contrast, while the early dinosaurian communities of the Late Triassic included a few small ornithischian herbivores, they also included very large, high-feeding, prosauropod herbivores such as *Plateosaurus* and *Melanorosaurus,* setting the stage for the herbivore communities of the Jurassic and Early Cretaceous (Galton 1985). It is possible that the large size attained by these early saurischian herbivores was a direct result of the "poor" quality of the newly dominant vegetation.

Land Plants as Food and Habitat 363

Jurassic to Middle Cretaceous

While there were many species of "small" dinosaurian herbivores (tens of kilograms to hundreds of kilograms in size) from the Triassic through the Cretaceous, two points are of note. First, the smallest dinosaurian herbivores were still perhaps two orders of magnitude larger than the smallest vertebrate herbivores among living mammals and birds. Second, while these "small" dinosaurian herbivores existed, Mesozoic terrestrial ecosystems were very strongly influenced, if not dominated, by the large dinosaurian herbivores. This is particularly true of the whole Earth in the Jurassic and Early Cretaceous and, after smaller Ornithischia radiated in the later Cretaceous of the Northern Hemisphere, of the Late Cretaceous in the Southern Hemisphere.

These large herbivores could have fed at least to 8 to 10 meters off the ground, and possibly to 15 meters if they were able to rear back on their hind legs and balance on their tails, as Bakker (1978, 1986) has suggested, although this may have been uncommon behavior (Dodson 1990a). This feeding height suggests that large sauropods commonly browsed on tree foliage, which in this case would be almost entirely provided by conifers. Some have questioned this high-feeding model, citing the difficulty of maintaining a sufficient supply of blood to a consistently elevated head (Dodson 1990a). This suggests the alternative hypothesis that sauropods stood in one place and fed over a wide range of lower vegetation. Indeed, some workers have suggested the sauropods fed on vast "fern prairies" (Coe et al. 1987). While pteridophytes were common in moist areas in the Mesozoic (Krassilov 1981; Saward 1992; Spicer et al. 1993), it seems unlikely that they grew in vast stands in the arid interiors of Mesozoic continents like grasses do in the present day, as the term *prairie* implies. Again, we need to recognize that the past is not the same as the present, and that large areas of the pre-angiosperm world may have been without vegetation. Thus, while shrubby vegetation (cycadophytes, seed ferns, some conifers) was certainly available and undoubtedly consumed, I suspect that sauropods did not confine themselves to such "boom-feeding," but used resources at several levels, including trees. Indeed, the radiation of low-feeding Ornithischia *after* the origin of the angiosperms (see below) suggests that "low-level" fodder was not a significant resource before the appearance of the angiosperms.

These large sauropod herbivores lacked the ability to chew fodder, as they possessed peg-like teeth that did not occlude. It seems likely that the teeth were used as "rakes," the animal closing its mouth around a branch and pulling its head back, peeling foliage off as it went. The poor quality of most available gymnosperm foliage, particularly in low and middle paleolatitudes, its high content of indigestible chemicals, and the inability of these herbivores to chew effectively lead one to suspect that the breakdown of the plant material took place in the gut. The great size of these herbivores would have allowed for a large intake and long residence time of relatively indigestible food. This would permit microbial fermentation to break down tough cell walls to allow the extraction of sufficient energy to support the organism (Farlow 1987). This could be aided by mechanical breakdown, possibly in a gizzard or muscular crop, or in the stomach through abrasion by gastroliths. The large intake, long passage time, and microbial fermentation permitted by large body size might also have allowed the dinosaurs to substantially negate the effects of digestion-inhibiting chemicals (Farlow 1987). Presumably, the greater diversity and density of high-latitude plant communities (Saward 1992; Spicer et al. 1993; Ziegler et al. 1993) provided a better source of fodder, and one prediction from the foregoing hypothesis is that the largest herbivorous dinosaurs from the high latitudes should have been somewhat smaller than their lower-latitude counterparts.

If the herbivores became quite large as a result of their interactions with plants, it is not surprising that the coeval carnivores evolved large size in order to deal with the herbivores. Thus, it may be possible to ascribe the size of large dinosaur predators to the nature of the Jurassic plant community.

While the focus has been on the ecologically important large herbivores, the smaller ones must have fed closer to the ground (e.g., some prosauropods, *Stegosaurus*, and several smaller Ornithischia). Many of these would have to have contended with the same poor food quality of tough conifer and cycadophyte foliage, but others could have fed on seed ferns, pteridophytes, and other gymnosperms with softer foliage, and thus would not have faced quite the same requirements of maintaining a digestive volume. Some may even have fed on the large seeds of cycads and some conifers (Weishampel 1984). It is possible that the youngsters of sauropod herbivores fed on higher-quality food until their requirements for food volume forced them to seek lower-quality fodder, although if dinosaurs possessed a lower metabolism than mammals and birds, this pattern may not have been as pronounced as it is in mammals and birds (Farlow 1987).

The sheer size of the dominant sauropod herbivores, coupled with the relatively (compared to the present day) scattered nature and low food quality of the Jurassic vegetation, at least beyond the higher latitudes, raises some interesting questions. How many individual herbivorous dinosaurs were there at any one time in the Jurassic? As many as there are living mammalian herbivores in the present? In the modern world, it is observed that the larger the body size of an animal, the fewer the total numbers of individuals in the species (Peters 1983). It is also recognized that the poorer the quality of the available food, the smaller the number of individual organisms that can be supported. The great size of many Jurassic herbivores, the relatively poor quality of the available food, its restricted distribution, and its low diversity suggest three predictions. First, we might expect a lower number of species of giant dinosaurian herbivores relative to the number of species of present-day large mammalian herbivores. Second, we might expect relatively few individuals to be present within each sauropod species. Third, individuals within a species might have to migrate over very large areas in order to obtain sufficient fodder to live.

Several lines of circumstantial evidence do not disprove these hypotheses, and lead to the suggestion that the very nature of the Late Triassic through middle Cretaceous terrestrial ecosystem was quite different from that of the ecosystems of the Tertiary and present day. First, the taxonomic diversity of dinosaurs is remarkably low compared to that of fossil mammals. Dodson (1990b) estimated that there were between 900 and 1,200 genera and about 1,100 to 1,500 species of dinosaurs in the 160 million years of dinosaur dominance, or about 7 to 9 species per million years. By contrast, there are about 3,000 genera of fossil mammals reported from the Tertiary. Assuming about 4 species per genus of living mammals (Nowak 1991), this suggests that there were between 3,000 (most conservatively) and 12,000 species during this 65-million-year period, or between 46 and 185 species per million years. Even allowing for various preservational biases, these figures suggest that dinosaurs had a very much lower specific diversity per unit time than would be expected by comparison with mammals. Dinosaur trackway evidence in North America suggests both that the large sauropod herbivores may have migrated long distances (e.g., Dodson 1990a), and that the individual herds were not very large. This is weak evidence for a low number of individuals in a species, but stands in sharp contrast to clear evidence for the large numbers of individuals in some herds of dinosaurs from the later Cretaceous (see below).

However, these speculations rest on two (if not more) basic assumptions. First, what is the physiology (or is it physiologies?) of the dinosaurs? Were the herbivorous sauropods endotherms, inertial homeotherms, or ectotherms? If they were ectotherms, they may have had lower food requirements than do living mammals, which are endotherms. If they were inertial homeotherms, we have the difficulty that no large terrestrial, inertial homeotherms exist in the present day upon which we can base an estimate of necessary food intake. Second, what is the effect of size on caloric requirements for an organism the size of a large sauropod? From studies of living animals, it has been demonstrated that the basal metabolic rate decreases in a linear manner with increasing size, all the way up to an elephant (Peters 1983; Schmidt-Nielsen 1984). However, we do not have any living herbivores that are as large as the average herbivorous sauropod to check if this relationship continues to be linear at a size of ten to twenty times that of an elephant. But even if sauropod herbivores did not possess a physiology comparable to that of living mammals, it is still probable that they would have been a formidable force of deforestation in the Mesozoic.

Late Cretaceous

The nature of the herbivore-plant relationship changed dramatically with the appearance and diversification of the angiosperms. Indeed, Bakker (1986) and Wing and Tiffney (1987) observed that the disturbance provided by dinosaur herbivory may well have selected for the appearance of a disturbance-tolerant seed plant with rapid growth and abilities for vegetative reproduction—in other words, an angiosperm. By the Late Cretaceous, the angiosperms provided a far more energetically rewarding and diverse source of food than the gymnosperms and ferns: one that grew in a wider range of habitats and a wider range of heights and shapes, that recovered rapidly from damage, and that possessed fewer mechanical and chemical impediments to digestion. The exact nature of the vegetation created by the angiosperms is still being discovered. It is clear that the angiosperms became the most taxonomically diverse plants on the earth's surface by the Late Cretaceous (Lidgard and Crane 1990), and that they occurred in plant communities from the equator to the poles (Saward 1992; Spicer et al. 1993). However, it is not clear how rapidly they actually came to dominate the global vegetation physically; it is possible that this high diversity of types was present in low individual numbers, at least in some areas of the globe (Wing et al. 1993).

The Late Cretaceous also witnessed a major radiation of the ornithischians, including the hadrosaurs and ceratopsians. The parallel between the radiation of the Ornithischia and angiosperms suggests the hypothesis that the angiosperms underwrote the ornithischian radiation (although some doubt this; cf. Insole and Hutt 1994). This hypothesis is circumstantially supported by several lines of evidence. First, the new ornithischian dinosaurs fed at a lower level than the preceding sauropod herbivores

(Bakker 1978), befitting the herbaceous, shrubby, or small-tree stature of many Cretaceous angiosperms. Second, these were relatively smaller herbivores (1–10 tons) than the preceding sauropods (10–50 tons), suggesting a more digestible food source requiring a less prolonged period of fermentation in the gut. Third, in a case where approximately contemporaneous faunas occurred in North America (Lehman 1987), areas dominated by conifers supported a fauna rich in herbivorous sauropods (*Alamosaurus*), while areas dominated by angiosperms supported an ornithischian herbivore fauna, suggesting an association of herbivore and food type.

Finally, two changes took place in dinosaur diversity. Dodson (1990b) observed that almost 50 percent of the described species of dinosaurs come from the last 20 million years of their existence, suggesting a substantial jump in species diversity in the Late Cretaceous. This throws the numbers of pre–Late Cretaceous dinosaur species into even more stark contrast to the numbers of mammals in the Tertiary. Furthermore, mass kill sites with large numbers of individuals of one species of dinosaur have been found only in the Late Cretaceous. In the most dramatic case, this involved a herd of up to 10,000 individuals of the duckbill herbivore *Maiasaura* (Weishampel and Horner 1990), and bonebeds consisting of large numbers of other herbivorous dinosaurs are known as well (e.g., Currie and Dodson 1984).

The appearance of angiosperms was not the sole factor in this radiation; evolutionary changes in the Ornithischia also played a role. Many of the hadrosaurs had teeth that occluded, and ceratopsians had shearing teeth, providing the ability to break down plant material prior to swallowing it (Norman and Weishampel 1985; Weishampel and Norman 1989). This increased the efficiency of digestion and the energy reward per unit of food consumed, favoring the greater success of these herbivores, and possibly their smaller size. However, as in the case of the saurischian herbivores, some questions remain open. For example, was the physiology of these ornithischian herbivores the same as the physiology of the large sauropod herbivores, or is part of the change in herbivore diversity and size due to an unrecognized change in physiology?

The appearance of angiosperms may have also affected the environment of the dinosaurs in other ways. The Jurassic vegetation of the earth probably included large areas in the low and mid latitudes where trees were widely spaced or occurred in patches, forming an open vegetation. Indeed, many well-known Jurassic dinosaur localities are associated with open vegetation (Krassilov 1981; Saward 1992; Ziegler et al. 1993). This parallels the modern day, when large herbivores are associated with open environments.

Following their appearance, the angiosperms are presumed to have created an increasingly dense vegetation, and one composed of a greater diversity of species than that provided by gymnosperms and pteridophytes alone. The nature of this transition is an active area of research. Wolfe and Upchurch (1987) suggest that Cretaceous angiosperm communities were "open woodlands," forming a more closed and widespread community than created by the preceding gymnosperms. Wing et al. (1993), however, describe fully fern-dominated communities in the Late Cretaceous, suggesting that, at least in one area of western North America in the Late Cretaceous, increasing moisture allowed ferns to form an at least locally dominant vegetation. If the radiation of angiosperms did change the three-dimensional environment of the Late Cretaceous, what effect did this change have upon the sauropod-ornithischian transition, the evolution of the Ornithischia in the Late Cretaceous, and the Cretaceous-Tertiary

Land Plants as Food and Habitat   367

extinction of dinosaurs? Regarding the last question, the development of a increasingly closed vegetation may have increasingly fragmented dinosaur populations, rendering them more susceptible to extinction.

The early Tertiary saw the first appearance of a dense angiosperm forest vegetation of modern aspect (Tiffney 1984). The early Tertiary also witnessed a massive radiation of birds and mammals. These herbivores were tiny in comparison to their dinosaurian predecessors (Fig. 25.5), and they fitted "within" the three-dimensional structure of the newly evolved angiosperm forests, rather than around its margins. Their small size suited them to feed on the organs of angiosperms, thus setting up a very different herbivore-plant interaction than prevailed in the age of dinosaurs. In fact, the most basic difference between the terrestrial ecosystems of the Mesozoic and those of the Cenozoic and the present is the scale of plant-animal interaction. The Mesozoic involved open vegetation and large herbivores that acted as "plant predators," preying on whole organisms. The early Cenozoic introduced a closed vegetation housing far smaller herbivores that preyed on plant organs. "Whole-plant" predation reappeared only with the evolution of more open angiosperm communities in the later Tertiary.

This account of the plants of the Mesozoic summarizes generally accepted information. However, the account of the interactions and interrelationships of Mesozoic plants and dinosaurs is much more speculative. Why? Ten years ago paleobotanists and vertebrate paleontologists tended to ignore the question of how these two important elements of the Mesozoic terrestrial ecosystem interacted. This is no longer so, but the field remains young. Many patterns have been recognized, and some hypotheses have been put forward to explain these observations. The hypotheses require testing, modification, and re-testing, however, and more patterns await observation. There is much here to entertain and frustrate the paleontologists of the future!

What is clear from these initial observations is that the dynamic interactions between plants and herbivores of the Late Triassic through the mid-Cretaceous were different from those of the Late Cretaceous, and that both were vastly different from the herbivore-plant interactions of the birds, mammals, and angiosperms in the Tertiary and present. The reader should not come away from this chapter thinking that the age of dinosaurs was "just like today," only with dinosaurs instead of birds and mammals. Rather, the age of dinosaurs should be viewed as a unique ecosystem that functioned in its own manner, and which offers an interesting look into one of the alternate worlds that have existed on our planet (Tiffney 1992).

References

Alvin, K. L. 1983. Reconstruction of a Lower Cretaceous conifer. *Botanical Journal of the Linnean Society* 86: 169–176.

Bakker, R. T. 1978. Dinosaur feeding behavior and the origin of flowering plants. *Nature* 274: 661–663.

Bakker, R. T. 1986. *The Dinosaur Heresies: New Theories Unlocking the Mystery of the Dinosaurs and Their Extinction.* New York: William Morrow.

Benton, M. J. 1983. Dinosaur success in the Triassic: A noncompetitive ecological model. *Quarterly Review of Biology* 58: 29–55.

Bond, W. J. 1989. The tortoise and the hare: Ecology of angiosperm dominance and gymnosperm persistence. *Biological Journal of the Linnean Society* 36: 227–249.

Coe, M. J.; D. L. Dilcher; J. O. Farlow; D. M. Jarzen; and D. A. Russell. 1987. Dinosaurs and land plants. In E. M. Friis, W. G. Chaloner, and P. R. Crane (eds.), *The Origins of Angiosperms and Their Biological Consequences,* pp. 225–258. Cambridge: Cambridge University Press.

Currie, P. J., and P. Dodson. 1984. Mass death of a herd of ceratopsian dinosaurs. In W.-E. Reif and F. Westphal (eds.), *Third Symposium on Mesozoic Terrestrial Ecosystems, Short Papers,* pp. 61–66. Tübingen: ATTEMPO Verlag.

Delevoryas, T., and R. C. Hope. 1971. A new Triassic cycad and its phyletic implications. *Postilla* 150: 1–14.

Dodson, P. 1990a. Sauropod paleoecology. In D. B. Weishampel, P. Dodson, and H. Osmólska (eds.), *The Dinosauria,* pp. 402–407. Berkeley: University of California Press.

Dodson, P. 1990b. Counting dinosaurs: How many kinds were there? *Proceedings of the National Academy of Sciences* (USA) 87: 7608–7612.

Doyle, J. A., and M. J. Donoghue. 1986. Seed plant phylogeny and the origin of angiosperms: An experimental cladistic approach. *Botanical Review* (Lancaster) 52: 321–431.

Doyle, J. A., and M. J. Donoghue. 1992. Fossils and seed plant phylogeny reanalyzed. *Brittonia* 44: 89–106.

Farlow, J. O. 1987. Speculations about the diet and digestive physiology of herbivorous dinosaurs. *Paleobiology* 13: 60–72.

Galton, P. M. 1985. Diet of prosauropod dinosaurs from the Late Triassic and Early Jurassic. *Lethaia* 18: 105–123.

Gillette, D. D. 1994. *Seismosaurus: The Earth Shaker.* New York: Columbia University Press.

Hallam, A. 1984. Continental humid and arid zones during the Jurassic and Cretaceous. *Palaeogeography, Palaeoclimatology, Palaeoecology* 47: 195–223.

Hallam, A. 1993. Jurassic climates as inferred from the sedimentary and fossil record. *Philosophical Transactions of the Royal Society of London* B 341: 287–296.

Insole, A. N., and S. Hutt. 1994. The palaeoecology of the dinosaurs of the Wessex Formation (Wealden Group, Early Cretaceous), Isle of Wight, southern England. *Zoological Journal of the Linnean Society* 112: 197–215.

King, G. 1996. *Reptiles and Herbivory.* London: Chapman and Hall.

Krassilov, V. A. 1981. Changes of Mesozoic vegetation and the extinction of dinosaurs. *Palaeogeography, Palaeoclimatology, Palaeoecology* 34: 207–224.

Lehman, T. M. 1987. Late Maastrichtian paleoenvironments and dinosaur biogeography in the western interior of North America. *Palaeogeography, Palaeoclimatology, Palaeoecology* 60: 189–217.

Lidgard, S., and P. Crane. 1990. Angiosperm diversification and Cretaceous floristic trends: A comparison of palynofloras and leaf macrofloras. *Paleobiology* 16: 77–93.

Mellett, J. S. 1982. Body size, diet, and scaling factors in large carnivores and herbivores. *Proceedings of the Third North American Paleontological Convention* 2: 371–376.

Meyen, S. V. 1987. *Fundamentals of Palaeobotany.* London: Chapman and Hall.

Midgley, J. J., and W. J. Bond. 1991. Ecological aspects of the rise of the angiosperms: A challenge to the reproductive superiority hypothesis. *Biological Journal of the Linnean Society* 44: 81–92.

Niklas, K. J.; B. H. Tiffney; and A. H. Knoll. 1985. Patterns in vascular land plant diversification: An analysis at the species level. In J. W. Valentine (ed.), *Phanerozoic Diversity Patterns: Profiles in Macroevolution,* pp. 97–128. Princeton, N.J.: Princeton University Press.

Norman, D. 1985. *The Illustrated Encyclopedia of Dinosaurs.* New York: Crescent Books.

Norman, D. B., and D. B. Weishampel. 1985. Ornithopod feeding mechanisms: Their bearing on the evolution of herbivory. *American Naturalist* 126: 151–164.

Nowak, R. M. 1991. *Walker's Mammals of the World.* Baltimore: Johns Hopkins University Press.

Land Plants as Food and Habitat 369

Peters, R. H. 1983. *The Ecological Implications of Body Size*. Cambridge: Cambridge University Press.

Saward, S. A. 1992. A global view of Cretaceous vegetation patterns. In P. J. McCabe and J. T. Parrish (eds.), *Controls on the Distribution and Quality of Cretaceous Coals,* pp. 17–35. Special Paper 267. Boulder, Colo.: Geological Society of America.

Schmidt-Nielsen, K. 1984. *Scaling: Why Is Animal Size So Important?* Cambridge: Cambridge University Press.

Spicer, R. A.; P. M. Rees; and J. L. Chapman. 1993. Cretaceous phytogeography and climate signals. *Philosophical Transactions of the Royal Society of London* B 341: 277–286.

Stebbins, G. L. 1981. Why are there so many species of flowering plants? *Bioscience* 31: 573–577.

Stewart, W. N., and G. W. Rothwell. 1993. *Paleobotany and the Evolution of Plants.* Cambridge: Cambridge University Press.

Stokes, W. L. 1987. Dinosaur gastroliths revisited. *Journal of Paleontology* 61: 1242–1246.

Taylor, T. N., and E. L. Taylor. 1993. *The Biology and Evolution of Fossil Plants.* Englewood Cliffs, N.J.: Prentice Hall.

Tiffney, B. H. 1984. Seed size, dispersal syndromes, and the rise of the angiosperms: Evidence and hypothesis. *Annals of the Missouri Botanical Garden* 71: 551–576.

Tiffney, B. H. 1989. Plant life in the age of dinosaurs. In K. Padian and D. J. Chure (eds.), *The Age of Dinosaurs,* pp. 35–47. Short Courses in Paleontology 2. Knoxville: Paleontological Society, University of Tennessee.

Tiffney, B. H. 1992. The role of vertebrate herbivory in the evolution of land plants. *The Palaeobotanist* 41: 87–97.

Vakhrameev, V. A. 1991. *Jurassic and Cretaceous Floras and the Climates of the Earth.* Cambridge: Cambridge University Press.

Weishampel, D. B. 1984. Interactions between Mesozoic plants and vertebrates: Fructifications and seed predation. *Neues Jahrbuch für Geologie Paläontologie Abhandlungen* 167: 224–250.

Weishampel, D. B., and J. R. Horner. 1990. Hadrosauridae. In D. B. Weishampel, P. Dodson, and H. Osmólska (eds.), *The Dinosauria,* pp. 534–561. Berkeley: University of California Press.

Weishampel, D. B., and D. B. Norman. 1989. Vertebrate herbivory in the Mesozoic: Jaws, plants and evolutionary metrics. In J. O. Farlow (ed.), *Paleobiology of the Dinosaurs,* pp. 87–100. Special Paper 238. Boulder, Colo.: Geological Society of America.

Wing, S. L.; L. J. Hickey; and C. J. Swisher. 1993. Implications of an exceptional fossil flora for Late Cretaceous vegetation. *Nature* 363: 342–344.

Wing, S. L., and B. H. Tiffney. 1987. The reciprocal interaction of angiosperm evolution and tetrapod herbivory. *Review of Palaeobotany and Palynology* 50: 179–210.

Wolfe, J. A., and G. R. Upchurch, Jr. 1987. North American nonmarine climates and vegetation during the Late Cretaceous. *Palaeogeography, Palaeoclimatology, Paleoecology* 61: 33–77.

Zavada, M. S., and M. T. Mentis. 1992. Plant-animal interaction: The effect of Permian megaherbivores on the glossopterid flora. *American Midland Naturalist* 127: 1–12.

Ziegler, A. M.; J. M. Parrish; Y. Jiping; E. D. Gyllenhaal; D. B. Rowley; J. T. Parrish; N. Shangyou; A. Bekker; and M. L. Hulver. 1993. Early Mesozoic phytogeography and climate. *Philosophical Transactions of the Royal Society of London* B 341: 297–305.

370   *Bruce H. Tiffney*

What Did Dinosaurs Eat? Coprolites and Other Direct Evidence of Dinosaur Diets

Karen Chin

26

What did the Mesozoic dinosaurs really eat? This question has spawned numerous hypotheses from scientists, dinosaur enthusiasts, and fantasy writers. Speculations about dinosaur diets are frequently based on indirect evidence that includes surveys of available food and theories about foraging abilities inferred from functional morphology. Such analyses are important tools that have suggested generalized dinosaur feeding strategies. Even so, indirect evidence cannot tell us which available foods were actually eaten. Did dinosaurs feast on certain ferns? Conifers? Mammals? Each other? We will never completely understand dinosaur food habits, but scrutiny of the fossil record has revealed a number of fortuitous traces of dinosaur feeding activities. These clues are usually rare and often controversial, but they provide us with paleobiological information which can help us better understand dinosaurs and their interactions with other organisms.

In order to look for direct evidence of dinosaur diets, we can consider all stages of feeding behavior, including search, capture, ingestion, digestion, and defecation (Bishop 1975). Although these activities generally leave little preservable evidence, animals spend a substantial proportion of their time seeking food, so it is not surprising that some traces of feeding activity have been preserved. Clues have been gleaned from a variety of trace fossils (fossils that indicate the activity of organisms) and from distinctive assemblages of skeletal material. These disparate sources of fossil evidence provide different perspectives on dinosaur feeding habits.

Trackways as Evidence of the Search for Food

The act of looking for food might seem to be untraceable, but dinosaurs occasionally left sets of tracks that suggest that they were actively seeking dinner entrées. At a famous Early Cretaceous site along the Paluxy River in Texas, tracks from one or more theropods appear to follow several sauropod trackways. It is apparent that the sauropods preceded the theropods because the carnivore footprints were superimposed on top of several of the sauropod tracks. The time differential between the passage of the two groups has not been ascertained, however. Dinosaur tracker Roland T. Bird suggested that the dovetailing trackways indicate that the theropods were hot on the heels of the sauropod herd (Farlow 1987). It is possible, however, that the theropods were stalking the sauropods at a distance (Lockley 1991). If the theropods were indeed pursuing the sauropods, this site may provide evidence that these particular theropods hunted in packs. Cooperative hunting is also suggested by another set of Upper Cretaceous theropod/sauropod trackways in Bolivia, where several theropod trackways parallel and overlap prints made by a group of sauropods (Leonardi 1984).

A different possible theropod hunting scenario is presented by a Cretaceous trackway assemblage in Australia. This site contains thousands of tracks and has been interpreted as a dinosaur "stampede" triggered by a single large theropod stalking a mixed group of small coelurosaurs and ornithopods (Thulborn and Wade 1979). Although the large theropod footprints do not actually follow the smaller dinosaur tracks, the long stride lengths and parallel trackways of the over one hundred small dinosaurs suggest that they were fleeing a significant threat.

Fossil tracks may also provide information about the foraging behavior of herbivorous dinosaurs. A set of intriguing Cretaceous footprints in the roof of a Utah coal mine were found clustered around fossil tree trunks that were preserved in growth position. The tracks are oriented toward the tree trunks and suggest the shuffling steps of browsing hadrosaurs (Parker and Rowley 1989; L. Parker, personal communication).

Fossil Assemblages That Indicate Predator/ Prey Interactions

Predator/prey interactions can occasionally be inferred from the associations of different organisms in exceptional fossil assemblages. One spectacular find from the Gobi Desert revealed the skeleton of a carnivorous *Velociraptor* entangled with a herbivorous *Protoceratops* (Fig. 26.1; Kielan-Jaworowska and Barsbold 1972). The relative positions of the two dinosaurs suggest that they were engaged in a struggle when they died, with the theropod's clawed feet extending into the *Protoceratops*'s throat and

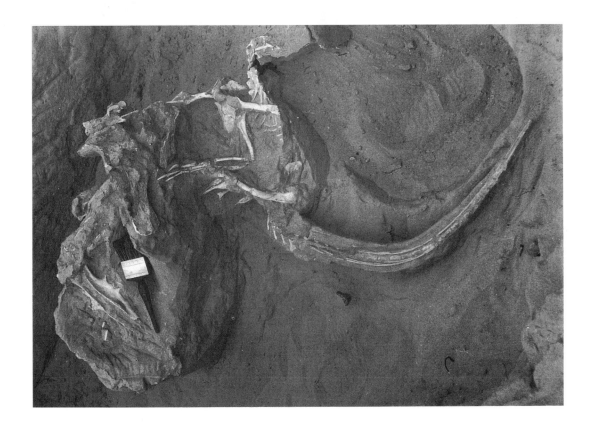

Figure 26.1. This Late Cretaceous Mongolian assemblage shows articulated skeletons of a *Velociraptor* (right) and a *Protoceratops* locked in an apparent struggle. The interactive nature of the encounter is indicated by the fact that the carnivore's right forelimb is caught in the herbivore's jaws. Photo by T. Jerzykiewicz.

belly. Although this association has often been cited as an example of fighting dinosaurs, one report disputes that view and suggests that the *Velociraptor* was simply feeding on a dead or dying animal (Osmólska 1993). This scenario portrays the *Velociraptor* as a scavenger that died of unknown causes while feeding. A more recent investigation (Unwin et al. 1995), however, argues that the taphonomic evidence supports the original predator/prey fight interpretation. Particularly telling is the fact that the theropod's arm is firmly locked in the herbivore's jaws—a position that could not have occurred accidentally. This study suggests that the struggling dinosaurs died simultaneously in a massive sandstorm.

The two different interpretations of the event recorded by this remarkable Upper Cretaceous Mongolian assemblage differ in their characterization of *Velociraptor* as a scavenger or as an active hunter. Both explanations, however, conclude that the *Velociraptor* fully intended to dine on the *Protoceratops*.

Other predator/prey relationships are suggested by associations of theropod teeth with bones from other animals. Dinosaur teeth were continually shed as new ones grew in, so we should expect to find them in feeding areas where vigorous biting accelerated tooth loss. One such probable theropod feeding site is indicated by the discovery of several theropod teeth with a partially articulated sauropod skeleton in the Upper Jurassic of Thailand (Buffetaut and Suteethorn 1989).

Even more compelling evidence for carnivory was found in the Lower Cretaceous of Montana, where fifteen different sites were found to have

What Did Dinosaurs Eat?   373

Deinonychus teeth associated with *Tenontosaurus* bones (Maxwell and Ostrom 1995). The frequent co-occurrence of these elements and the dearth of *Deinonychus* teeth in the vicinity of bones from other possible prey animals suggest that the herbivorous *Tenontosaurus* may have been the preferred prey of *Deinonychus*. At one particularly distinctive locality (Fig. 26.2), more than thirty-five *Deinonychus* teeth and skeletal elements from four *Deinonychus* individuals were found with the partial remains of one *Tenontosaurus*. The bones were found in fine overbank deposits and could not have been transported by fluvial processes. Thus the assemblage has been interpreted as the scavenged remains of a struggle between a large *Tenontosaurus* and a pack of the much smaller *Deinonychus*. The presence of both *Deinonychus* and *Tenontosaurus* bones at the site suggests that the prey animal and members of the attacking *Deinonychus* pack were killed during the struggle and were subsequently consumed (Ostrom 1990; Maxwell and Ostrom 1995).

These skeletal associations tell us much about interactions between different dinosaurs because both predator and prey organisms have been identified. Fossil assemblages suggesting clear examples of predatory behavior are rare, however, and must be carefully scrutinized so that inadvertent associations of fossil bones are not misinterpreted.

Tooth Marks on Bone: The Result of Fighting or Feeding Behavior

If theropods dined on other dinosaurs, we might expect to find numerous bite marks on dinosaur bones. While a number of researchers have reported tooth-damaged dinosaur bone (e.g., Jacobsen 1995), the incidence of such traces appears to be considerably lower than that of marks found on bones from communities with large mammalian carnivores (Fiorillo 1991). This discrepancy may reflect differences in carcass utilization patterns (Hunt 1987; Fiorillo 1991) or taphonomic biases (Erickson and Olson 1996).

Tooth-damaged dinosaur bone can be recognized by distinctive markings such as grooves or punctures. Although some damage may have been inflicted during intraspecific dominance fights (Tanke and Currie 1995), most bite marks probably indicate carnivory. Identification of damaged bone can tell us that a particular species of dinosaur was eaten, but it generally does not indicate whether the prey was hunted and killed or opportunistically scavenged. In some cases, however, it may be possible to associate different tooth marks with specific predator activities based on the types and distribution of damage. Multiple bite marks on the ends of sauropod limb bones, for example, are more likely to represent feeding traces than assault wounds (Hunt et al. 1994b).

The identity of the animal responsible for bite marks is usually difficult to determine because many Mesozoic vertebrates (including crocodiles) were capable of causing generalized tooth damage to bone. Fortunately, well-preserved tooth marks can occasionally exhibit distinctive shapes, spacing, and/or serration marks that allow comparisons with fossil jaws of contemporaneous carnivores. For example, the spacing of the teeth in an *Allosaurus* jaw was found to match the patterns of scoring found on bones of an *Apatosaurus* (Matthew 1908). A more definitive identification was made by using dental putty to make molds of puncture marks found in bone from the Hell Creek Formation of Montana. This clever technique revealed that marks in a *Triceratops* pelvis and an *Edmontosaurus* phalanx

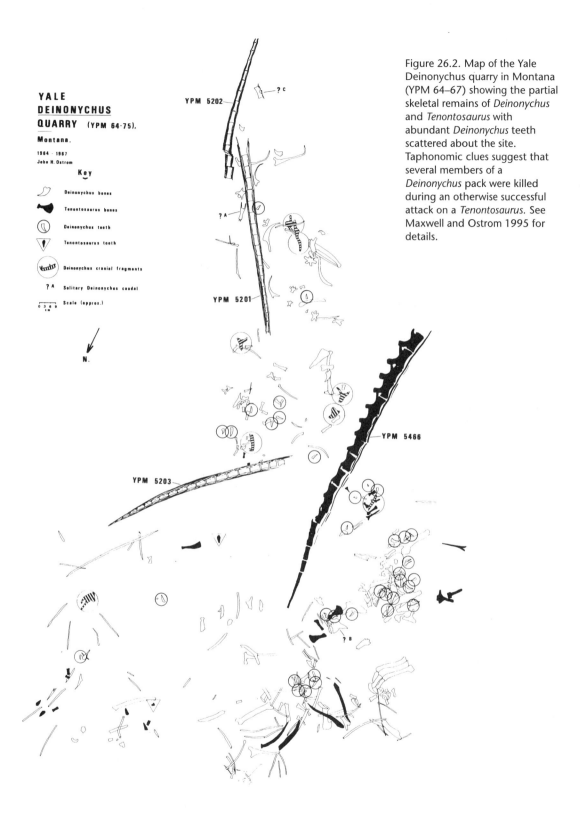

Figure 26.2. Map of the Yale Deinonychus quarry in Montana (YPM 64–67) showing the partial skeletal remains of *Deinonychus* and *Tenontosaurus* with abundant *Deinonychus* teeth scattered about the site. Taphonomic clues suggest that several members of a *Deinonychus* pack were killed during an otherwise successful attack on a *Tenontosaurus*. See Maxwell and Ostrom 1995 for details.

What Did Dinosaurs Eat?   375

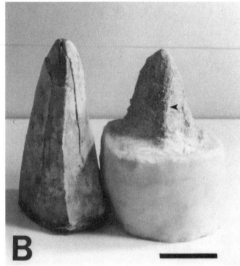

Figure 26.3. (A) Tooth puncture marks in a *Triceratops* pelvic bone (Museum of the Rockies specimen number 799) from the Late Cretaceous Hell Creek Formation of Montana. Fifty-eight distinctive tooth marks were found in the ilium and sacrum of this animal. (B) A mold of one of the deepest bite marks (*right*) was made by pressing dental putty into the puncture. The mold clearly matches a cast of a *Tyrannosaurus* tooth (*left;* cast of University of California Museum of Paleontology museum number 118742). Scale bar is 2 cm. Photos by G. Erickson.

had been inflicted by *Tyrannosaurus* teeth (Fig. 26.3; Erickson and Olson 1996).

Even more dramatic are the very rare examples of dinosaur teeth actually stuck in the bones of their prey. In Montana, a tyrannosaurid tooth was found embedded in a *Hypacrosaurus* fibula (J. R. Horner, personal communication), providing more indisputable evidence of carnivory.

Stomach Contents: Evidence of Food Already Ingested

If a dinosaur died with a full belly, it is possible that its partially digested last meal would be evident in the gut region. This would require the exceptional preservation of an articulated specimen that had been undisturbed by erosion or scavenging. One case of possible herbivore stomach contents was reported in the early 1900s (Kräusel 1922) when conifer needles, twigs, and seeds were found in the body cavity of an *Edmontosaurus*. A good taphonomic evaluation did not accompany this report, however, and another researcher (Abel 1922) suggested that the plant fragments might simply represent a mass of plant debris that had floated into the carcass. Unfortunately, reassessment of this find is not possible because the plant material was removed during preparation of the specimen (Weishampel and Horner 1990).

Reports of carnivore stomach contents are more convincing. The articulated holotype specimen of tiny *Compsognathus* from the Solnhofen lime-stone was found to have a lizard in its belly (Ostrom 1978). The partial lizard skeleton is clearly sandwiched within the ribcage of the *Compsognathus* and indicates that this dinosaur preyed on other reptiles (Fig. 26.4).

Another example of *in situ* stomach contents reveals a more surprising feeding behavior. Two articulated skeletons of the Triassic dinosaur *Coelophysis* were found to have skeletal material from other *Coelophysis* dinosaurs within their thoracic cavities. These finds raise the question of whether the enclosed bones are embryonic and indicate ovoviviparity (whereby hard-shelled eggs hatch inside the mother so the young appear to be born live). That possibility is effectively eliminated by the large size of the bones found in the gut regions. The included bones are up to two-thirds the size of the same elements from the enclosing skeletons and are therefore too large to have passed through the pelvic opening during birth. It thus appears that *Coelophysis* engaged in cannibalism—through either predation or scavenging (Colbert 1989).

Coprolites: The End Result of Feeding Activities

Reconstructions of the lives of dinosaurs often conveniently overlook the fact that dinosaurs produced fecal material—lots of it. Because feces are the unutilized waste products of digestion, they literally provide the end view of food habits. Studies on the diets of extant wild animals often rely on fecal analyses because many dietary components are still identifiable after passage through an animal's gut. It thus makes sense to look to dinosaur coprolites (fossil feces) for direct evidence of diet. Although few coprolites have been unequivocally attributed to dinosaurs, the number of reported specimens is growing.

Preservational biases help account for the paucity of known dinosaur coprolites, since the fossilization potential of fecal material is dependent on composition and the environment of deposition. Although plant-eating dinosaurs and other terrestrial herbivores undoubtedly outnumbered their meat-eating counterparts, carnivore coprolites are much more common—apparently because the high calcium phosphate content of a carnivorous diet helps facilitate mineralization (Bradley 1946). Depositional conditions are just as important; the great majority of described coprolites were probably produced by aquatic organisms that lived in environments that were subject to rapid sedimentation. In contrast, fecal matter deposited on land is less likely to be preserved because it is vulnerable to decomposition, desiccation, trampling, erosion, and coprophagy (consumption). Thus it is likely that the fortuitous preservation of dinosaur feces required rapid burial.

Recognizing possible dinosaur coprolites can be problematic, especially since many vertebrates produce similarly shaped feces. Spiral coprolites are known to have been produced by primitive fish (e.g., sharks) that had spiral intestinal valves, but most other fossil feces lack taxonomically distinct morphologies (Hunt et al. 1994a). Fortunately, large fecal volume can be a potentially distinctive characteristic, because large dinosaurs would have generated sizable fecal masses. This size factor, however, can both help and hinder the recognition of dinosaur coprolites. A sizable total fecal volume is informative because it indicates a large source animal; although small pelletoid feces can be excreted by large animals, small animals cannot

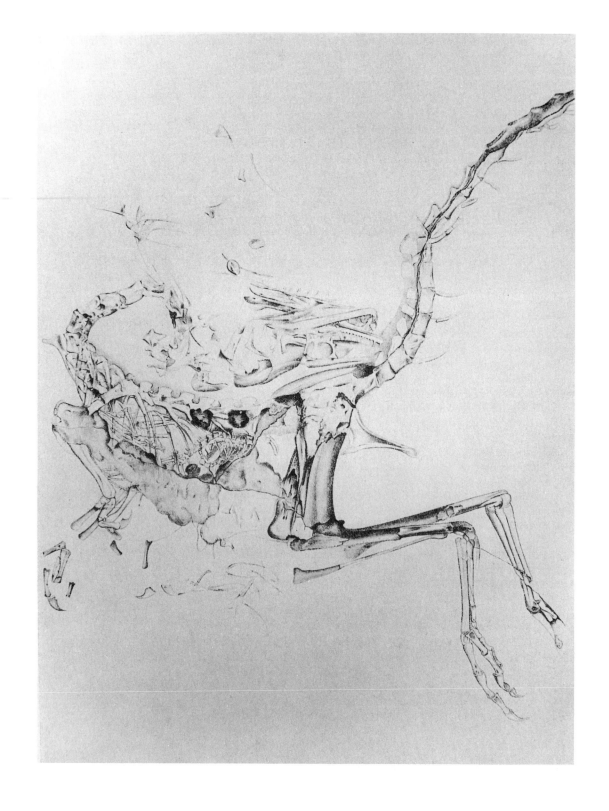

produce large quantities of dung. Yet large fecal masses are highly suscep-tible to breakage and deformation during deposition, exposure, and burial. As a result, dinosaur feces can be fragmented into numerous irregular pieces. Some of the pieces may still be larger than coprolites from other animals, but they might not bear familiar fecal shapes, and other criteria may be required to help establish fecal origins. Large size, then, may be a diagnostic feature of dinosaur coprolites, especially when used in conjunc-tion with other characteristics such as appropriate age, depositional envi-ronment, and contents.

Figure 26.4. A partial skeleton of the lizard *Bavarisaurus* can be seen in the gut region of the holotype specimen of *Compsog-nathus longipes* (Bayerische Staatssammlung für Paläon-tologie und historische Geologie specimen number AS I 563). The lizard bones are clearly enclosed by the ribs of *Compsognathus*, indicating that the dinosaur had ingested the lizard.

Because indisputable dinosaur coprolites are rare, the conspicuous absence of large quantities of dinosaur feces in Mesozoic sediments has led to claims that questionable material is dinosaur dung. There is speculation, for example, that numerous suspiciously shaped nodules are poorly pre-served dinosaur coprolites, even though these rocks contain no organic inclusions that provide positive evidence of a fecal origin. In other cases, bonafide coprolites have been ascribed to dinosaurs despite the fact that the evidence for a dinosaurian origin is weak (Thulborn 1991). Although most of the specimens that have been convincingly linked to dinosaurs do not have the familiar fecal shapes of other described coprolites, they provide valuable dietary information since they contain recognizable components that withstood both digestion and diagenesis.

One very unusual deposit of numerous flattened coprolites was found in the Jurassic sediments of Yorkshire, England (Hill 1976). The indi-vidual carbonaceous traces scarcely resemble animal droppings, but it is apparent that the collective deposit represents an assemblage of over 250 small (approximately 1 cm in diameter) pelletoid coprolites similar to the pellet groups produced by present-day deer. The coprolites contain large quantities of cycadeoid leaf cuticle and indicate that the source animal was a terrestrial herbivore. The dimensions of the flattened traces suggest that the original total fecal volume would have been approximately 130 cm^3, so this mass can be reasonably attributed to a dinosaur because other known Jurassic herbivores were too small to have generated so much fecal material.

Markedly different herbivorous dinosaur coprolites have been recov-ered from the Upper Cretaceous Two Medicine Formation of Montana (Fig. 26.5). Although these large, blocky specimens lack typical morpholo-gies, the presence of dung beetle burrows in and around the specimens helped to establish their fecal origin (Chin and Gill 1996). Most of the coprolites are primarily composed of woody conifer stem tissues, indicat-ing a highly fibrous diet. These contents may document a dietary preference for conifers, or they may reflect seasonal differences in the availability of browse. The specimens may also indicate a preservational bias that favored the fossilization of woody components. In any event, the fragmented nature of the wood indicates that the source animals were well equipped for processing tough food. This observation is consistent with the fact that these coprolites are often found in close association with fossils of the hadrosaur *Maiasaura*. This large herbivore had a battery of grinding teeth that could have managed a woody diet, and is the most likely producer of these coprolites.

Despite the transient nature of diet, direct evidence of dinosaur feeding activity has been gleaned from a surprising variety of fossil sources. Exceptional trackways, skeletal assemblages, tooth marks, stomach con-tents, and coprolites have provided bits and pieces of information that help reveal feeding traces or specific food items. Some of these finds help confirm

What Did Dinosaurs Eat? 379

Figure 26.5. (A) Fragments of conifer stems appear as dark linear inclusions in this blocky Cretaceous coprolite from Montana. Large coprolites such as this may have atypical shapes because they have been broken or deformed by trampling, coprophagy, erosion, and/or diagenesis. Ruler at base is 17 cm long. (B) Traces of dung beetle activity in sediment adjacent to a coprolite. Arrow indicates a burrow backfilled with fecal material which a beetle would have used for feeding or nesting purposes. (C) Thin section of coprolitic material showing fragments of conifer wood tissue. Millimeter scale. Museum of the Rockies specimen number 771; photos by K. Chin.

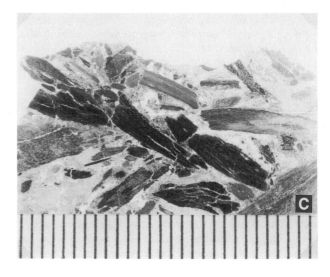

previous speculations about dinosaur herbivory or predator/prey interactions, while others bolster arguments for feeding strategies such as pack hunting or cannibalism.

Unfortunately, fossil dietary evidence is quite rare, and can be subject to multiple interpretations. Analysis of additional finds, however, will help resolve some of the ambiguities. As this information is combined with dietary predictions based on contemporaneous organisms and dinosaur morphology, a clearer picture of dinosaur food habits will emerge. This in turn will shed light on the interactions of dinosaurs with other organisms in Mesozoic ecosystems.

References

Abel, O. 1922. Diskussion zu den Vorträgen R. Kräusel and F. Versluys. *Paläontologische Zeitschrift* 4: 87.

Bishop, G. A. 1975. Traces of predation. In R. W. Frey (ed.), *The Study of Trace Fossils*, pp. 261–281. New York: Springer-Verlag.

Bradley, W. H. 1946. Coprolites from the Bridger Formation of Wyoming: Their composition and microorganisms. *American Journal of Science* 244: 215–239.

Buffetaut, E., and V. Suteethorn. 1989. A sauropod skeleton associated with theropod teeth in the Upper Jurassic of Thailand: Remarks on the taphonomic and palaeoecological significance of such associations. *Palaeogeography, Palaeoclimatology, Palaeoecology* 73: 77–83.

Chin, K., and B. D. Gill. 1996. Dinosaurs, dung beetles, and conifers: Participants in a Cretaceous food web. *Palaios* 11 (3): 280–285.

Colbert, E. H. 1989. The Triassic dinosaur *Coelophysis*. *Bulletin of the Museum of Northern Arizona* 57: 1–160.

Erickson, G. M., and K. H. Olson. 1996. Bite marks attributable to *Tyrannosaurus rex*: Preliminary description and implications. *Journal of Vertebrate Paleontology* 16: 175–178.

Farlow, J. O. 1987. *Lower Cretaceous Dinosaur Tracks, Paluxy River Valley, Texas.* Waco, Tex.: South Central Geological Society of America, Baylor University.

Fiorillo, A. R. 1991. Prey bone utilization by predatory dinosaurs. *Palaeogeography, Palaeoclimatology, Palaeoecology* 88: 157–166.

Hill, C. R. 1976. Coprolites of *Ptiliophyllum* cuticles from the Middle Jurassic of North Yorkshire. *Bulletin of the British Museum (Natural History), Geology* 27: 289–294.

Hunt, A. P. 1987. Phanerozoic trends in nonmarine taphonomy: Implications for Mesozoic vertebrate taphonomy and paleoecology. *South Central Section, Geological Society of America, Abstracts with Program* (Waco, Texas) 19: 171.

Hunt, A. P.; K. Chin; and M. G. Lockley. 1994a. The palaeobiology of vertebrate coprolites. In S. K. Donovan (ed.), *The Palaeobiology of Trace Fossils*, pp. 221–240. New York: Wiley.

Hunt, A. P.; C. S. Meyer; M. G. Lockley; and S. G. Lucas. 1994b. Archaeology, toothmarks and sauropod dinosaur taphonomy. *Gaia* 10: 225–231.

Jacobsen, A. R. 1995. Ecological interpretations based on theropod tooth marks: Feeding behaviour of carnivorous dinosaurs. *Journal of Vertebrate Paleontology* 15 (Supplement to no. 3): 37A.

Kielan-Jaworowska, Z., and R. Barsbold. 1972. Narrative of the Polish-Mongolian Palaeontological Expeditions 1967–1971. *Palaeontologia Polonica* 27: 5–13.

Kräusel, R. 1922. Die Nahrung von *Trachodon*. *Paläontologische Zeitschrift* 4: 80.

Leonardi, G. 1984. Le impronte fossili di Dinosauri. In J. F. Bonaparte, E. H. Colbert, P. J. Currie, A. de Ricqlès, Z. Kielan-Jaworowska, G. Leonardi, N. Morello, and P. Taquet (eds.), *Sulle Orme dei Dinosauri*, pp. 165–186. Venice: Erizzo Editrice.

Lockley, M. G. 1991. *Tracking Dinosaurs*. Cambridge: Cambridge University Press.

Matthew, W. D. 1908. *Allosaurus*, a carnivorous dinosaur, and its prey. *American Museum Journal* 8: 2–5.

Maxwell, W. D., and J. H. Ostrom. 1995. Taphonomic and paleobiological implications of *Tenontosaurus-Deinonychus* associations. *Journal of Vertebrate Paleontology* 15: 707–712.

Osmólska, H. 1993. Were the Mongolian "fighting dinosaurs" really fighting? *Revue de Paléobiologie,* vol. spéc. 7: 161–162.

Ostrom, J. H. 1978. The osteology of *Compsognathus longipes* Wagner. *Zitteliana* 4: 73–118.

Ostrom, J. H. 1990. Dromaeosauridae. In D. B. Weishampel, P. Dodson, and H. Osmólska (eds.), *The Dinosauria,* pp. 269–279. Berkeley: University of California Press.

Parker, L. R., and R. L. Rowley Jr. 1989. Dinosaur footprints from a coal mine in East-Central Utah. In D. D. Gillette and M. G. Lockley (eds.), *Dinosaur Tracks and Traces,* pp. 361–366. Cambridge: Cambridge University Press.

Tanke, D. H., and P. J. Currie. 1995. Intraspecific fighting behavior inferred from tooth mark trauma on skulls and teeth of large carnosaurs (Dinosauria). *Journal of Vertebrate Paleontology* 15 (Supplement to no. 3): 55A.

Thulborn, R. A. 1991. Morphology, preservation and palaeobiological significance of dinosaur coprolites. *Palaeogeography, Palaeoclimatology, Palaeoecology* 83: 341–366.

Thulborn, R. A., and M. Wade. 1979. Dinosaur stampede in the Cretaceous of Queensland. *Lethaia* 12: 275–279.

Unwin, D. M.; A. Perle; and C. Truman. 1995. *Protoceratops* and *Velociraptor* preserved in association: Evidence for predatory behaviour in dromaeosaurid dinosaurs? *Journal of Vertebrate Paleontology* 15 (supplement to no. 3): 57A–58A.

Weishampel, D. B., and J. R. Horner. 1990. Hadrosauridae. In D. B. Weishampel, P. Dodson, and H. Osmólska (eds.), *The Dinosauria,* pp. 534–561. Berkeley: University of California Press.

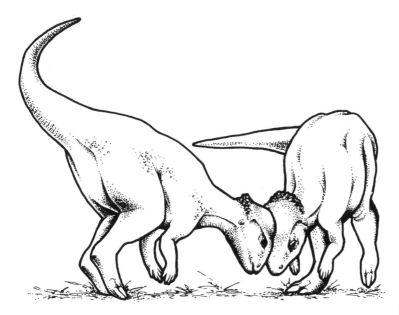

Dinosaur Combat and Courtship

Scott Sampson

During the "dinosaur renaissance" of the last three decades—which witnessed transformation of the sluggish, dim-witted beasts of old into sleeker, faster, more complex models—it has become common to ascribe social functions to many odd and often bizarre skeletal features of dinosaurs, such as crests, horns, plates, and spikes. Structures long thought to have functioned in defense against predators have been redefined as mating signals used to attract mates or compete with rivals. Thus, depictions of the age-old confrontation between *Triceratops* and *Tyrannosaurus* have been largely replaced by rutting ceratopsians, tooting hadrosaurs, and head-crunching pachycephalosaurs. Sexual dimorphism, differences between males and females of a given species, has been postulated for several kinds of dinosaurs, once again usually in connection with some social/sexual role. The aim of this chapter is twofold: to provide a brief overview of the current ideas relating to sexual dimorphism and social behaviors (particularly courtship and combat) in nonavian dinosaurs, and to assess the evidence for these contentions.

New Ideas for Old Bones

Animal communication includes behaviors for signaling species membership, for attracting mates, and for competing against rivals. In the latter category, aggression between members of the same species, or "agonistic" behavior, can take many forms, including threat displays as well as physical

27

conflict. Speculations on the behavioral functions of peculiar and distinctive morphologies have been made for numerous dinosaurs. For example, among saurischians, the heads, necks, and tails of sauropods have all been implicated as social structures (Bakker 1968, 1971, 1986; Coombs 1975). Their ample necks are thought to have been well suited for sparring in a manner akin to the fighting style of living giraffes, while the long whip-like tails of some species (e.g., *Diplodocus*) may also have served as agonistic weapons. Alternatively, some sauropods (e.g., *Apatosaurus*) may have reared up on their hind legs to use their large hand claws in combat. Sauropod necks and tails might also have performed admirably as display devices, signaling threats to members of the same sex, or courtship to members of the opposite sex. Finally, Bakker (1986) has put forth the possibility that some sauropods (e.g., *Camarasaurus, Brachiosaurus*) had an inflatable sac covering the nasal opening, which served as a display device similar to that in living elephant seals.

Among the carnivorous theropods, bony horn-like structures are present in many species, occurring variably over the eyes (*Albertosaurus*), the nose (*Ceratosaurus*), or much of the skull roof (*Dilophosaurus*). In most cases, these features are thin and fragile, suggesting that they were adaptations for display rather than combat. In at least one theropod taxon, *Spinosaurus*, the vertebral spines were extremely elongate, once again suggesting a possible display function. (A similar claim can be made for the long vertebral spines of certain ornithopods [*Ouranosaurus, Hypacrosaurus*] and sauropods [*Amargasaurus*].) Some theropods (e.g., dromaeosaurids) possessed enlarged claws on the hind foot that could have functioned like the spurs of fighting cocks, used to cause physical injury to rivals, while their reinforced tails might have acted like an additional limb during combat (Ostrom 1986; Coombs 1990).

Many well-known members of the plant-eating Ornithischia possess elaborate structures with possible social functions. For example, the plates and spikes of stegosaurs are good candidates for having been mating signals. These bony structures may have acted as agents of display, enhancing the apparent size of the animal, particularly in broadside view, and the tail spikes may have been formidable weapons to combat other stegosaurs as well as to ward off predators (Davitashvili 1961; Farlow et al. 1976).

Ankylosaurs are characterized by a variety of species-specific spikes, armor, and spines, as well as tail clubs and cranial horns in some forms. Members of one group, the ankylosaurids, possess massive tail clubs that could have been employed in mace-like fashion to combat rivals, while their broad, flattened, armored skulls could plausibly have served in head-to-head shoving matches. Members of the other major group, the nodosaurids, have narrower skulls and lack tail clubs. Many nodosaurids have enlarged spikes in the shoulder region, however, which could have been interlocked with those of an opponent in contests of dominance (Carpenter 1982; Coombs 1990).

Perhaps the most often depicted and colorful example of intraspecific combat postulated for dinosaurs is that of pachycephalosaurs. The thickened skull domes of these relatively rare animals have inspired some investigators to postulate their use as battering rams in head-butting contests against rivals (Galton 1970; Molnar 1977; Sues 1978), much as in the confrontations between males of extant bighorn sheep.

The diverse and abundant hadrosaurs, which are differentiated largely on the basis of skull shapes, are divided into two groups. The lambeosaurines, or crested duckbills (e.g., *Parasaurolophus, Lambeosaurus*), possess elaborate species-specific crests housing complex narial passages. Several authors have noted the potential for these crests to function as visual signals (Abel 1924; Davitashvili 1961; Dodson 1975; Hopson 1975), while others (Weishampel 1981) note that the enclosed narial passages could have acted as resonating chambers useful in vocal display. The second clade, known as hadrosaurines, or flat-headed duckbills (e.g., *Edmontosaurus*), lack the hollow crest but possess an enlarged depression in the narial region that stretches up to one-half the length of the skull. Hopson (1975) postulated that this remarkable "narial fossa" housed an inflatable sac or "diverticulum" that functioned in visual display, akin to that proposed later by Bakker for sauropods. Hopson (1975) speculated further that certain hadrosaurines with reinforced nasal platforms (e.g., *Kritosaurus*) may have engaged in head-to-head shoving contests.

The frills and horns of ceratopsians are likely candidates for social organs important in the competition for mates (Davitashvili 1961; Farlow and Dodson 1975; Sampson 1995a; Sampson et al. in press). The frills would have made these animals look bigger, particularly when viewed head-on. The long, paired orbital (eye) horns characteristic of one group, the chasmosaurines (e.g., *Triceratops*), were likely used to lock against those of an opponent in wrestling contests (Fig. 27.1). Among centrosaurines, with abbreviated horns above the eyes (e.g., *Centrosaurus*), the skull roof is modified into a variety of designs, ranging from large nose horns of variable orientation to thick, roughened "bosses" over the eyes and nose. This diversity in skull form suggests that centrosaurine species varied in their combat styles, using horns or bosses to lock heads in wrestling matches.

Finally, it is now often suggested, or even assumed, that many dinosaurs employed bright colors as mating signals. Striking color patterns undoubtedly would have enhanced the visual impact of the various display structures noted above (e.g., ceratopsian frills, hadrosaur crests). This new trend in thinking is best illustrated (literally) by looking at depictions of dinosaurs over the past one hundred years. The latest generation of artists has virtually transformed Jurassic and Cretaceous landscapes; the hulking, drab-colored behemoths that characterized dinosaur depictions early in the century have been replaced by more nimble, active, and riotously colored creatures.

The Evidence

Is there solid evidence to support these radical new ideas about dinosaur social behavior, or is it likely that future generations of paleontologists will look back on this period as a time of unchecked, misguided speculation? While it is beyond the bounds of this chapter to address all of the above ideas, the following discussion will address the study of dinosaur behavior generally, and consider several examples noted above. More than in any other area of dinosaur studies, contentions about dinosaurian social lives are often purely speculative, with little or no basis in direct evidence. The reason for this is simple, of course: we cannot observe directly the behavior of extinct animals. Thus, many of the above hypotheses are virtually untestable, and hence unscientific. As a rule, all inferences about dinosaur behavior should be greeted with skepticism, particularly when no evidence is cited. Unfortunately, it is often the most fanciful notions about dinosaurian lifestyles that receive the bulk of attention in the public and lay scientific press, often with little or no mention of supporting data!

Not all is lost, however. Indirect evidence of the behavior of extinct

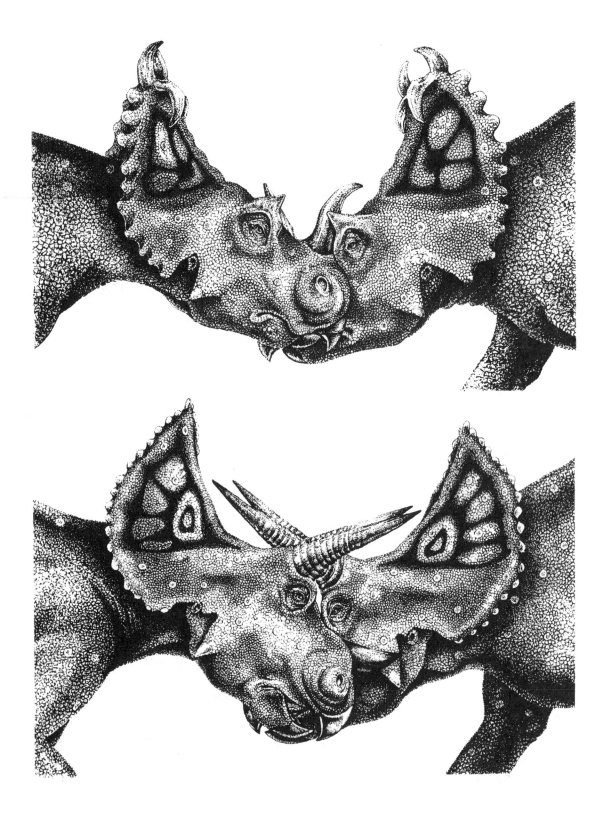

animals comes in many forms, from fossilized bones and teeth to trackways and nests. These ancient clues, combined with observations of living animals, enable paleontologists to reconstruct some aspects of behavior with reasonable confidence.

The Living Shed Light on the Dead

In many ways, the most powerful evidence in our understanding of dinosaur social behavior comes from animals living today. This is true in spite of the fact that the closest extant relatives of dinosaurs—birds and crocodiles—differ considerably in structure and size from most dinosaurs, and almost certainly in their behaviors as well. A prime example of how observations of the living help us make inferences about the extinct comes from the problem of color in dinosaurs. Unlike the vast majority of mammals, birds and crocodiles have color vision, and the mating signals of many birds are enhanced by spectacular coloration. Therefore, although we have no direct evidence of bright colors in extinct dinosaurs, they can be plausibly inferred from observations of living relatives.

Figure 27.1. Possible styles of grappling combat in *Centrosaurus* (above) and *Triceratops* (below). Drawing used by permission of the artist, Bill Parsons.

Information gleaned from the behavior of living animals is also crucial because this source provides the only direct evidence of how organisms use particular structures to survive and reproduce. Fortunately for paleontologists, convergence—the independent acquisition of similar structures and behaviors in unrelated organisms—is a common evolutionary theme, and is also the best evidence for evolutionary adaptation. Again and again we find that evolution solves similar problems in parallel fashion. For example, we have confidence that ornithomimid dinosaurs were fast runners, not only because they look so much like the swift ostriches, but because their hind legs are designed like those of most large animals adapted for running (e.g., horses and antelope): that is, long limbs with a relatively short upper portion (femur) and an elongate lower portion (tibia, ankle, and foot).

Similarly, it is compelling that among living animals, horns and horn-like organs (crests, frills, etc.) are found in numerous unrelated groups—from beetles and bovids to chameleons and cassowaries—and the primary function in virtually all instances involves sex: attracting mates and/or competing with rivals for reproductive success. Furthermore, closely related species of living animals are often differentiated on aspects of their mating signals, from the vocalizations of birds, frogs, and insects to the antlers of deer and horns of antelope. The same pattern holds true for the often bizarre dinosaurian structures reviewed above, which tend to be the primary means of identifying and distinguishing species (Dodson 1975; Horner et al. 1992; Sampson 1995b). Therefore, the current generation of paleontologists is probably quite accurate in ascribing social functions to horns and related features present in dinosaurs.

The real problem comes with testing specific hypotheses and assessing various alternatives. How are we to determine if hadrosaur crests were visual/vocal signals used to proclaim species membership, threaten rivals, combat rivals, and court mates, and/or were used in some other non-social capacity such as aiding thermoregulation? It is often remarkably difficult to determine the various functions of structures in living, breathing, behaving animals, let alone extinct forms. Indeed, in most cases, when it comes to the behavior of extinct organisms, it is simply not possible to eliminate all but one hypothesis, and one is left with the unhappy circumstance of several plausible explanations.

Dinosaur Combat and Courtship   387

Nonetheless, an important first step in discerning among alternative behavioral hypotheses is to reconstruct the anatomy of soft tissues as accurately as possible (Witmer 1995). By necessity, of course, the work of dinosaur paleontologists focuses largely on the remains of fossilized bones. Yet bones are but a single tissue type present in vertebrates, and many of the "soft tissues"—blood vessels, nerves, muscle, cartilage, skin, etc.—are often more informative with regard to anatomical function. Fortunately, soft tissues frequently leave telltale signatures on bones, from muscle scars to holes that transmit nerves and blood vessels. The problem comes with interpreting these features, and this is where it is essential to study the anatomy of living vertebrates, including both close relatives (such as birds and crocodiles) and more distantly related forms possessing features that might be analogous to those of dinosaurs. This kind of detailed comparative work has only just begun (Witmer 1995), and it is almost certain that major insights into behavior are still to come as we learn to interpret the messages preserved in dinosaur bones.

Biomechanical Considerations

Biomechanical analyses also have the potential to support or falsify functional hypotheses by demonstrating that a given structure is well designed or poorly designed for a specific use. With regard to dinosaur social behavior, biomechanical tests may be particularly useful to investigate contentions of intraspecific combat. For example, the skull and neck of *Triceratops* appear to be well designed for confrontations in which opponents lock horns and wrestle to establish dominance (Farlow and Dodson 1975; Ostrom and Wellnhoffer 1986). Possible adaptations for fighting include a pair of large horns above the eyes, a solid triangular skull with a double layer of bone over the brain, reinforced bony eye sockets, and fusion of the first three neck vertebrae. This view is further supported by the fact that living horned animals engage in similar contests (Geist 1966; Schaffer and Reed 1972).

Conversely, although a detailed analysis remains to be done, the common notion of head-butting in pachycephalosaurs seems highly improbable on biomechanical grounds. Among extant bovids that engage in violent head-to-head contests (e.g., bighorn sheep), the horns are designed to provide a broad platform of contact and thereby prevent injuries due to twisting of the neck. Thus, rounded skull domes are the opposite of what one would predict in animals that ram heads. It is much more likely that the rounded pachycephalosaur domes were used to butt the flanks of opponents (Goodwin 1995; cf. Alexander, chap. 30 of this volume). Among living animals, flank-butting occurs in numerous groups, whereas the kind of violent head-ramming envisioned for pachycephalosaurs is restricted to a handful of mammalian species.

Bonebeds and Growth Patterns

Mass death assemblages, or "bonebeds," are known for many kinds of dinosaurs (e.g., *Iguanodon, Coelophysis, Allosaurus, Maiasaura*). These sites are critical because they often preserve the remains of tens, hundreds, or even thousands of individuals of a single species, thereby providing firsthand insight into variation. Numerous bonebeds in western North America preserve the remains of short-frilled ceratopsids, or centrosaurines (e.g., *Styracosaurus, Einiosaurus*). A comparative study of several of these localities (Sampson et al. in press) revealed that each species has a characteristic pattern of hooks, horns, spikes, frills, and/or bosses. However, the presence of several growth stages further indicates that these features did not develop fully until the animals were virtually adult-sized, a pattern seen also in extant animals with elaborate secondary sexual characters (Jarman 1983). A direct corollary of this finding is that some taxa based solely on immature specimens (e.g., *Brachyceratops montanensis*) are invalid because we cannot say with certainty what the adults would have looked like. In other words, if these immature individuals had survived to full adulthood, they might have developed the horn and frill characteristics that we associate with some other species.

If horns and frills functioned primarily to fend off predators, one would expect them to develop early in life so they could be put to use as soon as possible. The fact that they occur at a later growth stage is more consistent with the hypothesis that they were employed primarily as mating signals. Mating signals among living animals frequently show delayed expression, developing fully late in maturity and often in association with the establishment of a dominance hierarchy, or "pecking order." Such a growth pattern allows individuals living in large groups to determine rank without risking life and limb in physical combat (Geist 1978). Based on this evidence, as well as the multiple occurrence of vast bonebeds indicative of sociality, it is certainly plausible, and perhaps likely, that some species of ceratopsids (and, on the basis of similar evidence, hadrosaurs) lived in complex, hierarchically organized herds (Sampson 1995a).

Sexual Dimorphism

Variation between the sexes of a species, or sexual dimorphism, occurs commonly among living animals, often in association with the competition for mates. Males of many species are often larger or more elaborately colored than their female counterparts, and males often possess social structures such as horns or crests that are reduced or absent in females. However, dimorphism between males and females may be due to factors unrelated to sex or mating signals. For example, females of a given species may be larger than males because of egg-laying requirements.

If extant vertebrates are any guide, it is probable that some dinosaur species were sexually dimorphic, but demonstrating this conclusively is—not surprisingly—problematic. Sexual dimorphism in skull shape has been postulated for several kinds of dinosaurs, including lambeosaurine hadrosaurs on the basis of crest variations (Dodson 1975), and ceratopsids based on variation in horns and frills (Dodson 1990; Lehman 1990; Sampson et al. in press). However, even with the increased sample sizes provided by bonebeds, the total number of skulls available for any one species (fewer than twenty) remains too small for statistical testing, and any conclusions about dimorphism must therefore be regarded cautiously. Without more specimens in hand, it is always possible that what we perceive to be sexual dimorphism is in reality variation due to one or more unrelated factors: age, geography, individual differences, or taxonomic differences (i.e., more than one species). The one exception to date is *Protoceratops andrewsi,* a small Asian ceratopsian known from literally dozens of complete skeletons. Morphometric analyses of skulls of *P. andrewsi* indicate that they are indeed sexually dimorphic, especially in the frill (Kurzanov 1972; Dodson 1976).

Returning to possible extant analogues, sexual dimorphism in both size and weaponry among large-bodied terrestrial mammals tends to be least in small-bodied forms (body mass less than 20 kg), greatest in medium-sized forms, and reduced in the largest-bodied species (body mass greater than 300 kg), particularly those inhabiting open environments (Jarman 1983). Females tend to mimic males in weaponry in the largest species, although sexual differences in horns are common. The reasons behind this pattern may relate to several factors, including predator defense and increased competition associated with life in social groups. On the basis of this modern analogy, we can predict minimal sexual dimorphism in body size and weaponry among ornithischian herbivores, many of which greatly exceeded 300 kilograms. Current evidence, although somewhat lacking, suggests that this prediction may hold true at least among ceratopsid dinosaurs, for which the only indication of sexual dimorphism occurs in details of the horns and frills (Dodson 1990; Lehman 1990; Sampson et al. in press).

Whereas differences between ornithischian males and females are most often associated with skulls, among theropods they are more commonly attributed to differences in body size and shape. Carpenter (1990) made a claim for sexual dimorphism in *Tyrannosaurus rex* on the basis of two skeletal types, one slender and the other more robust. With about fifteen specimens to compare, however, the problem of sample size applies here as well. Colbert (1990) and Raath (1990) posit the existence of sexual differences in the skeletal characters of the two small theropods *Coelophysis* and *Syntarsus*, respectively, both preserved in large mass death assemblages. Raath's study of *Syntarsus* is particularly noteworthy because he conducted a quantitative analysis of more than thirty individuals; the results support existence of two body types, or "morphs," one sturdier and more robust than the other.

Interestingly, both Raath (1990) and Carpenter (1990) independently suggest that the larger or more robust morph in each case represents the female. Raath bases this conclusion on sex differences among living raptorial birds, while Carpenter notes that the angle between the rear of the pelvis and the tail vertebrae is greater in the more robust form of *Tyrannosaurus*, perhaps as a result of constraints involved with egg-laying. Finally, Larson (1994) supports this viewpoint as well, claiming that the more robust specimens of *Tyrannosaurus* lack the small V-shaped chevron bone on the bottom of the first tail vertebra, once again supposedly because of egg-laying constraints. If Larson's technique of sexing individuals turns out to be of general utility among dinosaurs, it will enable paleontologists to determine aspects of sexual dimorphism with much greater confidence.

Mating Signals and Dinosaur Evolution

Given the assumption that many of the dinosaurian structures described above were used to signal mates and compete with rivals, they may have far greater evolutionary significance than previously supposed. It was noted previously that species within groups of living animals tend to be differentiated on the basis of mate signals, and that this pattern may well have applied to the elaborate features found in many groups of dinosaurs—including ceratopsians, hadrosaurs, pachycephalosaurs, ankylosaurs, and stegosaurs.

Recent work in theoretical biology strongly supports the notion that evolutionary processes, including sexual selection and selection for species recognition, may incidentally result in the origin of new species through the creation of novel mating signals (West-Eberhard 1983; Paterson 1985). If two populations of a given species become isolated and evolve distinct signals for mating, the animals of each may no longer recognize each other as potential mates, and two new species will have formed. Such processes may exert a powerful influence not only on the origin of new forms but also on the success of certain groups over others. Thus, the remarkable diversity of ceratopsids and hadrosaurs, for example, ultimately may be due to their mating behaviors. Ongoing investigations into both fossil and living organisms should provide further insight into this fascinating proposition.

Two humbling points should be kept in mind when considering the social behavior of dinosaurs. First, mating signals in living animals exploit the entire spectrum of sensory channels—visual, acoustic, tactile, and chemical—whereas fossils generally provide direct evidence for only a minute portion of one of these categories: visual signals preserved in skeletal tissues. (There are occasional exceptions; a recently discovered and exquisitely preserved skeleton of the hadrosaur *Maiasaura* includes the impression of what may be a dewlap, or large flap of skin below the neck [H.-D. Sues, personal communication].) Second, dinosaurs lived at a time very different from today, and undoubtedly engaged in behaviors completely unique to them, with no modern analogue. Thus, paleontologists have certain well-defined limits on the amount and quality of evidence potentially available, and, barring the invention of a tractable time machine, the study of dinosaurian social lives is destined to remain a highly speculative endeavor.

Given these constraints, some evolutionary biologists seem to think that the study of dinosaur behavior is approximately equivalent to the study of life on other planets (exobiology), with approximately equal amounts of available evidence. A number of paleontologists, however, would strongly disagree. Although many behavioral hypotheses are untestable and unscientific in the sense that we cannot witness that behavior or rule out all alternatives, it is nonetheless possible to make inroads into the understanding of how dinosaurs lived. Every species, living or extinct, represents an evolutionary experiment, and the search for recurring patterns in the history of life provides a means of testing many kinds of evolutionary hypotheses. Through comparisons with living animals, biomechanical analyses, studies of variation in mass death assemblages, and the application of new technologies, we have the capability to address ancient behavior, including certain aspects of social and mating behaviors. Of course, in order to provide appropriate rigor, paleontologists must actively address alternatives, and distinguish clearly between contentions supported by evidence and bald speculation. Finally, if we simply ignored interesting but problematic issues such as behavior and evolutionary processes, dinosaur research would be reduced to describing and categorizing newly discovered forms—a kind of glorified fossil stamp collecting—and a number of paleontologists undoubtedly would be pursuing alternative occupations.

References

Abel, O. 1924. *Die Stämme der Wirbeltiere*. Berlin: De Gruyter.
Bakker, R. T. 1968. The superiority of dinosaurs. *Discovery* 3: 11–12.
Bakker, R. T. 1971. The ecology of brontosaurs. *Nature* 229: 172–174.

Bakker, R. T. 1986. *The Dinosaur Heresies: New Theories Unlocking the Mystery of the Dinosaurs and Their Extinction.* New York: William Morrow.

Carpenter, K. 1982. Skeletal and dermal armor reconstruction of *Euoplocephalus tutus* (Ornithischia: Ankylosauridae) from the Late Cretaceous Oldman Formation of Alberta. *Canadian Journal of Earth Sciences* 19: 689–697.

Carpenter, K. 1990. Variation in *Tyrannosaurus rex*. In K. Carpenter and P. J. Currie (eds.), *Dinosaur Systematics: Approaches and Perspectives,* pp. 141–145. Cambridge: Cambridge University Press.

Colbert, E. H. 1990. Variation in *Coelophysis bauri*. In K. Carpenter and P. J. Currie (eds.), *Dinosaur Systematics: Approaches and Perspectives,* pp. 81–90. Cambridge: Cambridge University Press.

Coombs, W. P. Jr. 1975. Sauropod habits and habitats. *Palaeogeography, Palaeoclimatology, Palaeoecology* 17: 1–33.

Coombs, W. P. Jr. 1990. Behavior patterns of dinosaurs. In D. B. Weishampel, P. Dodson, and H. Osmólska (eds.), *The Dinosauria,* pp. 32–42. Berkeley: University of California Press.

Davitashvili, L. Sh. 1961. *Teoriia polovogo otbora (The theory of sexual selection).* Moscow: Izdatel'stvo Akademii Nauk (Academy of Sciences Press).

Dodson, P. 1975. Taxonomic implications of relative growth in lambeosaurine hadrosaurs. *Systematic Zoology* 24: 37–54.

Dodson, P. 1976. Quantitative aspects of relative growth and sexual dimorphism in *Protoceratops*. *Journal of Paleontology* 50: 929–940.

Dodson, P. 1990. On the status of the ceratopsids *Monoclonius* and *Centrosaurus*. In K. Carpenter and P. J. Currie (eds.), *Dinosaur Systematics: Approaches and Perspectives,* pp. 211–229. Cambridge: Cambridge University Press.

Farlow, J. O., and P. Dodson. 1975. The behavioral significance of frill and horn morphology in ceratopsian dinosaurs. *Evolution* 29: 353–361.

Farlow, J. O.; C. V. Thompson; and D. E. Rosner. 1976. Plates of *Stegosaurus*: Forced convection or heat loss fins? *Science* 192: 1123–1125.

Galton, P. M. 1970. Pachycephalosaurids: Dinosaurian battering rams. *Discovery* 6: 23–32.

Geist, V. 1966. The evolution of horn-like organs. *Behaviour* 27: 173–214.

Geist, V. 1978. On weapons, combat and ecology. In L. Krames, P. Pliner, and T. Alloway (eds.), *Aggression, Dominance and Individual Spacing,* pp. 1–30. New York: Plenum Press.

Goodwin, M. B. 1995. A new skull of the pachycephalosaur *Stygimoloch* casts doubt on head butting behavior. *Journal of Vertebrate Paleontology* 15 (Supplement to no. 3): 32A.

Hopson, J. A. 1975. The evolution of cranial display structures in hadrosaurian dinosaurs. *Paleobiology* 1: 21–43.

Horner, J. R.; D. J. Varricchio; and M. B. Goodwin. 1992. Marine transgressions and the evolution of Cretaceous dinosaurs. *Nature* 358: 59–61.

Jarman, P. 1983. Mating system and sexual dimorphism in large, terrestrial, mammalian herbivores. *Biological Review* 58: 485–520.

Kurzanov, S. M. 1972. Sexual dimorphism in protoceratopsians. *Palaeontological Journal* 1972: 91–97.

Larson, P. 1994. *Tyrannosaurus* sex. In G. D. Rosenberg and D. L. Wolberg (eds.), *Dino Fest,* pp. 139–155. Paleontological Society Special Publication no. 7. Knoxville: University of Tennessee.

Lehman, T. M. 1990. The ceratopsian subfamily Chasmosaurinae: Sexual dimorphism and systematics. In K. Carpenter and P. J. Currie (eds.), *Dinosaur Systematics: Approaches and Perspectives,* pp. 211–229. Cambridge: Cambridge University Press.

Molnar, R. E. 1977. Analogies in the evolution of combat and display structures in ornithopods and ungulates. *Evolutionary Theory* 3: 165–190.

Ostrom, J. H. 1986. Social and unsocial behavior in dinosaurs. In M. H. Nitecki and J. A. Kitchell (eds.), *Evolution of Animal Behavior,* pp. 41–61. Oxford: Oxford University Press.

Ostrom, J. H., and P. Wellnhoffer. 1986. The Munich specimen of *Triceratops*, with a revision of the genus. *Zitteliana* 14: 111–158.

Paterson, H. E. H. 1985. The recognition concept of species. In E. S. Vrba (ed.), *Species and Speciation*, pp. 21–29. Transvaal Museum Monograph 4. Pretoria, South Africa.

Raath, M. A. 1990. Morphological variation in small theropods and its meaning in systematics: Evidence from *Syntarsus rhodensis*. In K. Carpenter and P. J. Currie (eds.), *Dinosaur Systematics: Approaches and Perspectives*, pp. 91–105. Cambridge: Cambridge University Press.

Sampson, S. D. 1995a. Horns, herds, and hierarchies. *Natural History* 194 (6): 36–40.

Sampson, S. D. 1995b. Two new horned dinosaurs from the Upper Cretaceous Two Medicine Formation of Montana; with a phylogenetic analysis of the Centrosaurinae (Ornithischia: Ceratopsidae). *Journal of Vertebrate Paleontology* 15: 743–760.

Sampson, S. D.; M. J. Ryan; and D. H. Tanke. In press. The ontogeny of centrosaurine dinosaurs (Ornithischia: Ceratopsidae), with new information from mass death assemblages. To be published in *Zoological Journal of the Linnean Society*.

Schaffer, W., and C. A. Reed. 1972. The co-evolution of social behavior and cranial morphology in sheep and goats (Bovidae, Caprini). *Fieldiana Zoology* 61: 1–88.

Sues, H.-D. 1978. Functional morphology of the dome in pachycephalosaurid dinosaurs. *Neues Jahrbuch für Geologie und Paläontologie Monathefte* 1978: 459–472.

Weishampel, D. B. 1981. Acoustic analyses of potential vocalizations in lambeosaurine dinosaurs (Reptilia: Ornithischia): Comparative anatomy and homologies. *Journal of Paleontology* 55: 1046–1057.

West-Eberhard, M. J. 1983. Sexual selection, social competition, and speciation. *Quarterly Review of Biology* 58: 155–183.

Witmer, L. M. 1995. The extant phylogenetic bracket and the importance of reconstructing soft tissues in fossils. In J. J. Thomason (ed.), *Functional Morphology in Vertebrate Paleontology*, pp. 19–33. Cambridge: Cambridge University Press.

Dinosaur Eggs

*Karl F. Hirsch
and
Darla K. Zelenitsky*

The two most frequently asked questions about dinosaur eggs are, How do you know the specimen is or is not an egg? and What kind of animal laid this egg? To understand this subject better, we must consider what an egg is, and how it functions (Figs. 28.1A, B, 28.2).

What Is an Egg?

The egg is the house of an embryo. This house must provide everything the embryo needs to develop: shelter, protection, food, water, fresh air (oxygen), expulsion of bad air (carbon dioxide), a constant temperature, calcium for growing bones, and a place to store waste (Ar et al. 1979; Hirsch 1994). The eggshell is the wall of this house. This wall is composed of calcareous and/or organic matter, and possesses a variety of functional and morphological properties. The eggshell must be strong enough to support the weight of brooding parents or the burden of overlying nesting material, and at the same time weak enough to allow the embryo to hatch. The eggshell also supplies the additional calcium required by the developing embryo. It must permit the diffusion of gases and water vapor, and in some cases allow the absorption of liquid water (Rahn et al. 1979). It also must protect the embryo by keeping out bacteria and parasites. The egg and its shell exist in a variety of shapes, shell structures, and pore systems in order to cope with different environmental conditions. The eggshell may

28

be camouflaged by pigments that can also help to regulate the egg's temperature by absorbing or reflecting radiation. The pointed pear shape of a gull egg prevents it from rolling off cliffs. All the above factors are in delicate equilibrium (Taylor 1970; Ar et al. 1979).

Eggs are categorized by the physical properties of their eggshell into soft-, flexible-, and hard-shelled types. Hard-shelled eggs have the best chance of fossilization, because their eggshell is dominated by calcareous matter and is composed of tightly abutting, interlocking shell units (Fig. 28.2). In the shells of the other two egg types, the organic matter dominates, and the poorly organized calcareous matter is barely recognizable as an eggshell component after the organic matter decays. To date, only hard-shelled dinosaur eggs have been identified (Fig. 28.1B). While it is possible that some dinosaurs laid soft- or flexible-shelled eggs, no such specimens have been recovered (Hirsch 1994).

A Digression: What Isn't an Egg?

Mother Nature fabricates many egg-like objects, from pelloidal specimens and concretions produced within the sediments, to sophisticated egg-shaped calculi formed in the stomach of a ruminant mammal (Fig. 28.1C; Hirsch 1986). In order to prove that a specimen is an egg, there must be evidence of eggshell structure.

The Study of Fossil Eggshell

The study of fossil eggshell is based on the knowledge gained from the study of modern eggshell. Because the study of eggshell is a relatively young discipline (cf. Carpenter et al. 1994), there is no well-established terminology that has been universally accepted. Current terminology is based on the well-studied avian eggshell (Fig. 28.2).

The hard or rigid eggshell is composed of an inner, organic shell membrane and an outer calcareous layer of abutting and interlocking shell units (Figs. 28.2A, 28.4A). The shell membrane and the shell units are attached to one another by a basal cap that grows from a nucleation center into the shell membrane (Fig. 28.2A).

The description of eggs and eggshells is based on the external morphology (appearance) and eggshell structure of the specimen. These features are examined at three structural levels:

1. The macrostructure includes the egg size and shape, the shell thickness, the pore pattern, and the outer surface sculpture.

2. The microstructure deals with the organization of the eggshell and includes the morphology and arrangement of the shell units and pore canals.

3. The ultrastructure represents the texture or extremely fine details of the eggshell—the organization and composition of the crystalline zones (tabular or squamatic) and their interwoven organic network.

A number of techniques, adopted and modified from other disciplines, have proven invaluable for the identification and classification of fossil eggs. Eggshell structure is studied using polarizing light microscopy (PLM) and scanning electron microscopy (SEM). Both of these techniques have their advantages, and the two tend to complement one another. The microstructure and ultrastructure of the eggshell are observed in radial

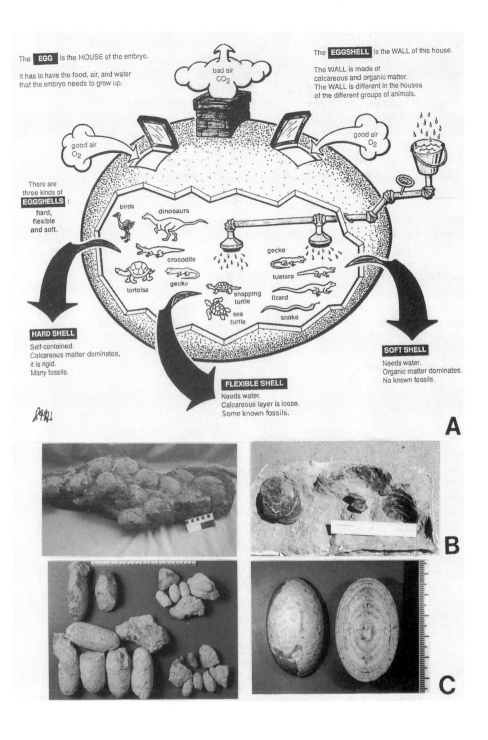

The **EGG** is the HOUSE of the embryo.

It has to have the food, air, and water that the embryo needs to grow up.

The **EGGSHELL** is the WALL of this house.

The WALL is made of calcareous and organic matter. The WALL is different in the houses of the different groups of animals.

bad air CO₂

good air O₂

good air O₂

There are three kinds of **EGGSHELLS**: hard, flexible and soft.

birds
dinosaurs
crocodile
gecko
tortoise
gecko
tuatara
snapping turtle
lizard
sea turtle
snake

HARD SHELL
Self-contained.
Calcareous matter dominates, it is rigid.
Many fossils.

FLEXIBLE SHELL
Needs water.
Calcareous layer is loose.
Some known fossils.

SOFT SHELL
Needs water.
Organic matter dominates.
No known fossils.

A

B

C

sections (Figs. 28.2B, 28.4A). Tangential views using SEM provide structural details of the outer and inner eggshell surfaces, and tangential sections using PLM reveal the shell's internal structure (Fig. 28.2B).

Additional techniques have been used to understand changes that occurred in the egg after burial. Mineralogical and elemental analysis and cathodoluminescence help us to recognize these changes. Egg size and shape can be estimated by measuring the radii of large shell fragments with the Geneva Lens Measure and engineering radius gauges. Computer-assisted tomography (CAT) scans can be used to detect embryonic bones inside an egg or to assist in the reconstruction of a compressed egg.

Classification of Eggshell

Rigid- or hard-shelled eggs of modern animals show a correlation between the basic structure of the eggshell and the egg-layer (Figs. 28.3,

Figure 28.1. (A) The Egg House and its inhabitants. (B) Dinosaur eggs. At left, a nest of spheroidal eggs from Henan Province, China; at right, a group of three ovoidal eggs, each of different preservation, from the North Horn Formation of Utah, U.S. (C) Pseudo-eggs. At left, pelloidal shaped objects formed within the sediment from the Eocene of France; at right, a stomach stone with a fragment of chalcedony in the center, from Utah, U.S.

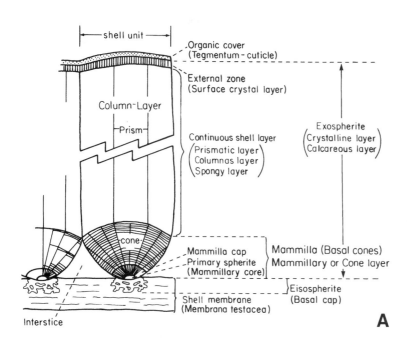

Figure 28.2. (A) A blueprint of avian eggshell structure and its nomenclature. (B) Egg geometry showing curvature components and different cuts and views.

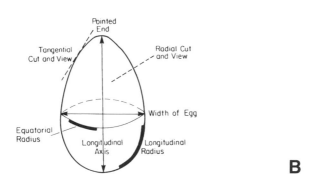

Dinosaur Eggs   397

BASIC TYPES OF EGGSHELL ORGANIZATION	STRUCTURAL MORPHOTYPES	PORE SYSTEM	PARATAXONOMIC FAMILIES	TAXONOMIC GROUPS
FOSSIL REMAINS OF EGGS OF MODERN AMNIOTES				
Testudoid	Spherurigidis Spheruflexibilis		Testudoolithidae Testudoflexoolithidae	CHELONIA
Geckonoid	Geckonoid	Rete-canaliculate	Gekkoolithidae	GEKKOTA
Crocodiloid	Crocodiloid		Krokolithidae	CROCODYLIA
Ornithoid	Prismatic ("neognathe")	Angusti-canaliculate		Gobipipus (embryos)
	Ratite	Angusti-canaliculate	Laevisoolithidae	?Enantiornithids
		Angusti-canaliculate	Medioolithidae	?
		Angusti-canaliculate		Struthionidae
		Angusti-canaliculate	Ornitholithidae	?Diatrymatidae
DINOSAUR EGGS				
Ornithoid	Ratite	Angusti-canaliculate	Elongatoolithidae	Theropoda (Oviraptor embryo)
Dinosauroid-spherulitic	Filispherulitic (Multispherulitic)	Multi-canaliculate	Faveoloolithidae	?Sauropoda
	Dendrospherulitic	Prolato-canaliculate	Dendroolithidae	?Sauropoda ?Ornithopoda
	?Dictospherulitic	Prolato-canaliculate	Dictyoolithidae	?Sauropoda
	Discretispherulitic (Tubospherulitic)	Tubo-canaliculate	Megaloolothidae	?Sauropoda ?Ornithischia
	Prolatospherulitic	Prolato-canaliculate	Spheroolithidae	Ornithopoda (some hadrosaur embryos)
	Angustispherulitic	Rimo-and angusti-canaliculate	Ovaloolithidae	?Ornithopoda
Dinosauroid-prismatic	Prismatic (Angustiprismatic)	Angusti-canaliculate	Prismatoolithidae	Theropoda (Troodon embryo) Protoceratopsids
	Prismatic (Obliquiprismatic)	Obliqui-canaliculate		?Ornithopoda

398   *Hirsch and Zelenitsky*

28.4). For example, turtles lay eggs with a common basic shell structure that differs from the basic eggshell structure of other groups of egg-layers. Therefore, we are able to identify different groups of egg-layers by their eggshell structure. The eggshells of modern turtles, geckos, crocodiles, and birds each exhibit a unique structure referred to as a "basic type of eggshell organization." All modern hard-shelled eggs are of one of the following four basic types of eggshell: testudoid (turtle), geckonoid (gecko), crocodiloid (crocodile), and ornithoid (bird). Through the study of fossil eggshells, these "basic types" can be traced back to Cretaceous and Jurassic times. The eggshell structure is stable within these larger groups of modern animals and therefore facilitates the classification of eggshell within certain extinct taxa.

Dinosaur eggs fit only partially into these modern basic types. One group of dinosaur eggs is of the ornithoid basic type, thus like that of birds.

Figure 28.3. Correlation chart of basic types of eggshell, their morphotypes and pore systems, and corresponding parataxonomic families and taxonomic groups. Question marks preceding taxonomic groups indicate correlations based on circumstantial evidence.

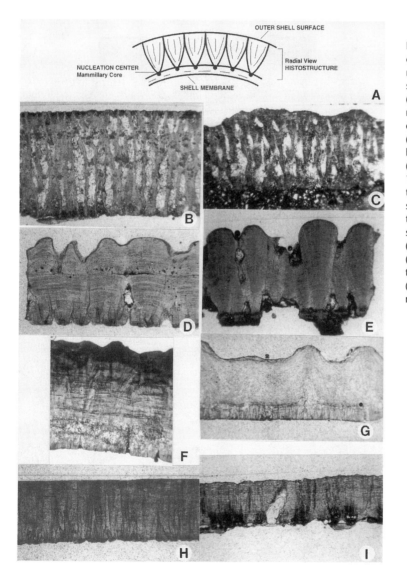

Figure 28.4. (A) Stylistic sketch of the important radial view needed for identification of the structural morphotypes. (B–I) The eight well-established morphotypes of dinosaurian eggshell in radial thin section. (B) Filispherulitic (multispherulitic); thickness 2.5 mm. (C) Dendrospherulitic; thickness 1.5 mm. (D) Prolatospherulitic; thickness 1.2 mm. (E) Discretispherulitic (tubospherulitic); thickness 1.6 mm. (F) Angustispherulitic; thickness 2.0 mm. (G) Ratite; thickness 1.4 mm. (H) Prismatic (angustiprismatic); thickness 0.9 mm. (I) Prismatic (obliquiprismatic); thickness 0.7 mm.

Dinosaur Eggs 399

For the others, two additional "basic types" have been established: dinosauroid-spherulitic and dinosauroid-prismatic (see Fig. 28.3; Hirsch and Packard 1987; Hirsch and Quinn 1990; Mikhailov 1991, 1992, 1997; Mikhailov et al. 1996).

The "basic types" are subdivided into structural morphotypes that express the general characteristics of the eggshell microstructure. Nine morphtoypes have been established for dinosaur eggshell (Figs. 28.3, 28.4B–I).

An egg is an object produced by an animal; thus it is a trace fossil and should be classified as such, even if it can be positively linked to an animal taxon by an embryo. The taxonomy or classification of the egg-layer is distinct from the parataxonomy (a separate, distinct taxonomy) of the eggs it laid. If these two classifications were combined, problems might occur if two different species of embryos were found in separate eggs having similar eggshell structure. The parataxonomy of the eggs is distinguished from the taxonomy of the egg-layer by using the prefix *oo-* in the nomenclature of the former. Therefore, the hierarchical classification of egg parataxonomy includes oofamilies, oogenera, and oospecies.

Until recently, the classification of dinosaur eggs was confusing because some eggs were classified as type 1, 2, 3 . . . , or type A, B, C . . . , and still others were given names. Fossil eggs from England were the first to be assigned to an oogenus and oospecies (Buckman 1860). This nomenclature for eggs was subsequently developed and expanded by the Chinese (Zhao 1979, 1993). The structural classification that includes the basic types and morphotypes, in conjunction with the system of naming eggs (oofamily, oogenus, oospecies), has produced a workable parataxonomic system for fossil egg classification (Fig. 28.3).

Figure 28.5. An amazing find: a skeleton of the small theropod *Oviraptor* preserved sitting on a clutch of its own eggs. This specimen was described by Norell et al. (1995); Dong and Currie (1996) described a similar specimen. The dinosaur is preserved sitting on its folded hind legs. The left and right hind feet project forward. The left and right forelimbs curl around the front edge of the block, surrounding the oval-shaped eggs, which are particularly easy to see adjacent to the right forelimb. The dinosaur probably was killed by a sandstorm that buried it while it was guarding or brooding its nest.

Hirsch and Zelenitsky

Identifying the Egg-Layer

An egg is a trace fossil that can be positively linked to an animal taxon if proof exists within the egg. The only definitive proof is the presence of an identifiable embryo. The next best evidence for correlating an egg with an animal taxon is eggs and/or eggshells found in a nest with hatchlings or embryos—or the remains of a parent. The presence of eggs and/or eggshells within the same stratigraphic horizon as skeletal remains is only circumstantial evidence and thus remains questionable.

Numerous cases exist where eggs and eggshells have been assigned to specific dinosaurs, but most of these assignments are based on circumstantial evidence. To date, there have been only three occurrences of identifiable embryos within eggs and one of hatchlings and eggshells in a nest documented in the literature:

- the theropod *Troodon formosus,* Egg Mountain, Montana, U.S. (these eggs were previously attributed to the hypsilophodontid *Orodromeus makelai;* see Varricchio et al. 1997 for details);
- the hadrosaurid *Hypacrosaurus stebingeri,* Devil's Coulee, Alberta, Canada;
- an unidentified oviraptorid, Ukhaa Tolgod, Gobi Desert, Mongolia; and
- the hadrosaurid *Maiasaura peeblesorum* (hatchlings), near Egg Mountain, Montana, U.S.

For these specimens, we are able to relate the oospecies to the egg-layer. If two eggs with similar form and structure are found, it does not necessarily mean that they were laid by the same species, especially if they are from different regions or stratigraphic horizons. If one egg contains an identifiable embryo and the other lacks it, we can be certain only of the taxonomic position of the egg with the embryo. If only eggshell fragments, similar to the egg with the embryo, are found in different regions or horizons, the correlation to the egg-layer is even more tenuous. Because eggs containing embryos are rare, we have limited information on the range of eggshell structure within certain dinosaur taxa.

Correlations between several eggshell morphotypes and the animals thought responsible for them are uncertain, and therefore are preceded by a question mark in the right-hand column of the correlation chart (i.e., ?Ornithopoda, ?Sauropoda, etc.; Fig. 28.3). These assignments are temporary and based on circumstantial evidence. Dinosaurs with these particular assignments were found in proximity to the egg localities. The presence of question marks is justified, as illustrated by the long-term mix-up between oviraptorid and protoceratopsid eggs (Norell et al. 1994; Mikhailov 1995). It was assumed that *Oviraptor* was stealing the eggs of *Protoceratops,* when in fact it was protecting or brooding its own eggs (Fig. 28.5). Such a scenario reveals that one should not be overly confident in suggesting the egg-layer if the assignment is based on circumstantial evidence.

References

Ar, A.; H. Rahn; and C. V. Paganelli. 1979. The avian egg, mass and strength. *Condor* 81: 331–337.

Buckman, J. 1860. Fossil reptilian eggs from the Great Oolite of Cirencester. *Quarterly Journal of the Geological Society of London* 16: 107–110.

Carpenter, K.; K. F. Hirsch; and J. R. Horner (eds.). 1994. *Dinosaur Eggs and Babies*. Cambridge: Cambridge University Press.

Dong, Z.-M., and P. J. Currie. 1996. On the discovery of an oviraptorid skeleton on a nest of eggs at Bayan Mandahu, Inner Mongolia, People's Republic of China. *Canadian Journal of Earth Sciences* 33: 631–636.

Hirsch, K. F. 1986. Not every "egg" is an egg. *Journal of Vertebrate Paleontology* 6: 200–201.

Hirsch, K. F. 1994. The fossil record of vertebrate eggs. In S. Donovan (ed.), *The Paleobiology of Trace Fossils*, pp. 269–294. London: John Wiley and Sons.

Hirsch, K. F., and M. J. Packard. 1987. Review of fossil eggs and their shell structure. *Scanning Microscopy* 1: 383–400.

Hirsch, K. F., and B. Quinn. 1990. Eggs and eggshell fragments from the Upper Cretaceous Two Medicine Formation of Montana. *Journal of Vertebrate Paleontology* 10: 491–511.

Mikhailov, K. E. 1991. Classification of fossil eggshells of amniote vertebrates. *Acta Palaeontologica Polonica* 36: 193–238.

Mikhailov, K. E. 1992. The microstructure of avian and dinosaurian eggshell: phylogenetic implications. In K. E. Campbell, Jr. (ed.), *Papers in Avian Paleontology*, pp. 361–373. Los Angeles: Natural History Museum of Los Angeles County.

Mikhailov, K. E. 1995. Theropod and protoceratopsian dinosaur eggs from the Cretaceous of Mongolia and Kazakhstan. *Paleontological Journal* 28 (2): 101–120.

Mikhailov, K. E. 1997. *Fossil and Recent Eggshell in Amniotic Vertebrates: Fine Structure, Comparative Morphology and Classification*. Special Papers in Palaeontology no. 56. London: Palaeontological Association.

Mikhailov, K. E.; E. S. Bray; and K. F. Hirsch. 1996. Parataxonomy of fossil egg remains (Veterovata): Principles and applications. *Journal of Vertebrate Paleontology* 16: 763–769.

Norell, M. A.; J. M. Clark; L. M. Chiappe; and Dashzeveg D. 1995. A nesting dinosaur. *Nature* 378: 774–776.

Norell, M. A.; J. M. Clark; Dashzeveg D.; Barsbold R.; L. M. Chiappe; A. R. Davidson; M. C. McKenna; Perle A.; and M. J. Novacek. 1994. A theropod dinosaur embryo and the affinities of the Flaming Cliffs dinosaur eggs. *Science* 266: 779–782.

Rahn, H.; A. Ar; and C. V. Paganelli. 1979. How bird eggs breathe. *Scientific American* 2: 46–56.

Taylor, T. G. 1970. How an eggshell is made. *Scientific American* 222 (3): 89–97.

Varricchio, D. J.; F. Johnson; J. J. Borkowski; and J. R. Horner. 1997. Nest and egg clutches of the dinosaur *Troodon formosus* and the evolution of avian reproductive traits. *Nature* 385: 247–250.

Zhao Z. 1979. Progress in the research of dinosaur eggs. In *Mesozoic and Cenozoic Redbeds of South China*, pp. 330–340. Beijing: Science Press.

Zhao Z. 1993. Structure, formation and evolutionary trends of dinosaur eggshells. In I. Kobayashi, H. Mutvei, and A. Sahni (eds.), *Structure, Formation and Evolution of Fossil Hard Tissues*, pp. 195–212. Tokyo: Tokai University Press.

How Dinosaurs Grew

R. E. H. Reid

Because nobody has ever been able to watch a dinosaur grow, or ever will be able to do so, any essay on dinosaur growth must be a study in inference. Luckily, two types of bone that formed as the animals grew tell us parts of the story; and because the same general growth processes are now common to all tetrapods, we can use information from modern forms to fill some of the gaps. But there is still a great deal we don't know, and may never know.

Logically, a study of growth should start with embryos and work through to the end of growth in adults. In recent years, various dinosaur embryos have been found (e.g., Norell et al. 1994), but few details have yet been published, and most seem to be specimens with bones already ossified. So here we need to start by looking at how bones arise in living animals.

Early Development and Further Growth

First, we can assume that most bones in a dinosaur's skeleton, apart from some in the skull, were at first formed from cartilage before ossification began. In its simplest form, cartilage is a tissue in which live cells called chondrocytes occur scattered within a gel-like matrix formed from the proteoglycan (a compound composed of protein and carbohydrate) chon-

29

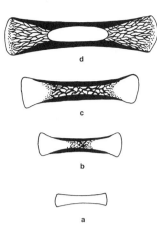

d

c

b

a

Figure 29.1. The early development of a dinosaurian limb bone. (a) Stage 1: A simple cartilage rod, coated by an external perichondrium. (b) Stage 2: Tissue coating the sides of the bone has become the periosteum and has started to lay down peri-osteal bone (*solid*). Cartilage in the shaft has calcified (*dots*), and is starting to be replaced by endochondral bone (*lines*). (c) Stage 3: Center of the shaft is now occupied by trabecular endochondral bone, with marrow-filled interspaces. Zones of calcification are moving toward the terminal surfaces. (d) Stage 4: Uncalcified cartilage is now restricted to parts underlying terminal surfaces, and is underlain by zones of calcification, with growing endochondral bone under them. A marrow cavity has formed and is expanding.

Figure 29.2. Drawing of one end of a bone. Terms on the left describe the parts of the bone. *From top to bottom:* the articulating terminal part (epiphysis); the expanded sub-terminal part (metaphysis); and the shaft (diaphysis). Terms on the right indicate processes characteristic of these regions in dinosaurian (but not mammalian) limb bones. *From top to bottom:* growth in length, by formation of new endochondral bone; external remodeling, by resorption of bone underlying the covering periosteum; growth in thickness, by accretion of periosteal bone.

droitin sulphate. There is usually also some content of collagen fibers. Cartilage thus resembles bone in being a cellular tissue; but it differs in that cells in it are able to divide, and to generate new matrix, thus allowing it to grow interstitially. These abilities are the basis of one of the two main processes by which bones grow.

Next we can look at how dinosaur limb bone would develop, as a model for all bones with cartilage prototypes (Fig. 29.1). At first it would be just a rod of cartilage (a), with an external coating of a fibrous tissue called the perichondrium, containing chondroblast cells, which were responsible for forming the cartilage. Early growth would be simply by formation of more cartilage at the sides and the ends, but two new processes would then start in the central parts. Bone-forming cells called osteoblasts would appear in the external perichondrium and start to coat the cartilage surface with bone, spreading out toward the two ends but not over them (b). In modern species, the external generative tissue after this change is called the periosteum, and bone formed by it is called periosteal. About the same time, a second process would begin with calcification of the cartilage at the center of the shaft, and invasion of this calcified cartilage by tissue of perichondral (or periosteal) origin, containing cartilage-resorbing cells termed chondroclasts (c). Osteoblasts appearing in this tissue, now called the endosteum, would next start lining cavities excavated by the chondroclasts with bone described as endochondral bone. This threefold process would then spread toward both ends of the developing bone, extending into the terminal parts (epiphyses; Fig. 29.2), but not as far as the terminal surfaces. These would remain cartilaginous and become the articular cartilages (d).

From this stage, further growth of the bone would proceed by two processes: growth in thickness, by accretion of more periosteal bone along the sides, and growth in length, by continued formation of new cartilage at the ends, and its progressive replacement by endochondral bone. In dinosaurs, periosteal bone formed after early life is dense (compact) bone;

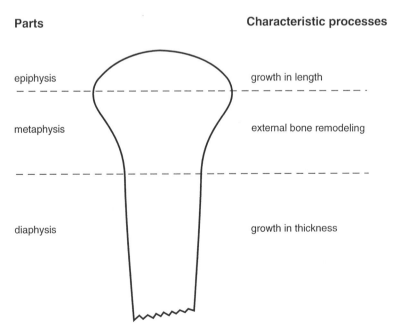

Parts

epiphysis

metaphysis

diaphysis

Characteristic processes

growth in length

external bone remodeling

growth in thickness

R. E. H. Reid

but endochondral bone, as in other tetrapods, is typically cancellous, and built from bony struts termed trabeculae with spaces called cancelli between them. In life, these would be lined by the endosteum, and otherwise filled with marrow tissues, including the hemopoietic tissue in which blood cells (erythrocytes, granular leukocytes, and thrombocytes) are formed. In addition, the bone would be subject to two further processes, not concerned with growth directly, but essential to its proper development. These are remodeling, in which bone formed earlier is resorbed, and internal reconstruction, in which it is replaced by new tissues. Bone resorption, involved in both, is carried out by cells called osteoclasts, which appear in the periosteum and endosteum, so that resorption can be either external or internal.

There are two main cases here. First, because bone cannot grow interstitially, a limb bone can grow in length only by addition of bone at its ends. But because the ends also become expanded to form articulatory structures, the cylindrical shaft (diaphysis) cannot be lengthened by simple terminal growth alone, and is extended by external bone resorption in the parts (metaphyses) just short of the ends until the right shape is reached (Fig. 29.3). Periosteal bone formation then takes over. Bone resorption is also involved when bones have to change their curvature during growth—for example, when a braincase is expanding. Enlow (1962) gives a good account of these processes. Second, as a limb bone grows, its marrow can expand only radially through resorption of bone which surrounds it. In hollow limb bones, seen in theropods and some ornithischians, endosteal osteoclasts would first produce a marrow (medullary) cavity in the central part of the shaft (Fig. 29.1d), and this would then expand both longitudinally and radially. In bones that lack a marrow cavity, as in sauropods, marrow expansion would occur through outward spread of a secondary cancellous (spongy) bone tissue, with marrow-filled interspaces. Internal bone reconstruction can also occur in other contexts, but these are the two most important here.

Endochondral Bone and Epiphyses

Endochondral bone has only one known pattern in dinosaurs. If you look at the seemingly articular surfaces of a well-preserved limb bone, or those of the centrum or zygapophyses of a vertebra, you will find that they are formed by a dense tissue, which may be featureless or show numerous closely spaced round pores (Figs. 29.4c, d). This is not bone but calcified cartilage, which once formed part of a growth zone underlying the soft articular cartilage. Its nature may not be apparent in surface views but is readily seen in sections, in which it appears as a tissue resembling plastic foam (Fig. 29.4b), containing numerous densely packed small rounded cavities which once contained chondrocytes. In life (Fig. 29.5) , the true articular surface would be formed by uncalcified cartilage, with a coating of perichondral and synovial tissues, and new cartilage would be added at the surface during growth. Some distance under it, however, would be a zone in which chondrocytes underwent enlargement (hypertrophy) and division until they became packed together in columns or without regular order. Deeper still would be a zone in which this hypertrophic cartilage was calcified, and under this a zone in which endochondral bone was replacing it. Thus, the seeming articular surfaces seen in the fossils (Figs. 29.4c, d) really represent the internal interface between uncalcified cartilage, which is lost, and the underlying calcified tissue. This style of longitudinal growth

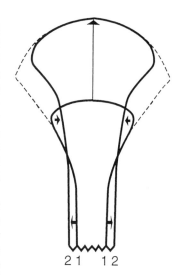

2 1   1 2

Figure 29.3. External bone remodeling. Overlapping solid outlines (1, 2) show the form of part of a limb bone at two successive stages of its growth. Because bone tissues cannot expand once they are formed, growth in length (*vertical arrow*) depends on the formation of new endochondral bone, and growth in thickness (*horizontal arrows, pointing outward*) depends on the accretion of new periosteal bone. But those processes alone would lead only to progressive expansion of the terminal parts (*see broken outlines*), without lengthening of the shaft (diaphysis); and this result is avoided by the resorption of bone (*horizontal arrows, pointing inward*) in parts below the growing terminal (epiphyseal) surface, allowing the shaft to be lengthened and the bone to maintain its proper shape.

Figure 29.4. Endochondral bone and calcified cartilage. (a) Both tissues as seen at the top end of an *Allosaurus* fibula. The upper half shows calcified cartilage containing spaces excavated by advancing marrow processes, with early endochondral bone (*dark tissue*) in places. The lower half shows trabecular endochondral bone (zones e and f of Figure 29.5). (b) Enlarged view of the calcified cartilage, in an area near top right in a, showing the rounded form of the spaces once occupied by cartilage cells. (c) The "articular" surface of an ornithomimid metatarsal, formed from a thin sheet of calcified cartilage with small perforating pores. (d) The "articular" surface of an *Iguanodon* limb bone, showing pores surrounded by thin rings of dark endochondral bone, which formed tubules like those described by Haines (1938) from young Nile crocodiles.

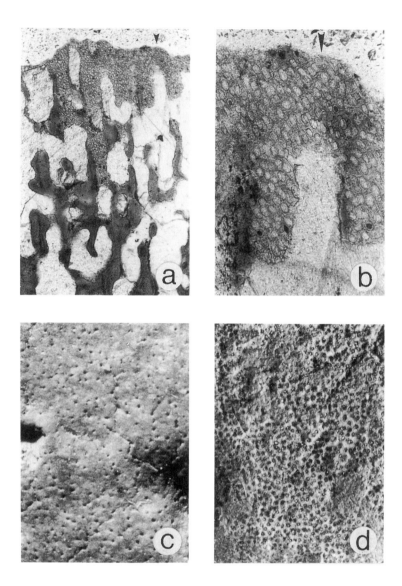

is now seen in turtles and crocodiles, from which the missing details have been taken here, and is also the primitive pattern for tetrapods in general (Haines 1938, 1942).

Details of the growth zone have been little studied, but some information is available (e.g., Reid 1984, 1996). The zone of calcified cartilage varies in thickness, from a few cells to many cells thick, while the cartilage-marrow contact ranges generally from more or less flat, with bone trabeculae simply abutting it, to deeply sculptured, with early endochondral bone extending into the zone of calcified cartilage. The endochondral trabeculae may be formed from bone only, or may contain cores or "islands" of calcified cartilage for some distance away from the growth zone. These were later destroyed by remodeling of the trabeculae. The "articular" surfaces may show no sign of what was happening under them; but in various ornithopods (e.g., *Iguanodon* [Reid 1984]), it appears to have been usual for numerous small cylindrical marrow processes to extend outward through the zone of calcified cartilage, so that the surface shows the pattern of small pores mentioned earlier. These pores may be surrounded by cartilage only or by thin rings of bone, which sections show to represent thin layers of endochondral bone, formed on the walls of hollow tubules that the marrow processes occupied (Fig. 29.4d). This pattern is also known from young crocodiles (Haines 1938), and so could be a common inheritance from primitive archosaurian ancestors.

Here we need to note a difference between dinosaurs and various other tetrapods, including mammals (Fig. 29.6). In dinosaurs, the processes leading to the replacement of cartilage with endochondral bone took place under the articular surfaces, as in crocodiles (Fig. 29.6a); but in mammals, lepidosaurs (lizards plus *Sphenodon*), and some amphibians, the epiphyses themselves develop centers of calcification, which may ossify as in mammals (Fig. 29.6b). Growth in length then takes place at a separate internal growth plate, in the form of a sheet of cartilage extending transversely between the calcified or ossified epiphysis and the expanded metaphysis at the end of the shaft, while cartilage extending across the end of the bone has a purely articular function. In mammals, moreover, the ossified epiphyses fuse with the metaphyses when growth ceases at adult size, and no further growth in length is then possible. Because of this, and because we know the ages at which different epiphyses fuse, their condition can be used to identify adults and even determine ages; but we cannot do either in dinosaurs, whose sub-articular style of endochondral growth would in-

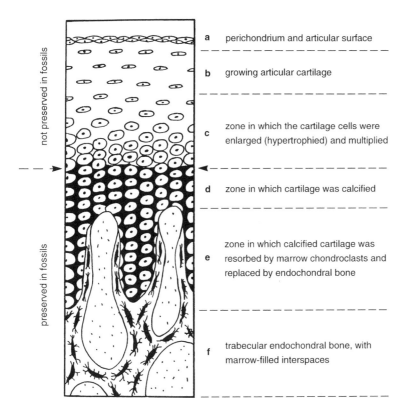

a perichondrium and articular surface

b growing articular cartilage

c zone in which the cartilage cells were enlarged (hypertrophied) and multiplied

d zone in which cartilage was calcified

e zone in which calcified cartilage was resorbed by marrow chondroclasts and replaced by endochondral bone

f trabecular endochondral bone, with marrow-filled interspaces

not preserved in fossils

preserved in fossils

Figure 29.5. How dinosaurian limb bones grew in length: reconstructed section through the "growth plate" in the terminal part (epiphysis; see Figs. 29.2 and 29.6a) of a growing bone. (a) Perichondrium, under which new cartilage was formed, coating the true articular surface. (b) Growing articular cartilage, formed progressively under the perichondrium. (c) Zone in which the cartilage cells became enlarged and multiplied, and arranged in vertical columns. In modern examples, such columns are formed by cell division at the top of each column. (d) Zone in which the cartilage matrix became calcified (*black shading*). The interface (*arrowed*) forms the "articular" surfaces of fossil bones. (e) Zone in which calcified cartilage was resorbed in intruding marrow processes and replaced by endochondral bone. (f) Trabecular endochondral bone, with marrow-filled interspaces (*dotted shading*). Note that bone cells differ from chondrocyte in emitting branching processes.

How Dinosaurs Grew   407

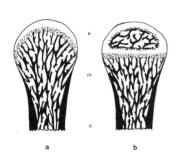

a　　　　　　　　b

Figure 29.6. How dinosaurs (a)
differed from mammals (b) in
the way in which their bones
grew in length. In dinosaurs (a),
the zones of hypertrophic and
calcified cartilage (*dotted
shading*) formed a "growth
plate" which underlay the
articular cartilage (*clear, at top*)
directly, and had endochondral
trabeculae (*black, reticulating*)
extending up to it. The "growth
plate" thus followed the form of
the articular surface; and the
seeming articular surfaces of
dinosaurian bones follow the
form of the interface between
uncalcified and calcified
cartilage. This style of growth in
length is now seen in turtles and
crocodiles, as well as in plesiosaurs, ichthyosaurs, and all
tetrapods older than Triassic. In
mammals (b), in contrast, the
epiphyses (*top*, e) develop
secondary centers of calcification, which then ossify by
endochondral replacement; and
growth in length by formation of
endochondral bone takes place
under an internal growth plate,
extending transversely between
the epiphysis (e) and the
metaphysis (m); d = diaphysis.
At maturity, this growth plate
disappears as the epiphyses fuse,
preventing any further growth in
length, whereas dinosaurs could
potentially have grown throughout their lives, like turtles and
crocodiles.

stead have allowed them to grow throughout their lives, like modern
reptiles. And de Ricqlès (1980) believed that they did, presumably because
many dinosaurian bones show no sign that growth had ceased before
death—although how far such evidence should be trusted is uncertain,
because it does not prove what it seems to show.

We can finish here with a puzzle. Most of the limb bones of birds show
the same style of growth in length as those of dinosaurs, and it is tempting
to see this as simply an inheritance from dinosaur ancestors. But is it? In
1931, it was shown experimentally by Landauer that any of the limb bones
may form calcified epiphyses in a condition described as chondrodys-
trophy; and as Haines (1938) saw seven years later, this must mean that the
mechanism for their production exists in all these bones, although it is
normally not used. Or, in genetic terms, the genes for their production must
exist, but are normally "switched off." And if that is so, then the usual sub-
articular style of growth must be secondary, whereas nothing is known to
suggest that it is anything but primitive in dinosaurs. Perhaps future work
on early birds will throw light on this conundrum.

Periosteal Bone

Having seen how bones grew in length during increase in stature, we can
now look at how they grew in thickness. In outline (Figs. 29.7, 29.8),
periosteal growth was often continuous, as in mammals and birds, but
sometimes periodic, as in reptiles, with some dinosaurs following both
styles in different parts of their skeletons (Reid 1984, 1987). Because of a
lack of systematic sampling, we do not know how common the latter
condition was. Variations in growth rates can also be detected, but only on
a relative basis.

Continuous periosteal growth in dinosaurs seems commonly to have
been rapid, because the tissue produced is of a type seen when modern
forms grow quickly. This tissue (Fig. 29.9), called fibro-lamellar bone (de
Ricqlès 1974), is formed initially as finely cancellous bone, and then is
compacted by internal deposition of more bone to form structures called
primary osteons. The initial cancellous framework is built from a fast-
growing tissue called woven bone, in which collagen fiber bundles are
arranged without order; and its cancellous form allows a given volume of
bone to produce a higher rate of radial growth than it would if it were laid
down in compact form. Dinosaurs with bone of this type seem to have
grown at rates comparable to those of large fast-growing mammals, in
which the same type occurs, and could do so up to brachiosaur sizes. Some,
however, have similar-looking bone, in which the use of crossed polarizers
shows the periosteal bone to have a layered structure, implying slower
growth than true woven bone, or the whole of the bone to be formed from
such tissue, without osteons, implying slower growth still (Fig. 29.10). But
what the actual rates were cannot be measured, because of the absence of
a time scale.

In contrast, the periosteal bone of some dinosaurs (Figs. 29.7b, c, 29.8b)
shows conspicuous "growth rings," or zones, like those seen in the bones
of modern reptiles, which result from the slowing or periodic cessation of
growth. Slowed growth is shown by bands of dense tissue known as annuli,
and arrested growth by features called resting lines (or arrest lines, or LAGs
[lines of arrested growth]), marking where the surface was during growth
pauses (see also Figs. 29.4b, c). Bone in the zones may be fibro-lamellar
bone, implying rapid growth while it was forming, or of some slower-

408　　*R. E. H. Reid*

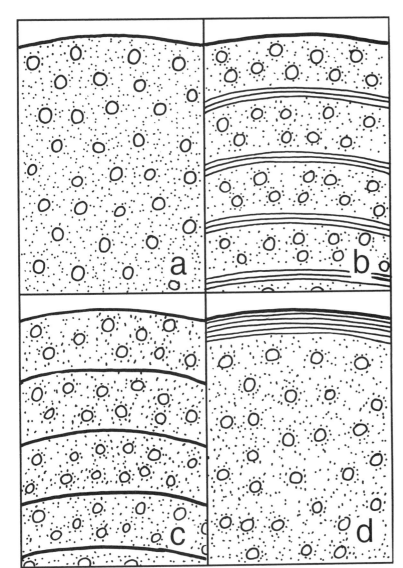

Figure 29.7. How different styles of growth are recorded in periosteal bone formed as bones grew in thickness. (a) Bone formed without interruptions or textural variations, implying continuous growth at a constant rate, with no sign of growth's having slowed before it ceased. This style of growth is by far the commonest in dinosaurs. (b) Bone divided into major "growth rings," or zones, by thin bands of lamellated tissue termed annuli, implying a regular alternation of periods of normal growth (the zones) and slowed growth (the annuli). (c) Bone divided into "growth rings" by resting lines, which mark periods in which growth was interrupted. (d) Bone like that seen in a, coated externally by a layer of lamellated tissue, implying a later switch from active growth to slow accretion. This pattern is rare in dinosaurs, although it is known from the sauropod *Brachiosaurus* and several small theropods (e.g., *Syntarsus*). Circular features are vascular canals.

growing type, including finely lamellated tissues, which are usually formed slowly. Bone with "growth rings" should potentially be able to yield data on ages and growth rates, if the rings are assumed to be annual as in crocodilians; but few such studies have been made, and the interpretation of growth rings involves problems. The main one is that, even when present, growth rings often record only the last part of a dinosaur's growth, because most of the periosteal bone produced earlier was resorbed as the marrow expanded or was replaced by secondary tissues. But if enough is left, it may then be possible to estimate how many rings have been lost, by extrapolation from the thicknesses of those which remain. By this method, used by Ferguson et al. (1982) to "age" alligators, a sauropod took twenty-eight or twenty-nine years to reach roughly the size of an adult modern elephant weighing five to six tons (Reid 1987, 1990), while the iguanodont *Rhabdodon* reached the size of a riding horse in about sixteen years (Reid

Figure 29.8. The contrasting appearances of periosteal bone when formed continuously (a) and discontinuously (b), as seen in samples from ribs of *Tyrannosaurus* (a) and a large Early Cretaceous theropod (b). The *Tyrannosaurus* rib shows no growth interruptions apart from the vascular canals that in life housed blood capillaries. The rib in b is divided into "growth rings" (zones) by circumferential resting lines, some of which are double as a result of aborted growth resumption.

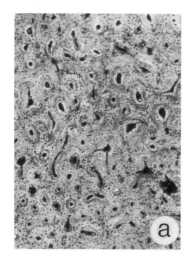
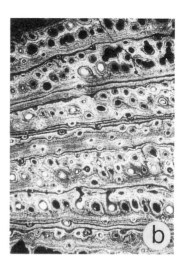

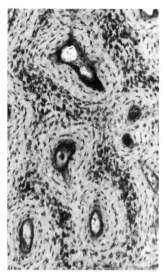

Figure 29.9. Fibro-lamellar bone, from a limb bone of the sauropod *Cetiosaurus*. In this figure, the large rounded spaces are vascular canals, which contained blood capillaries, and the small dark bodies are mineral-filled spaces (lacunae) which once contained bone cells (osteocytes). The tissue was formed initially as finely cancellous (spongy) bone, represented by the dark tracts in which the osteocyte lacunae are largest, and then was compacted by internal deposition of more bone, which grew inward toward enclosed blood vessels. This bone formed structures called primary osteons, in which the osteocyte lacunae are smaller than in the initial framework, and show a rough concentric arrangement around the vascular canals.

1990). David Varricchio (1993) suggests three to five years for the small theropod *Troodon* to reach 50 kilograms in weight, although he warns that the resting lines relied on may not have been annual; and another, *Saurornitholests*, seems to have taken six or seven years to reach about the same size (Reid 1993). But none of these estimates tells us anything about dinosaurs whose growth was continuous; and the rates of bone accretion again cannot be quantified, because we do not know the length of the periods through which growth was active.

Last, we need to note two problems. First, modern crocodilians, including American alligators, can grow fast enough to form fibrolamellar bone when young; but they lose that ability later, and in old age form only thin layers of avascular bone. But dinosaurs could form fibrolamellar bone continuously, up to sizes several times larger than those of the largest known terrestrial mammals, and why they could needs explaining. Various authors have expressed conflicting views, of which none has yet gained general acceptance. Second, while many dinosaurian bones show no sign of cessation of active growth before death, a few show a late periosteal switch to formation of thin superficial layers of dense lamellated bone (Figs. 29.7d, 29.11). This condition was first described from *Brachiosaurus* (Gross 1934), near the top of the size range, and has recently been rediscovered in three small theropods, *Syntarsus* (Chinsamy 1990), *Troodon* (Varricchio 1993), and *Saurornitholestes* (Reid 1993). As was seen by Chinsamy (1990), such tissues resemble accretionary bone, which is sometimes formed in fast-growing mammals after active growth has ceased; and in *Saurornitholestes*, a sequence of resting lines suggests that this dinosaur lived longer after active growth ceased than before (Reid 1993). In these respects, these genera seem more "mammal-like" than typical dinosaurs; on the other hand, *Syntarsus* and *Saurornitholestes*, and less certainly *Troodon*, have the bone formed during active growth divided into "growth rings," and in this seem more "reptilian." These conflicting resemblances to contrasting kinds of modern animals are yet another puzzle for speculators on dinosaurian physiology.

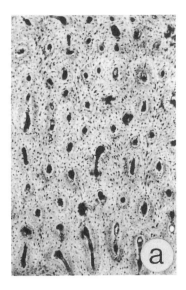

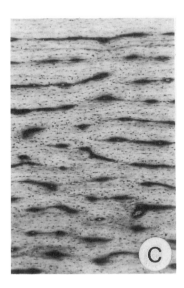

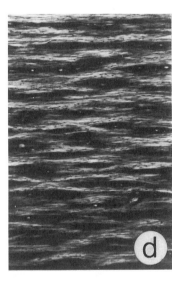

Figure 29.10. Apart from contrasting arrangements of vascular canals (the large dark perforations), these bone samples from *Tyrannosaurus* (a) and *Ceratosaurus* (c) appear similar when viewed with normal lighting, but are strikingly different (b, d) when viewed with polarizers crossed. The *Tyrannosaurus* bone (b) is a specialized fibro-lamellar tissue, containing numerous small primary osteons marked by dark four-armed "axial crosses"; while the intervening periosteal bone shows an unusual pattern of light patches and dark extinction lines. In contrast, the *Cerato-saurus* bone (d) is built entirely from roughly stratified periosteal bone, with no unstratified woven bone framework, and no osteon system. The *Tyrannosaurus* bone can be presumed to have been the faster-growing, although how fast the two tissues grew cannot be measured.

This chapter uses data from both dinosaurs and modern forms, to allow a more complete picture than would otherwise be possible. The main points made are as follows:

Most dinosaurian bones must have been formed from cartilage proto-types, in the same ways as those of modern tetrapods. During further growth, other processes (remodeling, internal reconstruction) also oper-ated.

The formation of new endochondral bone as limb bones grew in length took place under the articular cartilages, as in crocodiles and turtles, and not under separate sub-epiphysial plates as in mammals.

Periosteal growth was commonly continuous, as in mammals and birds, but sometimes instead periodic. Fast and slower styles of growth can be recognized histologically in both cases, but only on a relative basis. Many dinosaurian bones show no sign of growth's having ceased before death, but may simply not record its having done so.

How Dinosaurs Grew 411

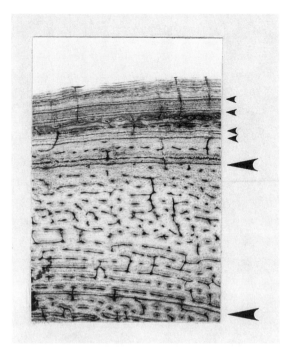

Figure 29.11. Apparent determinate growth in a small Cretaceous theropod, *Saurornitholestes*. This figure shows the outer parts of a bone built mainly from a highly vascular tissue, divided into major "growth rings" or zones by resting lines marking pauses in growth. The last of these zones is seen between the large arrows at right. This is then followed by bone showing several closely spaced resting lines (*small arrows*), and this in turn by almost avascular bone showing fine circumferential lamellation. Active growth thus seems to have been followed by a switch to slow accretion, as can happen when modern endotherms stop growing at a maximum size.

References

Chinsamy, A. 1990. Physiological implications of the bone histology of *Syntarsus rhodesiensis* (Saurischia: Theropoda). *Palaeontologia africana* 27: 77–82.

de Ricqlès, A. J. 1974. Evolution of endothermy: Histological evidence. *Evolutionary Theory* 1: 51–80.

de Ricqlès, A. J. 1980. Tissue structure of dinosaur bone: Functional significance and possible relation to dinosaur physiology. In R. D. K. Thomas and E. C. Olson (eds.), *A Cold Look at the Warm-blooded Dinosaurs,* pp. 103–139. American Association for the Advancement of Science Selected Symposium 28. Boulder, Colo.: Westview Press.

Enlow, D. H. 1962. A study of the post-natal growth and remodelling of bone. *American Journal of Anatomy* 110: 79–102.

Ferguson, M. W. J.; L. S. Honig; P. Bringas, Jr.; and H. C. Slavkin. 1982. *In vivo* and *in vitro* development of first branchial arch derivatives in *Alligator mississippiensis.* In A. D. Dixon and B. Sarnat (eds.), *Factors and Mechanisms Influencing Bone Growth,* pp. 275–296. New York: Alan R. Liss.

Gross, W. 1934. Die Typen des mikroskopischen Knochenbaues bei fossilen Stegocephalen und Reptilien. *Zeitschrift für Anatomie* 103: 731–764.

Haines, R. W. 1938. The primitive form of the epiphysis in the long bones of tetrapods. *Journal of Anatomy* 72: 323–343.

Haines, R. W. 1942. The evolution of epiphyses and endochondral bone. *Biological Reviews* 17: 267–292.

Landauer, W. 1931. Untersuchungen über der Krüperkuhn. II. Morphologie und Histologie des Skelets, inbesondere de Skelets der langen Extremitätenknochen. *Zeitschrift für mikroskopische-anatomische Forschung* 25: 115–141.

Norell, M. A.; J. M. Clark; Dashzeveg D.; Barsbold R.; L. M. Chiappe; A. R. Davidson; M. C. McKenna; Perle A.; and M. J. Novacek. 1994. A theropod dinosaur embryo and the affinities of the Flaming Cliffs dinosaur eggs. *Science* 266: 779–782.

Reid, R. E. H. 1984. The histology of dinosaurian bone, and its possible bearing on dinosaurian physiology. In M. W. J. Ferguson (ed.), *The Structure, Development and Evolution of Reptiles,* pp. 629–663. Orlando, Fla., and London: Academic Press.

Reid, R. E. H. 1987. Bone and dinosaurian "endothermy." *Modern Geology* 11: 133–154.

Reid, R. E. H. 1990. Zonal "growth rings" in dinosaurs. *Modern Geology* 15: 19–48.

Reid, R. E. H. 1993. Apparent zonation and slowed late growth in a small Cretaceous theropod. *Modern Geology* 18: 391–406.

Reid, R. E. H. 1996. Bone histology of the Cleveland-Lloyd dinosaurs, and of dinosaurs in general. Part I: Introduction: Introduction to bone tissues. *Brigham Young University Geology Studies* 41: 25–71.

Varricchio, D. J. 1993. Bone microstructure of the Upper Cretaceous theropod dinosaur *Troodon formosus. Journal of Vertebrate Paleontology* 13: 99–104.

Engineering a Dinosaur

R. McNeill Alexander

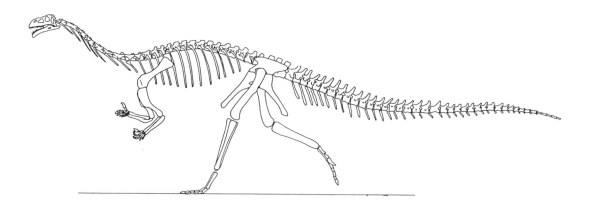

This chapter looks at dinosaurs from an engineering point of view, considering the strengths of their skeletons and the stresses that must have acted on them when the animals moved. It asks what problems arise from being large, and how well these problems were solved in the course of dinosaur evolution. Were the largest dinosaurs lumbering giants, barely able to support their great weight, or were they formidable and fast-moving? How did the sauropods support their long necks, and could they have reared up on their hind legs to feed from high branches? What can engineering analysis tell us about dinosaur weapons such as ceratopsian horns and the reinforced skulls of dome-headed dinosaurs?

Sizes and Size Limits

The large dinosaurs were enormous, much larger than any modern land animal. *Diplodocus*, with its long neck and its very long tail, had an overall length of about 25 meters (82 feet; an American football field is about four *Diplodocus* long by two *Diplodocus* wide). *Brachiosaurus*, carrying its neck erect like a giraffe, was 13 meters (43 feet) tall and could have looked over the roofs of four-story buildings. Giraffes grow to heights of only about 5.5 meters (18 feet).

For the questions asked in this chapter, the weights of dinosaurs are generally more important than their heights or lengths. We cannot weigh living dinosaurs, and have to rely instead on estimates based on their fossil remains. Such estimates usually depend on scale models of the animals as they are thought to have appeared in life. The dimensions of the skeleton are known, but the modeler must rely on her or his judgment in deciding how much flesh and skin to put around the bones. The volume of the model is measured, preferably by weighing it in air and in water (see Alexander 1989 for an explanation). The volume of the living animal is calculated by scaling up, and is multiplied by the presumed density of the animal to obtain its mass. An alternative approach depends on measurements of the circumferences of leg bones (Anderson et al. 1985). It seems fine for some purposes, but not for the engineering analyses of this chapter. If we estimated a dinosaur's weight from the thicknesses of its bones and then asked whether the bones were thick enough to support its weight, we would be indulging in circular argument.

Unfortunately, there is a great deal of uncertainty about dinosaur masses. Using the model method, Colbert (1962) estimated the mass of *Brachiosaurus brancai* as 78 tonnes; I estimated it as 47 tonnes (Alexander 1989), and Paul (1988) calculated only 32 tonnes. (The metric tonne used by scientists is almost the same as the ton used in commerce.) The reasons for the difference, for this and other species, are that different authors have based their estimates on different individual animals, some have used relatively skinny models and others more portly ones, and there is some doubt about dinosaur densities.

Brachiosaurus is the largest dinosaur for which model-based mass estimates are available. Whichever estimate is nearest the truth, *Brachiosaurus* was many times heavier than the largest modern land animals. Male African elephants grow to about 6 tonnes, Asiatic elephants are a little smaller, and the heaviest rhinoceros are about 3 tonnes (Nowak 1991). *Tyrannosaurus rex*, the biggest carnivorous dinosaur known from reasonably complete skeletons, has been estimated from models to have had masses in the range of 5.7 to 8.0 tonnes (Farlow 1990), far more than polar bears (up to 0.8 tonnes) and tigers (0.3 tonnes; Nowak 1991). Two other carnivorous dinosaurs, *Giganotosaurus* and *Carcharodontosaurus*, although not known from skeletons as complete as those of *Tyrannosaurus*, may have been even heavier (Coria and Salgado 1995; Sereno et al. 1996).

It has often been suggested that large dinosaurs were close to an upper limit of size, which land animals cannot exceed. The commonest suggestion has been that the limit is mechanical: larger animals could not support their weight. This problem of large size was recognized by Galileo (1638), who lived long before dinosaurs had been discovered. Imagine two animals of different sizes that are geometrically similar—that means they are exact scale models of each other. If, for example, the larger is twice as long as the smaller, it is also twice as wide and twice as high, and has bones of twice the diameter. The strength of each bone is proportional to its cross-sectional area and so to the square of the animal's linear dimensions, four times as strong in an animal of twice the length. The weight the skeleton has to support, however, is proportional to the animal's volume, and so to the cube of its length, eight times as heavy in an animal of twice the length. Thus the larger animal in our example has to carry eight times the weight on a skeleton of four times the strength. As size increases (while geometric similarity is maintained), a limit must be reached above which the animal could not stand. The limit could be shifted upward if larger animals had relatively thicker bones (and buffalo do have relatively thicker leg bones than gazelles), but there must be some limit to leg bone thickness if a large animal is not to be hopelessly cumbersome.

Engineering a Dinosaur   415

The only modern animals that are heavier than large sauropods are whales (blue whales grow to well over 100 tonnes; Nowak 1991). It used to be thought that large sauropods could not have supported their weight on land and must have lived in lakes, wading in water deep enough for them to have been supported largely by buoyancy. However, Bakker (1971) argued that they seem better adapted for terrestrial than semi-aquatic life, and Hokkanen (1986) calculated that animals could be larger even than the largest whales and still support themselves on land.

Another possibility is that animal size may be limited, because excessively large animals would overheat. The argument is more complicated than the one about strength because metabolic heat production is not quite proportional to body mass, and because the rate of heat loss depends on the thickness of the skin (which serves as heat insulation) as well as on its area. Also, there is controversy about dinosaurs' metabolic rates: Did they produce heat at the rate expected for a reptile of the same size, or at the rate expected for a mammal, or at some other rate? Rough calculations based on the cooling rates of modern reptiles of different sizes suggest that if they had mammal-like metabolism, they would have had to allow a great deal of water to evaporate from their bodies to prevent overheating (Alexander 1989). It is possible that the largest dinosaurs were near an upper limit of size, above which they could not have avoided overheating in warm climates, but this is uncertain.

Farlow (1993) suggests that the sizes of land animals may be limited by problems neither of support nor of heat loss, but by the problem of maintaining a viable population. The larger the members of a species are, the fewer individuals can be supported by the resources of their geographic range. An excessively large species could have had so few individuals that chance extinction would be highly likely.

Limits to Athleticism

These arguments suggest that the large dinosaurs may not have been near any absolute size limit based on mechanical considerations. It nevertheless seems interesting to consider the mechanical consequences of their being so large. The largest modern land animals are not very athletic. Elephants can run at moderate speeds but do not gallop and do not jump. Were dinosaurs still less athletic? Among modern mammals, the fastest runners and the strongest jumpers seem to be in the middle range of sizes (Garland 1983; but note that many of the speed records he quotes are subjective estimates, not measurements). Was this true also of dinosaurs?

The most direct available evidence of dinosaur movement comes from fossil footprints, from which speeds can be estimated by the method of Alexander (1976; see also Farlow and Chapman, chap. 36 of this volume). There are a few tracks of medium-sized dinosaurs that seem to have been running faster than humans can sprint, at around 12 meters per second (Farlow 1981), but no records of fast running by very large dinosaurs. All known sauropod tracks seem to show human walking speeds, well under 2 meters per second, and large theropod tracks show only slightly higher speeds (Thulborn 1990).

Fossil footprints seem to tell us that large dinosaurs usually walked, but do not exclude the possibility that they may sometimes have run. To judge how athletic (or otherwise) dinosaurs may have been, we must look for other evidence. It would help greatly if we knew how big their muscles were, but we do not. In the circumstances, the best approach seems to be to study their skeletons. Were their skeletons strong enough for running and jumping?

The more athletically a person or animal behaves, the larger (in general) are the forces that act on the leg skeleton. When a person walks, the maximum forces exerted by each foot on the ground are about equal to body weight; in jogging, peak ground forces are about 2.5 times body weight; in sprinting, about 3.6 times body weight (Nigg 1986); and in jumping, even more. Athletic animals need strong skeletons.

Calculating the forces on the skeleton of a running or jumping animal is complicated, and fraught with uncertainty if there is no possibility of observing the animal in action. Also, there are uncertainties in relating skeleton strength to athletic performance, because it would not be satisfactory for a bone to be *just* strong enough for the animal's most strenuous activity. A factor of safety is needed to allow for the unexpectedly large forces that may sometimes occur accidentally, for example if the animal stumbles. Similarly, engineers build factors of safety into their structures: if a bridge is expected to have to bear ten-ton loads, it will probably be designed to be strong enough to carry twenty tons.

Alexander (1985) tried to assess dinosaur athleticism, using a method designed to overcome these difficulties. It depends on comparisons of dinosaur skeletons with those of living animals and depends on several assumptions:

- That dinosaur bone is about as strong as bone from modern mammals. Little is known about the strength of reptile bone, but bird and mammal bone are about equally strong, and there is no reason to expect reptile bone to be different.
- That safety factors for the skeletons of dinosaurs were about the same as for the skeletons of modern mammals. This assumption seems reasonable, but may lead to error because it seems likely that the best safety factor (i.e., the one that will be preferred by natural selection) will depend on circumstances. For example, a herbivore that is not threatened by predators has little need to run, so it may not matter if its leg bones are so thick as to be cumbersome—and it may be advantageous to have exceptionally strong bones for extra safety. The leg bones of rhinoceros seem to be built to unusually higher factors of safety (Alexander and Pond 1992).
- That dinosaurs moved pretty much like modern mammals. More precisely, to use technical terms that will be explained in the next few paragraphs, dinosaur movements were fairly nearly dynamically similar to the movements of modern mammals running at equivalent speeds. The comparison here is with mammals, not reptiles, because fossil footprints show that dinosaurs walked and ran like mammals, placing their feet under the body (see Farlow and Chapman, chap. 36 of this volume). Unlike lizards and other modern reptiles, they did not use sprawling gaits with the feet placed far out on either side of the body.

We have already encountered the concept of geometric similarity, as applied to shapes. Dynamic similarity is the equivalent for comparisons between movements. Two shapes are geometrically similar if one could be made identical to the other by a uniform change in the scale of length. Two movements are dynamically similar if they could be made identical by uniform changes of the scales of length, time, and force. If you had films of a large animal and a small one running in dynamically similar fashion, you could make the projected images identical by showing them at different magnifications (adjusting the scale of length) and running the projectors at different speeds (adjusting the scale of time).

A basic principle of mechanics tells us that animals of different sizes cannot move in dynamically similar fashion unless they are traveling with equal Froude numbers.

$$\text{Froude number} = \frac{(\text{speed})^2}{\text{leg length} \times \text{gravitational acceleration}}$$

The principle can be expressed in various ways. One is that movements over land cannot be similar unless the animals have equal ratios of kinetic energy ($\frac{1}{2}$ mass $\times$ speed^2) to potential energy (mass $\times$ gravity $\times$ height). The Froude number is twice this ratio. For example, if a horse has legs four times as long as those of a dog, the two animals can be expected to move similarly only when the horse is traveling twice as fast as the dog (giving them equal values of (speed)^2/leg length). They can then be expected to use the same gait (walk, trot, or gallop); to have stride lengths proportional to the length of their legs; to exert forces on the ground that are equal multiples of body weight; and so on. A study of mammals ranging from small rodents to rhinoceros showed that these predictions from theory are remarkably close to the truth (Alexander and Jayes 1983). Obviously a dog is not the same shape as a horse, so it cannot move in precisely similar fashion—but it is strikingly close to dynamic similarity when their Froude numbers are equal.

When animals of different sizes move in dynamically similar fashion (both trotting or both cantering, for example), the forces on their skeletons are proportional to their body weights. That means that if two animals have skeletons whose strengths are in proportion to the weights they have to carry, and if one of the animals can perform a particular athletic activity, the other's skeleton is strong enough for it to do the same thing, to move in dynamically similar fashion.

Alexander (1985) defined a "strength indicator" as a measure of bone strength in relation to body weight. It indicates ability to withstand forces acting at right angles to the bone shaft, because these set up much larger stresses in bones than forces along the shaft. To appreciate its meaning, imagine a leg bone fixed horizontally, with the proximal end (the end that was nearer the trunk in the living animal) firmly held. Now suspend the animal from the bone's other end (Fig. 30.1). If the maximum stress in a cross-section of the shaft of the bone is S, the strength indicator is $1/S$. Thus if the bone is thick, the stress S is low and the strength indicator is high. The explanation just given applies to bipeds. For quadrupeds the weight used in the calculation is not the weight of the whole body but only that of the forequarters (in the case of foreleg bones) or hindquarters (for hind legs). Alexander (1985) explains how the calculations are performed, using body weight (estimated from models, in the case of dinosaurs) and detailed measurements of the bones.

Results are shown in Table 30.1. Strength indicators are expected to be lower for larger animals because weight is proportional to the cube of length and strength only to the square (see above). Strength indicators are larger for the buffalo than for the elephant, reflecting the buffalo's better athletic ability—buffalo can gallop, but elephants can manage only a slow run. Values for the sauropod *Apatosaurus* are close to those for elephant, indicating that its skeleton was strong enough for it to be as athletic as an elephant despite being so much larger: it seems likely to have been capable

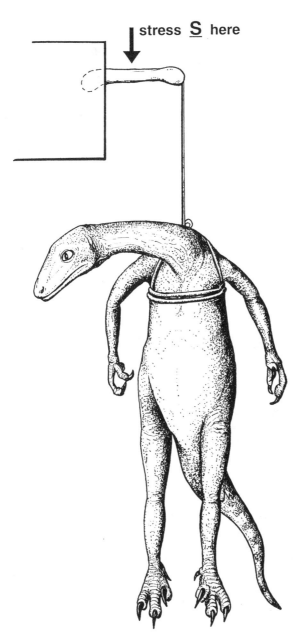

stress **S** here

Figure 30.1. A diagram illustrating the meaning of the term *strength indicator*. See text for additional explanation.

of a slow run. Values for *Triceratops* are higher (though still lower than those for the buffalo). Alexander (1985) suggested that this dinosaur might have been able to gallop slowly, but against that it has been claimed that *Triceratops* and related dinosaurs held their forefeet well out to either side of the body, rather like modern reptiles, and could not have galloped (Farlow 1990; Johnson and Ostrom 1995; but see Lockley and Hunt 1995 for a contrary view). Some crocodiles gallop (Webb and Gans 1982), showing that reptilian posture does not necessarily make galloping impossible, but it would probably increase the stresses on the bones.

The legs of sauropods and ceratopsians are sufficiently like those of

Engineering a Dinosaur   419

Table 30.1

**Strength Indicators for Large Modern Mammals
and for Dinosaurs**

	Body mass (tonnes)	Strength indicator (m²/GN) femur	tibia	humerus
African elephant	2.5	11	7	9
Young white rhinoceros	0.75	31	26	33
African buffalo	0.50	21	22	27
Apatosaurus	34	9	6	14
Triceratops	6–9*	15–21	12–20	
Tyrannosaurus	8	9		

*A range of values is given because body mass estimates vary.
Source: Alexander 1989 and Alexander and Pond 1992.

large modern mammals for us to assume that they moved much like modern mammals; thus we can draw conclusions from comparisons of strength indicators. *Tyrannosaurus*'s legs, however, had proportions unlike those of any modern biped. It could not have run like an ostrich (which has a long tarsometatarsal segment, between the ankle and the foot, and a short femur) nor like people (who put the whole sole of the foot on the ground). Its low (elephant-like) strength indicators suggest poor athletic ability, but leave some room for doubt.

There are doubts in any case for all the dinosaurs, because of uncertainty about body weight. If they were less heavy than we have supposed, they could have been more athletic. In particular, Paul (1988) has claimed that *Apatosaurus* had a mass of only 18 tonnes. If he is right, its strength indicators should be much higher than those given in Table 30.1. Also, the strength indicator of *Tyrannosaurus* has recently been reassessed (Farlow et al. 1995). The specimen in question seems to have had a mass of only 6 tonnes, but its femur is more slender than the one previously measured, and its strength indicator is confirmed as 7.5–9 m²/GN.

In the same paper, Farlow et al. (1995) argued that *Tyrannosaurus* could not have run fast because of the danger of severe injury if it fell, with no adequate forelimbs to save itself. An admittedly rough calculation suggested that forces of six or more times body weight might act on the torso in a fall. Alexander (1996) used data from car crashes to confirm that such forces would be dangerous but argued that many modern animals risk serious injuries from falls. A giraffe that trips while galloping or a gibbon that falls from a high branch is likely to break bones.

Dinosaurs as Giraffes

Following his argument that sauropod dinosaurs were not semi-aquatic (Bakker 1971), Bakker (1978) proposed that they used their long necks to feed like giraffes from trees. Some, such as *Brachiosaurus,* had long been reconstructed with their necks in an erect, giraffe-like posture. Others, such as *Apatosaurus* and *Diplodocus,* with differently shaped neck vertebrae, seem to have carried their necks horizontal. Bakker made the astonishing claim that these latter sauropods also fed like giraffes, rearing up on their hind legs to do so. Could so large an animal have stood on its hind legs?

The question is not one of strength but one of balance: the forces that act on the feet in running are so much greater than those in standing that an animal strong enough to run on all fours should also be strong enough to stand on one pair of legs. The critical question is whether the hind legs could be placed directly under the animal's center of gravity, enabling it to balance on them.

Alexander (1985) set out to estimate the positions of dinosaurs' centers of gravity. The first step was to locate the center of gravity of a plastic model of the living dinosaur. Then a correction had to be made to allow for the lungs' being full of air, not flesh (the models were solid plastic). A more sophisticated calculation might have taken account of the different proportions of bone in different parts of the body and of the possibly air-filled cavities around sauropod neck vertebrae, but such refinement seemed unnecessary in calculations that could not be made precise. The calculations showed a clear difference between *Brachiosaurus,* whose center of gravity was midway between the fore and hind legs, and *Diplodocus,* whose center of gravity was near its hips (Fig. 30.2). If the *Diplodocus* in the illustration had moved its right hind foot forward, beside the left one, both hind feet would have been under the center of gravity, and it would have been easy for the animal to rear up on them. It would have been much harder for *Brachiosaurus* to rear up, and it seems unlikely to have done so.

Dodson (1990) agrees that sauropods may have reared up on their hind legs on special occasions, for example in a fight or when all but the highest leaves had been eaten, but he thinks it unlikely that they reared up often. He also questions how much time they spent with their necks raised high, even with all four feet on the ground. One reason for these doubts is that when the head was high above the heart, very high blood pressure would be needed to get blood to the brain. Giraffes carry their brains about 3 meters above the heart and have blood pressure about double that of other mammals. If *Brachiosaurus* stood with its neck held high, its brain would be about 8 meters above its heart, and it would need blood pressure twice as high as in giraffes (Seymour 1976). Such high blood pressure would be remarkable but does not seem impossible.

Much higher blood pressure is needed when the head is held high, but

Figure 30.2. The positions of the centers of gravity of *Diplodocus* and *Brachiosaurus,* determined by experiments on models. Modified from Alexander 1989.

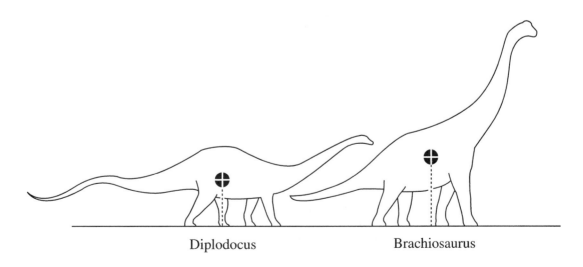

Diplodocus   Brachiosaurus

Engineering a Dinosaur   421

the forces needed to support the head and neck are greater when it is low. The reason is that the further the neck's center of gravity is in front of the shoulders, the greater the leverage of its weight. A calculation of the kind that engineers make to work out forces in girders led to the conclusion that the tension in the back muscles of *Diplodocus* must have been 13 tonnes of force when the neck and tail were held horizontal (Alexander 1985).

Diplodocus and some other sauropods have V-shaped spines projecting upward from the vertebrae of the neck and thorax. It seems possible that a ligament ran along the top of the neck, filling these Vs. Tension in it could have supported the weight of the head, saving the animal the fatigue of muscle tension—and if the ligament was elastic, the animal could easily have raised and lowered its head. Cattle and other hoofed mammals have neck ligaments like this to support their heavy heads. They are made principally of elastin, a rubber-like protein. My calculations confirm that a ligament of similar composition filling the V-shaped vertebral spines would have been strong enough to support the head of *Diplodocus* (Alexander 1985).

Dinosaur Weapons

Many dinosaurs had structures that seem to have been weapons. Some of these are so different from any weapons found on modern animals that we can only guess how they were used. No modern animals have weapons like the half-meter (20-in.) spikes on the tail of *Stegosaurus* or the heavy bony club on the tail of some ankylosaurs. Were these defensive weapons used to drive off predators, or were they used in fights between rival males?

Some other weapons have close modern parallels and seem likely to have been used in the same manner as their modern equivalents. The horns of ceratopsians, for example, resemble the horns of antelopes, which are used in ritualized fights between males competing for mates (cf. Lehman 1989; Godfrey and Holmes 1995; Sampson 1995; Forster 1996a). It seems likely that male *Triceratops* interlocked horns and wrestled, as stags and male antelopes do today. If so, we would expect to find bigger horns on males: in many antelope species the females have no horns, and in those that do, their horns are markedly more slender than those of males (Packer 1983). Lehman (1989, 1990) and Farlow (1990) point out that differences in horn thickness and shape between *Triceratops* fossils may be differences between thick-horned males and thin-horned females (it should be noted, however, that at least some of the putative sexual differences could instead indicate the existence of different species of *Triceratops*; Forster 1996b).

Triceratops horns are considerably shorter and more slender, however, than one would expect to find on an antelope of the same body mass. The horns of a 6-tonne *Triceratops* are about the same length as those of an eland of one-tenth its mass, and the bases have only about half the diameter that would be predicted for a 6-tonne male antelope, by scaling up from existing species. In making this comparison, Alexander (1989) wondered whether *Triceratops* horns were strong enough for fights between such large animals. There is no obvious way to calculate the forces, but Farlow (1990) points out that elephant tusks are also slender, though males use them for wrestling. According to his data, the bony cores of the horns of *Triceratops* have about the same cross-sectional areas as those of African elephants of similar body mass.

The remarkable domes on the skulls of pachycephalosaurs also seem to have been weapons. For example, a *Pachycephalosaurus* skull 62 centimeters long has a roof 22 centimeters thick, of spongy bone (Sues 1978). The older view was that males fought like bighorn rams, running at each other head down and colliding head to head (Galton 1971). Sues (1978) questioned this view, pointing out that whereas rams' horns interlock, the domed skulls of pachycephalosaurs would be apt to bounce off each other at unpredictable angles. This might impose dangerous torques on the neck. His alternative suggestion was that pachycephalosaurs butted the sides of rivals.

Let us think about the mechanics of head-butting and see whether the torques would really be likely to be so dangerous. We do not know how fast pachycephalosaurs ran, so we will use data for bighorn rams. Films that have been analyzed show rams running at each other at speeds up to 6 meters per second (14 miles per hour; Kitchener 1988). That looks impressive but is not actually very fast: for humans, it would be a middle-distance speed, rather than a sprint.

At that speed, the kinetic energy of the ram's head is 18 joules per kilogram of head mass. We will apply this estimate to a pachycephalosaur. The worst that would be likely to happen in a glancing collision is that the head would be deflected sideways, with this same kinetic energy. A kilogram of muscle can absorb about 170 joules of energy when stretched rapidly, so a pachycephalosaur's neck muscles could safely halt this sideways movement if the muscle on each side of the neck amounted to 11 percent of head mass.

The more rapidly a body is halted in an impact, the bigger the forces involved. That is why it is more painful to fall on rigid concrete than on a soft mattress. Animals' flanks are relatively soft, so the forces in flank-butting would not be enormous, and would not require any extraordinary thickening of the skull roof. Further, if flank-butting were used to drive off predators, a rounded dome would seem a surprising weapon: sharp horns would do much more damage. We cannot exclude the possibility of ritualized flank-butting contests between rival males, but I know no modern examples of this.

The bone of the pachycephalosaur dome seems to have had a spongy texture, so it probably deformed enough in an impact to provide useful cushioning. If the whole body were brought to a halt in head-butting, in the few centimeters that deformation of the dome might allow, the forces would be enormous. More probably, only the head decelerated so abruptly, while the back functioned like the crumple zone of a car, reducing the forces by decelerating the body over a longer distance. Kitchener (1988) analyzed films of colliding bighorn rams and estimated the decelerations of the body as no more than 34 meters per second squared, implying a force of only 3.4 times body weight. A body decelerating at this rate from a speed of 6 meters per second would come to a halt in 0.17 second, with a deceleration distance of 0.5 meter. If this calculation is accurate (and it is admittedly difficult to measure accelerations from films), the rams' backs must have bent enough for the center of mass of the body to travel half a meter forward after the initial impact. This would have allowed the back muscles to absorb a great deal of the energy of the impact. The same may have been true of pachycephalosaurs.

Engineering is the science of structure and movement, devised for the analysis of human-made structures but equally applicable to animals, living and extinct. In this chapter it has helped us to evaluate ideas about

Engineering a Dinosaur   423

the lives of large dinosaurs: how they could support their great weight, how agile they were, how sauropods may have used their long necks, and how some other dinosaurs may have fought.

References

Alexander, R. McN. 1976. Estimates of speeds of dinosaurs. *Nature* 261: 129–130.

Alexander, R. McN. 1985. Mechanics of posture and gait of some large dinosaurs. *Zoological Journal of the Linnean Society* 83: 1–25.

Alexander, R. McN. 1989. *Dynamics of Dinosaurs and Other Extinct Giants*. New York: Columbia University Press.

Alexander, R. McN. 1996. *Tyrannosaurus* on the run. *Nature* 379: 121.

Alexander, R. McN., and A. S. Jayes. 1983. A dynamic similarity hypothesis for the gaits of quadrupedal mammals. *Journal of Zoology* 201: 135–152.

Alexander, R. McN., and C. M. Pond. 1992. Locomotion and bone strength of the white rhinoceros, *Ceratotherium simum*. *Journal of Zoology* 227: 63–69.

Anderson, J. F.; A. Hall-Martin; and D. A. Russell. 1985. Long bone circumference and weight in mammals, birds and dinosaurs. *Journal of Zoology* (A) 207: 53–61.

Bakker, R. T. 1971. The ecology of the brontosaurs. *Nature* 229: 172–174.

Bakker, R. T. 1978. Dinosaur feeding behaviour and the origin of flowering plants. *Nature* 274: 661–663.

Colbert, E. H. 1962. The weights of dinosaurs. *American Museum Novitates* 2076: 1–16.

Coria, R. A., and L. Salgado. 1995. A new giant carnivorous dinosaur from the Cretaceous of Patagonia. *Nature* 377: 224–226.

Dodson, P. 1990. Sauropod paleoecology. In D. B. Weishampel, P. Dodson, and H. Osmólska (eds.), *The Dinosauria*, pp. 402–407. Berkeley: University of California Press.

Farlow, J. O. 1981. Estimates of dinosaur speeds from a new trackway site in Texas. *Nature* 294: 747–748.

Farlow, J. O. 1990. Dynamic dinosaurs [book review]. *Paleobiology* 16: 234–241.

Farlow, J. O. 1993. On the rareness of big, fierce animals: Speculations about the body sizes, population densities, and geographic ranges of predatory mammals and large carnivorous dinosaurs. *American Journal of Science* 293A: 167–199.

Farlow, J. O.; M. B. Smith; and J. M. Robinson. 1995. Body mass, bone "strength indicator," and cursorial potential of *Tyrannosaurus rex*. *Journal of Vertebrate Paleontology* 15: 713–725.

Forster, C. A. 1996a. New information on the skull of *Triceratops*. *Journal of Vertebrate Paleontology* 16: 246–258.

Forster, C. A. 1996b. Species resolution in *Triceratops*: Cladistic and morphometric approaches. *Journal of Vertebrate Paleontology* 16: 259–270.

Galilei, G. 1638. *Dialogues Concerning Two New Sciences*. English translation. New York: Dover Publications, 1954.

Galton, P. M. 1971. A primitive dome-headed dinosaur (Reptilia. Pachycephalosauridae) from the Lower Cretaceous of England and the function of the dome of pachycephalosaurids. *Journal of Paleontology* 45: 40–47.

Garland, T., Jr. 1983. The relation between maximal running speed and body mass in terrestrial mammals. *Journal of Zoology* 199: 157–170.

Godfrey, S. J., and R. Holmes. 1995. Cranial morphology and systematics of *Chasmosaurus* (Dinosauria: Ceratopsidae) from the Upper Cretaceous of western Canada. *Journal of Vertebrate Paleontology* 15: 726–742.

Hokkanen, J. E. I. 1986. The size of the largest land animal. *Journal of Theoretical Biology* 118: 491–499.

Johnson, R. E., and J. H. Ostrom. 1995. The forelimb of *Torosaurus* and an analysis of the posture and gait of ceratopsian dinosaurs. In J. J. Thomason (ed.), *Functional Morphology in Vertebrate Paleontology*, pp. 205–218. Cambridge: Cambridge University Press.

Kitchener, A. 1988. An analysis of the forces of fighting of the blackbuck (*Antilope cervicapra*) and the bighorn sheep (*Ovis canadensis*) and the mechanical design of the horns of bovids. *Journal of Zoology* 214: 1–20.

Lehman, T. M. 1989. *Chasmosaurus mariscalensis*, sp. nov., a new ceratopsian dinosaur from Texas. *Journal of Vertebrate Paleontology* 9: 137–162.

Lehman, T. M. 1990. The ceratopsian subfamily Chasmosaurinae: Sexual dimorphism and systematics. In K. Carpenter and P. J. Currie (eds.), *Dinosaur Systematics: Perspectives and Approaches,* pp. 211–229. Cambridge: Cambridge University Press.

Lockley, M. G., and A. P. Hunt. 1995. Ceratopsid tracks and associated ichnofauna from the Laramie Formation (Upper Cretaceous: Maastrichtian) of Colorado. *Journal of Vertebrate Paleontology* 15: 592–614.

Nigg, B. M. 1986. *Biomechanics of Running Shoes.* Champaign, Ill.: Human Kinetics Publishers.

Nowak, R. M. 1991. *Walker's Mammals of the World.* 5th ed. Baltimore: Johns Hopkins University Press.

Packer, C. 1983. Sexual dimorphism: The horns of African antelopes. *Science* 221: 1191–1193.

Paul, G. S. 1988. The brachiosaur giants of the Morrison and Tendaguru with a description of a new subgenus, *Giraffatitan,* and a comparison of the world's largest dinosaurs. *Hunteria* 2 (3): 1–14.

Sampson, S. D. 1995. Two new horned dinosaurs from the Upper Cretaceous Two Medicine Formation of Montana; with a phylogenetic analysis of the Centrosaurinae (Ornithischia: Ceratopsidae). *Journal of Vertebrate Paleontology* 15: 743–760.

Sereno, P. C.; D. B. Dutheil; M. Iarochene; H. C. E. Larsson; G. H. Lyon; P. M. Magwene; C. A. Sidor; D. J. Varricchio; and J. A. Wilson. 1996. Predatory dinosaurs from the Sahara and Late Cretaceous faunal differentiation. *Science* 272: 986–991.

Seymour, R. S. 1976. Dinosaurs, endothermy and blood pressure. *Nature* 262: 207–208.

Sues, H.-D. 1978. Functional morphology of the dome in pachycephalosaurid dinosaurs. *Neues Jahrbuch für Geologie und Paläontologie Monatshefte* 1978 (8): 459–472.

Thulborn, R. A. 1990. *Dinosaur Tracks.* London: Chapman and Hall.

Webb, G. J. W., and C. Gans. 1982. Galloping in *Crocodylus johnstoni:* A reflection of terrestrial activity. *Records of the Australian Museum* 34: 607–614.

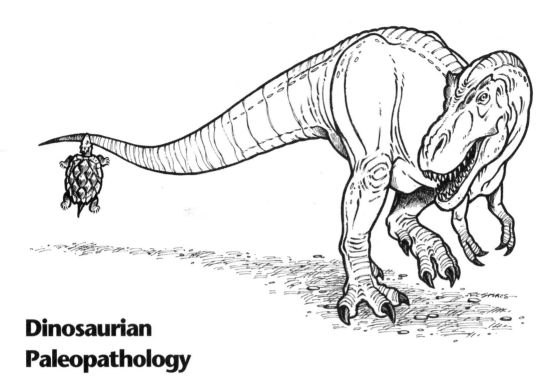

Dinosaurian Paleopathology

Bruce M. Rothschild

Dinosaurs were relatively healthy Mesozoic animals. While the title of this chapter refers to paleopathology, it perhaps more properly could be called paleo-health (Rothschild and Martin 1993). The isolated nature of most examples of pathology is compatible with three hypotheses: (1) The animals were actually quite healthy; (2) any pathologic conditions present were so catastrophic that the animal did not survive long enough to develop osseous (bony) changes (which are recognizable as disease); and (3) their health was so good that any bone pathology healed so fully as not to be recognizable.

Alterations in bony remains are easiest to identify in the fossil record. While other diseases affecting soft tissues may have been present, they are as yet unidentified. Infectious diseases which in people can rapidly cause death (e.g., influenza) cause no bony changes and would not be directly recognizable. Genetic abnormalities would be equally difficult to identify. Those factors limiting life span or increasing susceptibility to infectious disease usually do not affect the skeleton and will not be recognized on gross examination.

31

Recognition of Bone Paleopathology: Problems and Procedures

Paleopathology, the study of ancient disease (Parks 1922; Moodie 1923, 1930; Swinton 1981; Sawyer and Erickson 1985; Monastersky 1990; Rothschild and Martin 1993), originally described the study of curiosities, sometimes with quite ridiculous conclusions (Anonymous 1934). The field was initially given full recognition in 1895, when it first appeared in the dictionary. Sir Marc Armond Ruffer was the first paleopathologist (Swinton 1981). He reported "spondylitis" in "sacred monkeys of the ancient temples near Thebes" and fusions of vertebrae in the lower Miocene crocodilian *Tomistoma dawsoni,* and in the cave bear *Ursus spelaeus* (Ruffer 1921).

The first dinosaur reports dealt predominantly with injuries and fusions (Moodie 1923; Brown and Schlaikjer 1937; Sternberg 1940; Tyson 1977). Current approaches allow exploration of details of dinosaur lifestyle and behavior (Rothschild and Martin 1993), similar to the insights provided by analysis of dinosaur trackways (Lockley 1986, 1989). The latter have concentrated on the gait patterns of apparently normal individuals and groups. While isolated pathologic theropod (possible *Tyrannosaurus*) footprints have been reported (Currie, Lockley, and Tanke, personal communication; Ishigaki 1986), I am unaware of any reports specifically analyzing gait disturbance in such individuals. Tucker and Burchette (1977) reported a digit III malformation in the trackway with footprints given the name *Anchisauripus,* a theropod, while Thulborn (1990) commented on apparent absent toes.

The field of paleopathology became a science with the introduction of hypothesis testing (Rothschild and Martin 1993). Recognition that bony tissues survive the process of fossilization (Dollo 1887; Broili 1922), along with our augmented ability to recognize artifacts introduced by diagenesis, has allowed such testing. Electron microscopic examination revealed intact microstructures in Triassic specimens (Isaacs et al. 1963). Structural integrity is not just limited to histology. Wyckoff (1980) reported detection of bone collagen in dinosaur bone, dated at 200 million years before the present. The potential for application of immunologic and perhaps even DNA techniques to dinosaurs is only now being explored (Rothschild and Martin 1993). Collaborative efforts by physicians, paleontologists, veterinarians, botanists, geographers, meteorologists, agronomists, entomologists, and geologists, first recommended by Sigerist in 1951, are essential to this process.

Critical to recognition of pathology is the ability to recognize changes in bone which are not disease-related, and which did not occur while the animal was alive. These changes are referred to as diagenetic, and the processes responsible for them are collectively termed diagenesis. Simple pressure of overlying sediment may result in postmortem bone breakage. As water percolates through buried bones, conditions of relative acidity and alkalinity may dissolve minerals from them. The pressure of overlying sediment on such softened bone may alter its shape or "bend" it. The results resemble the bowing of long bones (from weight-bearing) that can occur in humans with vitamin D deficiency (i.e., rickets or osteomalacia), even though such changes in human bones obviously are not diagenetic.

Additionally, minerals (leached from overlying sediment) may be deposited in buried bone. Thus bone density assessment by chemical or physical analysis of mineral content may not be valid. Diagenesis includes the damage from the burial process itself. Fractures (broken bones) are especially difficult to interpret. Only fractures that have evidence of healing can unequivocally be considered pathology. Fractures occurring within several weeks of death may not be distinguishable from sediment-induced breakage, or from damage from the effects of trampling on the bones by large animals. Diagenetic changes may occur from tumbling of the bone in a river stream, which smooths and grinds down articular surfaces.

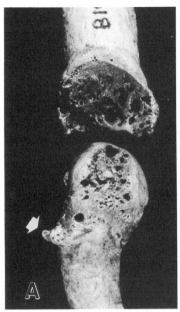
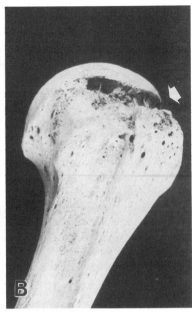

Figure 31.1. Effects of erosive arthritis. (A) Anterio-lateral view of the metacarpal-phalangeal (between hand and finger bones) joint in a gorilla with spondyloarthropathy. Erosive arthritis produced holes/erosions in the subchondral bone (*arrow:* bone originally covered by cartilage at the site of new bone formation). (B) Posterio-lateral view of the proximal humerus of a human with rheumatoid arthritis. Erosive arthritis produced erosion of margin of articular surface (*arrow*).

As rich sources of phosphate, both unfossilized and fossilized bones may "attract" plant roots. Insects may bore holes in and animals may gnaw on the bones of recently dead animals. The presence and pattern of such diagenetic changes provide insights into the speed of burial events after death. If bones are not buried immediately, their exposure to extremes of heat and cold may cause the periosteum (the outer layer of the bone cortex) to separate from the cortex. This creates a periosteal elevation (separation) image that mimics the effects of bleeding into the bone during the animal's life, a condition that occurs in vitamin C deficiency (scurvy). Scurvy might be diagnosed in a dinosaur on the basis of an isolated finding, if this were not considered. Because scurvy is a systemic disease (affecting many long bones), periosteal elevations in one or a few bones of an animal are more likely to represent diagenesis.

One intriguing form of bone pathology is erosive disease (Fig. 31.1). Areas of bone resorption may be caused by infection, but microfossil studies offer an interesting caveat. Studies of regurgitated owl pellets often reveal partially digested rodent bones. The latter contain areas of bone resorption (digestion) that may be very difficult to distinguish from osteoclastic bone resorption caused by disease.

Complementing routine x-ray techniques is the application of computerized tomographic (CT) scanning. Digital (computer) reconstruction techniques now allow us to generate three-dimensional images of a bone, or even to non-destructively "dissect out" the areas of interest or allow the area of interest to be visualized from almost any angle or perspective (Artzy et al. 1980; Skinner and Sperber 1982; Conroy and Vannier 1985; Virapongse et al. 1986; Woolson et al. 1986; Hadley et al. 1987; Farrell and Zappulla 1989). Though limited in resolution to only a 1-millimeter-thick slice, the three-dimensional reconstruction image is occasionally so reliable that the actual fossil preparation requirements may be reduced, and the preparator may even be directed in the process. The encoded data can even be utilized to physically produce three-dimensional models (Roberts et al. 1984; Sundberg et al. 1986). Perhaps the most revolutionary finding is that Magnetic Resonance Imaging (MRI) techniques can be applied to fossils (Sebes et al. 1991).

Current MRI is based on the magnetic properties of hydrogen. Atoms with odd numbers of protons produce a small magnetic field. Because of hydrogen's odd number of protons, compounds containing hydrogen can be polarized. A powerful magnet aligns the individual hydrogen atoms. Short exposure to additional magnetism changes the direction of the polarization. The MRI measures the extent of change and the time needed for the change in polarity to return to baseline normal. For polarization to occur, however, the protons of the hydrogen atom must not be constrained. Current MRI consequently does not alter or measure the proton fluxes (changes) in all hydrogen-containing substances. Hydrogen in cortical bone, for example, is tightly bound, and bone is not visualized (i.e., it appears black) on MRI. Air also contains insufficient hydrogen for it to be visualized. Air therefore also appears black. Lipid and water, however, have sufficient hydrogen and, more important, "mobile protons." Mobile protons are required to visualize a structure with MRI. Therefore, lipid- and water-containing structures can be visualized.

Fossils, lacking many "mobile" protons, cannot be visualized by MRI. Simply placing a fossil in water or a light oil, however, sufficiently "proto-nates" it to allow its internal structure to be visualized (Rothschild and Martin 1993).

Exostoses

Perhaps the simplest form of recognizable pathology is that related to abnormalities of bone growth. While bone size is one manifestation of growth, growth is not always symmetrical. Asymmetry may take the form of overgrowth of a portion or even a fragment of bone. If the outer layer of

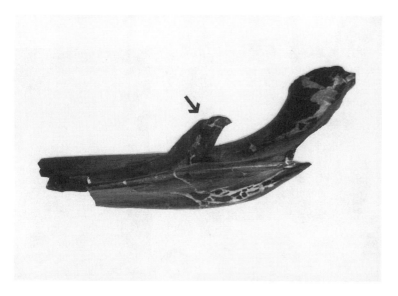

Figure 31.2. Ventro (inferio)-medial view of a scapula (shoulder blade) of *Triceratops* (National Museum of Natural History [USNM] specimen number 8013). Note anteriorly directed, hook-like exostosis (*arrow*).

Figure 31.3. Lateral view of the chest region of a skeleton of *Allosaurus* (United States National Museum [USNM] 4734). Note exostoses on the scapula, which normally is a long, thin bone.

bone (periosteum) is responsible in part for bone growth, any process that results in areas of partial periosteal detachment will produce isolated bone asymmetry. If one end of the periosteum is pulled free (e.g., by a tendon avulsion—tendon ripping off the bone), the remaining bone may have a ragged appearance with projecting bone spicules. Continued growth results in an area of new bone formation. The new bone is called an exostosis. If the bone contains a fragment of cartilage, growth may occur, producing an exostosis with a cartilage cap. The latter is referred to as an osteochondroma. Overgrowth of bone with spicule formation is manifest in exostoses of a *Triceratops* mandible and scapula (Fig. 31.2) and in an *Allosaurus* scapula (Fig. 31.3; Gilmore 1919; Rothschild 1989b).

Figure 31.4. "Divots" in articular surfaces of hadrosaur phalanges. (A) Superior (dorsal) view of Museum of the Rockies (MOR) specimen number 553. The "scooped-out" area is prominent. (B) Radiograph of another hadrosaur phalanx, Royal Tyrrell Museum (RTM) specimen number P.67.9.61. Note focal excavation/indentation, compared with normal phalanx (C).

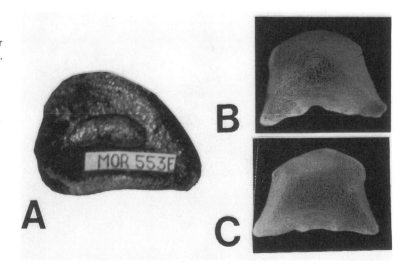

Divots

Erosions of bone articular surfaces have not been observed in dinosaurs. Other surface defects, however, have been observed. In appearance these defects (Fig. 31.4) are quite similar to the divot of grass which an inconsiderate golfer fails to replace. This condition has been observed so far only in lambeosaurine hadrosaurs. These divots may represent developmental cartilage defects, that is, cartilage islands which never ossified.

Another divot-like structure was noted in a specimen of *Tyrannosaurus* (Fig. 31.5). It consists of a defect in the humerus. Localized to the site of tendon attachment, it appears to represent a tendon avulsion (the site where a tendon was ripped off the bone). It appears similar to occasional lesions found in human distal femora, which are likewise attributed to tendon avulsion injuries (S. Murphy, personal communication).

Fractures

Evidence of trauma in dinosaurs is represented by fractures in various stages of healing. Only a healed or partially healed fracture can be distinguished from postmortem damage. Thus fractures less than a few weeks old would probably not be recognizable as having occurred during life. Given this caveat, there is very little evidence of clearly survived injuries which involved osseous structures in dinosaurian herbivores. Only isolated records of injuries exist, such as sauropod rib fractures (Riggs 1903), a wound of the ilium and fractures of the basal tail of *Camptosaurus* (Gilmore 1909, 1912), and an ischial fracture in *Iguanodon* (Blows 1989).

Fractures and injuries are fairly common in many theropods. Forelimb and foot fractures, perhaps received while struggling with prey, were present in 27 percent and 25 percent, respectively, of theropod specimens examined (Fig. 31.6; cf. Molnar and Farlow 1990; Molnar in press). Vance (1989), however, suggested an alternative hypothesis, that forelimb fractures might have occurred during mating activities. Madsen

Figure 31.5. Anterior (A) and lateral (B) views of casts of the humerus of a *Tyrannosaurus* specimen informally known as "Sue." The "divot" shown in A (*arrow*) was probably caused by tendon avulsion. Note the spicule of bone in B (*arrow*) probably related to that process. (C) Lateral radiograph of a human foot. Note similar divot (*arrow*).

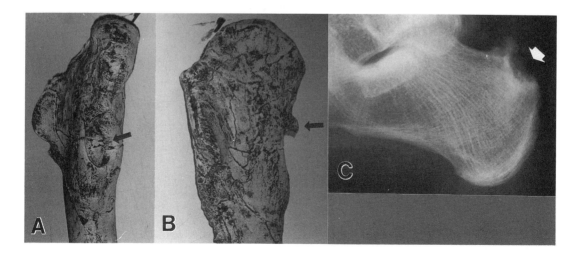

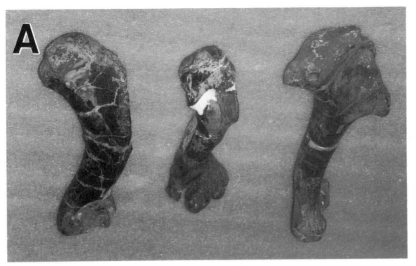

Figure 31.6. (A) Lateral view of MOR 379, a humerus of *Albertosaurus* (middle bone), compared to normal humeri of this dinosaur. Note shortening and deviation of distal portion. (B) Ventro-medial view of MOR 379.

(1976) reported a fractured *Allosaurus* radius (forearm bone), and Molnar (in press) noted metatarsal fractures in *Syntarsus* and a phalangeal fracture in *Deinonychus*. Lambe (1917) reported fractured ribs and fibula (lower leg) in *Albertosaurus* (*Gorgosaurus*), and Russell (1970) reported distal humeral pathology in *Albertosaurus*. Curiously, most theropod forelimb fractures involve the more proximal bone. Tyrannosaurid humeral fractures have been reported by Vance (1989), Larson (1991), and Molnar and Farlow (1990), complementing Larson (1991) and Petersen et al.'s (1972) report in *Allosaurus*. While survival of fractures of major weight-bearing bones may seem unlikely, Molnar (in press) reported a healed tibial (lower leg) fracture in *Syntarsus* and a healed fibula fracture in *Allosaurus;* fractured fibulae occur in *Albertosaurus* and *Tyrannosaurus* (my own observations and D. Tanke, personal communication). Molnar (in press) noted that 13 percent of observed theropod pathologies affected the skull or mandible and 27 percent the vertebral column or ribs (cf. Larson 1991).

The only systematic, quantitative studies of dinosaur bone fractures have been carried out on Late Cretaceous ceratopsians and hadrosaurs from Alberta (Tanke in press a, b). These were made possible by detailed field examination and collection of all specimens from dinosaur bonebeds. An approach that collected only perfect, complete specimens would underestimate the frequency of pathology, while special attention to collection of pathologic specimens would bias the frequency estimate in the opposite direction. Tanke's (in press b) studies of some 30,000 bones are especially enlightening. He found fractures in 0.025 to 0.5 percent of *Centrosaurus* and 0.2 to 1.0 percent of *Pachyrhinosaurus* bones. Most were localized to mid- and posterior dorsal ribs. Fractures in centrosaurines are quite similar in distribution to the mid- and posterior dorsal rib fractures noted in adult male American bison (McHugh 1958). The flank-butting behavior of the latter is by inference suspected in the former (Tanke in press b). Cases of horn and frill disruptions provide further evidence for intraspecific fights. These include *Triceratops* (Gilmore 1919; Erickson 1966; Czerkas and Czerkas 1990), *Torosaurus* (Tokaryk 1986; Czerkas and Czerkas 1990;

Johnson 1990), *Pentaceratops* (Czerkas and Czerkas 1990), and *Diceratops* (Gilmore 1906).

In hadrosaurs, fractured proximal caudal neural spines and caudal fusions predominate (Fig. 31.7; Tanke in press a). Tanke has speculated that these injuries are related to mating activities (related to the effect of the male's weight resting on the female's tail and delicately constructed neural spine tips). Gilmore (1909, 1912) has previously blamed such injuries in *Camptosaurus* on copulation, as has Blows (1989) for *Iguanodon*.

Stress Fractures

Fractures generally occur as the direct effect of a specific trauma or injury or when the mechanical integrity of bone is suddenly overcome (Rothschild and Martin 1993). Repetitive stresses, below the threshold for acute fracture induction, may also leave their imprint on bone. Unconditioned stresses (novel stresses, those to which the bone has not been previously exposed) are usually responsible for this condition in humans. A military recruit who suddenly marches twenty miles subjects his or her metatarsals (foot bones) to repetitive stress/trauma. This contrasts with "building endurance" by walking a mile a day for a week, then a few miles a day for a week, thus gradually conditioning the body/bones to the stresses/trauma. The recruit instead plunges right into an activity. Particularly if an exercising individual fails to warm up and stretch to get the limbs ready, the stress of activity results in sudden shocks to vulnerable bones. If an activity is performed at a rate or to an extent that overcomes the physical strength of the bone, a stress fracture may occur. These occur in the metatarsals, tibia, femur, and even the lumbar spine pedicles of ballet dancers. Cross-country running can produce stress fractures in the midfoot, tibia, and femur, javelin throwing can result in stress fractures of the distal humerus and proximal ulna, and jumping can cause such injuries to the calcaneum (heel) and proximal fibula.

When unconditioned, repetitive stress forces overcome the intrinsic strength of bone, microfractures and accelerated bone remodeling result in stress or fatigue fractures (Hartley 1943; Daffner 1978; Worthen and Yanklowitz 1978; Rothschild 1982). These occur when the rate of stress-induced bone remodeling (buttressing of the stressed area) does not

Figure 31.7. Left lateral view of a hadrosaur proximal tail vertebra (MOR 7.20.92.37). Note healed spinous process fracture (*arrow*).

Dinosaurian Paleopathology   433

"outpace" the rate and extent of stress application (Goodship et al. 1979). The most common sites of stress fractures in humans are the metatarsals, but phalangeal stress fractures also have been noted (Morris and Blickenstaff 1967; Wilson and Katz 1969; Daffner 1978; Orava et al. 1978; Rothschild 1982).

Stress fractures are recognized in x-rays because the actual defect may cause a partial separation. The separation appears black in comparison to the white x-ray image. It is therefore described as radiolucent, because it does not block the x-ray photons which cause the film to turn black. Intact bone covers part of the x-ray film, limiting exposure to the x-ray photons and resulting in the bone's appearing relatively white or clear. Stress fractures in otherwise normal bone are typically oblique, compared to the long axis of the bone. When the stress fracture occurs in abnormal bone (e.g., Paget's disease), it is more commonly perpendicular to the long axis of the bone. Both types of stress fractures typically appear as thin lucent areas.

Oblique radiolucent knife-slice-like clefts are diagnostic of this condition, but periosteal overgrowth may be the only visible bone alteration. The latter is due to the callus/bone healing response (Wilson and Katz 1969). Apart from these radiolucent clefts, the most recognizable signs of stress fractures are indistinct overgrowth of the outer layer (periosteum) of bone and a focal increase in bone density (sclerosis; Wilson and Katz 1969).

Bone infection, tumors, and even trauma-induced bone contusion may mimic features of stress fractures, but the radiolucent clefts present in stress fractures adequately distinguish the latter from these other conditions. Unfortunately, there are features of another ailment that resemble the radiolucent clefts of stress fractures. These are the so-called Looser lines of osteomalacia (vitamin D deficiency). Looser lines are pseudofractures that appear in x-rays as clear zones, 1 to 10 millimeters thick, of poorly mineralized bone. While stress fractures may have associated thickening of the outer layer of bone cortex (periosteal reaction), the pseudofractures of osteomalacia do not.

Figure 31.8. (A) Lateral view of a ceratopsian phalanx (National Museums of Canada 40741). Bump (*arrow*) indicates a stress fracture. (B) Lateral radiograph of the same bone. Note knife-slice-like cleft (*arrow*) at the site of the stress fracture.

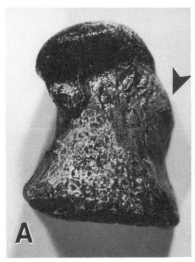
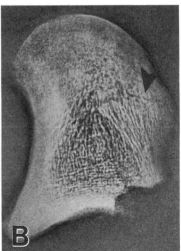

Linear stress fractures, unassociated with bone destruction, have been found in ceratopsian phalanges (Fig. 31.8) (Rothschild 1988a). They are localized to the proximal phalanges of digits II through IV of *Centrosaurus*, *Styracosaurus*, *Pachyrhinosaurus*, and *Triceratops* (Rothschild and Martin 1993). The presence of associated periosteal reaction allows confident diagnosis of stress fractures and eliminates consideration of osteomalacia (vitamin D deficiency). It is suspected that the ceratopsian stress fractures occurred because of foot stamping or sudden accelerations (a possible predator response), or perhaps during long-distance migrations.

Dental Pathology

Dental pathology also appears to have been quite rare. There are no examples among thousands of hadrosaur teeth observed in the field in Alberta or in collections of the Royal Tyrrell Museum of Palaeontology (Rothschild and Tanke 1991; D. Tanke, personal communication). A dental abscess was, however, reported by Moodie (1930) in a cf. *Lambeosaurus* jaw (Olshevsky 1978). The bulging deep pocket was associated with tooth-row groove destruction and tooth loss. Other dental pathologies that have been reported include occasional broken teeth in theropods (Farlow et al. 1991; Farlow and Brinkman 1994), lost teeth in a specimen of *Allosaurus* (formerly known as *Labrosaurus ferox*; Fig. 31.9), and malocclusion in an uncataloged *Tyrannosaurus* maxilla (Fig. 31.10). Gilmore (1930) also reported an "injured right premaxilla" in the nodosaur *Edmontonia* (*Palaeoscincus*).

Infectious Disease

Joint (articular) infections have not yet been positively identified in a dinosaur (Rothschild and Martin 1993), although that possibility exists in a specimen of *Tyrannosaurus* informally known as "Sue." Infection of

Figure 31.9. Lateral view of USNM 2315, lower jaw of *Allosaurus* (originally described under the name *Labrosaurus*). Note the prominent groove, with missing teeth, adjacent to distal portion of the bone.

Dinosaurian Paleopathology 435

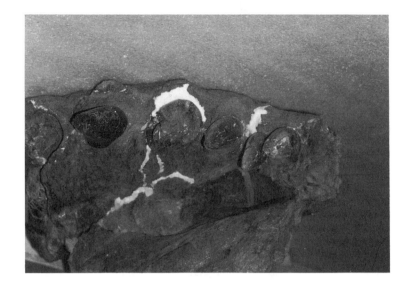

Figure 31.10. Ventral view of an uncatalogued *Tyrannosaurus* maxilla in the collection of the Museum of the Rockies. The second tooth from the right is malaligned.

bone (osteomyelitis) is also rare. These infections were either rare or rarely survived events (Rothschild and Tanke 1991). Most infections appear secondarily related to trauma, such as Moodie's (1926) report of a fractured and infected hadrosaur manus. A fracture would likely have penetrated the thin overlying skin. The resulting injury, termed a compound fracture, probably became infected, in the same manner as compound fractures in human bones. A *Camptosaurus* ilium developed osteomyelitis at the site of a possible bite mark, documenting its probable origin (Moodie 1917). While isolated reports have been identified, evidence for infection, or at least for survival of acute infections, is rare. They are infrequently observed, even in large samples of bones from the same taxon. Isolated reports include an abscessed *Dilophosaurus* humerus (Molnar in press), an *Allosaurus* phalanx (Madsen 1976), and a *Troodon* parietal (Molnar in press). Infection of the outer layer of bone (infectious periostitis) in a ceratopsian scapula (Rothschild and Martin 1993), and an isolated *Triceratops* coronoid exhibiting a destructive process (compatible with infection) have been noted (D. Russell, personal communication).

Osteoarthritis

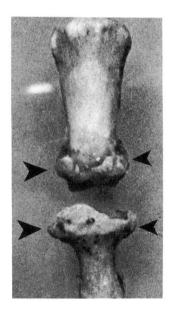

Figure 31.11. Anterior view of a human distal interphalangeal joint. Osteophytes (*arrows*) are prominent.

The most common form of arthritis in *Homo sapiens* is osteoarthritis. Almost universal in people older than seventy-five (Lawrence et al. 1966), it is recognized in defleshed bone on the basis of bony articular surface overgrowths, called spurs or osteophytes (Fig. 31.11). This is in contrast to vertebral centrum marginal spurs (also called osteophytes), which are indicative of a phenomenon called spondylosis deformans (Fig. 31.12). The latter spurs parallel the flat ends of the vertebrae, and are possibly caused by tears in the anulus fibrosus and disk prolapse. Intervertebral disks have two components, the inner nucleus propulsus and the surrounding anulus fibrosus. Weakness or tears in the anulus fibrosus allow the nucleus propulsis to extrude, referred to as protrusion. While osteoarthritis is frequently a source of pain or disability, spondylosis deformans is an

436   *Bruce M. Rothschild*

asymptomatic phenomenon. It is found in approximately 60 percent of women and 80 percent of men over age fifty.

Osteoarthritis has traditionally been considered a common disorder in dinosaurs (Abrams 1953; Jurmain 1977; Norman 1985). I was always intrigued by that statement, because I had been unable to identify a single instance among the remains of 10,000 dinosaurs (Rothschild 1990). It was only after a visit to the Royal Museum of Natural History in Brussels, Belgium, that two examples (Fig. 31.13) were finally identified, in *Iguanodon* (see below). The belief that osteoarthritis was common in dinosaurs is erroneous. One of the major signs of osteoarthritis is growth of osteophytes or spurs. While the centrum of a vertebral body is often the site of spur formation, that is a disk space, not a joint (Rothschild 1989a, 1989b), and so spur formation here cannot be regarded as osteoarthritis. Because vertebral body osteophytes are not a cause of pain or morbidity, they are far removed from what is usually recognized as osteoarthritis. Such lesions represent spondylosis deformans.

Osteophytes on joints, however, are sufficient to establish the diagnosis of osteoarthritis (Rothschild 1982; Resnick and Niwayama 1988; Rothschild and Martin 1993). Eburnation (grooving of articular surfaces) occurs with severe osteoarthritis. Other signs of osteoarthritis include increase in bone density and cyst formation. These occur just under articular surfaces and can be recognized on cross-section or by x-ray. In addition to showing remodeling with spur formation, x-rays reveal increase in density of the bone underlying the articular surface (subchondral bone) and, occasionally, formation of cysts in that area. Microfractures appear to produce such holes (cysts) (Meachim and Brooke 1984). The quality of preservation of many fossil bones is so excellent that the actual trabecular pattern is visible on cross-section or x-ray examination (Rothschild and Martin 1993). Thus, x-ray of dinosaur bones is a viable technique for identifying pathology.

Figure 31.12. Lateral view of a vertebra of *Edmontonia* (USNM 11868). The marginal overgrowth of the vertebral centrum (*arrow*) is called an osteophyte, indicative of spondylosis deformans. This is not osteoarthritis.

Osteoarthritis has not been found in hadrosaurs, ankylosaurs, theropods, stegosaurs, ceratopsians, pachycephalosaurs, or sauropods (Rothschild 1990). Neither examination of intact bones nor x-rays of those bones have revealed any evidence of osteophytes, eburnation, cysts, or subchondral (below articular surfaces) sclerosis. The absence of osteoarthritis in sauropods such as *Camarasaurus, Apatosaurus, Diplodocus, Haplocanthosaurus,* and *Barosaurus* indicates that weight was not the major determinant of osteoarthritis development in dinosaurs. While osteoarthritis has not been found in one genus of iguanodont, *Camptosaurus* (Rothschild 1990), the only instances of dinosaurian osteoarthritis in weight-bearing bones were in a related form, *Iguanodon*. Two specimens of *Iguanodon* in a sample of thirty-nine individuals had ankle osteophytes (Fig. 31.13; Rothschild and Martin 1993).

While excess weight (obesity) has been suggested as important in the development of human osteoarthritis (Silberberg and Silberberg 1960; Sokoloff et al. 1960; Saville and Dickson 1968; Leach et al. 1973), the stability of the joint appears to be more important (O'Donoghue et al. 1971; Jurmain 1977). Highly stabilized joints are those which have movement essentially limited to one plane of motion (e.g., elbow-hinge joints). Such highly stabilized joints appear to be protected from osteoarthritis (Harrison et al. 1953; Puranen et al. 1975; Funk 1976; Cassou et al. 1981). In contrast, the least constrained joints (with complicated movement) are most susceptible to osteoarthritis (Radin 1978). The human knee, for example, is not simply a hinge joint, but actually has a rotational component to its motion. It is therefore minimally constrained and highly

Figure 31.13. Anterior view of the ankle of *Iguanodon* (Royal Museum of Natural History, Brussels, Belgium). Note overgrowth of articular margins forming osteophytes (*arrows*).

Dinosaurian Paleopathology   437

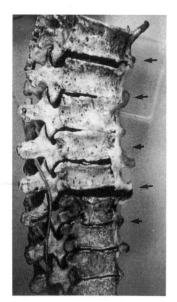

Figure 31.14. Lateral view of human thoracic vertebrae. Notice ossification (*arrows*) giving the appearance of candle wax having "dripped down the spine." This condition is diffuse idiopathic skeletal hyperostosis (DISH).

susceptible to injury and osteoarthritis. The low frequency of osteoarthritis in dinosaurs suggests that they generally had highly constrained joints, most probably hinge type, with little rotatory movement.

Inflammatory Arthritis

Gout is a painful inflammation of the joints, particularly those of the hands and feet, that is associated with excessively high levels of uric acid in the blood. Bones affected by gout develop characteristic spheroidal erosion surfaces, often with new bone growth at their borders.

Gout has been identified in two individual carnivorous dinosaurs (Rothschild et al. 1997). Two hand bones of a *Tyrannosaurus* from the Hell Creek Formation of South Dakota show the osteological signature of gout, as does a tyrannosaurid toe bone from the Dinosaur Park Formation (Judith River Group) of Alberta. In humans, at least, gout is associated with the consumption of food with a high purine content, such as red meat. Although the incidence of gout in tyrannosaurids seems to have been low, perhaps they, too, occasionally suffered the consequences of too rich a diet.

Diffuse Idiopathic Skeletal Hyperostosis

The normal alignment of the vertebral column is partially maintained by ligaments that run between bones. Chief among these are the longitudinal ligaments. Ossification of vertebral longitudinal ligaments defines a condition known as diffuse idiopathic skeletal hyperostosis, or DISH (Oppenheimer 1942; Forestier and Lagier 1971; Resnick et al. 1978), also called Forestier's disease. A tortuous paravertebral calcification, anterior and lateral to the vertebral centra, gives the appearance that candle wax has dripped longitudinally along the spine (Fig. 31.14; Rothschild and Berman 1991). This is a ligamentous phenomenon, sparing the zygapophyseal (facet) and sacroiliac joints. DISH rarely occurs in humans under age forty, but is present in approximately 20 percent of men and 4 percent of women over age fifty (Rothschild 1985).

While the earliest sign of DISH appears to be new bone formation adjacent to the midportion of the vertebral body (Fornasier et al. 1983), the ligamentous calcification is mainly separate from the actual vertebral centra. Radiologically, this is recognized as a radiodense line (from several millimeters to more than a centimeter thick) paralleling the long axis of the vertebral centra, but separated from it by a clearly defined space (Resnick et al. 1978; Rothschild 1985; Rothschild and Martin 1993). Such ligamentous ossification is not always limited to vertebrae, and may also occur at sites of tendon, ligament, or joint capsule insertion.

Its pathological appearance notwithstanding, DISH appears not to be a disease. No associated pathology has been identified in humans, with the exception of a possible association with diabetes (glucose intolerance) (Rothschild 1985; Schlapbach et al. 1989). In fact, back and neck injuries and complaints are more common in the general population than in people with DISH. It is a phenomenon rather than a disease. There are no proved hazards from its presence (Rothschild 1985; Schlapbach et al. 1989), and it may even protect one from back pain (Rothschild 1989b; Rothschild and Martin 1993).

DISH must be distinguished from infectious spondylitis. Coalescence of vertebrae with spongy, reactive new bone formation often characterizes the latter. This produces a filigree appearance, very similar to the intricate goldwork seen in Faberge eggs. It is a minute lacy pattern of new bone.

Two other conditions can cause ligamentous ossification: fluorosis and hypervitaminosis A. Ingestion of excessive quantities of fluoride produces fluorosis, while excess ingestion of vitamin A produces hypervitaminosis. Hypervitaminosis A has been noted as an acute poisoning phenomenon in Arctic explorers (who ate polar bear liver), and also as a chronic intoxication. The latter results in ossification at sites of tendon, ligament, and joint capsule insertion. Hypervitaminosis A has characteristic bone histology. While the ligamentous ossification of DISH has the appearance of Haversian (lamellar) bone, fluorosis and hypervitaminosis A are characterized by a disorganized, non-lamellar bone pattern of calcification. Exuberant periosteal reaction (overgrowth of the outer layer of cortical bone) is found in individuals with hypervitaminosis A (Seawright and English 1965; Pennes et al. 1985), while irregular periosteal thickening is reported in fluorosis (Singh et al. 1962).

Ossification of spinal ligaments actually is the rule, rather than the exception, in most dinosaurs. Just as in human DISH, Haversian systems are found on cross-section (Rothschild and Martin 1993), confirming the osseous transformation of the ligaments. Ceratopsians, hadrosaurs, iguanodonts, and pachycephalosaurs had ossified ligaments, apparently acquired as juveniles (Moodie 1926, 1928; Rothschild 1985, 1987a, 1987b; Rothschild and Tanke 1991). Because it is now recognized that dinosaurs were not "tail-draggers," but actually held their tails off the ground, the presence of such a mechanism to stiffen the tail is not surprising. The exceptions perhaps prove the point. Those dinosaurs likely to use their tails as weapons (e.g., stegosaurs) did not have such generalized spinal ligamentous fusion.

The distribution of ligamentous fusion in ankylosaurs is even more dramatic. The only fusion occurring in ankylosaurs is apparently in those genera with tail clubs, where it is limited to the distal caudal vertebrae. This would appear to provide greater stability to the tail appendage, making it a more effective weapon. Gilmore (1930) also reported a "kinked tail"

Figure 31.15. Diffuse idiopathic skeletal hyperostosis. (A) Lateral view of Carnegie Museum of Natural History specimen number 94, *Diplodocus,* showing fusion of two caudal vertebrae (*white arrow*). Note normal zygapophyseal (facet) joints. (B) Computerized tomogram of fused vertebrae of *Diplodocus* (American Museum of Natural History [AMNH] 655). Note separation (*clear black spaces*) between the ossified ligaments and the vertebral centra.

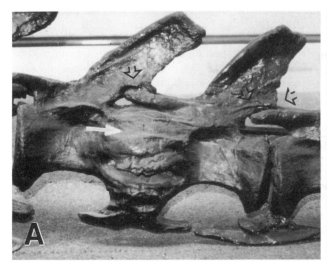
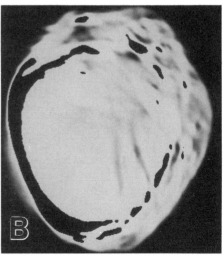

with fused caudals in the nodosaur *Edmontonia* (*Palaeoscincus*), which also had an "injured right premaxilla." Trauma is suspected as the source of pathology in this individual.

The occurrence of DISH in sauropods is even more interesting. Fusion of two to four caudal vertebrae in *Diplodocus* has been the source of much speculation. While some investigators originally considered the phenomenon post-traumatic (perhaps related to stresses associated with rearing up on hind legs), the preservation of intact zygapophyseal (facet) joints did not fit that hypothesis. The absence of recognized stress fractures in sauropods also suggests that the animals were not subjected to the stresses of such behavior (Rothschild and Martin 1993). Routine x-rays of affected vertebrae were inconclusive. Computerized tomography of such specimens, however, revealed separation between the ossified ligaments and the vertebral centra (Fig. 31.15). DISH limited to caudal vertebrae 17 to 23 is the predicted site for buttressing a cantilever structure. Its presence helps to explain why sauropod trackways don't show tail drag marks (Thulborn 1984). Such fusion (by DISH) facilitated the tail's being kept in the air, as suggested by Bakker (1968) on the basis of the trackway record. Although DISH may have facilitated defensive use of the tail (Hatcher 1901; Holland 1915), I speculate that its function was more likely related to intraspecific territorial or courting behavior or that it was even integral to the mating act (Rothschild 1987b; Rothschild and Berman 1991). Because the phenomenon was present in 50 percent of *Diplodocus* and *Apatosaurus* and 25 percent of *Camarasaurus* specimens examined (Rothschild and Berman 1991), I believe it to represent sexual dimorphism. Tail stiffening in the female would maintain the cloaca in a position more accessible for mating.

Vertebral Fusion

Varieties of arthritis in humans that produce fusion of peripheral or axial joints (such that the bones become conjoined/co-ossified) are categorized as forms of spondyloarthropathy (Bywaters 1960; Martel 1968; McEwen et al. 1971; Rothschild 1982; Ortner and Putschar 1985; Resnick and Niwayama 1988). That term encompasses several subgroups: ankylosing spondylitis, Reiter's syndrome, psoriatic arthritis, and the arthritis associated with inflammatory bowel disease (ulcerative colitis or Crohn's disease) (Rothschild 1982; Resnick and Niwayama 1988). Reiter's syndrome or reactive arthritis is a disorder related to indirect effects of infectious-agent diarrhea (food poisoning) or is sexually transmitted (by *Chlamydia*). Spondyloarthropathy classically affects the vertebral centra and posterior (zygapophyseal or facet) joints with erosions or fusion (Rothschild 1982; Kelly et al. 1985; Resnick and Niwayama 1988; Katz 1989; Rothschild and Martin 1993).

Calcification of the anulus fibrosus (outer layer of the intervertebral disc) produces vertebral bridgings, referred to as syndesmophytes (Fig. 31.16). Syndesmophytes are overgrowths of bone paralleling the longitudinal axis of the animal, typically bridging between vertebral body centra. With full fusion, the adjacent centra may look like a piece of bamboo. The term *bamboo spine* is used to describe this phenomenon. Such vertebral centra and zygapophyseal joint fusion are specific for spondyloarthropathy in humans. These lesions are easily distinguished from the destructive lesions with exuberant new bone formation characteristic of vertebral infection (infectious spondylitis) (Rothschild 1982; Resnick and

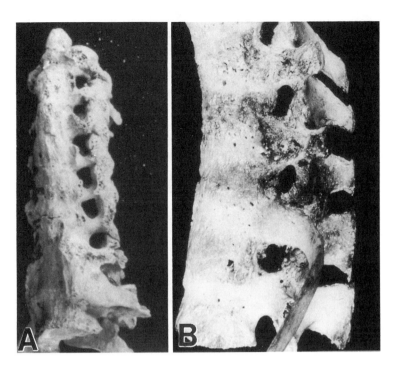

Figure 31.16. Calcification of the human anulus fibrosus. Involvement of adjacent vertebral centra mimics the appearance of a piece of bamboo. Zygapophyseal (facet) joint fusion is also present. (A) Fused cervical vertebrae. (B) Fused lumbar vertebrae. Note also costovertebral (rib) fusion.

Niwayama 1988). Diffuse idiopathic skeletal hyperostosis (DISH) may mimic spondyloarthropathy, but does not cause zygapophyseal joint fusion.

Fusion of the first three cervical (neck) vertebrae has been noted in ceratopsians (Fig. 31.17) (Rothschild 1987a; Tanke in press b). Although the diagnosis of spondyloarthropathy was originally entertained, the fact that fusion is limited to the first three cervical vertebrae and the lack of peripheral skeletal involvement made that diagnosis highly unlikely. Recognition of this phenomenon in all adults of several genera raised another possibility: Is this a developmental phenomenon? I concluded that fusion of the first three cervical vertebrae is not pathologic in ceratopsians. It may have an important mechanical function, support of the massive skull. Smaller, presumably younger ceratopsians have incomplete fusion of this cervical bar, in contrast to larger, presumably older ceratopsians (belonging to the genera *Chasmosaurus, "Monoclonius," Pachyrhinosaurus,* and *Triceratops*) (Rothschild 1987a; Tanke in press b). This controversial hypothesis will be testable only when adequate numbers of cervical bars (first three cervical vertebrae, identified as to species) become available for evaluation. *Leptoceratops* clearly has segmented cervical bars. While some *Protoceratops* cervical bars appear fused on visual examination, radiologic evaluation of all available specimens reveals persistence of segmentation. *Pachyrhinosaurus* manifests a different pattern from that of other large ceratopsians, as variable fusion is noted in large, adult specimens. If fusion is an ontogenetic (growth-related) event, absence of fusion in what appear to be small ceratopsians raises the possibility that they might actually represent juveniles of another species.

Dinosaurian Paleopathology   441

Figure 31.17. Lateral view of ceratopsian cervical vertebrae. (A) Fusion of the first three cervical vertebrae (*arrow*) in AMNH 4842. (B) Fusion of the first three cervical vertebrae in MOR 571, an unidentified ceratopsian.

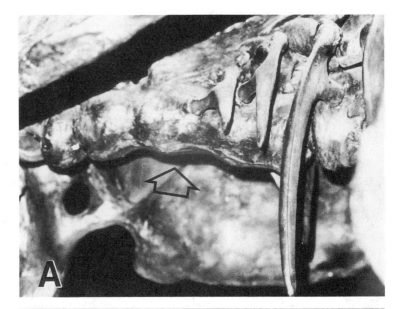

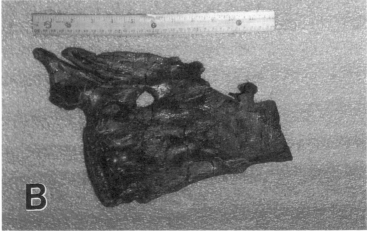

A specimen of *Tyrannosaurus* (American Museum of Natural History 5027; Carpenter 1990) shows a peculiar fusion of the centra (bodies) of two vertebrae. This pathology, involving the last cervical and first thoracic vertebrae, produced a significant postural abnormality. The neck is suddenly bent backward and upward at this site because of the fusion. The cause is unclear. However, the peculiar nature of the fusion and the associated deformity raise the possibility of healing after a bite. Proof may lie in close examination. Such proof could come in the form of the presence of a predator's tooth embedded in the vertebrae, as has been observed in a mosasaur (marine lizard) specimen.

Vertebral fusion may also occur after injury or infection. The latter is typically associated with a very disorganized bony architecture (Rothschild 1982; Resnick and Niwayama 1988). Such trauma was the source of tailtip fusions noted in hadrosaurs (Tanke in press a) and *Allosaurus* (Rothschild and Martin 1993).

442   *Bruce M. Rothschild*

Tumors

Examples of tumors in dinosaurs are problematic. Moodie (1923) suggested that a sauropod caudal fusion represented a hemangioma. Hemangiomas are vascular tumors which have a very characteristic appearance on x-ray. The trabecular pattern is accentuated (prominent) in the longitudinal axis (Resnick and Niwayama 1988). A bubbly or honeycombed trabecular pattern may also be noted in these tumors, which are commonly present in human vertebrae (allegedly in 10 percent of some populations).

The original specimen that Moodie described has unfortunately been lost, a fact that precludes diagnostic assessment. The published drawing (Moodie 1923), however, does not show longitudinal trabecular pattern prominence or honeycombing, thus making the diagnosis of hemangioma highly unlikely. The evidence from other sauropod vertebral fusions (see above) also challenges his interpretation. What Moodie described as a tumor is more likely an example of DISH.

Wade Miller of Brigham Young University and Leon Goldman of the San Diego Naval Hospital suggested that a "cauliflower-like growth" on the humerus of a theropod 135 to 150 million years old (probably *Allosaurus* or *Torvosaurus*) might represent a type of cartilage cancer called a chondrosarcoma. Details of this case have not yet been published, to my knowledge. Chondrosarcoma in humans appears as a lobulated mass with speckled internal calcification (Copeland 1956). Associated elevation of periosteum (the outer layer of the cortex) and wide margins of irregular thickness and density are highly characteristic of this condition in *Homo sapiens*. I have not had the opportunity to examine this theropod specimen, its x-rays or histology. It will be intriguing to learn if this find contributes to our understanding of the antiquity of cancer.

Dinosaur Eggs

Dinosaur eggs were not immune to pathology (Dughi and Sirugue 1957, 1958). Erben et al. (1979) reported increasing frequencies of multilayered eggshells in *Hypselosaurus*. Hirsch et al. (1989) reported finding a Jurassic dinosaur egg shell with two layers of calcification. As the pores of the layers were not aligned, embryonic respiration would have been impaired, and the embryo probably did not survive. An apparently normal embryo was, however, recognized on computerized tomographic (CT) examination of an intact egg.

The study of paleopathology opens up new vistas. While the "time line" of disease is of interest to some, paleopathology affords other important opportunities. Unique insights into lifestyle, habitat, and physiology can be gained though study of the fossil record of dinosaur mishaps and disease.

References

Abrams, N. R. 1953. Etiology and pathogenesis of degenerative joint disease. In J. Hollander (ed.), *Arthritis and Allied Conditions,* p. 691. 5th ed. Philadelphia: Lea and Febiger.

Anonymous. 1934. Glands may have caused evolution of freak dinosaurs. *Science News Letter* 25: 182.

Artzy, E.; G. Frieder; and G. T. Herman. 1980. The theory, design, implementation and evaluation of a three-dimensional surface detection algorithm. *Computer Graphics and Image Processing* 15: 1–24.

Bakker, R. T. 1968. The superiority of dinosaurs. *Discovery* 3: 11–22.

Blows, W. T. 1989. A pelvic fracture in *Iguanodon*. *Archosaurian Articulations* 1: 49–50.

Broili, F. 1922. Über den feineren Ban der "Verknocherten Sehnen" (= Verknocherten Muskeln) von Trachodon. *Anatomischer Anzeiger* 55: 464–475.

Brown, B., and E. M. Schlaikjer. 1937. The skeleton of *Styracosaurus* with the description of a new species. American Museum *Novitates* 955: 1–12.

Bywaters, E. 1960. The early radiologic signs of rheumatoid arthritis. *Bulletin of the Rheumatic Diseases* 11: 231–234.

Carpenter, K. 1990. Variation in *Tyrannosaurus rex*. In K. Carpenter and P. J. Currie (eds.), *Dinosaur Systematics: Perspectives and Approaches*, pp. 141–145. Cambridge: Cambridge University Press.

Cassou, B.; J. P. Camus; and J. G. Peyron. 1981. Recherche d'une arthrose primitive de la cheville chez les sujets de plus de 70 ans. In J. G. Peyron (ed.), *Epidemiologie de l'Arthrose*, pp. 180–184. Paris: Geigy.

Conroy, G. C., and M. W. Vannier. 1985. Endocranial volume determination of matrix-filled fossil skulls using high-resolution computed tomography. In K. D. Lawrence (ed.), *Evolution: Past, Present and Future*, pp. 419–426. Lawrence, Kans.: Alan R. Liss.

Copeland, M. M. 1956. Tumors of cartilaginous origin. *Clinical Orthopedics and Related Research* 7: 9–26.

Czerkas, S. J., and S. A. Czerkas. 1990. *Dinosaurs: A Complete World History*. New York: B. Mitchell.

Daffner, R. H. 1978. Stress fractures: Current concepts. *Skeletal Radiology* 2: 221–229.

Dollo, L. 1887. Note sur les ligaments ossifies des dinosauriens de Bernissart. *Archives of Biology* 7: 249–264.

Dughi, R., and F. Sirugue. 1957. Les oeufs de dinosauriens du Bassin d'Aix-en-Provence. *Compte Rendu de l'Academie Sciences, Paris* 245: 707–710.

Dughi, R., and F. Sirugue. 1958. Observations sur les oeufs de dinosaures du bassin d'Aix-en-Provence: Les oeufs a coquilles biostratifiées. *Compte Rendu de l'Academie Sciences, Paris* 246: 2271–2274.

Erben, H. K.; J. Hoefs; and K. H. Wedepohl. 1979. Paleobiological and isotopic studies of eggshells from a declining dinosaur species. *Paleobiology* 5: 380–414.

Erickson, B. R. 1966. Mounted skeleton of *Triceratops prorsus*. *Science Publications Science Museum, St. Paul, Minneapolis* 1: 1–16.

Farlow, J. O., and D. L. Brinkman. 1994. Wear surfaces on the teeth of tyrannosaurs. In G. D. Rosenberg and D. L. Wolberg (eds.), *Dino Fest*, pp. 165–175. Paleontological Society Special Publication 7. Knoxville: University of Tennessee.

Farlow, J. O.; D. L. Brinkman; W. L. Abler; and P. J. Currie. 1991. Size, shape, and serration density of theropod dinosaur lateral teeth. *Modern Geology* 16: 161–198.

Farrell, E. J., and R. A. Zappulla. 1989. Three-dimensional data visualization and biomedical applications. *Critical Reviews in Biomedical Engineering* 16: 323–363.

Forestier, J., and R. Lagier. 1971. Ankylosing hyperostosis of the spine. *Clinical Orthopedics and Related Research* 74: 65–83.

Fornasier, V. L.; G. Littlejohn; and M. B. Urowitz. 1983. Enthesial new bone formation: The early changes of spinal diffuse idiopathic skeletal hyperostosis. *Journal of Rheumatology* 10: 939–947.

Funk, F. J., Jr. 1976. Osteoarthritis of the foot and ankle. In American Academy of Orthopedic Surgeons (eds.), *Symposium on Osteoarthritis*, pp. 287–301. St. Louis: C. V. Mosby.

Gilmore, C. W. 1906. Notes on some recent additions to the exhibition series of fossil vertebrates. *Proceedings of the U.S. National Museum* 30: 607–611.

Gilmore, C. W. 1909. Osteology of the Jurassic reptile *Camptosaurus*, with a revision of the species of the genus, and descriptions of two new species. *Proceedings of the U.S. National Museum* 36: 197–332.

Gilmore, C. W. 1912. The mounted skeletons of *Camptosaurus* in the United States National Museum. *Proceedings of the U.S. National Museum* 41: 687–696.

Gilmore, C. W. 1919. A new restoration of *Triceratops* with notes on the osteology of the jaws. *Proceedings of the U.S. National Museum* 55: 97–112.

Gilmore, C. W. 1930. On dinosaurian reptiles from the Two Medicine Formation of Montana. *Proceedings of the U.S. National Museum* 77: 1–39.

Goodship, A. E.; L. E. Lanyon; and H. McFie. 1979. Functional adaptation of bone to increased stress. *Journal of Bone and Joint Surgery* 61A: 539–546.

Hadley, M. N.; V. K. Sonntag; M. R. Amos; J. A. Hodak; and L. J. Lopez. 1987. Three-dimensional computed tomography in the diagnosis of vertebral column pathological conditions. *Neurosurgery* 21: 186–192.

Harrison, M. H.; F. Schajowicz; and J. Trueta. 1953. Osteoarthritis of the hip: A study of the nature and evolution of the disease. *Journal of Bone and Joint Surgery* 35B: 598–626.

Hartley, B. J. 1943. "Stress" or "fatigue" fractures of bone. *British Journal of Radiology* 16: 225–262.

Hatcher, J. B. 1901. *Diplodocus* (Marsh), its osteology, taxonomy and probable habits, with a restoration of the skeleton. *Memoirs of the Carnegie Museum* 1: 1–63.

Hirsch, K. F.; K. L. Stadtman; W. E. Miller; and J. H. Madsen, Jr. 1989. Upper Jurassic dinosaur egg from Utah. *Science* 243: 1711–1713.

Holland, W. J. 1915. Heads and tails: A few notes relating to the structure of the sauropod dinosaurs. *Annals of the Carnegie Museum* 9: 273–278.

Isaacs, W. A.; K. Little; J. D. Currey; and L. B. Tarlo. 1963. Collagen and cellulose-like substance in fossil dentine and bone. *Nature* 197: 192.

Ishigaki S. 1986. *The Dinosaurs of Morocco.* Tokyo: Tsukiji Shokan.

Johnson, R. E. 1990. Biomechanical analysis of forelimb posture and gait in *Torosaurus. Journal of Vertebrate Paleontology* 10 (Supplement 3): 29A–30A.

Jurmain, R. 1977. Stress and the etiology of osteoarthritis. *American Journal of Physical Anthropology* 80: 229–237.

Katz, W. A. 1989. *Diagnosis and Management of Rheumatic Disease.* 2nd ed. Philadelphia: Lippincott.

Kelly, W. N.; E. D. Harris, Jr.; S. Ruddy; and C. B. Sledge. 1985. *Textbook of Rheumatology.* 2nd ed. Philadelphia: Saunders.

Lambe, L. 1917. The Cretaceous theropodous dinosaur *Gorgosaurus. Geological Survey of Canada Memoir* 100: 1–84.

Langston, W., Jr. 1961. News from members. *Society of Vertebrate Paleology News Bulletin* 63: 7–9.

Larson, P. L. 1991. The Black Hills Institute *Tyrannosaurus:* A preliminary report. *Journal of Vertebrate Paleontology* 11 (Supplement 3): 41A–42.

Lawrence, J. S.; J. M. Bremner; and F. Bier. 1966. Osteo-arthrosis: Prevalence in the population and relationship between symptoms and x-ray changes. *Annals of the Rheumatic Diseases* 25: 1–24.

Leach, R. E.; S. Baumgard; and J. Broom. 1973. Obesity: Its relationship to osteoarthritis of the knee. *Clinical Orthopedics and Related Research* 93: 271–273.

Lockley, M. J. 1986. *A Guide to Dinosaur Tracksites of the Colorado Plateau and American Southwest.* University of Colorado at Denver, Geology Department Magazine, Special Issue 1: 1–56.

Lockley, M. J. 1989. Tracks and traces: New perspectives on dinosaurian behavior, ecology, and biogeography. In K. Padian and D. J. Chure (eds.), *The Age of Dinosaurs,* pp. 134–145. Short Courses in Paleontology no. 2. Knoxville: Paleontological Society, University of Tennessee.

Madsen, J. H., Jr. 1976. *Allosaurus fragilis:* A revised osteology. Utah Geological and Mineral Survey Bulletin 109. Salt Lake City.

Martel, W. 1968. Radiologic signs of rheumatoid arthritis with particular reference to the hand, wrist, and foot. *Medical Clinics of North America* 52: 655–665.

McEwen, C.; D. DiTata; and J. Lingg. 1971. Ankylosing spondylitis and spondylitis accompanying ulcerative colitis, regional enteritis, psoriasis, and Reiter's disease: A comparative study. *Arthritis and Rheumatism* 14: 291–318.

McHugh, T. 1958. Social behavior of the American Bison (*Bison bison bison*). *Zoologica* 43: 1–40.

Meachim, G., and G. Brooke. 1984. The pathology of osteoarthritis. In R. W. Moskowitz, D. S. Howell, V. M. Goldberg, and H. J. Mankin (eds.), *Osteoarthritis: Diagnosis and Management*, pp. 29–42. Philadelphia: Saunders.

Molnar, R. E. In press. Theropod paleopathology: A literature survey. To be published in B. M. Rothschild and S. Shelton (eds.), *Paleopathology*. London: Archetype Press.

Molnar, R. E., and J. O. Farlow. 1990. Carnosaur paleobiology. In D. B. Weishampel, P. Dodson, and H. Osmólska (eds.), *The Dinosauria*, pp. 210–224. Berkeley: University of California Press.

Monastersky, R. 1990. Reopening old wounds: Physicians and paleontologists learn new lessons from ancient ailments. *Science News* 137 (3): 40–42.

Moodie, R. L. 1917. Studies in paleopathology. Part I: General consideration of the evidences of pathological conditions found among fossil animals. *Annals of Medical History* 1: 374–393.

Moodie, R. L. 1923. *Paleopathology: An Introduction to the Study of Ancient Evidences of Disease*. Urbana: University of Illinois Press.

Moodie, R. L. 1926. Excess callus following fracture of the fore foot in a Cretaceous dinosaur. *Annals of Medical History* 8: 73–77.

Moodie, R. L. 1928. The histological nature of ossified tendons found in dinosaurs. *American Museum Novitates* 311: 1–15.

Moodie, R. L. 1930. Dental abscesses in a dinosaur millions of years old, and the oldest yet known. *Pacific Dental Gazette* 38: 435–440.

Morris, J. M., and L. D. Blickenstaff. 1967. *Fatigue Fractures: A Clinical Study*. Springfield, Ill.: Charles C. Thomas.

Norman, D. 1985. *The Illustrated Encyclopedia of Dinosaurs*. New York: Crescent Books.

O'Donoghue, D. J.; G. R. Frank; and G. L. Jeter. 1971. Repair and reconstruction of the anterior cruciate ligament in dogs: Factors influencing long-term results. *Journal of Bone and Joint Surgery* 53A: 710–718.

Olshevsky, G. O. 1978. The Archosaurian taxa. *Mesozoic Meanderings* 1: 1–50.

Oppenheimer, A. 1942. Calcification and ossification of vertebral ligaments (spondylitis ossificans ligamentosa): Roentgen signs of pathogenesis and clinical significance. *Radiology* 38: 160–173.

Orava, S.; J. Puranen; and L. Ala-Ketoal. 1978. Stress fractures caused by physical exercise. *Acta Orthopaedica Scandinavica* 49: 19–27.

Ortner, D. J., and W. G. Putschar. 1985. *Identification of Pathological Conditions in Human Skeletal Remains*. Washington, D.C.: Smithsonian Institution Press.

Parks, W. A. 1922. *Parasaurolophus walkeri*: A new genus and species of crested trachodont dinosaur. *University of Toronto Studies, Geological Series* 13: 1–32.

Parks, W. A. 1935. New species of trachodont dinosaurs from the Cretaceous formations of Alberta: With notes on other species. *University of Toronto Studies, Geological Series* 37: 1–45.

Pennes, D. R.; W. Martel; and C. N. Ellis. 1985. Retinoid-induced ossification of the posterior longitudinal ligament. *Skeletal Radiology* 14: 191–193.

Petersen, K.; J. I. Isakson; and J. H. Madsen, Jr. 1972. Preliminary study of paleopathologies in the Cleveland-Lloyd dinosaur collection. *Utah Academy of Sciences Proceedings* 49: 44–47.

Puranen, J.; L. Ala-Ketola; and P. Peltokallio. 1975. Running and primary osteoarthritis of the hip. *British Medical Journal* 2: 424–425.

Radin, E. L. 1978. Our current understanding of normal knee mechanics and its implications for successful knee surgery. In *American Association of Orthopedic Surgeons Symposium on Reconstructive Surgery of the Knee*, pp. 37–46. St. Louis: Mosby.

Resnick, D., and G. Niwayama. 1988. *Diagnosis of Bone and Joint Disorders*. Philadelphia: Saunders.

Resnick, D.; R. F. Shapiro; and K. B. Wiesner. 1978. Diffuse idiopathic skeletal hyperostosis (DISH) (ankylosing hyperostosis of Forestier and Rotes-Querol). *Seminars in Arthritis and Rheumatism* 7: 153–187.

Riggs, E. S. 1903. Structure and relationships of opisthocoelian dinosaurs. Part 1: *Apatosaurus* Marsh. *Publications of the Field Columbian Museum, Geological Series* 2 (4): 165–196.

Roberts, E. D.; G. B. Baskin; E. Watson; W. G. Henk; and T. C. Shelton. 1984. Calcium pyrophosphate deposition disease (CPDD) in nonhuman primates. *American Journal of Pathology* 166: 359–361.

Rothschild, B. M. 1982. *Rheumatology: A Primary Care Approach.* New York: Yorke Medical Press.

Rothschild, B. M. 1985. Diffuse idiopathic skeletal hyperostosis (DISH): Misconceptions and reality. *Clinical Rheumatology* 4: 207–212.

Rothschild, B. M. 1987a. Diffuse idiopathic skeletal hyperostosis as reflected in the paleontologic record: Dinosaurs and early mammals. *Seminars in Arthritis and Rheumatism* 17: 119–125.

Rothschild, B. M. 1987b. Paleopathology of the spine in Cretaceous reptiles. In T. Appelboom (ed.), *Art, History and Antiquity of Rheumatic Disease,* pp. 97–99. Brussels: Elsevier.

Rothschild, B. M. 1988a. Stress fracture in a ceratopsian phalanx. *Journal of Paleontology* 62: 302–303.

Rothschild, B. M. 1988b. Diffuse idiopathic skeletal hyperostosis. *Comprehensive Therapy* 14: 65–69.

Rothschild, B. M. 1989a. Skeletal paleopathology of rheumatic diseases: The subprimate connection. In D. J. McCarty (ed.), *Arthritis and Allied Conditions,* pp. 3–7. 11th ed. Philadelphia: Lea and Febiger.

Rothschild, B. M. 1989b. Paleopathology and its contributions to vertebrate paleontology: Technical perspectives. *Journal of Vertebrate Paleontology* 9 (Supplement 3): 36A–37A.

Rothschild, B. M. 1990. Radiologic assessment of osteoarthritis in dinosaurs. *Annals of the Carnegie Museum* 59: 295–301.

Rothschild, B. M., and D. Berman. 1991. Fusion of caudal vertebrae in Late Jurassic sauropods. *Journal of Vertebrate Paleontology* 11: 29–36.

Rothschild, B. M., and L. Martin. 1993. *Paleopathology.* Montclair, N.J.: Telford Press.

Rothschild, B. M., and D. Tanke. 1991. Paleopathology: Insights to lifestyle and health in prehistory. *Geoscience* 19: 73–92.

Rothschild, B. M.; D. Tanke; and K. Carpenter. 1997. Tyrannosaurs suffered from gout. *Nature* 387: 357.

Ruffer, M. A. 1921. *Studies in the Paleopathology of Egypt.* Chicago: University of Chicago Press.

Russell, D. A. 1970. Tyrannosaurs from the Late Cretaceous of western Canada. National Museum of Natural Sciences, *Publications in Paleontology* 1: 1–34.

Saville, P. D., and J. Dickson. 1968. Age and weight in osteoarthritis of the hip. *Arthritis and Rheumatism* 11: 635–644.

Sawyer, G. T., and B. R. Erickson. 1985. Injury and diseases in fossil animals: The intriguing world of paleopathology. *Encounters* (May/June): 25–28.

Schlapbach, P.; C. Beyeler; N. J. Gerber; S. van der Linden; U. Burgi; W. A. Fuchs; and H. Ehrengruber. 1989. Diffuse idiopathic skeletal hyperostosis (DISH) of the spine: A cause of back pain? A controlled study. *British Journal of Rheumatology* 28: 299–303.

Seawright, A. A., and P. B. English. 1965. Hypervitaminosis A and hyperostosis of the cat. *Nature* 206: 1171–1172.

Sebes, J. I.; J. W. Langston; M. L. Gavant; and B. M. Rothschild. 1991. Magnetic resonance imaging of growth recovery lines in fossil vertebrae. *American Journal of Roentgenology* 157: 415–416.

Sigerist, H. E. 1951. *A History of Medicine.* New York: Oxford University Press.

Silberberg, M., and R. Silberberg. 1960. Osteoarthritis in mice fed diets enriched with animal or vegetable fat. *Archives of Pathology* 70: 385–390.

Singh, A.; R. Dass; S. Singhhayreh; and S. S. Jolly. 1962. Skeletal changes in endemic fluorosis. *Journal of Bone and Joint Surgery* 44B: 806–815.

Skinner, M. F., and G. H. Sperber. 1982. *Atlas of Radiographs of Early Man*. New York: Liss.

Sokoloff, L.; O. Mickelsen; E. Silverstein; G. E. Jay, Jr.; and R. S. Yamamoto. 1960. Experimental obesity and osteoarthritis. *American Journal of Physiology* 198: 765–770.

Sternberg, C. M. 1940. Ceratopsidae from Alberta. *Journal of Paleontology* 14: 468–480.

Sundberg, S. B.; B. Clark; and B. K. Foster. 1986. Three-dimensional reformation of skeletal abnormalities using computed tomography. *Journal of Pediatric Orthopedics* 6: 416–419.

Swinton, W. E. 1981. Sir Marc Armand Ruffer: One of the first palaeopathologists. *Canadian Medical Association Journal* 124: 1388–1392.

Tanke, D. H. In press a. Paleopathologies in Late Cretaceous hadrosaurs (Reptilia: Ornithischia): Behavioral implications. To be published in B. M. Rothschild and S. Shelton (eds.), *Paleopathology*. London: Archetype Press.

Tanke, D. H. In press b. The rarity of paleopathologies in "short-frilled" ceratopsians (Reptilia: Ornithischia: Centrosaurinae): Evidence for non-aggressive intraspecific behavior. To be published in B. M. Rothschild and S. Shelton, (eds.), *Paleopathology*. London: Archetype Press.

Thulborn, R. A. 1984. Preferred gaits of bipedal dinosaurs. *Alcheringa* 8: 243–252.

Thulborn, R. A. 1990. *Dinosaur Tracks*. London: Chapman and Hall.

Tokaryk, T. T. 1986. Ceratopsian dinosaurs from the Frenchman Formation (Upper Cretaceous) of Saskatchewan. *Canadian Field-Naturalist* 100 (2): 192–196.

Tucker, M. E., and T. P. Burchette. 1977. Triassic dinosaur footprints from South Wales: Their context and preservation. *Palaeogeography, Palaeoclimatology, Palaeoecology* 22: 195–208.

Tyson, H. 1977. Functional craniology of the Ceratopsia (Reptilia: Ornithischia) with special reference to *Eoceratops*. Master's thesis, University of Alberta.

Vance, T. 1989. Probable use of the vestigial forelimbs of the tyrannosaurid dinosaurs. *Bulletin of the Chicago Herpetological Society* 24: 41–47.

Virapongse, C.; M. Shapiro; A. Gmitro; and M. Sarwar. 1986. Three-dimensional computed tomographic reformation of the spine, skull, and brain from axial images. *Neurosurgery* 18: 53–56.

Wilson, E. S. Jr., and F. N. Katz. 1969. Stress fractures: An analysis of 250 consecutive cases. *Radiology* 92: 480–486.

Woolson, S. T.; P. Dev; L. L. Fellingham; and A. Vassiliadis. 1986. Three-dimensional imaging of bone from computerized tomography. *Clinical Orthopaedics and Related Research* 202: 239–244.

Worthen, B. M., and B. A. Yanklowitz. 1978. The pathophysiology and treatment of stress fractures in military personnel. *Journal of the American Podiatric Association* 68: 317–325.

Wyckoff, R. W. 1980. Collagen in fossil bones. In P. E. Hare (ed.), *Biogeochemistry of Amino Acids,* pp. 17–22. New York: Wiley.

Dinosaurian Physiology: The Case for "Intermediate" Dinosaurs

R. E. H. Reid

In 1976, a chance find of typical reptilian "growth rings" in a dinosaurian bone led me to set up a program of bone research, which led me in turn to the opinion that dinosaurian physiology was probably somehow intermediate between the "cold-blooded" and "warm-blooded" types now seen in tetrapods (Reid 1984a, 1984b, 1987, 1990). This chapter was written to show you why I formed that opinion, and how I relate the bone evidence to other considerations.

Dinosaurian physiology can currently be seen as a puzzle with three main possible solutions:

1. Dinosaurs were "cold-blooded" reptiles, wholly comparable with modern ones, except that many grew to sizes at which sheer bulk protected them from day-to-day temperature variations. In its modern form, this classic view of dinosaurs is based mainly on the thermal inertia work of Colbert et al. (1946) and Spotila et al. (1973).

2. Dinosaurs were "warm-blooded" animals, essentially comparable with modern ones. The main exponent of this view has been R. T. Bakker (e.g., 1972, 1975, 1986), although others held it before him (e.g., L. S. Russell [1965]).

3. Dinosaurs were a third kind of animal, neither "cold-" nor "warm-blooded" by modern standards, but between the two and different from both. Although hinted at earlier by de Ricqlès (1974), this concept first emerged fully in a study by Regal and Gans (1980).

To these may be added the further possibility that dinosaurs included animals of two or all three of these kinds.

32

This high level of uncertainty is due partly to authors' approaching the problem in different ways, or holding different views on what is relevant evidence, or on the meaning of a given piece of evidence. Bakker (1986), for instance, argues persuasively from various kinds of evidence that fit his views, but pays little attention to evidence or arguments that do not. Conversely, some authors have ignored valid points made by Bakker. Spotila et al. (1991) see the oceangoing turtle *Dermochelys* as providing a useful physiological model for dinosaurs, but I see its environment and lifestyle as far too different from those of dinosaurs for their physiological adaptations to be comparable (Box 32.1). But the real problem is that how animals regulate temperature can be determined only in living ones, and not in fossils by any technique now available or foreseeable. Even, for instance, if isotope chemistry could tell us the temperature at which a dinosaur's bones were formed (e.g., Barrick and Showers 1994), this could not tell us certainly how that temperature was maintained (Box 32.2). Accordingly, all that can be done is to try to decide which solution fits the evidence best, without making claims that go beyond the competence of that evidence.

My own reading of the riddle is based primarily on an argument from the dynamics of blood circulation (a.k.a. hemodynamics) developed by Hohnke (1973), Seymour (1976), and Ostrom (1980), with thermal inertia studies of Colbert et al. (1946), Spotila et al. (1973), and McNab and Auffenberg (1976) as its second main component, and with bone providing evidence against both the "cold-" and "warm-blooded" solutions. The picture that emerges is essentially that given by Regal and Gans (1980), with some modifications which I see as implied by the type of circulation they envisaged.

The First Step: Heights and Hearts

First, modern reptiles other than crocodiles have three-chambered hearts (Fig. 32.1A), with a single ventricle and no complete separation of the pulmonary circulation to the lungs from the systemic circulation to the body. This primitive arrangement is adequate because the vertical distance through which blood has to circulate is not great, even in the largest reptiles. Dinosaurs, however, not only reached much larger sizes, they also stood upright, and sometimes held their heads high above their backs on long necks. The head of the giant sauropod *Brachiosaurus*, for instance, if its pose has been interpreted correctly, was carried as much as roughly 11 meters (36 feet) above the ground, and 7.5 meters (24.6 feet) above its heart.

The pressure required to circulate blood through such vertical distances would be greater than in any modern reptile, and almost certainly too great to be applied to blood vessels in the lungs without causing a fatal leakage of blood fluid. This points to dinosaurs' having some form of "double-pump" circulation, as in mammals and birds, with four-chambered hearts and with complete separation of the pulmonary and systemic circulations. Crocodiles (Fig. 32.1B) have a primitive form of double-pump circulation, with four-chambered hearts; but they do not stand upright, and the right- and left-hand sides of the heart still communicate. Large dinosaurs, walking upright, would have needed a fully evolved double-pump system; and this would probably have been of the type now seen in birds (Figs. 32.1C, 32.2), which appears to have arisen from the crocodilian type by suppression of the left systemic arch (or left aorta; compare Figs. 32.1B and C), as Russell (1965) saw.

Box 32.1.

Dermochelys
and
Dinosaurs

In a study picturing dinosaurs as simply "gigantotherms," Spotila et al. (1991: 207) held that the leatherback turtle *Dermochelys* "may provide the best model for the study of gigantothermy in living reptiles." Perhaps so; but *Dermochelys* is also likely to be very different physiologically from any terrestrial reptile, let alone from any dinosaur. Here's why.

1. As a highly specialized marine deep diver, *Dermochelys* has an environment and a lifestyle not shared with any known dinosaur, and must have evolved physiologically under adaptive selection pressures to which no known dinosaur was ever subjected. This applies especially to essentially permanent immersion in a medium—seawater—with very different thermal properties from those of air. Its high tolerance to cold conditions is also probably related to its deep-diving habit, and may have evolved over a long enough time for progressive Cenozoic cooling of deep oceanic water to be a further factor. None of this would apply to any dinosaur. Further, like all marine forms, *Dermochelys* is essentially insulated from the effects of wind chill and rain chill, to which dinosaurs would need to be fully adapted.

2. As a medium-supported animal, *Dermochelys* has no use for weight-supporting energetics, which must instead have been highly developed in dinosaurs, which walked upright and weighed 30 tons or more.

3. As an animal known to dive at least as deep as sperm whales, *Dermochelys* must be highly adapted to maintaining activity in the absence of pulmonary respiration. The release of muscular energy will need to be almost wholly anaerobic, and other special adaptations are likely. Judging from what is known of other divers, these are likely to include (a) a high capacity for storing oxygen in blood and muscles before diving, (b) a high tolerance to accumulated carbon dioxide and lactic acid, produced by activity while diving, and (c) a means of giving the nervous system preferential access to available oxygen. There might also be a means of shunting blood away from the lungs when they are not in use, as in crocodiles. None of these adaptations would be needed by any known dinosaur, despite occasional dips in lakes or rivers, and their activity metabolism is also fairly likely to have been fully aerobic (see above).

4. Other minor adaptations, not needed by dinosaurs, must include a skin tolerant to permanent immersion in salt water, and arrangements for dealing with salt ingested with food.

To put this differently, it needs to be realized that thermal physiology does not exist in isolation from other physiological systems, but is simply one segment of an integrated complex of interdependent systems, which all need to be considered together in picturing an animal's physiology. In that light, *Dermochelys* is likely to be very different physiologically from any modern terrestrial reptile, let alone from any dinosaur. Further, if its spectacular cold-tolerance is essentially a product of its lifestyle, as seems almost certain, it was evolved by a route which is not open to any terrestrial reptile. Crocodiles, large tortoises, and the Komodo monitor are then the real guides to what such reptiles can achieve as "gigantotherms."

If this conclusion is correct, as I think, there was a major anatomical difference between dinosaurs and all modern reptiles, which could well have opened the way to further physiological differences; and since this type of circulatory system is now seen only in "warm-blooded" animals, its postulated presence implies that dinosaurs could have been "warm-blooded." The question then is not whether they differed from modern reptiles physiologically, but how much they differed.

Dinosaurian Physiology   451

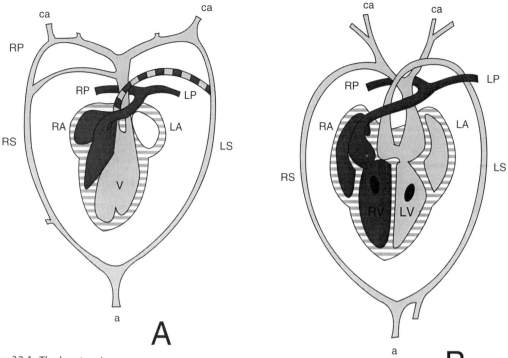

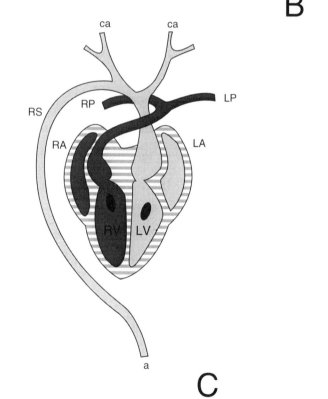

Figure 32.1. The heart and associated arteries in (A) a lizard, (B) a crocodile, and (C) a dinosaur, assumed to have complete double-pump circulation. Abbreviations: a = aorta (dorsal aorta); ca = carotids; LA = left auricle (or atrium); LP = left pulmonary artery; LS = left systemic arch; LV = left ventricle; RA = right auricle (or atrium); RP = right pulmonary artery; RS = right systemic arch; RV = right ventricle; V = ventricle. In the lizard (A), deoxygenated and reoxygenated blood are both pumped by the same single ventricle, with only some separation by incomplete ventricular septa. In the crocodile (B), the ventricles are separate, but the right and left systemic arches communicate through an aperture (the Foramen of Panizza), through which deoxygenated blood is shunted away from the lungs during diving. The pattern suggested for dinosaurs (C) is supposed to be derived from the crocodilian pattern by loss of the left systemic arch, as suggested by L. S. Russell (1965).

"Warm" Blood, "Cold" Blood, and Other Notions

What Is "Warm" Blood?

Next, we need to look at the technical terminology used in defining "cold" and "warm" blood scientifically, and at some associated concepts that are relevant to discussion of dinosaurs. First, it needs to be understood clearly that the definitions of "cold" and "warm" blood involve more than just temperature. We call amphibians and reptiles "cold-blooded" because many, which have lower temperatures than our own, feel cold when they are handled; but some reptiles can have temperatures approaching our own or even higher. In optimum conditions, the body temperatures of American alligators can reach 33° to 34°C (Coulson and Hernandez 1983), close to the human level (37°C), while the desert iguana *Dipsosaurus* has a temperature of 40° to 42°C when active.

Scientifically, three pairs of contrasting terms are used to describe animals. They are called (1) *homoiothermic* if they are able to maintain a steady temperature in their normal environments, irrespective of daily variations in atmospheric (ambient) temperature, or *poikilothermic* if they cannot; (2) *endothermic* if they maintain an activity temperature by means of internally generated heat, or *ectothermic* if they depend on external heat sources; and (3) *tachymetabolic* if their body chemistry (metabolism) runs at a high rate, or *bradymetabolic* if it runs at a low one.

Ideally, "warm-blooded" animals are homoiothermic, endothermic, and tachymetabolic, with their rapid metabolism providing an internal heat source, while "cold-blooded" animals are poikilothermic and bradymetabolic. The conditions they show are described as, for example, homoiothermy, endothermy, and tachymetabolism in the "warm-blooded" case. In practice, however, not all "warm-blooded" forms conform to the ideal pattern; some birds, for instance (e.g., hummingbirds, chickadees), save energy by lowering their temperatures overnight. Such forms, in which this habit appears to be secondary, are distinguished from typical endotherms as heterotherms. The naked mole-rat *Heterocephalus* of East Africa, which spends its life in burrows, is almost completely poikilother-

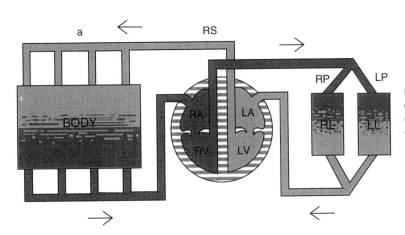

Figure 32.2. How blood would circulate in a dinosaur with complete double-pump circulation. LL = left lung; RL = right lung; other abbreviations as in Figure 32.1.

Dinosaurian Physiology   453

mic. The temperatures maintained by modern endotherms also vary, being generally between 28° and 30°C in sloths and monotremes, between 33° and 36°C in marsupials, between 36° and 38°C in most placentals, and between 40° and 41°C in birds.

A mistake which needs to be avoided here is the notion that poikilotherms have no control over their temperatures, which will simply correspond with those of their environments. Ectothermic lizards which cannot maintain an optimum temperature overnight can often maintain it quite precisely during the daytime, at levels above or below the ambient temperature, by various behavioral and physiological means; and even if dinosaurs were simply normal reptiles, they would probably have had similar abilities.

"Warm" Blood through Bulk

In living tetrapods, homoiothermy is restricted to endotherms with high metabolic rates; but some authors have argued that an ectotherm can become homoiothermic if it simply grows large enough, without needing a high metabolic rate. This supposed condition has been called inertial homoiothermy (McNab and Auffenberg 1976), on the basis of its being the thermal inertia of large animal bodies, or simply mass homoiothermy (de Ricqlès 1974). The basic idea was derived from experiments on alligators made by Colbert et al. (1946), who found that large examples take longer to change temperature than small ones. This is explicable as being due to body volume increasing by cubes of linear dimensions, whereas surface area increases by squares only. For any given temperature change, more heat must thus pass through a given surface area in a large form than in a small one. Extrapolating upward from their alligator data, they concluded that a ten-ton "cold-blooded" dinosaur would take eighty-six hours to change its temperature by 1°C. Since there are only twenty-four hours in a day, such an animal would be able to maintain an essentially steady temperature despite normal daily temperature variations.

In its modern form, however, the concept dates from a mathematical study by Spotila et al. (1973). Assuming a model dinosaur to have a body diameter of 1 meter, a 5-centimeter layer of insulating fat beneath its skin, and the energy budget of an alligator, they found that in a climate such as that of Florida, with air temperatures cycling between 22° and 32°C daily, its core temperature (Fig. 32.3, Tc) would remain between 28.5° and 29.6°, and that the time taken for a change between any two equilibrium temperatures would be forty-eight hours. Many dinosaurs had larger bodies, and some were much larger. Small forms and juveniles, however, would still be poikilothermic, and would have to use other means of regulating temperature. Halstead (1975), for instance, took the minimum size for homoiothermy through bulk as that of the prosauropod *Plateosaurus,* and pictured the small ornithopod *Fabrosaurus,* only 1 meter long, as moving in and out of sunlight to regulate its temperature, in the manner described as shuttling heliothermy when practiced by modern lizards.

Some authors, including Halstead (1975), have applied the term *warm-blooded* to dinosaurs pictured as mass homoiotherms, but this equates "warm-bloodedness" with homoiothermy only. Since metabolic rates rise with temperature, an ectothermic dinosaur large enough for mass homoiothermy would admittedly gain a stabilized metabolic rate in the upper part of whatever the range for such animals was; but a high metabolic rate maintained by a high steady temperature depending on bulk is clearly a different matter from a high steady temperature maintained by a high

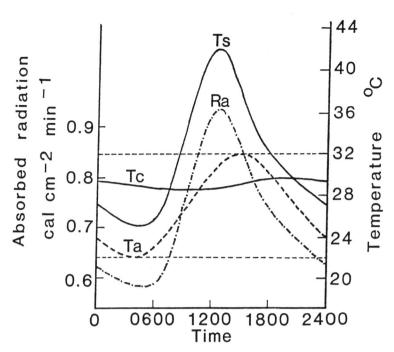

Figure 32.3. Inertial (or ectothermic) homoiothermy in a 1-meter-diameter model dinosaur devised by Spotila et al. (1973), whose Figures 2a and 2b are here combined. In Florida-like conditions, with air temperatures cycling daily between 22° and 32°C, the internal (core) temperature (Tc) of a 1-meter-diameter animal, resembling an American alligator physiologically, would vary only between 28.5° and 29.6°C, and thus by only 1.1°C daily despite a 10°C daily variation in air temperature (Ta). Surface temperature (Ts) would, however, vary more widely, with a pattern paralleling that of absorbed radiation (Ra). From Reid 1987: Fig. 1, by permission; recaptioned.

metabolic rate independently of bulk. So dinosaurs envisaged in this manner should not be called "warm-blooded," if we want the term to mean the same as in mammals and birds. The terms *endothermy* and *endotherm* also need to be used with the same meanings. Some authors have treated endotherms as simply animals in which most of the heat in the body is of internal origin; but because all metabolic activity produces heat, this could apply to a large enough ectotherm with a low metabolic rate, again as a result of the "cube and square effect." Such "endothermy," depending on bulk and termed "mass endothermy" by de Ricqlès (1983), is again clearly different from the tachymetabolic endothermy of mammals and birds.

Two other features of modern reptiles can be pictured as likely to have supplemented mass effects in dinosaurs. First, McNab and Auffenberg (1976) found that heat exchange through the surface occurs more quickly in lizards than in mammals below about 100 kilograms in weight, but less quickly above that weight (Fig. 32.4). If dinosaurs resembled lizards here, this would reinforce the mass effect of large size; and again, most reached much larger sizes. Second, Regal and Gans (1980) pictured mass effects as supplemented by vascular control over heat exchanges, through contraction or dilation of superficial blood vessels as in some modern reptiles. This combination of mass effects with physiological control has been called gigantothermy by Spotila et al. (1991).

"High" Metabolic Rates

Although it is commonly said that endotherms have high metabolic rates while ectotherms have low ones, the true picture is really much more complex. First, the metabolic rates of animals are not constant, but vary

Dinosaurian Physiology   455

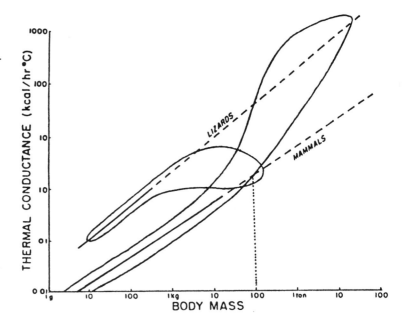

Figure 32.4. How thermal conductance changes with size in lizards and mammals according to McNab (1978), whose Figure 1 is here simplified. At sizes under 1 kilogram, the thermal conductance of the skin of a lizard can be as much as ten times that of the skin of a mammal of similar weight. Above 1 kilogram, however, the conductance values converge, and they reverse their relationship above 100 kilograms. If the conductance of dinosaurian skin was similar to that of large lizards, or even lower at larger sizes, this would contribute substantially to the maintenance of inertial homoiothermy. From Reid 1987: Fig. 2, by permission; recaptioned.

with their level of activity. Every animal has a basic metabolic rate, termed its resting or standard metabolic rate (SMR), and defined by the minimum level of oxygen consumption required to sustain life at a standard temperature and pressure. Any form of activity, even digestion, requires a higher rate, and an active ectotherm can have a higher metabolic rate than a resting endotherm the same size. When endotherms are said to have high metabolic rates, this means that their standard metabolic rates are generally six or more times higher than those of comparable ectotherms. At 28°C, for instance, the SMR of a 70-kilogram American alligator is said to be under 4 percent of that of a 70-kilogram human (Coulson and Hernandez 1983). Thus, if high metabolic rates are claimed as grounds for seeing dinosaurs as endotherms, this must refer to their basic metabolic rates, and not to activity metabolism. Some small dinosaurs were probably highly active, with high levels of activity metabolism and resultant heat production; but the endothermy of mammals and birds is not based on activity metabolism, but on high basic metabolic rates which do not depend on activity. This is why they can maintain high steady temperatures while sleeping, even at small sizes, which a dinosaur maintaining a high temperature through activity could not do.

Second, the statement that endotherms have higher basic metabolic rates than ectotherms applies only to animals of similar size, because of the two ways in which SMRs change with size. In terms of total oxygen consumption, they are inevitably highest in large forms and lowest in small ones; but minimum oxygen consumption per unit mass, termed the mass-specific SMR, is instead highest in small forms (Box 32.3), to the extent that small ectothermic lizards can have higher mass-specific SMRs than large endothermic mammals (Fig. 32.5). These changes in SMRs with size have two implications for dinosaurs.

456   R. E. H. Reid

Box 32.2.

Isotopes
and
Endothermy

Barrick and Showers (1994) have introduced a method of using isotopes to assess dinosaurian physiology, by comparing depositional temperatures recorded in bones from core parts and extremities. On the basis of finding only a 4°C difference between these parts in *Tyrannosaurus rex,* they describe it as an endotherm. Their approach is attractive, because of the relative objectivity of isotopic data, but still involves problems.

First, while it allows comparisons between the temperature ranges at which bone was deposited in different parts of a skeleton, it yields no direct evidence of the animal's metabolic rate, which is left as a matter for inference. So while it can indicate homoiothermy, it cannot prove endothermy; and there is no certain way of distinguishing between a true endothermic homoiotherm and a large "intermediate" mass homoiotherm (or "gigantotherm") with a metabolic rate be-

tween those of modern ectotherms and endotherms. Their result could fit either, and Barrick and Showers say themselves (p. 224) that *T. rex*'s metabolic rate may not have matched that of modern endotherms.

Second, and more seriously, such isotopic data can yield reliable results only if bone deposition has been equally continuous in the parts that are compared. Barrick and Showers do not say whether this was so in their *Tyrannosaurus;* and it certainly was not so in allosaurs the writer has studied (Reid 1990, 1996). In them, resting lines (LAGs), marking pauses in growth, are usually sparse or lacking in large limb bones, but more numerous in bones from near extremities (Reid 1990: Fig. 14, distal caudal; 1996: Fig. 61, phalanx) and very numerous in the claw-bearing unguals (1990: Fig. 13; 1996: Fig. 27). At the very least, this means that isotopic data from extremities would be less complete

than those obtained from core bones; and while there is no proof that growth pauses in extremities were temperature-related, they could obviously mark occasions when the temperature fell below a level at which bone was not deposited. Variations in temperature between core parts and extremities could thus have been considerably greater than isotopic data would record, and such data would then make the animal seem more homoiothermic than it was. Such results should therefore be treated with caution, unless bone deposition is shown to have been equally continuous in all parts sampled.

One is that small forms would automatically have higher mass-specific SMRs than large ones, without this implying any movement toward endothermy—although, lacking the temperature stabilization of mass effects, they could have been under greater selection pressure to move toward it. Assuming parallels, for instance, with Eckert et al.'s cat and elephant figures (1988: Table 16–2), the total SMR of a 3,833-kilogram (3.833-tonne) tyrannosaur would be about 158 times that of a 2.5-kilogram compsognathid; but the latter's mass-specific SMR would be 9.7 times that of the tyrannosaur, and thus almost a whole order higher. The contrast between it and a 30-tonne sauropod, for example, would have been even greater.

In addition, mass-specific SMRs can fall considerably during growth if adults are much larger than juveniles, as appears to have been generally the case in large dinosaurs. Walter Coombs (1980), for

Dinosaurian Physiology   457

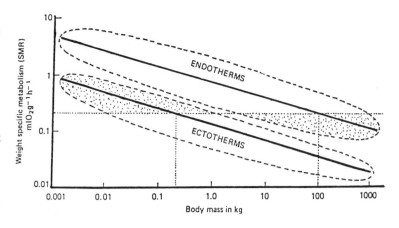

Figure 32.5. How mass-specific standard metabolic rates change with size in ectotherms and endotherms. In both, weight-specific SMR declines from the smallest to the largest, with the averaged regression lines for their scatters following slopes of about 0.25. As a result, small ectotherms in the shaded area at left have *higher* metabolic rates than large endotherms in the shaded area at right. This is why the terms *low* and *high* metabolic rates, when used comparatively, must always be understood as referring to animals of similar size. Modified from McFarland and Heiser 1979: Fig. 7.21, lower part; this version from Reid 1987: Fig. 4, by permission; recaptioned.

instance, judged *Psittacosaurus* hatchlings to weigh only 0.7 percent of adult weight, and Ted Case's (1978) estimates of 2.9 kilograms and 5,300 kilograms (5.3 tonnes) for hatchlings and adults of *Hypselosaurus* give the still lower figure of only 0.055 percent. For comparison, values obtained by Coulson and Hernandez (1983) show the mass-specific SMR of a 7-kilogram alligator as only 17 percent of that of a 35-gram hatchling, falling further to 7.66 percent at 70 kilograms and only 3.6 percent at 700 kilograms (Box 32.3). A considerable fall in mass-specific SMR during growth is hence a predictable feature of all large dinosaurs, irrespective of whether they were ectotherms, endotherms, or intermediates.

A Digression: Why Regulate Temperature?

Regulation of temperature (thermoregulation) by reptiles, mammals, and birds has been noted above without comment on why this should happen. Explained simply, regulation of temperature within narrow limits in terrestrial species is usually interpreted (e.g., Pough et al. 1989) as allowing maximum efficiency in the coordination of the thousands of enzyme-catalyzed and temperature-sensitive biochemical processes on which life depends. High optimum temperatures are said to allow faster nervous reactions than lower ones, and a permanently stabilized temperature allows activity to be independent of external heat sources. An ectothermic dinosaur growing large enough for full mass homoiothermy would gain the same independence in activity, plus a raised and stabilized metabolic rate, without the energetic cost of true endothermy, in which 90 percent of energy expenditure is devoted to heat production.

In addition, some thermoregulation is related to a need to keep temperature within limits outside which life processes will break down irreversibly. Because of their dependence on external heat sources, many ectothermic reptiles can tolerate cooling to well below their activity temperatures; but endotherms are killed by hypothermia if heat losses outstrip their ability to compensate by increased heat production. At the other extreme, both ectotherms and endotherms have little tolerance to core temperatures above their normal ranges, and adopt various methods of avoiding this problem. Panting, for instance, is a physiological method of dispersing excess heat by increasing respiratory cooling; and many forms living in warm deserts take refuge in shade or underground during the hottest hours. How tolerant small dinosaurs would have been to lowered temperatures would depend on what type of physiology they had; but overheating would potentially be a problem for all large dinosaurs, in warm conditions and strong sunlight especially, irrespective of their thermal physiology.

A Model for Dinosaurs

Although de Ricqlès (1974, 1980) hinted at a physiological difference between dinosaurs and all modern tetrapods, the details of how they could have differed first emerged in a model proposed by Philip Regal and Carl Gans (1980), designed to allow large animals to make optimum use of a limited food supply. Their ideal dinosaur was an animal with complete double-pump circulation, a temperature stabilized by bulk, and a low metabolic rate (SMR), with aerobic activity metabolism, vascular control over heat exchanges, and tolerance of temperature instability as further features. This combination, they thought, could give such animals "extraordinary abilities to be active and grow rapidly on a limited food supply," as well as explain how giants such as the sauropods could avoid overheating, and feed themselves with only small heads.

Besides having double-pump circulation, such dinosaurs would differ from modern reptiles in a second major way if their activity metabolism was aerobic. In modern forms, reliance on anaerobic glycolysis as an energy source restricts high activity to short periods, producing "oxygen debts" which then have to be repaid, whereas energy release by direct oxidation, as in mammals and birds, would give dinosaurs a similar capacity for continuous activity. Although monitor lizards (varanids or goannas) show partial aerobic activity, complete double-pump circulation would presumably allow its full development in dinosaurs.

A low basic metabolic rate was specified because the higher this is, the more food animals of similar size need to sustain it: modern mammals, for instance, need ten to thirteen times as much as modern reptiles (Pough 1979). Whether dinosaurs were really adapted to limited food supplies is debatable (cf. Farlow et al. 1995), but a low SMR would also let more of the energy derived from their food be devoted to growth. Vascular control over heat exchange, by dilation or contraction of superficial blood vessels, is known from various modern reptiles, and would add an active control over heat loss to the passive insulation of Spotila et al.'s (1973) model. Some attempts to dismiss the concept of mass homoiothermy (e.g., Desmond 1975) have assumed that dinosaurs had no active means of controlling heat exchanges; but since they were live animals and not inanimate objects, this is very unlikely.

Since full double-pump circulation is now restricted to endotherms, it needs to be asked whether its possession would have converted dinosaurs into endotherms automatically, thus not allowing the existence of animals of the type that Regal and Gans envisaged. This is possible but does not seem likely. Crocodiles have a form of double-pump circulation, in which systemic arteries carry only oxygenated blood except during diving, when the lungs are out of use; yet they show no sign of moving toward becoming endotherms. On the other hand, I think it unlikely that the three-way combination of full double-pump circulation, full aerobic activity, and a capacity for fast continuous growth could be evolved without at least some upward movement of basic metabolic rates resulting; and I have hence pictured Regal-Gans dinosaurs as likely to show at least some movement toward endothermy. The question then becomes how far they would move toward it.

Box 32.3.

Metabolic
Rates and
Size

A characteristic feature of vertebrates (and many invertebrates) is that while their total oxygen consumption increases with size, their consumption per unit of body mass instead decreases. If the mass-specific standard metabolic rates (SMRs) of modern vertebrates are plotted logarithmically, those of ectotherms and endotherms cluster around parallel regression lines, with –0.25 slopes, which fall at levels corresponding with their average difference in metabolic rates (Fig. 32.5). We have hence to expect a similar pattern in dinosaurs of different sizes, irrespective of whether they were ectotherms, endotherms, or somehow intermediate between the two. This, of course, assumes that one basic type of physiology was common to all dinosaurs; but this assumption is justified in showing what would then be expected.

In the following tables, data from Coulson and Hernandez (1983) for SMRs in mammals and alligators are analyzed in terms of percentages, for ease of comparison. First, figures for representative mammals from a very small shrew to the largest mammal known (the blue whale) provide a picture of the range of variation among members of one class. (See Table One, facing page.)

If you multiply the figures in the first two columns, you will find that the total minimum oxygen consumption (= total SMR) for the whale is more than 75,000 times that for the shrew; but at the mass-specific level, that of the shrew is nearly 660 times that of the whale, and 60 times that of an average (70 kg) man.* This is why the difference between the total figures does not simply correspond with the differences in weight. To the nearest simple fraction, the human figure is roughly two-thirds that of the 10-kilogram dog, one-fifth that of the 200-gram rat, one-sixteenth that of a 20-gram mouse, and one-sixtieth that of a 2-gram shrew. If we assume the same pattern for dinosaurs, the mass-specific SMR of a 100-tonne sauropod would be under one-tenth (9.1%) that of a 70-kilogram deino-

nychid, and about one-sixteenth that of a 10-kilogram hatchling dinosaur.

The same authors' figures for American alligators give a striking picture of how much SMR can change during growth, when adults are much larger than juveniles. (See Table Two, facing page.)

From these figures, a 20,000-times mass increase leads SMR to fall by a factor of nearly 28. How far dinosaurs would parallel alligators is conjectural, but the figures give an indication of the scale of change to be expected in large forms up to 700 kilograms in mass, assuming that hatching masses were between 0.5 and 7 kg.

---

*Read the comparative SMR percentage vertically by column. For example, in the first column the shrew = 100%, and thus the SMR of the blue whale compared to the shrew is 0.15%. In the next column, the mouse = 100%, and the SMR of the blue whale compared to the mouse is 0.58%.

This then leads to the question, How do endotherms evolve? Nobody knows; but one theory, proposed by Brian McNab (1978), is that animals must first become inertial (mass) homoiotherms, and then progressively become smaller, raising SMRs higher and higher as this happens in order to maintain homoiothermy. In fact, the mammal-like reptiles (therapsids) on which this picture was based were not an evolutionary series; but the earliest-known mammals were mouse-sized, and

TABLE ONE (mammals)

	Mass in kilograms	O$_2$ used in liters/kg/day	Comparative SMRs by mass			
Shrew	0.002	322.8	100%			
Mouse	0.02	84.0	26%	100%		
Rat	0.2	27.7	8.6%	32.9%	100%	
Dog	10.0	8.0	2.5%	9.5%	28.9%	100%
Man	70.0	5.38	1.7%	6.4%	19.4%	67.3%
Blue whale	100,000.0	0.49	0.15%	0.58%	1.8%	6.1%

TABLE TWO (alligators)

Mass in kilograms	O$_2$ used in liters/kg/day	Relative change in SMR during growth				
0.035	2.35	100%				
0.050*	2.07*	88.1%				
0.12	1.56	66.4%				
0.5*	1.05*	44.7%	100%			
1.0	0.77	32.8%	73.3%	100%		
3.5*	0.56*	23.8%	53.3%	72.7%	100%	
7.0	0.40	17.0%	38.1%	51.9%	71.4%	100%
70.0	0.18	7.66%	17.1%	23.4%	32.1%	45%
700.0	0.084	3.6%	8.0%	10.9%	15.0%	21%

*From Coulson and Hernandez 1983: Fig. 2.1; the rest is from their table 2.1.

the first known bird, *Archaeopteryx*, was no bigger than a pigeon. This at least suggests that "warm-bloodedness," as developed in mammals and birds, was evolved at small sizes. But dinosaurs were typically medium- to large-sized animals from their earliest appearances; and while growth to large sizes would enhance mass effects, it would lead them away from true endothermy if Brian McNab was right. This led me to suggest (Reid 1984b, 1987) that we should see them as "failed endotherms," which took two of the critical steps toward true endothermy (double-pump circulation and aerobic activity) but then took a "wrong turn" which led to the point where most of them were relying on bulk as their main means of stabilizing temperature. And in making this suggestion, I was influenced by evidence from bone.

Evidence from Bone

In 1907, Adolf Seitz showed that a tissue now called Haversian bone can be developed in dinosaurs to a level now seen chiefly in large mammals. Early in the present controversy, it was regarded as evidence of endothermy by both Bakker (1972) and de Ricqlès (1974); but others have preferred to relate it to mass homoiothermy (e.g., McNab 1978) or to mechanical effects of large size (e.g., Hotton 1980; Ostrom 1980). It can have no strict causal connection with high SMR levels, since it does not occur in the smallest birds and mammals despite their having the highest metabolic rates; and as a secondary tissue, replacing others, it is known to be influenced by various non-thermal factors (e.g., Enlow 1962), some of which (e.g., muscular attachments; Reid 1984a) have been shown to have affected it in dinosaurs. It is also not restricted to endotherms, as Marianne Bouvier (1977) saw, and was even developed as extensively as in humans in the tortoise *Geochelone triserrata* (Reid 1987: Fig. 3b), which is not a likely endotherm. Because of these problems, my own work has been centered on a different tissue, which yields direct evidence of how dinosaurs grew.

Periosteal bone is a tissue formed on the surfaces as bones grow in thickness, whose significance in this context was first recognized by Armand de Ricqlès (1974). When formed slowly, it is typically a dense, finely lamellated tissue, with few or none of the channels for blood capillaries known as vascular canals. When formed quickly, however, it is instead highly vascular, and at first forms a finely cancellous (spongy) framework, within which slowly formed lamellar bone grows inward toward blood vessels to form structures called primary osteons (Fig. 32.6a). As de Ricqlès saw, the periosteal bone of modern reptiles, described

Figure 32.6. Fibro-lamellar and zonal bone. (a) Fibro-lamellar bone, showing its initial formation as finely cancellous bone (*top*), and subsequent compaction with formation of primary osteons (*lower half*). (b) Zonal bone, with lamellated annuli between zones with small primary osteons in a non-lamellated matrix. (c) Zonal bone formed wholly from lamellated tissue, with zones defined by resting lines.

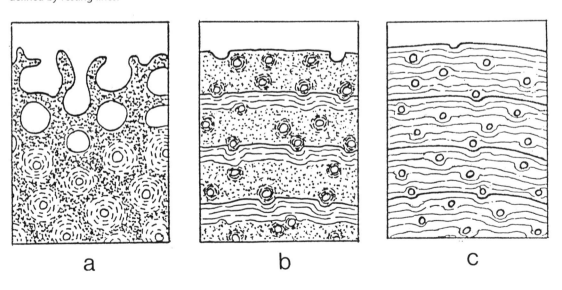

a   b   c

R. E. H. Reid

as zonal (Gross 1934) or lamellar-zonal bone (de Ricqlès 1974), is most commonly built partly or entirely from bone of the first type (Figs. 32.6b, c), and is often also divided into major "growth rings," or zones, through the slowing or periodic cessation of growth.

In various modern reptiles, including crocodiles, the "growth rings" have been shown to be annual. But the periosteal bone of large, fast-growing mammals (e.g., cattle) and birds (e.g., ostriches) is instead bone of the second type, termed fibro-lamellar bone, and is characteristically formed continuously, without annual interruptions, although non-cyclical "growth rings" may occur if, for example, disease or starvation causes interrupted growth. Such bone is widely distributed in dinosaurs (cf. de Ricqlès 1980: Table 10) and was originally thought by de Ricqlès (1974, 1976) to be the only type that they produced. As he warned, its only strict implication is that dinosaurs could grow quickly and continuously, in the manner of large mammals and birds; but as such growth is now restricted to "warm-blooded" animals, it could be thought to be possible only for such animals.

This view has been followed by Bakker (e.g., 1975, 1986), but has three major flaws:

First, as shown earlier by Donald Enlow (1969), bone with the structure of fibro-lamellar bone is formed intermittently by both turtles and croco-diles, which show periodic rapid growth, and then forms the major parts of their growth rings. In young American alligators, it can even be the only tissue formed (Figs. 32.7a, b); and in their size range, figures given by Coulson and Hernandez (1983: Table 2.1) show the mass-specific SMR of a 7-kilogram specimen (0.40 liters O_2/kg/day) as only 5 percent of that of a slightly larger (10 kg) dog (8.0 liters). Hence, such bone cannot be taken as implying endothermic metabolic rates. Paul (1991) has tried to avoid this conclusion by claiming that fast growth is seen only in animals raised under optimum conditions; but the example figured here was both wild and from North Carolina, where these animals experience winter freeze-ups, as in Louisiana (cf. Joanen and McNeese 1980).

Second, continuous growth of fibro-lamellar bone is unsafe as an index of endothermy without knowledge of the thermal physiology of the oldest forms to show it (Reid 1984a, 1984b). Further, evidence that it need not be

Figure 32.7. Bone from two modern reptiles. (a) Bone from a young American alligator, showing obvious fibro-lamellar structure and four weakly defined zones. (b) The same tissue enlarged, showing the non-lamellated character of the periosteal framework (the dark trabeculae). (c) Vascular bone without "growth rings," from the femur of a half-grown Galapagos tortoise.

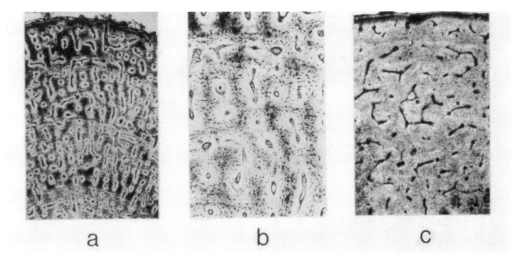

a   b   c

is seen in forms leading to mammals, in which it occurs first in primitive therapsids (deinocephalians, titanosuchians, eotheriodonts) showing no other signs of being endotherms (Kemp 1982). Some dinosaurs, moreover, formed vascular bone without "growth rings" which is not fibro-lamellar, but a uniform tissue without osteons (Currey 1962; Reid 1984a); and this can be matched in the Galapagos tortoise, *Geochelone elephantopus* (Fig. 32.7c), which is plainly not an endotherm. This, however, may be a reflection of growth rates rather than metabolic rates.

Third, de Ricqlès's argument for endothermy (1974, 1976) involved the assumption that typical reptilian "growth rings" do not occur in dinosaurs, and was withdrawn when it was shown that they do (Reid 1981; de Ricqlès 1983). They are now known from a wide range of genera (Reid 1990), from bones other than limb bones (e.g., Figs. 32.8, 32.9), and from limb bones as well (e.g., Fig. 32.10).

Despite these problems, however, fibro-lamellar bone has three things to tell us about dinosaurs. First, as de Ricqlès (1974, 1980) saw, it shows many of them as able to grow quickly and continuously up to large and very large sizes, at rates comparable with those now seen in large fast-growing mammals and birds. No modern reptiles can grow in this manner, yet dinosaurs could do so up to the sizes of huge sauropods such as *Brachiosaurus* (Gross 1934), weighing 30 tons or more. Second, such growth must mean that their circulatory systems were at least as efficient as those of large terrestrial mammals, thus confirming the argument from vertical distances. Animals can be programmed genetically to grow more slowly than they could, as, indeed, we are ourselves; but none can grow faster than a maximum potential rate determined by the rate at which the vascular system can supply substrates and energy to growing tissues. Rapid growth to large sizes hence means that dinosaurs must have had hearts and blood that could support such growth, which would not have been possible otherwise. Third, continuous formation of fibro-lamellar bone in small forms and juveniles shows that it cannot have been a product of mass effects, as some writers (e.g., Hotton 1980) have supposed. If it were, it would not be seen in, for example, the small ornithopods *Hypsilophodon* (Reid 1984b: Fig. 1j) and *Dryosaurus* (Chinsamy 1995); and J. R. Horner has traced it back to pigeon-sized hatchlings, with bone like that of young birds. And since its presence in this size range cannot be a mass effect, it must have some different explanation.

Bone with zonal "growth rings" (Figs. 32.8–10), in contrast, shows dinosaurs as able to grow in the manner of ectothermic reptiles as well as that of endothermic mammals, and as sometimes also growing in both styles in different parts of their bodies. De Ricqlès (1968), for instance, found good uninterrupted fibro-lamellar bone in limb bones of the large prosauropod *Euskelosaurus;* but I found equally good zonal bone in its ribs, and a similar contrast between the femur and the pubis in the theropod *Megalosaurus* (Reid 1990). This in turn shows up a further failure of modern data as a basis for understanding dinosaurs. At present, these types of bone are seen characteristically in endotherms and ectotherms respectively; but they cannot reflect different types of thermal physiology when both are found in one animal, and when what determined the development of one or the other must have been something different. Luckily, dinosaurs themselves provide evidence of what this could have been. As first shown by de Ricqlès (1983), bones that grew asymmetrically can show cyclically interrupted growth in parts where growth was slowest, with the resulting "growth rings" clearly related to whether the local growth rate was below or above

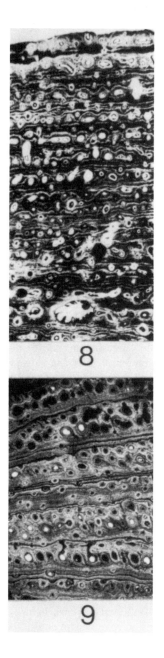

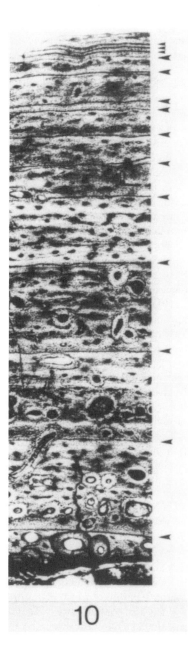

Figure 32.8. Zonal bone from the pubis of a sauropod dinosaur, from the Lower Bajocian of Northamptonshire, England. This tissue is strikingly similar to that figured by Gross (1934: Fig. 4) from *Nothosaurus* as typical of reptilian zonal bone.

Figure 32.9. Bone with zonal "growth rings" from a rib of an allosaur-sized carnosaur, from the Wessex Formation (Lower Cretaceous, Wealden, ?Barremian) of the Isle of Wight, England.

Figure 32.10. Bone with zonal "growth rings" from the femur of the ornithopod *Rhabdodon*, from the Maastrichtian of Szentpeterfalva, Romania. Progressive thinning of the rings toward the external surface (*at top*) implies progressively slowing growth, like that seen in modern crocodiles, although the thickness of the rings shows growth as faster than in any known crocodile. The specimen was collected by Baron Ferenc Nopcsa, before World War I.

some critical level, and this could potentially apply to all occurrences in dinosaurs.

Bone with zonal "growth rings" also yields useful evidence in two other ways. First, it adds further evidence that continuous growth in dinosaurs was not a mass effect, by occurring in forms large enough to show high levels of mass homoiothermy. If, for instance, this could not eliminate it in the sauropod *Camarasaurus* (cf. Reid 1990: Fig. 7), it is unlikely to have done so in anything smaller. Second, if the rings are assumed to be annual, they can sometimes yield data on the rates at which dinosaurs grew, which cannot be measured without annual markers. For instance, although the sauropod whose bone is shown in Figure 32.8 is known only from hip bones (the pubis and ischium), these indicate an animal at least as big as any modern elephant; and one of them (the pubis) shows twenty-three rings in a shaft with room for only twenty-eight or twenty-nine at most (Reid 1987, 1990). So while this bone was formed in a "reptilian" manner, the animal grew at a rate that no modern reptile can match, and little short of that of modern elephants, which can take twenty to twenty-five years to reach full size.

Last, the end of growth in dinosaurs can follow various patterns. Many bones with uninterrupted fibro-lamellar bone show no sign of growth's having ceased, or even slowed, before the animals died (e.g., *Dryosaurus* [Chinsamy 1995]); and this led de Ricqlès (1980) to picture dinosaurs as combining the fast continuous growth of large endotherms with the indefinite growth of reptiles. This, he thought, could point to their being unique physiologically. Forms with growth rings may again show no sign that growth slowed before ceasing (Fig. 32.8; cf. *Massospondylus* [Chinsamy 1993]), or they may show that it slowed progressively (Fig. 32.10), as in crocodiles, although much less pronouncedly. The giant sauropod *Brachiosaurus* (Gross 1934) and the small theropod *Troodon* (Varricchio 1993) show thin superficial layers of avascular lamellated bone, resembling bone sometimes formed by slow accretion in endotherms after growth has ceased at maturity, but possibly simply marking old age. The small theropods *Syntarsus* (Chinsamy 1990) and *Saurornitholestes* (Reid 1993) show similar superficial tissue in bones in which earlier bone shows "growth rings," so that features now typical of endotherms and ectotherms respectively occur together.

These various mixtures of "endothermic" and "ectothermic" growth styles have been my main reason for envisaging an intermediate style of physiology for dinosaurs. Before discussing this, however, we need to look at a further kind of evidence.

Cavernous Bones

In saurischian dinosaurs, although not ornithischians, it is common for neck and body vertebrae to show deep lateral hollows, termed pleurocoels, or to show small lateral openings leading into a labyrinth of internal cavities separated by thin plates of bone (Fig. 32.11). The upper parts of ribs may also contain cavities, again with small external openings. In birds, very similar hollowed vertebrae contain extensions of a system of air sacs into which air is drawn through the lungs; and this has led various authors (e.g., Swinton 1934; Romer 1945; Janensch 1947) to see saurischians as having a similar air sac system. Strictly, there is no way of proving this without having soft parts preserved; but the histological structure of such cavernous bones is essentially identical with that seen in modern pneumatic bones, apart from differences due to dinosaurs' sometimes reaching massively large sizes (Reid 1996). As in modern forms, the internal cavities were produced by internal bone resorption, and then lined with a tissue like that which may line marrow cavities; but unresorbed cancellous bone was also typically converted into compact bone, which is not the case in marrow-filled bones. In some forms, the resulting tissue can be quite indistinguishable histologically from bone from bird vertebrae or from the pneumatic parts of elephant skulls. This makes it very likely that saur-

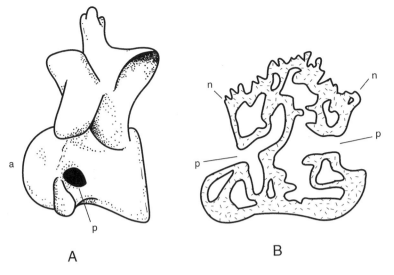

Figure 32.11. (A) Late cervical vertebra of the theropod *Allosaurus,* showing a pleurocoel (p) in the form of a small lateral perforation; a = anterior. (B) Transverse section of a detached centrum through the pleurocoels, showing how they open into internal cavities in the manner of pneumatopores. The neural arch has been detached along the line of the neurocentral suture (n -). A, X 2/3. B, X 1.

ischians did indeed have a bird-like air sac system; and this style of respiration, which is more efficient than that of mammals, is reasonable grounds for presuming aerobic activity.

An air sac system could also have made a substantial contribution to respiratory cooling, on which dinosaurs probably relied extensively. As various authors have recognized (e.g., Spotila 1980), excess heat would increasingly be a problem for dinosaurs the larger they grew, and this suggests that the respiratory system would be increasingly important correspondingly. Various factors would be involved here. First, dinosaurian skin when it is known was apparently a hide like that of crocodiles, whereas respiratory airways would be lined by thin membranes, partly thin enough for gaseous exchange in the lungs. Second, because of this, the respiratory system would probably be the main avenue for evaporative cooling. Third, while cooling through the skin would at times have had to operate in still air, and could simply have been impossible in bright sunshine, respiratory cooling would operate in a permanent airflow situation, and hence be a more constant factor. Fourth, because the growth of lung tissues involves the multiplication of respiratory elements, and not simply their progressive enlargement, the ratio of the internal surface area of the lungs to the volume of the body does not follow the same square-to-cube rule as the external surface. To these factors, which apply to all dinosaurs, an air sac system would add at least extra internal cooling surfaces (Fig. 32.12), and potentially the extra efficiency of a one-way airflow system like that of birds. Both could have been especially important in large sauropods, in which the problem of dispersing excess heat would have been greatest.

Lack of cavernous structures in ornithischians points to air sacs' being either less developed or simply absent; but ornithischians do not seem to have been physiologically disadvantaged compared with saurischians, except for not reaching the sizes of large sauropods, and mammals have not needed air sacs to become fully aerobic.

Dinosaurian Physiology   467

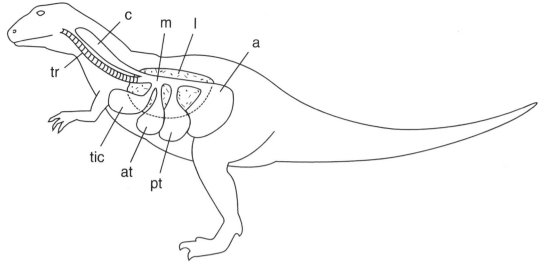

Figure 32.12. A theropod dinosaur envisaged as having an air-sac system like that of a bird. Figure shows structures of the right side as seen from the left. tr = trachea; l = lung; c = cervical sac; tic = interclavicular sac; at = anterior thoracic sac; pt = posterior thoracic sac; a = abdominal sac; m = meso-bronchus.

Last Steps: Matching the Pieces

It remains to assess the possibilities set out at the start of this chapter in terms of the principles and evidence now discussed. First, I think it clear that we must reject the notion that dinosaurs were simply mass homoio-therms (or gigantotherms), not differing otherwise from modern reptiles. Three points are critical here: (1) recognition by Hohnke (1973), Seymour (1976), and Ostrom (1980) that blood circulation in large dinosaurs would require systemic pressures which only fully evolved double-pump hearts could provide; (2) recognition by de Ricqlès (1974) that fibro-lamellar bone shows dinosaurs as able to grow quickly and continuously, up to the largest known sizes; and (3) the further implication that such growth would be possible only for animals with cardiovascular systems that could support it.

In saurischians, there is also the evidence for a bird-like air sac system, very probably also implying aerobic activity. There is simply no way in which animals with this combination of characters can be seen as no more than large "good reptiles." Mass homoiothermy also cannot explain fast continuous growth in small species and juveniles too small to show mass effects, or alone explain ability to grow quickly to large sizes; and it also could not release "good reptile" dinosaurs from the limitations of anaero-bic activity metabolism, which would be increasingly a problem the larger they grew. According to Coulson (1984), for instance, a 100-ton "reptil-ian" dinosaur forced to use up all the energy available to it anaerobically would have needed at least three weeks to replace it. With large predators watching for animals in trouble, its survival would be doubtful.

Taking next the possibility that dinosaurs were endotherms, we can first reject the view (e.g., Spotila 1980) that heat stress at large sizes would have made this impossible, except perhaps for large sauropods. As was seen by de Ricqlès (1980), only sauropods grew much larger than the Tertiary rhinoc-eros *Paraceratherium* (a.k.a. *Indricotherium* or *Baluchitherium),* said to have weighed 16 tons (Halstead 1979); and even the African elephant, reaching 6 tons or more, is heavier than was likely for most non-sauropod dinosaurs.

Rejection of this notion, however, is not an argument for endothermy. To show dinosaurs as endotherms in the same sense as mammals and birds, two things need to be proved beyond a reasonable doubt: the possession of high *basal* metabolic rates, in the same general range of those of mammals and birds, and the ability to maintain high steady temperatures at small sizes, independently of bulk and activity. On the first point, it is critically important to distinguish between basic and activity metabolism, and to discount claims or arguments that fail to distinguish between them. In particular, since small gracile dinosaurs such as *Hypsilophodon* or *Troodon* were probably highly active at times, they have been said to have had "high metabolic rates"; but the high metabolic rates characteristic of endotherms are not based on activity metabolism, and a high metabolic rate achieved during maximum activity is not a basic metabolic rate.

Bakker (e.g., 1972) has argued that such animals would need to be endotherms to sustain the speeds they seem to be built for; but whether their maximum speeds could be sustained for an appreciable time or were reached only briefly would have depended on whether their basis was aerobic or anaerobic, and not on endothermy. The high metabolic rates that were formerly claimed by de Ricqlès (1974, 1976; but not 1983) were basal metabolic rates; but when the fibro-lamellar bone they depended on can be formed by young American alligators (Fig. 32.7a, b) at an SMR of only one-twentieth that of a comparable mammal, there is no obvious reason why dinosaurs should need endothermic SMRs to do so. Admittedly, they commonly also grew continuously, but so can the Galapagos tortoise (Fig. 32.7c), without being an endotherm.

This leaves only Bakker's (1972) concept of predator-prey ratios as providing what could be real evidence of high basic metabolic rates, since he seems to show carnivorous dinosaurs as having similar food requirements to mammalian carnivores. But as J. O. Farlow (1976) has shown, Bakker's results are hard to duplicate, even when very careful allowance is made for every possible biasing factor; and even if confirmed, they would still apply only to the carnivores (i.e., theropods) and not other dinosaurs.

Moreover, there is currently no evidence whatsoever for the ability to maintain high steady temperatures at small sizes, and quite probably there never will be. A domestic cat or a chicken, for instance, can maintain a high temperature while sleeping, but there is no way of showing that the chicken-sized *Compsognathus* could have done this, however active it may have been while awake. Further, avian and mammalian endothermy are based partly on possession of external insulation, which also is not simply passive but is used with active neuromuscular control of individual feathers and hairs, triggered physiologically. In contrast, dinosaurian skin, when it is known, shows no sign of external insulation; and this problem cannot be avoided by calling *Archaeopteryx* a dinosaur, or by drawing imaginary feathers on theropods, but only by showing all dinosaurs to have been either feathered or derived from feathered ancestors. Short of that, there is nothing to show that dinosaurs had any means of regulating temperature beyond those now seen in reptiles.

This leaves us with the possibility that dinosaurs were "intermediate" animals, combining some of the features of endotherms with reptilian styles of thermoregulation. No true evolutionary intermediates between ectotherms and endotherms now exist, despite various overlaps; but they must once have existed, for endotherms to have evolved from ectotherms, and their absence now is due to this process having taken place long ago.

Mammalian endothermy, for instance, appears to be an inheritance from the pre-mammalian cynodonts (Kemp 1982; Ruben 1995), and so to date from the Late Permian. In addition, while large mammals have played the same part in post-Cretaceous faunas that dinosaurs did in Mesozoic faunas, they inherited a small-animal style of physiology from ancestors that spent the dinosaur period in the size range of modern rats and mice, and so could be endotherms more by inheritance than by necessity. Without a comparable pre-commitment to endothermy, the dinosaurs might have been able to evolve a less energetically wasteful way of being large active animals. But is there any evidence that they did so?

In my opinion there is, if some of the evidence from periosteal bone is accepted as counting against their being endotherms. This applies especially to the various mixtures of "endothermic" and "reptilian" growth styles, to which de Ricqlès (1980) first drew attention, and especially the occurrence of zonal "growth rings" when these are extensively developed (Figs. 32.8–32.10). Some caution is needed here, because annual "growth rings" occur in some mammals, but they are typically seen either in marine forms that encounter cold conditions (e.g., the porpoise *Phocoena*: Buffrenil 1982) or in small forms from cold regions (e.g., beavers, muskrats, hamsters; Klevezal and Kleinenberg 1969), some of which also hibernate. They are also seen mainly in slowly formed avascular bone, in which "resting lines" marking growth pauses occur close together. In contrast, dinosaurs with "growth rings" could form highly vascular tissues, and the examples figured here all lived under warm conditions. The sauropod discussed earlier (Fig. 32.8), for instance, lived near a sea in which corals built patch reefs, and the ornithopod *Rhabdodon* (Fig. 32.10) lived in lands lapped by the tropical ocean Tethys. We have thus to deal with animals that could form bone in the manner of typical reptiles under even tropical conditions; and with similar tissues unknown from any comparable modern mammal, or any bird, such animals are not likely endotherms.

It would, of course, be possible for dinosaurs to have had more than one type of physiology; but *Rhabdodon,* which shows "growth rings" in its limb bones (Fig. 32.10), is otherwise just as typical an ornithopod as *Iguanodon,* which does not. No physiological difference between them hence seems likely. Different thermal physiologies also cannot be reflected when both main growth styles occurred in one animal, and asymmetrical bones can then show them as related instead to local growth rates. The simplest, or "most parsimonious," reading of such evidence is that all dinosaurs shared one type of physiology, which allowed them to form bone continuously when or where growth was rapid, but also allowed them to form "growth rings" when or where it was slower (Reid 1990). This then suggests, in turn, that while more like birds and mammals physiologically than modern reptiles are, they were also more like modern reptiles than mammals and birds are. And if this is correct, it fits their being intermediate animals of the kind that Regal and Gans (1980) envisaged.

The case for intermediate dinosaurs rests here. It cannot be proved conclusively, but it fits current evidence better than the view that they were endotherms. Hemodynamics and fast growth to large sizes both demand more efficient circulation than any modern reptile possesses, and the evidence of air sacs in saurischians points strongly to aerobic activity. Both can be expected to have led to SMRs shifting upward; but the evidence from bone does not require this to have been to endothermic levels, which also cannot be inferred from high activity, and have yet to be proved by any other means. Dinosaurs did have the ability to grow in the manner of fast-growing endotherms, but could also still grow in the manner of normal reptiles. Nothing shows dinosaurs in general as having any means of regulating temperature but those now seen in reptiles; but even doubling average modern reptilian SMR levels would have given them more capacity for exploiting mass homoiothermy than any comparable modern reptile. That bulk was especially important to them physiologically is strongly suggested by the fact that so many were large, and especially by their failure to radiate as small animals, as both the ectothermic lizards and the endothermic birds and mammals have. Small forms would have been pre-adapted for evolving into endotherms if they could also develop external insulation; but judging from current evidence, this seems likely to have happened only in the stock that led to birds. All these points are consistent with most dinosaurs' being "failed endotherms," which started on the road to endothermy but never attained it through opting for bulk as their main means of stabilizing temperature.

In short, it seems fairly likely that the great success of dinosaurs in their time need not mean that they were endotherms. They could instead have been sub-endothermic "super-reptiles," or "super-gigantotherms," with a more advanced circulatory system than any modern reptile, and no true modern physiological counterparts. Together with inherited upright limbs, which would have allowed them high mobility and pre-adapted them to support massive weights, this could have made them superior as large terrestrial animals to all possible Mesozoic competitors. In the end, a specialized physiology which worked best at large adult sizes, and not at small ones, could in turn have been a factor in their downfall; but if so, that is part of another story.

References

Bakker, R. T. 1972. Anatomical and ecological evidence of endothermy in dinosaurs. *Nature* 238: 81–85.

Bakker, R. T. 1975. Dinosaur renaissance. *Scientific American* 232 (4): 58–78.

Bakker, R. T. 1986. *The Dinosaur Heresies: New Theories Unlocking the Mystery of the Dinosaurs and Their Extinction.* New York: William Morrow.

Barrick, R. E., and W. J. Showers. 1994. Thermophysiology of *Tyrannosaurus rex*: Evidence from bone isotopes. *Science* 265: 222–224.

Bouvier, M. 1977. Dinosaur Haversian bone and endothermy. *Evolution* 31: 449–450.

Buffrenil, V. de. 1982. Données préliminaire sur la présence de lignes de'arrêt de croissance périostiques dans la mandibule du marsouin commun, *Phocoena phocoena* (L.), et leur utilisation comme indicateur de l'âge. *Journal canadien de zoologie* 60: 2557–2567.

Case, T. J. 1978. Speculations on the growth rate and reproduction of some dinosaurs. *Paleobiology* 4: 320–328.

Chinsamy, A. 1990. Physiological implications of the bone histology of *Syntarsus rhodesiensis* (Saurischia: Theropoda). *Palaeontologia africana* 27: 77–82.

Chinsamy, A. 1993. Bone histology and growth trajectory of the prosauropod dinosaur *Massospondylus carinatus. Modern Geology* 118: 319–329.

Chinsamy, A. 1995. Ontogenetic changes in the bone histology of the Late Jurassic ornithopod *Dryosaurus lettowvorbecki. Journal of Vertebrate Paleontology* 15: 96–104.

Colbert, E. H.; R. B. Cowles; and C. M. Bogert. 1946. Temperature tolerances in the American alligator, and their bearing on the habits, evolution and extinction of the dinosaurs. *Bulletin of the American Museum of Natural History* 86: 327–373.

Coombs, W. 1980. Juvenile ceratopsians from Mongolia: The smallest known dinosaur specimens. *Nature* 283: 380–381.

Coulson, R. A. 1984. How metabolic rate and anaerobic glycolysis determine the habits of reptiles. In M. W. J. Ferguson (ed.), *The Structure, Development and Evolution of Reptiles,* pp. 425–441. Orlando, Fla., and London: Academic Press.

Coulson, R. A., and T. Hernandez. 1983. Alligator metabolism: Studies on chemical reactions *in vivo. Comparative Biochemistry and Physiology* 74: i–iii, 1–182.

Currey, J. D. 1962. The histology of the bone of a prosauropod dinosaur. *Palaeontology* 5: 238–246.

de Ricqlès, A. J. 1968. Recherches paléohistologiques sur les os longs des tétrapods. I. Origine du tissue osseux plexiforme des dinosauriens sauropodes. *Annales de Paléontologie (Vertébrés)* 54: 133–145.

de Ricqlès, A. J. 1974. Evolution of endothermy: Histological evidence. *Evolutionary Theory* 1: 51–80.

de Ricqlès, A. J. 1976. On bone histology of fossil and living reptiles, with comments on its functional and evolutionary significance. In A. d'A. Bellairs and C. B. Cox (eds.), *Morphology and Biology of Reptiles,* pp. 123–150. London: Academic Press.

de Ricqlès, A. J. 1980. Tissue structure of dinosaur bone: Functional significance and possible relation to dinosaur physiology. In R. D. K. Thomas and E. C. Olson (eds.), *A Cold Look at the Warm-Blooded Dinosaurs,* pp. 103–139. American Association for the Advance of Science Selected Symposium 28. Boulder, Colo.: Westview Press.

de Ricqlès, A. J. 1983. Cyclical growth in the long limb bones of a sauropod dinosaur. *Acta palaeontologia Polonica* 28: 225–232.

Desmond, A. J. 1975. *The Hot-Blooded Dinosaurs.* London: Blond and Briggs.

Eckert, R.; D. Randall; and G. Augustine. 1988. *Animal Physiology: Mechanisms and Adaptations.* 3rd ed. New York: W. H. Freeman.

Enlow, D. H. 1962. Functions of the Haversian system. *American Journal of Anatomy* 110: 268–306.

Enlow, D. H. 1969. The bone reptiles. In C. Gans and A. d'A. Bellairs (eds.), *Biology of the Reptilia,* vol. 1, pp. 45–80. London and New York: Academic Press.

Farlow, J. O. 1976. A consideration of the trophic dynamics of a Late Cretaceous large-dinosaur community (Oldman Formation). *Ecology* 57: 841–857.

Farlow, J. O.; P. Dodson; and A. Chinsamy. 1995. Dinosaur biology. *Annual Review of Ecology and Systematics* 26: 445–471.

Gross, W. 1934. Die Typen des mikroskopischen Knochenbaues bei fossilen Stegocephalen und Reptilien. *Zeitschrift für Anatomie* 103: 731–764.

Halstead, L. B. 1975. *The Evolution and Ecology of the Dinosaurs.* London: Peter Lowe.

Halstead, L. B. 1979. *The Evolution of the Mammals.* London: Book Club Associates.

Hohnke, L. A. 1973. Haemodynamics in the Sauropoda. *Nature* 244: 309–310.

Hotton, N. III. 1980. An alternative to dinosaur endothermy: The happy wanderers. In R. D. K. Thomas and E. C. Olson (eds.), *A Cold Look at the Warm-Blooded Dinosaurs,* pp. 311–350. American Association for the Advancement of Science Selected Symposium 28. Boulder, Colo.: Westview Press.

Janensch, W. 1947. Pneumatizität bei Wirbeln von Sauropoden und anderen Saurischiern. *Palaeontographica,* Suppl. VII: 1–25.

Joanen, T., and L. McNeese. 1980. The effects of a severe winter freeze on wild alligators in Louisiana. In *Crocodiles: Proceedings of the 9th Working Meeting of the Crocodile Specialist Group,* pp. 21–32. Gland, Switzerland: World Conservation Union.

Kemp, T. S. 1982. *Mammal-Like Reptiles and the Origin of Mammals.* London and New York: Academic Press.

Klevezal, G. A., and S. E. Kleinenberg. 1969. *Age Determination of Mammals from Layered Structures in Teeth and Bone.* Jerusalem: Israel Program for Scientific Translations.

McFarland, W. N.; F. H. Pough; T. J. Cade; and J. B. Heiser (eds.). 1979. *Vertebrate Life.* London and New York: Collier Macmillan International.

McNab, B. K. 1978. The evolution of endothermy in the phylogeny of mammals. *American Naturalist* 112: 1–21.

McNab, B. K., and W. Auffenberg. 1976. The effect of large body size on the temperature regulation of the Komodo dragon, *Varanus komodoensis. Comparative Biochemisty and Physiology* 55A: 345–350.

Ostrom, J. H. 1980. The evidence for endothermy in dinosaurs. In R. D. K. Thomas and E. C. Olson (eds.), *A Cold Look at the Warm-Blooded Dinosaurs,* pp. 15–54. American Association for the Advancement of Science Selected Symposium 28. Boulder, Colo.: Westview Press.

Paul, G. S. 1991. The many myths, some old, some new, of dinosaurology. *Modern Geology* 16: 69–99.

Pough, F. H. 1979. Modern reptiles. In W. N. McFarland, F. H. Pough, T. J. Cade, and J. B. Heiser (eds.), *Vertebrate Life,* pp. 455–513. London and New York: Collier Macmillan International.

Pough, F. H.; J. B. Heiser; and W. N. McFarland. 1989. *Vertebrate Life.* 3rd ed. London and New York: Collier Macmillan and Macmillan Publishing.

Regal, P. J., and C. Gans. 1980. The revolution in thermal physiology. In R. D. K. Thomas and E. C. Olson (eds.), *A Cold Look at the Warm-Blooded Dinosaurs,* pp. 167–188. American Association for the Advancement of Science Selected Symposium 28. Boulder, Colo.: Westview Press.

Reid, R. E. H. 1981. Lamellar-zonal bone with zones and annuli in the pelvis of a sauropod dinosaur. *Nature* 292: 49–51.

Reid, R. E. H. 1984a. The histology of dinosaurian bone, and its possible bearing on dinosaurian physiology. In M. W. J. Ferguson (ed.), *The Structure, Development and Evolution of Reptiles,* pp. 629–633. Orlando, Fla., and London: Academic Press.

Reid, R. E. H. 1984b. Primary bone and dinosaurian physiology. *Geological Magazine* 121: 589–598.

Reid, R. E. H. 1987. Bone and dinosaurian "endothermy." *Modern Geology* 11: 133–154.

Reid, R. E. H. 1990. Zonal "growth rings" in dinosaurs. *Modern Geology* 15: 19–48.

Reid, R. E. H. 1993. Apparent zonation and slowed late growth in a small Cretaceous theropod. *Modern Geology* 18: 391–406.

Reid, R. E. H. 1996. Bone histology of the Cleveland-Lloyd dinosaurs and of dinosaurs in general. Part I: Introduction: Introduction to bone tissues. *Brigham Young University Geology Studies* 41: 25–71.

Romer, A. S. 1945. *Vertebrate Paleontology.* Chicago: University of Chicago Press.

Ruben, R. 1995. The evolution of endothermy in mammals and birds: From physiology to fossils. *Annual Review of Physiology* 57: 69–95.

Russell, L. S. 1965. Body temperature of dinosaurs and its relationship to their extinction. *Journal of Paleontology* 39: 497–501.

Seitz, A. L. L. 1907. Vergleichender Studien über den mikroskopischen Knochenbau fossiler und rezenter Reptilien und dessen Bedeutung für das Wachstum und Umbildung des Knochengewebes in allgemeinen. *Nova Acta, Abhandlungen der kaiserlichen Leopold-Carolingischen deutschen Akademie der Naturforscher* 87: 230–370.

Seymour, R. S. 1976. Dinosaurs, endothermy and blood pressure. *Nature* 262: 207–208.

Spotila, J. R. 1980. Constraints of body size and environment on the temperature regulation of dinosaurs. In R. D. K. Thomas and E. C. Olson (eds.), *A Cold Look at the Warm-Blooded Dinosaurs,* pp. 233–252. American Association for the Advancement of Science Selected Symposium 28. Boulder, Colo.: Westview Press.

Spotila, J. R.; P. W. Lommen; G. S. Bakken; and D. M. Gates. 1973. A mathematical model for body temperature of large reptiles: Implications for dinosaur ecology. *American Naturalist* 107: 391–404.

Spotila, J. R.; M. P. O'Connor; P. Dodson; and F. V. Paladino. 1991. Hot and cold running dinosaurs: Body size, metabolism and migration. *Modern Geology* 16: 203–227.

Swinton, W. E., 1934. *The Dinosaurs.* London: Thomas Murby.

Varricchio, D. J. 1993. Bone microstructure of the Upper Cretaceous theropod dinosaur *Toodon Formosus. Journal of Vertebrate Paleontology* 13: 99–104.

Dinosaurian Physiology   473

Oxygen Isotopes in Dinosaur Bone

Reese E. Barrick,
Michael K. Stoskopf,
and
William J. Showers

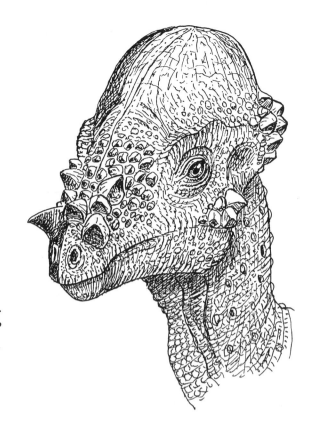

33

Thermoregulation, Metabolic Rates, and Oxygen Isotopes

Thermoregulation in animals is affected by their metabolic rate, and so by understanding the thermoregulation of dinosaurs, we can make more accurate interpretations of their metabolic strategy. A better understanding of dinosaur thermoregulation and metabolism provides important insights into dinosaurs' life history and evolution.

Oxygen isotopic ratios in dinosaur bone provide a direct means of determining the thermoregulatory strategy of the dinosaurs. Oxygen is one of the most abundant elements in the earth's crust, atmosphere, and hydrosphere, as well as in animal blood, tissue, and bone. Atoms of oxygen may contain between sixteen and eighteen neutrons in their nucleus, but the number of protons remains constant. The differences in the mass of these isotopes of oxygen result in their having slightly different physicochemical properties. Differences in chemical properties can lead to isotope effects in chemical reactions.

One of the most interesting of these mass-related isotope effects is the dependence upon temperature of the oxygen isotope exchange between oxygen atoms in fluids and those in minerals. This means that during formation of the mineral component of bone, the ratio of the oxygen–18 to oxygen–16 ($^{18}O/^{16}O$) atoms in bone phosphate [$Ca_{10}(PO_4)_6(OH)_2$] will vary, depending upon (1) the animal's body temperature, and (2) the $^{18}O/$

^{16}O value of the body water with which the oxygen atoms of the bones are exchanging. Therefore, if the isotopic composition of both the bone phosphate and the body water is known, it is possible to calculate the body temperature at which the bone was formed. The equation that is used in this calculation is

$$T°C = 111.4 - 4.3(\partial^{18}O_p - \partial^{18}O_w)$$

where T°C is body temperature in degrees centigrade, $\partial^{18}O_p$ (that is, delta $^{18}O_{phosphate}$) is the isotopic composition ($^{18}O/^{16}O$) of the bone phosphate, and $\partial^{18}O_w$ is the isotopic composition ($^{18}O/^{16}O$) of the body water.

Delta values for oxygen isotope ratios require a bit of explanation if their significance for interpreting the temperatures at which bones grow is to be understood. Because there are very few atoms of ^{18}O compared to the number of atoms of ^{16}O in the natural environment, differences in the $^{18}O/^{16}O$ ratio between any two test samples are extremely small. By convention, geochemists do a numerical manipulation to make differences in this ratio among different samples look bigger.

This manipulation involves comparing the ratio of the heavier to the lighter isotope of the test sample (such as a bone) with the same ratio measured in accordance with some arbitrary standard. In the case of oxygen isotope ratios, the standard used is known as standard mean ocean water, or SMOW. Because the difference in the oxygen isotope ratio between a test sample and SMOW will be extremely small, it is multiplied by 1,000 to make the difference (that is, delta) in isotopic ratios between the test sample and SMOW a bigger number. This procedure is entirely equivalent to the everyday practice of expressing the difference between two numbers as a percentage (otherwise known as parts per hundred), but because in the case of isotope ratio differences we are multiplying by 1,000—not 100—the resulting number is said to be in parts per thousand.

If the $^{18}O/^{16}O$ ratio of the sample is larger than that of SMOW, the $\partial^{18}O_{sample}$ will be a positive number. On the other hand, if the oxygen isotope ratio of the sample is less than that of SMOW, the delta value will be negative.

Inspection of the equation presented above indicates that if the $\partial^{18}O_w$ values are the same for all samples, then more positive bone isotopic values (more ^{18}O) reflect colder temperatures, and negative bone isotopic values (more ^{16}O) reflect warmer temperatures (Longinelli and Nuti 1973; Luz and Kolodny 1989).

Body water composition depends upon the $^{18}O/^{16}O$ value of ingested water as well as on the rate of metabolism. Without knowing the body water $^{18}O/^{16}O$ values for different dinosaur species, it is not possible to calculate individual body temperatures. All that remains of the dinosaurs are the bones. However, body water isotopic composition is in equilibrium throughout every individual (Pflug et al. 1979; Schoeller 1986; Wong et al. 1988). Thus the body water, with which all the bones of any given individual exchange oxygen atoms, contains essentially identical $^{18}O/^{16}O$ values throughout the animal's body. Any variation in the $^{18}O/^{16}O$ values from different bones within any individual should correlate to differences in the temperature at which those bones were formed. This means that the $^{18}O/^{16}O$ values of the bone phosphate of all skeletal elements from the body core of an individual endothermic animal should have little variation, because endotherms are characterized by a narrow range of body core temperatures. Modern reptiles, however, do not maintain constant temperatures (Cloudsley-Thompson 1971; Greenberg 1980). Seasonal tem-

Oxygen Isotopes in Dinosaur Bone   475

perature fluctuations are recorded by the variability of the oxygen isotopic compositions in the bones.

In living animals, the terms *endothermy* and *ectothermy* describe two hypothetical extremes of a continuum of adaptive strategies to optimize species productivity. Pure application of either strategy throughout a species' entire life history is extremely rare. Nevertheless, in species that are primarily ectothermic, metabolic rate rises and falls in a relatively consistent exponential relationship with environmental temperature (Keeton 1967). Most extant ectotherms have adopted various behavioral, biochemical, and physiological strategies to minimize the effects of varying ambient temperature on their metabolic processes. For example, they may limit themselves to relatively narrow environmental conditions through active relocation (Johnson et al. 1976; Pearson and Bradford 1976), sacrifice peak enzymatic efficiencies for broader ranges of moderate enzyme activity (Bouche 1975; Staton et al. 1992), or employ complex hormonally mediated metabolic modifications in response to environmental temperature cues (Paxton and Umminger 1979). Nevertheless, food supply and environmental temperature are usually the primary factors controlling growth efficiency in ectotherms (Lillie and Knowlton 1897; Brett et al. 1969). These in turn can affect both an animal's survival and its reproductive potential.

Metabolic rates of primarily endothermic species are also closely linked to temperature. These species have adopted one or many strategies for generating heat within their tissues to maintain body temperature in a relatively narrow range (homeothermy), generally above environmental temperatures (Bligh and Johnson 1973). This is usually combined with mechanisms to retard heat loss to the environment, which conserves metabolic heat (Whittow 1976). Endotherms must consistently eat large amounts of food to support high rates of metabolism and maintain a constant body temperature, but the efficiency of maintaining metabolic enzymes and other functional molecules in narrow optima allows them to successfully exploit a much greater range of environmental conditions than ectotherms can. Ectothermic metabolism, on the other hand, adjusts to take advantage of a plentiful food supply or to prolong survival in periods of scarce resources. The two energetic strategies dictate important contrasts in life strategies.

It is often noted that the distinction between the metabolic strategies of endothermy and ectothermy are greatest at small body masses, and most obscure at large body masses. However, it is also incorrectly assumed that the masses typical of dinosaurs are in the range where endothermy and ectothermy are obscure. This is perhaps true of large sauropods, but most dinosaurs had adult masses between those of a chicken and an African elephant. Determining the thermal physiology of dinosaurs and other extinct vertebrates is therefore an important aspect of deciphering their life histories.

Intrabone and Interbone Isotopic Variability

There are two ways of assessing oxygen isotopic variation within an animal: intrabone variation (different places within the same bone) and interbone variation (among different bones). Multiple samples taken sequentially across a bone constitute a time series of temperature and/or body water $^{18}O/^{16}O$ variability during the growth of that bone. This pattern of isotopic variability, however, is modified by processes of bone remodeling (the resorption and reprecipitation of bone for maintaining its mechanical and physiological competence). Therefore, the variation across a bone cannot be cyclically correlated to specific daily, seasonal, or annual variations in body temperature or body water $^{18}O/^{16}O$ values as can be done, for example, with bivalve mollusk shells. Remodeling rates in human bone, for instance, vary between 2 and 10 percent per year depending on the bone type and the skeletal element (Marshall et al. 1973; Frost 1980; Francillon-Vieillot et al. 1990). These rates of bone turnover are slower than those causing monthly or seasonal isotopic variability within an individual. As a result, seasonal isotopic variability will appear in the isotopic composition of the bone phosphate, but the pattern of variability will depend upon the pattern of bone remodeling.

Each difference of 1 part per thousand (‰) in the $^{18}O/^{16}O$ ratio of bone phosphate between samples within a skeletal element is the result of a 4.3°C difference in temperature between the times of bone formation. This is calculated using the slope of the phosphate–water temperature equation (Longinelli and Nuti 1973):

$$\Delta T°C = 4.3(\Delta \partial^{18}O_p)$$

where $\Delta T°C$ is the difference in temperature between two samples, and $\Delta \partial^{18}O_p$ is the difference in the $^{18}O/^{16}O$ ratio between two bone samples. Remember that more positive isotopic values (more ^{18}O) represent bone deposited at colder temperatures, and more negative isotopic values (more ^{16}O) represent bone deposited at warmer temperatures. A homeotherm by definition is an animal whose body temperature varies by less than ±2°C (Bligh and Johnson 1973). Consequently a homeotherm should have a temperature-related isotopic variability of less than 1‰ in any bone (corresponding to a 4°C range of temperature variability). A heterotherm (poikilotherm), on the other hand, will often have greater than 1‰ variability in its bones, corresponding to changes in body temperature greater than 4°C. The amount of temperature-related variability in a heterothermic ectotherm will depend upon its size and the magnitude of the seasonal climate changes over which it is actively growing or remodeling bone.

Bone Isotope Ratios in Modern Animals

Several modern animals were analyzed to test the utility of the bone isotope ratio method to detect intra- and interbone temperature variability (Figs. 33.1, 33.2). The first four animals (cow, deer, pheasant, and opossum) all come from Iowa, where, on a thirty-year average, the normal mean daily temperature between January and August differs by more than 30°C (Statistical Abstracts of the United States 1994). The domestic cow has very low intrabone isotopic variability in a rib and dorsal vertebra, and only slightly greater variability in the tibia (all less than 0.5‰). Its interbone variability suggests only a 0.5°C mean temperature difference between the tibia, on one hand, and the rib and vertebra, on the other. The wild deer has greater intra- and interbone isotopic variability, reflecting a degree of regional heterothermy (that is, differences in temperature in different parts of the body). Yet the mean temperature difference between the tibia and the dorsal vertebra is only 0.75‰ or 3.2°C. Temperature variability in the tibia during bone deposition is again only 4°C, though the mean ambient temperature varies more than 20°C during an annual cycle.

Perhaps the method lacks the resolution to detect regional heterothermy. Analyses of the bones of an opossum and a pheasant, however, suggest that

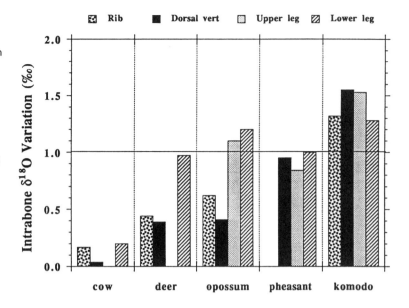

Figure 33.1. Intrabone isotopic variability from bones of modern animals: a cow, deer, opossum, pheasant, and Komodo dragon lizard (*Varanus komodoensis*). Each 1‰ of isotopic variability represents about 4°C of temperature variation. "Dorsal vert" represents thoracic and lumbar vertebrae. The upper leg is represented by the femur and the lower leg by the tibia.

this is not so. Interbone isotopic variability within the opossum is very low, except for the tail. Opossums have long, naked tails, which through heat loss or countercurrent heat exchange (a system in which vessels carrying warm blood from the animal's interior are closely appressed to vessels carrying cool blood from the animal's periphery back toward its interior, with the result that the chilled blood is warmed by its proximity to the warmer flow, helping the animal to retain its body heat) should more closely reflect ambient temperature. This is indeed the case; the tail vertebra is 3.5‰ more positive than the other skeletal elements, which suggests that it was deposited at about 15°C below body core temperature. The opossum also exhibits increased intrabone variability in the legs—more than 1‰, suggesting a 4 to 5°C temperature variability.

Figure 33.2. Maximum inter-bone temperature variability in modern animals. Negative temperatures indicate the number of degrees centigrade colder the most isotopically positive bone from the limbs or tail is relative to the most isotopically negative (i.e., warmest) bone from the body core.

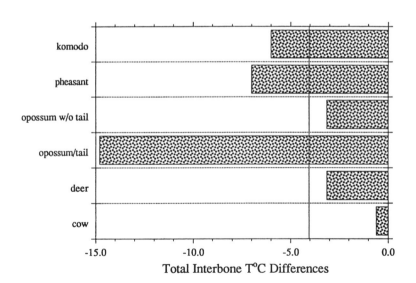

A pheasant from Iowa also exhibits regional heterothermy. The femur and tibiotarsus reflect temperatures about 5°C and 7°C, respectively, colder than the vertebra and furcula. The ulna is about 4.5°C colder than the body core temperature. Thus the technique is capable of picking up signals of regional heterothermy. This regional heterothermy is prominent in bird legs and feet (especially in waterfowl). The mean temperature difference in the ungulates, however, is unexpectedly small, suggesting homeothermy for the legs as well as the body core, albeit with the legs maintained at cooler temperatures than the body core. Even in the pheasant, however, the intrabone variability is much smaller than the annual range of mean ambient monthly temperatures.

What about an ectotherm? Samples from a large lizard, the Komodo dragon (*Varanus komodoensis*), were also analyzed. This 2.3-meter-long individual was raised at the San Diego Zoo, where the maximum mean temperature variability to which it was exposed was approximately 8 to 10°C. Maximum intrabone variability for the ribs and vertebrae is about 1.4‰, and for the legs and tail about 1.5‰. The lizard's body temperature during bone deposition fluctuated by 6 to 8°C. There is little difference in the intrabone isotopic variability between the core body and extremities. This may be due to the relatively controlled environment in which the dragon was kept, or to the fact that there was little countercurrent heat exchange occurring, thus allowing both the body core and extremities to fluctuate outside the limits of homeothermy. Interbone variability in the lizard, however, does suggest that its extremities experienced temperatures some 4 to 7°C cooler than the body core. Because the intrabone variability is similar for bones from the core body and the extremities, it is likely that the interbone variability results simply from greater heat loss in the extremities, rather than from countercurrent heat exchange.

This preliminary analysis of several modern animals suggests that both the intra- and interbone ranges in a terrestrial ectotherm with a low metabolic rate will fluctuate outside the defined limits of homeothermy, and will rather closely track changes in mean ambient temperature, as long as these are not outside the range in which growth occurs. Endotherms exhibiting regional heterothermy may have large interbone variability between the body core and extremities, but low intrabone variability in both the extremities and body core, as in the pheasant. Alternatively, like the deer, they may have increased intrabone temperature variability (≤4°C) while also having interbone temperature differences ≤4°C. Endothermic homeotherms in more equable climates should show low intra- and interbone variability. The next step in our modern animal studies will be to analyze wild elephants from South Africa, alligators from Florida, and farm-raised ostriches from North Carolina.

A Necessary Digression: Diagenesis and Bone Isotope Ratios

Diagenesis is a term that encompasses all of the chemical changes that occur to a fossil bone after it is buried. Diagenetic processes include dissolution, recrystallization, and the replacement of bone by other minerals. An understanding of the diagenetic processes that have altered the bone is important, because meaningful isotopic analyses of fossil bone require that the bone contain the original oxygen atoms in bone phosphate that were there when the dinosaur was alive.

Oxygen Isotopes in Dinosaur Bone   479

Bone mineral in its simplest form is composed of the mineral dahllite, $Ca_{10}(PO_4)_6(OH,F,CO_3)_2$. After burial, bone mineral generally recrystallizes to the more stable form of fluorapatite, in which F substitutes for OH and CO_3 and many trace element cations (e.g., Sr, U, Mg) from the soil substitute for Ca (Hubert et al. 1996). Such substitutions during recrystallization do not affect the PO_4 oxygen atoms. If a significant number of PO_4 ions are lost from the bone (Stuart-Williams et al. 1996) or added to the bone from the groundwater, the oxygen isotopic composition of the bone phosphate can be altered. Oxygen atoms from the original bone phosphate do not exchange with groundwater unless they are digested and used for metabolic processes by bacteria and later re-precipitated. Thus, in some cases (where much bone has been lost from the system or much phosphate secondarily added), dinosaur bone isotopic values may be altered. Indeed, with the large variety of burial and post-burial environments to which dinosaur bones have been subjected, it is likely that some bones have been so altered as to not be of use for paleobiologic (isotopic) studies. Hubert et al. (1996) have suggested that secondary precipitation of phosphate did occur in some Jurassic dinosaurs from Dinosaur National Monument, Utah. They recognized this by utilizing transmission electron microscopy (TEM). If these PO_4 ions are indeed secondarily precipitated from groundwater rather than recrystallized PO_4 ions from bone or bone collagen, then they will affect bone isotopic composition. Whether there is enough secondary phosphate to significantly alter the original bone isotopic composition or mask intra- or interbone isotopic trends will depend upon the volume of the secondary phosphate relative to the original bone phosphate. Fortunately, there are several methods which can be used to determine whether bones have been isotopically altered or dominantly retain their original biologic signal.

Bone mineral occupies only 10 to 30 percent of the total bone volume in cancellous or spongy bone, but occupies 95 percent of the total volume in compact bone (Francillon-Vieillot et al. 1990). This results in much greater

Figure 33.3. Mean isotopic values and standard deviation (a measure of the variability within the data) of the cancellous and dense bone samples from six of the dinosaurs analyzed in this study. Because in nature approximately 998 of 1000 atoms of oxygen are ^{16}O and only 2 are ^{18}O, it is easier to report the isotopic composition of a sample as the difference in its isotopic composition in parts per thousand (‰) relative to a standard, rather than as the sample's absolute isotope abundance. The isotopic composition of the standard is arbitrarily set to 0‰. The oxygen isotopic composition of phosphates is reported relative to the isotopic composition of standard mean ocean water (SMOW). For example, the *Camarasaurus* sample, which has a mean $\partial^{18}O$ value of +13.9‰, has on average 13.9 more ^{18}O atoms per 1000 oxygen atoms than does SMOW.

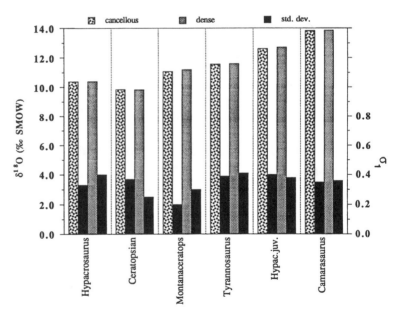

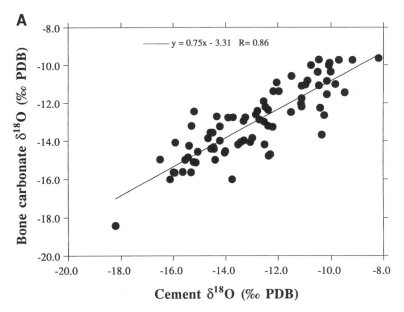

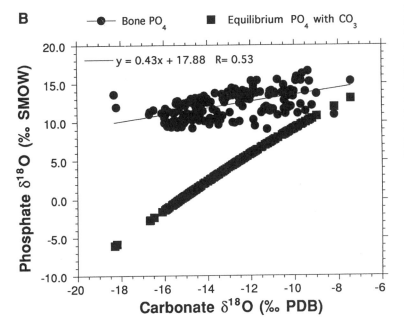

Figure 33.4. (A) Covariation of the bone carbonate isotopic values ($\partial^{18}O_c$) with those of the corresponding carbonate cements ($\partial^{18}O_{cc}$). Carbonate isotopic compositions are reported relative to the PDB standard rather than to SMOW. The PDB standard was derived from the shell of a belemnite (an extinct mollusk) from the Peedee Formation in South Carolina. (B) Filled circles represent covariation of bone phosphate and carbonate isotopic values for the Cretaceous dinosaurs in our sample. Filled squares represent equilibrium values of phosphate with diagenetic carbonate. This is the line expected if the bone phosphate had been isotopically altered along with the carbonate.

bone crystal surface area being exposed to groundwater (and thus possible diagenetic alteration of the isotopic composition) in spongy bone than in compact bone. One would therefore expect that diagenetic alteration of $\partial^{18}O$ values by isotopic exchange with groundwater, precipitation of secondary phosphate, or dissolution and loss of phosphate would be greatest in spongy bone, and less in compact bone.

One way of testing this is to compare mean $\partial^{18}O$ values of cancellous and compact bone samples for fossilized bones of the same animal (Fig.

Oxygen Isotopes in Dinosaur Bone   481

33.3). Differences in mean $\partial^{18}O$ values between cancellous and compact bone are less than 0.3‰ in all cases. The standard deviation (a routine statistical measurement of variability in a data set) is also very similar for samples of cancellous and compact bone for each animal. With the vast difference in the volume occupied by bone mineral in spongy and compact bone, this could occur only if the $\partial^{18}O$ values are unaltered, or if there was complete re-equilibration of all phosphate oxygen with groundwater.

One way of determining which of these alternatives is correct is to analyze the bone carbonate and co-occurring calcite cement (calcite which has precipitated in the open spaces in the bone after burial) $\partial^{18}O$ values. Although mediated by different enzymes as an animal deposits bone, the carbonate (CO_3) and phosphate (PO_4) systems should behave analogously with respect to their isotopic compositions. If diagenesis (after death and burial) has isotopically altered the bone carbonate but not the phosphate, there will be a covariation between the $\partial^{18}O$ of the calcite cement (CO_3) and the bone carbonate (CO_3), but no covariation between these and the bone phosphate $\partial^{18}O$. Alternatively, if the phosphate has been completely re-equilibrated with groundwater, there will be a covariation among the bone phosphate, bone carbonate, and calcite cement. Finally, no diagenetic alteration of the bone isotopes will be indicated if the bone phosphate and bone carbonate are in equilibrium, and there is no covariation with the cement isotopic values.

Figure 33.4 shows the covariation of the $\partial^{18}O$ values of the bone carbonate, bone phosphate, and calcite cements. The structural bone carbonate appears to have been partially altered when correlated with the calcite cements, but the bone phosphate appears unaltered. This indicates that the bone phosphate preserved its original biologic signal.

Kolodny et al. (1996) suggest that recrystallization of dahllite to fluorapatite causes isotopic alteration due to the addition of organic phosphate to the bone crystal structure. They recognize that dissolution-precipitation will not affect the oxygen isotopic record of phosphates in an inorganic environment. However, Stuart-Williams et al. (1996), in their study of human bones in archaeological sites, and Person et al. (1996), in a consideration of the effects of heating on modern bone, concluded that organic materials are removed from bone prior to bone recrystallization. This conclusion is supported by our unpublished data for *Tyrannosaurus rex* that show no correlation between bone crystallinity (recrystallization) and the isotopic composition of fossil bones. The only way to determine if there has been isotopic alteration of bone phosphate is to correlate apatite isotopic values with known diagenetic values and bone density, as described above.

One final line of evidence arguing against isotopic alteration is that of five dinosaurs and one lizard from the Late Cretaceous Two Medicine Formation, there is evidence of heterothermy in the lizard and homeothermy in the dinosaurs (see below). If diagenetic alteration had occurred, then the bones from all of the animals would have homogeneous isotopic ratios, and even the lizard would appear to have been a homeotherm.

Bone Isotope Ratios in Fossil Species

The bones of several Late Cretaceous and Late Jurassic dinosaurs and a Cretaceous varanid lizard (a relative of the Komodo dragon) have been analyzed for intrabone isotopic variability (Fig. 33.5). The intrabone variability displayed by the varanid lizard for those bones with three or four samples ranged from 1.7‰ for a proximal caudal vertebra to 3.4‰

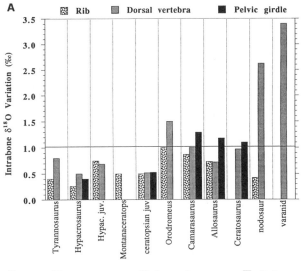

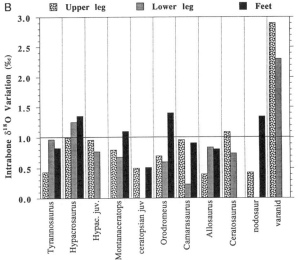

Figure 33.5. Intrabone isotopic variability from bones of (A) the body core, (B) the limbs, and (C) the tails for ten dinosaurs and a varanid lizard. Variability less than 1‰ suggests less than 4°C total temperature variation during bone growth.

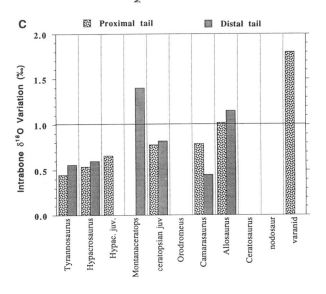

Oxygen Isotopes in Dinosaur Bone   483

for a dorsal vertebra. The limb intrabone variabilities were between 2.0 and 2.5‰ for the humerus, radius, and tibia, and 3.2‰ for the femur. A lizard of this size had little capability for mass homeothermy, and so all of its bones were likely deposited within the same range of temperatures. The isotopic data from this varanid suggest that body temperatures fluctuated seasonally by at least 10 to 15°C (Barrick et al. 1996). It must be pointed out that this relates to body temperatures during which bone deposition occurred, and body temperatures could have ranged more widely (i.e., become significantly cooler) during times when bone deposition was not occurring, during seasons when the animal was neither active nor growing.

The total range of intrabone isotopic variability of bones from the body core (ribs and dorsal vertebrae) of some Cretaceous dinosaurs (*Tyrannosaurus*, *Hypacrosaurus*, *Montanoceratops*, juvenile *Achelousaurus*) are near or below 1‰, indicating that they were formed under homeothermic conditions (±2°C). This is especially notable for the small (roughly 1.5 meters long; Varricchio and Horner 1993) *Hypacrosaurus* juveniles (Barrick and Showers 1995).

The rib of the small hypsilophodont *Orodromeus*, however, shows greater than 1‰ (4°C) variability, and is near the limit of homeothermy. This dinosaur seems to have had slightly weaker temperature regulation capabilities than the other homeothermic Cretaceous dinosaurs. Similarly, the dorsal vertebra of a nodosaurid ankylosaur has an isotopic range of 2.5‰, suggesting a within-bone temperature variability of nearly 11°C. The pelvic girdles of the Jurassic dinosaurs show intrabone variations suggesting a heterothermic temperature range (±2.2 to 2.8°C).

Intrabone variability in the extremities of nearly all of the Cretaceous dinosaur specimens falls near the range of homeothermy. The feet of four of the ten dinosaurs exhibit increased intrabone variability, suggestive of regional heterothermy, as do the distal portions of the tails of *Montanoceratops* and *Allosaurus*. The femur of the Jurassic *Ceratosaurus* slightly exceeds 1‰.

It is easiest to see the effect of temperature on interbone variability by assuming that it was the only cause of the isotopic differences (Fig. 33.6).

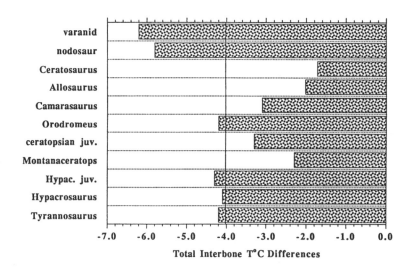

Figure 33.6. Maximum interbone temperature variability in fossil bone samples. Negative temperatures indicate the number of degrees centigrade colder the most isotopically positive bone from the limbs or tail is relative to the most isotopically negative (i.e., warmest) bone in the body core. All dinosaurs except the nodosaur fall near or within the definition of homeothermy.

Mean values for each skeletal element were used in determining the inter-bone temperature differences. $^{18}O/^{16}O$ values for each skeletal element are then compared to the bone from the body core with the most negative isotopic value (i.e., the warmest temperature). The mean interbone temperature differences suggest that the extremities remained within about 4°C of the core body temperature.

On the basis of intrabone and interbone isotopic variability, nearly all of the dinosaurs studied thus far appear to have been homeotherms that experienced some regional heterothermy. The exception is the nodosaur, in which the dorsal plate and metapodial are 5 to 6°C colder than the dorsal vertebra, and the body core underwent about a 10°C range of variability.

The interbone temperature differences in the varanid lizard must be viewed with caution, because the number of samples per bone from which the means were calculated varied from one to four, and thus they are not directly comparable. The proximal caudal vertebra indicates temperatures 2 to 6°C colder than the dorsal vertebrae, and the radius and tibia temperatures about 2°C cooler than the humerus and femur, respectively. The dermal scutes of the varanid suggest temperatures as much as 8°C colder than the mean body core temperature. However, each sample was a composite of two to three scutes.

Putting It All Together: What Was Dinosaur Physiology Like?

The heterothermy seen in the Cretaceous varanid is similar to that observed in its modern relative, the Komodo dragon; both lizards show large intrabone isotopic variability in both the body core and the extremities. In both, the extremities are cooler than the core body by a mean temperature difference of 5 to 6°C.

Compared to the lizards, most of the dinosaurs appear to have been much more sophisticated thermoregulators. Is the homeothermy seen in these dinosaurs the result of large mass (Spotila et al. 1991), or does it indicate true endothermy?

The Cretaceous dinosaurs range in size from a body mass of about 60 kilograms to the roughly 6000-kilogram *Tyrannosaurus* (similar in size to a bull African elephant; Horner and Lessem 1993). All but the nodosaur (which is from Texas) are from Montana, which during the Late Cretaceous was situated at 53° north latitude. The climate in Montana during the Late Cretaceous was similar to that of present-day Louisiana or North Carolina, although seasonal temperature changes may not have been quite as large as in those modern climatic counterparts. Sloan and Barron (1990), however, suggest that seasonal climate change was likely to have been significant.

A biophysical model of bradymetabolic (having a low metabolic rate) hadrosaurs ranging in mass from 2 to as much as 5000 kilograms (i.e., from nestlings to large adults) in such a simulated Cretaceous climate suggests that these dinosaurs would have undergone annual variations in body core temperatures of between 20 and 30°C (Dunham et al. 1989). Given that there are thirteen weeks each of spring, summer, fall, and winter, there would have been ample time for core body temperatures of bradymetabolic animals to adjust to seasonal averages, and so such fluctuations in body temperature are not an unreasonable expectation.

Temperature fluctuations of this magnitude recorded in bone phosphate would require intrabone isotopic variabilities of about 4‰. With the exception of the nodosaur, this is not seen in any of the Cretaceous dinosaurs.

One would expect the smaller dinosaurs, *Montanoceratops* and the juvenile hadrosaurs, particularly to show greater isotopic variability if they had had the low metabolic rates typical of an ectotherm—such as the varanid lizard from the same geologic formation. Interestingly, it is the nodosaur from a much lower latitude that indicates the temperature variability suggested by the biophysical models to represent ectothermy. While the limbs of this animal were kept cooler than the body core, it was the body core that underwent the greatest temperature fluctuations. This indicates that seasonally the nodosaur could not maintain its preferred high body temperature, unlike the other dinosaurs studied. This suggests that nodosaurs had much lower metabolic rates than the other Cretaceous dinosaurs. One other possible explanation is that nodosaurs seasonally hibernated, during which time their metabolic rates dropped significantly, allowing their body temperatures to drop (G. Paul, personal communication, 1996). Nodosaurs do appear to be poorer thermoregulators than other Cretaceous dinosaurs.

It is apparent that the homeothermy seen in most Cretaceous dinosaurs could have been supported only by relatively high metabolic rates (Barrick and Showers 1994, 1995; Barrick et al. 1996). Metabolic rates for these dinosaurs may or may not have been as high as those of modern endotherms, but they were elevated enough that body temperatures were maintained largely by a controlled metabolic rate, in contrast to modern ectotherms.

The question must then be asked, How low could metabolic rates have been in dinosaurs and still have maintained homeothermy? Obviously the answer will differ for dinosaurs of different sizes living in various climates. Interestingly, the pattern of isotopic variation in the dinosaurs is not exactly the same as that observed in any of the modern endotherms. The dinosaurs exhibit both increasing intrabone variability (more than 4°C) in the distal extremities, and interbone temperature differences near or slightly greater than 4°C. The deer and nocturnal opossum have larger intrabone ranges in their extremities, but also have interbone temperature differences (excluding the opossum tail) well below 4°C. The pheasant has pronounced interbone temperature gradients toward the limbs, but the intrabone variability remains below 1‰ in all bones.

The Cretaceous dinosaurs lived in a climate much less extreme than that presently found in Iowa (north-central U.S.), yet the isotopic variability of the dinosaurs is similar to that of the modern endotherms in our sample, or slightly greater. This suggests that the dinosaurs had elevated metabolic rates, but that those metabolic rates remained lower than those found in modern mammals. Thus these reptiles may have had a unique metabolic strategy that might be termed "dinosaurian" endothermy.

The Jurassic dinosaurs analyzed thus far also appear to have been homeotherms. Few bones show intrabone isotopic variability beyond expectations for homeotherms, and interbone variability indicates homeothermy between body regions. These dinosaurs could also have been endotherms. However, the sauropod *Camarasaurus*, because of its large size (roughly 15 tons; G. Paul, personal communication, 1993), might instead have been a mass homeotherm. Only after a smaller, juvenile specimen is analyzed will it be possible to determine if the sauropods grew as endotherms or ectotherms. Similarly, more Jurassic theropods need to be analyzed before more conclusive interpretations can be made regarding the effect of climate variability on bone isotopic values in these dinosaurs.

Skeptics of the idea of dinosaur endothermy (e.g., Spotila et al. 1991) have suggested that 15-ton sauropods could have maintained homeothermy (even in their necks and tails) with very low metabolic rates. However, these authors failed to consider the matter of how such dinosaurs could have maintained homeothermy (or even *if* they maintained homeothermy) before reaching adult size. Other critics (Ruben et al., chap. 35 of this volume) have suggested that 5-to-6-ton endothermic dinosaurs should have had hetero-thermic limbs and tails, and so any isotopic studies suggesting little variabil-ity are faulty.

These authors presume that any mention of endothermy in dinosaurs *assumes* mammalian-style metabolic rates in dinosaurs, or that if one dinosaur is presumed to have been an endotherm, then all of them were. However, as we have already argued, dinosaurs probably did not have mammalian-style endothermy. They did not inherit elevated metabolic rates from small mammals, or from the cynodont ancestors of mammals. Dino-saurs that exhibited endothermy inherited this physiological feature from their dinosaurian or ornithosuchian ancestors, or else evolved it indepen-dently themselves. Endothermy is simply defined as the maintenance of homeothermy by means of an elevated metabolic rate (Bligh and Johnson 1973)—and does not presume an elevated metabolic rate at the level seen in mammals! Endothermy *does* mean that the metabolic rates must be elevated above those found in ectotherms.

Perhaps as adults the large sauropods maintained homeothermy with low mass-specific metabolic rates—what has been termed gigantothermy (Paladino et al. 1990; Spotila et al. 1991). However, a large body mass and a compact, tank-like body did not keep a nodosaur from showing heterothermy in core body as well as limb temperatures over the annual cycle. Nodosaurs have been hypothesized to have had lower metabolic rates than other dinosaurs based on their simple dentition and lesser running ability (Coombs and Maryańska 1990). Suggestions that adult bradymetabolic, ectothermic dinosaurs would undergo body temperature variations of as much as 20°C (Dunham et al. 1989) are often met with ridicule, and with the observation that a heating and cooling curve would take weeks. Yet seasonal changes in temperature occur on the order of several months, which would be ample time for mean body temperature shifts in truly bradymetabolic dinosaurs. Again, our data indicate that the nodosaur's core temperature apparently shifted by roughly 11°C, and this dinosaur's extremities were significantly cooler than its body core.

Perhaps the term *gigantothermy* should be reserved for the truly gigan-tic, adult sauropods. If so, the interesting question then becomes how juvenile sauropods thermoregulated while growing to their adult sizes: Did they use dinosaurian endothermy or ectothermy? This more intriguing question remains to be answered, and has implications for the possible role of endothermy in the evolution of large size.

Ruben et al. (chap. 35 of this volume) claim to have found a "Rosetta stone" for endothermy in nasal turbinates, which in mammals were inherited from some "mammal-like reptiles." Thus they have found direct evidence for mammalian endothermy (the highly elevated rates found in small mammals). However, a few mammals, notably elephants, and also some birds lack or have very poorly developed respiratory turbinals (Bang 1971; Sikes 1971), suggesting that other methods for water conservation in the face of the rapid ventilation rates required by endothermy are possible. Highly derived nasal turbinates may or may not have been necessary to maintain metabolic rates elevated enough to support dinosaurian endo-thermy.

Some dinosaurs did maintain homeothermy at sizes that necessitate metabolic rates greater than those of modern reptiles. Thus it must be possible to elevate metabolic rates, with or without respiratory turbinals. Ruben and his associates do not account for the likelihood of intermediate metabolic strategies. However, such studies are important, as are studies of bone histology (see Reid, chap. 32 of this volume), and do play a role in understanding the physiology of dinosaurs and other extinct vertebrates.

Dinosaurs apparently displayed a range of thermoregulatory and metabolic strategies. Some evolved dinosaurian-style endothermy in maintaining homeothermy, while others remained heterotherms, an indication of lower metabolic rates. This keeps the future study of dinosaur thermoregulation and metabolism wide open. Continued study of dinosaur bone isotopic composition will be one of the important keys to unlocking the pattern and history of dinosaur physiology.

References

Bang, B. G. 1971. Functional anatomy of the olfactory system in 23 orders of birds. *Acta Anatomica* 79: 1–71..

Barrick, R. E., and W. J. Showers. 1994. Thermophysiology of *Tyrannosaurus rex:* Evidence from oxygen isotopes. *Science* 265: 222–224.

Barrick, R. E., and W. J. Showers. 1995. Oxygen isotope variability in juvenile dinosaurs (*Hypacrosaurus*): Evidence for thermoregulation. *Paleobiology:* 21 450–459.

Barrick, R. E.; A. G. Fischer; and W. J. Showers. 1996. Comparison of thermoregulation of four ornithischian dinosaurs and a varanid lizard from the Cretaceous Two Medicine Formation: Evidence from oxygen isotopes. *Palaios* 11: 295–305.

Bligh, J., and K. G. Johnson. 1973. Glossary of terms for thermal physiology. *Journal of Applied Physiology* 35: 941–961.

Bouche, G. 1975. Researches on the nucleic acids and protein synthesis during prolonged starvation and refeeding in carp. Thesis Docteur D'Etat, mention sciences, Université Paul Sabatier de Toulouse, France.

Brett, J. R.; J. E. Shelbourn; and C. T. Shoop. 1969. Growth rate and body composition of fingerling sockeye salmon in relation to temperature and ration size. *Journal of the Fisheries Research Board of Canada* 26: 2363–2394.

Cloudsley-Thompson, J. L. 1971. *The Temperature and Water Relations of Reptiles.* Watford, England: Merrow Technical Library.

Coombs, W. P. Jr., and T. Maryańska. 1990. Ankylosauria. In D. B. Weishampel, P. Dodson, and H. Osmólska (eds.), *The Dinosauria,* pp. 456–483. Berkeley: University of California Press.

Dunham, A. E.; K. L. Overall; W. P. Porter; and C. A. Forster. 1989. Implications of ecological energetics and biophysical and developmental constraints for life-history variation in dinosaurs. In J. O. Farlow (ed.), *Paleobiology of the Dinosaurs,* pp. 1–19. Special Paper 238. Boulder, Colo.: Geological Society of America.

Farlow, J. O. 1990. Dinosaur energetics and thermal biology. In D. B. Weishampel, P. Dodson, and H. Osmólska (eds.), *The Dinosauria,* pp. 43–55. Berkeley: University of California Press.

Frost, H. M. 1980. Skeletal physiology and bone remodeling. In M. R. Urist (ed.), *Fundamentals and Clinical Bone Physiology,* pp. 208–241. Philadelphia: J. B. Lippincott.

Francillon-Vieillot, H.; V. de Buffrénil; J. Castanet; J. Géraudie; F. J. Meunier; J. Y. Sire; L. Zylberberg; and A. de Ricqlès. 1990. Microstructure and mineralization of vertebrate skeletal tissues. In J. G. Carter (ed.), *Skeletal Biomineralization: Patterns, Processes and Evolutionary Trends,* vol. 1, pp. 471–529. New York: Van Nostrand Reinhold.

Greenberg, N. 1980. Physiological and behavioral thermoregulation in living reptiles. In R. D. K. Thomas and E. C. Olson (eds.), *A Cold Look at the Warm-Blooded Dinosaurs,* pp. 141–166. American Association for the Advancement of Science Selected Symposium 28. Boulder, Colo.: Westview Press.

Horner, J. R., and D. Lessem. 1993. *The Complete T. rex.* New York: Simon and Schuster.

Hubert, J. F.; P. T. Panish; D. J. Chure; and K. S. Prostak. 1996. Chemistry, microstructure, petrology, and diagenetic model of Jurassic dinosaur bones, Dinosaur National Monument, Utah. *Journal of Sedimentary Research* 66: 531–547.

Johnson, C. R.; W. G. Voigt; and E. N. Smith. 1976. Thermoregulation in crocodilians. Part III: Thermal preferenda, voluntary maxima and heating and cooling rates in the American alligator, *Alligator mississippiensis. Zoological Journal of the Linnean Society* 62: 179–188.

Keeton, W. T. 1967. Body temperature and metabolic rate. In W. T. Keeton (ed.), *Biological Science,* pp. 145–149. New York: W. W. Norton.

Kolodny, Y.; B. Luz; M. Sander; and W. Clemens. 1996. Dinosaur bones: Fossils or pseudomorphs? The pitfalls of physiology reconstruction from apatitic fossils. *Palaeogeography, Palaeoclimatology, Palaeoecology* 126: 161–171.

Lillie, F. R., and F. P. Knowlton. 1897. On the effect of temperature on the development of animals. *Zoological Bulletin* 1: 179–193.

Longinelli, A., and S. Nuti. 1973. Revised phosphate-water isotopic temperature scale. *Earth and Planetary Science Letters* 19: 373–376.

Luz, B., and Y. Kolodny. 1989. Oxygen isotope variation in bone phosphate. *Applied Geochemistry* 4: 317–324.

Marshall, J. H.; J. Liniecki; E. L. Lloyd; G. Marotti; C. W. Mays; J. Rundo; H. A. Sissons; and W. S. Snyder. 1973. Alkaline earth metabolism in adult man. *Health Physics* 24: 125–221.

Paladino, F. V.; M. P. O'Connor; and J. R. Spotila. 1990. Metabolism of leatherback turtles, gigantothermy and thermoregulation of dinosaurs. *Nature* 344: 858–860.

Paxton, R., and B. L. Umminger. 1979. Role of hexokinase in the maintained hyperglycemia of cold-acclimated goldfish. *American Zoologist* 19 (3): 974.

Pearson, O. P., and D. F. Bradford. 1976. Thermoregulation of lizards and toads at high altitudes in Peru. *Copeia* 1976: 155–170.

Person, A.; H. Bocherens; A. Mariotti; and M. Renard. 1996. Diagenetic evolution and experimental heating of bone phosphate. *Palaeogeography, Palaeoclimatology, Palaeoecology* 126: 135–149.

Pflug, K. P.; K. D. Schuster; J. P. Pichotka; and H. Forstel. 1979. Fractionation effects of oxygen isotopes in mammals. In E. R. Klein and P. D. Klein (eds.), *Stable Isotopes: Proceedings of the Third International Conference,* pp. 553–561. New York: Academic Press.

Schoeller, D.; C. Leitch; and C. Brown. 1986. Doubly labeled water method: In vivo oxygen and hydrogen isotope fractionation. *American Journal of Physiology* 251: 1137–1143.

Sikes, S. K. 1971. *The Natural History of the African Elephant.* New York: Elsevier.

Sloan, L., and E. Barron. 1990. "Equable" climates during Earth history? *Geology* 18: 489–493.

Spotila, J. R.; M. P. O'Connor; P. Dodson; and F. V. Paladino. 1991. Hot and cold running dinosaurs: Body size, metabolism and migration. *Modern Geology* 16: 203–227.

Staton, M. A.; H. M. Edwards; I. L. Brisbin; T. Joanen; and L. McNease. 1992. The influence of environmental temperature and dietary factors on utilization of dietary energy and protein in purified diets by alligators, *Alligator mississippiensis* (Daudin). *Aquaculture* 107: 369–381.

Stuart-Williams, H. LeQ.; H. Schwarcz; C. White; and M. Spence. 1996. The isotopic composition and diagenesis of human bone from Teotihuacan and Oaxaca, Mexico. *Palaeogeography, Palaeoclimatology, Palaeoecology* 126: 1–14.

Varricchio, D. J., and J. R. Horner. 1993. Hadrosaurid and lambeosaurid bone beds from the Upper Cretaceous Two Medicine Formation of Montana: Tapho-

nomic and biologic implications. *Canadian Journal of Earth Sciences* 30: 997–
1006.

Whittow, G. C. 1976. Regulation of body temperature. In P. D. Sturkie (ed.), *Avian
Physiology*, pp. 154–184. New York: Springer Verlag.

Wong, W. W.; W. J. Cochran; W. J. Klish; E. O. Smith; L. S. Lee; and P. D. Klein.
1988. In vivo isotope-fractionation factors and the measurement of deuterium
and oxygen–18 dilution spaces from plasma, urine, saliva, respiratory water
vapor, and carbon dioxide. *American Journal of Clinical Nutrition* 47: 1–6.

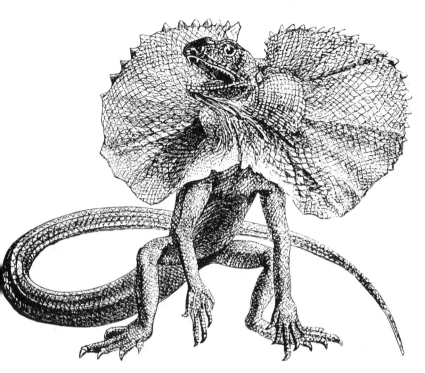

A Blueprint for Giants: Modeling the Physiology of Large Dinosaurs

Frank V. Paladino,
James R. Spotila,
and
Peter Dodson

What are the selection pressures that affect the physiology of animals? Why is it that many different kinds of animals can evolve and live in similar habitats, yet their physiology operates under very different design constraints? In the past thirty years there has been much debate about the basic design—the physiological blueprint—of dinosaurs. Was their physiology more like that of reptiles, or were dinosaurs physiologically similar to mammals and birds? This fundamental question of which blueprint best describes the physiology of dinosaurs cannot be answered definitively without a live animal to test. An extinct animal's physiology is difficult to ascertain from fossils; how an animal actually functions cannot fully be characterized on the basis of bones or other fossils.

However, we can investigate some aspects of this question when we look at the physiology of extant vertebrates. Dinosaur paleontologists have recognized this, but too often studies of the physiology of small lizards, mammals, and birds have been uncritically used to make interpretations of how medium-sized to gigantic ectothermic or endothermic dinosaurs would have functioned. Consider, for example, the unsubstantiated statements made by Bakker (1986) about running speeds of more than 40 miles per hour attained by large theropods such as *Tyrannosaurus*. Not only does a biomechanical analysis of the dinosaur's skeleton not support this claim (Alexander 1985; Farlow et al. 1995b), but the physiological demands on both the respiratory and cardiovascular systems for such strenuous activity would have been enormous and prohibitive for animals the size of *Tyrannosaurus*. In addition, data on the physiology of large living reptiles, such

34

as crocodilians, marine turtles, giant tortoises, and elephants, are probably more useful for setting constraints on the physiological designs possible for extinct reptilian giants than are measurements made on small lizards (Paladino et al. 1990), many of which may have become specialized for lifestyles as small ectotherms (Gans and Pough 1982).

Paleontologists also may have been overly influenced by systematics when thinking about paleophysiology. The consensus among systematists that taxonomic groupings of organisms should be based upon recency of common ancestry, and not on overall morphological similarity, has revolutionized vertebrate classification (Benton 1988). However, such cladistic thinking has perhaps had an unfortunate side effect on attempts to reconstruct dinosaur physiology. Recognition of natural, monophyletic groups depends upon the identification of derived, specialized features, and totally disregards primitive characters (even if two taxa have more primitive than derived features in common). Paleontologists may therefore be predisposed to ascribe derived features of physiology (seen in extant members of lineages) to the earliest members of those lineages (clades), and at the same time predisposed not to consider the possibility that the earliest members of clades were physiologically more like members of different, but related, clades than like their own descendants.

Consider an extreme case. In terms of the recency of their common ancestors, a lungfish is more closely related to a cow than to a trout. If one were therefore to use the cow as a model for interpreting the physiology and ecology of a Devonian lungfish, the likelihood of success would be rather low. The trout and the lungfish have more primitive characters in common than the number of derived characters shared by the cow and the lungfish—which is, after all, why the traditional Linnean classification regards both the trout and the lungfish as fishes.

Most (but not all) paleontologists now think that birds are derived from small carnivorous dinosaurs. In the cladistic sense, this means that extinct dinosaurs are more closely related to birds than to crocodilians and other non-dinosaurian archosaurs, because dinosaurs and birds had a common ancestor that lived more recently than the common ancestor of all archosaurs.

Many paleontolologists may therefore have a (perhaps unconscious) tendency to read bird-like features of physiology into extinct dinosaurs. However, we think that in many features, typical dinosaurs were more like crocodilians (and even large lizards and turtles) than their avian descendants. All known birds have feathers; as yet there is no evidence that any dinosaurs did, but there is evidence that many did not. Modern living birds do not have teeth (some extinct forms did), and modern living birds do not have a long bony tail (again some Mesozoic birds did)—two other features in which dinosaurs more closely resemble crocodilians than modern birds. Few modern birds approach the body sizes attained by typical dinosaurs, unlike many turtles, crocodilians, and even some extinct marine lizards (mosasaurs). Finally, most birds are highly modified for flight, perhaps the most distinctive way in which birds differ from typical dinosaurs—and in which dinosaurs, to the contrary, were more like living reptiles.

We therefore think that it may be more reasonable to interpret dinosaur physiology on the basis of what is seen in large living reptiles than to focus on the physiological traits of living birds. Crocodilians and turtles originated about the same time as the dinosaurs, and have changed very little since the Mesozoic Era. As in the trout-lungfish-cow comparison, comparisons across different (in this case reptilian) clades may be more informative than comparisons of early and later members of the same clade for understanding the physiology of the non-avian members of the dinosaurian clade.

This chapter considers how a large mammal or large bird, on the one hand, or a large reptile, on the other, would function physiologically at the gigantic sizes of some dinosaurs, using physiological data from living species. On the basis of these comparisons, we will describe the physiological features most likely to have characterized large dinosaurs.

Issues in the Interpretation of Dinosaur Physiology

Reconstructing dinosaur paleophysiology requires consideration of several fundamental questions: What is the original amniote physiological blueprint upon which the subsequent physiological evolution of turtles, crocodiles, dinosaurs, birds, and mammals was based? What insights do the physiological designs (respiratory and cardiovascular features, and size-related metabolic rates) of present-day reptiles, mammals, and birds provide about the physiology of giant dinosaurs? How does size affect the physiology of an animal, and what are the selection pressures for or against gigantism? and Which physiological design would be most likely to produce a gigantic animal living in a tropical to subtropical environment?

Tetrapod Physiological Blueprints

Studies of the physiology of living reptiles indicate some common features that presumably characterized the ancestral reptiles, and some of which probably typified the earliest amniotes as well:

1. Reptiles typically do not have the fast oxidative glycolytic (FOG) muscle fiber types typical of endotherms that rely upon high activity ATP (adenosine triphosphate, the "fuel" used by organisms during cellular metabolism) and high sustained oxidative metabolic rates. Reptiles consequently tend to have lower citrase synthetase activity (the enzyme/chemical necessary to make this high fuel use possible) in the muscle fibers that they do have.

2. Reptiles can have complex multicameral (many-chambered) lungs, some with primitive air sacs (Perry 1989). Reptiles were among the first vertebrates to develop the negative-pressure thoracic pump. In this kind of respiratory system, movement of air into the thoracic (chest) cavity results from expansion in volume of the thoracic cavity and/or internal movements of the viscera, which creates a negative internal chest air pressure compared with the air pressure outside the animal's body. Expansion and contraction of the chest cavity permits rapid movement of air into and out of the respiratory system without the use of a diaphragm, the sheet of muscle that in mammals separates the chest/thorax cavity from the abdomen and is used like a bellows to expand and contract the volume of the chest cavity.

3. Reptiles can have four-chambered hearts and complex circulation and pressure changes, but most modern forms have three-chambered hearts. Since all living archosaurs have four-chambered hearts, we will assume that dinosaurs did also.

4. Reptiles are mostly uricotelic, which means that nitrogen waste products formed as a result of protein breakdown occur in the form of concentrated uric acid. This adaptation is probably a consequence of the fact that the majority of the lineage reproduce by laying cleidoic eggs,

which are enclosed in a shell and membranes (Romanoff 1967; Carey 1996).

5. Most reptiles have seasonal growth, which is reflected in the typical pattern of growth rings formed in their bone. In addition, living reptiles show indeterminate growth; they do not stop growing throughout their lifetime.

The sophistication of the reptilian blueprint is often not appreciated by paleontologists, but it greatly impresses physiologists. For example, Grigg (1989) concluded that crocodilians have the most anatomically complex heart and outflow chambers of any vertebrate, combining the best features of reptilian, mammalian, and avian designs.

In contrast to the reptilian blueprint, the physiology of modern mammals and birds has the following features:

1. Mammals and birds have a multitude of complex muscle fiber types, and employ fast oxidative glycolytic and aerobic pathways (using oxygen) that have high ATPase activity and citrate synthetase activity. These produce the excess internal heat involved in endothermic thermoregulation. Interestingly, similar biochemical cellular traits are found in the muscles of endothermic fishes such as tuna.

2. Mammals and birds have complex multicameral lungs (with cul-de-sac alveoli in mammals), and birds have parabronchial lungs (parallel tubes less than 1 mm in diameter which pass through the gas-exchange tissues and are punctured by thousands of tiny air capillaries that branch off into the extensive capillary networks of the rigid avian lung) and air sacs (Scheid 1979). These permit the highly efficient removal of oxygen from the environment that is needed to support the high oxygen demands of endothermy and high aerobic scope for activity. Both groups breathe by means of negative-pressure-pump pulmonary designs; a diaphragm is used for this purpose in mammals, but not in birds.

3. All birds and mammals have complex four-chambered hearts.

4. Placental mammals are ureotelic (urea is the nitrogen waste product from protein metabolism); birds are uricotelic.

5. Birds and mammals have determinate growth (they stop growing in their adult stage). In addition, growth in these animals is less influenced by seasonality than is the growth of reptiles, and so their bones seldom show pronounced seasonal growth rings.

6. Birds and mammals have a specialized outer plumage or pelage, which serves an important thermoregulatory function. The feathers of birds obviously function as well during flight.

So which of these very different physiological blueprints, the reptilian or the avian/mammalian, would have best served gigantic dinosaurs? Before this question is addressed, it will be helpful to consider the physiology of some living giants: elephants.

Elephant Physiology

We studied the respiratory and metabolic physiology of the largest living land mammals, the Indian (*Elephas maximus*) and African (*Loxodonta africana*) elephants, to better understand the physiological mechanisms and adaptation of a large terrestrial, endothermic mammal (Table 34.1). The first comprehensive work on elephant physiology was published by Benedict (1936). Our findings more than fifty years later (Paladino et al. 1981) confirmed much of his work with respect to the metabolic, respiratory, and thermoregulatory physiology of these animals.

Table 34.1.
Respiration and Metabolic Data from Resting and Active Elephants

From Paladino et al. 1981; Benedict 1936. See Table 34.2 for a key to variable abbreviations.

Animal*	State	Age (yrs)	Mass (kg)	V_T (1)	f (BPM)	V_E (1/min)	O_2 (% ↓)	V_{O2} (ml/min)	M.R. (W/kg)	CO_2 (% ↑)	V_{CO2} (ml/min)	R.Q.
Roxie	Rest	12	1955	6.5	10.6	63.89	4.22	2.7	0.452	4.09	2.61	.97
	Predicted	N/A	1955	14.7	5.6	82.08	3.86	3.17	0.531	N/A	N/A	N/A
	Exercised	12	1955	13.5	13.5	129	5.69	7.34	1.23	6.24	8.05	1.09
Gee Gee	Rest	10	2046	6.0	8.6	58.9	4.01	2.36	0.378	3.98	2.34	.99
	Predicted	N/A	2046	15.4	5.5	84.93	3.86	3.28	0.526	N/A	N/A	N/A
	Exercised	10	2046	21.9	7.5	164.25	5.15	8.61	1.38	5.34	8.77	1.02
Irma	Rest	5	1273	8.1	8.2	67	2.72	1.8	0.469	2.63	1.76	.98
	Predicted	N/A	1273	9.55	6.2	59.42	3.85	2.29	0.845	N/A	N/A	N/A
	Exercised	5	1273	12.66	15.4	195	4.22	8.23	2.12	4.7	9.17	1.11
Jap	Rest	37	3672	26.5	8.8	209	2.78	5.81	0.518	N/A	N/A	N/A
	Predicted	N/A	3672	27.5	4.8	131.5	3.89	5.12	0.457	N/A	N/A	N/A
Gretchen (African)	Rest	16	2182	18.5	6.2	114.75	2.04	2.34	0.351	2.4	2.75	1.18
	Predicted	N/A	2182	16.37	5.44	89.02	3.88	3.45	0.518	N/A	N/A	N/A
	Exercised	16	2182	15.3	12.2	186.14	4.89	9.1	1.37	5.1	9.49	1.04
Nemo (male)	Rest	10	1360	19.9	6.6	131.3	N/A	N/A	N/A	N/A	N/A	N/A
	Predicted	N/A	1360	10.2	6.12	62.43	3.86	2.41	0.581	N/A	N/A	N/A
Mean Values Rest		15	2081	14.3	8.2	107.5	3.15	2.3	0.434	3.28	2.37	1.03

*Animals were Asian females except as noted. Roxie and Gee Gee were from a circus, Irma and Gretchen from a zoo, and Jap and Nemo from the 1936 Benedict study.

We determined elephant metabolic rates using standard procedures as described in Prange and Jackson (1976) and Jackson and Prange (1979). These experiments involved the collection of respiratory gas samples from elephants and the subsequent analysis of those gas samples for the amount of oxygen removed by the elephant and the amount of carbon dioxide given off. These two respiratory gas samples provide an indirect measure of the level of energy metabolism (fuel consumption) of these animals (cf. Kleiber 1975).

End expiratory gas samples were collected in Douglas bags (similar to a weather balloon) and analyzed for oxygen content, carbon dioxide content, and total volume from the breaths of both Indian and African elephants at the Cincinnati Zoo and the Ensinger Circus. Measurements were made on elephants at rest and after ten minutes of exercise (Paladino et al. 1981).

Our measurements were used to calculate respiratory CO_2 production (V_{CO2}) from the product of expired volume and the difference between fractional concentrations (the absolute amount of CO_2 in inspired and expired air). Computation of V_{O2} was corrected for the calculated difference between inspired and expired volumes, based on the difference in O_2 concentration in inspired and expired air. All gas measurements were corrected to STPD (standard temperature and pressure, dry air) using standard methods (Depocas and Hart 1957).

A Blueprint for Giants   495

The Respiratory Quotient (RQ), the ratio of CO_2 produced or released by the animal compared to the O_2 consumed or removed from the air sample by the animal, was calculated from matched subsamples of respiratory gases for elephants. Our experimental data were pooled with respiratory data from both male and female elephants taken from Benedict (1936). RQ and V_E (air ventilation values, the amount of air moved into and out of the respiratory system per unit time) were then calculated, using methods described in Jackson 1985.

The average elephant in these studies was fifteen years old and had a mass of 2081 kilograms, resting tidal volumes (see Table 34.2 for an explanation of this and other parameters) of 14.3 liters, an average of 8.2 breaths per minute, and a respiratory minute volume (the amount of air moved into the respiratory system in one minute) of 107.5 liters per minute. Exercised male and female elephants exhibited a typical mammalian response to mild activity with a two- to threefold increase in metabolic rate and respiration.

Measurements of physiological parameters in elephants can be compared with predicted values of those same parameters for elephant-sized animals, using allometric equations that link the relevant variables to animal body mass (Table 34.2), to gain insight into the physiology of large and gigantic animals. There is remarkable agreement between the measured physiological parameters of elephants and the values predicted for mammals using allometric (predictive) equations (Table 34.1). This agreement is especially impressive when one considers that the allometric equations are primarily derived from data obtained from small to medium-sized animals (Calder 1984).

Elephants, like other mammals, have a typically high level of metabolism and a large-capacity, rapid-turnover respiratory system which requires diaphragmatic breathing, especially while active. In the tropical to subtropical environments in which elephants are found, there are many evolutionary adaptations that aid in heat dissipation. The large ears of African elephants and the absence of pelage are just two of the very obvious adaptations in this regard (Benedict 1936; Wright 1984; Williams 1990; Phillips and Heath 1992).

Designing a Dinosaur: Cardiovascular Constraints

The close agreement between predicted and measured values of physiological parameters encourages us to use allometric equations in an effort to model the physiological performance of dinosaurs. First, though, we must consider the likely nature of the cardiovascular system that supported dinosaur metabolism.

The dinosaur cardiovascular system probably employed a four-chambered heart not dissimilar to that found in present-day archosaurs (crocodilians and birds), but without the diving adaptations (the foramen of Panizzae and other shunts from the right to the left side of the heart) that are typical of crocodilians (see Reid, chap. 32 of this volume). Small to medium-sized alligators with four-chambered hearts are capable of generating systemic blood pressures as high as 80 mmHg (Johansen and Burggren 1980; Lillywhite 1988). Even reptiles that have three-chambered hearts, such as sea snakes and varanid lizards, can generate systemic blood pressures in excess of 100 mmHg (Johansen and Burggren 1980; Lillywhite 1988), with pulmonary blood pressures maintained at 20 to 45 mmHg. Although this suggests that a four-chambered heart would not be necessary

Table 34.2.

Allometric Scaling Equations Used to Relate Body Mass to Various Physiological Parameters in Reptiles, Mammals, and Birds

Equations from Dejours 1981; Dubach 1981; Nagy 1982; Calder 1984; and Paladino et al. 1990.

RMR (for all reptiles, T_b = 30°C, in W kg^{-1}) = 0.378(M)$^{-.17}$
RMR (for mammals, T_b = 39°C, in W kg^{-1}) = 3.35(M)$^{-.25}$
RMR (for birds, T_b = 41°C, in W) = 0.047(m)$^{.72}$
T.L.V. (for reptiles in ml) = 1.237(m)$^{.75}$
T.L.V.^{m} (for mammals in ml) = 0.035(m)$^{1.06}$
T.L.V.^{b} (for birds in ml) = 0.034(m)$^{0.97}$
V_T (for reptiles in ml) = 0.020(m)$^{.80}$
V_T^m (for mammals in ml) = 0.0075(m)$^{1.0}$
V_T^b (for birds in ml) = 0.0076(m)$^{1.08}$
f (for reptiles in breaths min^{-1}) = 20.6(m)$^{-.04}$
f (for mammals in breaths min^{-1}) = 209(m)$^{-.25}$
f (for birds in breaths min^{-1}) = 146(m)$^{-0.04}$
HM (for reptiles) = 0.0051(M)
HM (for mammals in g) = 5.8(M)$^{0.98}$
HM (for birds in g) = 8.6(M)$^{0.94}$
HR (for mammals in bpm) = 241(M)$^{-0.25}$
HR (for birds in bpm) = 156(M)$^{-0.23}$
SV (for mammals in ml) = 0.78(M)$^{1.06}$
SV (for birds in ml) = 1.83(M)$^{1.01}$
wf (for reptiles in ml H$_2$O day^{-1}) = 45(M)$^{0.66}$
wf (for mammals in ml H$_2$O day^{-1}) = 123(M)$^{0.80}$
wf (for birds in ml H$_2$O day^{-1}) = 115(M)$^{0.75}$

Abbreviations: M = mass in kg and m = mass in g; RMR = resting metabolic rate (the lowest sustained energy consumption of an animal at rest, in the dark, at least ten hours after eating, in a thermally non-stressful environment, calculated at T_b = 30°C for reptiles); TLV = total lung volume; V_T = tidal volume (amount of air moved into or out of the respiratory system in one breath); f = respiratory rate (in breaths per minute); wf = water flux (water loss or gain per unit time); HM = heart mass in g or kg; HR = heart rate in beats per minute; SV = cardiac stroke volume, or the volume of blood pumped by the ventricles to the body.

to generate the blood pressures necessary to provide ample circulation throughout the body, the fact that four-chambered hearts are characteristic of all living archosaurs, both birds and crocodilians, makes it highly probable that the same was true of dinosaurs.

The absence of a diaphragm in birds and crocodilians likewise suggests that large dinosaurs had to move huge volumes of air without a diaphragm (Hengst and Rigby 1994). Some dinosaurs (e.g., sauropods) had long necks that would have created large dead-space volumes (air left after each breath in the mouth, trachea, or air tubes leading into the lungs), necessitating a fair amount of air turnover to provide the lungs with sufficient oxygen to survive. The allometrically predicted tidal volume for an *Apatosaurus* if it had a lizard-like respiratory system would be only about 19 liters (Table 34.3). This amount would be totally insufficient and would not even move enough

A Blueprint for Giants   497

air to replace the dead-space volumes predicted for these long-necked giants (Hengst and Rigby 1994). Mammals with a muscular diaphragm can create large pressure differences to move such large masses of air. What did sauropods do?

There are primitive air sacs and multicameral lungs in the most primitive reptiles today (Perry 1989). We suspect that complex lungs, and in saurischian dinosaurs air sacs (Reid 1996), were a necessity for large dinosaurs to enhance oxygen uptake. This was especially true of sauropods, with their huge dead-space volumes (Daniels and Pratt 1992; Hengst and Rigby 1994). Air sacs and a one-way flow-through lung (Reid, chap. 32 of this volume) would have allowed a large sauropod such as *Apatosaurus,* with a total lung volume estimated at about 1400 liters and a dead-space volume of 184 liters (Hengst and Rigby 1994), to have maintained a reasonable respiratory rate in the absence of a muscular diaphragm.

In reviewing our allometric calculations for a "mammal"-, "bird"-, or "reptile"-like respiratory system (Table 34.3), we predict that a 30,000-kilogram *Apatosaurus* would have needed an avian-type lung and air sacs just to supply the gas exchange necessary for a relatively low reptilian metabolism. The exchange requirements for an even greater oxygen demand resulting from avian or mammalian endothermic metabolism are well above the allometric predictions. If *Apatosaurus* had a total thoracic volume of 1700 liters and the lung tissue occupied 232 liters (lung mass = $0.013(M)^{0.95}$; Calder 1984), and the heart added an additional 500 liters of volume, this would have left a lung capacity of about 900 liters. This volume is very close to the 600-liter volume calculated for an avian-type parabronchial or rigid lung. It may seem surprising that a 30,000-kilogram dinosaur would have had a calculated tidal volume that was greater than the predicted lung volume, but this could be easily accounted for by air in the hypothesized air sac system. This large predicted tidal volume would have been necessary to dilute the used air remaining in the tracheal dead space of the long necks of giant dinosaurs. To then add to this requirement for an efficient respiratory system due to giantism and huge dead-space volumes the additional demand for oxygen associated with endothermy seems highly improbable.

The Evolution of Giants

For most of their evolutionary history, mammals were small-bodied animals. Large and gigantic mammals did not become prominent in terrestrial faunas until the later Cenozoic Era, in response to cooling and

Table 34.3.
Physiological Performance of a Gigantic Dinosaur (*Apatosaurus*), Using Predictions from Allometric Physiological Equations for Mammals, Birds, and Reptiles

Dinosaur species	Animal blueprint	Mass (kg)	Lung volume (l)	Tidal volume (l)	Resp. rate (f)	Heart rate (BPM)	Stroke volume (l)	Heart mass (kg)
Apatosaurus	Mammal	30,000	2949.96	225	2.82	18.31	43.45	141.56
	Bird	30,000	608.54	903.87	0.70	14.57	60.86	138.99
	Reptile	30,000	501.43	19.18	10.35	*	*	153.00

*No appropriate allometric equation is available.

drying climatic trends (Behrensmeyer et al. 1992; Prothero 1994). Even since large forms evolved, the majority of mammalian species at any given time have been small in size (Brown 1995; Farlow et al. 1995a).

Although large-bodied tropical mammals (such as living elephants) obviously do exist, there are reasons for thinking that the tropics do not constitute the best thermal environment for very large endotherms. Under warm environmental conditions, large endotherms would have difficulty dumping the huge amounts of heat that their bulk would produce (Spotila et al. 1991). This may account for the tendency of very large tropical mammals to forage more at night than during the daytime, and for the lack of a thick pelage in these animals (Owen-Smith 1988), and also for the previously mentioned heat-dumping role of elephant ears.

A diversity of gigantic mammals does exist in the oceans, however. Their huge bulk is sustained by the high productivity of the small-bodied prey (e.g., krill) upon which they feed, and the high heat conductance of water enables them to dissipate the huge thermal loads of their bodies. Even so, the biggest whales spend much time in cold, polar waters; they rarely enter warm waters except to calve.

The majority of extinct dinosaur species were large to enormous animals (Peckzis 1994; Farlow et al. 1995a). Dinosaurs lived during a time when warm thermal environments were characteristic of most of the planet (Behrensmeyer et al. 1992; Farlow et al. 1995a). Large dinosaurs would have had little need for elevated resting metabolic rates. Like large modern reptiles, big dinosaurs could have employed a suite of physiological features collectively termed gigantothermy (the use of large body size, circulatory adjustments, and layers of body insulation to maintain constant high body temperatures with low ectothermic metabolic rates, even in relatively cool thermal environments; Paladino et al. 1990). Furthermore, if very large dinosaurs had been endotherms, they might have experienced a lethal inability to dump excess body heat (Spotila et al. 1991).

Presently available fossil evidence (Carpenter et al. 1994) indicates that dinosaurs were egg-layers. As with their modern relatives, crocodilians and birds, the constraints of development in a cleidoic (closed system) egg necessitated uricotelism (the formation of uric acid as a less toxic waste product of nitrogen metabolism than ammonia). The fact that many modern birds and reptiles have evolved extrarenal (non-kidney) methods of salt removal (such as nasal/ocular salt glands) provides further evidence that the lineage that led to modern mammals was quite different from that which led to present-day reptiles and birds. Allometric calculations of the water flux for different-sized reptiles, mammals, and birds indicate that an *Apatosaurus*-sized sauropod built following the mammalian blueprint would be closely tied to water resources, needing at least 469 liters of external water per day to survive (equation in Table 34.2). Dinosaurs constructed along the lines of a reptilian blueprint would have been best able to cope with the dry conditions that often characterized dinosaur environments (Behrensmeyer et al. 1992; Farlow et al. 1995a). However, an apatosaur with a bird-like respiratory system (one-way flow through lungs with air sacs), a reptilian resting metabolism (using the avian equation and substituting a reptilian metabolism), and extrarenal salt glands would require only about 56 percent of the mammalian/avian water requirements, some 262 liters per day, to survive. A dinosaur with a reptile-like respiratory system and kidney/salt glands would need only about 41 liters of external water to survive. Water conservation, along with selective pressure to evolve a less toxic form for nitrogenous waste storage in a cleidoic egg, may have led to the development of uricotelism in dinosaurs.

A Blueprint for Giants   499

Ectothermic dinosaurs would have required considerably less food than endotherms of comparable size (Farlow 1993; Farlow et al. 1995a), which would have permitted dinosaur populations to survive on a much lower resource base than is possible for populations of large mammals. It may be difficult for gigantic mammals to maintain large enough populations to avoid chance extinction (Farlow et al. 1995a), which could be an additional reason (along with problems of heat stress) why there are no whale-sized terrestrial mammals, even in polar regions, where overheating would presumably not be a problem. In contrast, species of ectothermic dinosaurs would have been able to attain considerably larger individual-animal body sizes, for a given population size, than is possible for endotherms. Consequently dinosaurs were well positioned physiologically to evolve gigantism, were environmental conditions to make this possible.

The beginning of the Mesozoic Era seems to have coincided with an increase in levels of atmospheric CO_2 over those of the later Paleozoic Era (Berner 1994; Graham et al. 1995). In addition to warming the planet, and thus creating an ideal environment in which ectotherms could opt for gigantothermy, elevated carbon dioxide levels might have stimulated plant productivity. This in turn could have provided a greater food resource base for dinosaurs than had been available to their therapsid predecessors of the later Paleozoic—thus fueling the evolution of gigantism in dinosaurs, making it possible for them to evolve to sizes at which gigantothermy would have been most beneficial. The pattern of indeterminate growth typical of most reptiles, including at least some dinosaurs (Farlow et al. 1995a), would have permitted many dinosaurs to continue to grow throughout their entire lifetime if sufficient nutrition was available.

In recent years the controversy about whether dinosaurs were endotherms or ectotherms has received considerable attention in the paleophysiological literature. Chinsamy (1990) and Chinsamy et al. (1994) used bone histology studies to determine that the bone growth pattern of extinct dinosaurs more closely resembles the periodic seasonal growth pattern of typical ectotherms. Additionally, data from studies on reptilian muscle power physiology (Ruben 1991) indicate that flapping flight probably evolved in primitive birds before the evolution of endothermy and constantly high regulated body temperatures. It appears that endothermy probably evolved well after birds split from the typical reptilian dinosaur lineage and well after the evolution of flight.

Endothermy was not a necessity for the evolution of flight (contrary to the increased aerobic capacity argument of Bennett and Ruben [1979]), and the muscle physiology of a reptile is fully capable of launching a primitive chicken-sized or even turkey-sized bird into the air from the ground (Ruben 1991). A bird of this size does not use the constant highly aerobic flapping flight we see in the small migratory birds of northern temperate climates. Larger birds use short bursts of their wings to propel them up where they can then glide, and also use intermittent bursts, or thermals, to fly long distances. Vultures, eagles, and hawks can migrate great distances without the need for continuous flapping flight. We suspect that the primitive birds of the same size that evolved from a chicken- to a turkey-sized bipedal dinosaur probably could do the same thing. The evolution of endothermy allowed birds to expand their niches into the more seasonal northern and southern latitudes, which became much colder as climates cooled in the Maastrichtian. These primitive birds would have then needed to maintain constant high body temperatures to ensure that their high-power, reptilian muscles were kept warm in the cold. Keeping these muscles warm was necessary to generate the thrust to get airborne. Chicken- to turkey-sized birds would have required endothermy to keep their muscles warm in colder climates. Once endothermy evolved in these medium-sized to large primitive birds, smaller birds with an even greater endothermic capacity could then also expand their aerobic scope to support the typical high-energy flapping flight that we see in the small species of today.

This scenario of thermoregulatory selection pressures and thermal niche expansion may have been the driving force in the evolution of endothermy in birds.

These arguments provide additional support for the hypothesis of Block et al. (1993) that endothermy in fishes evolved as a result of niche expansion, and not of the need to support higher aerobic capacities. In the monophyletic fish group comprising tunas and their allies, the Scombroidei, endothermy has evolved at least three separate times (Block 1991; Block et al. 1993). The scombroid phylogeny in these studies was generated by genetic comparison of the cytochrome b gene in muscles; outgroups used for comparison in these phylogenetic studies all are fishes with laterally placed red muscle (*Sphyraena*, Serranidae, and *Coryphaena*).

Four genetic modifications were identified at points that are inferred mutations that resulted in the evolutionary development of endothermic red muscle in four different genera of the Scombroidei. In each case the modification involved a different mechanism. For example, in the billfish group, the genetic modification of the superior rectus muscle into a thermogenic organ (countercurrent heat exchanger formed from the carotid artery) led to the development of regional endothermy. For the butterfly mackerel, a modification of the lateral rectus muscle into a thermogenic organ (heat exchanger derived from the lateral dorsal aorta) was a different route leading to endothermy. For tunas, a systemic endothermy using vascular countercurrent heat exchangers in the muscle, viscera, and brain (heat exchanger in the brain formed from the carotid artery) represents the adaptations that led to endothermy, with some internalization of red muscle along the horizontal septum.

Close relatives of the tuna, such as bonitos, also swim very rapidly but are ectothermic. Tunas, on the other hand, are true endotherms, with thermally isolated and specialized red muscle that has exceptional aerobic capacities. The evolution of endothermy allowed tunas to expand their thermal niches into the colder northern and southern oceans as well as into deeper, colder water than is possible for their ectothermic relatives.

In contrast to endothermic fishes, leatherback turtles are gigantotherms that use large body size and a constant low-wattage power supply to fuel their very active long-distance migrations, and to dive to depths of 1000 meters in the cold ocean while maintaining constant high body temperatures (Paladino et al. 1990).

The scombroid fish data, data from a gigantothermic turtle, reptilian muscle-power calculations, and the fossil bone histology of dinosaurs and primitive birds all suggest paths toward the evolution of endothermy that are consistent with the interpretation made here: that there were no selection pressures that would have led to the evolution of endothermy in typical dinosaurs. Large extinct dinosaurs were ectotherms.

The Dinosaurian Blueprint

The rapid evolution of large body size in early archosaurs would have served as a strong selective force in the evolution of a four-chambered heart. Such a heart may have been necessary to meet increased demands for

A Blueprint for Giants   501

more efficient nutrient and oxygen delivery as archosaurs became increasingly larger. The circulatory system that evolved to meet these needs had all the necessary features that could be refined to provide the physiological framework of avian endothermy.

For dinosaurs, however, the same kind of heart could be used to support evolution in an entirely different direction: toward even bigger body sizes. In saurischian dinosaurs, the evolution of a one-way-flow parabronchial lung with air sacs may have been a further response to selection pressures associated with the evolution of large body size.

Once such a respiratory system was in place, however, it was ideally suited for dealing with the huge dead-space volumes associated with the long necks of sauropods. This may have been one of the key adaptations that enabled these dinosaurs to attain such huge sizes.

References

Alexander, R. McN. 1985. Mechanics of posture and gait of some large dinosaurs. *Zoological Journal of the Linnean Society* 83: 1–25.

Bakker, R. T. 1986. *The Dinosaur Heresies: New Theories Unlocking the Mystery of the Dinosaurs and Their Extinction.* New York: William Morrow.

Behrensmeyer, A. K.; J. D. Damuth; W. A. DiMichele; R. Potts; H.-D. Sues; and S. L. Wing (eds.). 1992. *Terrestrial Ecosystems through Time: Evolutionary Paleoecology of Terrestrial Plants and Animals.* Chicago: University of Chicago Press.

Benedict, F. G. 1936. *Physiology of the Elephant.* Washington, D.C.: Carnegie Institution Publication no. 474.

Bennett, A., and J. Ruben. 1979. Endothermy and activity in vertebrates. *Science* 206: 649–654.

Benton, M. J. (ed). 1988. *The Phylogeny and Classification of the Tetrapods.* Vol. 1. Systematics Association Special Volume 35A: 289–332.

Berkson, H. 1966. Physiological adjustments to prolonged diving in the Pacific green turtle. *Comparative Biochemistry and Physiology* 18: 101–119.

Berner, R. A. 1994. Geocarb II: A revised model of atmospheric CO_2 over Phanerozoic time. *American Journal of Science* 294: 56–91.

Block, B. 1991. Evolutionary novelties: How fish have built a heater out of a muscle. *American Zoologist* 31: 726–742

Block, B.; J. Finnerty; A. Stewart; and J. Kidd. 1993. Evolution of endothermy in fish: Mapping physiological traits on a molecular phylogeny. *Science* 260: 210–214.

Brown, J. H. 1995. *Macroecology.* Chicago: University of Chicago Press.

Calder, W. A. 1984. *Size, Function, and Life History.* Cambridge, Mass.: Harvard University Press.

Carey, C. 1996. Reproductive energetics. In C. Carey (ed.), *Avian Energetics and Nutritional Ecology,* pp. 324–374. New York: Chapman and Hall.

Carpenter, K.; K. F. Hirsch; and J. R. Horner (eds.). 1994. *Dinosaur Eggs and Babies.* Cambridge: Cambridge University Press.

Chinsamy, A. 1990. Physiological implications of the bone histology of *Syntarsus rhodesiensis* (Saurischia: Theropoda). *Palaeontologia Africana* 27: 77–82.

Chinsamy, A.; L. Chiappe; and P. Dodson. 1994. Growth rings in Mesozoic birds. *Nature* 368: 196–197.

Daniels, C. B., and J. Pratt. 1992. Breathing in long necked dinosaurs: Did the sauropods have bird lungs? *Comparative Biochemistry and Physiology* 101A: 43–46.

Dejours, P. 1981. *Principles of Comparative Respiratory Physiology.* Amsterdam: Elsevier.

Depocas, F., and J. S. Hart. 1957. Use of the Pauling oxygen analyzer for measurement of oxygen consumption in open-circuit systems and in short lag, closed-circuit apparatus. *Journal of Applied Physiology* 10: 388–392.

Dixon, D.; B. Cox; R. J. G. Savage; and B. Gardiner. 1992. *The Macmillan Illustrated Encyclopedia of Dinosaurs and Prehistoric Animals*. New York: Collier Books, Macmillan.

Dubach, M. 1981. Quantitative analysis of the respiratory system of the house sparrow, budgerigar, and violet eared hummingbird. *Respiratory Physiology* 46: 43–60.

Farlow, J. O. 1993. On the rareness of big, fierce animals: Speculations about the body sizes, population densities, and geographic ranges of predatory mammals and large carnivorous dinosaurs. *American Journal of Science* 293-A: 167–199.

Farlow, J. O.; P. Dodson; and A. Chinsamy. 1995a. Dinosaur biology. *Annual Review of Ecology and Systematics* 26: 445–471.

Farlow, J. O.; M. B. Smith; and J. M. Robinson. 1995b. Body mass, bone "strength indicator," and cursorial potential of *Tyrannosaurus rex*. *Journal of Vertebrate Paleontology* 15: 713–725.

Gans, C., and H. Pough. 1982. Physiological ecology: Its debt to reptilian studies, its value to students of reptiles. In C. Gans and F. H. Pough (eds.), *Biology of the Reptilia*, vol. 12, pp. 1–11. New York: Academic Press.

Gleeson, T. T. 1991. Patterns of metabolic recovery from exercise in amphibians and reptiles. *Journal of Experimental Biology* 160: 187–207.

Graham, J. B.; R. Dudley; N. M. Aguilar; and C. Gans. 1995. Implications of the late Palaeozoic oxygen pulse for physiology and evolution. *Nature* 375: 117–120.

Grigg, G. C. 1989. The heart and pattern of cardiac outflow in Crocodilia. *Proceedings of the Australian Physiological and Pharmacological Society* 20: 43–57.

Hengst, R., and J. K. Rigby, Jr. 1994. *Apatosaurus* as a means of understanding dinosaur respiration. In G. D. Rosenberg and D. L. Wolberg (eds.), *Dino Fest*, pp. 199–212. Special Publication 7. Knoxville: The Paleontological Society, University of Tennessee.

Jackson, D. C. 1985. Respiration and respiratory control in the green turtle, *Chelonia mydas*. *Copeia* 1985: 664–671.

Jackson, D. C., and H. D. Prange. 1979. Ventilation and gas exchange during rest and exercise in adult green turtles. *Journal of Comparative Physiology* 134: 315–319.

Johansen, K., and W. W. Burggren. 1980. Cardiovascular function in the lower vertebrates. In G. H. Bourne (ed.), *Hearts and Heart-like Organs*, vol. 1, pp. 61–117. New York: Academic Press.

Kleiber, M. 1975. *The Fire of Life*. New York: R. E. Krieger.

Lillywhite, H. B. 1988. Snakes, blood circulation and gravity. *Scientific American* 259: 92–98.

Nagy, K. A. 1982. Field studies of water relations. In C. Gans and F. H. Pough (eds.), *Biology of the Reptilia*, vol. 12, pp. 483–501. New York: Academic Press.

Owen-Smith, R. N. 1988. *Megaherbivores: The Influence of Very Large Body Size on Ecology*. Cambridge: Cambridge University Press.

Paladino, F. V., and J. R. King. 1984. Thermoregulation and oxygen consumption during terrestrial locomotion by white-crowned sparrows, *Zonotrichia leucophrys gambelii*. *Physiological Zoology* 57: 226–236.

Paladino, F. V.; M. P. O'Connor; and J. R. Spotila. 1990. Metabolism of leatherback turtles, gigantothermy, and thermoregulation of dinosaurs. *Nature* 344: 858–860.

Paladino, F. V.; J. R. Spotila; and D. Pendergast. 1981. Respiratory variables of Indian and African elephants. *American Zoologist* 21: 143.

Peckzis, J. 1994. Implications of body-size estimates for dinosaurs. *Journal of Vertebrate Paleontology* 14: 520–533.

Perry, S. F. 1989. Structure and function of the reptilian respiratory system. In S. C. Wood (ed.), *Comparative Pulmonary Physiology: Current Concepts*, pp. 193–236. Lung Biology in Health and Disease, vol. 39. New York: Marcel Dekker.

Phillips, P. K., and J. E. Heath. 1992. Heat exchange by the pinna of the African elephant (*Loxodonta africana*). *Comparative Biochemistry and Physiology* 101A: 693–699.

Prange, H. D., and D. C. Jackson. 1976. Ventilation, gas exchange and metabolic scaling of a sea turtle. *Respiratory Physiology* 27: 369–377.

Prothero, D. R. 1994. *The Eocene-Oligocene Transition: Paradise Lost*. New York: Columbia University Press.

Reid, R. E. H. 1996. Bone histology of the Cleveland-Lloyd dinosaurs and of dinosaurs in general. Part I: Introduction: Introduction to bone tissues. *Brigham Young University Geology Studies* 41: 25–71.

Romanoff, A. L. 1967. *Biochemistry of the Avian Embryo*. New York: John Wiley and Sons.

Ruben, J. 1991. Reptilian physiology and flight capacity of *Archaeopteryx*. *Evolution* 45: 1–17.

Scheid, P. 1979. Mechanism of gas exchange in bird lungs. *Reviews of Physiology, Biochemistry and Pharmacology* 86: 137–186.

Spotila, J. R.; M. P. O'Connor; P. Dodson; and F. V. Paladino. 1991. Hot and cold running dinosaurs: Body size, metabolism and migration. *Modern Geology* 16: 203–227.

Williams, T. M. 1990. Heat transfer in elephants: Thermal partitioning based on skin temperature profiles. *Journal of Zoology, London* 222: 235–245.

Wood, S. C. 1989. *Comparative Pulmonary Physiology: Current Concepts*. Lung Biology in Health and Disease, vol. 39. New York: Marcel Dekker.

Wright, P. S. 1984. Why do elephants flap their ears? *South African Journal of Zoology* 19: 266–269.

New Insights into the Metabolic Physiology of Dinosaurs

*John Ruben,
Andrew Leitch,
Willem Hillenius,
Nicholas Geist,
and
Terry Jones*

Endothermy, or "warm-bloodedness," is one of the major evolutionary developments of vertebrates, and among the most significant features that distinguish existing birds and mammals from reptiles, amphibians, and fishes. Endothermy, which has clearly evolved independently in birds and mammals, provides organisms with distinct physiological and ecological benefits, and appears to be largely responsible for the present success of birds and mammals in aquatic and terrestrial environments (Ruben 1995). Elevated rates of lung ventilation, oxygen consumption, and internal heat production (via aerobic metabolism), which are the hallmarks of endothermy, enable birds and mammals to maintain thermal stability over a wide range of ambient temperatures. As a result, these animals are able to thrive in environments with cold or highly variable thermal conditions, and in nocturnal habitats generally unavailable to ectothermic vertebrates.

Furthermore, the increased aerobic (oxygen-consuming) capacities of endotherms allow them to sustain activity levels well beyond the capacity of ectotherms (Bennett 1991). With some noteworthy exceptions, ectotherms, such as reptiles, typically rely on non-sustainable anaerobic metabolism for all activities beyond relatively slow movements. Although capable of often spectacular bursts of intense exercise, ectotherms generally fatigue rapidly as a result of lactic acid accumulation. In contrast, endotherms are able to sustain even relatively high levels of activity for extended periods of time. This enables these animals to forage widely and to migrate over extensive distances. The capacity of bats and birds to

35

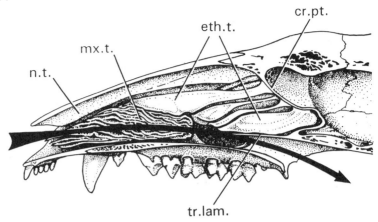

A

cr.pt.

eth.t.

mx.t.

n.t.

tr.lam.

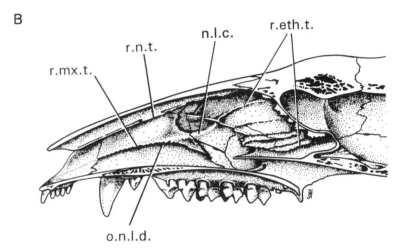

B

n.l.c.

r.eth.t.

r.n.t.

r.mx.t.

o.n.l.d.

Figure 35.1. The nasal turbinates of mammals. The arrow describes the path of air flow through the nasal region into the oral cavity. (A) Right longitudinal section of the skull of the opossum, *Didelphis*. (B) Similar to A, but with turbinates removed to reveal turbinate attachment ridges. (C) Cross-sections through the anterior nasal turbinates of several mammalian taxa, including the African buffalo (*Syncerus*), badger (*Meles*), and seal (*Phoca*). Not to scale; after Hillenius 1994. Abbreviations: cr.pt. = cribiform plate; eth.t. = ethmoturbinates (olfactory); mx.t. = maxilloturbinate, or respiratory concha; n.l.c. = nasolacrimal canal; n.t. = nasoturbinate (olfactory); o.n.l.d. = opening of nasolacrimal canal; r.eth.t. = ridges for olfactory ethmoturbinates; r.mx.t. = ridge for maxilloturbinate; r.n.t. = ridge for nasoturbinate (olfactory).

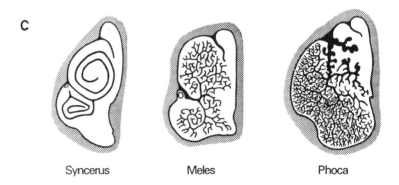

C

Syncerus   Meles   Phoca

sustain long-distance powered flight is far beyond the capabilities of modern ectotherms (Ruben 1991).

Considering the significance of endothermy to the biology of extant vertebrates, it is hardly surprising that the evolution of endothermy has received so much attention (e.g., Hopson 1973; Bennett and Ruben 1979; Bennett and Huey 1990). Additionally, in the past few decades there has been considerable speculation about the possible presence of endothermy, or the lack of it, in extinct reptilian and avian lineages, such as pterosaurs and dinosaurs, which dominated the terrestrial and aerial environments for most of the Mesozoic Era (e.g., Farlow 1990; Padian 1983).

Until very recently, endothermy has been virtually impossible to demonstrate clearly in extinct forms. Endothermy is almost exclusively an attribute of the "soft anatomy," which leaves a poor, or usually nonexistent, fossil record. Physiologically, endothermy is achieved through prodigious rates of cellular oxygen consumption: In the laboratory, mammalian basal, or resting, metabolic rates are typically about six to ten times greater than those of reptiles of the same body mass and temperature; avian resting rates are often greater still—up to fifteen times reptilian rates. In the field, metabolic rates of mammals and birds often exceed those of equivalent-sized ectotherms by about twenty times. To support such high oxygen-consumption levels, endotherms possess profound structural and functional modifications to facilitate oxygen uptake, transport, and delivery. Both mammals and birds have greatly expanded rates of lung ventilation, fully separated pulmonary and systemic circulatory systems, and expanded cardiac output. They also have greatly increased blood volume and blood oxygen-carrying capacities, as well as increased tissue aerobic enzymatic activities (Ruben 1995). These key features of endothermic physiology are unlikely to have ever been preserved in fossils, mammalian, avian, or otherwise.

Consequently, previous conjecture concerning the possible presence of endothermy in a variety of extinct vertebrates has relied primarily on hypothesized correlations of metabolic rate with a variety of weakly supported criteria, including, but not limited to, predator-prey ratios (Bakker 1980), fossilized trackways (Bakker 1986), and correlations with avian or mammalian posture (Bakker 1971). Close scrutiny, however, has revealed that virtually all of these correlations are, at best, equivocal (Farlow et al. 1995). Attempts have also been made to associate supposedly high overall growth rates in endotherms with hypothesized fast growth and endothermy in some dinosaurs. However, new information (as well as reanalysis of old data) suggests that growth rates in a variety of extant endotherms and ectotherms are not broadly different from one another (Chinsamy 1990, 1993; Owerkowicz and Crompton 1995). In addition, growth rate in at least one dinosaur (*Troodon*) has been shown to have overlapped that of the American alligator (Ruben 1995).

Most recently, relative quantities of fossilized bone oxygen isotope ($O^{16}:O^{18}$) were purported to demonstrate relatively little *in vivo* variation between extremity and deep-body temperature in some large dinosaurs (e.g., *Tyrannosaurus*) (Barrick and Showers 1994). This was assumed to signify that these large dinosaurs were endothermic because living endotherms, unlike ectotherms, were presumed to maintain relatively uniform extremity vs. core temperatures. Unfortunately, there are abundant data demonstrating that many birds and mammals often maintain extremity temperatures well below deep-body, or core, temperatures. Additionally, fossil bone oxygen isotope ratios may be strongly influenced by groundwater temperatures (Kolodny et al. 1996). Consequently, fossilized bone

New Insights into Metabolic Physiology   507

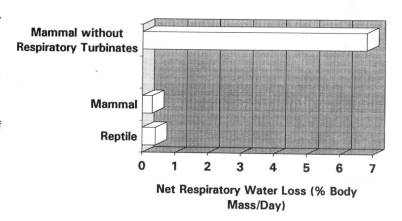

Figure 35.2. Daily net respiratory evaporative water loss rates (respiratory evaporative water loss minus metabolic water production) for a free-living 1-kg reptile and a 1-kg mammal, and probable net respiratory evaporative water loss for a free-living mammal lacking the use of respiratory turbinates (that is, with a reptile-like nasal anatomy and reptile-like net respiratory water loss rates per cubic centimeter of O_2 consumed). Without the water-conserving function of respiratory turbinates, daily water flux rates in mammals and birds would be out of balance by about 30 percent. Calculations based on field metabolic and water flux rates for lizards and eutherian mammals (regressions provided by Nagy [1987] and Nagy and Peterson [1988]) and observed rates of net respiratory evaporative water loss in lizards and intact and experimentally altered mammals (Hillenius 1992). For thermoregulating lizards (body temperature = 37°C), net respiratory water loss (at ambient temperature = 15°C) approximates 1.5 mg of H_2O per cc of O_2 consumed; for intact mammals, net respiratory water loss per cc of O_2 consumed is negligible or slightly positive at ambient temperatures = 15°C (see Hillenius 1992).

oxygen isotope ratios in dinosaurs are likely to reveal little, if any, definitive information about dinosaur metabolic physiology.

Perhaps more to the point, virtually all of the arguments used previously were based predominantly on apparent similarities to the mammalian, or avian, condition, without a clear functional correlation to endothermic processes per se. Until recently, no empirical studies were available which described an unambiguous, and exclusive, functional relationship to endothermy of a preservable morphological characteristic.

This situation changed with the demonstration that the respiratory turbinates (described below) of almost all species of living mammals are essential to, and have a tight functional correlation with, maintenance of high lung ventilation rates and endothermy. It has also been discovered recently that respiratory turbinates, and, presumably, elevated lung ventilation and metabolic rates, also occurred in at least two groups of Permo-Triassic mammal-like reptiles (Hillenius 1994). Consequently, the respiratory turbinates represent the first direct morphological indicator of endothermy that can be observed in the fossil record.

Complex respiratory turbinates are also found in all existing birds. Although independently derived in avians, these structures are remarkably similar to their mammalian analogues. Limited, but reliable, data suggest that avian respiratory turbinates have a similar functional association with high lung ventilation rates and endothermy. Consequently, as in the therapsid-mammal lineage, the occurrence or absence of these structures may well serve as a "Rosetta stone" for revealing patterns of lung ventilation rate and metabolism in early birds and their ancestors, the dinosaurs.

Respiratory Turbinates and Their Relationship to Endothermy in Mammals and Birds

Turbinate bones, or cartilages, are scroll- or baffle-like elements located in the nasal cavity of virtually all reptiles, birds, and mammals. In most mammals, these usually consist of two sets of mucous membrane–lined structures which protrude directly either into the main nasal airway, or into blind "alleyways" immediately adjacent to the main respiratory airway (Fig. 35.1). Those situated within the main anterior nasal air passageway (i.e., the nasal passage proper), the maxilloturbinates, or respiratory turbinates, are thin, complex structures, lined with moist respiratory epithelia. Olfactory turbinates (also known as lateral sphenoids and naso- or ethmoturbinates)

INHALATION

$T_{ambient} = 15°C$

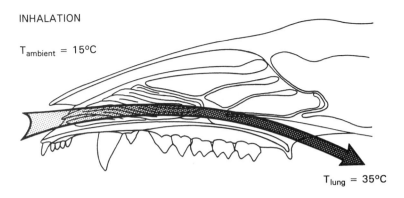

$T_{lung} = 35°C$

EXHALATION

$T_{exhaled} = 17.9°C$

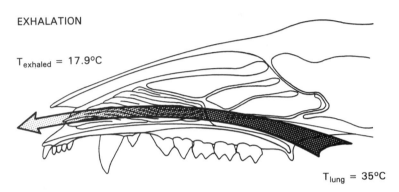

$T_{lung} = 35°C$

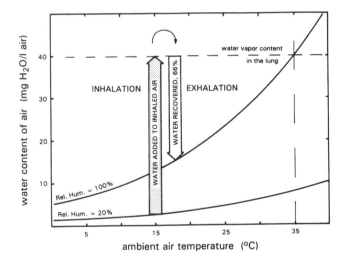

Figure 35.3. The water-recovery mechanism of respiratory turbinates. *Top:* During inhalation, ambient air passes over the respiratory turbinates and is warmed to body temperature. As a result, inhaled air is saturated with water vapor, and the turbinates are cooled by evaporative heat loss. *Center:* During exhalation, warm air from the lungs returns through the nasal passage, and is cooled as it passes outward over the turbinates. As a consequence, exhaled air becomes supersaturated with water vapor, and excess moisture condenses on the turbinate lamellae. *Bottom:* Graph depicting water vapor added to inhaled air (*shaded bar*) and water (condensate) recovered on exhalation (*open bar*) in an opossum. The curves represent water vapor content of saturated air (relative humidity [RH] = 100 percent), and air at 20 percent RH. In this example, inhaled air is modified from ambient air temperature and humidity (15°C, 20 percent RH) to lung air conditions (35°C, RH = 100 percent). Subsequent to cooling by the turbinates, exhaled air is still saturated, but only 17.9°C. Consequently, of the water vapor originally added to inhaled air, about 66 percent is "reclaimed." Graph after Schmidt-Nielsen et al. 1970; data from Hillenius 1992.

New Insights into Metabolic Physiology   509

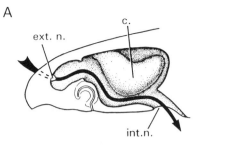

A

c.

ext. n.

int.n.

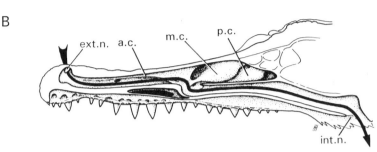

B

ext.n. a.c.   m.c.   p.c.

int.n.

Figure 35.4. The nasal turbinates of reptiles. All turbinates in these animals are exclusively olfactory in function. The arrow describes the path of air flow through the nasal region. (A) Right sagittal section of the nasal cavity of a lizard, *Lacerta*. (B) Right sagittal section through the nasal cavity of *Alligator*. Not to scale. After Hillenius 1994. Abbreviations: ext.n. = external nares; int.n. = internal nares; a.c., c., m.c., p.c. = olfactory turbinates.

are located just out of the main path of respired air, usually dorsal and posterior to the respiratory turbinates. Olfactory turbinates are lined with olfactory (sensory) epithelia and are the primary centers for the sense of smell. They occur in all reptiles, birds, and mammals and have no particular association with the maintenance of endothermy.

Only the respiratory turbinates have a strong functional association with endothermy. In both birds and mammals, endothermy is tightly linked to high levels of oxygen consumption and elevated rates of lung ventilation. In the laboratory, lung ventilation rates of birds and mammals exceed those of similar-sized reptiles by three and a half to five times, and avian and mammalian pulmonary ventilation rates in the field undoubtedly exceed reptilian rates by about twenty times. In mammals, the respiratory turbinates are essential for alleviating potentially unacceptably high rates of water loss associated with elevated lung ventilation rates (Fig. 35.2).

Respiratory turbinates facilitate an intermittent countercurrent exchange of respiratory heat and water between respired air and the moist epithelial linings of the turbinates (Fig. 35.3). Briefly, as cool external air is inhaled, it absorbs heat and moisture from the turbinate linings. This prevents desiccation of the lungs, but it also cools the respiratory epithelia and creates a thermal gradient along the turbinates. During exhalation, this process is reversed: Warm air from the lungs, now fully saturated with water vapor, is cooled as it once again passes over the respiratory turbinates. The exhaled air becomes supersaturated as a result of this cooling, and "excess" water vapor condenses on the turbinate surfaces, where it can be reclaimed and recycled. Over time, a substantial amount of water and heat can thus be saved, rather than lost to the environment. In the absence of respiratory turbinates, continuously high rates of oxidative metabolism and endothermy might well be unsustainable because respiratory water-loss rates would frequently exceed tolerable levels, even in species living in non-desert environments (Hillenius 1992, 1994).

Mammalian respiratory turbinates have no analogues or homologues

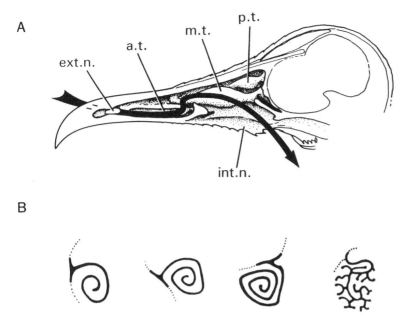

A

ext.n.

a.t.

m.t.

p.t.

int.n.

B

Coragyps   Fulmarus   Rhea   Dromaius

Figure 35.5. The nasal turbinates of birds. The arrow describes the path of air flow through the nasal cavity. Anterior and middle turbinates are respiratory in function; posterior turbinates are olfactory. (A) Right sagittal section through the nasal cavity of a gull, *Larus*. (B) Cross-sections through the middle (respiratory) concha of a vulture (*Coragyps*), a fulmar (*Fulmarus*), a rhea (*Rhea*), and an emu (*Dromaius*). Not to scale. After Hillenius 1994. Abbreviations: ext.n. = external nares; int.n. = internal nares or choanae; a.t., m.t. = anterior and middle (respiratory) turbinates; p.t. = posterior (olfactory) turbinates.

among any living reptiles or amphibians. In living reptiles, one to three simple nasal turbinates are present, but these are exclusively olfactory in function (Fig. 35.4). Like the mammalian olfactory turbinates (nasal or ethmoturbinates), these are typically located in the posterodorsal, olfactory portion of the nasal cavity. There are no structures in the reptilian nasal cavity (or the nasal cavity of any extant ectotherm) specifically designed for the recovery of respiratory water vapor, nor are they as likely to be needed. Reptilian lung ventilation rates are sufficiently low that pulmonary water loss rates seldom create significant problems, even for desert species (Fig. 35.2).

Complex turbinates comparable to those of mammals are also found in almost all birds (Bang 1971), and it is likely that they share a similar function as well (e.g., Schmidt-Neilsen et al. 1970). The extent and complexity of the nasal cavity of birds vary widely with the shape of the bill, but in general, the avian nasal passage is elongate with three cartilaginous, or sometimes ossified, turbinates in succession (Fig. 35.5). The anterior turbinate is often relatively simple, but the others, particularly the middle turbinates, are often more highly developed into prominent scrolls with multiple turns. Sensory (olfactory) epithelium is restricted to the posterior turbinate. Like mammalian olfactory turbinates, this structure is situated outside the main respiratory air stream, often in a separate olfactory chamber. Embryological and anatomical studies indicate that only the posterior turbinate is homologous to those of reptiles; the anterior and middle turbinates are new features that have evolved independently in birds from the respiratory turbinates of mammals (Witmer 1995).

The anterior and middle turbinates of birds, like the respiratory turbinates of mammals, are situated directly in the respiratory passage, and are covered primarily with respiratory epithelium. The position of these turbinates leaves them well positioned to modify bulk respired air. Reliable data

New Insights into Metabolic Physiology 511

suggest that these turbinates function as well as, or better than, mammalian respiratory turbinates in the recovery of water vapor contained in exhaled air (Schmidt-Neilsen et al. 1970). Consequently these structures in birds probably represent an adaptation to high lung ventilation rates and endothermy, fully analogous to respiratory turbinates of mammals.

To summarize, physiological data imply that independent selection for endothermy in birds, mammals, and/or their ancestors was, by necessity, tightly associated with the convergent evolution of respiratory turbinates in these taxa. In the absence of these structures, unacceptably high rates of pulmonary water loss would probably always have posed a chronic obstacle to maintenance of bulk lung ventilation consistent with endothermy, or with metabolic rates approaching endothermy.

Mere coincidence is unlikely to account for the striking anatomical and functional similarity of the independently derived avian and mammalian turbinate systems. This is because maintenance of the requisite alternating thermal countercurrent system at any point on the respiratory tree other

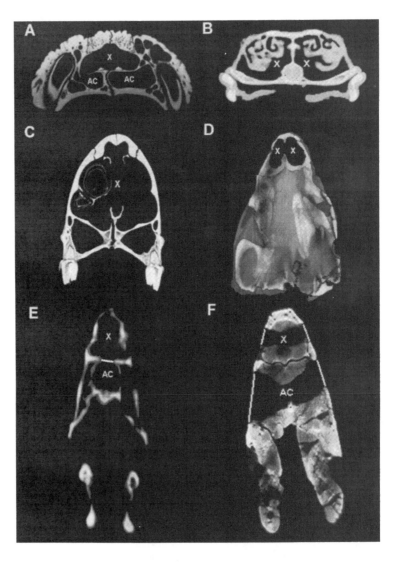

Figure 35.6. Cross-sectional, computed axial tomography (CT scans) of the nasal passage in (A) crocodile (*Crocodylus*), (B) ostrich (*Struthio*), (C) bighorn sheep (*Ovis*), (D) the tyrannosaurid theropod dinosaur *Nanotyrannus*, (E) the "ostrich-like" theropod dinosaur *Ornithomimus*, and (F) the lambeosaurine "duckbill" dinosaur *Hypacrosaurus*. Respiratory turbinates in mammals and birds are housed in voluminous nasal passages; note the complex structures of bone located in the chambers labeled "x" in parts B and C of this figure (see also Fig. 35.3). As in the alligator, the tube-like nasal passage (cavum nasi proprium) in the theropod dinosaurs appears to have been housed primarily within the maxillary and nasal bones. The main nasal passage in duckbill dinosaurs was probably an elongated nasal vestibulum, contained largely within the nasal bone. Some minimal post-depositional distortion of these fossils is evident. Nevertheless, the relatively narrow nasal passages in the dinosaurs indicate that, as suggested by their fossils, respiratory turbinates were probably absent in the living animals. Scale bar = 1 cm. Abbreviations: AC = accessory cavity; X = nasal passage proper.

than the nasal cavity would be untenable. Intermittently cool and warm countercurrent exchange sites in the body cavity would necessarily preclude deep-body homeothermy; an efficient tracheal ("windpipe") countercurrent system would inevitably result in chronic oscillation of brain temperature because of the proximity of the trachea to arterial blood bound for the brain (via the carotid circulation). Consequently the confirmed absence of respiratory turbinates, or similar structures, is likely to be strongly indicative of ectothermic, or near-ectothermic, rates of lung ventilation and metabolism, in any taxa, living or extinct.

The Metabolic Status of Some Dinosaurs

The confirmed presence or absence of respiratory turbinates is likely a bellwether indicator of lung ventilation and metabolic rates in virtually all terrestrial taxa, living or extinct. Although complex respiratory and olfactory turbinates are virtually ubiquitous among extant birds, these structures remain poorly known in fossil birds or their presumed ancestors the dinosaurs (but see Feduccia 1996 for a different view of avian origins). The widespread occurrence of respiratory turbinates among living birds suggests that these structures may predate the last common ancestor of extant avian taxa, a form that perhaps lived during the Late Cretaceous (Feduccia 1995, 1996), but it is uncertain how much older, or how widespread, these structures might be in the dinosaurian-avian lineage.

Several problems complicate the study of the evolutionary history of turbinates. Although they are occasionally preserved in extinct taxa (e.g., olfactory turbinates in phytosaurs [Camp 1930]; very early mammals [Lillegraven and Krusat 1991]), turbinates are highly fragile structures in living taxa, and are generally poorly preserved or absent in fossilized specimens. Furthermore, although they ossify or calcify in many extant taxa, these structures often remain cartilaginous in birds, which further decreases their chances for preservation. Nevertheless, we have determined

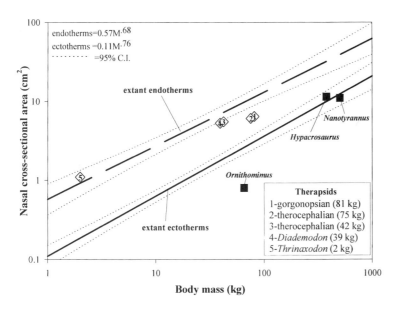

Figure 35.7. The relation of nasal passage proper cross-sectional area to body mass in modern endotherms (birds and mammals) and ectotherms (lizards and crocodilians), three genera of Late Cretaceous dinosaurs, and five genera of therapsids (values for dinosaurs and therapsids were not included in regression calculations). Dinosaur masses (estimated from head and/or body skeletal length): ostrich dinosaur, *Ornithomimus* (Theropoda: Ornithomimidae), 70 kg (Campanian Stage; Royal Tyrrell Museum of Palaeontology specimen 95.110.1); juvenile "duckbill" dinosaur, *Hypacrosaurus* (Ornithischia: Hadrosauridae), 375 kg (Maastrichtian Stage; American Museum of Natural History specimen 5461); tyrannosaurid dinosaur, *Nanotyrannus* (Theropoda: Tyrannosauridae), 500 kg (Maastrichtian Stage; Cleveland Museum of Natural History specimen 7541. Modified from Ruben et al. 1996.

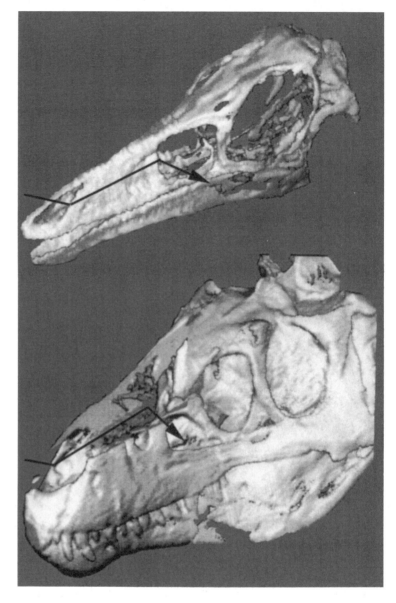

Figure 35.8. Three-dimensional CT scan representation (left lateral view) of the skulls of the *Ornithomimus* (above) and the tyrannosaurid dinosaur *Nanotyrannus* (below). Respiratory turbinates appear to have been absent. Arrows indicate the path of airflow through the nasal cavities of the skulls.

that the presence of respiratory turbinates in extant endotherms is inevitably associated with marked expansion of the proportionate cross-sectional area of the nasal cavity proper (Ruben et al. 1996) (Figs. 35.6, 35.7). Increased nasal passage cross-sectional area in endotherms probably serves to accommodate elevated lung ventilation rates as well as to provide increased rostral volume to house the respiratory turbinates. Significantly, relative nasal-passage diameter in a sequence of successively more recent, increasingly mammal-like therapsids approaches and, in the very mammal-like *Thrinaxodon*, even attains mammalian/avian nasal-passage cross-sectional proportions (Fig. 35.7).

The recent application of computed axial tomography, or CT scans, to paleontological specimens has greatly facilitated non-invasive study of fine

Ruben, Leitch, Hillenius, Geist, and Jones

details of the nasal region in fossilized specimens, especially those which have been "incompletely" prepared. In some cases, CT scans of particularly well-preserved specimens have revealed delicate remnants of calcified, cartilaginous, and/or lightly calcified cartilaginous structures. In the tyrannosaurid *Nanotyrannus* (Figs. 35.6–35.8), CT scans clearly demonstrate that in life this animal boasted particularly well-developed olfactory turbinates, but it was unlikely to have possessed respiratory turbinates: They are absent from the fossil, and, most important, nasal-passage cross-sectional dimensions are virtually identical to those in extant ectotherms (Figs. 35.6, 35.7). Additionally, CT scans of the nasal region of another theropod dinosaur, the ornithomimid *Ornithomimus,* as well as of the ornithischian dinosaur *Hypacrosaurus,* also indicate the presence of narrow, ectotherm-like nasal cavities, unlikely to have housed respiratory turbinates (Figs. 35.6, 35.7). In this way it is strikingly similar to the nasal region of many extant reptiles (e.g., *Crocodylus;* Figs. 35.6, 35.7) and is strong evidence for low lung ventilation rates and ectothermy, or near-ectothermy, in these dinosaurs.

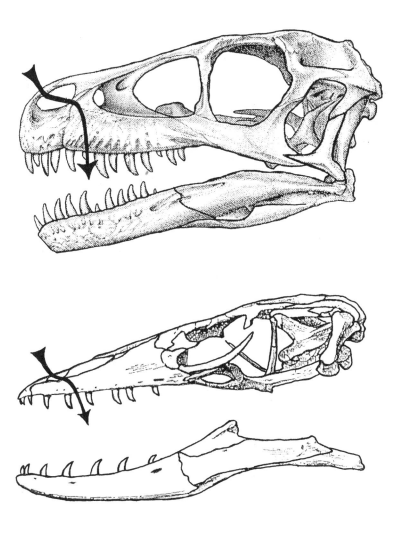

Figure 35.9. Lateral views of the skulls of the maniraptoran theropod dinosaur *Dromaeosaurus* (Theropoda: Dromaeosauridae) (*above*) and the monitor lizard *Varanus* (Squamata: Varanidae) (*below*). The arrow describes the likely path of air flow through the nasal region into the oral cavity. In *Dromaeosaurus,* and probably in some other dromaeosaurid dinosaurs (e.g., *Deinonychus*), the short, direct path of air flow into the oral cavity was similar to that in *Varanus* (and many other extant lizards) and almost certainly precluded sufficient space to have housed respiratory turbinates. Dromaeosaurid dinosaurs are often assumed to be the sister group of birds. Modified from Ruben et al. 1996.

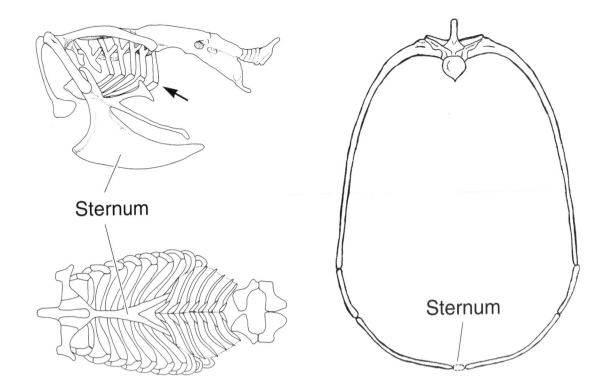

Figure 35.10. Portions of the axial skeletons of an alligator (*bottom left, ventral view*), a pigeon (*top left, lateral view*), and the maniraptoran dinosaur *Deinonychus* (*right, cross-sectional view*). Ventilation of the avian "high-performance," flow-through septate lung is tightly linked to highly mobile sternal rib–costal rib articulations (*arrow*) and a posteriorly elongated sternum, the posterior end of which moves dorsoventrally during ventilation. Dinosaurs, like alligators and many lizards, appear to have possessed relatively unmodified ribs and sternum, consistent only with a relatively simple, "bellows-like," septate lung. Such a lung would likely have been incapable of supporting mammal- or bird-like rates of oxygen consumption during extended periods of exercise.

In addition, the almost varanid lizard–like arrangement of the nostrils and choanae (openings of the nasal tract in the roof of the mouth) in some dromaeosaurs (e.g., *Deinonychus, Dromaeosaurus*) strongly suggests that respiratory turbinates were unlikely to have been present in these theropod taxa (Fig. 35.9).

Some Further Observations

The absence of respiratory turbinates in the varied taxa presented here suggests that ectothermic, or near-ectothermic, rates of lung ventilation and metabolism were possibly widespread in dinosaurs during periods of routine activity. These data provide little insight into dinosaurian capacities for oxygen consumption during periods of intense activity. However, the presence of a relatively unmodified thoracic skeleton (e.g., crocodile-like ribs; relatively short, immobile sternum) in many dinosaurs is consistent with a lizard- or crocodile-like septate lung that was probably incapable of avian or mammalian rates of gas exchange during periods of intense exercise (Perry 1983) (Fig. 35.10). In contrast, a high-performance, avian-style, "flow-through" septate lung is associated with highly flexible sternal rib–costal rib articulations, and a strongly modified sternum capable of extensive dorso-ventral movement (Fig. 35.10).

Together, these observations suggest that many dinosaurs lacked basic anatomical attributes causally linked to endothermic modes of existence in modern taxa. Nevertheless, it would be erroneous to conclude that dinosaurs were therefore necessarily similar in lifestyle to most modern temperate-latitude reptiles (i.e., sluggish herbivores or "sit-and-wait" predators). The dynamic skeletal structure of many dinosaurs strongly suggests that they possessed a bird- or mammal-like capacity for at least burst activity. Moreover, even if they were fully ectothermic, if dinosaurs possessed aerobic metabolic capacities and predatory habits equivalent to those of some modern tropical-latitude varanid lizards (e.g., *Varanus komodoensis*), they may well have maintained large home ranges, actively pursued and killed large prey, and defended themselves fiercely when cornered.

References

Bakker, R. T. 1971. Dinosaur physiology and the origin of mammals. *Evolution* 25: 636–658.

Bakker, R. T. 1980. Dinosaur heresy-dinosaur renaissance. In R. D. K. Thomas and E. C. Olson (eds.), *A Cold Look at the Warm-Blooded Dinosaurs*, pp. 351–462. American Association for the Advancement of Science, Selected Symposium 28. Boulder, Colo.: Westview Press.

Bakker, R. T. 1986. *The Dinosaur Heresies: New Theories Unlocking the Mysteries of the Dinosaurs and Their Extinction*. New York: William Morrow.

Bang, B. 1971. Functional anatomy of the olfactory system in 23 orders of birds. *Acta Anatomica* 79: 1–71.

Barrick, R. E., and W. J. Showers. 1994. Thermophysiology of *Tyrannosaurus rex*: Evidence from oxygen isotopes. *Science* 265: 222–224.

Bennett, A. F. 1991. The evolution of activity capacity. *Journal of Experimental Biology* 160: 1–23.

Bennett, A. F., and B. Dalzell. 1973. Dinosaur physiology: A critique. *Evolution* 27: 170–174.

Bennett, A. F., and R. B. Huey. 1990. Studying the evolution of physiological performance. In D. Futuyma and J. Antonovics (eds.), *Oxford Surveys in Evolutionary Biology*, vol. 7, pp. 251–284. Oxford: Oxford University Press.

Bennett, A. F., and J. A. Ruben. 1979. Endothermy and activity in vertebrates. *Science* 206: 649–654.

Camp, C. L. 1930. *A Study of the Phytosaurs*. Berkeley: University of California Press.

Chinsamy, A. 1990. Physiological implications of the bone histology of *Syntarsus rhodesiensis* (Saurischia: Theropoda). *Paleontologia Africana* 27: 77–82.

Chinsamy, A. 1993. Bone histology and growth trajectory of the prosauropod dinosaur *Massospondylatus carinatus* (Owen). *Modern Geology* 18: 319–329.

Farlow, J. O. 1990. Dinosaur energetics and thermal biology. In D. B. Weishampel, P. Dodson, and H. Osmólska (eds.), *The Dinosauria*, pp. 43–55. Berkeley: University of California Press.

Farlow, J. O.; P. Dodson; and A. Chinsamy. 1995. Dinosaur biology. *Annual Review of Ecology and Systematics* 26: 445–471.

Feduccia, A. 1995. Explosive evolution in Tertiary birds and mammals. *Science* 267: 637–638.

Feduccia, A. 1996. *The Origin and Evolution of Birds*. New Haven, Conn.: Yale University Press.

Hillenius, W. J. 1992. The evolution of nasal turbinates and mammalian endothermy. *Paleobiology* 18: 17–29.

Hillenius, W. J. 1994. Turbinates in therapsids: Evidence for Late Permian origins of mammalian endothermy. *Evolution* 48: 207–229.

Hopson, J. A. 1973. Endothermy, small size and the origin of mammalian reproduction. *American Naturalist* 107: 446–452.

Kolodny, Y.; B. Luz; M. Sander; and W. A. Clemens. 1996. Dinosaur bones: Fossils or pseudomorphs? The pitfalls of physiology reconstruction from apatitic fossils. *Palaeogeography, Palaeoclimatology, Palaeoecology* 126: 161–171.

Lillegraven, J. A., and G. Krusat. 1991. Craniomandibular anatomy of *Haldanodon exspectatus* (Docodontia; Mammalia) from the Late Jurassic of Portugal and its implications to the evolution of mammalian characteristics. *Contributions to Geology, University of Wyoming* 28: 39–138.

Nagy, K. A. 1987. Field metabolic rate and food requirement scaling in mammals and birds. *Ecological Monographs* 57: 111–128.

Nagy, K., and C. C. Peterson. 1988. Scaling of water flux in animals. *University of California Publications in Zoology* 120: 1–172.

Owerkowicz, T., and A. W. Crompton. 1995. Bone of contention in the evolution of endothermy. *Journal of Vertebrate Paleontology* 15 (Supplement to no. 3): 47A.

Padian, K. 1983. A functional analysis of flying and walking in pterosaurs. *Paleobiology* 9: 218–239.

Perry, S. F. 1983. Reptilian lungs: Functional anatomy and evolution. *Advances in Anatomy, Embryology and Cell Biology* 79: 1–81.

Ruben, J. A. 1991. Reptilian physiology and the flight capacity of *Archaeopteryx*. *Evolution* 45: 1–17.

Ruben, J. A. 1995. The evolution of endothermy: From physiology to fossils. *Annual Review of Physiology* 57: 69–95.

Ruben, J. A.; W. J. Hillenius; N. R. Geist; A. Leitch; T. D. Jones; P. J. Currie; J. R. Horner; and G. Espe III. 1996. The metabolic status of some Late Cretaceous dinosaurs. *Science* 273: 1204–1207.

Schmidt-Nielsen, K.; F. R. Hainsworth; and D. E. Murrish. 1970. Counter-current heat exchange in the respiratory passages: Effect on heat and water balance. *Respiratory Physiology* 9: 263–276.

Witmer, L. M. 1995. Homology of facial structure in extant archosaurs (birds and crocodilians), with special reference to paranasal pneumaticity and nasal turbinates. *Journal of Morphology* 225: 269–327.

The Scientific Study of Dinosaur Footprints

*James O. Farlow
and
Ralph E. Chapman*

Perhaps the most vivid impression of a dinosaur as a living creature comes not from seeing a mounted skeleton, but instead from the examination of a well-preserved trackway. Looking at footprints made, one after the other, by a dinosaur going about its business millions of years ago gives one an almost palpable sense of the trackmaker as a real animal—as opposed to some analytical construct created from the ruminations and speculations of an imaginative paleontologist.

Edward Hitchcock must have experienced similar feelings. He was an early-nineteenth-century American geologist who conducted pioneering studies of the fossil footprints of the Connecticut Valley of the eastern United States (see Colbert, chap. 3 of this volume), and whose scientific work influenced the thinking of such literary figures of his day as Herman Melville, Emily Dickinson, Henry David Thoreau, and Henry Wadsworth Longfellow (Dean 1969). In his monumental description of this early Mesozoic footprint fauna (Fig. 36.1), the *Ichnology of New England* of 1858, Hitchcock wrote:

> What a wonderful menagerie! Who would believe that such a register lay buried in the strata? . . . At first men supposed that the strange and gigantic races which I had described, were mere creatures of imagination, like the Gorgons and Chimeras of the ancient poets. But now that hundreds of their footprints, as fresh and distinct as if yesterday impressed upon the mud, arrest the attention of the sceptic on the ample slabs of our cabinets, he might as reasonably doubt his own corporeal existence as that of these enormous and peculiar races. (P. 190)

36

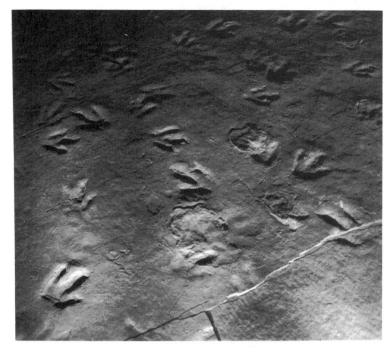

Figure 36.1. Early Jurassic (East Berlin Formation) dinosaur footprints, Dinosaur State Park, Rocky Hill, Connecticut. The footmarks are about 30 to 40 cm long. The makers of these prints are here regarded as theropods, but Weems (1987, 1992) argues that large early Mesozoic dinosaur footprints such as these were made by prosauropods. Similar three-toed footmarks, both large and small, occur at many tracksites in early Mesozoic rocks of eastern North America. Prints like these were studied by the pioneering ichnologist Edward Hitchcock.

Dinosaur footprints provide a source of information about dinosaurs that is at least partly independent of evidence from skeletal material. This chapter describes how dinosaur footprints are studied, how we identify the kinds of dinosaurs that made fossilized tracks, how and why dinosaur footprints are given scientific names of their own, and what such trace fossils tell us about how their makers walked and ran. Lockley (chap. 37 of this volume) will give more information about how dinosaur footprints were preserved, and describe the use of footprint assemblages in reconstructing dinosaur faunas.

Before we begin, however, a brief note about terminology is in order. In everyday English, the word *footprint* has an unambiguous meaning. The word *track*, however, can refer either to an individual footprint or to a sequence of footprints made by the same animal. To avoid confusion, we will use the terms *footprint, print,* and *footmark* here as synonyms to refer to a mark made by an individual foot. The word *trackway* refers to a sequence of footprints made by the same animal. Finally, the term *track* can be used more generically, when it isn't necessary to specify whether a single footprint or a trackway is being discussed.

How Dinosaur Footprints Are Studied: Field Work

One of the most important tasks in describing a dinosaur footprint site is to record the positions of the footprints relative to each other. Footprints made by the same animal need to be recognized, and the direction of travel of each animal at the site recorded. The best way to preserve such information is to make a map of the site (Fig. 36.2).

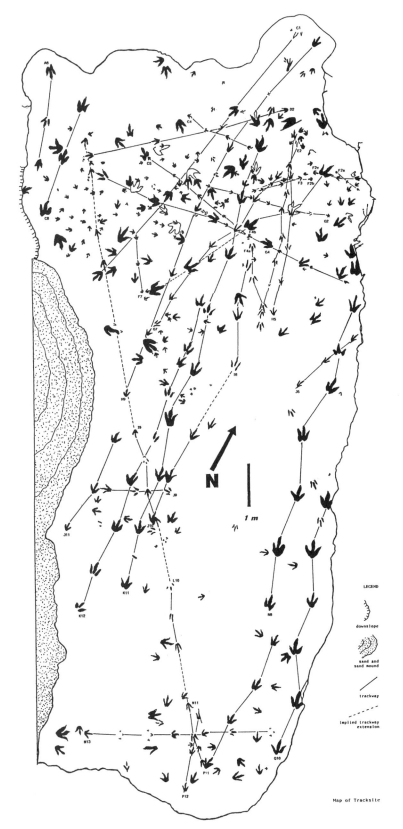

Figure 36.2. Map of an Early Jurassic dinosaur footprint site from the Moenave Formation, northeastern Arizona. The linear scale is 1 m long. Reproduced from Irby 1996a.

The Scientific Study of Dinosaur Footprints 521

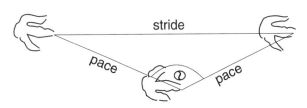

Figure 36.3. Typical measurements made on dinosaur trackways. The pace is the distance between two successive footprints made by the opposite feet, and the stride is the distance between two successive marks of the same foot. The pace angulation Θ measures how nearly prints of the left and right foot are in a single line; the pace angulation is 180 degrees if the animal puts its left and right feet directly in front of each other as it walks. The pace angulation also serves as an indirect indication of how wide the trackway is, and therefore may provide information about the breadth of the trackmaker's body, or whether it walked in an erect or sprawling fashion. Other measurements commonly made include the length of each separate toemark, the angles between toes, and the angle made between individual footprints and the trackmaker's overall direction of travel. For quadrupedal dinosaurs, the position of prints of the front and hind feet can be used to estimate the distance between the trackmaker's shoulders and its hips. For more information about footprint and trackway measurements, see Leonardi 1987.

This can be done in many different ways. One simple method is to lay a square grid on the tracksite surface, with subdivisions of the grid made at fixed intervals. A similar grid is drawn—obviously at a much smaller scale!—in a field notebook; arithmetic graph paper also works well for this. The individual footprints can then be sketched in the notebook at the proper distance from, and orientation relative to, intersections in the grid, and at the proper scale to provide information about their size. A more elaborate version of this method is to photograph each square on a square grid drawn on the tracksite, prepare a photomosaic of the entire track surface, and then trace the position of each print from the photomosaic onto a sheet of paper or plastic.

An alternative way of making a map is to measure the distance of each footprint from some fixed landmark on the tracksite, along with the compass bearing from the landmark to the print. The size and compass heading of each footprint are also measured. These measurements can then be used to draw the position, orientation, and size of each track on a map.

Some ichnologists (scientists who study footprints) make relatively simple, diagrammatic maps; other, more artistically talented ichnologists prepare maps in which the features of each individual footprint are beautifully rendered. Regardless of how the map is constructed, as many photographs of the tracksite should be taken as possible. These can be used to check the accuracy of the map, and also provide a permanent record of the tracksite.

For detailed studies of the movements of trackmakers, additional measurements are useful (Fig. 36.3). The distance between successive prints of the animal's left and right feet (the pace) and the distance between two successive footprints of the same foot (the stride) obviously tell us something about how the animal was moving; the longer the animal's steps, the more quickly it was moving. The extent to which left and right footprints of the animal are aligned, or widely separated from one another, may indicate how wide-bodied or narrow-bodied the trackmaker was, and whether it had an erect or a sprawling carriage. The angle made by the longest dimension of the footprint with respect to the animal's direction of travel indicates whether the trackmaker walked with its feet angling outward or pointed straight ahead, or with its toes turned inward.

In order for the makers of footprints to be identified as accurately as possible, the shapes of individual prints must be recorded as faithfully as possible. Measurements of the overall size of each footprint, of the lengths of the toes (and of segments of each toe), and of the angles between toes are often made. This can be done either in the field, from the footmarks themselves, or later in the laboratory, from casts or photographs of the prints. Photographs of individual footprints are best taken from directly overhead, and not obliquely; photographing footprints from an angle distorts their shape (Fig. 36.4). Another confounding factor is the lighting;

522   *Farlow and Chapman*

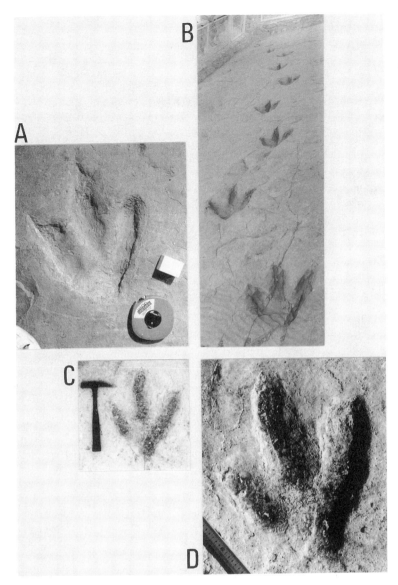

Figure 36.4. Different ways of photographing dinosaur footprints in the field. (A) An Early Jurassic theropod print (East Berlin Formation), Dinosaur State Park, Rocky Hill, Connecticut. Footmark length about 35 cm. The print is viewed from directly overhead, to provide as accurate an impression of its shape as possible. The tape indicates the size of the footmark, and the shadow of the little cardboard box shows the direction of lighting in the photograph. (B) Oblique view of an Early Cretaceous theropod trackway (Enciso Group), Los Cayos, La Rioja, Spain. The prints are about 45 cm long. Such an oblique view shows the arrangement of the footprints in the trackway, but it greatly distorts the shape of individual prints. (C) Overhead view of an Early Cretaceous theropod footmark (Fort Terrett Formation), F[6] Ranch site, Kimble County, Texas, as seen when the sun is directly overhead. The footprint is about 37 cm long. (D) Overhead view of the same footprint seen in C, but here photographed at dusk, under low-angle illumination.

a footmark can look quite different at dusk, when the sun's light is at a low angle to the tracksite, than it does at noon, when the sun is directly overhead.

Because of this, it is a good idea to make as many casts of the footprints as is practical, to take back to the lab for further study. The medium used for casting tracks depends on the preservation of the footprints, and also on the ichnologist's budget. Some fossil tracks are preserved as depressions in the surface across which the trackmaker walked; others are preserved as natural casts from sediments that later filled the footprints. If the footmarks are simple depressions, with no undercuts (where toes or edges of the track extend farther into the rock, below the surface of the track-bearing layer, than they do at the surface), a simple, quick, and cheap way to make a cast is to grease the rock surface with some separator (such as petroleum jelly),

The Scientific Study of Dinosaur Footprints   523

Figure 36.5. Different ways of
illustrating the shape of a
dinosaur footprint. (A) Photograph of a print of the right foot
of a large carnivorous dinosaur
in situ in the limestone bed of
the Paluxy River at Dinosaur
Valley State Park, Glen Rose,
Texas. The footmark is about 50
cm long. (B) Photograph of a
cast of the same footprint (cast
made by Peggy Maceo). Because
the cast was made by filling the
actual print with the casting
medium, the cast reverses the
topography of the real footprint.
Note the "notch" along the
inner edge of the footmark (on
the right of the print in this cast,
arrow); this indentation occurs
because the inner digit did not
impress over its entire length.
(C) Artist's drawing of a cast of
the same footprint (drawing by
Jim Whitcraft). (D) Computer-
drawn topographic map of the
same footprint cast. The surface
of the cast was read into a
computer file using a three-
dimensional digitizer, after which
software converted the raw data
into the form shown here.
(E) and (F) Wire-net diagrams
showing two perspective views
of the footprint. Each of the
methods of illustrating the
footprint has its desirable
features, and also its shortcom-
ings. Photographs and computer
images are more objective
portraits of the footmark than
drawings are, but may not
emphasize features that the
ichnologist considers important.
Drawings can be prepared in
such a way as to emphasize
features not obvious in photographs or computer images, but
may not accurately portray
the overall shape of the print.
Whenever possible, it is desirable
to illustrate a footprint by more
than one method.

and then pour a batter of plaster of Paris into it. Once the plaster cures, you have your cast.

If there are undercuts, however, it is better to make a latex rubber cast of the footprint. This involves putting down thin coats of liquid or paste latex on the print, one after another, allowing time for each coat to dry before the next is added. After several such layers of latex have cured, strips of burlap or cheesecloth can be pressed into a new coat of latex while it is still wet, and then still more coats of latex are painted onto the cloth. The cheesecloth or burlap makes the cast resistant to tears, and also gives it a bit of rigidity. Working with latex in this manner can obviously be rather time-consuming.

Once several layers of latex have been applied to the footprint and allowed to cure, the cast can be peeled away from the rock surface. It should then be put back against the print, and a rigid backing of fiberglass, burlap, or cheesecloth dipped in plaster is made for the cast, so that it will retain its shape and not collapse under its own weight. Alternatively, the latex cast can be filled with expandable foam, or even with a mushy mixture of latex and tissue paper that will later harden. Making a support for a latex cast is especially important for casts of very large footprints. Latex eventually deteriorates (as the existence of innumerable unplanned children—including the first author of this chapter—proves). It is consequently a good idea to make permanent copies of latex casts of footprints in some other medium.

When footmarks are preserved as natural casts, they also can be copied, although it is often a trickier procedure. If, for example, natural casts of footprints project from the undersurface of a rock ledge, it is pretty hard to make plaster copies of them.

Latex and plaster are not the only suitable materials for copying foot-prints in the field or lab, but they are relatively cheap and readily available. Other media, such as silicone rubber, can also be used, but these are often fairly expensive.

Sometimes ichnologists even collect the actual footprints themselves, although this can be a labor-intensive and time-consuming enterprise. Probably the most ambitious project of this kind was the collection of some 40 tons of dinosaur footprint–containing rock of Early Cretaceous age from the bed of the Paluxy River near Glen Rose, Texas, by Roland T. Bird of the American Museum of Natural History (Bird 1985). Trackways of a sauropod and large theropod were removed in pieces, and later reassembled as an exhibit in the museum.

Laboratory Work

Field work is enjoyable, but it is only the start. The real work begins when maps, measurements, photographs, and casts are brought back to the lab for more detailed study. Casts of footprints can be photographed under more controlled conditions in the lab than in the field, and other ways of illustrating the footmarks are employed as well (Fig. 36.5). Individual footprints from the same trackway are compared, to see how variable tracks made by the same animal are. Measurements of the tracks in one trackway are then compared with those of other trails, to see how similar or different they are. Footprint shapes are compared with foot skeletons of various kinds of dinosaurs, to determine the kind(s) of dinosaur likely to have made the prints, and with previously described footprints from other tracksites, to try to determine if the same or

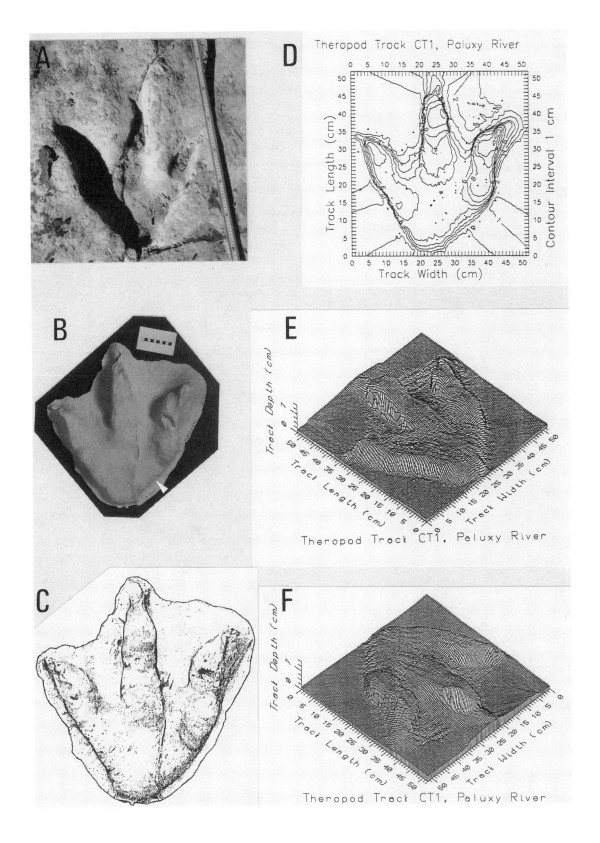

A

B

C

D Theropod Track CT1, Paluxy River

Track Length (cm)

Contour Interval 1 cm

Track Width (cm)

E Theropod Track CT1, Paluxy River

Track Depth (cm)

Track Length (cm)

Track Width (cm)

F Theropod Track CT1, Paluxy River

Track Depth (cm)

Track Length (cm)

Track Width (cm)

The Scientific Study of Dinosaur Footprints   525

different kinds of animals are likely to have been making tracks at the various sites.

If the footprints from a new site are significantly different from those that have previously been named in the scientific literature, it is necessary to prepare a formal description of the new footprints, and to give them their own name. Frequently, too, an ichnologist may find it necessary to re-study previously described footprints and tracksites, and to rename previously described trace fossils in some way. In any case, the final result of field and lab work is the preparation of a report for publication that describes the tracksite and its footprints, so that scientists can add this information to the overall database about dinosaurs.

Identifying the Makers of Dinosaur Footprints: The Remarkable Birds of Professor Hitchcock

At the time that Edward Hitchcock described the Early Jurassic dinosaur footprints of the Connecticut Valley, he had no reason to suppose that his trackmakers had been dinosaurs. He knew that animals such as *Iguanodon* had existed during the Mesozoic Era (Hitchcock 1858: 175), but his image of what dinosaurs had looked like undoubtedly reflected Richard Owen's interpretation of dinosaurs as huge, four-footed (quadrupedal) animals (see Torrens, chap. 14 of this volume). The three-toed footmarks of the Connecticut Valley had obviously been made by two-legged creatures. Hitchcock concluded that most of these three-toed prints could have been made only by big, flightless birds, like living emus, ostriches, rheas, or cassowaries. Hitchcock's confidence in his interpretation of the nature of the Connecticut Valley trackmakers was bolstered (Hitchcock 1858: 76–79, 178) by one of Owen's scientific triumphs: Owen's recognition that New Zealand had once been the home of several species of large flightless birds, the moa (Gruber 1987). The former existence of moa (the word *moa* is used as both a singular and a plural), and of the elephantbird on Madagascar, combined with the continued existence of modern flightless birds, seemed ample reason for attributing the Connecticut Valley footprints to similar creatures. Ironically enough, Hitchcock's studies of the New England footprints had a minor reciprocal influence on Owen's research, for Owen named some of his moa species after specific names of Connecticut Valley footprints (Anderson 1989).

Figure 36.6. Footprints known as *Gigandipus* (described by Edward Hitchcock under the name *Gigantitherium*) in the collection of the Pratt Museum, Amherst College, Massachusetts. Note the groove running the length of the slab, from footprint to footprint, interpreted as a tail drag mark. The tracks are about 40 to 45 cm long.

Hitchcock's reconstruction of a vanished world populated by a diversity of huge birds gripped the imagination of his literary contemporaries. In an essay from *The Encantadas*, Herman Melville described himself as being like "an antiquary of a geologist, studying the bird-tracks and ciphers upon the exhumed slates trod by incredible creatures whose very ghosts are now defunct." In a poem titled "To the Driving Cloud," Henry Wadsworth Longfellow (whose father-in-law helped fund Hitchcock's work; Dean 1969) imagined an ancient race of Native Americans who "once by the margin of rivers stalked those birds unknown, that have left us only their footprints."

Even so, Hitchcock had lingering doubts about the avian nature of some of the Connecticut Valley trackmakers. One of Hitchcock's footprint taxa, which he called the *Gigantitherium* (now known as *Gigandipus*; Lull 1953), combined bird-like footprints with a most un-bird-like tail drag mark (Fig. 36.6). Hitchcock speculated (1858: 93, 179–180) that the *Gigantitherium*-maker had been a huge, bipedal, bird-like lizard or amphibian; "if a biped, its body must have had somewhat the form of a bird, in order to keep it properly balanced . . . how very strange must have been the appearance of a lizard, or batrachian [amphibian in modern terms], with feet and body like those of a bird, yet dragging a veritable tail!" (180). Another of Hitchcock's footprint taxa, *Anomoepus,* combined a three-toed, bird-like foot with a long hind-foot "heel" and a five-toed forefoot that reminded Hitchcock very strongly of the forefeet of marsupial (pouched) mammals such as kangaroos (Fig. 36.7). Hitchcock found *Anomoepus* very hard to interpret:

> If a phrase so compound could be used as should imply a participation in the characters of marsupials, birds, batrachians, and even lizards, it would better than any other express my present convictions; for the longer I study these tracks, the stronger my impressions become that some of these ancient animals possessed characters now more exclusively belonging to two, three, or even four [vertebrate] classes. (1858: 60).

To be fair to Hitchcock's thinking, and to see it in its proper historic context, it must be emphasized that he did not interpret the mixture of characters of different vertebrate classes in his footprints in an evolutionary framework; Hitchcock remained committed throughout his life to the concept of special creation (Guralnick 1972; Lawrence 1972).

Ironically, although Hitchcock was not an evolutionist, the discovery of *Archaeopteryx,* the earliest known bird, gave him considerable satisfaction. Here was a bird of very "low grade" whose skeleton showed the very features that Hitchcock had identified, and found troubling, in the Connecticut Valley footprints. In a posthumous *Supplement* to his *Ichnology* of 1858, Hitchcock concluded that the *Anomoepus*-maker must have been a "low grade" bird like *Archaeopteryx,* and "if we can presume that the *Anomoepus* was a bird, it lends strong confirmation to another still more important conclusion, which is that all the fourteen species of thick-toed

Figure 36.7. (A) Hind-foot prints (*Anomoepus*) of a small ornithischian dinosaur of early Mesozoic age in the collection of the Pratt Museum at Amherst College. The small scale bars are 1 cm long. The specimen shown here is unusual, in that it appears to have been made by a sitting animal, thus giving the prints long "heel" marks (not the anatomical heel but the foot bones proper: dinosaurs walked "tiptoe"). Even though the *Anomoepus*-maker usually walked bipedally, footmarks of this kind are occasionally associated with five-toed impressions of the hand. (B) Trackway of a small kangaroo, Idalia National Park, Queensland, Australia. The photograph shows two sets of prints. Paired hand-shaped manus prints are positioned behind each pair of two-toed pes marks. Edward Hitchcock initially thought that the *Anomoepus*-maker had been a bird-like marsupial, but the discovery of *Archaeopteryx* convinced him that it had instead been a very primitive bird. See Figure 36.10A–F for the hand and foot skeletons of animals similar to the *Anomoepus*-maker.

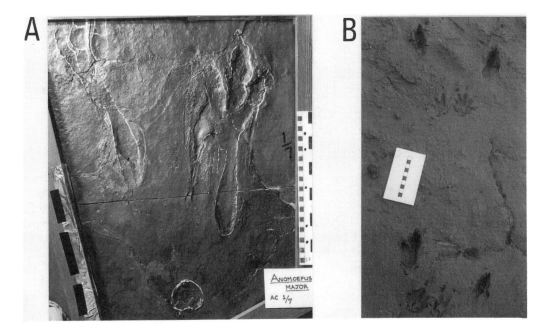

bipeds which I have described in the Ichnology, and in this paper, were birds" (Hitchcock 1865: 32).

In 1858, the same year that Hitchcock's *Ichnology* was published, Joseph Leidy of the Academy of Natural Sciences of Philadelphia described the skeleton of an ornithopod dinosaur that he named *Hadrosaurus*. Noting that the hind limbs of this animal were considerably longer than its forelimbs, Leidy suggested that *Hadrosaurus* and other dinosaurs might have been bipeds (Glassman et al. 1993), an idea that had actually been anticipated by Gideon Mantell for *Iguanodon* (see Torrens, chap. 14 of this volume). This conclusion was amply confirmed in the decades that followed, with the discovery of a diversity of bipedal theropods, ornithopods, and other dinosaurs.

It didn't take long for naturalists to realize the implications of the existence of bipedal dinosaurs for the interpretation of the Connecticut Valley three-toed footprints. In 1867, a note in the *Proceedings of the Academy of Natural Sciences of Philadelphia* reported that Edward Drinker Cope "gave an account of the extinct reptiles which approached the birds" in their structure (Anonymous 1867: 234); Cope concluded that "the most bird-like of the tracks of the Connecticut sandstone" had been made by dinosaurs. The following year, in a spirited discussion of the implications of *Archaeopteryx* and the small theropod dinosaur *Compsognathus* for the correctness of the theory of evolution in general, and the evolutionary derivation of birds from dinosaurs in particular, Thomas Henry Huxley noted that the Connecticut Valley footprints proved that "at the commencement of the Mesozoic epoch, bipedal animals existed which had the feet of birds, and walked in the same erect or semi-erect fashion. These bipeds were either birds or reptiles, or more probably both" (Huxley 1868: 365). The implication was that many of these trackmakers might have been dinosaurs, rather than birds in the strict sense. In 1877, Othniel Charles Marsh of Yale University went even further, asserting that all of the Connecticut Valley three-toed footprints had been made by dinosaurs rather than birds.

Hitchcock had come very close to deducing the body form of bipedal dinosaurs from his study of the early Mesozoic footprints of New England. He had called the trackmakers birds and not dinosaurs, but given the little that was known about dinosaurs when he was studying his tracks, "bird" was as accurate an identification of the trackmakers as was possible for him to have made.

Identifying Trackmakers

Although modern paleontologists have a much clearer understanding of what dinosaurs were like than Hitchcock did, deciding exactly what kind of dinosaur was responsible for a particular footprint still can be tricky. Except for those rare instances in which dinosaur "mummies" preserve replicas of desiccated soft tissues of the foot (such as the *Corythosaurus* described by Brown [1916]), we do not know what the soft parts of the feet of dinosaur taxa—which are described on the basis of skeletal material—looked like when those animals were alive. Consequently, when we say that a given footprint is a theropod footmark, for example, what we are really saying is that the print has a shape consistent with what we think would have been made by the soft tissues around the bones of a theropod's foot when pressed into sediment. This conclusion is based on what we know about the skeletal structure of the theropod foot, and the relationship between the foot skeleton and its enveloping soft tissues in modern vertebrates, particularly crocodilians and birds, the closest living relatives of dinosaurs.

Some aspects of trackmaker identification are easier than others. One important determination in identifying the maker of a trackway is whether the animal walked quadrupedally or bipedally. Some dinosaurs, including most theropods and pachycephalosaurs, were habitual bipeds, while other dinosaurs, such as sauropods, stegosaurs, ankylosaurs, and ceratopsians, probably walked on all fours nearly all the time. Ornithopods and prosauropods may have used both quadrupedal and bipedal gaits with some regularity.

The shape and size of the individual footprints in a trail are also important. The various groups of quadrupedal dinosaurs differed among themselves in the size of the forefoot compared with that of the hind foot, and also in the number and shape of toes on each foot (Fig. 36.8). The many

Figure 36.8. Manus (forefoot) and pes (hind foot) skeletons of quadrupedal dinosaurs. (A) and (B) Right manus and pes of the prosauropod *Plateosaurus*. (C) and (D) Right manus and pes of the sauropod *Apatosaurus*. (E) and (F) Right manus and pes of the stegosaur *Kentrosaurus*. (G) and (H) Right manus and pes of the ankylosaur *Talarurus*. (I) and (J) Left manus and pes of the ceratopsian *Centrosaurus*. Scale bar is 10 cm long. Figures redrawn from Lull 1933; Gilmore 1936; Galton 1990; and Thulborn 1990.

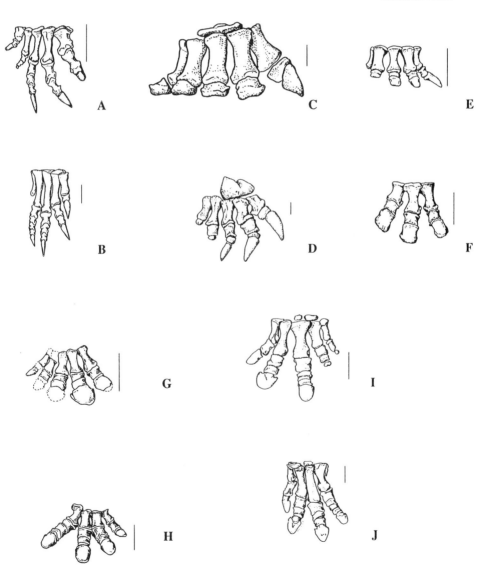

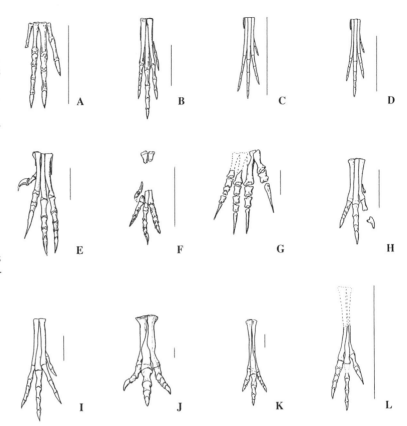

Figure 36.9. Manus and pes skeletons of theropod dinosaurs. (A) and (B) Right manus and pes of the ceratosaur *Coelophysis*. (C) Right pes of the small theropod *Compsognathus*. (D) Right pes of the ceratosaur *Procompsognathus*. (E) Left pes of the dromaeosaur *Deinonychus*. (F) Left pes of the bird(?) *Mononykus*. (G) Right pes of the therizinosaur (segnosaur) *Erlikosaurus*. (H) Right pes of the oviraptorosaur *Ingenia*. (I) Right pes of the elmisaur *Chirostenotes*. (J) Left pes of the tyrannosaur *Tyrannosaurus* (*Tarbosaurus*). (K) Left pes of the ornithomimosaur *Struthiomimus*. (L) Right pes of the troodont *Borogovia*. Scale bar is 10 cm long. Figures redrawn from Ostrom 1969, 1978; Maleev 1974; Colbert 1989; Barsbold and Maryańska 1990; Barsbold et al. 1990; Barsbold and Osmólska 1990; Currie 1990; Osmólska and Barsbold 1990; and Perle et al. 1994.

kinds of four-footed dinosaurs also differed among themselves in overall adult body size. A well-preserved trackway consisting of very large, rather elephant-like footprints, in which the hind-foot footmarks are noticeably bigger than those of the forefoot, is most likely to have been made by a sauropod. On the other hand, a trackway consisting of prints of more modest size, in which the forefoot impressions have five toes and the hind-foot footmarks four, could very well have been made by an ankylosaur.

Alas, nature isn't always that cooperative. Footprints of quadrupedal dinosaurs are often poorly preserved, and sometimes can be recognized as prints only because they are arranged in an obvious trackway pattern. Sometimes interpreting the major group to which the trackmaker belonged is little more than an educated guess. Even if one is fairly certain that a particular trackway was made by, say, a sauropod, it is usually impossible to say what kind of sauropod known from skeletal material the trackmaker was. Indeed, in many cases the actual trackmaker may have belonged to a species whose skeletal remains have not yet been discovered.

The same problems apply to the identification of three-toed (tridactyl) prints of bipedal dinosaurs (Figs. 36.9, 36.10). Most (but perhaps not all; Harris et al. 1996) theropods made three-toed prints during normal locomotion. Many bipedal ornithischians were likewise tridactyl, but some forms had a long first toe that might or might not generally have touched the ground. Bipedally walking prosauropods may have made a four-toed footprint (Lockley 1991). However, Weems (1987, 1992) suggested that some of the larger three-toed tracks of the early Mesozoic (such as those shown in

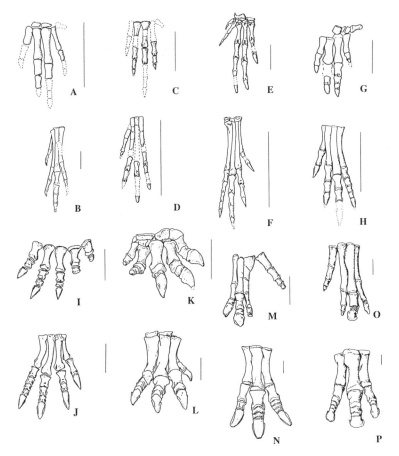

Figure 36.10. Manus and pes skeletons of ornithopods and other potentially bipedal ornithischians. (A) and (B) Left manus and pes of the basal ornithischian *Lesothosaurus* (*Fabrosaurus*). (C) and (D) Left manus and pes of the basal thyreophoran *Scutellosaurus*. (E) and (F) Right manus and pes of the basal ornithopod *Heterodontosaurus*. (G) and (H) Left manus and pes of the hypsilophodontid *Hypsilophodon*. (I) and (J) Left manus and right pes of the tenontosaur *Tenontosaurus*. (K) and (L) Right manus and pes of the iguanodont *Camptosaurus*. (M) and (N) Left manus ("thumb" not shown) and pes of the iguanodont *Iguanodon*. (O) and (P) Right manus and pes of the hadrosaur *Edmontosaurus*. Scale bar 2.5 cm in A, B, C, E, and G, 10 cm in all other cases. Figures redrawn from Gilmore 1909; Thulborn 1972; Galton 1974; Santa Luca 1980; Colbert 1981; Norman 1986; Forster 1990; and Thulborn 1990.

Fig. 36.1) were made by prosauropods, either specialized tridactyl forms whose skeletal remains have yet to be found, or *Plateosaurus*-like prosauropods that walked with the inner toe of the hind foot held above the ground. Are there any criteria that could serve to distinguish among tridactyl footprints of theropods, bipedal ornithischians, and prosauropods (if any three-toed prosauropod footmarks exist)?

Well-preserved three-toed prints often have conspicuous swellings, or digital nodes, at places along the lengths of the toemarks. At least the larger of these nodes, particularly those nearest the bases of the toes, often correspond to joints between adjacent foot and toe bones, or between toe bones (Heilmann 1927; Peabody 1948; Baird 1957; Thulborn 1990; Fig. 36.11). This makes it possible to estimate the lengths of individual toe bones (formally known as phalanges) of the foot skeleton of a trackmaker from the pattern of the digital nodes, although this should be done only with caution. The pattern of digital nodes does not always reflect the pattern of the foot skeleton with perfect fidelity (notice the outer toes of the rhea foot shown in Fig. 36.11). Once the foot skeleton is reconstructed from the digital node pattern, this reconstruction can be compared with real foot skeletons of three-toed bipedal dinosaurs.

It is therefore interesting to look at the proportions of foot skeletons of various kinds of dinosaurs that may have made tridactyl footprints, to see

The Scientific Study of Dinosaur Footprints   531

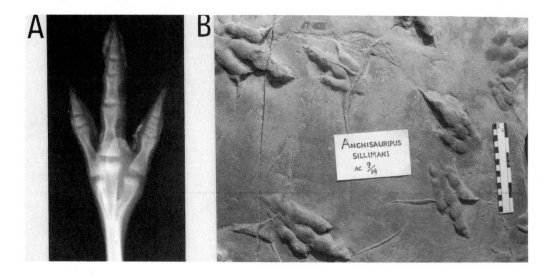

Figure 36.11. (A) X-ray of the right foot of a young greater rhea (*Rhea americana*), a flightless bird from South America. Joints between toe bones (phalanges) are clearly visible. In the middle toe (digit III), swellings (digital nodes or pads) in the soft tissues of the toe correspond to joints between phalanges. This correspondence is less obvious in the toes of the outer digits (II and IV), which have rather short phalanges, all of which are incorporated in the same digital pad of each toe. (B) Natural cast of beautifully preserved small theropod footprints (*Anchisauripus* or *Grallator*—see Weems 1992) from the collection of the Pratt Museum. Notice the well-developed nodes in the footprints' toemarks.

if there are any differences in foot skeletal structure among groups of potential trackmakers that might be discernible in footprints with well-preserved digital nodes. A few such distinctive features do show up (Figs. 36.12, 36.13). Theropods tend to have relatively longer toe bones in the middle portion of each toe, compared with the bones at either end of the toe, than do other potential three-toed trackmakers. Theropods also tend to have relatively shorter toe claws than do ornithischians or prosauropods (Fig. 36.13; cf. Farlow and Lockley 1993).

There are even suggestions that subgroups within the major bipedal dinosaur categories can be distinguished from one another on the basis of foot proportions. Ceratosaurs, tyrannosaurids, and ornithomimids plot in different regions of morphological space in a principal-components analysis (Fig. 36.12; see Chapman, chap. 10 of this volume, for a discussion of this analytical technique). Hadrosaurids and *Iguanodon* plot well away from other groups, because of the exaggeratedly short middle phalanges of their toes. *Iguanodon* and *Tenontosaurus* plot higher along the Factor 3 axis than do other ornithopods, because of their disproportionately long toe claws.

However, there is overlap among groups. The graphs suggest that one would probably not be able to distinguish footprints of tyrannosaurids from those made by *Allosaurus* and its kin. The region of morphological space occupied by hypsilophodontids overlaps or is adjacent to that of *Psittacosaurus* and several small theropods. Note, too, that hypsilophodontids and *Tenontosaurus* plot close to the region occupied by prosauropods. It might therefore be difficult to discriminate among footprints of some of these dinosaurs on the basis of the proportions of the three main toes of the hind foot alone (although dromaeosaurids would probably not ordinarily leave impressions of the long claw of the second digit). One would have to hope for the presence of additional criteria, such as the occasional impressions of digit I of the foot, or tracks made by the hand, to identify trails made by members of some of these dinosaur groups. Even with this additional information, it might be hard to tell trackways of some of these kinds of dinosaurs apart (e.g., the various groups of small ornith-

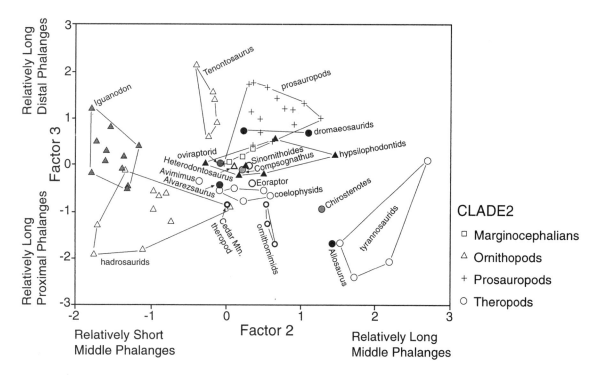

Figure 36.12. Principal-components analysis (PCA) of foot skeletons of potential makers of three-toed dinosaur tracks. The first factor, related to the overall size of each specimen, has been removed. Factor 2 compares the lengths of bones (phalanges) from the middle portion of a toe with the lengths of proximal (closest to the leg) and distal (farthest from the leg) bones of the foot. Factor 3 compares the lengths of the most distal bones of the foot, including the claw bones (unguals), relative to the most proximal bones of the toes. "Marginocephalians" here is *Psittacosaurus*. Theropods tend to have relatively short distal phalanges, while those of prosauropods and many ornithopods are relatively long. Bones from the middle portions of toes are relatively long in theropods, while those of hadrosaurids, *Tenontosaurus,* and *Iguanodon* are extremely short. There is considerable overlap between hypsilophodontids and certain small theropods.

ischians). Although this is not depicted in the graphs shown here, foot skeletons of members of the same genus sometimes plot more closely to each other than to foot skeletons of other members of the same group. In other cases, however, foot skeletons of members of the same genus plot as far away from each other as they do from foot skeletons of other genera in the group. Even when it is possible to assign a footprint to a particular group of bipedal dinosaurs on the basis of toe-bone proportions estimated from the print, it may not be possible to be certain about which genus within the group was the trackmaker.

If tridactyl prosauropod footprints exist, the way that prosauropod points plot in morphological space (Figs. 36.12–36.14) suggests that three-toed prosauropod prints would more likely be mistaken for tracks of tridactyl ornithischians than for tracks made by theropods. In fact, the hind foot of *Plateosaurus* is remarkably similar to that of *Tenontosaurus* (Figs. 36.8B, 36.10J), which suggests that confusing tracks made by bipedal prosauropods with those made by (for now hypothetical) early Mesozoic large, bipedal ornithischians could in theory be a real problem, regardless of the number of toes on the hind foot! Interestingly enough, one of Hitchcock's footprint types, *Otozoum,* was made by a four-toed biped that has recently been interpreted as a prosauropod (Lockley 1991; Farlow 1992), an ornithopod (Thulborn 1990), and a thyreophoran ornithischian (Gierliński 1995).

It must be emphasized that the preceding comments about the possibility of identifying the makers of tridactyl dinosaur footprints from foot proportions apply only to those instances where footprints are so well preserved that they clearly show digital pads, from which the relative lengths of individual toebones can be estimated. Quite often three-toed footmarks do not show distinct toe pads. Are there any criteria by which

The Scientific Study of Dinosaur Footprints   533

the makers of tridactyl footprints can be identified on the basis of gross print shape?

One set of potential characters involves comparisons of the lengths of the three toemarks (often made in descriptions of dinosaur footprints; cf. Moratalla et al. 1988, 1992; Demathieu 1990; Gierliński 1988, 1994; Casanovas Cladellas et al. 1993; Gierliński and Ahlberg 1994). Many tridactyl footprints show a conspicuous indentation along the inner margin of the footprint, near the rear of the print (Fig. 36.5B; cf. Thulborn 1990), which can be useful for identifying whether a footmark was made by the left or the right foot of the dinosaur. This indentation exists because the inner toe, digit II, did not touch the ground over its entire length. The back end of the impression of digit II was usually made by the joint between the first and second phalanges of that toe. In like manner, the most proximal indication of the middle toe of the foot (digit III) also represents the joint between phalanges 1 and 2. In contrast, digit IV usually did impress over its entire length, and so the back end of its impression—and often the back of the footprint itself—was made by the joint between the first phalanx of the toe and the fourth metatarsal bone of the foot.

Consequently the toemarks of digits II and III will generally consist of the combined lengths of phalanges 2–3 and 2–4, respectively, while the toemark of digit IV will be made by the combined lengths of all five phalanges of that toe. Comparing the relative lengths of the three main toes of the foot (Fig. 36.14) results in some separation of theropods from other major kinds of potential tridactyl footprint-makers, but there is a lot of overlap among groups.

Large theropods tend to have relatively narrower phalanges than do big ornithopods (Figs. 36.9, 36.10), and so a big, three-toed print with long, skinny toemarks (Fig. 36.5) was more likely made by a theropod than an ornithopod, while a big footmark with three relatively short, thick toe-

Figure 36.13. Plot of the ratio of the length of phalanx III2 (the second bone of the third toe of the foot) to the length of phalanx IV1 (the first phalanx of the fourth toe) against the ratio of the length of the ungual of digit III (III4) to the length of phalanx III2. In tridactyl dinosaurs, the second toe is the innermost of the three large walking toes, the third toe is the middle digit, and the fourth toe is the outer digit. Phalanx III2 is near the midpoint of the third digit, and phalanx IV1 is the most proximal bone of the fourth toe. Consequently the ratio of the lengths of phalanx III2/phalanx IV1 is related to factor 2 in the PCA shown in Figure 36.12. Similarly, the III4/III2 ratio is related to factor 3 in the PCA. "Basal Ornithischians" here is Lesothosaurus. Theropods tend to have a relatively long III2 in comparison with IV1 and III4, while prosauropods and ornithischians tend to do just the opposite.

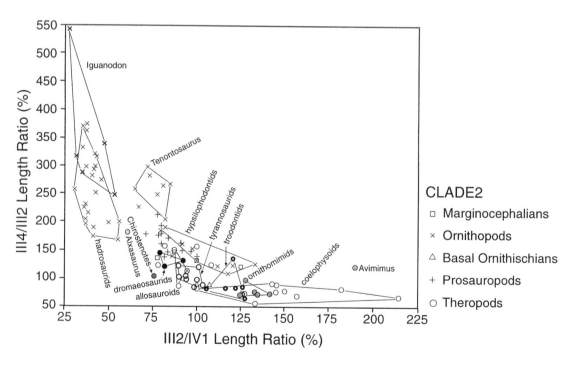

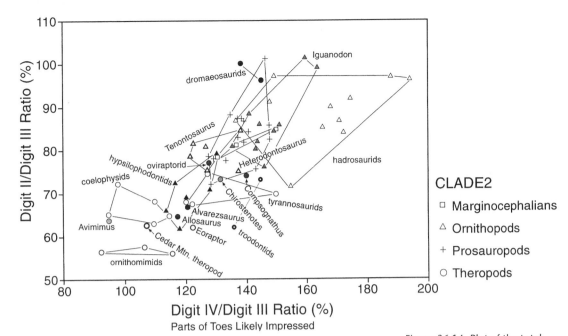

Figure 36.14. Plot of the total length of the portion of digit II (phalanges 2 and 3) that was likely to have touched the ground during normal walking against the total length of digit IV, both as a percentage of the length of that portion of the middle toe (digit III) likely to be recognizably impressed in a footprint (phalanges III2-III4). Theropods tend to have a relatively long middle toe, but do show overlap with other groups, such as hypsilopho-dontids. The huge ungual on digit II in dromaeosaurids puts them well away from other theropods.

marks (Fig. 36.15) probably had an ornithopod maker. However, many of the smaller bipedal ornithischians had relatively long, narrow toes (Fig. 36.10), and so toemark slenderness might not be very helpful in distinguishing prints of small theropods from those of small ornithischians. It is not surprising, for example, that the maker of an early Mesozoic footprint known as *Atreipus* has been identified as an ornithischian and as a theropod by different paleontologists (Thulborn 1990).

Many tridactyl dinosaurs, both theropods and ornithischians, have an outer toe (digit IV) that is conspicuously narrower than digits II and III (Fig. 36.10). However, the troodontid *Borogovia* and the early bird(?) *Mononykus* have a very stout digit IV and a relatively slim central toe (digit III). Such differences in digit thickness might be recorded in very well-preserved footprints.

Ichnologists have proposed additional criteria for identifying the makers of tridactyl dinosaur footprints that are less readily related to the skeletal structure of the foot (Thulborn 1990, 1994; Lockley 1991; Weems 1992). These include such features as the overall width/length ratios of footprints, the angles between toemarks, the angles made between footprint long axes and the trackmaker's direction of travel, and trackway pace angulations (Fig. 36.3). Such parameters may well be useful, but probably do not permit as confident identification of the trackmaker as do parameters more directly related to skeletal structures.

The job of identifying dinosaurian trackmakers can be made somewhat easier by a consideration of the stratigraphic occurrences of dinosaur footprints. For example, the Early Cretaceous Glen Rose Limestone of Texas has yielded beautifully preserved sauropod footprints (Fig. 36.16). We cannot say for sure what kind of sauropod made those tracks. However, the fact that skeletons of a brachiosaurid known as *Pleurocoelus* have been found in rocks of the same age, in the same region, and that what is known

The Scientific Study of Dinosaur Footprints   535

of the foot skeleton of *Pleurocoelus* is consistent with the shape of the footprints, suggests that *Pleurocoelus* was very likely the Glen Rose Limestone's sauropod footprint-maker. Similarly, large theropod footprints from the Glen Rose Limestone are quite likely to have been made by a big allosaurid, *Acrocanthosaurus*, whose skeletal remains are found in rocks of the appropriate age from the same region. Using the same reasoning, an enormous tridactyl footprint with relatively narrow clawmarks from latest Cretaceous sediments of New Mexico could well have been made by the gigantic Maastrichtian theropod *Tyrannosaurus* (Lockley and Hunt 1994).

We can also make inferences about what dinosaurs are unlikely to have been the trackmakers of any particular stratigraphic unit. Because skeletal remains of tyrannosaurids are known only from rocks of Late Cretaceous age, it is unlikely that any of the Early Jurassic large theropod footprints of the Connecticut Valley were made by those particular large theropods. Tyrannosaurids are not presently known from Late Cretaceous rocks of South America, where abelisaurids were the common later Mesozoic large theropods; it is therefore more likely that Late Cretaceous large theropod footprints from South America were made by abelisaurids than by tyranno-saurids. For similar reasons we can be skeptical that any Late Cretaceous footprints of quadrupedal dinosaurs were made by prosauropods (a group that is presently known only from the Triassic and Jurassic) walking on all fours. These are, of course, statements only of what is likely, and not of what is known to be true with complete certainty. It is possible that someone will someday find the skeletal remains of an Early Jurassic tyrannosaurid or a

Figure 36.15. Computer-drawn topographic map of a large ornithopod footmark in the collection of the Royal Tyrrell Museum of Palaeontology, Alberta, Canada. Note the relatively short, thick toemarks; compare this shape with that of the large theropod print shown in Figure 36.5.

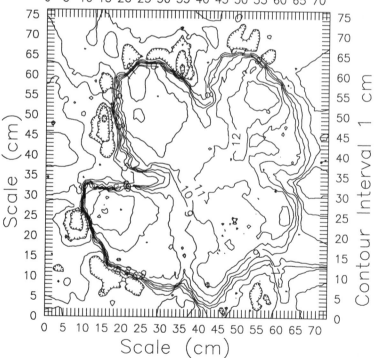

Cretaceous prosauropod—but it seems very unlikely. It is somewhat less improbable that, in rocks of Late Cretaceous age, remains of a tyrannosaurid will someday turn up in South America.

Sometimes, however, stratigraphic considerations make the identification of trackmakers more complicated. Three-toed footprints similar to typical theropod footmarks have been found in Middle Triassic rocks that are older than the most ancient skeletal fossils of carnivorous dinosaurs (Haubold 1984; Demathieu 1989; Arcucci et al. 1995—but see King and Benton 1996). Either theropods evolved rather earlier than we presently think, or these footprints were made by non-dinosaurian archosaurs with very theropod-like feet and gaits. If the latter is true, it raises the question of whether any Late Triassic or even Early Jurassic footprints that we presently regard as having been made by theropods could instead have been made by archosaurs that were closely related to, but were not quite, dinosaurs.

Figure 36.16. Sauropod trackway from the bed of the Paluxy River, Dinosaur Valley State Park, Glen Rose, Texas. Tracks of this kind are known by the name *Brontopodus*. Superimposed on many of the sauropod prints are tridactyl footmarks of a big theropod that may have been following the sauropod. Photograph by R. T. Bird.

The Scientific Study of Dinosaur Footprints   537

Identifying the makers of dinosaur tracks in any particular stratigraphic unit will always involve an element of uncertainty. We will probably never be able to reconstruct the composition of dinosaur faunas on the basis of dinosaur footprint assemblages with as much confidence as is possible for well-preserved skeletal assemblages (but see Lockley, chap. 37 of this volume, for a less pessimistic view). On the other hand, for stratigraphic units with few skeletal remains, footprints will remain our only window on those otherwise lost dinosaur communities.

Naming Dinosaur Footprints

Hitchcock's *Ichnology* of 1858 was preceded by several shorter articles about the Connecticut Valley tracks. In some of these earlier works, Hitchcock coined names that he applied specifically to the footprints themselves. By 1845, however, Hitchcock had concluded that names such as *Gigantitherium* and *Anomoepus* were suitable tags not just for the footprints, but also for the creatures that made them. This belief reflects the confidence that early-nineteenth-century naturalists had in the predictive powers of comparative anatomy. The great French anatomist Georges Cuvier had correctly predicted, from the scant skeletal evidence initially visible at the surface of the enclosing rock, that a certain fossil mammal would turn out to be a marsupial once its bones were more fully exposed (Rudwick 1976: 113–116). Richard Owen had similarly been able to deduce the former existence of moa as huge flightless birds on the basis of a fragmentary femur. Hitchcock therefore felt confident in asserting,

> The grounds on which I propose to name and describe the animals that made the fossil footmarks, are derived from comparative anatomy and zoology. These sciences show a mathematically exact relation to exist, not only between different classes and families of animals, but between different parts of the same animals. . . . Why then, in view of such facts, should we not name and describe an unknown animal, though nothing but its tracks remain from which to judge of its nature? For in truth a track . . . scarcely differs from the foot petrified; and if Cuvier's principle be true, it will generally give us a tolerably correct idea of the other parts of the body. Why can we not construct the whole animal from this petrified foot, as well as the anatomist can from a single bone belonging to some other part of the frame? (1858: 23–24)

The unfortunate answer to Hitchcock's rhetorical question is contained in the discussion of the previous section of this chapter. Although it may be possible to get a "tolerably correct idea" of the kind of animal responsible for a particular footprint, this may seldom be accurate enough to allow us to say exactly which kind of beast made the track. We would probably not be able to distinguish footprints of one species and perhaps even genus of hadrosaurid, or tyrannosaurid, from another.

Ichnologists have therefore reverted to Hitchcock's initial approach, in which scientific names are created to describe distinctive kinds of footprints themselves, rather than the animals that made them. Footprints are named as ichnogenera and ichnospecies, to distinguish them from genera and species named, in the case of fossil vertebrates such as dinosaurs, on the basis of skeletal material (cf. Baird 1957; Demathieu 1970; Haubold 1984; Sarjeant 1989, 1990; Thulborn 1990; Lockley 1991; Leonardi 1994; Sarjeant and Langston 1994). There are, however, differences in opinion as to the criteria by which the footprints of dinosaurs and other fossil vertebrates should be named.

Vertebrate footprints are only one kind of trace fossil; other vertebrate trace fossils include such things as eggs, coprolites, and bite marks (see Chin, chap. 26, and Hirsch and Zelenitsky, chap. 28, of this volume). Furthermore, vertebrates are not the only kinds of organisms that left traces of their activities in ancient sediments; a diversity of invertebrates, plants, and even microorganisms did so as well. Bottom-living (benthic) invertebrates, both marine and freshwater, create marks on or in sediments (or even hard rock) as they crawl, bore, burrow, and feed on the the floors of seas and lakes. Frequently ichnologists cannot identify even the phylum of the animal responsible for a particular invertebrate trace fossil. On the other hand, the kinds of animal activities reflected in invertebrate trace fossils that occur in sedimentary rocks can provide very useful information about the environmental conditions under which the sediments accumulated (Osgood 1987; Maples and West 1992). Consequently ichnologists who study invertebrate traces name these sedimentary structures on the basis of the kind of tracemaker behavior recorded in the trace, and not on the basis of the generally unknowable taxonomic affinities of the tracemaker (Sarjeant and Kennedy 1973; Basan 1979; Magwood 1992). A marine invertebrate can make one kind of trace while sitting motionless on the sea floor, and an entirely different trace while crawling about. Following the procedure of ichnologists who study invertebrate traces, such different traces of the same animal are given different names.

Some ichnologists (e.g., Sarjeant 1990; Sarjeant and Langston 1994) advocate using the same approach when it comes to recognizing and naming vertebrate ichnotaxa. Footprints of a given characteristic shape should be given a name regardless of whether they can be correlated with a particular category of trackmaker. Furthermore, if footprints made by the same kind of animal differ in shape under different circumstances (such as when the beast is running, as opposed to walking), those different footmarks made by the same kind of animal should be given different names.

Most ichnologists agree that vertebrate footprints that have a characteristic morphology indicative of a distinctive foot structure should be named, regardless of whether the taxonomic affinities of their maker can be ascertained. The controversy arises over whether differences in footprint shape caused by differences in behavior should also be used as a basis for naming footprints, as is done with invertebrate traces.

We are skeptical about the usefulness of doing this for two reasons. First of all, even though, as already discussed, there can be uncertainty as to the maker of a particular vertebrate track, this level of uncertainty is at a much lower taxonomic level for vertebrates than for most benthic invertebrates (Schult and Farlow 1992). Being unable to tell footmarks of theropods of one family from those of another family, or even theropod prints from ornithopod tracks, is a rather different thing from being unable to distinguish burrows made by "worms" of one phylum from those of a different phylum. Tracks of dinosaurs and other vertebrates will usually preserve a greater amount of information about the taxonomic affinities of their makers than invertebrate traces will.

Secondly, differences in benthic invertebrate traces, reflecting as they do differences in the behavior of the tracemakers, preserve useful information about ecological conditions (such as the distribution of food in sediments) of the environments in which the traces were made (Osgood 1987). In contrast, differences in the shape of footprints of a particular kind of vertebrate caused by differences in the creature's behavior, or differences in the way the prints were formed and preserved, will seldom reveal anything particularly interesting about the environments (apart from substrate conditions) in which the tracks were made (Schult and Farlow 1992). Whether a dinosaur was walking, running, hopping, sitting, or limping may tell us nifty things about what the animal was up to, but these will not reflect anything ecologically significant about the environment in which the animal was doing those things. We therefore don't see anything useful to be gained from naming vertebrate ichnotaxa in exactly the same way that invertebrate trace fossils are named.

All of this leads to a basic question: What are ichnotaxa for? Or, what is the goal, and what is to be gained, by giving names to vertebrate trace fossils?

As noted by Lockley (1991, and chap. 37 of this volume), in some stratigraphic units there is a better record of dinosaurs and other Mesozoic vertebrates in the form of footprints than in body fossils. The early Mesozoic rocks of eastern North America (such as the rocks of the Connecticut Valley that Hitchcock studied) are a good example.

Suppose that we had studied such a rock unit that was rich in dinosaur footprints but had few dinosaur bones. Suppose, too, that we wanted to compare the dinosaur fauna inferred from the footprints of our rock unit with dinosaur faunas inferred from footprints, or known from skeletons, from some other geographic area or stratigraphic interval. If in naming ichnotaxa we had used footprint shape differences created by differences in trackmaker behavior, and not just track shape differences that reflected differences in foot structure of the trackmakers, then a list of ichnotaxa from our rock unit might show very little correspondence to the kinds of animals, and the diversity of animal taxa, that had actually inhabited the region. We would not be able to use the ichnotaxa to say very much about whether the dinosaur fauna represented by our footprint assemblage was similar to, or different from, dinosaur faunas of other places and times.

Many ichnologists therefore prefer an approach to naming vertebrate footprints (cf. Baird 1957; Olsen and Galton 1984; Farlow et al. 1989; Lockley et al. 1994; Lockley and Hunt 1995a) that is somewhere in the middle of the positions represented by Hitchcock's view of 1858 and the approach adopted by students of invertebrate traces. Footprints should be given names only when they display a distinctive shape that is unlikely to be an artifact of their formation, a shape that to some extent reflects the skeletal structure of the trackmaker. Furthermore, the foot structure inferred from the footprint should be different from the foot structure inferred from previously named ichnotaxa. The names given to footprints are not the same as those given to the creatures that made them, however. For example, the name *Brontopodus* was assigned to the sauropod tracks from the Glen Rose Formation that very likely (but not certainly) were made by *Pleurocoelus* (Farlow et al. 1989).

This approach has the advantage that a list of vertebrate ichnotaxa will bear a closer relationship to the fauna of animals that made the tracks than will a list compiled on the basis of the approach used by students of invertebrate traces. There will be a correlation between skeletal taxa and ichnotaxa at some taxonomic level. What that level will be will probably vary from case to case. As noted by Olsen and Galton (1984: 94), "The ichnogenus *Grallator* [another of Hitchcock's footprint names] could have been made by any of the conservative members of the . . . Theropoda from any part of the Mesozoic. The ichnogenus *Anomoepus* could have been made by cursorial members of the families Fabrosauridae, Hypsilophodontidae, some Iguanodontidae, Psittacosauridae, and some Leptoceratopsidae."

Inspection of Figures 36.12–36.14 gives some support to that view. The presently available data on tridactyl dinosaur hind-foot proportions suggest that one ichnotaxon (perhaps an ichnogenus) could be justified for footprints made by *Tenontosaurus*, should such tracks ever be discovered. Another ichnotaxon might well incorporate all tracks of hadrosaurids. Yet another ichnotaxon could be created for tyrannosaurid (and other large theropod?) footprints, and still other ichnotaxa erected for tracks made by ornithomimids and ceratosaurs. An ichnotaxon that would likely include all the shapes of tracks made by different kinds of hypsilophodontids would probably also have to include prints made by *Heterodontosaurus* and psittacosaurs, and possibly other small bipedal ornithischians as well. Thus a list of ichnotaxa from a rock unit named on the basis of the approach advocated in this chapter will probably underestimate the diversity of animals in the living fauna, but it will also be as realistic a reflection of the number and relationships of animal taxa in that fauna as can be obtained from footprint evidence.

Unfortunately, Figures 36.12–36.14 also suggest a more disturbing possibility: that a particular ichnotaxon might have to include prints made by unrelated groups of dinosaurs. Note once more the overlap between small ornithischians and some small theropods in the graphs. Although this concern is raised by comparisons of foot skeletons, and not footprints, there is another reason for thinking that we are dealing with more than a hypothetical problem.

Over the course of the Cenozoic Era, numerous groups of ground-living (often flightless) birds evolved in many parts of the world (Feduccia 1996): ratites (ostriches, rheas, cassowaries, emus, kiwis, moa, and elephantbirds); tinamous; gastornithiforms (*Diatryma* and its kin); gruiforms (bathornithids, seriemas, phorusrhacoids, Messel rails, eogruids, ergilornithids, the adzebill, bustards, cranes, and rails); large flightless pigeon-like birds (the dodo and the solitaire); big flightless ducks and geese; galliforms (pheasants, turkeys, megapodes, and their kin); secretarybirds and similar birds of prey; and mihirungs (large to gigantic flightless birds of uncertain relationships from the Australian region), to name some of the more impressive forms.

The evolutionary relationships of these various groups of ground birds are not completely understood. Most notably, it is still uncertain whether the various living ratite species are more closely related to each other (as suggested by molecular evidence; Cooper et al. 1992; Cooper and Penny 1997) than to other, non-ratite bird groups (Feduccia 1996).

That being the case, comparing footprints made by different kinds of ground-living birds (Fig. 36.17) raises some interesting questions. If the ratites do constitute a monophyletic group, it is intriguing to note that the two-toed ostrich makes a print very different from that of other ratites—a footmark that in gross appearance might resemble prints made by certain extinct gruiform birds, the similarly two-toed ergilornithids, more closely than it does the footprints of other ratites. Emus and cassowaries are universally believed to be close relatives, but the enormous claw on digit II of the cassowary foot makes cassowary prints differ more in gross appearance from emu footprints than the latter do from prints of rheas, moa, mihirungs, and bustards.

Footprints of two taxa of trackmakers can resemble each other for one of the following reasons: (1) both taxa retain the primitive foot shape of a common ancestor, a primitive foot shape that they also share with other taxa; (2) both taxa have evolved a similar foot shape that they share with no other taxa because they are part of a monophyletic group (cf. Holtz and

The Scientific Study of Dinosaur Footprints 541

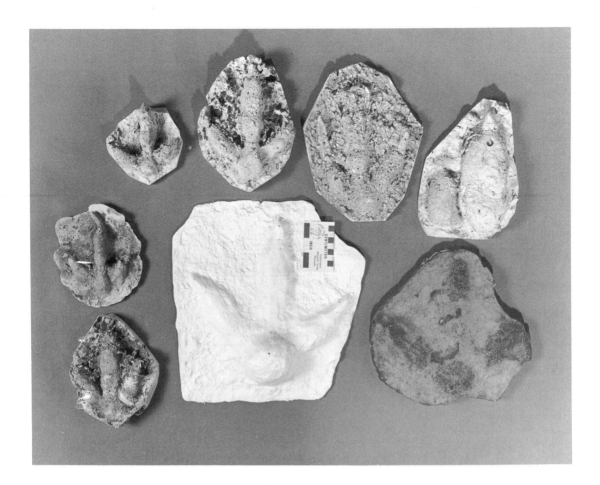

Figure 36.17. Casts of footprints of living and extinct species of large ground-living birds. *Clockwise from the upper left-hand corner:* Right footprint of a young emu (*Dromaius novaehollandiae*); right footprint of an adult emu; left footprint of a cassowary (*Casuarius casuarius*—note the enormous clawmark on the inner toe, digit II [on the left side of the cast in this view]; the same digit in the adjacent emu footprint is on the right side of the cast); right footprint of an adult ostrich (*Struthio camelus*—note the absence of the digit II, which would have been to the right of the large toemark for digit III in this view); right(?) footprint of a mihirung (dromornithid) from Tasmania; right(?) footprint of a moa from New Zealand; left footprint of a lesser

(continues)

Brett-Surman, chap. 8 of this volume); (3) the two taxa are not close relatives, but have independently evolved a similar foot shape because of functional similarities associated with a comparable mode of life; (4) one taxon retains the primitive foot form, and the other taxon has reverted to that foot shape from a more derived condition, perhaps because it has returned to the way of life of its ancestors.

Which of these is the correct explanation for similarities in footprint form probably varies from case to case, whether we are talking about dinosaur or bird footprints. The key point, however, is that most systematists now contend that natural phylogenetic groups should be recognized only on the basis of shared derived features (see Holtz and Brett-Surman, chap. 8 of this volume). Consequently there will be congruence between groupings of taxa on the basis of foot shapes and groupings of taxa in monophyletic clades only if foot shapes consistently change at the same time as do the characters used to define monophyletic groups (characters that could include, in the case of dinosaurs, the rest of the skeleton as well as the foot). Figures 36.12–36.14 and 36.17 suggest that this is frequently not the case.

Footprint taxa are usually distinguished from one another, or considered to be similar to each other, on the basis of overall differences or similarities in shape. Whether similarities in print shape are due to the second reason given above (the only reason that would involve monophyletic groups; Olsen 1995) or to one of the other three explanations cannot always (or perhaps even often) be determined.

That being the case, there will inevitably be an unavoidable lack of resolution in correlations between dinosaur skeletal and footprint taxa (although we once again emphasize that this will not be as great a problem as in invertebrate ichnotaxonomy). This will significantly affect the resolution we can hope for in reconstructing the composition of ecological communities, making biostratigraphic correlations, or interpreting biogeographic patterns on the basis of dinosaur ichnotaxa.

For example, Lockley (chap. 37 of this volume) suggests that the presence of the same theropod ichnotaxa in Eurasia and North America indicates the presence of land connections between these landmasses throughout the Jurassic Period. We do not question the idea that it was possible for terrestrial vertebrates to move relatively easily from one landmass to another at that time. However, the presence of the same ichnogenus on two or more continents will provide convincing evidence for intercontinental connections only if footprints from the different regions are placed in the same ichnogenus because similarities in foot shape paralleled shared derived characters that would indicate membership of the trackmakers from the different regions in a monophyletic clade. In the present example, we would have to be confident that the large theropods in North America responsible for the relevant ichnogenera were more closely related to the Eurasian makers of the same ichnogenera than either group was to other kinds of large theropods. Whether footprint shapes can yield this much taxonomic resolution is arguable. As we have already seen, closely related birds and dinosaurs do not necessarily have similar foot shapes, while distantly related bird and dinosaur taxa can have foot shapes hard to tell apart. On a brighter note, though, if footprint assemblages from different regions share many of the same ichnotaxa (cf. Olsen and Galton 1984), the chances that all of these similarities are misleading should be greatly reduced.

Regardless of just how much useful taxonomic information can be extracted from well-preserved footprints, we suspect that many—perhaps even most—dinosaur footprints do not preserve enough information about the skeletal foot structures of their makers to warrant their being given names. We doubt, for example, that many of the tridactyl dinosaur ichnotaxa that have been named over the years are really different enough from each other to be very useful proxies for the kinds of animals that made them, apart from such generalized identifications as "theropod footprint" or even "bipedal dinosaur print." We think it best to err on the side of caution. Giving a formal name to a vertebrate trace fossil creates the impression that there is something anatomically distinctive about that fossil. We advocate creating such names only in those (rare?) instances where this really is true.

(or Darwin's) rhea (*Rhea pennata*); right footprint of a kori bustard (*Ardeotis kori*). Although the emu is more closely related to the cassowary than to the other birds, its footprint is as much or more like that of the kori bustard. Similarly, the rhea, emu, cassowary, and moa may be more closely related to each other and to the ostrich than to the kori bustard or the mihirung, but one would be hard-pressed to see this in comparisons of footprint shapes. This suggests that footprints belonging to the same ichnotaxon will not necessarily be made by animals that are closely related, and that animal species that are close relatives might make footprints that would be placed in different ichnotaxa.

Footprints and Dinosaur Locomotion

Much of our understanding about what dinosaurs were like as living animals is based on interpretations of how dinosaur skeletons functioned as living machines (see Alexander, chap. 30 of this volume). One aspect of such interpretations has to do with how dinosaurs walked and ran. Reconstructions are made of the kinds of movements possible at joints between bones (based on the shapes of the surfaces of those bones). The postures of living dinosaurs are inferred from the way that limb bones articulate with each other and with the shoulder or hip girdles. Interpretations of how strenuous the activities of dinosaurs were can be made from calculations of the mechanical strengths of the bones involved.

As useful as such functional morphological studies are, it is nice to have an independent source of information for testing hypotheses about dinosaur biomechanics. Dinosaur trackways provide such an independent line of evidence.

In the early 1900s, a bitter controversy arose over the body carriage of sauropod dinosaurs (Desmond 1976). For some time, American paleontologists, following the lead of O. C. Marsh, had reconstructed the limbs of sauropods as held in an erect, elephant-like fashion. When casts of the skeleton and models of the life appearance of *Diplodocus* were prepared by the Carnegie Museum of Natural History in Pittsburgh at the behest of the museum's patron, Andrew Carnegie, and distributed to museums around the world, several paleontologists objected to the erect posture given the dinosaur. It was argued that *Diplodocus* instead was more likely to have adopted a sprawling carriage, with its belly close to the ground—as living crocodiles were said to do.

This anatomical revisionism prompted vigorous counterattacks from W. J. Holland of the Carnegie Museum, among others. It was noted that giving *Diplodocus* and other sauropods a sprawling posture would have required major dislocations of joints between limb bones, and would have positioned the deep (from top to bottom) body in such a way that a sprawling sauropod would have gouged a trench in the ground with its belly as it moved!

The case for erect sauropods was further strengthened by the discovery of a nearly complete skeleton of a juvenile *Camarasaurus*. "The articulated right hind limb furnishes indisputable evidence in favor of those who have supported the view that these animals walked in an upright quadrupedal attitude, and it should quiet for all time those who advocate a crawling, lizard-like posture for the sauropod dinosaurs" (Gilmore 1925: 349).

Support for Gilmore's view came from the discovery of the sauropod trackways of the Glen Rose Limestone by R. T. Bird of the American Museum of Natural History (Fig. 36.16). The distance between left and right footprints in these trackways was much too narrow for the trackmakers to have had anything other than an erect posture, with their limbs directly beneath their bodies. The trackway evidence thus provided the clinching evidence in a controversy that had previously been argued entirely on the basis of skeletal anatomy.

Trackways have figured in a more recent controversy as well, this one involving the forelimb carriage of ceratopsid dinosaurs. Most paleontologists of the early twentieth century reconstructed *Triceratops* and its kin with fully erect hind limbs, but with forelimbs held in a sprawling fashion (Dodson 1996; Dodson and Farlow, in press). This interpretation was challenged by R. T. Bakker (1986 and references therein), who argued that ceratopsids moved in a rhinoceros-like fashion, their forelimbs held just as vertically as their hind limbs.

The discovery of what were probably ceratopsid trackways (named *Ceratopsipes* by Lockley and Hunt [1995a]) added an ichnological dimension to the controversy. Although the width across left and right forefoot (manus) impressions was greater than the width across left and right hindfoot (pes) prints, Lockley and Hunt concluded that this was no more than would be expected in other quadrupedal dinosaurs that were felt to have had an erect forelimb carriage. Lockley and Hunt concluded that the *Ceratopsipes*-maker held its forelimbs either in a semi-erect or (more likely, in their opinion) erect fashion. Paul (1991) likewise argued that the *Ceratopsipes* trackway pattern was most consistent with an erect forelimb carriage.

Although agreeing with Paul (1991) and Lockley and Hunt (1995a) that *Ceratopsipes* does not suggest a sprawling forelimb carriage in ceratopsids, Dodson and Farlow (in press) found it harder to say whether the forelimbs were held in an erect or semi-erect fashion. They noted problems in estimating the body size of the *Ceratopsipes*-maker, and in reconstructing the width across the ceratopsid shoulders as compared with the width across the hips, all of which would influence the interpretation of forelimb carriage from the trackway pattern. A solution to this controversy that will satisfy all participants is not yet apparent.

In primitive land vertebrates, the humeri and femora project outward from the body, and there is a sharp downward bend of the limb at the elbow or knee (Bakker 1971; Charig 1972). This puts footprints of the left and right side well apart from each other. As already described, an indirect way (but not the only way; see Leonardi 1987) of describing this in quantitative terms is by the pace angulation (also called the step angle) of a trackway (Fig. 36.3). Trackways of sprawling amphibians and reptiles often have pace angulations less than 100 degrees (Peabody 1948; Haubold 1984; Padian and Olsen 1984a; Lucas and Heckert 1995).

Crocodilians walking on land generally do not use a sprawling, lizard-like gait, but rather move in what is called a "high walk," with the limbs held in a more nearly vertical, semi-erect fashion (Cott 1961; Bakker 1971; Charig 1972; Webb and Manolis 1989; Parrish, chap. 15 of this volume). Ironically, crocodilians do not usually walk on land with the sprawling gait that was attributed to them by advocates of sprawling sauropods! Dinosaurs and certain other archosaurs had a fully erect gait, at least for the hind limb (Bakker 1971; Charig 1972). Trackways of semi-erect and fully erect quadrupedal reptiles have pace angulations greater than 100 degrees (Peabody 1948; Demathieu 1970; Padian and Olsen 1984b), sometimes reaching 180 degrees; trackways of quadrupedal dinosaurs have pace angulations between 100 and 120 degrees (Farlow et al. 1989; Thulborn 1990).

Of course, the pace angulation of a trackway depends on more than how the trackmaker holds its forearms and thighs during walking. A wide-bodied, short-legged quadruped will make a trail with a lower pace angulation than a narrow-bodied, long-legged quadruped will, even if both of them have an erect stance. The trackways of two erect animals with legs of equal length may show different pace angulations if the legs of one are directed straight downward from the shoulder and hip, but the legs of the other animal slant slightly inward, as well as downward, from the shoulder and hip. Finally, the pace angulation of the trail of a running animal can be greater than that of the same animal moving more slowly.

Trackways of bipedal dinosaurs have higher pace angulations (sometimes reaching 180 degrees, with tracks of the left and right foot falling on a straight line) than do the trackways of quadrupedal dinosaurs (Farlow 1987; Thulborn 1990). A quadruped has four feet supporting its body, so the animal has a broad base of support as it walks. It isn't very likely to tip over sideways, or to fall forward or backward. Not so a biped: Consider a typical theropod, with its head and body on the one side and its tail on the other, both balanced over the beast's hips (Fig. 36.18). With only two feet constituting its base of support, our theropod constantly risks falling on its side, face, or rump (or arse, if one prefers).

Of course, if its body were wide compared with the length of its legs, with its left and right feet far apart, our theropod would be fairly stable from side to side. However, the dinosaur's head, torso, and tail would extend well to the front and rear of the creature's center of balance; even

The Scientific Study of Dinosaur Footprints   545

standing still would involve a rather precarious balancing act to keep from tilting forward or backward. If that weren't difficulty enough, with every step the animal would diminish its stability of support in the very direction in which it was already risking a fall.

On the other hand, if—like real theropods—our biped is relatively narrow-bodied and long-legged, its long legs will act to check falls of its body in the forward direction; the animal will be as stable as is possible during forward motion. It will now be unstable in the sideways direction, but it partly mitigates this risk by having a narrow body, so that its weight remains directly in line above its feet. It can also avoid moving sideways any more than is absolutely necessary! All of these considerations are reflected in the way theropods and other bipedally walking dinosaurs moved, with the result that they made rather narrow trackways, with high pace angulations.

Countless old monster movies to the contrary notwithstanding, dinosaurs usually did not walk with their tails dragging on the ground, as lizards and crocodilians often do. Trackways of both bipedal and quadrupedal dinosaurs seldom show tail drag marks (Hitchcock's *Gigantitherium* [Fig. 36.6] is a notable exception). As one would infer from skeletal functional morphology (Molnar and Farlow 1990), trails confirm that theropods moved like animated see-saws, the weight of their tails counterbalancing that of their heads and bodies over their hips (Fig. 36.18). Quadrupedal dinosaurs did not need their tails for balance, but trackways suggest that even long-tailed quadrupeds (like some sauropods) rarely allowed their tails to drag passively across the ground.

From observations of the movements of a variety of living animals (humans, horses, jirds, elephants, and ostriches), the British zoologist R. McNeill Alexander (1976) devised a formula that analyzed the length of an animal's legs (its hip height), its stride length, and its speed of movement in terms of a dimensionless parameter known as the Froude number (Alexander,

Figure 36.18. Restoration of the huge theropod *Tyrannosaurus* in its normal walking pose, as suggested by the typical pattern of theropod dinosaur trackways. The dinosaur puts one foot down nearly in front of the other, and the tail is carried well off the ground. Drawing by Jim Whitcraft, based on a model by Matt Smith.

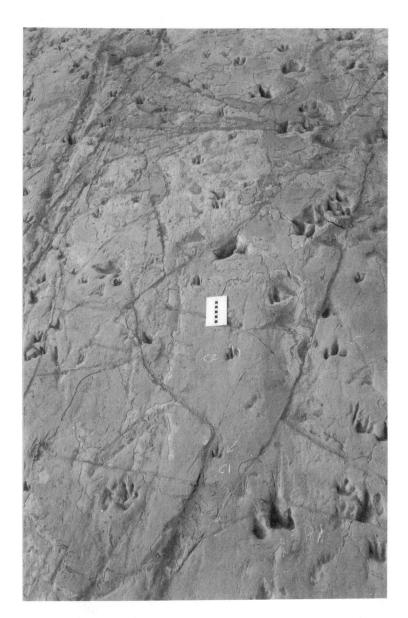

Figure 36.19. Footprints of small bipedal dinosaurs at the Lark Quarry (Winton Formation, Mid-Cretaceous), Queensland, Australia. The trackmakers were moving away from the viewer. The long stride lengths of the numerous little dinosaurs at this tracksite suggest that the trackmakers were running flat out.

chap. 30 of this volume). Alexander's formula could be recast in a form from which an animal's speed could be estimated from its trackway:

speed (meters/second)
= 0.25 × gravitational constant$^{0.5}$ × stride length$^{1.67}$ × hip height$^{-1.17}$

To estimate the speed of a dinosaur from its trackway, one first measures the stride directly from the trail. The hip height obviously cannot be obtained from a trackway. However, Alexander noted that, based on skeletal reconstructions, the hip height of dinosaurs was roughly four times the length of the portion of the foot that touched the ground; one could therefore estimate a trackmaker's hip height from the length of its footprint. (Thulborn [1990] went on to provide more refined equations for estimating hip heights from footmark lengths for individual dinosaur groups.)

The Scientific Study of Dinosaur Footprints   547

Figure 36.20. Three footprints in a trackway of a running theropod dinosaur, F[6] Ranch site, Kimble County, Texas. The middle footprint in this sequence (next to the meterstick) is shown in Figure 36.4C and D.

Many paleontologists have applied Alexander's equation to estimating dinosaur speeds (see Thulborn 1990 for a review). Unsurprisingly, most speed estimates suggest that the dinosaur trackmakers were walking, at speeds between 2 and 10 kilometers per hour. Like living animals, dinosaurs probably didn't go tearing across landscapes any more than they had to, but 5 to 10 kilometers per hour will get you a fair distance if you keep at it long enough. It is therefore not unreasonable to speculate that some dinosaurs might have migrated over fairly long distances (Hotton 1980; Lockley, chap. 37 of this volume).

Some trackways, however, seem to record running on the part of their makers (Thulborn 1990; Irby 1996b). In such trails, the stride length is roughly three or more times the trackmaker's estimated hip height (Thulborn 1990). The most impressive collection of trackways of apparently running dinosaurs at any one place is the Lark Quarry, a middle Cretaceous site in western Queensland, Australia (Thulborn and Wade 1984). At this locality, numerous small ornithopods and theropods dashed across a mudflat, possibly having been spooked by the approach of a large meat-eating dinosaur (Fig. 36.19). The estimated speeds of these little dinosaurs were about 12 to 16 kilometers per hour, which may not seem that fast. However, because the trackmakers were small animals, this was probably about as fast as their short legs could carry them.

The fastest speeds estimated for dinosaurs are based on trackways of medium-sized theropods from a site in Texas (Farlow 1981). Most of the trackmakers at this site were walking, but three of the theropods seem to have been running (Fig. 36.20), with estimated speeds of 30 to 40 kilometers per hour. A comparable speed has been estimated for the trackway of a medium-sized theropod from the Early Cretaceous of Spain (Viera and Torres 1995).

To date, trackways that appear to have been made by running dinosaurs are limited to those made by bipedal forms of small to medium size. There are as yet no trackways that suggest rapid running on the part of large dinosaurs, quadrupeds or bipeds. Consequently, interpretations of the peak running abilities of big dinosaurs are presently restricted to speculation based on the functional morphology and limb strengths of dinosaur skeletons (Coombs 1978; Bakker 1986; Paul 1988; Alexander 1989, and chap. 30 of this volume; Thulborn 1990; Farlow et al. 1995). Some paleontologists think that even large dinosaurs such as *Triceratops* and *Tyrannosaurus* were capable of speeds as fast as those estimated from the trackways of smaller dinosaurs at the above-mentioned tracksite in Texas (an interpretation that figured prominently in the movie *Jurassic Park*!). Other scientists doubt that large dinosaurs could move that quickly. As in the case of sauropod postures, trackway evidence could provide the telling evidence in this controversy. It would take only one trackway of a galloping *Triceratops* or a sprinting *Tyrannosaurus* to prove that these dinosaurs were capable of very fast locomotion. On the other hand, if—after enough tracksites are studied—no trails made by rapidly moving big dinosaurs ever turn up, it will be hard to reject the conclusion that large dinosaurs either were incapable of running, or did so very rarely.

References

Alexander, R. McN. 1976. Estimates of speeds of dinosaurs. *Nature* 261: 129–130.
Alexander, R. McN. 1989. *Dynamics of Dinosaurs and Other Extinct Giants*. New York: Columbia University Press.

Anderson, A. 1989. *Prodigious Birds: Moas and Moa-Hunting in Prehistoric New Zealand*. Cambridge: Cambridge University Press.

Anonymous. 1867. Untitled report of Professor Cope's "account of the extinct reptiles which approached the birds." *Proceedings of the Academy of Natural Sciences of Philadelphia* 1867: 234–235.

Arcucci, A. B.; C. A. Forster; F. Abdala; C. L. May; and C. A. Marsicano. 1995. "Theropod" tracks from the Los Rastros Formation (Middle Triassic), La Rioja Province, Argentina. *Journal of Vertebrate Paleontology* 15 (Supplement to no. 3): 16A.

Baird, D. 1957. Triassic reptile footprint faunules from Milford, New Jersey. *Bulletin of the Museum of Comparative Zoology, Harvard University* 117: 449–520.

Bakker, R. T. 1971. Dinosaur physiology and the origin of mammals. *Evolution* 25: 636–658.

Bakker, R. T. 1986. *The Dinosaur Heresies: New Theories Unlocking the Mystery of the Dinosaurs and Their Extinction*. New York: William Morrow.

Barsbold R. and T. Maryańska. 1990. Segnosauria. In D. B. Weishampel, P. Dodson, and H. Osmólska (eds.), *The Dinosauria*, pp. 408–415. Berkeley: University of California Press.

Barsbold, R.; T. Maryańska; and H. Osmólska. 1990. Oviraptorosauria. In D. B. Weishampel, P. Dodson, and H. Osmólska (eds.), *The Dinosauria*, pp. 249–258. Berkeley: University of California Press.

Barsbold R. and H. Osmólska. 1990. Ornithomimosauria. In D. B. Weishampel, P. Dodson, and H. Osmólska (eds.), *The Dinosauria*, pp. 225–244. Berkeley: University of California Press.

Basan, P. B. 1979. Trace fossil nomenclature: The developing picture. *Palaeogeography, Palaeoclimatology, Palaeoecology* 28: 143–167.

Bird, R. T. 1985. *Bones for Barnum Brown: Adventures of a Dinosaur Hunter*. Fort Worth: Texas Christian University Press.

Brown, B. 1916. *Corythosaurus casuarius*: Skeleton, musculature, and epidermis. *Bulletin of the American Museum of Natural History* 35: 709–716.

Casanovas Cladellas, M. L.; R. Ezquerra Miguel; A. Fernández Ortega; F. Pérez-Lorente; J. V. Santafé Llopis; and F. Torcida Fernández. 1993. Icnitas de dinosaurios. Yacimientos de Navalsaz, Las Mortajeras, Peñaportillo, Malvaciervo y la Era del Peladillo 2 (La Rioja, España). *Zubía* 5: 9–133.

Charig, A. J. 1972. The evolution of the archosaur pelvis and hindlimb: An explanation in functional terms. In K. A. Joysey and T. S. Kemp (eds.), *Studies in Vertebrate Evolution*, pp. 121–155. New York: Winchester Press.

Colbert, E. H. 1981. *A Primitive Ornithischian Dinosaur from the Kayenta Formation of Arizona*. Bulletin 53. Flagstaff: Museum of Northern Arizona Press.

Colbert, E. H. 1989. *The Triassic Dinosaur Coelophysis*. Bulletin 57. Flagstaff: Museum of Northern Arizona Press.

Coombs, W. P. Jr. 1978. Theoretical aspects of cursorial adaptations in dinosaurs. *Quarterly Review of Biology* 53: 393–418.

Cooper, A.; C. Mourer-Chauviré; G. K. Chambers; A. von Haeseler; A. C. Wilson; and S. Pääbo. 1992. Independent origins of the New Zealand moas and kiwis. *Proceedings of the National Academy of Sciences U.S.A.* 89: 8741–8744.

Cooper, A., and D. Penny. 1997. Mass survival of birds across the Cretaceous-Tertiary boundary: Molecular evidence. *Science* 275: 1109–1113.

Cott, H. B. 1961. Scientific results of an inquiry into the ecology and economic status of the Nile crocodile (*Crocodilus niloticus*) in Uganda and northern Rhodesia. *Transactions of the Zoological Society of London* 29: 211–357.

Currie, P. J. 1990. Elmisauridae. In D. B. Weishampel, P. Dodson, and H. Osmólska (eds.), *The Dinosauria*, pp. 245–248. Berkeley: University of California Press.

Dean, D. R. 1969. Hitchcock's dinosaur tracks. *American Quarterly* 21: 639–644.

Demathieu, G. R. 1970. *Les Empreintes de Pas de Vertébrés du Trias de la Bordure Nord-Est du Massif Central*. Paris: Centre National de la Recherche Scientifique.

Demathieu, G. R. 1989. Appearance of the first dinosaur tracks in the French Middle Triassic and their probable significance. In D. D. Gillette and M. G.

Lockley (eds.), *Dinosaur Tracks and Traces*, pp. 201–207. Cambridge: Cambridge University Press.

Demathieu, G. R. 1990. Problems in discrimination of tridactyl dinosaur footprints, exemplified by the Hettangian trackways, the Causses, France. *Ichnos* 1: 97–110.

Desmond, A. 1976. *The Hot-Blooded Dinosaurs: A Revolution in Palaeontology.* New York: Dial Press.

Dodson, P. 1996. *The Horned Dinosaurs.* Princeton, N.J.: Princeton University Press.

Dodson, P., and J. O. Farlow. In press. The forelimb carriage of ceratopsid dinosaurs. To be published in D. Wolberg (ed.), *Dinofest II.*

Farlow, J. O. 1981. Estimates of dinosaur speeds from a new trackway site in Texas. *Nature* 294: 747–748.

Farlow, J. O. 1987. *Lower Cretaceous Dinosaur Tracks, Paluxy River Valley, Texas.* Waco, Tex.: South Central Section, Geological Society of America, Baylor University.

Farlow, J. O. 1992. Sauropod tracks and trackmakers: Integrating the ichnological and skeletal records. *Zubía* 10: 89–138.

Farlow, J. O., and M. G. Lockley. 1993. An osteometric approach to the identification of the makers of early Mesozoic tridactyl dinosaur footprints. In S. G. Lucas and M. Morales (eds.), *The Nonmarine Triassic*, pp. 123–131. Bulletin no. 3. Albuquerque: New Mexico Museum of Natural History and Science.

Farlow, J. O.; J. G. Pittman; and J. M. Hawthorne. 1989. *Brontopodus birdi*, Lower Cretaceous sauropod footprints from the U.S. Gulf Coastal Plain. In D. D. Gillette and M. G. Lockley (eds.), *Dinosaur Tracks and Traces*, pp. 371–394. Cambridge: Cambridge University Press.

Farlow, J. O.; M. B. Smith; and J. M. Robinson. 1995. Body mass, bone "strength indicator," and cursorial potential of *Tyrannosaurus rex. Journal of Vertebrate Paleontology* 15: 713–725.

Feduccia, A. 1996. *The Origin and Evolution of Birds.* New Haven, Conn.: Yale University Press.

Forster, C. A. 1990. The postcranial skeleton of the ornithopod dinosaur *Tenontosaurus tilletti. Journal of Vertebrate Paleontology* 10: 273–294.

Galton, P. M. 1974. The ornithischian dinosaur *Hypsilophodon* from the Wealden of the Isle of Wight. *Bulletin of the British Museum (Natural History) Geology* 25: 1–152c.

Galton, P. M. 1990. Basal Sauropodomorpha—Prosauropoda. In D. B. Weishampel, P. Dodson, and H. Osmólska (eds.), *The Dinosauria*, pp. 320–344. Berkeley: University of California Press.

Gierliński, G. 1988. New dinosaur ichnotaxa from the Early Jurassic of the Holy Cross Mountains, Poland. *Palaeogeography, Palaeoclimatology, Palaeoecology* 85: 137–148.

Gierliński, G. 1994. Early Jurassic theropod tracks with the metatarsal impressions. *Przeglad Geologiczny* 42: 280–284.

Gierliński, G. 1995. Thyreophoran affinity of *Otozoum* tracks. *Przeglad Geologiczny* 43:123–125.

Gierliński, G., and A. Ahlberg. 1994. Late Triassic and Early Jurassic dinosaur footprints in the Höganäs Formation of southern Sweden. *Ichnos* 3: 99–105.

Gilmore, C. W. 1909. Osteology of the Jurassic reptile *Camptosaurus*, with a revision of the species of the genus, and descriptions of two new species. *Proceedings of the U.S. National Museum* 36: 197–332.

Gilmore, C. W. 1925. A nearly complete articulated skeleton of *Camarasaurus*, a saurischian dinosaur from the Dinosaur National Monument, Utah. *Memoirs of the Carnegie Museum* 10: 347–384.

Gilmore, C. W. 1936. Osteology of *Apatosaurus*, with special reference to specimens in the Carnegie Museum. *Memoirs of the Carnegie Museum* 11: 175–298.

Glassman, S.; E. A. Bolt, Jr.; and E. E. Spamer. 1993. Joseph Leidy and the "Great Inventory of Nature." *Proceedings of the Academy of Natural Sciences of Philadelphia* 144: 1–19.

Gruber, J. W. 1987. From myth to reality: The case of the moa. *Archives of Natural History* 14: 339–352.

Guralnick, S. M. 1972. Geology and religion before Darwin: The case of Edward Hitchcock, theologian and geologist (1793–1864). *Isis* 63: 529–543.

Harris, J. D.; K. R. Johnson; J. Hicks; and L. Tauxe. 1996. Four-toed theropod footprints and a paleomagnetic age from the Whetstone Falls Member of the Harebell Formation (Upper Cretaceous: Maastrichtian), northwestern Wyoming. *Cretaceous Research* 17: 381–401.

Hatcher, J. B. 1901. *Diplodocus* (Marsh): Its osteology, taxonomy, and probable habits, with a restoration of the skeleton. *Memoirs of the Carnegie Museum:* 1: 1–63.

Haubold, H. 1984. *Saurierfährten.* Wittenberg Lutherstadt, Germany: A. Ziemsen.

Heilmann, G. 1927. *The Origin of Birds.* New York: D. Appleton. Reprint, New York: Dover, 1972.

Hitchcock, E. 1858. *Ichnology of New England: A Report on the Sandstone of the Connecticut Valley, Especially Its Fossil Footmarks.* Boston: William White. Reprint, New York: Arno Press.

Hitchcock, E. 1865. *Supplement to the Ichnology of New England.* Boston: Wright and Potter.

Hotton, N. III. 1980. An alternative to dinosaur endothermy: The happy wanderers. In R. D. K. Thomas and E. C. Olson (eds.), *A Cold Look at the Warm-Blooded Dinosaurs,* pp. 311–350. American Association for the Advancement of Science Selected Symposium 28. Boulder, Colo.: Westview Press.

Huxley, T. H. 1868. On the animals which are most nearly intermediate between birds and reptiles. *Geological Magazine* 5: 357–365.

Irby, G. V. 1996a. Paleoichnology of the Cameron Dinosaur Tracksite, Lower Jurassic Moenave Formation, northeastern Arizona. In M. Morales (ed.), *The Continental Jurassic,* pp. 147–166. Bulletin 60. Flagstaff: Museum of Northern Arizona.

Irby, G. V. 1996b. Paleoichnological evidence for running dinosaurs worldwide. In M. Morales (ed.), *The Continental Jurassic,* pp. 109–112. Bulletin 60. Flagstaff: Museum of Northern Arizona.

King, M. J., and M. J. Benton. 1996. Dinosaurs in the Early and Mid-Triassic? The footprint evidence from Britain. *Palaeogeography, Palaeoclimatology, Palaeoecology* 122: 213–225.

Lawrence, P. J. 1972. Edward Hitchcock: The Christian geologist. *Proceedings of the American Philosophical Society* 116: 21–34.

Leonardi, G. (ed.). 1987. *Glossary and Manual of Tetrapod Footprint Palaeoichnology.* Brazilia: Brazilian Department of Mines and Energy.

Leonardi, G. 1994. *Annotated Atlas of South America Tetrapod Footprints (Devonian to Holocene).* Brazilia: República Federativa do Brasil, Ministério de Minas e Energia.

Lockley, M. G. 1991. *Tracking Dinosaurs: A New Look at an Ancient World.* Cambridge: Cambridge University Press.

Lockley, M. G.; J. O. Farlow; and C. A. Meyer. 1994. *Brontopodus* and *Parabrontopodus* ichnogen. nov. and the significance of wide- and narrow-gauge sauropod trackways. *Gaia* 10: 135–145.

Lockley, M. G., and A. P. Hunt. 1994. A track of the giant theropod dinosaur *Tyrannosaurus* from close to the Cretaceous/Tertiary boundary, northern New Mexico. *Ichnos* 3: 1–6.

Lockley, M. G., and A. P. Hunt. 1995a. Ceratopsid tracks and associated ichnofauna from the Laramie Formation (Upper Cretaceous: Maastrichtian) of Colorado. *Journal of Vertebrate Paleontology* 15: 592–614.

Lockley, M. G., and A. P. Hunt. 1995b. *Dinosaur Tracks and Other Fossil Footprints of the Western United States.* New York: Columbia University Press.

Lucas, S. G., and A. B. Heckert (eds.). 1995. *Early Permian Footprints and Facies.* Bulletin 6. Albuquerque: New Mexico Museum of Natural History and Science.

Lull, R. S. 1933. *A Revision of the Ceratopsia or Horned Dinosaurs.* Memoirs of the Peabody Museum of Natural History 3: 1–135.

The Scientific Study of Dinosaur Footprints   551

Lull, R. S. 1953. *Triassic Life of the Connecticut Valley.* Bulletin 81. Hartford: Connecticut State Geological and Natural History Survey.

Magwood, J. P. A. 1992. Ichnotaxonomy: A burrow by any other name . . . ? In C. G. Maples and R. R. West (eds.), *Trace Fossils,* pp. 15–33. Paleontological Society Short Course no. 5. Knoxville: University of Tennessee.

Maleev, E. A. 1974. [Gigantic carnosaurs of the family Tyrannosauridae]. *Sovm. Sov.-Mong. Paleontol. Eksped. Trudy* 1: 132–191. (In Russian.)

Maples, C. G., and R. R. West (eds.). 1992. *Trace Fossils.* Paleontological Society Short Course no. 5. Knoxville: University of Tennessee.

Marsh, O. C. 1877. Introduction and succession of vertebrate life in America. *American Journal of Science,* Third Series, 14: 337–378.

Molnar, R. E., and J. O. Farlow. 1990. Carnosaur paleobiology. In D. B. Weishampel, P. Dodson, and H. Osmólska (eds.), *The Dinosauria,* pp. 210–224. Berkeley: University of California Press.

Moratalla, J. J.; J. L. Sanz; and S. Jiménez. 1988. Nueva evidencia icnologica de dinosaurios en el Cretacico Inferior de La Rioja (España). *Estudios Geológicos* 44: 119–131.

Moratalla, J. J.; J. L. Sanz; S. Jiménez; and M. G. Lockley. 1992. A quadrupedal ornithopod trackway from the Lower Cretaceous of La Rioja (Spain): Inferences on gait and hand structure. *Journal of Vertebrate Paleontology* 12: 150–157.

Norman, D. B. 1986. On the anatomy of *Iguanodon atherfieldensis* (Ornithischia: Ornithopoda). *Bulletin de l'Institut Royal des Sciences Naturelles de Belgique* 56: 281–372.

Olsen, P. E. 1995. A new approach for recognizing track makers. *Geological Society of America Abstracts with Program* 27 (1): 72.

Olsen, P. E., and P. M. Galton. 1984. A review of the reptile and amphibian assemblages from the Stormberg of southern Africa, with special emphasis on the footprints and the age of the Stormberg. *Palaeontologia africana* 25: 87–110.

Osgood, R. G. Jr. 1987. Trace fossils. In R. S. Boardman, A. H. Cheetham, and A. J. Rowell (eds.), *Fossil Invertebrates,* pp. 663–674. Palo Alto, Calif.: Blackwell Scientific Publications.

Osmólska, H., and R. Barsbold. 1990. Troodontidae. In D. B. Weishampel, P. Dodson, and H. Osmólska (eds.), *The Dinosauria,* pp. 259–268. Berkeley: University of California Press.

Ostrom, J. H. 1969. *Osteology of Deinonychus antirrhopus, an Unusual Theropod from the Lower Cretaceous of Montana.* Bulletin 30. New Haven, Conn.: Peabody Museum of Natural History, Yale University.

Ostrom, J. H. 1978. The osteology of *Compsognathus longipes* Wagner. *Zitteliana* 4: 73–118.

Padian, K., and P. E. Olsen. 1984a. Footprints of the Komodo monitor and the trackways of fossil reptiles. *Copeia* 1984: 662–671.

Padian, K., and P. E. Olsen. 1984b. The fossil trackway *Pteraichnus:* Not pterosaurian, but crocodilian. *Journal of Paleontology* 58: 178–184.

Paul, G. S. 1988. *Predatory Dinosaurs of the World: A Complete Illustrated Guide.* New York: Simon and Schuster.

Paul, G. S. 1991. The many myths, some old, some new, of dinosaurology. *Modern Geology* 16: 69–99.

Peabody, F. E. 1948. Reptile and amphibian trackways from the Lower Triassic Moenkopi Formation of Arizona and Utah. *Bulletin of the Department of Geological Sciences,* University of California, Berkeley 27: 295–468.

Perle A.; L. M. Chiappe; Barsbold R.; J. M. Clark; and M. A. Norell. 1994. Skeletal morphology of *Mononykus olecranus* (Theropoda: Avialae) from the Late Cretaceous of Mongolia. American Museum *Novitates* 3105: 1–29.

Rudwick, M. J. S. 1976. *The Meaning of Fossils: Episodes in the History of Palaeontology,* 2nd ed. New York: Neale Watson Academic Publications.

Santa Luca, A. P. 1980. The postcranial skeleton of *Heterodontosaurus tucki* (Reptilia, Ornithischia) from the Stormberg of South Africa. *Annals of the South African Museum* 79: 159–211.

Sarjeant, W. A. S. 1989. "Ten paleoichnological commandments": A standardized procedure for the description of fossil vertebrate footprints. In D. D. Gillette and M. G. Lockley (eds.), *Dinosaur Tracks and Traces*, pp. 369–370. Cambridge: Cambridge University Press.

Sarjeant, W. A. S. 1990. A name for the trace of an act: Approaches to the nomenclature and classification of fossil vertebrate footprints. In K. Carpenter and P. J. Currie (eds.), *Dinosaur Systematics: Approaches and Perspectives*, pp. 299–307. Cambridge: Cambridge University Press.

Sarjeant, W. A. S., and W. J. Kennedy. 1973. Proposal of a code for the nomenclature of trace-fossils. *Canadian Journal of Earth Sciences* 10: 460–475.

Sarjeant, W. A. S., and W. Langston, Jr. 1994. *Vertebrate Footprints and Invertebrate Traces from the Chadronian (Late Eocene) of Trans-Pecos Texas*. Bulletin 36. Austin: Texas Memorial Museum.

Schult, M. F., and J. O. Farlow. 1992. Vertebrate trace fossils. In C. G. Maples and R. R. West (eds.), *Trace Fossils*, pp. 34–63. Paleontological Society Short Course no. 5. Knoxville: University of Tennessee.

Thulborn, R. A. 1972. The post-cranial skeleton of the Triassic ornithischian dinosaur *Fabrosaurus australis*. *Palaeontology* 15: 29–60.

Thulborn, R.A. 1990. *Dinosaur Tracks*. London: Chapman and Hall.

Thulborn, R. A. 1994. Ornithopod dinosaur tracks from the Lower Jurassic of Queensland. *Alcheringa* 18: 247–258.

Thulborn, R.A., and M. Wade. 1984. Dinosaur trackways in the Winton Formation (Mid-Cretaceous) of Queensland. *Memoirs of the Queensland Museum* 21: 413–517.

Viera, L. I., and J. A. Torres. 1995. Análisis comparativo sobre dos rastros de Dinosaurios Theropodos: Forma de marcha y velocidad. *Munibe* 47: 53–56.

Webb, G., and C. Manolis. 1989. *Australian Crocodiles: A Natural History*. Chatswood, New South Wales: Reed Books.

Weems, R. E. 1987. A Late Triassic footprint fauna from the Culpeper Basin, northern Virginia (U.S.A.). *Transactions of the American Philosophical Society* 77: 1–79.

Weems, R. E. 1992. A re-evaluation of the taxonomy of Newark Supergroup saurischian dinosaur tracks, using extensive statistical data from a recently exposed tracksite near Culpeper, Virginia. In P. C. Sweet (ed.), *Proceedings of the 26th Forum on the Geology of Industrial Minerals, May 14–18, 1990*, pp. 113–127. Charlotte: Commonwealth of Virginia, Division of Mineral Resources.

The Scientific Study of Dinosaur Footprints 553

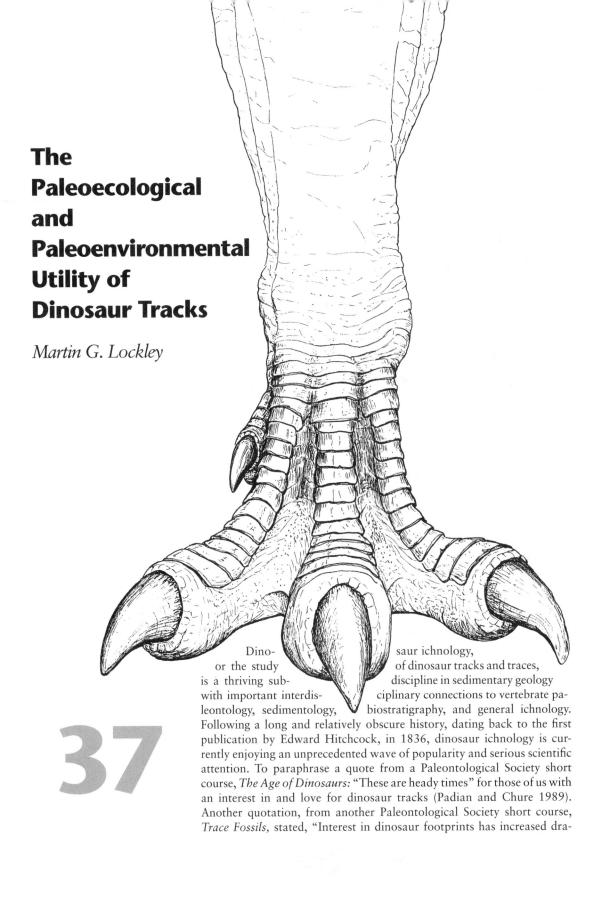

The Paleoecological and Paleoenvironmental Utility of Dinosaur Tracks

Martin G. Lockley

37

Dino- saur ichnology, or the study of dinosaur tracks and traces, is a thriving sub- discipline in sedimentary geology with important interdis- ciplinary connections to vertebrate pa- leontology, sedimentology, biostratigraphy, and general ichnology. Following a long and relatively obscure history, dating back to the first publication by Edward Hitchcock, in 1836, dinosaur ichnology is currently enjoying an unprecedented wave of popularity and serious scientific attention. To paraphrase a quote from a Paleontological Society short course, *The Age of Dinosaurs:* "These are heady times" for those of us with an interest in and love for dinosaur tracks (Padian and Chure 1989). Another quotation, from another Paleontological Society short course, *Trace Fossils,* stated, "Interest in dinosaur footprints has increased dramatically over the last decade, along with attempts to integrate the dinosaur trace fossil record with the skeletal record of the great reptiles" (Schult and Farlow 1992: 47). To underscore this point, we need only look at the growing number of new books on the subject (Gillette and Lockley 1989; Thulborn 1990; Lockley 1991a; Lockley and Hunt 1995) and the proliferation of articles dealing with fossil footprints in volumes devoted to dinosaurs and ichnology (Lockley 1987, 1989a, 1991a; Schult and Farlow 1992; Lockley et al. 1994a; Lockley and Hunt 1994a).

It is worth stressing that there is a sound scientific basis for the renewed interest in this field. Ever since John Ostrom (1969) and Robert Bakker (1975) spearheaded the so-called "Dinosaur Renaissance" by redescribing dinosaurs as agile, smart, successful, and dynamic rather than as stupid, cumbersome, and defunct, we have become much more aware that tracks are the epitome of dynamic evidence, because of the obvious fact that they were produced by living animals. The classic example of the utility of tracks in this regard is the debate about the speeds attained by dinosaurs (see Alexander 1989; Thulborn 1990; Lockley 1991a, 1991b, 1991c; and Farlow and Chapman, chap. 36 of this volume). Another classic example is the utility of tracks for revealing evidence of social behavior, or herding, which is now known to have been quite common, especially amongst sauropods and large ornithopods (Bird 1944; Lockley 1991a, 1991b, 1991c, 1995).

However, the utility of tracks goes far beyond the realms of locomotor studies and interpretation of individual and social behavior. As shown by the author and his colleagues (e.g., Lockley and Hunt 1994a, 1994b; Lockley et al. 1994a), tracks are very useful for paleoecological census studies, and can be considered representative of ancient animal communities to various degrees as discussed below. Moreover, evidence that particular track assemblages (ichnocoenoses) are repeatedly associated with particular ancient environments allows us to define distinctive ichnofacies and go some way toward demonstrating the environmental preferences of particular animal communities. Synthesis of such data within an appropriate sedimentological, stratigraphical, and paleobiogeographical framework adds much to our knowledge of the spatial and temporal distribution of various dinosaur groups, and in at least one case sheds important light on the long-standing debate about dinosaur extinction at the Cretaceous/Tertiary (K/T) boundary.

The lack of articulated skeletal remains of dinosaurs for at least 3 meters below the K/T boundary has generated debate about whether dinosaurs died out before the boundary. This debate was almost impossible to resolve because isolated bones in or above this 3-meter gap could be the result of reworking rather than evidence of the presence of dinosaurs. The discovery of hadrosaur and ceratopsian tracks only 37 centimeters below the boundary and at several other horizons within the top 2 meters of this gap (Lockley 1991a; Lockley and Hunt 1995) makes most of the debate about the significance of isolated bone redundant, and is clear proof that at least two families of dinosaurs survived until the last minute.

Recent work has also allowed dinosaur ichnologists to demonstrate that tracks are far more abundant than previously supposed. For example, megatracksites represent single surfaces, or thin packages of beds, that are track-rich over areas on the scale of tens of thousands of square kilometers, and so literally contain millions—even billions—of tracks. Such an abundance of fossil footprints obviously provides a readily accessible database that can substantially supplement the body fossil record.

Although the body and trace fossil records provide different levels of taxonomic resolution, what the track record may lack in resolution, it often makes up for in astonishing abundance. This does not mean that a large record of poorly preserved tracks is better than a sparse record of well-preserved bones. But, on average, the extensive track record provides a significant number of well-preserved tracks, which forces us to consider just how many individuals are represented by footprints as compared to bones. The track record also deals consistently with the same type of morphological information pertaining to feet (foot bones and flesh), and so we are able to compare the same information from sample to sample and from site to site, whereas with bones we may have to compare ribs with teeth, vertebrae, or other skeletal elements.

Compilation of available data allows us to establish important points of comparison between the trace fossil and body fossil record, and demonstrates some of the fundamental biases in the fossil record. When this is done, it can be shown that in many cases, the track record is far more complete than the bone record in terms of number of specimens and number of taxa documented. However, the ultimate purpose of making such comparisons, as done in this chapter, is not to argue that the track record is better or worse than the bone record—a case of comparing apples and oranges—but rather to pool available data for a better understanding of dinosaurs and other fossil vertebrates.

One of the paradoxes of paleontology is that trace fossils and body fossils are fundamentally different, yet they are often subject to similar treatment in paleobiological studies. For example, as outlined by Farlow and Chapman in chapter 36, body fossils and trace fossils both require detailed systematic description if their morphology is to be adequately understood. Both are labeled using the Linnean bionomial system, though the systematics of tracks and traces is referred to as ichnotaxonomy or parataxonomy rather than taxonomy. Moreover, the level of taxonomic resolution achieved by the two systems is somewhat different. In theory, well-defined ichnospecies, based on well-preserved material, are more or less equivalent to osteological species, as they are in the modern world. In reality, however, with fossil tracks we don't usually know precisely which species made the tracks, except in rare cases; for example, the track *Tyrannosauripus pillmorei* almost certainly represents the species *Tyrannosaurus rex* (Lockley and Hunt 1994b). In such ideal cases the ichnospecies is just as useful and informative from a morphological point of view as an osteological species that is based on only a few bones, although it is not nearly as informative in this regard as a species based on a complete or reasonably complete skeleton.

In practice, however, many tracks are referred to the broader categories of ichnogenus, which in some cases represent very broad "catchall" categories that probably accommodate many species of trackmakers from larger familial or ordinal-level taxonomic groups. There are several reasons for such lumping, including real and perceived similarities between the morphology of footprints from different animals of different geologic ages. Such similarities result from conservatism and convergence in foot morphology, masking or blurring of true foot morphology by poor preservation and the dynamics of foot emplacement, and the reluctance or inability of ichnologists to engage in detailed systematic work that could discriminate valid from invalid ichnotaxa.

This reluctance to straighten out taxonomic problems is understandable in the light of the confused history of ichnotaxonomic research on the tracks of dinosaurs and other fossil vertebrates, and the relatively small number of specialists in the field. However, this does not mean that significant progress cannot, or should not, be made in rectifying this situation. Basic principles of systematics and ichnology hold that we must describe morphological variation wherever we encounter it in suitably well preserved samples (Baird 1957). As progress is made in this direction, ichnologists can hope to dispense with ill-conceived names and establish an ichnotaxonomy that is a true reflection of morphological variation in the fossil footprint record. Such an objective should be regarded as a simple return to basics, rather than an unrealistic or unobtainable goal. Continued progress in applied ichnological fields such as biostratigraphy (palichnostratigraphy), paleoecological census studies, and ichnofacies analysis ultimately relies on a sound ichnotaxonomic basis.

Track Preservation: Problems and Solutions

Body fossils are generally preserved within sedimentary layers, while tracks are preserved on the surfaces between layers. This difference generally reflects the fact that tracks are made during breaks in sedimentation, whereas bones are buried during sedimentation events. In the broadest sense, tracks are much more commonly preserved than bones. For example, Dodson (1990: 7608) noted that only about 2,100 "generically determinate, articulated specimens" of dinosaurs are preserved in the world's major museums, after more than 150 years of worldwide exploration. By contrast, in a single region, such as the western United States, thousands of trackways, most representing different individual dinosaurs, have been recorded in less than a decade (Lockley and Hunt 1994a, 1995).

The biases that affect bone preservation are similar, in many instances, to those that affect track preservation. For example, large tracks, like large bones, are much more likely to be preserved and recognized (Lockley 1991a). Such biases should be taken into account in the compiling of osteological and trackway census data.

Despite the abundance of tracks in many sedimentary deposits, their quality of preservation varies considerably. Like skeletons and bones, trackways may be incomplete, and may have suffered from erosion and weathering in the pre-burial phase (taphonomic alteration), in the post-exhumation phase, or during both phases. Tracks, however, unlike bones, may also have been poorly preserved from the time that they were first formed, as a result of sub-optimal substrate conditions.

A classic example of this is the so-called trackway of a running ornithomimid described by Welles (1971) from the Lower Jurassic of Arizona, and named *Hopiichnus*. Apart from the fact that ornithomimids are not known from the Lower Jurassic, (Haubold 1984; Thulborn and Wade 1984), and that the trackway is evidently attributable to the well-known Lower Jurassic ichnogenus *Anomoepus* (Lockley and Hunt 1994a, 1995), the trackway is so poorly preserved that the impression of long steps, taken as a sign of running, is probably the result of missing footprints due to poor preservation of segments of the trackway! Despite this simple explanation, poor preservation, several authors have apparently accepted the step measurements, without looking at the material in question, and so have published speed estimates ranging from 26.4 to 82.5 kilometers per hour (Haubold 1984; Thulborn and Wade 1984). Given that the fastest of these speed estimates exceeds any derived from fossil trackways (Farlow 1981; Thulborn 1990) and exceeds even theoretical estimates (Thulborn

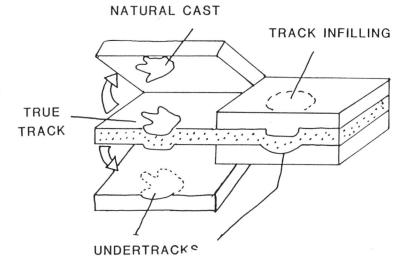

NATURAL CAST

TRACK INFILLING

TRUE TRACK

UNDERTRACKS

Figure 37.1. Preservation of tracks. True footprints form in the surface across which the animal actually walks. Undertracks form in sedimentary layers beneath the one across which the trackmaker moves, as underlying layers are themselves deformed by the animal's weight. Natural casts form from sediments that fill the footprint after it forms. Redrawn from Lockley 1991a.

1982, 1990), the estimates derived from the poorly preserved *Hopiichnus* trackway should be regarded as dubious at best.

One very important aspect of track preservation pertains to the phenomenon of undertracks, also known as underprints or ghost prints (Fig. 37.1). These are formed as a result of the transmission of footprints into underlayers, below the actual surface on which the trackmaker progressed. Because there are always layers of sediment, of various thickness, between the surface with the true tracks and the underlayers, it follows, as a general rule of thumb, that the undertracks are less well defined. Undertracks also reveal diffuse margins that give the impression that the tracks are larger than they really are. Ideally, true tracks preserve fine details of preservation, such as skin and pad impressions, that are usually not seen in undertracks. Trackers should be aware of these differences and look out for them so as not to read too much morphological meaning into undertracks.

We should note that exposed track-bearing surfaces typically erode along primary planes of weakness such as clay partings. For this reason, surfaces with undertracks are just as likely to be exposed as surfaces with "real" tracks. Although we can cite many examples of surfaces with undertracks, no one has attempted to record the proportion of known surfaces with undertracks relative to those revealing true tracks. At a number of sites, true tracks and the corresponding undertracks can be observed on two or more successive surfaces, thus allowing us to recognize what undertracks of a particular ichnotaxon look like.

There is at least one classic example of the misinterpretation of undertracks leading to incorrect interpretation of dinosaur behavior. In 1944, Roland Bird described an incomplete trackway from the Cretaceous Glen Rose Formation that apparently consisted mainly of front-foot impressions, with a single partial hind-foot impression. He interpreted this as the trackway of a "swimming" or partially buoyant animal (Bird 1944), and in several of his publications revealed his predisposition to believe that sauropods were aquatic or semi-aquatic animals. Subsequent work has revealed that the trackway consists of undertracks on an underlayer, and that faint hind footprints are also visible on the surface (Lockley and Rice 1990; Pittman 1990; Lockley 1991a). Thus the trackway represents an animal walking on an emergent surface, and not a sauropod swimming in a shallow marine environment.

Despite this reinterpretation of the data in terms of preservation rather than unusual behavior, several recent publications have continued to refer to the swimming sauropod hypothesis without reference to the sedimentology of the track-bearing layers, or to the revised interpretation (e.g., Thulborn 1990; Czerkas and Czerkas 1990; Norman 1991; Gardom and Milner 1993). This adherence to the old hypothesis is in part a time lag in learning, but it also reflects a common inclination of paleontologists, and people in general, to assume that tracks must have a behavioral significance, and that dramatic activity can be inferred. As this and the previous example show, a much safer approach is to understand preservation first, then interpret behavior, and in addition to bear in mind that unusual behavior is the exception, not the rule, in the trackway record.

Collecting Trackway Data: Tracks, Trackways, and Tracksites

As with any paleontological study, one of the prime objectives of ichnology is to maximize the retrieval of useful information. Individual tracks or footprints (the terms are used here with the same meaning) provide information on foot morphology and size that helps with trackmaker identification and ichnotaxonomy. However, individual footprints are components of trackways (also called trails), which also provide important additional information on the posture, gait, and locomotion of trackmakers. Such paramaters as trackway width (or straddle), pace angulation, and relative size and placement of hind and front footprints are all features that are diagnostic of particular ichnotaxa (Thulborn 1990; Lockley 1991a; Farlow and Chapman, chap. 36 of this volume).

Thus tracks and trackway segments are the basic units of currency found at any tracksite, whether it be a site that yields only a single footprint, or a much larger tracksite yielding thousands of footprints. These tracks may be found on a single surface or on multiple track-bearing layers. A tracksite revealing a single layer can be considered a single track assemblage or ichnocoenosis, while a site revealing multiple layers provides evidence of multiple ichnocoenoses formed at various successive time intervals.

Census Studies, Biomass, Ichnocoenoses, and the Ichnofacies Concept

As with any other branch of paleontology, a census of tracks and trackways can be made simply by counting the number of individuals of different types represented in any assemblage. The most efficient way to do this is to produce a complete map of the track-bearing surface (ichnocoenosis). Such a map (Fig. 37.2) can be viewed as analogous to a quarry map at an osteological site, and has the advantage of recording the orientation of tracks and trackways, as well as the total number of these trace fossils. Once a map has been made, individual trackways can be counted and numbered. This allows us to estimate the number of individuals that crossed a particular area. If all the trackways represent animals of different types (different ichnotaxa) or at least of different sizes, we can be sure that no individual animal was represented more than once. If there are

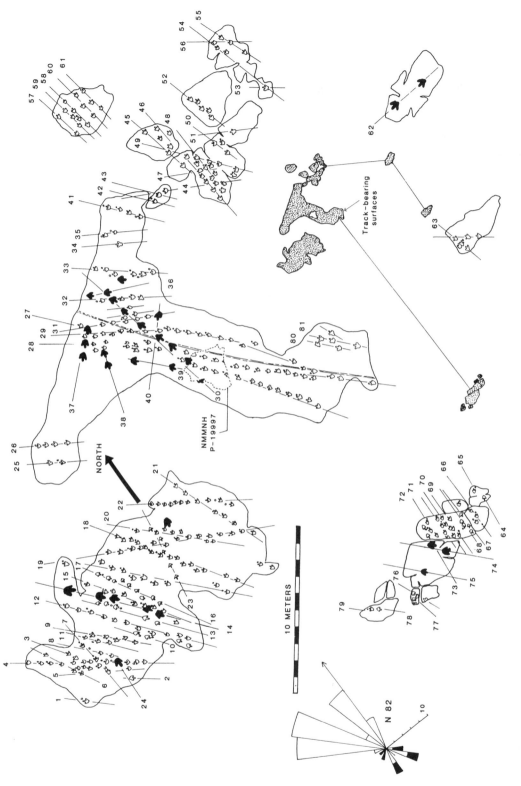

Mosquero Creek Tracksite

NORTH

10 METERS

N 82

NMMNH
P-19997

Track-bearing
surfaces

many trackways of the same type representing animals of the same size, then it is possible that particular individuals are represented more than once.

Size frequency data from a number of sites (e.g., Lockley 1994) suggest that most trackways represent different individuals, and so the simplest explanation is to assume that one trackway represents one individual. This "estimation" approach is not foolproof, but any other method of counting requires a number of more tenuous assumptions. Trackers can easily estimate the minimum number of track types (ichnotaxa) and individual animals (of different sizes) at any site. In most cases this minimum number of individuals represents a large percentage of the total number of trackways. (It would be possible to subtract from the total count any trackways that were of the same size and ichnotaxon as ones already counted. But this approach could lead to underestimates of the actual number of individuals—and, besides, has never been attempted.) Given the wide range of track sizes recorded in most published data sets, the one-individual-per-trackway approach must be considered the most parsimonious.

Once we have estimated the number of trackmakers (individuals) present at a particular site, we have an ichnological census of the track assemblage (ichnocoenosis) that can be expressed in terms of the relative abundance of different ichnotaxa. Such data can be expressed in terms of raw data, such as number of trackways in a pie diagram (Lockley 1991a; Lockley et al. 1994a; Fig 37.3), or modified to reflect estimates of biomass (the total amount of living material represented by the population in question). This latter form of synthesis has only recently been employed in ichnological studies (Lockley and Hunt 1995).

Although it is relatively easy to estimate the weight of a dinosaur (Colbert 1962) from skeletal dimensions, until now nobody has ever thought to convert numbers of tracks (trackways) into biomass estimates. To show the value of such an exercise, I have estimated the biomass of theropods and sauropods from footprint dimensions at the famous Late Jurassic Purgatoire tracksite in Colorado (Lockley and Hunt 1995), and shown how the results compare with simple numbers of trackways (Fig. 37.3). A count of trackways shows that there are about sixty theropod trackways and forty sauropod trackways associated with the main track level. Such a large number of theropod trackways is exactly the sort of

Figure 37.2. A map of a dinosaur tracksite summarizes a great deal of information in a very concise fashion. This map illustrates the orientation of dinosaur trackways at the Mosquero Creek site (Cretaceous, Dakota Group, New Mexico) and provides evidence of herding among ornithopods. From Lockley and Hunt 1995.

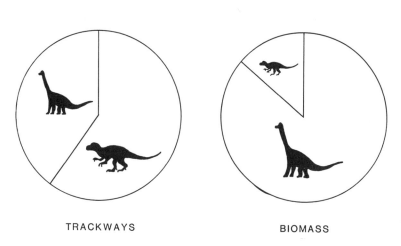

TRACKWAYS   BIOMASS

Figure 37.3. Pie diagrams illustrating different ways of summarizing information about the relative abundance of sauropods and theropods at the Late Jurassic Purgatoire site of eastern Colorado. The left diagram compares relative abundances in terms of simple counts of the numbers of trackways of the two trackmaker groups. The right diagram prorates abundance in terms of biomass by multiplying the number of trackways of each kind by the average individual body mass of that kind of trackmaker. From Lockley and Hunt 1995.

evidence that raises questions about bias in the track record. For example, because the tracks of theropods or carnivorous dinosaurs (predators) are often much more abundant than those of herbivorous prey species, it has been suggested that they were more active. While it is right to question why theropods appear to be so overrepresented in the track record, when in the bone record they make up only 3 to 5 percent, instead of 60 percent, the explanation may be simpler.

If we estimate biomass, however, we arrive at a very different conclusion. Based on average theropod track size (foot length of 37 cm) at the Purgatoire site, we can estimate that the average animal weighed about 1 ton (Thulborn 1990). By contrast, the average sauropod had a foot length of about 67 centimeters. Because volume (and mass) increases as the cube of the linear dimension, it is obvious that the sauropods had a much greater individual mass than did theropods. Estimates for the average sauropod at this site are in the range of 9 to 10 tons. Thus the 60:40 ratio becomes a 60:360 or 60:400 ratio. As shown in Figure 37.3, this produces a very different picture of the predator : prey ratio. The predator biomass is only 15 to 17 percent of the total, not 60 percent. Given that studies of bones suggest a small percentage of theropods (Coe et al. 1987), we can argue that the results obtained using this method are probably more representative because they are closer to those obtained from bone census data.

Because tracks are so abundant, they are potentially useful for shedding light on biomass, community dynamics, and predator:prey ratios. In the future we hope that we will be able to produce reliable estimates of biomass of different trackmaking groups from the abundant track census data that we are beginning to compile. Such an approach will require some standardization of how trackers estimate size and weight (biomass) of individual trackmakers. But as this example shows, the results are promising in terms of arriving at biomass estimates that reflect ancient ecology and community dynamics more accurately than numbers alone.

Experience reveals that various ichnotaxa recur, often in similar proportions in particular sedimentary deposits, thus reflecting the makeup, distribution, geologic age, and/or environmental preference of the trackmakers. Following the precedents set in the realm of invertebrate ichnology, recurrent associations of this type allow us to define ichnofacies (specific sedimentary deposits that repeatedly have distinctive track assemblages). By definition ichnofacies reflect, to some degree, animal communities that lived in particular environments Lockley et al. 1994a). It is now becoming clear that similar ichnocoenoses recur in particular sedimentary facies. Consequently it is possible to compile relative abundance data on track types from multiple ichnocoenoses to characterize the distribution of trackmakers on a regional rather than a local scale. For example, the repeated occurrence of sauropod tracks (*Brontopodus*) and large theropod tracks from a dozen tracksites from the Early Cretaceous carbonates of Texas proves that the patterns seen at individual sites are not merely random distributions. Studies of tracks in modern environments reveal that they reflect the rank order of relative abundance of trackmakers in local animal populations (Cohen et al. 1993). Schult and Farlow (1992) also found that the track assemblage from the famous hominid tracksite at Laeotoli in East Africa, which is between 3 and 4 million years old, compared favorably with the bone record from sediments of the same age in that area.

How representative the tracks are of animal communities in particular environments, in comparison with other forms of fossil evidence such as bones and coprolites, is debatable, but some degree of "representativeness" is undeniable. In short, representativeness can be defined as being recognized at three or more levels. First, any track must represent a minimum of one species that inhabited, or passed through, the ancient environment in which the track was made. Second, the evidence from modern and ancient studies that, at least in some cases, tracks reflect the animal communities (actual rank order of observed animals or their skeletal remains) is another level of representativeness showing that tracks reflect ancient animal communities. Third, recurrent track assemblages in particular sedimentary facies show that each one is representative of a particular fauna in a particular ancient environment. This latter phenomonon is the ichnofacies concept, which is well known in ichnology, though it was not applied in vertebrate paleontology until very recently (Lockley et al. 1994a). On this note, we actually do not know how representative the bone record is, only that sometimes it is sparse, and at other times abundant, and that some assemblages are common or typical and others rare. Tracks may sample the same portion of the ancient animal community as is represented by bones, but in other cases tracks may sample a different portion of the fauna. We will return to this topic later.

The term *ichnofauna* has also been used to describe a particular track assemblage (ichnocoenosis) or assemblages (either multiple ichnocoenoses or ichnofacies). In general, though, the term is rather vague (Lockley et al. 1994a) and is probably best used as a general descriptive term in cases where ichnocoenoses and ichnofacies have not been defined precisely.

Size-Frequency Data

Because tracks are so abundant, they provide a useful source of size-frequency data that is not normally so readily available from skeletal remains, except, perhaps, in the case of certain monospecific bone beds that represent mass mortality accumulations or other special taphonomic conditions. It is not unusual for large or moderately large sites to provide size-frequency data for several dozen trackways that belong to a single ichnotaxon. In cases where such trackways are found in parallel alignment, on a single surface, indicative of herd activity at a single instant in geologic time, it is possible to infer that the sample represents individuals from a single population. For example, Lockley (1994) summarized sauropod trackway data available from three sites, revealing a substanial amount of information not available from the skeletal record.

On the basis of several samples, including a large sample of predominantly small sauropod trackways from multiple levels in South Korea, it is possible to demonstrate that, in comparison with the track record, the skeletal record of sauropods is heavily biased toward large individuals. Dodson (1990), for example, noted that the skeletal remains of juvenile sauropods are very uncommon, and that the "overwhelming majority of sauropod specimens are at least 80 percent of adult size." Using the evidence from many tracksites that full-grown sauropods had hind feet up to about 1 meter in length, the trackway evidence presents a very different picture, providing evidence of a large proportion of small sauropods that are less than half adult size (Lockley 1994), and therefore only a fraction of full adult body mass.

Another interesting application of size-frequency data is to use it to estimate the age of trackmakers. This approach is problematic because we do not know the exact relationship between dinosaur age and size. Thus the reader should be warned that many of the inferences presented herein are based on preliminary work. Although estimating individual ages of extinct organisms is a perennial problem in paleontology, there is no shortage of work in this field, and in recent years some progress has been made in estimating the ages of dinosaurs. Work by Horner (1992) has produced the first tentative growth curves for large Cretaceous ornithopods. If we assume that these curves are correct, then it is a relatively straightforward procedure to calculate the approximate age of large ornithopods on the basis of track size (Lockley 1994). It is also possible to assume that the growth curves proposed for large ornithopods are applicable to other large dinosaurs, but further study is needed. Regardless of the reliability of available growth rate models, there is clearly a general relationship between size and age that allows track size to be used for estimates of relative (not absolute) age within a given sample.

To date, available evidence suggests that very small tracks of well-known dinosaur groups, such as ornithopods and sauropods, are rare. This observation is reminiscent of an earlier debate over the rarity of small dinosaurs (Richmond 1965). Current evidence, however, suggests that the rarity of small dinosaurs is in part a function of preservational bias and collection biases among previous generations of paleontologists. It is also in part due to previous neglect of the field—a situation that has been rectified to some degree in recent years (Weishampel et al. 1990; Carpenter et al. 1994). This scarcity of small individuals may also be due to the fact that dinosaurs had rapid early growth rates, and thus spent only a very short period of their trackmaking lives as small individuals. According to Horner (1992), dinosaurs could accomplish as much as 40 to 50 percent of their growth in the first year (but see Farlow et al. 1995 for a cautionary note).

The rarity of baby dinosaur tracks (cf. Leonardi 1981) is also due in part to preservational biases (Lockley 1991a, 1994), and may also be explained by rapid early growth rates. Thus it seems unlikely that ichnologists will ever find much in the way of tracks of very small juveniles of large species. Such an inference seems especially probable when we consider that many dinosaurs completed the first few months of growth in nests or nest colonies (but see Geist and Jones 1996 for an alternative view), where the ground was heavily trampled and compacted, and therefore unsuitable for the making or preservation of small footprints. However, on the basis of recorded track size ranges and preliminary age estimates (Horner 1992; Lockley 1994), the track record evidently reveals a reasonably broad range of dinosaur age groups, beginning with animals estimated to have been about one year old.

Finally we should note that a number of Late Triassic and Early Jurassic deposits reveal large numbers of trackways of small dinosaurs. Based on the known size of skeletal remains of potential trackmakers, and the preservational bias arguments outlined above, most of these are inferred to represent the trackways of adult individuals of small-bodied species.

Trackway Orientation Data

In any study of dinosaur tracksites that reveals multiple trackways, it is important to compile trackway orientation data. Such data can be very useful in determining the extent to which different trackmakers (ichnotaxa) were progressing in random or preferred directions. In some cases track-makers of different sizes, attributed to a particular ichnotaxon, moved in different directions. For example, at Cretaceous tracksites at Dinosaur Ridge in Colorado and at Mosquereo Creek in New Mexico, large and small ornithopods, assigned to the same ichnospecies (*Caririchnium leonardii*), were moving in different directions (Fig. 37.2). Such examples demonstrate the utility of including both size-frequency and orientation data in tracksite analysis.

As summarized by Ostrom (1972) and Lockley (1986b, 1991a), parallel trackways attributable to a particular trackmaker (ichnotaxon) can be indicative of purely "biologic" herd behavior, to the passage of individuals along a "physically controlled pathway" such as a shoreline, or to a combination of these two factors (Lockley 1991a). The most compelling evidence for shore-parallel trackways comes from ichnocoenoses that show bimodal distributions (from point A to point B and from B back to A) that are parallel to wave ripple crests or other indicators of shoreline trend (Lockley et al. 1986; Lockley 1991a). By contrast, the best evidence for gregarious behavior comes from ichnocoenoses that reveal unidirectional, equally spaced trackways of uniform depth that neither cross nor deviate substantially. Such patterns, referred to as regular "intertrackway spacing" (Lockley 1989a, 1991a), are reminiscent of soldiers marching in formation, and are hard to explain as movements of individuals at different times, especially in cases where all the parallel trackways are made by individuals that fall within a discrete size cluster within a particular ichnotaxon.

According to Lucas (1994: 210), such multiple parallel trackways are "regarded by many paleontologists as the strongest evidence . . . of . . . social behavior." This principle was first demonstrated by Bird (1944), who reported twelve parallel trackways of sauropods (ichnogenus *Brontopodus*) near the top of the Cretaceous Lower Glen Rose Formation of Texas, at what is now Dinosaur Valley State Park near the town of Glen Rose. Bird (1944) also reported and mapped twenty-three parallel *Brontopodus* trackways from a higher stratigraphic level, the top of the Upper Glen Rose Formation on West Verde Creek (see Pittman 1989, 1992 for stratigraphic locations). There is further evidence of parallel sauropod trackways at the top of the Upper Glen Rose Formation at a site on the South San Gabriel River, and at various other locations in Texas (Pittman and Lockley 1994).

The West Verde Creek site, originally known as the "Davenport Ranch" site, has become famous as a compelling example of trackways that almost certainly indicate the passage of a single herd, even though there is much overlap of trackways, rather than regular intertrackway spacing. The Davenport Ranch example has been cited many times (Ostrom 1985; Haubold 1984; Lockley 1987, 1991a, 1991b, 1995; Thulborn 1990; Pittman and Lockley 1994; Lockley and Hunt 1995) in reference to dinosaurian herding behavior, and even as an example of a "structured herd" with the "very largest footprints made only at the periphery of the herd; the very smallest . . . only in the center" (Bakker 1968: 20). Subsequent analysis of the footprints fails to support this hypothesis, although there is clearly evidence for a mixture of small and large individuals within the herd (Ostrom 1985; Lockley 1987, 1991a, 1991b, 1995).

A considerable amount of recent work has revealed further evidence for social behavior at sauropod tracksites from the Late Jurassic of North America (Lockley et al. 1986) and Europe (Lockley et al. 1994b) and from the Late Cretaceous of South America (Leonardi 1984). There is also abundant trackway evidence to suggest that large Cretaceous ornithopods (iguanodontids and hadrosaurids) were gregarious (see Currie 1983 and Lockley and Hunt 1995 for examples from North America, and Lim et al.

Dinosaur Tracks 565

1989 and Lockley 1991a for examples from South Korea). Thus trackway evidence for social behavior is quite common, especially with respect to trackways attributable to large herbivorous dinosaurs.

Relating Track Assemblages to Depositional Environments

Because tracks are *in situ* sedimentary structures, they should always be considered in the context of the sedimentary successions in which they occur. Considered first from the two-dimensional perspective of single surfaces, and on the local scale, of individual tracks, we can record track and trackway parameters in relation to the composition and texture of the sedimentary units in which they occur, and in relation to relative degree of water saturation and slope of sedimentary bed. Such approaches provide information on local sediment consistency at the time of trackmaking (cf. Allen 1989). In such cases tracks are "experiments in soil mechanics" (*sensu* Seilacher 1986), which mainly reveal information about preservation and very localized parts of the depositional environment.

At intermediate scales, on the order of magnitude of typical tracksite exposures (tens to hundreds of meters), tracks and trackway segments, in conjunction with other lines of sedimentological evidence, such as ripple marks and mud cracks, may provide evidence of local shorelines or other paleogeographic features of the landscape (Lockley 1986a, 1986b; Lockley et al. 1986). They may even provide evidence that substrates were temporarily emergent in successions that otherwise consist of subaqueous deposits.

Other intermediate-scale examples of tracks on two-dimensional or planar surfaces include reports of footprints on the foreset slopes of sand dunes, where they can be used to determine paleodepositional dip (Lockley 1986a, 1986b, 1991a, and references therein). There is currently debate about the paleoenvironmental conditions (wet, dry, subaerial, or subaqueous) that prevailed when such tracks were formed (Loope 1984, 1992; McKeever 1991; Brand and Tang 1991; Lockley 1992a; Lockley and Hunt 1995).

Trampling

Further examples of the relation between tracks and local paleoenvironments include the study of trampled biotas such as plant and invertebrate body fossils (Lockley 1986b; Lockley et al. 1986; Santos et al. 1992). Although this phenomenon has not been studied in detail, there is evidently considerable potential for learning about substrate conditions and the local biota by studying plant and animal remains that have been trampled. At several bone quarry sites there is evidence of trampling in and around skeletal remains. For example, at the Late Jurassic Howe Quarry of Wyoming, there is evidence of both large and small theropods' having made tracks among various carcasses. Such evidence is clearly a compelling example of scavenging behavior.

Although it is outside the scope of this chapter to discuss trampling in detail, there is abundant evidence that Mesozoic substrates were heavily trampled in some areas (Lockley 1991a, 1991b, 1991c, 1992b). In most cases the trampling appears to be relatively localized, but examples are known where trampled beds can be correlated for several kilometers. Because trampling, or "dinoturbation" as it has been called (Dodson et al.

1980; Lockley 1991a, 1991b) causes such disruption of the substrate, it is hard to measure in quantitative terms. However, a dinoturbation index has been proposed (Lockley and Conrad 1989; Lockley 1991a, 1991b) that distinguishes light, moderate, and heavy trampling as a measure of the surface area covered by tracks (Fig. 37.4). It is also evident that dinoturbation increases in the late Mesozoic (Late Jurassic to Late Cretaceous), and appears to be mainly associated with sauropod and large ornithopod tracksites (Lockley et al. 1988; Lockley 1991a, 1991b, 1991c; Lockley and Hunt 1995). This is not surprising when one considers that these animals were large, abundant, and gregarious.

It is worth adding that trampling intensity (the dinoturbation index) is not simply a measure of population density in the area, or of the level of dinosaur activity, although these factors clearly may come into play. The index is also a measure of time. Ichnologists who study bioturbation intensity in marine deposits have long known that there is a correlation between trace fossil intensity (ichnofabric) and depositional rates (e.g., Bromley 1990). The same principle applies to the study of trampled beds in terrestrial successions, where trampling intensity may be a measure of duration between depositional events as much as a measure of biological activity. Here again we see an example of a physical or preservational explanation being just as important as a biological or behavioral interpretation.

Regional-Scale Patterns

On a larger, regional scale, it is necesary to look at the distribution of tracks in space and time. Sedimentary facies have a three-dimensional

Figure 37.4. *Left:* The dinoturbation index, a simple way of characterizing the degree to which a track-bearing surface has been trampled by the passage of animals across it. After Lockley and Conrad 1989 and Lockley 1991a, 1992b. *Right:* Differences in trampling intensity by terrestrial vertebrates through time, expressed in terms of the number of trampled sites of different ages. After Lockley and Hunt 1995.

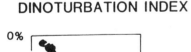

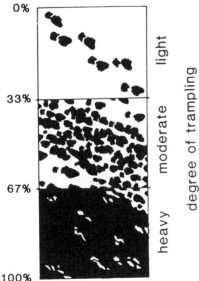

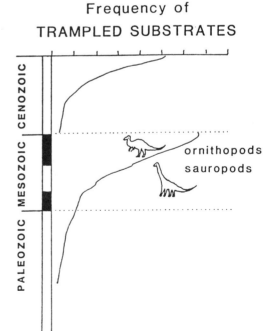

distribution in the geologic record, and as demonstrated by the existence of ichnofacies, tracksites very often mirror this three-dimensional geometry. Before the 1980s, little attention was paid to such phenomena—at best, most vertebrate ichnologists simply observed that tracks were associated with shorelines or mudflats. Subsequent work (e.g., Lockley 1989b: Fig. 50.6) has suggested that the distribution of tracks is predictable, and non-random in many cases.

For example, tracks are very common in association with crevasse splay deposits in fluvial systems (Nadon 1993) and interdune playa deposits in desert systems (Lockley 1991; Lockley and Hunt 1995). They are also common in association with lacustrine (lake) strandline deposits, where they may be stacked in multiple stratigraphic levels, depending on lake geometry. Tracks also appear to have a special relationship to aggradational deposits (located at sites where sediments accumulate), and so very often occupy specific stratigraphic positions in basinal sequences (Hunt and Lucas 1992; Lockley and Hunt 1995).

Tracks are also abundant in association with coastal plain deposits. Here they tend to occupy thin but laterally extensive zones associated with aggradational deposits that form the lower portions or early phases of transgressive systems tracts (aggradational deposits resulting from rising sea level). Such track-rich facies have in some cases been dubbed "mega-tracksites" or "dinosaur freeways."

Megatracksites

The phenomenon of regionally extensive track-bearing layers (Lockley et al. 1988) or megatracksites (Lockley and Pittman 1989; Lockley 1991a, 1991b, 1991c) was first noted in association with a number of Jurassic and Cretaceous coastal plain deposits in the western United States, but has since been observed also in Europe (Meyer 1993). As currently defined, a megatracksite is a regionally extensive single surface, or very thin package of beds, that is track-bearing or track-rich over a large area, on the order of hundreds to thousands of square kilometers.

The best examples currently known are the Middle-Late Jurassic "Moab Megatracksite" (Lockley 1991b: Lockley and Hunt 1994a, 1995), which extends for about 1,000 square kilometers in the Entrada–Upper Summerville transition zone in Utah; the Late Jurassic Solothurn Limestone megatracksite in Switzerland (Meyer 1993), which extends for about 400 square kilometers, and Lower to mid-Cretaceous megatracksite complexes associated with two levels in the Glen Rose limestone (Gulf of Mexico) and a complex of multiple levels in the Dakota Sandstone (Western Interior Seaway). Although estimates vary, these Cretaceous megatracksites have very large areal extents, on the order of tens of thousands of square kilometers (see Lockley 1991a, 1991b, 1991c, 1992b; Lockley and Hunt 1994a, 1994b; Lockley et al. 1992; Pittman 1992).

It is interesting to note that, although each is different in terms of track type and sedimentary facies, all known megatracksites consist of multiple sites that reveal similar ichnocoenoses associated with similar lithofacies. Thus they are all distinct ichnofacies as well as megatracksites. For example, the Moab megatracksite reveals multiple theropod ("*Megalosauripus*") ichnocoenoses in a siliciclastic (non-carbonate) transgressive facies, the Solothurn and Texas examples reveal theropod-sauropod track assemblages (*Brontopodus* ichnofacies) in platform carbonates (limestones), and the Dakota example reveals an ornithopod (*Caririchnium*)-

gracile theropod–crocodilian assemblage in a coal-bearing coastal plain facies assemblage.

With the exception of the Dakota megatracksite complex, which averages about 10 meters in thickness, all the other examples are associated with a single surface, or a very thin (about 1 m thick) track-bearing stratigraphic unit. At first sight this would suggest a useful biostratigraphic marker or time line. However, because the tracks are evidently associated with transgressive deposits at stratigraphic sequence boundaries, the surfaces or beds are time-transgressive, at least to a minor degree (Fig. 37.5). Initially I suggested that megatracksites form during falling sea level (Lockley 1989a, 1991a). However, further work indicates that megatracksites form during rising sea level, which causes sediments along the coastline to aggrade or accumulate (Fig. 37.5). This may seem counterintuitive, because we see tracks being made when the tide ebbs, not when it floods. But tidal cycles are not the same as long-term sea-level changes. Sequence stratigraphic studies show that rising sea level causes aggradation and leads to enhanced preservation of sediments and the tracks and other fossils they contain (Haubold 1990).

The Dakota megatracksite has been dubbed the "dinosaur freeway" and has been cited as an example of an ancient migration route. As explained by Lockley et al. (1992), the migration route hypothesis is debatable. However, it is known that the same ichnocoenoses occur in the megatracksite complex throughout an area of 80,000 square kilometers, and that large gregarious herbivores, in this case iguanodontids, were probably migratory. Thus, the Dakota "dinosaur freeway" does provide evidence that is consistent with the hypothesis of dinosaur migration.

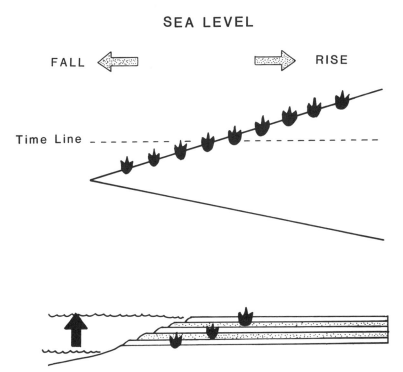

SEA LEVEL

FALL ⬅ ➡ RISE

Time Line

Figure 37.5. Schematic illustration showing how megatracksites represent accumulations of tracks through time. As sea level rises, shorelines move laterally across the land surface. Sedimentary rocks that form in coastal environments—a setting where vertebrate footprints are likely to be preserved—will differ in age from place to place, being older in the areas that are the first to be flooded as sea level rises. This idea is depicted in the upper diagram by having the track-bearing surface cut across a time line (modified from Lockley 1989b). The bottom diagram shows how such aggradation of sedimentary rock layers during sea level rise will result in the accumulation of a series of rock layers containing footprints, each layer younger than the one below it, and each formed later in the course of flooding of the continental margin.

Certainly similarly structured dinosaur faunas (or communities) ranged through this large area throughout the entire time that it took for the track beds to accumulate.

Tracks as Correlation Tools: The Science of Palichnostratigraphy

Traditionally, fossil organisms that were abundant and widespread are potentially useful for correlation or biostratigraphy—that is, for establishing the age relations of sedimentary rock layers. Most classic examples of fossils with great biostratigraphic utility come from the marine realm, and include such well-known groups as graptolites, conodonts, and ammonites. In the terrestrial realm, palynomorphs (fossilized pollen grains) have proved useful in correlation, but macrofloral remains (fossilized leaves, stems, and other plant parts), invertebrates, and vertebrates have proved of limited, or at best modest, utility. Mammals, whose teeth are relatively abundant in late Mesozoic and Cenozoic deposits, are useful for establishing biostratigraphic units known as land mammal ages. However, in pre-Cenozoic deposits, the establishment of land vertebrate ages, or biochrons (*sensu* Lucas 1991, 1993a, 1993b), is a relatively inexact science that rarely achieves the modest resolution of geologic stages (ages), and in many cases is accurate only at the broad level of geologic series (epoch).

It is therefore interesting to note that tracks have significant utility in biostratigraphy. Although their work is not widely known, European vertebrate ichnologists have proposed a series of Late Carboniferous through Early Jurassic footprint zones, and coined the term *palichnostratigraphy* for this branch of biostratigraphy (Haubold 1984, 1986; Haubold and Katzung 1978). The fact that such track zones are recognizable for this interval of time is a reflection of paleogeography. In the late Paleozoic and early Mesozoic, when the continents were coalesced as Pangaea, terrestrial vertebrate faunas were relatively cosmopolitan, and therefore of greater utility for correlation. The track zones defined by Haubold (1984, 1986) for the first two epochs of the age of dinosaurs (Late Triassic and Early Jurassic) are clearly of some utility on a global scale (see reviews in Lockley 1993a, 1993b; Lockley and Hunt 1994a, 1995; and Lockley et al. 1994a; and monographs in Ellenberger 1972, 1974).

For example, it is possible to identify latest Triassic track zones in Europe, North America, and Africa that contain the first abundant small *Grallator* tracks in association with probable prosauropod tracks assigned to the ichnogenera *Tetrasauropus* and *Pseudotetrasauropus*, and miscellaneous non-dinosaurian archosaur tracks. By contrast, younger Early Jurassic track assemblages contain larger tridactyl tracks (*Eubrontes*), purported ornithopod tracks (*Anomoepus*), and a different probable prosauropod track (*Otozoum*). Recent work demonstrates that distinctive track types (ichnotaxa) can be correlated between North America, Europe, and Asia during the Middle and Late Jurassic, after the break-up of Pangaea. For example Lockley et al. (in press) have shown that the distinctive Middle Jurassic theropod track *Carmelopodus* from the Carmel Formation of eastern Utah also occurs in deposits of exactly the same age in England. Similarly, the track type assigned to the ichnogenus *Megalosauripus* has been identified in Upper Jurassic deposits in North America, Europe, and Asia (Lockley et al., in press). Such evidence has implications for the existence of land bridges or connections between these continents throughout the Jurassic. Lockley et al. (1994a) have noted that like other fossils, tracks are facies-controlled to some degree. It is therefore important to exercise caution when correlating between different regions, and to expect greater similarites when correlating between like facies, and less similarity when correlating between different facies.

Comparison between Track and Skeletal Databases

It is clear from the landslide of new discoveries and the documentation of dinosaur tracks and other footprints that tracks are very abundant. Part of the reason for this is that any animal can potentially make many thousands of tracks in a lifetime, but it has only one skeleton. However, this is only one partial explanation, and studies at individual tracksites show that most tracks were made by different-sized individuals (as explained above). Thus the repeated activity explanation is not particularly compelling.

An alternative explanation is that tracks are an integral part of the sedimentary succession, and, though vulnerable to erosion and weathering, they are not susceptible to chemical disolution; nor are they subject to destruction by the activity of predators and scavengers. Moreover, they are most easily made and preserved in basinal settings where aggradational processes predominate, and for this reason have a relatively high preservation potential. Although individual tracks do not survive long on the surface (Cohen et al. 1991)—reminding us of the common and very resonable question, Why weren't these tracks washed away?—there is little scientific evidence that tracks *are* easily washed away. In fact, the surface expression, or two-dimensional planar view, of tracks is obscured by the very process that preserves them—by filling them in. Moreover, there is a high preservation potential for the undertrack expressions of these tracks, which are already buried, nestled in the substrate from the moment that they are made (Lockley 1991a).

Having established that tracks are common, and not rare as once supposed, we should address the widely held belief that they are associated only, or mainly, with certain types of deposits, particularly those that lack bones. Available evidence suggests that this is, to a large extent, a myth (Lockley 1991a; Lockley and Hunt 1994a). Tracks are known from a variety of different types of deposits, including those that contain abundant bones (though they are not usually found in exactly the sames beds as bones within a particular formation). For example, in the bone-rich Morrison Formation, from which only a handful of tracksites (about 5) were known prior to the 1980s, we now know of about forty sites—hardly a sparse record. Moreover, this record adds several hundred individual animals to our census of Morrison dinosaurs, and so compares favorably with the numbers on which compilations of articulated skeletal remains are based (Russell 1989; Schult and Farlow 1992).

It is perhaps more accurate to say that dinosaur skeletal remains are quite rare in many formations, thus giving the impression that tracks occur abundantly where bones do not. To emphasize this point, we can look at the distribution of dinosaurs throughout the Mesozoic, on a stage-by-stage basis, as reviewed by Dodson (1990). Such a compilation shows that dinosaurs are rare in many stages, and diverse and abundant in only a handful of Late Jurassic and Late Cretaceous deposits (Fig. 37.6). By contrast, dinosaur tracks are much more consistently abundant throughout the same sequences (Lockley and Hunt 1994a, 1995).

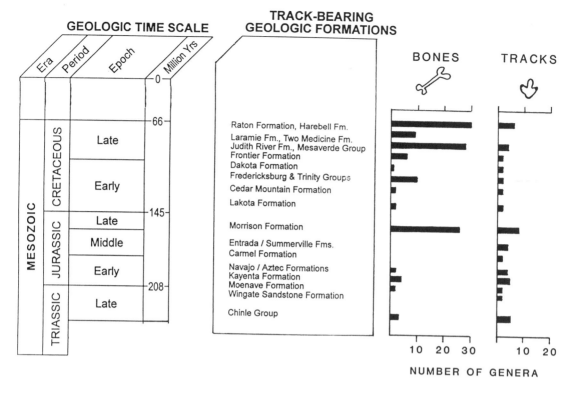

GEOLOGIC TIME SCALE

Era | Period | Epoch | Million Yrs

MESOZOIC

CRETACEOUS — Late, Early
JURASSIC — Late, Middle, Early
TRIASSIC — Late

0
66
145
208

TRACK-BEARING
GEOLOGIC FORMATIONS

BONES   TRACKS

Raton Formation, Harebell Fm.
Laramie Fm., Two Medicine Fm.
Judith River Fm., Mesaverde Group
Frontier Formation
Dakota Formation
Fredericksburg & Trinity Groups
Cedar Mountain Formation
Lakota Formation

Morrison Formation

Entrada / Summerville Fms.
Carmel Formation
Navajo / Aztec Formations
Kayenta Formation
Moenave Formation
Wingate Sandstone Formation

Chinle Group

10 20 30   10 20

NUMBER OF GENERA

Figure 37.6. A comparison of the skeletal and ichnological records of dinosaurs. The number of genera of dinosaurs described from skeletal material ("Bones" in the diagram; modified from Dodson 1990) varies greatly from one interval of Mesozoic time to another. In contrast, the ichnological record ("Tracks" in the diagram; modified from Lockley and Hunt 1995) is less spotty, with the number of ichnogenera more evenly distributed over time.

It is worth noting here that dinosaur track abundance can be measured in two ways: first as the number of trackways (taken as the number of individuals) in a given unit, and second as the number of sites that have yielded trackways. Ideally it is best to use both methods, because each site may represent a distinct stratigraphic level—again emphasizing that trackways from different sites (levels) in a formation represent different individuals. This is the method employed by Lockley and Hunt (1994a, 1995; Lockley et al. 1994a) to record the number of sites (ichnocoenoses) known from a particular formation, and it has the advantage of revealing how many times a particular track type appears in a given formation or facies.

For example, sauropod tracksites occur at eleven sites in the Gulf of Mexico Basin of Texas (Pittman 1992). Farlow (1992) advocated this type of approach, though he did not use it himself in his global compilation of sauropod tracksites, and so scored the Texas sauropod track occurrences as if they were a single site. When we record every level as a site (Lockley et al. 1994b), a much more comprehensive picture of the distribution of sauropod tracks around the world is obtained (but see McIntosh et al., chap. 20 of this volume, for a more skeptical view of this approach).

Using a compilation of number of tracksites in Mesozoic formations in the western United States, Lockley and Hunt (1994a, 1995) concluded that tracks are far more abundant than skeletal remains in the majority of formations. In many cases this conclusion is based on a bone-to-tracksite ratio of 0:15, as in the case of the Wingate Formation, or a ratio of 1:35, as in the case of the Entrada-Summerville zone. If the ratio of individuals (= minimum number of documented trackways), rather than tracksites, were calculated, the ratio disparity would be increased by an order of magnitude into the range of 0:150 and 1:350, respectively. Thus it can be seen that even using the conservative measure of counting sites, not trackways, demonstrates the quantitative superiority of trackway evidence in many cases.

The counting of individual trackways, as well as tracksites, is also important for characterizing ichnocoenoses, as well as for establishing the relative abundance of different trackmaking groups. As stated above, such track data are easy to compile, and lead rapidly to quantitatively substantial results. A three-year study at Dinosaur National Monument resulted in a trackway census of about 250 individuals, which compares favorably with the entire skeletal record excavated from this famous area (Lockley and Hunt 1993). Similarly, trackway censuses derived from the Dinosaur Freeway area (Dakota Group) run into hundreds of mapped and documented trackways (Lockley et al. 1992; Fig. 37.7).

In conjunction with the arguments presented above, it is reasonable to propose that track abundance is as much a real biologic and preservational phenomenon in the stratigraphic record as it is an artifact of repeated activity. It is therefore important not to overlook the importance of tracks

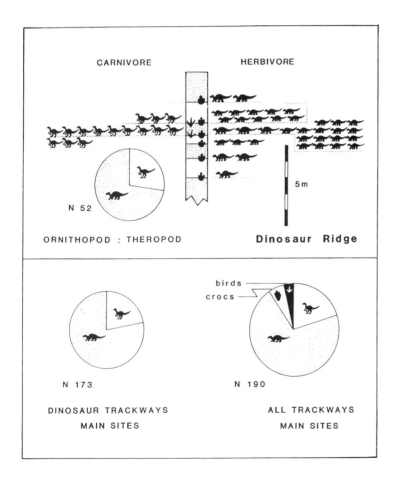

Figure 37.7. Information about the relative abundance of trackmakers of different kinds from a well-sampled set of tracksites. The upper diagram summarizes data for Dinosaur Ridge, a Cretaceous site (Dakota Group) near Denver, Colorado. Each dinosaur silhouette represents a single trackway attributed to either a small theropod (carnivore) or a large ornithopod (herbivore), out of a total of 52 trackways at the site. The pie diagram illustrates the relative abundance of the two kinds of trackmakers in a different way. The lower diagrams pool data on relative abundance of different kinds of trackmakers from numerous sites in the Dakota Group, including Dinosaur Ridge. On the left, a pie diagram compares the relative abundance of theropod and ornithopod trackways at all sites, based on a sample of 173 trackways. On the right, a pie diagram compares the relative abundance of dinosaur, bird, and crocodilians trackways, on the basis of 190 trackways. Information from Lockley et al. 1992 and Lockley and Hunt 1995.

Dinosaur Tracks   573

in filling in an otherwise very incomplete skeletal record. For too many years, tracks were regarded as a relatively unimportant line of paleontological evidence. Then, in the wake of the Dinosaur Renaissance, they were given more attention for their utility in understanding dinosaur behavior.

The approaches discussed in this chapter, with reference to the utility of tracks in understanding relationships between facies and faunas, the potential for paleoecological characterization of dinosaur populations and communities, the meaning of megatracksites, and the astonishing abundance of tracks at so many sites, allow us to ask a provocative question: Is the track record more complete and more representative than the body fossil record? (Lockley and Hunt 1994a: 95). Based on current knowledge, the answer appears to be yes!—at least in purely quantitative terms, and in terms of the frequency of occurrence of tracksites throughout the geologic record. This conclusion does not imply that it is a qualitatively better record, because the skeletal record is usually superior in terms of the taxonomic resolution that can be derived from relatively complete body fossils. The conclusion does, however, provide significant food for thought, and raises fundamental questions regarding the completeness of the fossil record. It is fair to conclude that dinosaur ichnology has come a long way in the last decade, and that the subdiscipline has found its rightful and legitimate place in the field of sedimentary geology. The future "looks bright and the wealth of available material promises to keep researchers active for many years" (Lockley 1989b: 447).

References

Alexander, R. McN. 1989. *Dynamics of Dinosaurs and Other Extinct Giants*. New York: Columbia University Press.

Allen, J. R. L. 1989. Fossil vertebrate tracks and indenter mechanics. *Journal of the Geological Society of London* 146: 600–602.

Baird, D. 1957. Triassic reptile footprint faunules from Milford, New Jersey. *Bulletin Museum of Comparative Zoology, Harvard University* 117: 449–520.

Bakker, R. T. 1968. The superiority of dinosaurs. *Discovery* 3: 11–22.

Bakker, R. T. 1975. Dinosaur renaissance. *Scientific American* 232 (4): 58–78.

Bird, R. T. 1944. Did *Brontosaurus* ever walk on land? *Natural History* 53: 60–67.

Brand, L. R., and T. Tang. 1991. Fossil vertebrate footprints in the Cocinino Sandstone (Permian) of northern Arizona: Evidence for underwater origin. *Geology* 19: 1201–1204.

Bromley, R. G. 1990. *Trace Fossils: Biology and Taphonomy*. London: Unwin Hyman.

Carpenter, K.; K. F. Hirsch; and J. R. Horner (eds.) 1994. *Dinosaur Eggs and Babies*. Cambridge: Cambridge University Press.

Coe, M. J.; D. L. Dilcher; J. O. Farlow; D. M. Jarzen; and D. A. Russell. 1987. Dinosaurs and land plants. In E. M. Friis, W. G. Chaloner, and P. R. Crane (eds.), *The Origins of Angiosperms and Their Biological Consequences*, pp. 225–258. Cambridge: Cambridge University Press.

Cohen, A.; J. Halfpenny; M. G. Lockley; and E. Michel. 1993. Modern vertebrate tracks from Lake Manyara, Tanzania and their paleobiological implications. *Paleobiology* 19: 443–458.

Cohen, A.; M. G. Lockley; J. Halfpenny; and E. Michel. 1991. Modern vertebrate track taphonomy at Lake Manyara, Tanzania. *Palaios* 6: 371–389.

Colbert, E. H. 1962. The weights of dinosaurs. American Museum *Novitates* 2076: 1–16.

Currie, P. J. 1983. Hadrosaur trackways from the Lower Cretaceous of Canada. *Acta Palaeontologica Polonica* 28: 63–73.

Czerkas, S. J., and S. A. Czerkas. 1990. *Dinosaurs: A Global View*. Limpsfield, U.K.: Dragon's World.

Dodson, P. 1990. Counting dinosaurs: How many kinds were there? *Proceedings of the National Academy of Sciences, U.S.A.* 87: 7608–7612.

Dodson, P.; A. K. Behrensmeyer; R. T. Bakker; and J. S. McIntosh. 1980. Taphonomy and paleoecology of the dinosaur beds of the Jurassic Morrison Formation. *Paleobiology* 6: 208–232.

Ellenberger, P. 1972. Contribution à la classification des pistes vertébrés du Trias: Les types du Stormberg d'Afrique du Sud (I). *Palaeovertebrata* Memoire Extraordinaire.

Ellenberger, P. 1974. Contribution à la classification des pistes de vertébrés du Trias: Les types du Stormberg d'Afrique du Sud (II partie: le Stormberg superieur- I. Le biome de la zone B/1 ou niveau de Moyeni: ses biocénoses). *Palaeovertebrata* Memoire Extraordinaire.

Farlow, J. O. 1981. Estimates of dinosaur speeds from a new trackway site in Texas. *Nature* 294: 747–748.

Farlow, J. O. 1992. Sauropod tracks and trackmakers: Integrating the ichnological and skeletal records. *Zubía* 10: 89–138.

Farlow, J. O.; P. Dodson; and A. Chinsamy. 1995. Dinosaur biology. *Annual Review of Ecology and Systematics* 26: 445–471.

Gardom, T., and A. Milner. 1993. *The Book of Dinosaurs: The Natural History Museum Guide.* London: Prima Publishing.

Geist, N. R., and T. D. Jones. 1996. Juvenile skeletal structure and the reproductive habits of dinosaurs. *Science* 272: 712–714.

Gillette, D. D., and M. G. Lockley (eds.). 1989. *Dinosaur Tracks and Traces.* Cambridge: Cambridge University Press.

Haubold, H. 1984. *Saurierfährten.* Wittenberg Lutherstadt: Die Neue Brehm-Bucheri.

Haubold, H. 1986. Archosaur footprints at the terrestrial Triassic-Jurassic transition. In K. Padian (ed.), *The Beginning of the Age of Dinosaurs,* pp. 189–201. Cambridge: Cambridge University Press.

Haubold, H. 1990. Dinosaurs and fluctuating sealevels during the Mesozoic. *Historical Biology* 4:176–206.

Haubold, H., and G. Katzung. 1978. Paleoecology and paleoenvironments of tetrapod footprints from the Rotliegend (Lower Permian) of central Europe. *Palaeogeography, Palaeoclimatology, Palaeoecology* 23: 307–323.

Hitchcock, E. 1836. Ornithichnology, description of the footmarks of birds (Ornithoidichnites) on New Red Sandstone in Massachusetts. *American Journal of Science* 29: 307–340.

Hitchcock, E. 1858. *Ichnology of New England: A report on the Sandstone of the Connecticut Valley, especially its Fossil Footmarks.* Boston: W. White. Reprint, New York: Arno Press, 1974.

Horner, J. 1992. Dinosaur behavior and growth. In R. S. Spencer (ed.), *Fifth North American Paleontological Convention, Abstracts and Program,* p. 135. Paleontological Society Special Publication 6.

Hunt, A. P., and S. G. Lucas. 1992. Stratigraphic distribution and age of vertebrate tracks in the Chinle Group (Upper Triassic), western North America. *Geological Society of America, Abstracts with Program* 24: 19.

Leonardi, G. 1981. Ichnological data on the rarity of young in North East Brazil dinosaurian populations. *Anais da Academia Brasileira de Ciências* 53: 345–346.

Leonardi, G. 1984. Le impronte fossili de dinosauri. In J. F. Bonaparte, E. H. Colbert, P. J. Currie, A de Ricqlès, Z. Kielan-Jaworowska, G. Leonardi, N. Morello, and P. Taquet (eds.), *Sulle Orme dei Dinosauri,* pp. 165–186. Venice: Erizzo Editrice.

Lim, S-Y.; S.-Y.Yang; and M. G. Lockley. 1989. Large dinosaur footprint assemblages from the Cretaceous Jindong Formation of South Korea. In D. D. Gillette and M. G. Lockley (eds.), *Dinosaur Tracks and Traces,* pp. 333–336. Cambridge: Cambridge University Press.

Lockley, M. G. 1986a. *A Guide to Dinosaur Tracksites of the Colorado Plateau and American Southwest.* Denver: University of Colorado at Denver, Geology Department Magazine Special Issue 1.

Dinosaur Tracks   575

Lockley, M. G. 1986b. The paleobiological and paleoenvironmental importance of dinosaur footprints. *Palaios* 1: 37–47.

Lockley, M. G. 1987. Dinosaur trackways. In S. J. Czerkas and E. C. Olsen (eds.), *Dinosaurs Past and Present*, pp. 80–95. Seattle: Natural History Museum of Los Angeles County/University of Washington Press.

Lockley, M. G. 1989a. Tracks and traces: New perspectives on dinosaurian behavior, ecology and biogeography. In K. Padian and D. J. Chure (eds.), *The Age of Dinosaurs*, pp. 134–145. Short Course 2. Knoxville: Paleontological Society, University of Tennessee.

Lockley, M. G. 1989b. Summary and prospectus. In D. D. Gillette and M. G. Lockley (eds.), *Dinosaur Tracks and Traces*, pp. 441–447. Cambridge: Cambridge University Press.

Lockley, M. G. 1991a. *Tracking Dinosaurs: A New Look at an Ancient World.* Cambridge: Cambridge University Press.

Lockley, M. G. 1991b. The dinosaur footprint renaissance. *Modern Geology* 16: 139–160.

Lockley, M. G. 1991c. The Moab Megatracksite: A preliminary description and discussion of millions of Middle Jurassic tracks in eastern Utah. In W. R. Averett (ed.), *Guidebook for Dinosaur Quarries and Tracksites Tour, Western Colorado and Eastern Utah*, pp. 59–65. Grand Junction, Colo.: Grand Junction Geological Society.

Lockley, M. G. 1992a. Comment: Fossil vertebrate footprints in the Coconino Sandstone (Permian) of northern Arizona—Evidence for underwater origin. *Geology* 20: 666–667.

Lockley, M. G. 1992b. La dinoturbación y el fenómeno de la alteración del sedimento por pisadas de vertebrados en ambientes antiguos. In J. L. Sanz and A. D. Buscalioni (eds.), *Los Dinosaurios y su Entorno Biotico,* pp. 269–296. Cuenca, Spain: Actas del Segundo de Paleontologia en Cuenca, Instituto "Juan de Valdes."

Lockley, M. G. 1993a. *Auf der Spuren der Dinosaurier.* (Translation of Lockley 1991a with additional chapter [no. 11] and prologue.) Berlin: Birkhauser.

Lockley, M. G. 1993b. *Siguiendo las Huellas de los Dinosaurios.* (Translation of Lockley 1991a with additional material.) Madrid: McGraw-Hill.

Lockley, M. G. 1994. Dinosaur ontogeny and population structure: Interpretations and speculations based on footprints. In K. Carpenter, K. F. Hirsch, and J. R. Horner (eds.), *Dinosaur Eggs and Babies*, pp. 347–365. Cambridge: Cambridge University Press.

Lockley, M. G. 1995. Track records. *Natural History* 104 (6): 46–50.

Lockley, M. G., and K. Conrad. 1989. The paleoenvironmental context and preservation of dinosaur tracksites in the western USA. In D. D. Gillette and M. G. Lockley (eds.), *Dinosaur Tracks and Traces*, pp. 121–134. Cambridge: Cambridge University Press.

Lockley, M. G., and V. F. dos Santos. 1993. A preliminary report on sauropod trackways from the Avelino Site, Sesimbra Region, Upper Jurassic, Portugal. *Gaia:* 6: 38–42.

Lockley, M. G., and A. P. Hunt. 1993. Fossil footprints: A previously overlooked paleontological resource in Utah's National parks. In V. L. Santucci (ed.), *National Park Service Paleontological Research Abstract Volume*, p. 29. Denver: U.S. Department of the Interior, Natural Resources Publication Office.

Lockley, M. G., and A. P. Hunt. 1994a. A review of vertebrate ichnofaunas of the Western Interior United States: Evidence and implications. In M. V. Caputo, J. A. Peterson, and K. J. Franczyk (eds.), *Mesozoic Systems of the Rocky Mountain Region, United States*, pp. 95–108. Denver: Society of Economic Paleontologists and Mineralogists (Rocky Mountain Section).

Lockley, M. G., and A. P. Hunt. 1994b. A track of the giant theropod dinosaur *Tyrannosaurus* from close to the Cretaceous/Tertiary boundary, northern New Mexico. *Ichnos* 3: 213–218.

Lockley, M. G., and A. P. Hunt. 1995. *Dinosaur Tracks and Other Fossil Footprints from the Western United States.* New York: Columbia University Press.

Lockley, M. G., and J. G. Pittman. 1989. The megatracksite phenomenon: Implications for paleoecology, evolution and stratigraphy. *Journal of Vertebrate Paleontology* 9 (Supplement to no. 3): 30A.

Lockley, M. G., and A. Rice. 1990. Did *"Brontosaurus"* ever swim out to sea? Evidence from brontosaur and other dinosaur footprints. *Ichnos* 1: 81–90.

Lockley, M. G.; K. Conrad; and M. Jones. 1988. Regional scale vertebrate bioturbation: new tools for sedimentologists and stratigraphers. *Geological Society of America Abstracts with Program* 20: 316.

Lockley, M. G.; J. Holbrook; A. Hunt; M. Matsukawa; and C. Meyer. 1992. The dinosaur freeway: A preliminary report on the Cretaceous Megatracksite, Dakota Group, Rocky Mountain Front Range and Highplains, Colorado, Oklahoma and New Mexico. In R. Flores (ed.), *Mesozoic of the Western Interior,* pp. 39–54. Denver: Society of Economic Paleontologists and Mineralogists Midyear Meeting Fieldtrip Guidebook.

Lockley, M. G.; K. J. Houck; and N. K. Prince. 1986. North America's largest dinosaur trackway site: Implications for Morrison paleoecology. *Bulletin Geological Society of America* 97: 1163–1176.

Lockley, M. G.; A. P. Hunt; and C. Meyer. 1994a. Vertebrate tracks and the ichnofacies concept: Implications for paleoecology and palichnostratigraphy. In S. Donovan (ed.), *The Paleobiology of Trace Fossils,* pp. 241–268. New York: Belhaven Press.

Lockley, M. G.; A. P. Hunt; M. Paquette; S. A. Bilbey; and A. Hamblin. In press. Dinosaur tracks from the Carmel Formation, northeastern Utah: Implications for Middle Jurassic paleoecology. *Ichnos.*

Lockley, M. G.; C. A. Meyer; and V. F. dos Santos (eds.). 1994b. Aspects of sauropod biology. *Gaia* 10: 1–279.

Lockley, M. G.; C. A. Meyer; and V. F. dos Santos. 1996. *Megalosauripus, Megalosauropus* and the concept of megalosaur footprints. In M. Morales (ed.), *The Continental Jurassic,* pp. 113–118. Bulletin 60. Flagstaff: Museum of Northern Arizona.

Lockley, M. G.; V. Novikov; V. F. dos Santos; L. A. Nessov; and G. Forney. 1994c. "Pegadas de Mula": An explanation for the occurrence of Mesozoic traces that resemble mule tracks. *Ichnos* 3: 125–133.

Loope, D. 1984. Eolian origin of Upper Paleozoic sandstones, southeastern Utah. *Journal of Sedimentary Petrology* 54: 563–580.

Loope, D. 1992. Comment on fossil vertebrate footprints in the Coconino sandstone (Permian) of northern Arizona: Evidence for underwater origin. *Geology* 20: 667–668.

Lucas, S. G. 1991. Sequence stratigraphic correlation of nonmarine and marine Late Triassic biochronologies, western United States. *Albertiana* 9: 11–18.

Lucas, S. G. 1993a. The Chinle Group: Revised stratigraphy and biochronology of Upper Triassic nonmarine strata in the western United States. *Museum of Northern Arizona Bulletin* 59: 27–50.

Lucas, S. G. 1993b. Vertebrate biochronology of the Jurassic-Cretaceous boundary, North America Western Interior. *Modern Geology* 18: 371–390.

Lucas, S. G. 1994. *Dinosaurs: The Textbook.* Dubuque, Iowa: William C. Brown.

Maples, C. G., and R. R. West (eds.). 1992. *Trace Fossils.* Short Course 5. Knoxville: Paleontological Society, University of Tennessee.

McKeever, P. 1991. Trackway preservation in eolian sandstones from the Permian of Scotland. *Geology* 19: 726–729.

Meyer, C. 1993. A sauropod dinosaur megatracksite from the Late Jurassic of northern Switzerland. *Ichnos* 3: 29–38.

Nadon, G. C. 1993. The association of anastomosed fluvial deposits and dinosaur tracks, eggs and nests: Implications for the interpretation of floodplain environments and a possible survival strategy for ornithpods. *Palaios* 8: 31–44.

Norman, D. 1991. *Dinosaur.* London: Boxtree.

Ostrom, J. H. 1969. Terrestrial vertebrates as indicators of Mesozoic climates. *North American Paleontological Convention, Chicago Proceedings* D: 347–376.

Ostrom, J. H. 1972. Were some dinosaurs gregarious? *Palaeogeography, Palaeoclimatology, Palaeoecology* 11: 287–301.

Dinosaur Tracks   577

Ostrom, J. H. 1985. Social and unsocial behavior in dinosaurs. *Bulletin Field Museum of Natural History* 55: 10–21.

Padian, K., and D. Chure (eds.). 1989. *The Age of Dinosaurs*. Short Course 2. Knoxville: Paleontological Society, University of Tennessee.

Pittman, J. G. 1989. Stratigraphy, lithology, depositional environment, and track type of dinosaur track-bearing beds of the Gulf Coastal Plain. In D. D. Gillette and M. G. Lockley (eds.), *Dinosaur Tracks and Traces*, pp. 135–153. Cambridge: Cambridge University Press.

Pittman, J. G. 1990. Dinosaur tracks and trackbeds in the middle part of the Glen Rose Formation, western Gulf Basin, USA. In G. R. Bergan and J. G. Pittman (eds.), *Nearshore Clastic-Carbonate Facies and Dinosaur Trackways in the Glen Rose Formation (Lower Cretaceous) of Central Texas*, pp. 47–83. Field Trip Guide no. 8. Dallas: Geological Society of America.

Pittman, J. G. 1992. Stratigraphy and vertebrate ichnology of the Glen Rose Formation, Western Gulf Basin, USA. Ph.D. thesis, University of Texas at Austin.

Pittman, J. G., and M. G. Lockley. 1994. A review of sauropod dinosaur tracksites of the Gulf of Mexico Basin. *Gaia* 10: 95–108.

Richmond, N. D. 1965. Perhaps juvenile dinosaurs were always scarce. *Journal of Paleontology* 39: 503–505.

Russell, D. A. 1989. *An Odyssey in Time: Dinosaurs of North America*. Toronto: University of Toronto Press/National Museum of Natural Sciences.

Santos, V. F.; M. G. Lockley; J. J. Moratalla; and A. M. Galopim de Carvalho. 1992. The longest dinosaur trackway in the world? Interpretations of Cretaceous footprints from Carenque, near Lisbon, Portugal. *Gaia* 5: 18–27.

Schult, M. F., and J. O. Farlow. 1992. Vertebrate trace fossils. In C. G. Maples and R. R. West (eds.), *Trace Fossils*, pp. 34–63. Short Course 5. Knoxville: Paleontological Society, University of Tennessee.

Seilacher, A. 1986. Dinosaur tracks as experiments in soil mechanics. In D. D. Gillette, D. D. (ed.), *First International Symposium on Dinosaur Tracks and Traces, Abstracts with Program*, p. 24. Albuquerque: New Mexico Museum of Natural History.

Thulborn, R. A. 1982. Speeds and gaits of dinosaurs. *Palaeogeography, Palaeoclimatology, Palaeoecology* 38: 227–256.

Thulborn, R.A. 1990. *Dinosaur Tracks*. London: Chapman and Hall.

Thulborn, R. A., and M. Wade. 1984. Dinosaur trackways in the Winton Formation (mid-Cretaceous) of Queensland. *Memoirs of the Queensland Museum* 21: 413–517.

Weishampel, D.; P. Dodson; and H. Osmólska (eds.). 1990. *The Dinosauria*. Berkeley: University of California Press.

Welles, S. P. 1971. Dinosaur footprints from the Kayenta Formation of northern Arizona. *Plateau* 44: 27–38.

Dinosaur Evolution in the Changing World of the Mesozoic Era

I can visualize the entire scene . . . the huge pterodactyls soaring through the heavy air . . . the mighty dinosaurs moving their clumsy hulks beneath the dark shadows of preglacial forests...

—Edgar Rice Burroughs, *The People That Time Forgot*

The general public often thinks of all the dinosaurs as having lived at the same time (possibly with sabertoothed "tigers," wooly mammoths, and cavepersons as neighbors). There is little appreciation of the fact that dinosaur faunas were constantly changing, that they differed from place to place and time to time.

The amount of time that separates the reader from the huge Late Cretaceous carnivorous dinosaur *Tyrannosaurus* is about the same amount of time by which *Tyrannosaurus* was preceded on the earth by *Allosaurus*, a big predator of the Late Jurassic. And the same stretch of time separates *Allosaurus* from one of the earliest carnivorous dinosaurs, little *Coelo-*

physis of the Late Triassic. When *Tyrannosaurus* stalked the Late Cretaceous wilds of western North America, the bones of *Allosaurus* had been fossilized for 70 million years. It is philosophically interesting to speculate about which dinosaur fossils *Tyrannosaurus* might have seen eroding from the ground upon which it stepped—and what, if anything, the huge predator would have made of them!

This section of the book puts dinosaur faunas into a geographic and chronological framework, and also describes some of the other animals that shared the world with the dinosaurs.

The first chapter serves as an introduction to those that follow by summarizing the principles of biogeography, the scientific study of the geographic distribution of living things. The tectonic forces that shape the surface of the earth, resulting in the changing configurations of continents and oceans, are described. The chapter then considers several topics germane to the study of dinosaur biogeography, such as different theoretical approaches to biogeography, methods of reconstructing ancient climates, and the impact of climatic change on animal evolution. The chapter ends by examining the implications of all of this material for understanding the geographic distribution of dinosaur faunas.

Dinosaurs were not the only Mesozoic vertebrates, and to understand their world one must know something about their contemporaries. We therefore have included a chapter that consists of a Mesozoic non-dinosaurian bestiary.

Chapters 40 and 41 describe changes in dinosaur faunas over time. The first of these focuses on the early part of the Mesozoic Era, when the major landmasses of the world remained in close proximity. We then explore the effects of increasing fragmentation of the continents during the later Mesozoic on the makeup of dinosaur faunas in different regions.

Although dinosaurs successfully weathered significant environmental changes over the course of their history, the Cretaceous-Tertiary (K-T) boundary event(s) proved to be more than they could handle. Nor were they alone: Numerous other terrestrial and marine organisms likewise perished. The last decade and a half has seen vigorous debate between "gradualists," who argue for slowly acting terrestrial causes of extinction, and "catastrophists," who explain the K-T extinctions in terms of the collision of a comet or asteroid with the earth. The final chapter of this section is a dialogue between a gradualist and a catastrophist, with each offering his own interpretation of the evidence pertaining to the Cretaceous extinctions, and the two authors together searching for areas of common ground between their positions.

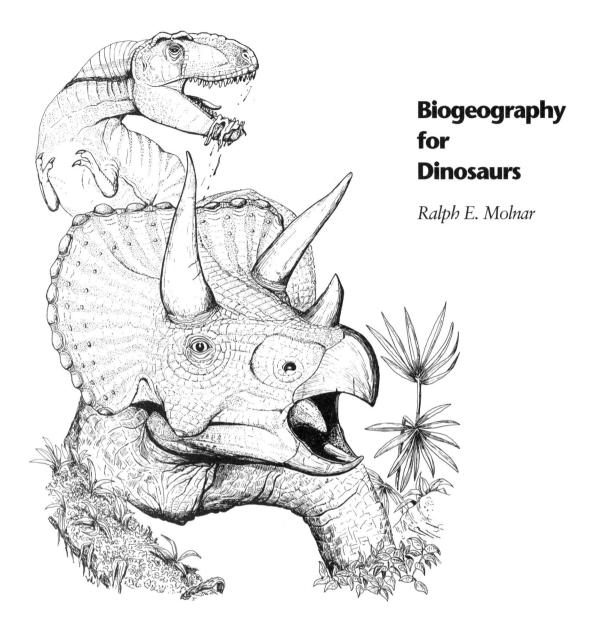

Biogeography for Dinosaurs

Ralph E. Molnar

The past was different; otherwise there could be no history. It has even been said that the past is a different country. The actual quotation (Hartley 1967: 3) is very much to the point: "The past is a foreign country: they do things differently there." We don't take this seriously, and simply imagine that the world of the past was very much like the present. But our imaginations are weak, for the past was literally a foreign world, and a "trip" into the Mesozoic would take us to a place unrecognizable except to specialists in the evolution and history of the earth.

We are raised on the notion that life has evolved, and so we realize that the creatures and plants of the Mesozoic were different from those now alive. But the climate and geography of the earth have also changed. To understand the distribution and ecology of dinosaurs, we need to understand how these aspects of the environment differed from those familiar

38

today and, more important, how we can discern this from the raggle-taggle remains of ancient organisms and the sedimentary detritus deposited when they lived and died.

Continental Drift and Plate Tectonics: A Brief History

The great voyages of European discovery of the sixteenth century—or rather the maps derived from them—revealed that the east and west coastlines of the Atlantic Ocean more or less matched. Sir Francis Bacon remarked in 1610 that this match could be no coincidence, and M. François Placet suggested that the ocean had been created by the biblical deluge, separating what previously had been a single landmass. In the less religious and more rational early twentieth century, the idea arose that the continents on both sides of the ocean had slipped apart, creating the Atlantic in their wake.

Although not the first to propose such a theory, the German meteorologist Alfred Wegener stated the hypothesis in the most detail and was first vilified, then hailed as the father of continental drift. Although the idea was supported by some geologists, mostly from the Southern Hemisphere, it did not gain general acceptance in either Europe or North America. Much has been made of the fact that Wegener was not a geologist but an "outsider"; however, be that as it may, there were two good reasons for rejecting continental drift—at first. One was that Wegener didn't propose any mechanism that could propel the massive continents around the surface of the earth, through solid rock; and the other was that geological evidence for drift, although clear in South America and southern Africa, was hard to find in the northern continents. Most geologists of the time were not wealthy enough to travel about the world simply to examine the local geology in detail. So most geologists regarded continental drift as a novel, but unlikely, hypothesis.

The change in attitude occurred in the 1960s, and it came about precisely from the discovery of convincing solutions to these two problems.

The Evidence for Plate Tectonics

The exploration of the Atlantic floor, particularly near the Mid-Atlantic Ridge, revealed that it was composed of solidified lavas. These lavas retained a record, a signature, of the direction of the earth's magnetic field when they solidified. These signatures formed clear bands parallel to the ridge (Fig. 38.1). The earth's magnetic field has repeatedly reversed polarity in the past, interchanging North and South poles, and this banding showed that the ages of the lavas varied from new, adjacent to the ridge, to older, away from it. It was as if the sea floor had formed at the ridge and slowly spread away, and this process—appropriately termed sea-floor spreading—is just what happened. As the sea floor spread, it pushed the continents away, sometimes driving them over the floors of other oceans. In the case of North America, as the Atlantic floor spread, the continent was driven over the Pacific floor to the west, elevating the Cordilleran Mountains (including the Rockies, Sierra Nevada, and others) in the process.

This evidence for sea-floor spreading was far from all the evidence for drift. Geographical features, geological formations, and the ranges of fossil animals and plants could be matched across oceans, most notably across

582   *Ralph E. Molnar*

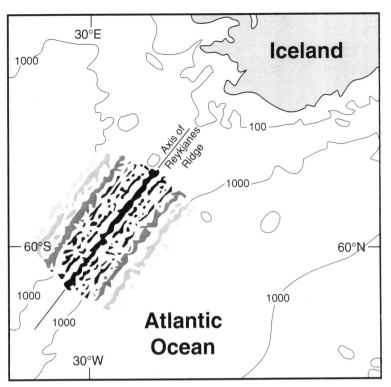

30°E

1000

Iceland

Axis of Reykjanes Ridge

100

1000

60°S

60°N

1000

1000

1000

1000

Atlantic Ocean

30°W

Figure 38.1. Magnetic "stripes" of the sea floor surrounding the Mid-Atlantic Ridge southwest of Iceland. New oceanic crust forms from volcanic eruptions at the crest of the ridge. As the lava solidifies, iron-bearing minerals align with the prevailing magnetic field of the earth. As the sea floor splits and moves to either side away from the ridge, a given magnetic band is torn in two to form a mirror-image pattern across the Mid-Atlantic Ridge. New molten rock rises from below to replace the solid crust that has been carried away. Should the earth's magnetic field reverse while this is happening, the newly crystallized ocean crust will have a polarity that is just the opposite from that prior to the reversal. In this diagram, dark-colored bands correspond to regions of oceanic crust that are magnetized with "normal" polarity, like that which prevails in the modern world. Light-colored bands correspond to "reversed" polarity. Redrawn from Press and Siever 1994.

the South Atlantic in both South Africa and southern Brazil (Colbert 1973). In both places, local geologists had long been convinced of the reality of continental drift. Most dramatically, the development of satellites and of methods of precisely measuring small distances and very short periods of time allowed actual measurement of the movement of the continents at several centimeters per year—roughly the rate at which your fingernails grow.

The Mechanism of Continental Movement

At first, the mechanism was not so convincingly obvious as the evidence. Geophysicists had long known that small pieces of rock (small as compared to the masses that make up the continents) are rigid and brittle on the surface of the earth, but at depth, under great pressure and heat, they become plastic and flow. In the earth's mantle (the thick region of the earth's interior separating the outer crust from the inner core) they move in great slow convection currents, rising from below, spreading along the surface of the mantle (beneath the crust), and then sinking back when they have cooled sufficiently. It is their slow spread along the surface that drags along the lighter, floating continents (and sea floors), enabling the continents to override the ocean floors, where the currents of the mantle have cooled, and so descend again into the depths of the planet (Fig. 38.2). So to geophysicists, the salient features of geography are not the continents and oceans but the crustal plates—including both continents and sea floors—

Biogeography for Dinosaurs 583

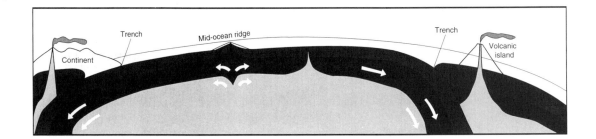

Figure 38.2. Diagram of the mechanism of plate tectonics. The earth's crust is shown in dark gray, the rigid uppermost part of the mantle in black, and the hotter, less rigid mantle below the outermost mantle in light gray. Molten rock from the mantle rises at a mid-ocean ridge (or its equivalent on land, a rift valley), solidifies, and then moves away to either side of the ridge. Eventually the lithospheric plate on which oceanic crust rides sinks into the mantle at a subduction zone, which is expressed at the earth's surface as a deep-ocean trench. The sinking plate melts, and some of the molten material rises to the surface to form a volcanic island arc or a chain of volcanic mountains at the edge of a continent. Isolated plumes of molten material that erupt onto the ocean floor create chains of volcanic islands, like the Hawaiian chain, as the plate moves over them.

that float on the mantle (Fig. 38.3), borne along by its currents. It is these plates that give this theory its name, plate tectonics.

So far, there is no direct evidence for these currents, and not much indirect evidence either. But there is some. Seismic studies show inclined, plate-like masses of (relatively) cool rock far beneath the west coasts of Oregon and Peru (cf. Vidale 1994). These masses are believed to be parts of sinking, or subducting, oceanic crust. How do these masses relate to convection currents in the mantle? Simple—there is no other known or hypothesized mechanism for dragging these masses of cool rock down into the mantle, except by being caught up in the convection currents (although the weight of the descending slabs may also play a role; Press and Siever 1994). In addition, other seismic studies show that the rock in the upper mantle is anisotropic—that is, upper mantle rocks show differences in their physical properties in different directions.

Dissent from Plate Tectonics and Continental Drift

Continental drift and the theory of its mechanism, plate tectonics, are widely, but not universally, accepted (cf. Chatterjee and Hotton 1992). Some geologists accept continental drift but propose other mechanisms for the motion. Others still believe that the continents have always remained in the same positions.

Most prominent among the theories that accept continental displacement, or drift, but not plate tectonics is the hypothesis that the continents have moved because the earth itself is expanding. So far, however, no convincing mechanism for this expansion has been put forth, nor is there any convincing evidence for it. Other geologists suggest that the continents rotated rather than moved in straight lines (Fig. 38.4).

Another hypothesis, "surge tectonics," suggests that the features of the earth's surface explained by plate tectonics are better explained by varying, rather than constant, flows of the mantle. Yet others doubt that the continents moved at all, at least laterally. One school, reminiscent of believers in "lost continents," contends that the major continental motions have been vertical. They believe that since the Jurassic, large regions have sunk into the oceans, mainly the western Pacific. And many geologists of the former Soviet Union simply adhered to pre-drift concepts of geology and geography.

All—well, almost all—of these theories were put forward by scientists to resolve what they saw as difficulties in the theory of plate tectonics. All have some evidence to support them. The adherents of plate tectonics, who

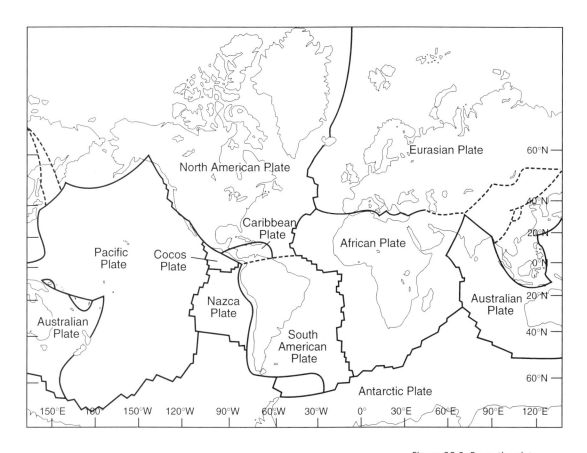

Figure 38.3. From the plate-tectonic viewpoint, the primary features of the earth's surface are the lithospheric plates, not the more obvious but superficial continents and oceans. This figure illustrates the modern pattern of lithospheric plates. Redrawn from Duxbury and Duxbury 1994.

are the majority, believe that the past success of this concept indicates that further research will resolve these difficulties, real or apparent, in a way supporting plate tectonics. And so far they have been right—but we cannot be sure that new discoveries will not generate another Wegener. Several of these alternative views, however, have influential supporters in places such as Australia, New Zealand, and Russia, and this must be remembered when reading paleobiogeographical papers from these countries.

Theories that the continents have moved laterally, from place to place, give different interpretations of biogeography from those that hypothesize that the continents have remained in place. Some views, especially the vertical motions school, may be consistent with plate tectonics, and hence have no special paleozoogeographical implications. The sinking of the Lord Howe Rise, east of Australia, seems to be associated with the fragmenting of the eastern part of the Australasian plate on its northward drift. The notion of sunken lands in the western Pacific (e.g., Pacifica or the Darwin Rise) may be relevant to the distribution of Asiamerican dinosaurs, such as tyrannosaurs or ceratopsids. Unlike some modern zoogeographers, no dinosaurian zoogeographers have taken it seriously—yet.

Plate tectonics is the most popular interpretation because of its success in linking together seemingly disparate kinds of geological observations. This theory relates paleobiogeography, position and structure of mountains and other topographic features, and location of mineral deposits—among other things—in a way not done by the alternatives (not yet,

Biogeography for Dinosaurs   585

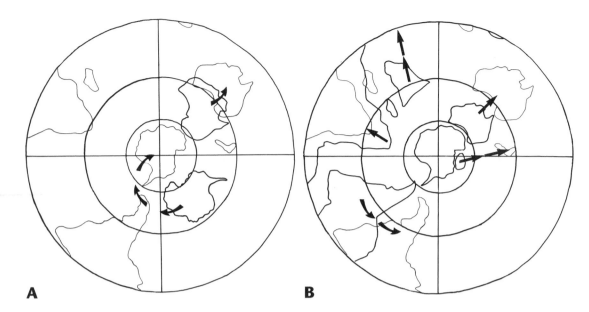

A   **B**

Figure 38.4. Continental rotation
(A) vs. continental drift (B) as
alternative explanations for the
movements of the southern
continents since the Cretaceous.
Cretaceous and modern
positions of the continents are
indicated, with arrows pointing
from the Cretaceous to the
modern postitions. Part A shows
the changes in position of
Australia and Antarctica
according to the rotation school
of thought (information about
the putative rotations of South
America and Africa are unavailable). Part B shows the generally
accepted view of continental
drift for all of the southern
continents. Here the landmasses
move in lines that are as straight
as is possible on the spherical
surface of the earth. (India, of
course, has disappeared over the
"horizon" into the Northern
Hemisphere.)

anyway), and certainly not by any previous geological theory. In other
words, so far it has more explanatory power than other theories. It has
proved quite influential in dinosaurian studies. Under the previous paradigm of static continents, no one seriously considered *Dryosaurus* from the
Late Jurassic of the Rocky Mountain region and *Dysalotosaurus* from the
Late Jurassic of southern Tanzania to be closely related. They lived too far
apart. But we now know that in the Late Jurassic, Africa was adjacent to
South America, and not far from North America, so considering them as
two species of a single genus (*Dryosaurus*) no longer seems unlikely on
geographical grounds.

Figures 38.5 through 38.12 summarize current thinking about the
changing shape of world geography during the Mesozoic Era. The positions of the continents had a profound effect on the distribution of dinosaurian faunas.

Theories of Biogeography

Having some notion of the geographical background for biogeography, we should now think about the aims of biogeography. Biogeography
attempts to explain the distributions of animals and plants—no one
seems to worry much about bacteria and fungi—in terms of four factors:
climate, barriers to dispersal (such as mountains, deserts, and seas), the
geographical distribution of resources and underlying soils or rock types,
and the evolutionary history of the organisms in the area of interest.
There seems to be a feeling among some biogeographers that biogeo-
graphical processes occur faster than evolutionary ones—that, for example, if the animals of an island are exterminated by a Krakatoa-like
volcanic eruption, those that reappear on the island when it cools may be
drawn from the same populations of the same species as were the original
inhabitants.

In this chapter we are interested in much longer times. Paleobiogeog-

586   *Ralph E. Molnar*

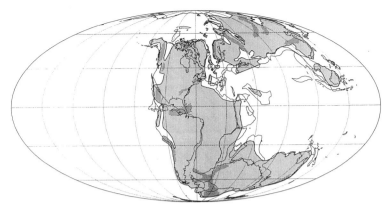

Figure 38.5. Paleogeographic reconstruction of continental positions during the Late Triassic (Carnian through Rhaetian Ages). Light stipple shows landmasses, with highlands in heavier stipple. The modern outlines of the continents are indicated by light lines. The major landmasses of the world have come together to form the supercontinent Pangaea (or Pangea), although a shallow, narrow seaway separates North America from Europe. Western Europe consists of a series of small islands, a configuration that will remain through the remainder of the Mesozoic Era. Map from Smith et al. 1994.

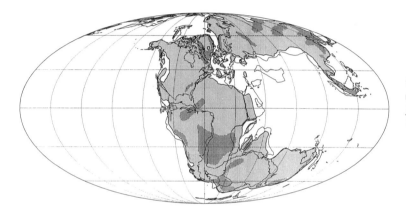

Figure 38.6. Paleogeographic reconstruction for the Early Jurassic (Pliensbachian Age). From Smith et al. 1994.

raphy attempts to make sense of the distributions of plants and animals in terms of their evolution, of the positions and past movements of continents, islands, and oceans, and of past climates.

Vicariance Biogeography

Before the acceptance of continental drift, no one realized that continents had histories. Mountains had risen and eroded, and epicontinental seas flooded the lowlands, but there were only two kinds of important changes, biogeographically speaking. These were the establishment and loss of land bridges linking different continents, and changes in the climate—and that was climatic, not continental, history. Thus paleobiogeography was simply biogeography on a grander, or at least longer, scale. With the recognition of continental drift, it became clear that continents had histories. For example, during the early Paleozoic Era, North America had been adjacent to Australia, but in the Mesozoic Era it was adjoined to

Biogeography for Dinosaurs   587

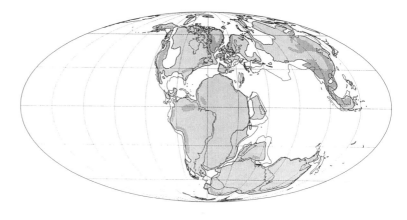

Figure 38.7. Paleogeographic reconstruction for the Middle Jurassic (Callovian Age). The Atlantic Ocean is in its infancy, and a shallow seaway has invaded western North America. From Smith et al. 1994.

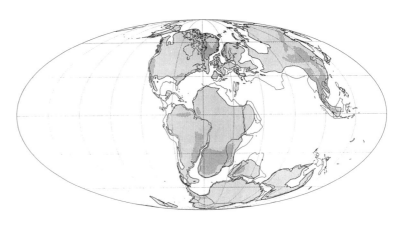

Figure 38.8. Paleogeographic reconstruction for the Late Jurassic (Tithonian Age). India, Antarctica, and Australia (East Gondwanaland) are separating from Africa and South America (West Gondwanaland). From Smith et al. 1994.

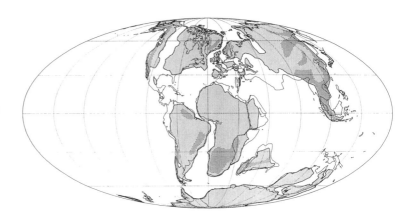

Figure 38.9. Paleogeographic reconstruction for the Early Cretaceous (Hauterivian-Barremian Ages). The North Atlantic continues to grow, and the South Atlantic is starting to become a significant body of water. From Smith et al. 1994.

Europe, and now it is connected to South America. Such different geographical relationships obviously had a great effect on the potential paths of migration and dispersal.

Furthermore, the movement of the continents suggested that, like a ship, they had carried with them complements of (evolving) plants and animals. Where continents had long been in contact, such as Europe and

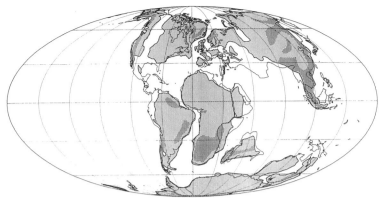

Figure 38.10. Paleogeographic reconstruction for the later Early Cretaceous (Albian Age). Antarctica and India are widely separated from Africa, India has lost contact with Australia and become an island, and shallow seas flood much of Australia and North America. From Smith et al. 1994.

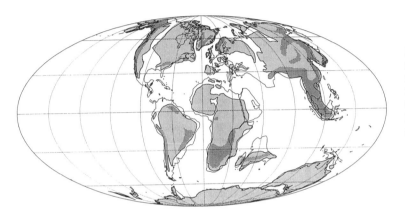

Figure 38.11. Paleogeographic reconstruction for the early Late Cretaceous (Turonian Age). Shallow seaways cover many of the continents, and Madagascar has separated from India. From Smith et al. 1994.

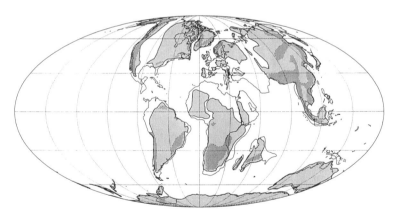

Figure 38.12. Paleogeographic reconstruction for the Late Cretaceous (Campanian Age). Continental flooding continues, and the continents are starting to move into positions reminiscent of modern geography. By the end of the Cretaceous Period (Maastrichtian Age), North America will have made a transient contact with South America, and the seas will have largely withdrawn from the continents. From Smith et al. 1994.

Asia, there would have been time for extensive intermingling of faunas, but where continents had been isolated, such as Australia and South America during the Tertiary Period of the Cenozoic Era, unique faunas would be expected. Vicariance biogeography (Fig. 38.13) considers the history and distributions of organisms in relation to the history and movements of continents and islands. It also explicitly concerns itself with the evolution-

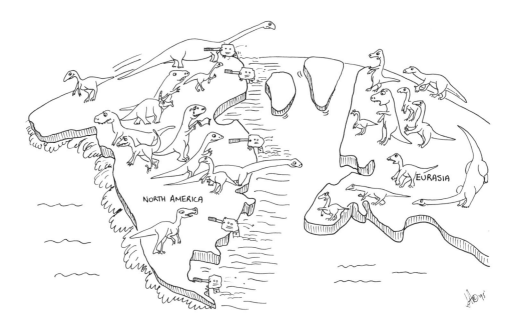

Figure 38.13. A cartoon illustrating the vicariance model of animal distributions: Whole continents move, carrying along their faunas (and floras). Drawing by S. Hocknull.

ary history of organisms, because understanding the phylogenetic relationships of organisms is essential to understanding their distribution, and how that distribution came about. Vicariance biogeography accepts that the main mechanism for distributing animals and plants across the globe is the movement of continents and smaller landmasses.

Note that vicariance biogeography applies in only a limited way to marine organisms. It does seem relevant to shallow-water, benthic (bottom-living) organisms—and perhaps to those of deep waters as well—but the swimmers and drifters of the high seas pretty much ignore it. Although the oceans and their basins may, in a sense, be said to move, they cover so much of the earth's surface that it is mainly currents, temperature, salinity, and physical barriers, such as continents, that determine the distributions of pelagic organisms.

Dispersalist Biogeography

The school of biogeography, and paleobiogeography, that prevailed before the acceptance of continental drift—and therefore didn't realize that it was just a school rather than the whole science—is now termed dispersalist (or Wallacean) biogeography (Fig. 38.14). The name follows from the preferred method of explaining the distributions of organisms by the movement of individuals (or populations) across continents and seas—in other words, the dispersal of organisms. Dispersalist biogeography accepts that the main mechanism for distributing animals and plants is their own dispersal, whether by flying, swimming, walking, drifting, or even being blown away by a hurricane. (Its alternative name honors Alfred Russel Wallace, who pioneered the science of biogeography.) Dispersal was aided by land bridges for large terrestrial animals, and by winds and ocean currents for small buoyant animals and seeds. It was inhibited or prevented

590   *Ralph E. Molnar*

Figure 38.14. The dispersalist model of animal distributions: Movement of animals is significant, not that of continents. This may be because the continents don't move (an idea no longer generally accepted), or because animals move from place to place much more rapidly than continents drift. Cartoon by S. Hocknull.

altogether by oceans, mountains, and deserts, as well as by other regions with unsuitable climates.

Dispersalist biogeographers took to heart the concept of allopatric speciation, that species usually or always arose in small, geographically isolated populations. Thus in order to acquire the broad geographical ranges often seen, animals and plants would have to emigrate outward from the small areas where they had originated. To get to islands, or different continents, they would have to migrate or disperse, by intent or by accident, across oceans or land bridges—unless, of course, they could fly. But even then, because most animals stick to some home territory, they would still disperse only when seeking a new territory or as the result of some unusual event, such as a hurricane.

Dispersalist paleobiogeography thus considered it important to discover where species or evolving lineages of organisms originated, and how they arrived (both literally and figuratively) at their present distributions. Although there were notable exceptions, such as the American paleontologist William Diller Matthew, some dispersalist biogeographers thought that dispersal occurred almost instantaneously compared with evolution. Thus distributions were the results of evolution, not components of it. Although vicariance biogeographers have criticized and played down the role of this school in understanding the changing distributions of plants and animals, some explanations by "dispersalists" have stood the test of time. The lineages of horses, for example, are widely accepted to have first appeared in North America, and then dispersed to Eurasia and Africa, eventually becoming extinct in their "homeland," North America. Similarly, we think that ceratopsians, tyrannosaurs, and hadrosaurs originated in Asia before dispersing into North America sometime in the middle of the Cretaceous.

Panbiogeography

There is a third school popular in some parts of the world (e.g., New Zealand). This is panbiogeography (Fig. 38.15), founded by Leon Croizat.

Biogeography for Dinosaurs   591

THAT MOUNTAIN WASN'T THERE WHEN MY FATHER'S FATHER'S ...... FATHER WAS HERE! SEE!

Figure 38.15. The panbiogeography model of animal distributions: Animals and continents both slowly evolve together. Cartoon by S. Hocknull.

This school is significant in that it is often taken to have been the forerunner of vicariance biogeography—but not by Croizat himself. Croizat, originally a botanist, was a prolific and engaging writer, but one who preferred painting a picture in words of his ideas to stating them explicitly. Basically he took the view that biogeography was intimately related to evolution and to time in geologically long periods. Organisms—that is, lineages of organisms—were older than the geography of today, and so modern biogeography is the result of paleobiogeography (and evolution). Panbiogeography is basically a way of looking at and studying the patterns of distributions of animals and plants, without—at least according to its proponents—presupposing how they came to be. Croizat did not accept that speciation necessarily took place at one point, geographically speaking, and then organisms spread from that point of origin. Instead, many species originated over large regions, including where they now live, and their ranges subsequently changed, either by extinction or through the emergence or submergence of lands: he did not accept plate tectonics. He also did not reject dispersal out of hand, but instead felt that dispersal "fine-tuned" the distributions we see today from those of the ancient past.

The "Modern Synthesis" of Biogeography

We can characterize, almost parody, dispersalist biogeography as assuming that animals move, but not continents, and vicariance biogeography as assuming that continents move, but not animals. Most biogeographers now accept that both occur. There is an important difference in scale here that isn't always obvious. Dispersalist biogeography treats the movements of organisms, exemplified by individuals, which may be adults or young. Vicariance biogeography treats the movements of entire faunas and floras. Dispersalist biogeography is concerned with relatively short times, geologically speaking, and vicariance with long periods. In this light we would expect most of the distributions at large scales to be explained by vicariance biogeography, while dispersalist biogeography accounts for the details of and the exceptions, mostly at small scales, to vicariance theory. No one would use vicariance biogeography to explain why lions, zebras, hyenas and hippos but not giraffes, crocodiles, and okapis live within Ngorongoro Crater, although all of them inhabit the Serengeti Plain (or other parts of Africa). Likewise, we no longer appeal to dispersalist theory to explain why (until recently, geologically speaking) the large tetrapods in Australia were marsupials and birds, and in New Zealand large ground-dwelling birds, rather than eutherian mammals as in most other places.

Predictions in Biogeography

From our knowledge of continental positions throughout the history of the earth, we should be able to work out what kinds of animals inhabited areas and times for which the fossil record is unknown. For example, Antarctica is covered by a massive ice cap that makes collecting fossils very difficult—although certainly not impossible. If we wish to get some idea of the Cretaceous dinosaurs of Antarctica, we can look to the distributions of dinosaurs of the other southern continents, which are reasonably well known. From these distributions we can work out what dinosaurs lived in Cretaceous Antarctica. In fact, this was done in 1989 (Molnar 1989), and one of the "predictions," that small ornithopods lived in Antarctica, was later verified (Hooker et al. 1991).

The discovery of dinosaurian fossils in Antarctica, together with the finding of remains of forests, shows that Antarctica was not always as cold as it is now. Because Antarctica has been in the vicinity of the South Pole since the beginning of the Cretaceous, this is an impressive example of climatic change over geological time, a topic which we shall now consider.

Paleoclimatology

Changes in the weather are more than well known; in fact, in some parts of the world "weather" is almost synonymous with swift and unexpected change. But the weather changes more than from day to day; it changes from month to month, with the seasons, and even over historical time, not to mention geological time. Long-term changes—say, those that take longer than a month—are considered climate, rather than weather, so in understanding dinosaurian biogeography, it is the Mesozoic climate, rather than the Mesozoic weather, that is important. Climate is controlled by well-known factors such as the input of heat from the sun and the position of the earth in its orbit, among others. Both of these have changed, but the former so slowly that we needn't worry about it here, even for climate, and the latter—at the moment—seems not to have had as great an influence on Mesozoic climates as it has since, especially during the ice ages.

But climate is also linked to geography, and hence to continental drift. As continents move north or south, they stray into regions of different climates. A prime example of this has been Australia. Early in the Cenozoic Era, this continent was situated in the southern temperate zone, with its southern coast near the Antarctic Circle. The climate was cool and moist.

Today Australia is farther north, almost straddling the Tropic of Capricorn, and it is basically a desert continent. It has crept north from the region where ascending air masses dropped their moisture as rain, into the region where descending air masses heat as they fall and so absorb any water available on the ground. So a cool, moist climate came to be replaced by a warm, dry one. This instance shows how the concept of drift can explain some apparent changes in climate seen in the geological record. In fact, here the global climate has not changed much—what changed was the position of Australia.

Some geologists believe that the continents, and their drift, can influence climate in a more profound way. Continental collisions can create mountain ranges, and mountains influence both how the air flows and how it is heated. It has been suggested, for example, that the rise of the Himalayas profoundly affected the climate during the Cenozoic (Raymo and Ruddiman 1992).

Evidence of Past Climate in Sedimentary Rocks

Mud cracks and raindrop impressions are obvious indicators of the weather of the past. But they do indicate weather rather than climate, unless there is layer after layer of them. Geochemical changes that occur in sediments during and just after their deposition are better indicators of climate. Several different kinds of continental sediments are laid down only, or usually, under certain climatic conditions. Coals, for example, form when the climate is moist enough to support abundant plant growth, often in swamps or similarly moist environments: there are no desert coals. Red beds are oxidized sediments, and were once thought to indicate arid conditions. But this view was unfounded, and they are now thought to form under several different kinds of climates. When the air is dry, water evaporates, leaving behind any dissolved salts it may contain. Seawater carries lots of dissolved substances, so when it evaporates, specific minerals known as evaporites are left behind, often in specific sequences. These, when found in thick beds, indicate a dry climate—if not really hot, at least warm enough to evaporate large amounts of seawater. Similarly some minerals may be deposited by evaporating groundwater. Such occurrences, especially carbonates known as caliches, also indicate aridity.

Of the various sedimentary paleoclimatic indicators, the most controversial are tillites. These are rocks formed from tills, unsorted deposits of the rocks, gravel, sand, and silt that result from glacial action. The controversy involves how rocks can be conclusively identified as tillites. Marine sediments, although not discussed here, can also provide paleoclimatic evidence: coral reef deposits, for example, indicate warm climates.

None of these methods gives specific measurements, such as temperatures, of past climatic conditions, but from the ratio of oxygen isotopes (^{16}O to ^{18}O), temperatures can be calculated. The rate at which these isotopes are taken up by various compounds, being a chemical process, depends upon the temperature, so with appropriate mathematical treatment, the ratio can indicate a temperature. These are often sea or groundwater temperatures, rather than air temperatures, but they usually indicate at least roughly what the air temperatures were. However, one must be wary. In lowland New Guinea, groundwater temperatures near rivers flowing down from the mountains reflect the cold air temperatures high in the mountains rather than the tropical conditions of the lowland. Furthermore, oxygen isotope ratios may be altered by diagenesis long after the sediments bearing them were deposited. Even so, with careful interpretation, they provide important information.

Paleobotanical Evidence for Past Climate

Fossil plants provide another way to gain some idea of past climates. Animals can shelter from the climate, but plants usually can't (unless they are annuals, or die back to roots or tubers). Thus the kinds of plants that lived together in some place indicate what the climate of that place was like—in a general way. There is, however, always some uncertainty about whether the fossil plants had the same taste in climate as do their modern descendants.

The weather is such an important factor to plants in some places that they develop specific adaptations to deal with it. Thus climates can be deduced from plant morphology (Dilcher 1973). Leaf size and the nature of leaf margins are often used. Large leaves and leaves with entire (i.e., smooth) margins (Fig. 38.16) indicate warm or moist conditions, and small leaves and leaves with notched ("non-entire") edges suggest cool or dry conditions. It is the proportion of the large leaves or those with entire margins among those fossilized at a site that is used as the indicator, not simply the occurrence of these kinds of leaves. And the inferences are necessarily a bit vague: conditions are interpreted as warm or moist (or both), but temperatures or rainfall amounts are not given as such. Drip tips, elongate tips to leaves (Fig. 38.16), are taken to imply that there was so much rain—at least seasonally—that specific structures were needed to

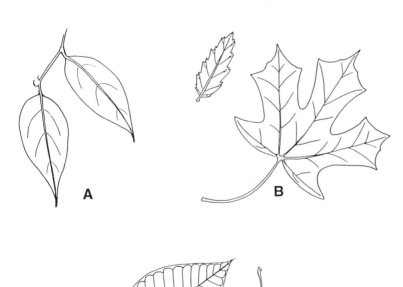

Figure 38.16. Features of leaves useful for inferring past climates. (A) Drip tips at the ends of leaves indicate high rainfall. (B) Notched leaf margins suggest cool or dry climates. (C) Unnotched, entire margins suggest wet or warm conditions. See Dilcher 1973 for details.

facilitate the shedding of the water. These correlations between climate and leaves were recognized in angiosperms, but some paleobotanists have applied them to other plants as well. Seasonality may be indicated by growth rings ("tree rings"). And Eocene (early Cenozoic) *Banksia* seeds, very similar to those of today that require fire to germinate, indicate seasonal dry conditions, conducive to fires.

Climatic conditions can be inferred both from features of the plant communities and from morphological features of individual plants. As mentioned above, the proportion of leaves (from all of the plants) with serrate margins is related to mean annual temperature. However, as with the oxygen isotope methods, caution must be used in inferring the temperature. The proportions of leaves preserved may be altered during the processes of fossilization, and soil types and amounts of rainfall also affect this proportion (Wing and Greenwood 1993).

Simulations of Climate

The basic factors that seem to determine the climate are well known—these include the amount of sunlight, the position of the earth in its orbit, the amount of solar radiation reflected off clouds, seas, and ice, and such (Fig. 38.17)—so it should be possible to use a computer to simulate the weather and the climate as well as to predict their changes. This use of computers, like many others, began in the 1940s with the Hungarian mathematician John von Neumann. He quickly discovered, however, that it wasn't a simple problem, and that many calculations were needed to simulate, hence to forecast, the weather. In fact, calculating the weather twenty-four hours in advance took very nearly twenty-four hours to do. As computers became more sophisticated and faster, simulating the global climate became feasible. Simulations in the 1970s showed the potential importance of the greenhouse effect and so encouraged the development of better models. Run "backward," these models have been used to simulate climates of the past.

The basics of the simulations for predicting the weather (and climate)

Figure 38.17. Simple model of the major features that influence weather and climate. These parameters and their interrelationships can be expressed as equations, which in turn can be used to simulate weather and climate. Redrawn from Casti 1991. © John Wiley & Sons Limited. Reproduced with permission.

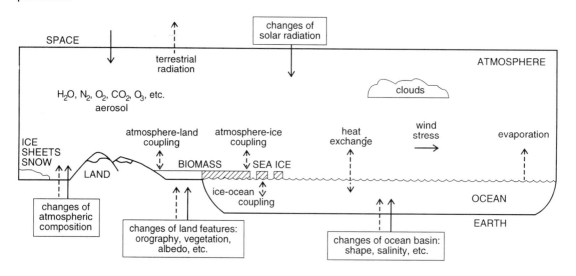

596   *Ralph E. Molnar*

are straightforward. Both can be expressed in numbers. These numbers are the values of certain properties of the weather, such as wind speed, barometric pressure, amount of rainfall per year or per season, etc. All are interrelated (see Fig. 38.17): Pressure changes cause winds, for example. The values of these at any one time can be calculated from their values at some previous time—in other words, tomorrow's weather depends on the weather today. This is simply saying that there is causation even in the weather.

The interrelationships among these variables are complex, but well understood, so they can be expressed as a series of equations that can be handled by a computer. The equations describe the dynamics of a fluid, the gases of the atmosphere, in four dimensions that include the familiar three dimensions of space and the fourth of time. They relate the input of heat from the sun to evaporation, changes in pressure, and hence air movement, both horizontally (winds) and vertically (updrafts and downdrafts). Vertical movements may result in precipitation and hence rain or snowfall, and winds can transport both moisture and heat. Specifically, five equations describe (1) horizontal and (2) vertical movement of the air, (3) conservation of mass in going from one to the other (so that, for example, no simulated air vanishes when a downdraft hits the simulated ground and transforms into a wind), (4) heating of the air, and (5) conservation of mass for the moisture in the air (as either humidity or precipitation). Each equation is then calculated for a given part of the atmosphere, imagined (i.e., simulated) as a three-dimensional grid (Fig. 38.18; cf. Casti 1991 for details).

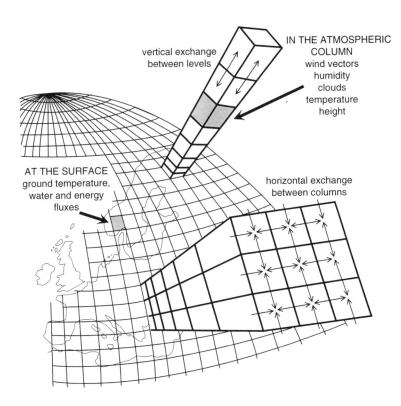

Figure 38.18. In quantitative models of climate (and weather), the atmosphere is imagined as a three-dimensional grid. Calculations are then carried out for each cell in this grid. The large size of these cells necessary to keep the calculations from being prohibitively complex is one source of error in such simulations. Redrawn from Casti 1991. © John Wiley & Sons Limited. Reproduced with permission.

Biogeography for Dinosaurs   597

But climate cannot be simulated merely by simulating the weather for a long time. Although the one stems from the other, they involve distinct processes, with obviously different time frames, and their simulations require different information. For example, in simulating the weather, we wish to know—among other things—the temperature at various times during the day. For climatic processes it is the average temperature that is important: for these processes the differences in temperature at different times of the day are just trivial fluctuations. Variations in the amount of heat arriving at the earth that occur over long periods of time have little (direct) effect on the weather, but they are influential in determining the climate.

There are two basic kinds of numerical models. Those starting with values of the variables—temperature, pressure, etc.—for some specific time are used for forecasting. Thus they can calculate (in theory, anyway!) tomorrow's weather from information about today's. The second type, which calculates from general, or average, values of variables such as the amount of heat arriving at the earth, are used in reconstructing climates of the past (and future). The most widely used of these are the GCMs, general circulation models.

It was generally hoped that such climatic models would allow us to accurately re-create, mathematically, the climates that prevailed at various times of geological history. This proved easier to hope than to do. Paleoclimatic models have encountered several problems and difficulties, and even simulations of the modern climate are not exact. First, there are problems with simulating the interactions among the atmosphere and other components of the planet. The oceans, ice caps, topography of the land, and even clouds all affect the atmosphere and weather but have proved difficult to simulate in detail. Second, there are problems with the actual computation of the models. Heated air moves much more quickly than heated ocean water, and the difference in rates of movement creates problems for the models because the rate of computation, and hence simulated movement, is the same whether simulating air or water. The simulations are also rather coarse-grained, which in recent models means that the weather is assumed to be uniform over patches about 500 kilometers across (Fig. 38.18). This is like assuming that the weather is always the same from New York City to North Carolina, or in both San Francisco and Los Angeles.

The simulation represents only large-scale features. However, the most extreme computational problems are those related to dynamic chaos, lightly referred to as the "butterfly effect." This occurs when imperceptibly small changes in the values of the variables can lead to substantially different results—in other words, when very small differences at first are greatly amplified later on into large or even very large ones—the proverbial butterfly who flaps her wings in Rio one week, which leads to a hurricane in London (or wherever) the next. The situation is exaggerated in paleoclimatic models because there are no real (i.e., measured) values to work with in the first place. All the values—temperatures, pressures, and such—are inferred, and so are doubtless less reliable than those we can measure today.

Viewed in this light, the last problem is perhaps no surprise. Simulated past climates do not always agree with climatic implications drawn from fossils. Simulations of the Permian (late Paleozoic) climate of central Gondwanaland (now South Africa) indicate extreme temperatures, to 60°C in the summer and –40°C in the winter. Yet a wide variety of fossil plants and vertebrates (largely therapsids) suggest a much more temperate climate. Perhaps the plants and therapsids were adapted to such extremes? But the sedimentary indicators and even oxygen-isotope paleotemperatures don't support these extremes either. It seems the problem is one of resolution and topography (Yemane 1993). There is evidence for large lakes in the region, with the water acting to cool the climate during the summers and warm it in winter. There is a similar lack of agreement between simulation and evidence for the Eocene (early Cenozoic) climate of central North America (Wing and Greenwood 1993). When considering claims for the ability of dinosaurs to survive where simulations indicate seasonally freezing temperatures, we must ask if there may have been similar ameliorating factors (such as the presence of high-latitude plant cover; Otto-Bliesner and Upchurch 1997) for the climate.

But these problems all seem to concern the details of the simulations rather than the basics. Unless there is something important that we don't know about the climate, climatic simulation looks to be a promising tool for interpreting paleobiogeography. It may be, however, that there are important things we don't know about the climate, for recently there have been two such surprises. Clouds have turned out to absorb from three to five times as much sunlight as modelers had thought (Kerr 1995). And the climate (and sea ice) around Antarctica has a tendency to run in cycles of eight to ten years in duration (Yuan et al. 1996). Nonetheless, the use of simulations helps point up the importance of such discoveries, so that they can be incorporated into later generations of simulations.

Geography, Climate, and Evolution

Geography and climate have certainly had some effect on the course of evolution, even if by and large they have attracted little attention from many evolutionary biologists. But they are also related to the processes of evolution, if only because they are factors in natural selection. Animals and plants survive, mate and reproduce in some environment, and the conditions of that environment play a role in determining their success at these activities. Climate and geography are major features of the environment, and so may be expected to have a significant effect on survival, mating and reproduction. The effect of natural selection on any organism is its success or failure at these three activities, so geography and climate are factors in this. Geography and climate may enter into selection in four basic ways: by the effects of land area; by the degree of separation or isolation between regions; by regions where species tend to originate ("cradles"); and by the effects of climatic change.

The basic, and still most widely accepted, mechanism for the origin of new species involves geography. This is the process of allopatric speciation. Very briefly, it involves the separation of a species into two parts, with no possibility of mating between any members of one part and those of the other. Allopatric speciation comes about by the implacement of some barrier, usually geographical such as a sea, between the two parts. And if the barrier is in place long enough, the two populations are thought to diverge genetically to the point that reproduction between them ceases, and new species are formed. Allopatric speciation was an integral part of dispersalist biogeography and remains so in vicariance biogeography.

Another component of the dispersalist school was the importance of land bridges. These were proposed in pre-continental drift times as a major means of dispersal of terrestrial animals from one continent to another. Although they were more a means of getting creatures around than a factor in evolution, they did play this role. When land bridges appeared, contact was possible between previously isolated faunas, and this "end of isolation" resulted in the evolution of new taxa and the extinction of old. But this leads us to another relationship, one between area and faunal diversity. Simply stated, it can be shown that, other things being equal, the landmass with the greater area will have the greater number of animal species (cf. Brown 1995). Islands have less diverse faunas than continents not only because they are difficult to reach, but also because their areas are smaller than those of continents.

This effect suggested to paleontologists, such as Henry Fairfield Osborn, early in this century that large landmasses were the "cradles" where new species and higher taxa originated. And it was this hypothesis that led him to fund the American Museum Mongolian expeditions of the 1920s under Roy Chapman Andrews (see Lavas, chap. 4 of this volume). Now that we accept continental drift, it seems that Osborn was right—many dinosaurian and mammalian groups did originate in Asia—but for the wrong reasons. If Osborn's idea were correct, we would expect the greatest rates of diversification and appearance of new forms in land animals to have occurred when all the continents were together in the supercontinent of Pangaea—but as far as we can tell, this didn't happen.

The idea of great landmasses' being "cradles" for speciation was never really controversial, but it led to a notion that was: that those animals that evolved on the largest landmasses, including modern Eurasia, were exposed to the most intense selection. In consequence it was thought that they could outcompete corresponding types of animals that had evolved on smaller landmasses. Thus it was no surprise to discover that North American/Eurasian mammals drove many of the native South American mammals to extinction when the former reached South America in the late Cenozoic. Well, maybe. The situation isn't really all that clear, because some of the southerners, such as glyptodonts and ground sloths, prevailed against the tide of northern mammals to the extent that ground sloths made it all the way to Alaska. In addition, armadillos have settled down and are surviving well in North America even against that most competitive mammal, humans.

The real problem in working out just what happened during the great American faunal exchange of the Late Cenozoic, and why, is twofold. First of all, you can't run experiments on intercontinental faunal exchanges— the scale is too big, the process takes too long—and above all think of the expense! And second, we have a good record of only one of these "invasions"—that of South America, as just mentioned.

The last of the effects noted in dispersalist biogeography, and again recognized by the vicariance school, is the effect of geographic isolation. This is not the kind of small-scale isolation involved in allopatric speciation, but isolation on the grand scale: the kind now "enjoyed" by Australia, New Zealand, and Antarctica. This grand isolation allowed the evolution over the Cenozoic Era of faunas and floras unlike those found anywhere else. The difficulty of reaching these places is shown by their biotas—prior to the coming of humans, especially Europeans, of course. In Australia the major terrestrial herbivores were marsupials, and the predators reptiles. In New Zealand the plant-eaters were flightless birds (moa), and the predators flying birds. On little Cocos Island (in the Indian Ocean), the main land animals were (and still are) crabs. Isolation also permitted the survival of animals and plants all of whose close relatives elsewhere had become extinct, the relicts or so-called living fossils. Australia, isolated throughout much of the Cenozoic Era, still retains a strong complement of such creatures.

The coming of vicariance biogeography did little to change our appreciation of these relationships between geography and the course of evolution—they are, after all, fairly obvious. This school did add some concepts, though, such as the carrying of whole faunas and floras across oceans on large or small fragments of continents, semi-facetiously termed "Noah's arks." And it did sharpen our understanding of the interaction of geography and evolution more generally. But its major contribution came from reconsidering isolation. During the course of the Mesozoic Era, a single large, contiguous landmass broke into several smaller (though still large) isolated or semi-isolated continents. This necessarily split up the populations of the land-dwelling and freshwater animals. In effect, it imposed allopatric speciation on a large scale.

Bjorn Kurtén (1969) proposed that the breaking up of the supercontinent Pangaea increased the diversity of land-dwelling animals, and this seems to be the explanation for the origin of those dinosaurian and mammalian groups in central Asia that Osborn thought originated as a result of the large area of that continent. It has recently been suggested, largely on what might be called circumstantial evidence, that the origins not only of some major groups of dinosaurs, but also of major groups of mammals, birds, and frogs, were related to the breakup of Pangaea (Hedges et al. 1996). If correct—and the agreement between the dates of origin of the groups and of the breakup of Pangaea is a bit vague—this indicates that continental drift can have evolutionary effects that reverberate up the taxonomic hierarchy, and produce new families and orders as well as just new species.

Recently more novel relationships have been proposed. First, and probably most important, is the role of climate. Elisabeth Vrba has proposed that climatic changes tend to occur more or less in synchrony around the world (cf. Tudge 1993). These changes, not surprisingly, drive changes in evolution. Vrba has analyzed the fossil records of late Cenozoic African mammals in detail and claims that such data provide support for her concept. It is not simply that the climate changes and the animals respond; her idea is more subtle. With changes in climate, small animals or specialist feeders may become extinct and be replaced by new species, while those with a more eclectic appetite simply change their diet. In Africa, the record shows extinctions and replacements in the specialist bovids (bovids include cattle, sheep, and antelope) around 2.25 million years ago, but no change in generalist feeders such as impala (which are also bovids) and pigs. This was a period of climatic change at least in Africa, northern South America, and China. Vrba's idea seems plausible, and much of the criticism of it is actually over whether the climatic and faunal changes are worldwide, or even Africa-wide, and not over their evolutionary impact (Tudge 1993).

Down under, the influence of climate on evolution has been taken up by paleomammalogist Tim Flannery (1994), who has argued that Aussie ecosystems are significantly different from those of the northern continents because of climatic conditions. In Australia the climate is very unpredictable: there may be rain, lots or little, in a given wet season, or it may be dry for a decade. Meteorologists can't tell, and neither can the local flora and fauna. Flannery uses the example that animals such as the North America's grizzly bears, which hibernate during the winter knowing that they can find sufficient food come spring, did not evolve in Australia. They could hibernate, all right, but depressingly often the food would come during the hibernation or not at all. Instead, most Australian animals (and plants) are geared to the rains—when the rains come, estivation breaks. For those at

Biogeography for Dinosaurs   601

the top of the food chain, it helps to be able to wait out the dry periods, which effectively means surviving famine. This implies a lower metabolic rate, and as mentioned before, many of the Australian top predators are (or were) lizards, snakes, and terrestrial crocodilians. Flannery claims more extensive evolutionary effects than Vrba, but for a less speculative reason, continental drift. He believes that the climatic change that occurred when Australia drifted north into the drier, and less predictable, subtropical climate has had a profound effect on evolution on the "island continent."

Since 1980, several rather speculative ideas have been proposed about interactions between evolution and geography. It is worthwhile to briefly consider these to appreciate recent thought in this area, although they are not yet established beyond doubt.

It has been proposed that several Cenozoic North American plants and vertebrates originated in the Arctic and later spread south over the continent (Hickey et al. 1983). No mechanism for this was given, but the notion of a geographic "cradle" that preferentially produces new species is in the air again. David Jablonski (1993) argued that ever since the Paleozoic, marine invertebrate species have tended to originate in tropical waters. Jablonski and Bottjer (1991) also proposed that marine invertebrates tend to originate in near-shore regions of the continental shelf. It has also been claimed that the Antarctic region was a "cradle" for marine invertebrates during the early Tertiary (Zinsmeister and Feldman 1984). Athough seemingly contradictory, this and the claim for tropical origins are not really in conflict. There is nothing mystical in these regions that churns out new species. Antarctica became separated from the other continents in the Eocene (early Cenozoic), and hence allopatric speciation came into play. The tropics have a very diverse biota and, if nothing else, provide lots of opportunities for new niches and lots of lineages to speciate.

Geerat Vermeij, who also works on marine invertebrates, maintains that the intensity of predation is reduced in polar regions (Vermeij 1987). These then are "safe places" where selection—that due to predation, at least—is reduced and relict forms can persist.

So far such notions have been applied to dinosaurs only in Australasia, which was near the South Pole through much of the Mesozoic, and which seems to have housed both relict and "precocious" dinosaurs. Perhaps Australian dinosaurs provide another example of the effects discussed in the last two paragraphs, but as we shall see, the Australian fossil record is poor and hence easily misinterpreted.

Probably the most significant new trend in biogeography is exemplified in the work of the expatriot Ugandan Jonathan Kingdon (1990). He has produced a detailed study of African biogeography, synthesizing modern ecology and climates with their changes over the past several million years. Kingdon's work integrates almost all of the themes mentioned here. In contrast to the grand sweep of Vrba's research, Kingdon's looks at the interactions of geography, topography, climate, and evolution at the fine scale, with an eye toward conservation. He has traced the changing fortunes of deserts and forests, as well as "cradles" where a great diversity of habitats in small areas has stimulated speciation, and regions that have remained stable and acted as refuges. The concept of refuges, regions where animals and plants survive times when most of their ranges are not suitable for them, has become important. It has obvious relevance for conservation, but also for our understanding of biogeography. The rainforests of Central America, for example, and possibly even those of the Amazon Basin, may be only about 10,000 years old (Lewin 1984). They may have "regenerated" from older rainforest species that survived drier or otherwise adverse times in refuges. An encouraging aspect of Kingdon's approach is that it allows prediction of where endemic and relict species may be found. This not only is useful to conservation efforts, but also shows that geographical understanding isn't restricted to any one or a few instances, but is applicable to all places over geological history. No application of this type of fine-scale biogeography to dinosaurian times and environments has been made, but with increasing knowledge of the Mesozoic and increasingly more precise dating methods, this should soon be feasible for certain areas.

How All This Applies to Dinosaurs

So finally we get around to dinosaurs. Like lions, pandas, wombats, and elk, dinosaurs were animals whose distributions were equally subject to climate and the locations of seas, mountains, and deserts. Looking at paleogeography and paleoclimates indicates that some forms, such as *Ammosaurus* in North America and *Protoceratops* in Asia, were desert dwellers, and others—especially the fin-bearing taxa, such as *Spinosaurus, Ouranosaurus,* and *Amargasaurus*—seem to have been adapted to tropical climates. Dinosaurs in Antarctica, Australia (e.g., *Atlascopcosaurus*), Alaska, and New Zealand lived in polar or near-polar regions.

This tells us more than just what environments were inhabited by dinosaurs, interesting though that may be. It also illuminates dinosaurian evolution and features of tetrapod evolution in general. The existence of dinosaurs in the tropics, near the poles, and in deserts shows that climate was not a limiting factor in dinosaurian distribution and evolution—for the group as a whole, if not individual species. The absence of oceanic, aquatic dinosaurs, however, suggests that some factor in their evolution, be it genetic or due to competition with the already existing marine saurians (or to something else), kept them from exploiting this environment. Near-polar dinosaurs lived in climates not suitable for large modern lepidosaurs or tortoises (Molnar and Wiffen 1994), or even, for that matter, contemporaneous ones (cf. Clemens and Nelms 1993). These dinosaurs had some physiological, or behavioral, method for coping with cool climates not possessed by modern large ectothermic tetrapods. The evidence that dinosaurs were endothermic in the same sense as are modern large mammals (being tachymetabolic endotherms) is not convincingly clear-cut—but the evidence is clear that they could and did survive in climates and places that modern large lizards, snakes, and tortoises do not (Rich 1996; Vickers-Rich 1996; Wiffen 1996).

Interpretations such as these are both enlightening and, simply, fun. But they are not, or at least not always, simple and straightforward. Not every species or even genus that ever lived is represented by fossils. In other words, the fossil record isn't complete. Look at dinosaurs: 40 percent are known from about the final 10 percent of Mesozoic time. Almost a third (126 out of 416) of the dinosaurian fossil localities listed by Weishampel (1990) are from the Late Cretaceous, which represents 23 percent (about one-fourth) of the period during which dinosaurs lived. Although the numbers of localities are not as biased toward the more recent as are the numbers of dinosaur specimens, both show that the more recent dinosaurs are better known than the more ancient forms. Any interpretations from the fossil record, of dinosaurs or anything else, have to contend with this loss of information, and it gets worse as the fossils get older.

This situation shows up in dinosaurian biogeography with fossils that seem to be "out of place." A single specimen reportedly of a pachycephalo-saur (*Majungatholus*) from Madagascar was the only representative of pachycephalosaurs from the Southern Hemisphere (but see below). Like-wise *Timimus* is the only ornithomimosaur from the Southern Hemisphere (*Elaphrosaurus*, previously thought to be an ornithomimosaur, now is regarded as a ceratosaur). In the same paper (Rich and Vickers-Rich 1994) in which *Timimus* was described, a neoceratopsian was also reported from the Early Cretaceous of southeastern Australia. Here the situation is even more problematic. The neoceratopsian not only is from Australia, where no other neoceratopsian material has been found, but also is the only Early Cretaceous neoceratopsian, so it is, in a sense, both "out of place" and "out of time" (Fig. 38.19). Furthermore, if the animal in question truly is a neoceratopsian, this implies that our understanding of the evolution of these dinosaurs is at best incomplete, because it is based on only part of the ceratopsian lineage, and perhaps not even a representative part.

These three examples emphasize how incomplete our knowledge of dinosaurian biogeography is. All of the animals described in the preceding paragraph are known from regions that are poorly known "dinosaurily" speaking, and all are known from very fragmentary specimens: a skull roof, a femur, and an ulna, respectively. Perhaps they indicate that some groups of dinosaurs were much more widespread (and older) than we realize. But they may also indicate convergence, in that the animals in question look like pachycephalosaurs, ornithomimosaurs, and neoceratopsians in those bones that are known, but only because of similar factors in natural selection, not because they are closely related. For example, recent dis-coveries in Madagascar show that *Majungatholus* is probably not a

Figure 38.19. A fossil that is "out of place": the reported Early Cretaceous neoceratopsian from Australia (if such it is). This figure illustrates just how far the Australian neoceratopsian is from the standard picture of ceratopsian evolution. Creta-ceous time is indicated as a heavy, dark vertical line along the left margin of the diagram (later is toward the top). Landmasses are represented by boxes along the base of the diagram, with arrows projected upward from the bases indicat-ing evolution of ceratopsians in those areas. As usually under-stood, ceratopsian evolution took place in Asia, beginning with a group of Late Jurassic dinosaurs (chaoyoungosaurs; Dong 1991: 94), and proceeded through the Early Cretaceous psittacosaurs to neoceratopsians. The earliest neoceratopsian is reported from the end of the Early Cretaceous in central Asia, and protoceratopsids and ceratopsids appear in the Late Cretaceous in Asia and North America. The "path" of ceratopsian evolution as usually understood is shown in this figure as solid arrows, and the extensive "detours" necessary to include the putative Australian neoceratopsian are represented by dashed arrows.

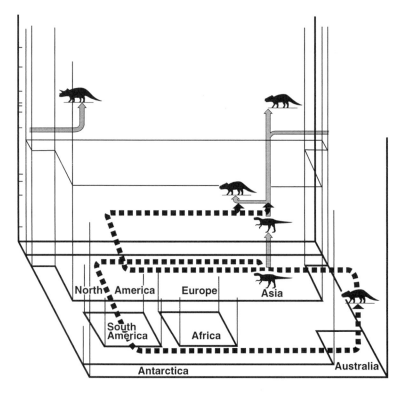

pachycephalosaur after all, but a theropod (Sampson et al. 1996). Like pachycephalosaurs, it too evolved thickened dome-like structures on its skull roof. This suggests that we should be skeptical about the identities of fragmentary, "out of place" fossils. We can be sure about them only when further, more complete fossils have been found.

Although we must be cautious regarding the details, the basic outlines—discussed in the following chapters—of dinosaurian biogeography seem to be pretty well known. Still there is plenty of room for exciting new discoveries to fill out, and maybe even substantially change, our understanding of dinosaurian biogeography and evolution.

References

Brown, J. H. 1995. *Macroecology.* Chicago: University of Chicago Press.

Casti, J. L. 1991. *Searching for Certainty: What Scientists Can Know about the Future.* New York: William Morrow.

Chatterjee, S., and N. Hotton III (eds.). 1992. *New Concepts in Global Tectonics.* Lubbock: Texas Tech University Press.

Clemens, W. A., and L. G. Nelms. 1993. Paleoecological implications of Alaskan terrestrial vertebrate fauna in latest Cretaceous time at high paleolatitudes. *Geology* 21: 503–506.

Colbert, E. H. 1973. *Wandering Lands and Animals.* London: Hutchinson.

Dilcher, D. L. 1973. A paleoclimatic interpretation of the Eocene floras of southeastern North America. In A. Graham (ed.), *Vegetation and Vegetational History of Northern Latin America,* pp. 39–59. Amsterdam: Elsevier.

Dong Z. 1991. *Dinosaurian Faunas of China.* Beijing: China Ocean Press.

Duxbury, A. C., and A. B. Duxbury. 1994. *An Introduction to the World's Oceans.* 4th ed. Dubuque, Iowa: Wm. C. Brown.

Flannery, T. F. 1994. *The Future Eaters.* Sydney, Australia: Reed Books.

Hartley, L. P. 1963. *The Go-Between.* New York: Stein and Day.

Hedges, S. B.; P. H. Parker; C. G. Sibley; and S. Kumar. 1996. Continental breakup and the ordinal diversification of birds and mammals. *Science* 381: 226–229.

Hickey, L. J.; R. M. West; M. R. Dawson; and D. K. Choi. 1983. Arctic terrestrial biota: Paleomagnetic evidence of age disparity with mid-northern latitudes during the Late Cretaceous and Early Tertiary. *Science* 221: 1153–1156.

Hooker, J. J.; A. C. Milner; and S. E. K. Sequeira. 1991. An ornithopod dinosaur from the Late Cretaceous of west Antarctica. *Antarctic Science* 3: 331–332.

Jablonski, D. 1993. The tropics as a source of evolutionary novelty through geological time. *Nature* 364: 142–144.

Jablonski, D., and D. J. Bottjer. 1991. Evironmental patterns in the origins of higher taxa: The post-Paleozoic fossil record. *Science* 252: 1831–1833.

Kerr, R. A. 1995. Darker clouds promise brighter future for climate models. *Science* 267: 454.

Kingdon, J. 1989. *Island Africa.* Princeton, N.J.: Princeton University Press.

Kingdon, J. 1990. The genesis archipelago. *BBC Wildlife* 8 (5): 296–302.

Kurtén, B. 1969. Continental drift and evolution. *Scientific American* 220 (3): 54–64.

Lewin, R. 1984. Fragile forests implied by Pleistocene pollen. *Science* 226: 36–37.

Molnar, R. E. 1989. Terrestrial tetrapods in Cretaceous Antarctica. In J. A. Crame (ed.), *Origins and Evolution of the Antarctic Biota,* pp. 131–140. Special Publication 47. London: Geological Society.

Molnar, R. E., and J. Wiffen. 1994. A Late Cretaceous polar dinosaur fauna from New Zealand. *Cretaceous Research* 15: 689–706.

Otto-Bliesner, B. L., and G. R. Upchurch, Jr. 1997. Vegetation-induced warming of high-latitude regions during the Late Cretaceous period. *Nature* 385: 804–807.

Press, F., and R. Siever. 1994. *Understanding Earth.* New York: W. H. Freeman.

Raymo, M. E., and W. F. Ruddiman. 1992. Tectonic forcing of late Cenozoic climate. *Nature* 359: 117–122.

Rich, T. 1996. Significance of polar dinosaurs in Gondwana. *Memoirs of the Queensland Museum* 39: 711–717.

Rich, T. H., and P. Vickers-Rich. 1994. Neoceratopsians and ornithomimosaurs: Dinosaurs of Gondwana origin? *National Geographic Research and Exploration* 10: 129–131.

Sampson, S. D.; D. W. Krause; C. A. Forster; and P. Dodson. 1996. Non-avian theropod dinosaurs from the Late Cretaceous of Madagascar and their paleobiogeographic implications. *Journal of Vertebrate Paleontology* 16 (Supplement to no. 3): 62A.

Smith, A. G.; D. G. Smith; and B. M. Funnell. 1994. *Atlas of Mesozoic and Cenozoic Coastlines*. Cambridge: Cambridge University Press.

Tudge, C. 1993. Taking the pulse of evolution. *New Scientist* 139 (1883): 32–36.

Vermeij, G. 1987. *Evolution and Escalation*. Princeton, N.J.: Princeton University Press.

Vickers-Rich, P. 1996. Early Cretaceous polar tetrapods from the Great Southern Rift Valley, southeastern Australia. *Memoirs of the Queensland Museum* 39: 719–723.

Vidale, J. E. 1994. A snapshot of whole mantle flow. *Nature* 370: 16–17.

Weishampel, D. B. 1990. Dinosaurian distribution. In D. B. Weishampel, P. Dodson, and H. Osmólska (eds.), *The Dinosauria*, pp. 63–139. Berkeley: University of California Press.

Wiffen, J. 1996. Dinosaurian paleobiology: A New Zealand perspective. *Memoirs of the Queensland Museum* 39: 725–731.

Wing, S. L., and D. R. Greenwood. 1993. Fossil and fossil climate: The case for equable continental interiors in the Eocene. *Philosophical Transactions of the Royal Society of London* B 341: 243–252.

Yemane, K. 1993. Contribution of Late Permian palaeogeography in maintaining a temperate climate in Gondwana. *Nature* 361: 51–54.

Yuan X.; M. A. Cane; and D. G. Martinson. 1996. Cycling around the South Pole. *Nature* 380: 673–674.

Zinsmeister, W. J., and R. M. Feldman. 1984. Cenozoic high latitude heterochroneity of Southern Hemisphere marine faunas. *Science* 224: 281–283.

606   *Ralph E. Molnar*

Major Groups of Non-Dinosaurian Vertebrates of the Mesozoic Era

Michael Morales

The Mesozoic Era, from 250 to 65 million years ago, is often called the "Age of Dinosaurs" because dinosaurs were the dominant land vertebrates of the time. Their supremacy is evidenced by their great abundance (number of individuals) and diversity (number of species), and by their enormous impact on contemporaneous environments and living communities. In one aspect, however, calling the Mesozoic the "Age of Dinosaurs" is incorrect, and in another it is misleading.

Dinosaurs were indeed the preeminent herbivores and carnivores on land during the Jurassic and Cretaceous periods of the Mesozoic, but they were not dominant during the Triassic. No bona fide dinosaur fossils are known from before the Late Triassic, and even after the dinosaurs appeared, they were not the commanding land vertebrates for many millions of years. During the Late Triassic, non-dinosaurian archosaurs were the main predators, and dicynodont synapsids (mammal relatives) were the primary herbivores. It was only after the close of the Triassic Period about 210 million years ago that dinosaurs became the dominant vertebrates on land.

The misleading aspect of the term *Age of Dinosaurs* is the implication that no other vertebrate groups were important during that time, and this is not true. Even during the Jurassic and Cretaceous, when dinosaurs were the top terrestrial vertebrates, many non-dinosaurian vertebrates played important roles on land. And in underwater marine environments of the Mesozoic, where no dinosaur ever lived, the dominant forms of life were either non-dinosaurian vertebrates or invertebrates. Thus, to truly under-

39

stand vertebrate life of the Mesozoic, one needs to know not only about dinosaurs, but about the other major groups as well. This chapter provides basic information about the lesser-known vertebrates of the Mesozoic Era.

Principal Divisions of Vertebrates

Animals that have a notochord sometime during their ontogeny (birth to death) belong to the phylum Chordata. The notochord is a rod of stiff tissue that runs along the back to give the body internal support. The vast majority of chordates are members of the subphylum Vertebrata, in which the notochord is surrounded by a string of separate but interconnected units of cartilage or bone called vertebrae. The whole string or column of vertebrae is called a backbone, spine, or spinal column, and all vertebrates have one.

Traditional Classes of Vertebrates

Vertebrates include a great diversity of body plans that were traditionally grouped into five main classes: Pisces (fishes), Amphibia (frogs, toads, salamanders, etc.), Reptilia (lizards, turtles, crocodiles, etc.), Aves (birds), and Mammalia (mammals). In more recent but still rather traditional classifications, fishes have been divided into five separate classes: Agnatha (jawless fishes), Placodermi (armored fishes), Chondrichthyes (cartilaginous fishes), Acanthodii (spiny fishes), and Osteichthyes (bony fishes). If these simplified Linnean classification systems are accepted for the moment, then the subphylum Vertebrata can be subdivided into different schemes of possible superclasses using three different criteria. One scheme is based on whether the paired appendages of the animals (or their more recent ancestors) are fins or legs, another on the presence or absence of jaws, and the third on the type of egg used in reproduction.

Fishes and Tetrapods

Vertebrates that swim using fins of various types are usually called fishes. Most other vertebrates move around on land using four legs rather than fins; hence they are called tetrapods ("four feet"). In many lineages of tetrapods, the legs have been modified into other structures or even lost through evolutionary history. Nevertheless, a vertebrate is still considered a tetrapod even if its legs have been lost (snakes) or have evolved into wings (birds and bats), flippers or paddles (seals and whales), or arms (humans).

Agnathans and Gnathostomes

All vertebrates can be divided into either jawed or jawless forms. Those that possess upper and lower jaws are called gnathostomes ("jaw mouths") and include most fishes and all amphibians, reptiles, birds, and mammals. The oldest vertebrates known, however, were fish-like forms that had a mouth but no true jaws at all. These agnathans ("without jaws"), often called jawless fishes, were the first members of the vertebrates. They were common during the Silurian and Devonian periods of the Paleozoic Era, about 440 to 360 million years ago. Though most agnathans were extinct by the end of the Devonian, there are two separate types still living today: lampreys and hagfish.

All vertebrates reproduce using some kind of egg. It may be laid in water, on or in land about the water line, or in trees, or the egg may not be laid at all, but instead may be kept inside the mother's body. All the various kinds of vertebrate eggs, and thus all vertebrates, can be grouped into two major categories, based on whether the egg is shelled or unshelled. These two groups are named on the basis of the presence or absence of a special membrane inside the shelled egg called the amnion.

Vertebrates with eggs lacking the special membrane are called anamniotes ("without amnion"). They include all fishes (jawless and jawed) and all amphibians. These animals usually lay their eggs in water or in moist environments outside water. The eggs are usually coated with a gelatin-like substance, but they lack any kind of shell. If the eggs are removed from water or moist conditions and become desiccated through contact with air, the developing embryos will dry up and die. Thus fishes and amphibians are intimately tied to water, or at least very moist conditions, for their reproductive success.

The eggs of reptiles, birds, and mammals, on the other hand, include a hard or leathery shell and an amnion; therefore these vertebrates are called amniotes. The amnion is one of several membranes that separate the area inside the shelled egg into different regions. Amniotes either lay their eggs outside of water or retain them inside the mother's body. In fact, if an amniote egg is laid in water, the embryo will suffocate! In this situation, the water surrounding the egg prevents oxygen from entering and carbon dioxide from leaving the egg through the shell.

Vertebrate Groups Used in This Chapter

Currently there are many different, often mutually exclusive classifications for vertebrates, so it would be impossible to give one scheme here that would satisfy everyone. This can be confusing for non-paleontologists who want to learn about fossil vertebrates. However, most modern classifications include many of the same groups of vertebrates, even though they show different conclusions concerning their interrelationships. In this chapter, therefore, non-dinosaurian vertebrates of the Mesozoic Era are surveyed using more traditional groups and their common names. The age range of each group that follows is given in parentheses. The main published sources for information contained in this chapter are listed in the bibliography.

Agnathans

Lampreys (late Carboniferous to the present) are one of two types of jawless fish still living today. They live in both freshwater and marine environments and have a long, eel-like body with a circular mouth possessing many tooth-like structures. Most of these agnathans live a parasitic life, attaching to other fishes with their sucker-like mouth and rasping the flesh of their host to obtain nutrients. Very few fossils of lampreys have been discovered, and none are known from Mesozoic rocks. However, fossils from the Carboniferous Period indicate that lampreys achieved their modern form, and presumably their current habits, very early in their evolutionary history.

Hagfishes (early Carboniferous to the present), too, have no Mesozoic fossil record, but like lampreys they existed then, because their fossils are known from before the Mesozoic, and living forms exist today. Hagfishes live in marine environments and have long, eel-like bodies like lampreys, but they are not parasitic. Instead they are predators of worms and other invertebrates and scavengers of rotting flesh. The hagfish mouth is not round in shape, and it lacks the lamprey's tooth-like structures.

Cartilaginous Fishes

The best-known members of the Chondrichthyes ("cartilage fishes") are sharks (Middle Devonian to the present) and rays (Early Jurassic to the present). The bodies of these fishes contain no bone; their internal skeletons are made of cartilage instead. Although sharks are known from as far back as the Middle Devonian, true rays appear for the first time in the Early Jurassic fossil record. Sharks and rays are usually associated with marine (salty) waters, but during the Mesozoic, some, such as *Hybodus* (Fig. 39.1a), lived in freshwater lakes and rivers on the continents.

Bony Fishes

The main group of fishes most people are familiar with is the Osteichthyes ("bone fishes"), which have internal skeletons made of bone. They live in fresh, salty, and brackish (mixed) waters, and often they move back and forth among the three. Bony fishes can be divided into two major

Figure 39.1. Fishes: (a) Hybodont shark *Hybodus,* 2 m long, Mesozoic. (b) Coelacanth (lobe-finned fish) *Diplurus,* 15 cm long, Triassic. (c) Lungfish (lobe-finned fish) *Neoceratodus,* 1 m long, Cretaceous to present. (d) Basal ray-finned fish *Perleidus,* 18 cm long, Triassic. (e) Intermediate ray-finned fish *Lepidotes,* 30 cm long, Jurassic. (f) Advanced ray-finned fish (teleost) *Ornategulum,* 25 cm long, Cretaceous. This and other drawings in this chapter by Susan Durning.

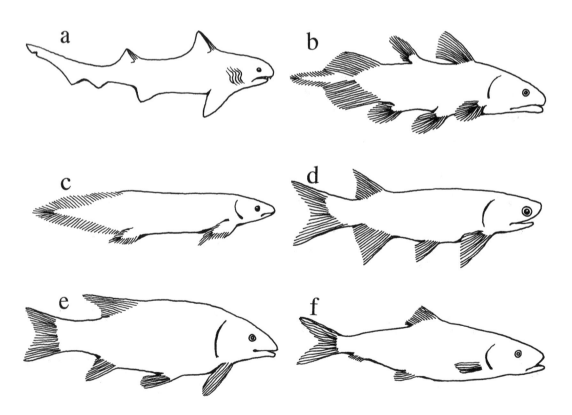

subgroups based on the structure of their fins: ray-finned forms and lobe-finned types.

The Actinopterygii ("ray fins") are bony fishes that have many internal rods or rays of bone supporting the fins, usually in a pattern radiating from the base of the fin. Ray-finned bony fishes are subdivided into three major groups. Basal ray-fins (Late Silurian to the present) include the early fossil forms (Fig. 39.1d) and the most primitive living ones, such as the paddlefish (*Polydon*) and sturgeon (*Acipenser*). Intermediate ray-fins (Early Carboniferous to the present) have more advanced mouth structures and generally more hydrodynamic bodies. They include many fossil forms (Fig. 39.1e) and the living gar (*Lepisosteus*) and bowfin (*Amia*). Advanced ray-fins are called teleosts (Late Triassic to the present), and they have very advanced mouths which can expand and project forward when the fishes seek food. Teleosts are the most abundant and diverse fishes today. Modern examples are the trout and salmon (*Salmo*) and perch (*Perca*). Teleosts were also common during the latter part of the Mesozoic (Fig. 39.1f).

The Sarcopterygii ("fleshy fins") or lobe-fins are bony fishes in which the base of the fins has a lobe of flesh internally supported by a few larger bones rather than many thinner rays. The outer part of the fin, however, does contain radiating bony rays. The lobe part of the fin allows much greater movement of the appendage than in ray-finned fishes. There are many different types of lobe-finned fish, some of which became extinct before the Mesozoic Era. Two groups of lobe-fins that lived during the Mesozoic include coelacanths (Middle Devonian to the present) and lungfishes (Early Devonian to the present). Both groups have living representatives that are not much different from their Mesozoic predecessors, such as the coelacanth *Diplurus* (Fig. 39.1b) and the lungfish *Neoceratodus* (Fig. 39.1c). Modern lungfishes live in fresh water, and Mesozoic forms seem to have done the same. In contrast, the modern coelacanth is a marine fish, whereas fossil members of this group are known from both marine and freshwater deposits.

Amphibians

Amphibians were the first vertebrates to develop legs with feet instead of fins for locomotion on the ground. Although they were tied to water for reproduction, they were able to venture onto land during at least part of their life cycle. The name Amphibia means "double life," a reference to this dual existence both on land and in water. These tetrapods had a long Paleozoic history of evolution before the Mesozoic Era, but many of the early amphibian groups were extinct by the beginning of the Triassic Period.

The only ancient amphibians that survived into the Mesozoic Era were labyrinthodonts (Late Devonian to Early Cretaceous), so named because of the highly convoluted infoldings of the enamel of their teeth, which in cross-section looks like a labyrinth or maze. Although several groups of labyrinthodonts existed during the Triassic Period, the Mesozoic as a whole was the time of their slow decline in abundance and diversity and eventual extinction. During the Paleozoic Era, some labyrinthodonts lived rather crocodile-like aquatic lives, but others were adapted to a more terrestrial existence, with stout legs that enabled them to move around on land rather well. During the Mesozoic, however, all surviving labyrinthodont groups were aquatic, usually living in freshwater environments, although a few were marine. The most common Mesozoic labyrinthodonts were capitosaurs (Early to Late Triassic) such as *Paracyclotosaurus* (Fig.

39.2a), plagiosaurs (Early to Late Triassic) such as *Gerrothorax* (Fig. 39.2b), metoposaurs (Middle to Late Triassic), trematosaurs (Early to Late Triassic), brachyopids (Late Permian to Late Jurassic), and the last group to become extinct, chigutisaurs (Early Triassic to Early Cretaceous).

Modern amphibians, which are usually united under the term *Lissamphibia* ("smooth double life"), include frogs and toads (Early Triassic to the present), salamanders and newts (Middle Jurassic to the present), and the little-known caecilians (Early Jurassic to the present). The oldest member of the frog group is *Triadobatrachus* from the Early Triassic. It is considered a link between true frogs with jumping locomotion (first known from the Early Jurassic) and the labyrinthodont ancestors of frogs. The oldest salamanders are from the Early Jurassic, and they probably looked very similar to modern forms. Living caecilians are legless, worm-like amphibians that evolved from legged ancestors. We know this because the earliest caecilian, from the Early Jurassic, had reduced limb and girdle bones.

Amniotes

Amniotes include reptiles, birds, and mammals, all of which reproduce by means of an amniote egg. The vast majority of fossil vertebrate material, however, is bones and teeth, not eggs. Therefore, fossil amniotes are generally characterized by their skeletal remains, with the skull (and its teeth) usually being the most informative part of the skeleton.

There are four basic types of amniote skulls, based on the number and position of holes or openings in the temporal region, behind the eyes sockets or orbits (see Parrish, chap. 15 of this volume). Amniotes with no holes in the temporal region of the skull are called anapsids ("without arch"). The name refers to the lack of a bony arch, formed by the postorbital and squamosal bones, that separates upper and lower skull openings when they are present. If there is no temporal opening in the skull, then there is, by definition, no bony arch, even though the two bones are present. The name may refer either to the structure of the skull (anapsid type) or to the group of amniotes possessing that type of skull. Turtles are modern anapsid reptiles.

Amniotes with a single temporal hole on each side of the skull, below the postorbital-squamosal arch, are labeled synapsids ("with arch"). Mammals and their reptile-like ancestors are synapsids. The latter used to be called mammal-like reptiles, but *reptile-like synapsids* is a better term. The term *reptiles* is now restricted to members of the other three amniote groups (except birds, as noted below).

If the single temporal opening is above the arch, then the name euryapsid ("wide arch") is used. There are no living euryapsids. The extinct forms are often referred to as marine reptiles, and include nothosasurs, plesiosaurs, and icthyosaurs.

Amniotes with two temporal holes in the skull, one above and one below the arch, are called diapsids ("two arch"). Modern diapsid reptiles include lizards, snakes, and crocodilians. Although birds have a modified diapsid skull structure, they are not referred to as reptiles (but see Holtz and Brett-Surman, chap. 8 of this volume, for the cladistic perspective on bird relationships).

Thus reptiles include all anapsids and euryapsids, and all diapsids except birds, but synapsids (mammals and their relatives) are not reptiles, even if some of them (non-mammal synapsids) may have had a reptile-like look to them. The earliest ancestral amniotes were anapsids. It appears that synapsids and diapsids evolved from them along separate evolutionary lines, and that euryapsids diverged from diapsids by growing together the bones surrounding the lower opening, thus closing this hole.

Anapsids

Anapsid amniotes include the oldest known fossil reptile, from the earliest Early Carboniferous. Of the several Paleozoic groups of anapsids, only turtles and procolophonids survived into the Mesozoic.

Turtles (including tortoises; Late Triassic to the present) are an ancient group of reptiles, dating back to the first period of the Mesozoic. Their distinctive shell is a hallmark of even the oldest forms. Turtles are divided into three major subgroups: proganochelydians (Late Triassic to Early Jurassic), pleurodirans (Late Triassic to the present), and cryptodirans (Early Jurassic to the present). The oldest subgroup, proganochelydians, were moderate to relatively large animals that had little or no ability to retract the neck inside the shell. In the two remaining subgroups, the neck is retracted in two different ways. In pleurodires, the neck vertebrae are flexed to the side, thus bringing the head under the lateral margin of the shell. In cryptodires, the neck vertebrae flex vertically and posteriorly, with the head pulled directly backward inside the shell.

Turtles are usually considered to be anapsids, which is why they are discussed at this point in the chapter. However, Rieppel and deBraga (1996) argued that the absence of the temporal opening in turtles is misleading, and that the true affinities of turtles are with diapsid reptiles. Wilkinson et al. (1997), however, suggested that traditional ideas about turtle relationships could still be correct, and that it may be difficult to determine who the closest relatives of turtles really are.

Procolophonids (Late Permian to Late Triassic) were rather lizard-like in their overall form and habits. They probably ate insects and other small prey animals, but some may have eaten plant material. It is likely they were active during the day and found places to hide during the night. Procolophonids and other small reptile groups filled the "lizard" role or niche during the early Mesozoic, before true lizards appeared in the Late Jurassic.

Non-Dinosaurian Vertebrates of the Mesozoic 613

Diapsids

Most diapsid reptiles are divided into two major groups or evolutionary lineages based on a variety of differences in skeletal structures resulting in a difference in posture and locomotion between the two. Lepidosauromorphs ("scale reptile forms"), which include lizards and snakes, usually have a sprawling body posture with the legs bent out to the sides, and movement of their legs is generally accompanied by lateral undulations of the body trunk. In contrast, archosauromorphs ("ruling reptile forms"), including dinosaurs and crocodilians, tend to have their legs held in a more upright position under the body, and reduce or eliminate lateral undulations of the trunk during locomotion. Another feature distinguishing the two groups is the presence of a large sternum in lepidosauromorphs; in archosaurs the sternum is relatively much smaller.

A few diapsid subgroups do not fit well into either the lepidosauromorphs or the archosauromorphs, so they will be discussed separately. Mesozoic examples of these enigmatic forms include aerial diapsids that glided or parachuted through the air and aquatic reptiles that swam in marine waters.

Lepidosauromorphs

Sphenodontians (Late Triassic to the present) today are represented by only one form, the tuatara (*Sphenodon*), which lives only in New Zealand. It is a relict of an old group that once had representatives on many other continents. Sphenodontians, like procolophonians, are another group of relatively small terrestrial reptiles that superficially looked and probably behaved much like lizards, although one subgroup, the pleurosaurids (Early Jurassic to Early Cretaceous), were adapted to an aquatic mode of life.

True lizards (Late Jurassic to the present) first appear in the fossil record during the Late Jurassic. During their entire geologic history, lizards have maintained a rather conservative but adaptable body form that has allowed them to adjust to a variety of habitats, from desert sands to jungle trees to near-shore rocky environments. In most cases the body did not become highly modified from the ancestral condition. One exception, the marine mosasaurs, will be discussed separately below.

Snakes (middle Cretaceous to present) are not as ancient a lepidosauromorph group as either sphenodontians or lizards. They have been thought to have evolved from lizard ancestors that became adapted to a fossorial or burrowing life; in the process they were presumed to have lost the front and hind limbs and the skull to have been highly modified. Recent work, however, suggests that snakes may have evolved as marine animals, and that they are closely related to the mosasaurs and their kin (Caldwell and Lee 1997).

Archosauromorphs

The group of reptiles that includes dinosaurs, crocodilians, pterosaurs (flying reptiles), and their relatives is called Archosauromorpha. These diapsids can be separated into archosaurs ("ruling reptiles") and all remaining archosauromorphs (see Parrish, chap. 15 of this volume). The former are characterized by a tendency toward better adaptations of the legs, feet, and hips for fast locomotion on land, and by the presence of yet another opening in each side of the skull, an antorbital hole in front of the eye socket.

Non-Archosaurs. Archosauromorph diapsids of the Mesozoic that are not archosaurs include trilophosaurs, rhynchosaurs, tanystropheids, and champsosaurids. Trilophosaurs (Late Triassic only) were small to medium-sized terrestrial reptiles that resembled modern lizards in form and probably habits too. *Trilophosaurus* (Fig. 39.3a) was a herbivore with broad teeth, each having three ridges, hence its name ("three ridge reptile"). It also had an unusually long tail for its body length.

Rhynchosaurs (Early to Late Triassic) were quadrupedal forms that ate plant material using powerful "beaks" to bite the vegetation (Fig. 39.3b). Seen from above, their skulls were triangular in shape, with the back being much broader than average for reptiles. Rhynchosaurs were geographically quite widespread during the Triassic and can be used to correlate deposits on different continents.

Tanystropheids (Middle to Late Triassic) were extremely odd reptiles that lived near and in marine waters. *Tanystropheus* (Fig. 39.3c) had a medium-sized body trunk, but its neck was almost unbelievably long,

Figure 39.3. Non-archosaur archosauromorphs: (a) *Trilophosaurus,* 2 m long, Triassic. (b) Rhynchosaur *Hyperodapedon,* 1.3 m long, Triassic. (c) *Tanystropheus,* 3 m long, Triassic. (d) *Champsosaurus,* 1.5 m long, Cretaceous.

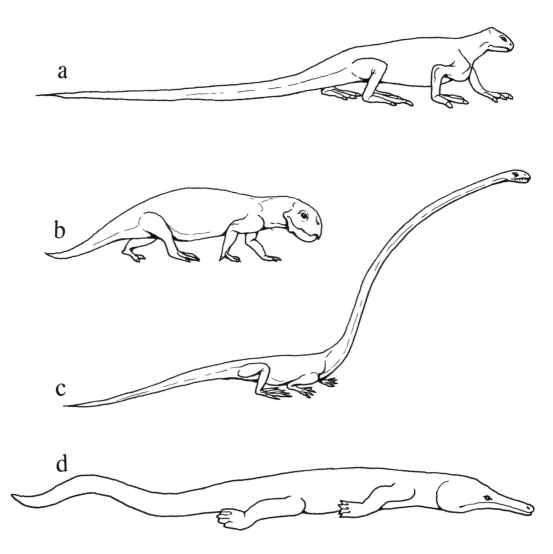

a

b

c

d

Non-Dinosaurian Vertebrates of the Mesozoic   615

ending in a small head. One idea is that while the body stood on the shore, the neck was stretched low over the water to catch fishes swimming by. Alternatively, the long neck could have been used to capture fishes while *Tanystropheus* swam underwater.

Champsosaurids (middle Cretaceous to early Eocene) were small to medium-sized aquatic reptiles that superficially looked like little crocodiles, although they are not closely related to them. *Champsosaurus* (Fig. 39.3d), known from freshwater deposits, had a long, slender snout, probably used to catch fishes and underwater invertebrates, and a long tail that was flattened side to side and was used to push the animal through the water by lateral undulations.

Archosaurs. The first known archosaurs are from the Late Permian, but this group quickly rose to prominence so that they were the dominant tetrapods on the continents during most of the Mesozoic Era (see Parrish, chap. 15 of this volume). In the Triassic Period, non-dinosaurian archosaurs were preeminent, even though dinosaurs were present by Late Triassic times. However, from the beginning of the Jurassic to the end of the Cretaceous, dinosaurs were the "rulers" of terrestrial environments worldwide.

The earliest archosaurs (Late Permian to Late Triassic) came in two main kinds. Both were relatively large and carnivorous, but one group was adapted to a more active life on land, and the other lived a more semiaquatic, crocodile-like existence in or near rivers and lakes. *Erythrosuchus* (Fig. 39.4a) is a typical example of the terrestrial kind of early archosaur.

Aetosaurs (Late Triassic only) were heavily armored archosaurs that lived in terrestrial environments and were herbivorous. To protect themselves from predators, they had body armor made of interlocking plates of bone embedded in the skin, sometimes with spikes sticking out. Their snouts were somewhat pig-like; hence one idea is that they rooted in the ground to search for plant food (Fig. 39.4b).

Imagine a big crocodile walking and running around on dry land, with rather upright front and hind legs well under the trunk of the body, and the tail held off the ground. That's roughly what rauischians (Middle to Late Triassic) looked like (Fig. 39.4c). In most regions of the world, they (and/or ornithosuchians) were dominant terrestrial predators during the Triassic Period, even when dinosaurian predators, which were smaller, inhabited the same area.

Phytosaurs (Late Triassic only) filled a niche now occupied by crocodilians. They even looked very much crocodiles and alligators (Fig. 39.4d), except that their nostrils were positioned in front of the eyes, rather than at the tip of the snout as in modern crocodilians. They lived in freshwater rivers and lakes, eating fishes and whatever else they could catch. Phytosaurs were probably the dominant predators of freshwater environments during the Late Triassic. Certainly they came out of the water to lay their eggs on land, but they probably spent the bulk of their time either in the water or on the shore basking in the sun.

Crocodylomorphs (Late Triassic to the present) are a group of archosaurs that includes crocodilians (crocodiles, alligators, caimans, and gavlals), known from the Late Triassic to the present, and their relatives. Some of the latter, such as sphenosuchians (Fig. 39.4e) and saltoposuchians (both Late Triassic to Early Jurassic), included rather small, fast-running forms with long and slender legs that lived on land rather than in water.

Ornithosuchian archosaurs (Late Triassic only—see Parrish, chap. 15 of this volume) were relatively large (10 feet or 3 meters long) terrestrial predators that may have walked on all four legs (like the rauisuchian in Fig.

39.4c), but probably ran rather fast bipedally, i.e., on the two hind legs only. Even though they lacked the ankle joint structure of true dinosaurs, ornithosuchians were among the most dinosaur-like of all the non-dino-saurian archosaurs.

Ornithodira (Middle and Late Triassic—see Parrish, chap. 15 of this volume) is the group of archosaurs to which dinosaurs belong, but it also includes pterosaurs (flying reptiles), birds, and some early forms either of uncertain affinities or closely related to dinosaurs and pterosaurs. The latter include *Lagosuchus* and *Lagerpeton* (Middle Triassic), long and

Figure 39.4. Early archosaurs: (a) Early archosaur *Erythrosuchus,* 4.5 m long, Triassic. (b) Aetosaur *Stagonolepis,* 3 m long, Triassic. (c) Rauisuchian *Ticinosuchus,* 3 m long, Triassic. (d) Phytosaur *Parasuchus,* 3 m long, Triassic. (e) Sphenosuchian crocodilo-morph *Terrestrisuchus,* 50 cm long, Triassic.

Non-Dinosaurian Vertebrates of the Mesozoic   617

slender-limbed terrestrial carnivorous or insectivorous reptiles of very small size that have many but not all the skeletal features that define primitive dinosaurs. These two genera, therefore, are often called pre- or proto-dinosaurs. Another early ornithodiran is *Sclermochlus* (Late Triassic), also with long and slender legs, which is sometimes mentioned in connection with the origin of pterosaurs. Two ornithodirans of uncertain relationships, *Sharovipteryx* and *Longisquama*, are discussed below, as are pterosaurs and birds. However, because dinosaurs are covered in considerable detail elsewhere in this book, they are not discussed in this chapter.

Aerial Diapsids

Among the diapsid reptiles, forms with adaptations to aerial locomotion evolved independently several times. Most examples can be categorized either as gliders or as true flyers that flapped their wings (front limbs) to propel themselves through the air.

The three best-known Mesozoic gliding reptiles (Late Triassic only) used rather different means to travel through the air, even though their basic body plan was similar. The ornithodiran *Sharovipteryx* (Fig. 39.5a) was a very small animal with skin membranes spread between its front legs and the front half of its trunk, and between its back legs and the front part of its tail. When both sets of legs were extended outward, the skin membranes formed an airfoil that allowed *Sharovipteryx* to glide through the air while being pulled down by gravity. The reptile was not able to flap its "wings" for powered flight.

Longisquama (Fig. 39.5b) was another ornithodiran about twice as big as *Sharovipteryx,* and instead of skin membranes it had elongated, hockey-stick-like scales that extended from its back. If these scales were held outward on either side of the animal, they could serve as an airfoil. When not in use, they might have been able to fold backward along the sides of the back.

The Kuehneosauridae is a family of small gliding lepidosauromorphs in which the ribs were extended laterally on both sides of the body, with a skin membrane stretched between them (Fig. 39.5c). When the "wing" was not in use, the ribs could fold back along the body, but when extended outward, the ribs and membrane formed a rather efficient airfoil with which the animal could glide downward from a high spot.

In pterosaurs (Late Triassic to end of Cretaceous) or flying reptiles (sometimes called pterodactyls, which actually refers to only one subgroup), the front legs (arms) were modified into true wings by elongation of the fourth finger, which supported a skin membrane that stretched to the body. Most pterosaurs probably flapped their wings to some degree to push themselves through the air. They lived both in shoreline marine areas, where they ate fish mainly, and within landmasses, where they probably ate insects and other small animals. At least one form, *Pterodaustro,* sieved tiny water creatures with a flamingo-like feeding mechanism. The flying reptiles are divided into two major subgroups: pterodactyloids (Late Jurassic to end of Cretaceous), with very short tails, and the more primitive rhamphorhynchoids (Late Triassic to end of Jurassic), with long tails (Fig. 39.5e). As a group, pterosaurs, like dinosaurs, became extinct at the end of the Cretaceous.

A supposed bird, provocatively named *Protoavis* ("first bird"), has been reported from the Late Triassic, but most vertebrate paleontologists who study bird origins do not consider it to be a bird at all. The earliest undisputed bird, therefore, is *Archaeopteryx* ("ancient wing") from the Late

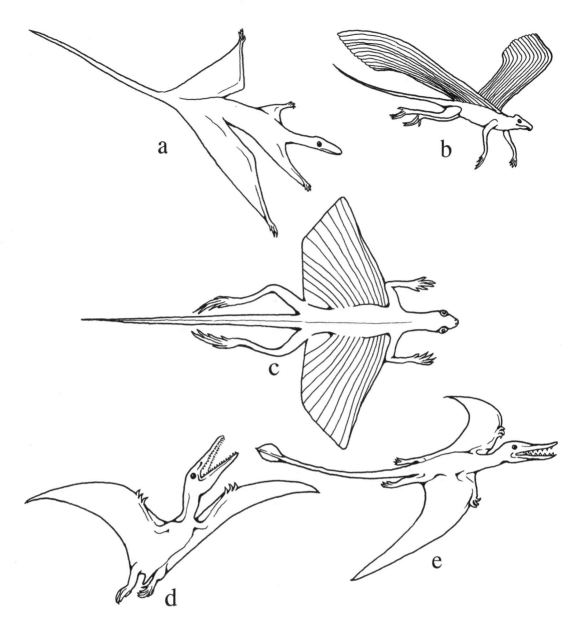

Figure 39.5. Aerial diapsids:
(a) Ornithodiran *Sharovipteryx* (formerly known as *Podopteryx*), wingspan 17 cm, Triassic.
(b) Ornithodiran *Longisquama*, "wingspan" 30 cm, Triassic.
(c) Lepidosauromorph *Kuehneosaurus*, wingspan 30 cm, Triassic.
(d) Pterodactyloid pterosaur *Pterodactylus*, wingspan 40 to 175 cm, Jurassic. (e) Rhamphorhynchoid pterosaur *Rhamphorhynchus*, wingspan 36 to 250 cm, Jurassic.

Jurassic. Its front legs were modified into true wings (Fig. 39.6a) that differed from the wings of pterosaurs. Instead of an airfoil made of a skin membrane, *Archaeopteryx* had feathers along its modified arms, as well as on its tail and body. Also, the whole hand, in a compact and concentrated form, supported wing feathers, rather than only the fourth finger bearing a skin membrane.

During the Cretaceous Period, birds became diversified into a variety of types, including forms with reduced wings, such as the flightless ground-dweller *Patagopteryx* (Fig. 39.6b) and the aquatic diving bird *Baptornis* (Fig. 39.6c). One controversial fossil, *Mononykus* ("one claw"), has been classified by some vertebrate paleontologists as a flightless bird, but by

Non-Dinosaurian Vertebrates of the Mesozoic   619

Figure 39.6. Birds: (a) Earliest bird, *Archaeopteryx,* 35 cm long, Jurassic. (b) An early flightless bird, *Patagopteryx,* 46 cm tall, Cretaceous. (c) Early aquatic bird (hesperornithiform) *Baptornis,* 1 m long, Cretaceous.

Figure 39.7. Euryapsids and other marine reptiles: (a) *Nothosaurus,* 3 m long, Triassic. (b) Long-necked plesiosaur *Elasmosaurus,* 14 m long, Cretaceous. (c) Short-necked plesiosaur *Kronosaurus,* 12.8 m long, Cretaceous. (d) Placodont *Placodus,* 2 m long, Triassic. (e) Ichthyosaur *Ophthalmosaurus,* 3.5 m long, Jurassic. (f) Marine crocodile (thallatosuchian) *Metriorhynchus,* 3 m long, Jurassic. (g) Marine lizard (mosasaur) *Platecarpus,* 4.25 m long, Cretaceous.

others as a dinosaur. It had reduced front limbs, which were not used as wings, and "hands" that ended with a single large claw.

Euryapsids and Other Mesozoic Marine Reptiles

Mesozoic euryapsids include nothosaurs, plesiosaurs, placodonts, and ichthyosaurs, all of which lived in marine seas and oceans. Some non-euryapsid reptiles lived in marine environments during the Mesozoic as well, including thalattosuchians (marine crocodilians), thallatosaurians (primitive archosaurmorphs), mosasaurs (marine lizards), and sea turtles.

Nothosaurs (Early to Late Triassic) were rather primitive euryapsids that were of small to moderate size (Fig. 39.7a) and probably lived a life something like that of a modern seal or otter. Nothosaurs had a mouth full of sharp, conical teeth that were probably used to catch fishes while either swimming underwater or standing at the water's edge. Their legs were modified into flippers, rather than the paddles of more advanced euryapsids.

Plesiosaurs (Early Triassic to end of Cretaceous) were medium-sized to large marine predators that came in two major body types: those with long necks (Fig. 39.7b), and those with short necks (Fig. 39.7c). Long-necked plesiosaurs had large body trunks, short tails, four legs modified into paddles, and a very long neck with a relatively small head. Their teeth were similar to those of nothosaurs, and they probably ate mainly fishes. They have been proposed as a model for what the Loch Ness monster (if it exists) is presumed to look like. Short-necked plesiosaurs had relatively much larger heads than the long-necked forms. They probably filled a niche in Mesozoic times similar to that of killer whales today. The structure of plesiosaur paddles indicates that they were not used not as oars that pushed

620   *Michael Morales*

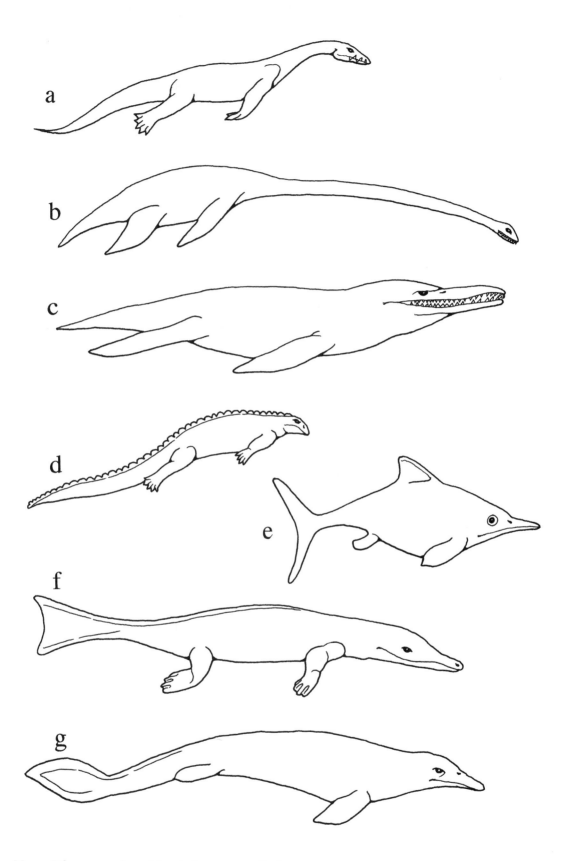

water from front to back and thereby moved the body forward. Instead, the paddles acted as "wings" that were flapped up and down, providing both forward propulsion and hydrodynamic lift, and allowing the animals to "fly" through the water, much as sea turtles and penguins do today.

Placodonts ("plate tooth") were rather sturdy euryapsids (Fig. 39.7d) that lived from the beginning to the end of the Triassic and had a long trunk and tail, limbs with feet that probably were webbed, and a head with large, plate-like crushing teeth. Propulsion during swimming was probably provided by lateral undulations of the tail, rather than movements of the legs and feet. Placodonts probably ate clams and other shelled invertebrates that they found while searching the sea floor on the continental shelf.

Ichthyosaurs ("fish reptiles") were highly modified euryapsids (Fig. 39.7e) that lived in the oceans from the beginning of the Triassic to the middle of the Cretaceous. They looked similar to and probably had habits much like modern dolphins, whales, and sharks. They were predators that ate fishes, marine invertebrates, and probably other marine reptiles as well. Interestingly, they became extinct before the end of the middle Cretaceous, before the great end-of-Cretaceous mass extinction. Perhaps ichthyosaurs were out-competed by mosasaurs (see below), first known from the middle Cretaceous.

Although some modern crocodilians spend at least part of their time in marine waters, none is truly adapted to an ocean existence. During the Mesozoic, however, one group of crocodilians, thalattosuchians ("sea crocodiles"—Early Jurassic to Early Cretaceous), became highly modified for a marine life. Advanced members of this group (Fig. 39.7f) had all four limbs changed to paddles, the far end of the bony tail bent downward to support a vertical fin, and long snouts presumedly to catch fishes. Thalattosuchians became extinct in the Early Cretaceous, even before ichthyosaurs did.

Thalattosaurians (Middle to Late Triassic) should not be confused with the similarly named marine crocodilians mentioned above. Thalattosaurians ("sea reptiles") were primitive archosauromorphs of medium size (6 feet or 2 meters long), with greatly elongated trunk and tail regions of the body. The four relatively short limbs probably had webbed feet, but were not modified into paddles. The snout had an adaptation often seen in aquatic reptiles (see phytosaurs above): placement of the nostrils well back from the tip of the snout.

Many lizards can swim, and even some modern forms, such as the marine iguana of the Galapagos Islands, spend a lot of time in the water. However, the most aquatic of all lizards were the mosasaurs (middle to end of the Cretaceous), which are relatives of monitor lizards. These medium-sized to large marine predators were highly adapted to living in seas and oceans. They had long bodies ending in a tail fin (Fig. 39.7g), four limbs altered to flippers, and very large jaws which were used to catch fishes, invertebrates, and other prey. They probably swam using a snake-like side-to-side motion, rather than the up-down motion of modern whales and dolphins.

Turtles that live exclusively in seas and oceans are known today, and they existed during the Mesozoic Era as well. In fact, the largest sea turtle ever known is probably *Archelon* from the Late Cretaceous, with a head-shell-tail length of more than 8 feet (2.5 meters). As with modern sea turtles, the four limbs of Mesozoic forms were modified into paddles, which were flapped as "wings" to "fly" through the water, as penguins do today.

Synapsids

The last major group of amniotes to be covered in this chapter are the synapsids, including mammals and their reptile-like relatives. The earliest synapsids, the pelycosaurs, first appeared in the Carboniferous Period and became extinct during the Late Permian. Thus there are no Mesozoic pelycosaurs. Therapsids are more advanced synapsids, which ranged in time from the Early Permian to the Middle Jurassic. One group of therapsids gave rise to mammals, known from the Late Triassic to today.

There are three major groups of therapsids that lived during the Mesozoic Era: anomodonts (Late Permian to Late Triassic), therocephalians (Late Permian to Middle Triassic), and cynodonts (Late Permian to Middle Jurassic). The most common anomodonts were a subgroup called dicynodonts ("two dog tooth"), large hippopotamus-like herbivores (Fig.

Figure 39.8. Therapsids and mammals: (a) Dicynodont therapsid *Kannemeyeria,* 3 m long, Triassic. (b) Cynodont therapsid *Thrinaxodon,* 50 cm long, Triassic. (c) Triconodont mammal *Megazostrodon,* 13 cm long, Triassic. (d) Early placental mammal *Zalambdalestes,* 20 cm long, Cretaceous. (e) Early marsupial mammal *Alphadon,* 10 cm long, Cretaceous. (f) Multituberculate mammal *Ptilodus,* 50 cm long, Jurassic.

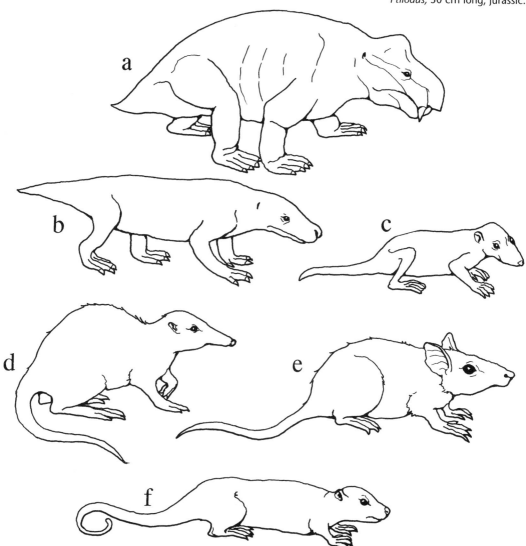

Non-Dinosaurian Vertebrates of the Mesozoic   623

39.8a) with two big canine-like teeth in the upper jaw, hence their name. During the Triassic, dicynodonts were the dominant herbivores in most regions of the world. Therocephalians were a relict group during the Mesozoic, their greatest days having already occurred in the Late Paleozoic. They were small to medium-sized quadrupedal therapsids that lived a terrestrial carnivorous or insectivorous life. Cynodonts were more advanced predatory therapsids, also of small to moderate size (Fig. 39.8b). Their skeletal structure and probably their physiology (cf. Ruben et al., chap. 35 of this volume) were more mammal-like than those of the therocephalians. Cynodonts had a more upright body posture, with limbs held more underneath the body, and probably a more advanced system for ventilating their lungs, thus improving their metabolism. Some cynodont skulls have small holes in the snout that have been interpreted as places where cat-like whiskers attached. Because whiskers are modified hairs, it is possible that some cynodonts had a covering of body hair. Neither anomodonts nor therocephalians left any descendants, but at least one group of cynodonts evolved into mammals.

It is interesting that the first mammals (Late Triassic to the present) and the first dinosaurs appeared at about the same time, during the Late Triassic. Although neither group was a major component of terrestrial ecosystems when they first evolved, dinosaurs rose to dominance by the Early Jurassic and were probably instrumental in keeping mammals restricted to less significant (or at least less showy) ecological niches throughout the rest of the Mesozoic Era. Indeed, mammals did not greatly diversify and rise to prominence until after dinosaurs became extinct at the end of the Mesozoic.

During the first two-thirds of their evolutionary history, therefore, mammals were not the dominant terrestrial vertebrates that they are today. Furthermore, if you compare the number of species and higher groups of mammals that appeared during the 65 million years after the Mesozoic with numbers from the 155 million years between the Late Triassic to the end of the Cretaceous, you will see that Mesozoic mammals were much less diverse than post-Mesozoic mammals, even though they existed for more than twice the amount of time.

Nearly all Mesozoic mammals were very small animals, approximately the size of a mouse or rat, with the largest ones about the size of a cat. It is believed that they were active mainly at night, when dinosaurs were asleep or at least less alert. Most early mammals probably ate insects or small vertebrates, but at least one group, the multituberculates, ate plant material. Mesozoic mammals almost certainly had the two features that define mammals in the common sense: hair and mammary glands, which secrete milk for the young. Although several Mesozoic mammals are known from complete or nearly complete skeletons, most forms are recognized mainly from their teeth.

The main groups of Mesozoic mammals include triconodonts (Late Triassic to Late Cretaceous), symmetrodonts (Late Jurassic to Late Cretaceous), docodonts (Middle to Late Jurassic), haramyoids (Late Triassic to Middle Jurassic), multituberculates (Late Jurassic to Late Eocene), monotremes (Early Cretaceous to the present), early marsupials (Middle Cretaceous to the present), and early placentals (Middle Cretaceous to the present).

Triconodonts and haramyoids are the oldest fossil mammals, known from the Late Triassic. Triconodonts ("three cone tooth"; Fig. 39.8c) had cheek teeth (molars and premolars) with top surfaces made of three cones or cusps lined in a straight row, hence their name. Haramyoids had teeth with many cusps in at least two parallel rows. Multituberculates (Fig. 39.8f), close relatives of haramyoids, also had cheek teeth with many cusps in more than one row. In addition, they had long, rodent-like incisor teeth at the front of the mouth. In general, multituberculates probably filled a "rodent" niche that had previously been filled by advanced cynodont therapsids, and was later filled by true rodents.

Symmetrodonts had upper and lower cheek teeth with many cusps that were arranged in a triangle pattern, which is an advance over teeth with cusps in a straight row. In docodonts, the cheek teeth were even more elaborated, with many cusps arranged in a T shape in the upper molars and a rectangular shape in the lower molars. In both symmetrodonts and docodonts, ridges of enamel often connected two or more cusps.

In early marsupials or pouched mammals (Fig. 39.8e), the upper and lower cheek teeth became more distinct, with large, roughly triangular uppers, and lowers made of two sections, a tall projecting part joined to a low basin. In early placentals (Fig. 39.8d), mammals in which the mother nourishes her developing fetus through a placenta, the cheek teeth are even more elaborated, with extra cusps and ridges. Finally, among the monotremes, true mammals (i.e., with hair and mammary glands) whose living forms lay eggs like reptiles, only the duckbilled platypus lineage goes back to the Mesozoic. It is represented by a single find of a lower jaw with three molars from the Early Cretaceous.

Although the Mesozoic Era is often considered to be the "Age of Dinosaurs" because of the great abundance and diversity of these magnificent creatures, many other vertebrates existed on land, in fresh and marine waters, and in the air as well. Many of these other vertebrates lived alongside and in many cases interacted with dinosaurs. They were an important part of the global ecosystem, of which dinosaurs were the "ruling" component on land. Therefore, to understand the range and complexity of vertebrate life of the Mesozoic, you need to know about not only the dinosaurs, but all the other vertebrates as well.

References

Benton, M. J. (ed.). 1990. *The Phylogeny and Classification of Tetrapods*. Vol. 1: *Amphibians, Reptiles, Birds*. Vol. 2: *Mammals*. The Systematic Association, Special Volumes nos. 35A and B. Oxford: Clarendon Press.

Benton, M. J. 1993. Reptilia. In M. J. Benton (ed.), *The Fossil Record 2*, pp. 681–715. London: Chapman and Hall.

Caldwell, M. W., and M. S. Y. Lee. 1997. A snake with legs from the marine Cretaceous of the Middle East. *Nature* 386: 705–709.

Cappetta, H.; C. Duffin; and J. Zidek. 1993. Chondrichthyes. In M. J. Benton (ed.), *The Fossil Record 2*, pp. 593–609. London: Chapman and Hall.

Carroll, R. L. 1988. *Vertebrate Paleontology and Evolution*. New York: W. H. Freeman and Company.

Colbert, E. H., and M. Morales. 1991. *Evolution of the Vertebrates*. 4th ed. New York: Wiley-Liss.

Feduccia, A. 1996. *The Origin and Evolution of Birds*. New Haven, Conn.: Yale University Press.

Gardiner, B. G. 1993. Osteichthyes: Basal Actinopterygians. In M. J. Benton (ed.), *The Fossil Record 2*, pp. 611–619. London: Chapman and Hall.

Halstead, L. B. 1993. Agnatha. In M. J. Benton (ed.), *The Fossil Record 2*, pp. 753–781. London: Chapman and Hall.

Long, J. A. 1995. *The Rise of Fishes*. Baltimore: Johns Hopkins University Press.

Maisey, J. G. 1996. *Discovering Fossil Fish*. New York: Henry Holt and Company.

Milner, A. R. 1993. Amphibian-grade Tetrapoda. In M. J. Benton (ed.), *The Fossil Record 2*, pp. 665–679. London: Chapman and Hall.

Milner, A. R. 1994. Late Triassic and Jurassic amphibians: Fossil record and phylogeny. In N. C. Fraser and H.-D. Sues (eds.), *In the Shadow of the Dinosaurs*, pp. 5–22. Cambridge: Cambridge University Press.

Norman, D. 1994. *Prehistoric Life: The Rise of the Vertebrates*. New York: Macmillan.

Patterson, C. 1993. Osteichthyes: Teleostei. In M. J. Benton (ed.), *The Fossil Record 2*, pp. 621–656. London: Chapman and Hall.

Rieppel, O., and M. deBraga. 1996. Turtles as diapsid reptiles. *Nature* 384: 453–455.

Schultze, H.-P. 1993. Osteichthyes: Sarcopterygii. In M. J. Benton (ed.), *The Fossil Record 2*, pp. 657–663. London: Chapman and Hall.

Stucky, R. K., and M. C. McKenna. 1993. Mammalia. In M. J. Benton (ed.), *The Fossil Record 2*, pp. 739–771. London: Chapman and Hall.

Unwin, D. M. 1993. Aves. In M. J. Benton (ed.), *The Fossil Record 2*, pp. 717–737. London: Chapman and Hall.

Wilkinson, M.; J. Thorley; and M. J. Benton. 1997. Uncertain turtle relationships. *Nature* 387: 466.

Continental Tetrapods of the Early Mesozoic

Hans-Dieter Sues

The Triassic Period represents a time of profound changes in the evolutionary history of continental tetrapods (four-footed vertebrates). Most of the major present-day groups (or their closest relatives) first appeared in the fossil record at this time: mammals, turtles, archosaurian reptiles including dinosaurs, squamate reptiles (lizards and snakes), and lissamphibians (frogs, salamanders, and caecilians). The dominant vertebrates on land during the preceding Permian Period (late Paleolozoic Era) had been the therapsids (often misleadingly called "mammal-like reptiles"), the majority of which became extinct during or at the end of the Permian. At the beginning of the Triassic, however, therapsids still constituted the majority of larger tetrapods on land. By the end of that period, though, dinosaurs were the most common large land animals. During the Middle Triassic, archosaurian reptiles became very diversified and abundant, while only a few therapsid groups (including the precursors of mammals) persisted in moderate diversities. The pattern of and possible cause(s) for this large-scale change in the composition of terrestrial vertebrate faunas are not yet understood, but the basic structure of continental ecosystems established during the early Mesozoic persists to the present day (Wing and Sues 1992). The only major subsequent faunal change was the replacement of most dinosaurs by mammals in many ecological roles, although, at least in terms of species diversity, one lineage of dinosaurs—birds—still vastly outnumbers mammals (about 9,000 species vs. 4,000 species).

During the entire length of the Triassic Period, the continents formed

40

a single large supercontinent, Pangaea (see Molnar, chap. 38 of this volume, for paleogeographic maps for different intervals of the Mesozoic Era). There were apparently few if any significant barriers to dispersal of land-living animals across this huge landmass. Late Triassic and especially Early Jurassic tetrapod assemblages show a remarkable absence of significant regional differences worldwide (see papers in Fraser and Sues 1994a).

This chapter briefly reviews the distribution of the principal assemblages of Triassic and Early Jurassic continental tetrapods in time and space, and discusses the profound changes among tetrapod communities during the early Mesozoic. Morales (chap. 39 of this volume) discusses the biology and diversity of the individual groups of reptiles and other land vertebrates that lived during the Triassic, and the reader is referred to that account for details.

Triassic Tetrapod Assemblages

Romer (1966) proposed an informal division of assemblages of Triassic continental tetrapods into three successive "faunas," labeled A, B, and C. "Fauna A" was still dominated by therapsids, while "Fauna C" was dominated by dinosaurs; the intermediate "Fauna B" is characterized by the first radiation of archosaurian reptiles. These three faunas only roughly coincide with the standard threefold division of the Triassic Period into Lower, Middle, and Upper Triassic (or, in terms of geological time, Early, Middle, and Late Triassic). Most assemblages assigned to "Fauna C" by Romer (1966) are now considered Early Jurassic in age (see below). Although somewhat oversimplified (Ochev and Shishkin 1989), Romer's scheme reflects the major steps of faunal succession among continental tetrapods during the early Mesozoic.

Most authors attempt a more precise chronological placement of the various early Mesozoic tetrapod assemblages in terms of the standard stages of the Triassic Period—Scythian (Lower Triassic), Anisian and Ladinian (Middle Triassic), and Carnian and Norian (Upper Triassic; the top of the Norian stage is sometimes considered a separate stage, the "Rhaetian"). Such attempts are fraught with tremendous difficulties, however, because those stages were originally based on sequences of marine sedimentary rocks in the European Alps; thus direct correlation to continental sedimentary rocks proves, for the most part, impossible. Fossil pollen and spores, as well as radiometric dates for volcanic rocks interbedded with marine sedimentary deposits, provide effective indirect means for correlating marine and continental strata. However, many details of intra- and intercontinental correlation between early Mesozoic tetrapod occurrences are controversial and are likely to remain so in the foreseeable future.

Early Triassic

The Early Triassic (Scythian) "Fauna A" is best documented from the *Lystrosaurus-Thrinaxodon* Assemblage Zone (formerly the *Lystrosaurus* Zone; Keyser and Smith 1978) of the Beaufort Group in South Africa (Fig. 40.1). It consists almost entirely of the herbivorous dicynodont therapsids, especially the extremely abundant *Lystrosaurus*, and to a lesser extent other therapsids such as early cynodonts (such as *Thrinaxodon*). The taxonomic diversity of these groups, however, is much reduced compared with what it had been in the preceding Late Permian. The Early Triassic assemblages also include a number of reptiles related to archosaurs, especially the superficially crocodile-like Proterosuchidae. *Lystrosaurus* and *Proterosuchus* are also known from India (Tripathi and Satsangi 1963; Chatterjee and Roy-Chowdhury 1974), Antarctica (Colbert 1982; Hammer 1989), and China (Sun 1989). *Lystrosaurus* (Kalandadze 1975) and forms closely related to *Proterosuchus* (Ochev et al. 1979) are known from the Lower Triassic of European Russia. However, the Early Triassic vertebrate communities from eastern Europe and Australia/Tasmania are characterized by a great abundance of temnospondyl amphibians and include very few therapsids (Ochev et al. 1979; Cosgriff 1984; Ochev and Shishkin 1989).

A stratigraphically slightly younger, possibly Middle Triassic fauna of the A type is that of the *Kannemeyeria* Assemblage Zone (formerly *Cynognathus* Zone; Keyser and Smith 1978) of South Africa. It is characterized by the abundance of the large dicynodont *Kannemeyeria* (with a skull length of up to 50 cm) and the large carnivorous cynodont *Cynognathus* (with a skull length of up to 40 cm). They occur with the presumably herbivorous cynodonts *Diademodon* and *Trirachodon* and with several relatives of archosaurian reptiles, such as the superficially hippopotamus-like, large-headed *Erythrosuchus* (with a skull length of up to 1 m) and the smaller *Euparkeria* (Keyser and Smith 1978). *Kannemeyeria* and *Cynognathus* are also known from the Lower Triassic of Argentina (Bonaparte 1978) and possibly Antarctica (Hammer 1989), and very closely related animals occurred in China (Young 1964; Sun 1989) and Russia (Ochev et al. 1979; Ochev and Shishkin 1989).

Middle Triassic

The Middle to early Late Triassic tetrapod assemblages of "Fauna B," most thoroughly documented from northwestern Argentina (Romer 1966, 1973; Bonaparte 1978) and southern Brazil (Huene 1935–42; Barberena et al. 1985), are distinguished by a great abundance of cynodonts with specialized (gomphodont) "cheek" teeth, and especially rhynchosaurs. Some large dicynodonts, such as *Stahleckeria* (skull length up to 60 cm), also occur in the B-type tetrapod assemblages. Most notable is the considerable diversity of archosaurian reptiles, which include the oldest known dinosaurs (Bonaparte 1978; Sereno et al. 1993).

The currently best known early B-type assemblage comes from the Chañares ("Ischichuca") Formation of La Rioja, Argentina (Fig. 40.2) (Romer 1966, 1973; Bonaparte 1978). It includes abundant small to medium-sized gomphodont cynodonts (*Massetognathus*) but apparently lacks rhynchosaurs. The Chañares Formation has also yielded a number of small carnivorous cynodonts (*Probainognathus, Probelesodon*), large dicynodonts (*Dinodontosaurus*), and a considerable diversity of archosaurs, including the large rauisuchian *Luperosuchus* (with a skull length of 60 cm), a possible early crocodile relative (*Gracilisuchus*), and several slender-limbed forms closely related to dinosaurs (*Lagerpeton, Marasuchus*; Sereno and Arcucci 1994a, 1994b). The therapsids *Massetognathus, Probelesodon,* and *Dinodontosaurus* also occur in the lower part of the Santa Maria Formation of Rio Grande do Sul (Brazil), which Barberena et al. (1985) called the *Dinodontosaurus* Assemblage Zone of the Santa Maria Formation, and appear to be Middle Triassic (Ladinian) in age.

Another diverse early B-type assemblage of tetrapods is known from the Middle Triassic Manda Formation of Tanzania (Attridge et al. 1964). In

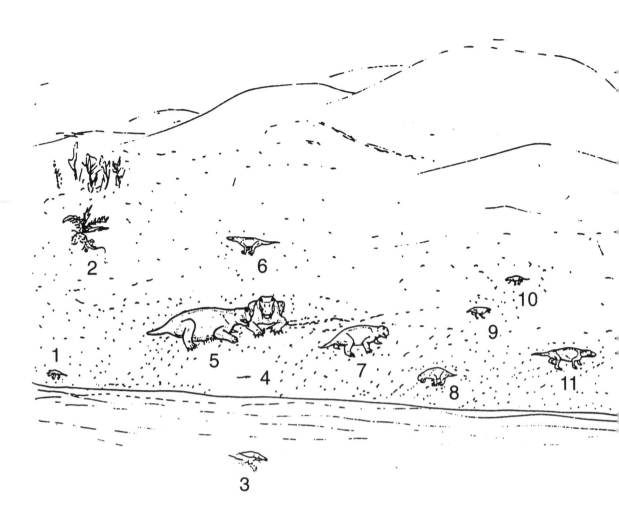

2   6   10   9   1   5   4   7   8   11   3

addition to the rhynchosaur *Stenaulorhynchus* and gomphodont cyno-donts, it includes rauisuchians and other archosaurian reptiles, but regrettably much of this important faunal material remains unpublished.

Late Triassic

Rhynchosaurs (*Scaphonyx*) and gomphodont cynodonts (e.g., *Exaeretodon*) constitute a large portion of the early Late Triassic (Carnian) tetrapod communities from the Ischigualasto Formation of northwestern Argentina (Bonaparte 1978) and the upper part (*Scaphonyx* Assemblage Zone) of the Santa Maria and Caturrita formations of southern Brazil (Barberena et al. 1985). Large (up to 6 m long) rauisuchian archosaurs such as *Prestosuchus* and *Saurosuchus* represented the top carnivores in these assemblages. Two of these formations have yielded the stratigraphically oldest dinosaurs known to date: *Staurikosaurus* from the upper part of the Santa Maria Formation and *Eoraptor, Herrerasaurus,* and *Pisanosaurus* from the Ischigualasto Formation (Sereno et al. 1993). The presence of the ornithischian dinosaur *Pisanosaurus* in the Ischigualasto Formation, re-

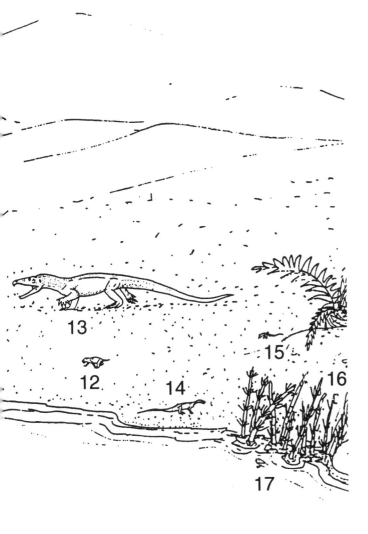

Figure 40.1. Early Triassic (*Lystrosaurus-Thrinaxodon* Assemblage Zone) vertebrates. Amphibians: 3, *Kestrosaurus;* 17, *Pneumatostega.* Procolophonids (primitive amniotes): 1, *Procolophon.* Therapsids: 5, 7, 8, *Lystrosaurus;* 6, *Tetracynodon;* 9, *Ericiolacerta;* 10, *Regisaurus;* 11, *Thrinaxodon;* 12, *Myosaurus;* 16, *Olivieria.* Rhynchosaurs: 2, *Noteosuchus.* Lizards: 4, *Colubrifer;* 15, *Paliguana.* Protorosaurs: 14, *Prolacerta.* Proterosuchians: 13, *Protero-suchus.* This and other drawings in this chapter by Tracy Ford.

cently radiometrically dated as middle Carnian in age, establishes that the evolutionary divergence of the Saurischia and Ornithischia predated the mid-Carnian (Rogers et al. 1993).

As medium-sized to large carnivores and herbivores, dinosaurs quickly came to dominate tetrapod assemblages of Romer's "Fauna C." Advanced therapsids, including the precursors of mammals, formed only a minor element of these communities. Various lines of geological evidence indicate that many of the allegedly Late Triassic C-type tetrapod assemblages discussed by Romer (1966) are in fact Early Jurassic in age (Olsen and Galton 1977; Olsen and Sues 1986).

An abundance of the crocodile-like phytosaurs and of metoposaurid amphibians, both groups constituting semi-aquatic or aquatic predators in lowland freshwater environments, characterizes the classical Late Triassic (Carnian-Norian) C-type assemblages from the Keuper of central Europe and the Chinle (Fig. 40.3) and Dockum groups of the southwestern United States. Other important faunal elements include the massively armored Stagonolepididae (Aetosauridae) and dinosaurs, especially the small thero-pod *Coelophysis* (see papers in Lucas and Hunt 1989). A rich tetrapod

Continental Tetrapods of the Early Mesozoic   631

assemblage, which includes abundant metoposaurs and phytosaurs as well as various dinosaurs, is known from the Carnian-age Argana Formation of Morocco (Dutuit 1972).

The mid- to late Norian tetrapod communities of the Stubensandstein and Knollenmergel (Keuper) of southern Germany and adjacent regions are famous for the abundance of the large (up to 7 m long) basal sauropodomorph *Plateosaurus* (Sander 1992). A virtually identical tetrapod assemblage occurs in the Norian-age Fleming Fjord Formation of East Greenland (Jenkins et al. 1993).

Figure 40.2. Middle Triassic (Chañares Formation) vertebrates. Therapsids: 1, *Jacheleria*; 7, *Probainognathus*; 8, *Massetognathus*; 9, *Dinodontosaurus*. The remaining animals are all archosaurs. Proterosuchians: 2, *Gualosuchus*; 4, *Chanaresuchus*. Lagosuchids: 5, *Marasuchus*; 10, *Lagosuchus*; 11, *Lagerpeton*. Rauisuchids: 12, *Luperosuchus*. Aetosaurs: 13, unidentifed form. Gracilisuchids: 3, *Gracilisuchus*; 6, *Lewisuchus*.

Metoposaurs and phytosaurs are very rare in or absent from most of the known Late Triassic communities from the Southern Hemisphere (Gondwana). A notable exception is the Carnian-age Maleri Formation of Deccan (India), which has yielded a "mixed" tetrapod assemblage that includes metoposaurs and phytosaurs as well as the gomphodont cynodont *Exaeretodon* (Chatterjee and Roy-Chowdhury 1974; Chatterjee 1982). It also includes the rhynchosaur *Hyperodapedon*, which is elsewhere known from the Lossiemouth Sandstone Formation of Scotland (Benton 1983). Similarly, fragmentary but diagnostic material of metoposaurs and phyto-

Continental Tetrapods of the Early Mesozoic   633

Figure 40.3. Late Triassic (Chinle Group) vertebrates. Amphibians: 14, *Buettneria.* Therapsids: 1, *Placerias;* 2, unidentified cynodont. Lizards: 7, *Kuehneosaurid.* Protorosaurians: 16, *Tanytrachelos.* Trilophosaurs: 4, *Trilophosaurus.* The remaining animals are all archosaurs. Rauisuchians: 10, *Postosuchus;* 13, *Chatterjea.* Aetosaurs: 3, *Desmatosuchus;* 6, *Stagonolepis.* Phytosaurs: 8, *Leptosuchus.* Sphenosuchians: 5, *Hesperosuchus.* Pterosaurs: 17, unidentified form. Dinosaurs: 9, *Anchisaurus;* 11, unidentified fabrosaurid; 12, unidentified coelophysid; 15, *Chindesaurus.*

saurs has been reported from the Upper Triassic of Madagascar (Dutuit 1978).

Sues and Olsen (1990) have suggested that faunal differences between northern (Laurasian) and southern (Gondwanan) tetrapod assemblages of Late Triassic age may reflect differences in geological age rather than geographic distribution. They regarded the B-type assemblages with rhynchosaurs and gomphodont cynodonts as slightly older than the oldest typical C-type assemblages with metoposaurs and phytosaurs. Rhynchosaurs and gomphodont cynodonts are associated with metoposaurs in the Wolfville Formation (Carnian) of Nova Scotia (Olsen et al. 1989). A gomphodont cynodont (*Boreogomphodon*) occurs in great abundance along with phytosaurs in the Turkey Branch Formation (Carnian) of Virginia (Sues and Olsen 1990). A rhynchosaur (*Hyperodapedon*) is common in the Lossiemouth Sandstone Formation (Carnian) of Scotland, which is noteworthy for the complete absence of metoposaurs and phytosaurs, although this may simply reflect a more arid environment (Benton and Walker 1985).

To date, the best-known Late Triassic tetrapod assemblage of "Fauna C" from the Southern Hemisphere comes from the Los Colorados Formation of northwestern Argentina (Bonaparte 1972, 1978). It includes the big (up to 11 m long) sauropodomorph *Riojasaurus* along with the huge rauisuchian *Fasolasuchus* (with a skull length of up to 95 cm), a sphenosuchian crocodylomorph (*Pseudhesperosuchus*), a crocodile precursor (*Hemiprotosuchus*), an ornithosuchid (*Riojasuchus*), a tritheledontid cynodont (*Chalimia*), and an indeterminate small theropod. In its faunal composition, the Los Colorados assemblage is intermediate between typically Late Triassic and Early Jurassic tetrapod communities, and more careful stratigraphic documentation is needed to determine whether all of these tetrapods actually occur in the same horizon.

A general trend toward wide—even cosmopolitan—geographic distribution of many taxa of continental tetrapods is apparent during the course of the Triassic, becoming most pronounced toward the end of that period and during the Early Jurassic (Olsen and Sues 1986; Sues and Reisz 1995).

Early Jurassic Tetrapod Assemblages

The redating of many allegedly Late Triassic occurrences of continental tetrapods worldwide by Olsen and his associates (Olsen and Galton 1977; Olsen and Sues 1986; Olsen et al. 1989) has invalidated the traditional view that there were few Early Jurassic tetrapod records because that time interval was allegedly represented mainly by marine sedimentary rocks. For a long time, the fossil record of Early Jurassic continental tetrapods seemed to be restricted to rare finds of dinosaurs and pterosaurs in marine strata from Europe. Subsequently, several bonebeds (southern Germany, southwest Britain) and fissure fillings with abundant disarticulated bones of small tetrapods (southwest Britain), spanning the Triassic-Jurassic boundary, were discovered (Clemens 1980; Evans and Kermack 1994). The Early Jurassic fissure fillings of southwest Britain are most famous for the large samples of the mammalian precursor *Morganucodon*; these occurrences apparently represent island biotas (Evans and Kermack 1994).

Some of the most diverse assemblages have been recovered from the Lower Lufeng Formation (Dark Red and Dull Purplish Beds) of Yunnan, China (Young 1951; Simmons 1965; Luo and Wu 1994). The Lower

Lufeng Formation includes the large sauropodomorph *Lufengosaurus* (Young 1951) and the large crested theropod *Dilophosaurus* (Hu 1993). However, the bulk of the assemblages comprises small to medium-sized animals: sphenodontian lepidosaurs (*Clevosaurus*), a sphenosuchian crocodylomorph (*Dibothrosuchus*), several small crocodile-like reptiles (e.g., *Platyognathus*), small ornithischian (e.g., *Tatisaurus*) and possibly small theropod dinosaurs, abundant tritylodontid cynodonts (e.g., *Bienotherium*), and various mammalian precursors (*Morganucodon, Sinoconodon*) (Luo and Wu 1994).

The well-known tetrapod assemblages of Early Jurassic age from Arizona (Moenave and Kayenta formations and Navajo Sandstone; Sues et al. 1994), Nova Scotia (McCoy Brook Formation; Shubin et al. 1994), and southern Africa (upper portion of Stormberg Group and its equivalents; Kitching and Raath 1984) share the presence of *Clevosaurus* and small crocodile-like archosaurs with the Lower Lufeng Formation. The mammal precursor *Morganucodon* is present in the Lower Jurassic fissure fillings of southwest Britain, the Kayenta Formation, the Lower Lufeng Formation, and possibly the upper Stormberg Group (*"Erythrotherium"*). The upper Stormberg also shares with the McCoy Brook Formation the presence of the tritheledontid cynodont *Pachygenelus* (Shubin et al. 1994) and with the Kayenta Formation the large (up to 4 m long) sauropodomorph *Massospondylus* and the small theropod *Syntarsus*. An early relative of crocodilians, *Protosuchus,* occurs in the upper Stormberg, the McCoy Brook Formation, and the Moenave Formation (Sues et al. 1994). The large (up to 6 m long) theropod *Dilophosaurus* occurs in the Kayenta Formation (Welles 1984) and in the Lower Lufeng Formation (Hu 1993). The small tritylodontid cynodont *Oligokyphus* is known from the Lower Jurassic of southern Germany and southwest Britain (Evans and Kermack 1994), the Kayenta Formation of Arizona (Sues et al. 1994), and the Lower Lufeng Formation of Yunnan (Luo and Sun 1994). This remarkable degree of cosmopolitanism among continental vertebrates during the Early Jurassic is puzzling in view of the fact that the disintegration of Pangaea was already under way (Sues and Reisz 1995).

Based on the currently available fossil record, Early Jurassic assemblages of continental tetrapods are primarily characterized by the absence of such characteristically Late Triassic forms as metoposaurid amphibians, phytosaurs, and procolophonids; apparently very few new groups of land vertebrates appeared during that time interval.

The Rise of Dinosaurs: Competition or Ecological Opportunism?

Several paleontologists have advanced scenarios to explain the spectacular diversification of archosaurian reptiles during the Triassic and their replacement of the therapsids as the dominant component of terrestrial tetrapod communities. Most of these scenarios invoke some form of competition, and assume that certain morphological and/or inferred physiological innovations provided these reptiles with selective advantages over potential competitors, especially therapsids. Many center on the differences in limb posture and gait between therapsids and other early tetrapods, on one hand, and archosaurian reptiles, on the other.

Therapsids were obligatory quadrupeds, with more or less sprawling, wide-track gaits (Hotton 1980). Archosaurs had more erect postures, with the limbs held and moving closer to the body. This supposedly permitted more rapid and, even in many early forms, at least occasionally bipedal locomotion.

Hotton (1980) further suggested that therapsids were replaced by archosaurian reptiles in part because the latter were able to excrete nitrogen as uric acid, with little loss of water. Present-day reptiles and birds excrete nitrogen in the form of uric acid, as a slurry (in birds) or nearly dry pellet (in lizards), whereas present-day mammals (and thus presumably the therapsids from which mammals arose) excrete nitrogen almost exclusively as urea, which requires copious amounts of water to be flushed out of the body. Given increasingly arid climates, especially in the later part of the Triassic, Hotton inferred that the water-saving metabolism of archosaurs, along with their locomotor specializations, gave them a competitive advantage over therapsids.

According to the competition scenario proposed by Charig (1980, 1984), the fully erect limb posture of dinosaurs resulted in "generally superior" locomotor abilities and supposedly accounted for the tremendous evolutionary success of this group. Thus the carnivorous dinosaurs became more efficient predators that in due course would have outcompeted other carnivorous archosaurs and carnivorous cynodonts and eliminated the various non-dinosaurian herbivores (dicynodonts, gomphodont cynodonts, and rhynchosaurs) of the time. The disappearance of those plant-eating forms could then have accelerated the evolutionary diversification of the various groups of herbivorous dinosaurs to exploit the plant resources (Charig 1984).

Benton (1983) questioned whether large-scale competition between dinosaurs with erect limb posture and synapsids and other tetrapods with sprawling limb posture over a period of several million years could have resulted in the former outcompeting the latter. Instead he claimed that the extinctions among non-mammalian synapsids and other major groups of tetrapods could be related to extrinsic factors, specifically floral changes, toward the end of the Triassic Period. However, his correlation of the extinction of the rhynchosaurs with the disappearance of floras dominated by the "seed-fern" *Dicroidium* is not supported by the available evidence (Rogers et al. 1993). Benton (1983) interpreted the early dinosaurs as mere ecological opportunists that succeeded by simply taking over the niches left vacant by their extinct predecessors.

The inherent incompleteness of the fossil record does not permit rigorous testing of the competing scenarios to explain the evolutionary success of dinosaurs. However, a recent survey of tetrapod diversity in the Ischigualasto Formation by Rogers et al. (1993) found no evidence for the kind of long-term gradual decline in the diversity of the earlier tetrapod groups predicted by Charig's and other models of competitive replacement. Furthermore, the "competitive superiority" of dinosaurs is less apparent than suggested by Charig. Various lineages of archosaurian reptiles independently evolved erect posture during the Late Triassic, yet most of these did not survive the end of the Triassic.

Early Mesozoic Extinctions in Continental Tetrapods

One or two major episodes of extinction among continental tetrapods occurred during the early Mesozoic. It has long been known that a mass extinction took place among marine invertebrates, especially ammonoid cephalopods, bivalves, and brachiopods, at the end of the Triassic Period

(Hallam 1981, 1990). In fact, this extinction event now ranks as one of the five biggest during the last 540 million years (Sepkoski 1986). Colbert (1958) first noted the disappearance of numerous tetrapod lineages at the Triassic-Jurassic boundary, and more recent surveys of early Mesozoic tetrapod biodiversity have confirmed his observations (Benton 1986, 1991, 1994; Olsen and Sues 1986). Despite continuing intensive research, considerable uncertainty persists in correlating the biotic changes on land and in the seas because of the problems in precise stratigraphic correlation discussed earlier in this chapter.

The stratigraphically well-constrained fossil record from the early Mesozoic sedimentary strata of the Newark Supergroup in eastern North America supports the hypothesis of a mass-extinction event in continental biotas at the end of the Triassic (Olsen and Sues 1986; Olsen et al. 1987, 1989, 1990). The Newark Supergroup represents the remnants of thousands of meters of sedimentary and volcanic rocks deposited in a chain of rift basins that formed during a 45-million-year episode of crustal extension and thinning preceding the Jurassic breakup of the supercontinent Pangaea and the opening of the northern Atlantic Ocean (Olsen et al. 1989). It has yielded a number of well-dated, diversified occurrences of skeletal remains and trackways of continental tetrapods that range in age from the Middle Triassic to the Early Jurassic.

The recent discovery of abundant tetrapod fossils of earliest Jurassic age in the lower part of the McCoy Brook Formation of Nova Scotia is of particular interest in the present context because these assemblages appear to postdate the radiometrically determined Triassic-Jurassic boundary by less than 1 million years (Olsen et al. 1987, 1989, 1990; Shubin et al. 1994). Although the fossil occurrences from the McCoy Brook Formation represent a number of different paleoenvironments, they all share the absence of certain amphibians and reptiles that characterized Late Triassic tetrapod communities in North America and elsewhere. These "losses" include metoposaurid amphibians, procolophonids, phytosaurs, stagonolepidids, and rauisuchians. The well-known assemblages of Early Jurassic tetrapods from the Lower Lufeng Formation of Yunnan (China) and the upper Stormberg Group of southern Africa show the same pattern of change in faunal composition.

The precise number of taxa disappearing at the end of the Triassic remains a matter of contention, but the extinctions do not appear to be taxonomically and/or ecologically selective (Olsen and Sues 1986; Benton 1991). All of the tetrapods found in the McCoy Brook Formation and elsewhere, on the other hand, represent taxa that were already present in the Late Triassic, and there appear to be very few originations of new taxa at the beginning of the Jurassic (Olsen and Sues 1986; Olsen et al. 1987; Benton 1991). Dinosaurs appear to have been essentially unaffected by the end-Triassic extinction event: To date, there is not a single recorded disappearance of a family-level taxon of dinosaurs across the Triassic-Jurassic boundary.

It is noteworthy in this context that the fossil record of pollen and spores from the sedimentary rock of the Newark Supergroup and elsewhere indicates a major floral change at or near the Triassic-Jurassic boundary (Olsen et al. 1990). The taxonomically diverse Late Triassic assemblages of pollen and spores are replaced by greatly impoverished Early Jurassic ones that are almost entirely composed of pollen of certain conifers (form-genus *Corollina* or *Classopollis*).

The causes for the large-scale biotic changes observed at the Triassic-Jurassic boundary are still not understood. Olsen et al. (1987) linked the changes to the giant Manicouagan impact crater in Quebec (Canada). Visible parts of that crater, located in remote terrain, have a diameter of about 70 kilometers, but more subtle features extend to about 100 kilometers. Unfortunately, the advance of huge glaciers during the Ice Age destroyed much of the original structure of the impact site. Silver (1982) has calculated that the crater was produced by an extraterrestrial object with a diameter of about 10 kilometers striking at a velocity of 25 kilometers per second and releasing an energy in excess of 100 million megatons upon impact. Hodych and Dunning (1992) have recently radiometrically redated the Manicouagan impact event at about 214 million years before the present, which makes it significantly older than the radiometrically determined Triassic-Jurassic boundary—and thus would rule it out as the "smoking gun" for the extinctions at the Triassic-Jurassic boundary.

Surprisingly, there exists as yet no evidence for a major extinction event at the time of the Manicouagan impact. Bice et al. (1992) reported several layers with shocked quartz in marine sedimentary rocks from the Triassic-Jurassic boundary in Italy, and hypothesized multiple impacts to account for their presence. However, that evidence is far less compelling than that for the terminal Cretaceous impact event (Benton 1994), and to date no obvious impact site of the appropriate age has been identified. Hallam (1981, 1990) favored terrestrial causes for the end-Triassic biotic crisis, specifically extensive volcanism in eastern North America and southern Africa and changes in sea level at the Triassic-Jurassic boundary. However, Olsen (in Olsen et al. 1989) argued that none of those scenarios can fully account for the currently available fossil evidence.

Benton (1991, 1994) has argued that an earlier, even more significant extinction of continental tetrapods occurred at the end of the Carnian stage of the Late Triassic. In his view, dinosaurs subsequently rapidly radiated during the Norian and continued to diversify in the Jurassic after the end-Norian extinction event. However, Rogers et al. (1993) and Fraser and Sues (1994b) have challenged the validity of the numbers of last appearances of taxa tabulated by Benton, which would significantly diminish the magnitude of the proposed end-Carnian event. Rigorous testing of Benton's intriguing hypothesis is currently difficult because there are few known tetrapod assemblages of undoubtedly early Norian age (contra Benton 1994).

After the extinction of all potential competitors among archosaurian reptiles (such as rauisuchians) at the end of the Triassic, dinosaurs became firmly established as the dominant large land animals worldwide during the Early Jurassic.

References

Attridge, J.; H. W. Ball; A. J. Charig; and C. B. Cox. 1964. The British Museum (Natural History)–University of London joint palaeontological expedition to northern Rhodesia and Tanganyika. *Nature* 201: 445–449.

Barberena, M. C.; D. C. Araújo; and E. L. Lavina. 1985. Late Permian and Triassic tetrapods of southern Brazil. *National Geographic Research* 1: 5–20.

Benton, M. J. 1983. Dinosaur success in the Triassic: A noncompetitive ecological model. *Quarterly Review of Biology* 58:29–55.

Benton, M. J. 1986. The Late Triassic tetrapod extinction events. In K. Padian (ed.), *The Beginning of the Age of Dinosaurs: Faunal Change across the Triassic-Jurassic Boundary,* pp. 303–320. Cambridge: Cambridge University Press.

Benton, M. J. 1991. What really happened in the Late Triassic? *Historical Biology* 5: 263–278.

Benton, M. J. 1994. Late Triassic and Middle Jurassic extinctions among continental tetrapods: Testing the pattern. In N. C. Fraser and H.-D. Sues (eds.), *In the Shadow of the Dinosaurs: Early Mesozoic Tetrapods,* pp. 366–397. Cambridge: Cambridge University Press.

Benton, M. J., and A. D. Walker. 1985. Palaeoecology, taphonomy, and dating of Permo-Triassic reptiles from Elgin, north-east Scotland. *Palaeontology* 28: 207–234.

Bice, D.; C. R. Newton; S. McCauley; P. W. Reiners; and C. A. McRoberts. 1992. Shocked quartz at the Triassic-Jurassic boundary in Italy. *Science* 255: 443–446.

Bonaparte, J. F. 1972. Los tetrápodos del sector superior de la formación Los Colorados, La Rioja, Argentina (Triásico superior). I Parte. *Opera Lilloana* 22: 1–183.

Bonaparte, J. F. 1978. El Mesozoico de América del Sur y sus tetrápodos. *Opera Lilloana* 26: 5–596.

Charig, A. J. 1980. Differentiation of lineages among Mesozoic tetrapods. *Mémoires de la Société Géologique de France,* n.s. 139: 207–210.

Charig, A. J. 1984. Competition between therapsids and archosaurs during the Triassic period: A review and synthesis of current theories. In M. W. J. Ferguson (ed.), *The Structure, Development and Evolution of Reptiles,* pp. 597–628. London: Academic Press.

Chatterjee, S. 1982. A new cynodont reptile from the Triassic of India. *Journal of Paleontology* 56: 203–214.

Chatterjee, S., and T. Roy-Chowdhury. 1974. Triassic Gondwana vertebrates from India. *Indian Journal of Earth Sciences* 1: 96–112.

Clemens, W. A. 1980. Rhaeto-Liassic mammals from Switzerland and West Germany. *Zitteliana* 5: 51–92.

Colbert, E. H. 1958. Tetrapod extinctions at the end of the Triassic. *Proceedings of the National Academy of Sciences,* U.S.A. 44: 973–977.

Colbert, E. H. 1982. Triassic vertebrates in the Transantarctic Mountains. In M. D. Turner and J. E. Splettstoesser (eds.), *Geology of the Central Transantarctic Mountains,* pp. 11–35. Antarctic Research Series, vol. 36. Washington, D.C.: American Geophysical Union.

Cosgriff, J. W. 1984. The temnospondyl labyrinthodonts of the earliest Triassic. *Journal of Vertebrate Paleontology* 4: 30–46.

Dutuit, J.-M. 1972. Introduction à l'étude paléontologique du Trias continental marocain. Description des premiers Stégocéphales, recueillis dans le couloir d'Argana (Atlas occidental). *Mémoires du Muséum National d'Histoire Naturelle* C, 36: 1–253.

Dutuit, J.-M. 1978. Description de quelques fragments osseux provenant de la région de Folakra (Trias supérieur malgache). *Bulletin du Muséum National d'Histoire Naturelle* 3 (69): 79–89.

Evans, S. E., and K. A. Kermack. 1994. Assemblages of small tetrapods from the Early Jurassic of Britain. In N. C. Fraser and H.-D. Sues (eds.), *In the Shadow of the Dinosaurs: Early Mesozoic Tetrapods,* pp. 271–283. Cambridge: Cambridge University Press.

Fraser, N. C., and H.-D. Sues. 1994a. *In the Shadow of the Dinosaurs: Early Mesozoic Tetrapods.* Cambridge: Cambridge University Press.

Fraser, N. C., and H.-D. Sues. 1994b. Comments on Benton's "Late Triassic to Middle Jurassic extinctions among continental tetrapods." In N. C. Fraser and H.-D. Sues (eds.), *In the Shadow of the Dinosaurs: Early Mesozoic Tetrapods,* pp. 398–400. Cambridge: Cambridge University Press.

Hallam, A. 1981. The end-Triassic bivalve extinction event. *Palaeogeography, Palaeoclimatology, Palaeoecology* 35: 1–44.

Hallam, A. 1990. The end-Triassic mass extinction event. Geological Society of America Special Paper 247: 577–583.

Hammer, W. R. 1989. Triassic terrestrial vertebrate faunas of Antarctica. In T. N. Taylor and E. L. Taylor (eds.), *Antarctic Paleobiology: Its Role in the Reconstruction of Gondwana,* pp. 42–50. New York: Springer-Verlag.

Hodych, J. P., and G. R. Dunning. 1992. Did the Manicouagan impact trigger end-of-Triassic mass extinction? *Geology* 20: 51–54.

Hotton, N. III. 1980. An alternative to dinosaur endothermy: The happy wanderers. In R. D. K. Thomas and E. C. Olson (eds.), *A Cold Look at the Warm-Blooded Dinosaurs,* pp. 311–350. American Association for the Advancement of Science Selected Symposium 28. Boulder, Colo.: Westview Press.

Hu S. 1993. [A new theropod (*Dilophosaurus sinensis* sp. nov.) from Yunnan, China.] *Vertebrata PalAsiatica* 31: 65–69. (In Chinese with English abstract.)

Huene, F. von. 1935–42. *Die fossilen Reptilien des südamerikanischen Gondwanalandes.* Munich: C. H. Beck'sche Verlagsbuchhandlung.

Jenkins, F. A. Jr.; N. H. Shubin; W. W. Amaral; S. M. Gatesy; C. R. Schaff; W. R. Downs; L. B. Clemmensen; N. Bonde; A. R. Davidson; and F. Osbæck. 1993. A Late Triassic continental vertebrate fauna from the Fleming Fjord Formation, Jameson Land, East Greenland. *New Mexico Museum of Natural History and Science, Bulletin* 3: 74.

Kalandadze, N. N. 1975. Pervaia nakhodka listrozavra na territorii evropeiskoi chasti SSSR. *Paleontologicheskii zhurnal* 1974 (4): 140–142.

Keyser, A. W., and R. M. H. Smith. 1978. Vertebrate biozonation of the Beaufort Group with special reference to the western Karoo Basin. *Annals of the Geological Survey of South Africa* 12: 1–35.

Kitching, J. W., and M. A. Raath. 1984. Fossils from the Elliot and Clarens formations (Karoo sequence) of the northeastern Cape, Orange Free State and Lesotho, and a suggested biozonation based on tetrapods. *Palaeontologia Africana* 25: 111–125.

Lucas, S. G., and A. P. Hunt (eds.). 1989. *Dawn of the Age of Dinosaurs in the American Southwest.* Albuquerque: New Mexico Museum of Natural History.

Luo, Z., and A. Sun. 1994. *Oligokyphus* (Cynodontia: Tritylodontidae) from the Lower Lufeng Formation (Lower Jurassic) of Yunnan, China. *Journal of Vertebrate Paleontology* 13: 477–482.

Luo, Z., and X. Wu. 1994. The small tetrapods from the Lower Lufeng Formation, Yunnan, China. In N. C. Fraser and H.-D. Sues (eds.), *In the Shadow of the Dinosaurs: Early Mesozoic Tetrapods,* pp. 251–270. Cambridge: Cambridge University Press.

Ochev, V. G., and M. A. Shishkin. 1989. On the principles of global correlation of the continental Triassic on the tetrapods. *Acta Palaeontologica Polonica* 34: 149–173.

Ochev, V. G.; G. I. Tverdokhlebova; M. G. Minikh; and A. V. Minikh. 1979. *Stratigraficheskoe i paleogeograficheskoe znachenie verknepermskikh i triasovykh pozvonochnykh Vostochno-evropeiskoi platformy i Priural'ia.* Saratov: Izdatel'stvo Saratovskogo universiteta.

Olsen, P. E.; S. J. Fowell; and B. Cornet. 1990. The Triassic/Jurassic boundary in continental rocks of eastern North America: A progress report. Geological Society of America Special Paper 247: 585–593.

Olsen, P. E., and P. M. Galton. 1977. Triassic-Jurassic extinctions: Are they real? *Science* 197: 983–986.

Olsen, P. E.; R. W. Schlische; and P. J. W. Gore (eds.). 1989. *Tectonic, Depositional, and Paleoecological History of the Early Mesozoic Rift Basins, Eastern North America.* 28th International Geological Congress, Field Trip Guidebook T351. Washington, D.C.: American Geophysical Union.

Olsen, P. E.; N. H. Shubin; and M. H. Anders. 1987. New Early Jurassic tetrapod assemblages constrain Triassic-Jurassic tetrapod extinction event. *Science* 237: 1025–1029.

Olsen, P. E., and H.-D. Sues. 1986. Correlation of continental Late Triassic and Early Jurassic sediments, and the Triassic-Jurassic tetrapod transition. In K. Padian (ed.), *The Beginning of the Age of Dinosaurs: Faunal Change across the Triassic-Jurassic Boundary,* pp. 321–351. Cambridge: Cambridge University Press.

Parrish, J. M.; J. T. Parrish; and A. C. Ziegler. 1986. Permian-Triassic paleogeography and paleoclimatology and implications for therapsid distribution. In N. Hotton III, P. D. MacLean, J. J. Roth, and E. C. Roth (eds.), *The Ecology and Biology of Mammal-like Reptiles,* pp. 109–131. Washington, D.C.: Smithsonian Institution Press.

Rogers, R. R.; C. C. Swisher III; P. C. Sereno; A. M. Monetta; C. A. Forster; and R. N. Martinez. 1993. The Ischigualasto tetrapod assemblage (Late Triassic, Argentina) and 40Ar/39Ar dating of dinosaur origins. *Science* 260: 794–797.

Romer, A. S. 1966. The Chañares (Argentina) Triassic reptile fauna. Part I: Introduction. *Breviora* 247: 1–14.

Romer, A. S. 1973. The Chañares (Argentina) Triassic reptile fauna. Part XX: Summary. *Breviora* 413: 1–20.

Sander, P. M. 1992. The Norian *Plateosaurus* bonebeds of central Europe and their taphonomy. *Palaeogeography, Palaeoclimatology, Palaeoecology* 93: 255–299.

Sepkoski, J. J. Jr. 1986. Phanerozoic overview of mass extinctions. In D. M. Raup and D. Jablonski (eds.), *Patterns and Processes in the History of Life,* pp. 277–296. Berlin: Springer-Verlag.

Sereno, P. C., and A. B. Arcucci. 1994a. Dinosaurian precursors from the Middle Triassic of Argentina: *Lagerpeton chanarensis. Journal of Vertebrate Paleontology* 13: 385–399.

Sereno, P. C., and A. B. Arcucci. 1994b. Dinosaurian precursors from the Middle Triassic of Argentina: *Marasuchus lilloensis* gen. nov. *Journal of Vertebrate Paleontology* 14: 53–73.

Sereno, P. C.; C. A. Forster; R. R. Rogers; and A. M. Monetta. 1993. Primitive dinosaur skeleton from Argentina and the early evolution of Dinosauria. *Nature* 361: 64–66.

Shubin, N. H.; P. E. Olsen; and H.-D. Sues. 1994. Early Jurassic small tetrapods from the McCoy Brook Formation of Nova Scotia, Canada. In N. C. Fraser and H.-D. Sues (eds.), *In the Shadow of the Dinosaurs: Early Mesozoic Tetrapods,* pp. 242–250. Cambridge: Cambridge University Press.

Silver, L. T. 1982. Introduction. In L. T. Silver and P. H. Schultz (eds.), *Geological Implications of Impacts of Large Asteroids and Comets on Earth,* pp. xiii–xix. Geological Society of America Special Paper 190.

Simmons, D. J. 1965. The non-therapsid reptiles of the Lufeng basin, Yunnan, China. *Fieldiana, Geology* 15: 1–93.

Sues, H.-D.; J. M. Clark; and F. A. Jenkins, Jr. 1994. A review of the Early Jurassic tetrapods from the Glen Canyon Group of the American Southwest. In N. C. Fraser and H.-D. Sues (eds.), *In the Shadow of the Dinosaurs: Early Mesozoic Tetrapods,* pp. 284–294. Cambridge: Cambridge University Press.

Sues, H.-D., and P. E. Olsen. 1990. Triassic vertebrates of Gondwanan aspect from the Richmond basin of Virginia. *Science* 249: 1020–1023.

Sues, H.-D., and R. R. Reisz. 1995. First record of the early Mesozoic sphenodontian *Clevosaurus* (Lepidosauria: Rhynchocephalia) from the Southern Hemisphere. *Journal of Paleontology* 69: 123–126.

Sun, A. 1989. *Before Dinosaurs.* Beijing: China Ocean Press.

Tripathi, S., and P. P. Satsangi. 1963. *Lystrosaurus* fauna from the Panchet Series of the Raniganj Coalfield. *Palaeontologia Indica,* n.s. 37: 1–53.

Welles, S. P. 1984. *Dilophosaurus wetherilli* (Dinosauria, Theropoda): Osteology and comparisons. *Palaeontographica* A, 185: 85–180.

Wing, S. L., and H.-D. Sues (rapporteurs). 1992. Mesozoic and early Cenozoic terrestrial ecosystems. In A. K. Behrensmeyer, J. Damuth, W. A. DiMichele, R. Potts, H.-D. Sues, and S. L. Wing (eds.), *Terrestrial Ecosystems through Time: Evolutionary Paleoecology of Terrestrial Plants and Animals,* pp. 327–416. Chicago: University of Chicago Press.

Young, C. C. 1951. The Lufeng saurischian fauna in China. *Palaeontologia Sinica,* n.s. C, 13: 1–96.

Young, C. C. 1964. The pseudosuchians in China. *Palaeontologia Sinica,* n.s. C, 19: 1–205.

Continental Tetrapods of the Early Mesozoic   643

Dinosaurian Faunas of the Later Mesozoic

*Dale A. Russell
and
José F. Bonaparte*

41

Although there are long traditions of interest relating to dinosaurian morphology, classification, geological age, and paleoecology, until recently relatively little attention was given to dinosaurian biogeography. One must assume that it would have been relatively easy to assess dinosaurian biogeography on the basis of living faunas. However, the fossil record of dinosaurs is notoriously incomplete, and estimates of the fraction of genera known, relative to the total number of genera that have existed, range between about 8 percent (Russell 1994, 1995) and 28 percent (Dodson 1990). It is not so widely appreciated that most of the known dinosaurian genera are represented by incomplete skeletal material. Thus, of the estimated total number of genera, the fraction that is currently represented by complete skeletons might range between 2 percent and 6 percent. Furthermore, those which are well known skeletally are very irregularly distributed in space and time (Table 41.1 and Dodson 1990; Russell 1994, 1995). Implications for the precision of resulting biogeographic inferences are obvious.

Molnar (1980a) compiled generic lists of tetrapods from various important late Mesozoic localities around the globe. He found that tetrapod assemblages from the Northern and Southern hemispheres were indistinguishable during Late Jurassic time, but had separated into well-defined Laurasian and Gondwanan biogeographic realms by Late Cretaceous time. Since Molnar's study, biogeographically significant remains of dinosaurs and other tetrapods have been described from the Jurassic of China (Sun et al. 1992, and references cited therein) and the Cretaceous of Argentina (Bonaparte 1990, 1991, and references cited therein). The fundamentally distinct nature of northern and southern faunas during Late Cretaceous time was confirmed by Bonaparte and Kielan-Jaworowska (1987). Holtz (1993), building on the work of Le Loueff (1991), subsequently also supported a model of cosmopolitan dinosaur distributions during the Late Jurassic, which began to fragment during the Early Cretaceous, and separated into "Asiamerican" and "Eurogondwanan" assemblages during the Late Cretaceous. Russell, on the basis of taxonomic (1994) and paleogeographic (1995) evidence, further proposed that central Asia was isolated from the remainder of Pangaea during Middle and Late Jurassic time. The more quantitative approach of Holtz (1996) also supports a Jurassic isolation for central Asia.

The purpose of this brief review is to offer our interpretations of the faunistic evidence that is currently available on dinosaurs, as it pertains to dinosaurian biogeography (see Molnar, chap. 38 of this volume, for paleogeographic maps of the various time intervals discussed here). In view of the incompleteness of the dinosaurian record, it is highly unlikely that our interpretations will be definitive. And in a few cases (noted below), each of us, in good humor and within the context of a larger consensus, is inclined to emphasize differing alternatives.

The Early Jurassic: Pangaea

Tetrapod assemblages from Early Jurassic localities in North America, Africa, and China strongly resemble each other, and early Jurassic tetrapod families were cosmopolitan (see Sues, chap. 40 this volume; Shubin and Sues 1991). Several well-known dinosaurian genera have intercontinental distributions (see Table 41.1). The presence of sauropods with *Camarasaurus*-like teeth, hypsilophodonts, and stegosaurs in China suggests that these lineages had already differentiated before central Asia became isolated near the beginning of Middle Jurassic time. Similarly, a peculiar Early Jurassic theropod (*Cryolophosaurus*) in an Early Jurassic assemblage from Antarctica suggests that allosaurids might have been distributed throughout Pangaea before Middle Jurassic time (Hammer and Hickerson 1993; Hammer in press). Cosmopolitanism continued to be characteristic of volant vertebrates through to the end of the Mesozoic (e.g., pterosaurs: Nessov 1991a; Bakhurina 1993; enantiornithine birds: Molnar 1986; Chiappe 1993; Sanz et al. 1993; Hutchison 1993; Lamb et al. 1993).

Middle and Late Jurassic: The Isolation of Central Asia

The dinosaurian assemblage from strata of probable Late Jurassic age (Shangshaximiao Formation) throughout central and northern China differs far more from those of eastern Africa (Tendaguru) and western North America (Morrison) than the latter differ from each other (Russell 1994; Russell and Zheng 1994). No yangchuanosaur or omeisaur-mamenchisaur remains have been definitively identified from the relatively well sampled North American and African assemblages (although a yangchuanosaur-like maxilla has been recovered from early Late Jurassic strata in England; Bakker et al. 1992). One of us (D.A.R.) has been unable to find diplodocid remains in any of the collections of Chinese sauropod material he has seen. Middle Jurassic (Xiashaximiao Formation) assemblages also resemble

Table 41.1.

Genera of Dinosaurs Known from Essentially Complete Skeletal Material, Including Skulls

After Weishampel et al. 1990, with referenced additions. Intercontinental distributions, often based on incomplete materials, are also noted (see references).

Early Jurassic: (10 genera)

North America:
Anchisaurus
Scutellosaurus
North America and Europe:
Scelidosaurus (North America; Padian 1989)
North America and Africa:
Syntarsus
Massospondylus (North America; Attridge et al. 1985)
Africa:
Lesothosaurus
Heterodontosaurus
North America and China:
Dilophosaurus (Hu 1993)
Asia:
Yunnanosaurus
Lufengosaurus

Middle Jurassic: (5 genera)

Asia:
Shunosaurus
Omeisaurus
Huayangosaurus
Agilisaurus (new genus; Peng 1992)
Yandusaurus

Late Jurassic: (17 genera)

North America:
Ceratosaurus
Allosaurus
Ornitholestes
Diplodocus
Apatosaurus
Stegosaurus
North America and Europe:
Camptosaurus (Raath and McIntosh 1987)
North America, Europe, and Africa:
Brachiosaurus
Camarasaurus
Europe:
Compsognathus
North America and Africa:
Dryosaurus
Africa:
Kentrosaurus
Dicraeosaurus

Asia:
Sinraptor (new genus; Currie and Zhao 1994)
Monolophosaurus (new genus; Currie and Zhao 1994)
Yangchuanosaurus (?Europe; Bakker et al. 1992)
Mamenchisaurus (new material; Russell and Zheng 1994)

Early Cretaceous: (10 genera)

North America:
Deinonychus
Tenontosaurus
Sauropelta
North America and Europe:
Hypsilophodon (North America; Galton and Jensen 1979)
Iguanodon
Asia:
Sinornithoides (new genus; Russell and Dong 1994)
Psittacosaurus
Africa:
Ouranosaurus
South America:
Carnotaurus
Amargosaurus

Late Cretaceous: (29 genera)

North America:
Ornithomimus
Struthiomimus
Dromiceiomimus
Albertosaurus
Daspletosaurus
Tyrannosaurus
Euoplocephalus
Thescelosaurus
Anatotitan
Edmontosaurus
Kritosaurus
Corythosaurus
Hypacrosaurus
Lambeosaurus
Parasaurolophus
Prosaurolophus
Maiasaura
Leptoceratops
Centrosaurus
Styracosaurus
Anchiceratops
Pentaceratops
Triceratops
North America and Asia:
Saurolophus
Asia:
Velociraptor
Gallimimus
Tarbosaurus
Pinacosaurus
Protoceratops

647

Shangshaximiao assemblages much more closely than those of the other two continents. Therefore, Russell (1994) postulated that central Asian dinosaurian assemblages were endemic, and essentially isolated from those of the remainder of the world.

As noted above, the isolation of central Asia would not have begun until near the beginning of Middle Jurassic time. Jerzykiewicz and Russell (1991), citing early Asian records of champsosaurs, dromaeosaurs, and iguanodonts, which are presumed to have originated beyond central Asia, postulated that the period of isolation ended sometime during the Early Cretaceous.

The Cretaceous: Formation of Laurasia

Excluding those from central Asia, assemblages of Late Jurassic age may be taken as ancestral to assemblages in both Laurasia and Gondwana during later Mesozoic time. The degree to which Laurasian (Morrison Formation) and Gondwanan (Tendaguru Formation) assemblages had already diverged has been variously interpreted by the two of us. Bonaparte (1990: 87) suggests that the presence of derived sauropods in the Tendaguru assemblage, which are absent in the Morrison, is evidence of endemism, while Russell (1994) suggests that differences between the assemblages may have an ecological basis. The recent documentation in the Morrison Formation of the U.S. of early records of discoglossid and pelobatid frogs (Evans and Milner 1993), and possibly of tyrannosaurids (Chure and Madsen 1993), small theropods allied to Cretaceous Asiamerican forms (Chure et al. 1993; Chure 1995), and polacanthids (Kirkland 1993), lends weight to Bonaparte's suggestion.

It seems clear that during the middle Cretaceous time, faunal exchange involving troodonts, dromaeosaurs, hypsilophodonts, iguanodonts, hadrosaurs, and polacanthids was taking place between North America and Europe and, later, Asia (Galton and Jensen 1979; Blows 1987; Weishampel and Bjork 1989; Jerzykiewicz and Russell 1991; Parrish and Eaton 1991; Pereda-Superbiola 1992, 1994; Howse and Milner 1993; Kirkland 1993; Kirkland et al. 1993; Russell and Dong 1994; Jacobs 1995; Norman 1995; Britt and Stadtman 1996; Kirkland 1996). Continental assemblages of the northern continents already differed significantly from those of Gondwana.

By Cenomanian time (Kirkland and Parrish 1995; Jacobs 1996), the Asian and North American continents were linked, probably through an emergent Bering isthmus. Dispersal flowed predominantly from Asia to North America, and the Cordilleran subcontinent or "peninsula" (Fig. 41.1) effectively became an appendage of Asia (Jerzykiewicz and Russell 1991; Nessov 1991b). However, according to patterns of ornithomimosaur, pachycephalosaur, and ceratopsian relationships, dispersal occurred only intermittently (Yaccobucci 1990; Sereno 1991; Forster and Sereno 1994). Ecological regionalism was present within the Asian-Cordilleran landmass. Deltaic environments in western Asia and Cordilleran North America contained similar dinosaurian assemblages (Nessov and Golovneva 1987; Nessov 1991b), although it should be noted that western Asian materials are for the most part incomplete and disarticulated. Northern and southern regions of the Cordilleran "peninsula" were inhabited by different varieties of ceratopsians (Rowe et al. 1992; Forster et al. 1993), and sauropods (probable immigrants from Gondwana) were apparently restricted to the southern region (Lehman 1987, in press).

Within the Asian heartland, dinosaurian assemblages contained an unusual variety of small ratite-like dinosaurs and birds (e.g., *Adasaurus, Anserimimus, Archaeornithodes, Avimimus, Borogovia, Conchoraptor, Elmisaurus, Gallimimus, Gobipteryx, Hulsanpes, Ingenia, Mononykus, Oviraptor, Saurornithoides* and *Velociraptor;* see Jerzykiewicz and Russell 1991, and references cited therein; Elzanowski and Wellnhofer 1993; Perle et al. 1993; Dashzeveg et al. 1995). The incomplete generotypic skull of *Archaeornithoides* exhibits similarities to those of large Afro-European theropods of middle Cretaceous age (Elzanowski and Wellnhofer 1993). The extent to which these similarities are indicative of relationship or due to convergence will be revealed by more complete material.

During much of Late Cretaceous time, the eastern portion of North America was separated from the Cordilleran "peninsula" by the Western Interior Seaway, although similarities between palynofloras (fossil pollen assemblages) on either side of the southern isthmus (Baghai 1994) suggest interchange between terrestrial biotas. Dinosaurian remains are infrequently collected east of the seaway, and are usually extremely fragmentary. Accordingly, the absence of records there for various groups of dinosaurs (e.g., ankylosaurs, pachycephalosaurs, and ceratopsians) is of moot biogeographic significance. Tyrannosaurs, in contrast, have been found on both sides of the seaway (Schwimmer in press). On the other hand, species-level distinctions in microvertebrate remains from New Jersey imply endemism (Grandstaff and Parris 1993), and the presence in the east of the fragmentary but peculiar generotypic skeleton of *Dryptosaurus,* a form unknown west of the Interior Seaway, does suggest an element of endemism on a subcontinental scale (Schwimmer et al. 1993).

The Cretaceous: Formation and Fragmentation of Gondwana

Southern Hemisphere dinosaurian assemblages are very poorly sampled, although the areal extent of the Cretaceous southern landmass implies that they may have been very diverse (Russell 1995). If such was the case, it might be expected that the attributes of the possibly less diverse northern dinosaurian assemblages (which were dominated by tyrannosaurs and diverse hadrosaurian and ceratopsian herbivores) would be defined by Gondwana standards (where assemblages are dominated by abelisaurs and diverse sauropod herbivores; Bonaparte et al. 1990; Calvo and Bonaparte 1991). This has, indeed, recently been done (Holtz 1996). The small, bird-like dinosaurs of the Southern Hemisphere (alvarezsaurids, velocisaurids, naosaurids; Bonaparte and Powell 1980; Bonaparte 1991) are unrelated to those in the Northern Hemisphere.

According to one concept (Sereno et al. 1994, 1996), Gondwana dinosaur faunas remained Pangaean in character well into Early Cretaceous time. Approximately simultaneously, by the beginning of Late Cretaceous time, the major Pangaean blocks had lost contact with each other. African dinosaur assemblages had accordingly become equally distinct from those of Asiamerica and South America. According to another concept (Forster 1996; Russell 1996; Sampson et al. 1996), dinosaur assemblages of the separating southern continents continued to form a zoogeographic entity distinguished from contemporary Asiamerican assemblages by the continued dominance of abelisaurs and titanosaurs.

During earlier Cretaceous time (Neocomian, perhaps, through Aptian; see Kellner 1994), terrestrial vertebrate distributions suggest that the South American landmass was in contact with Africa (Russell 1994 and references therein). In addition to abelisaurids and titanosaurs, both regions supported peculiar spinosaurid theropods (Kellner and Campos 1996; Russell 1996) and long-spined sauropods (McIntosh 1990; Calvo and Salgado 1991; Salgado and Bonaparte 1991; Bonaparte 1995; Russell 1996).

The two Australian dinosaurian taxa of this age which are known from relatively complete material include an iguanodontoid (*Muttaburrasaurus*; Bartholomai and Molnar 1981; Norman and Weishampel 1990) and an

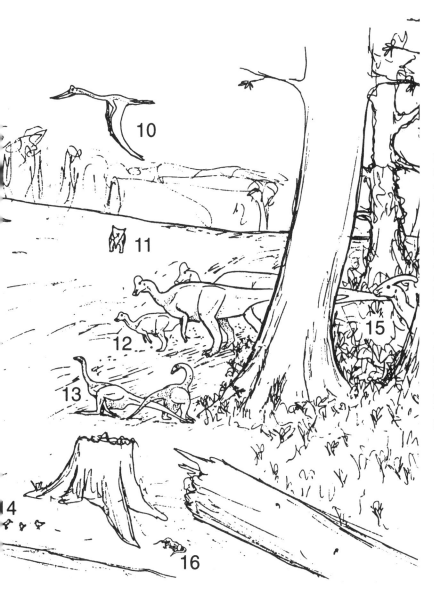

Figure 41.1. Late Cretaceous dinosaurs and other vertebrates of western North America (Dinosaur Park Formation, Alberta). Choristoderans: 16, *Champsosaurus* (eating a snake). Plesiosaurs: 6, *Leurospondylus.* Crocodilians: 1, cf. *Brachychampsa.* Pterosaurs: 10, *Quetzalcoatlus.* Birds: 14, indeterminate forms. Dinosaurs: Theropods: 3, *Chirostenotes;* 5, *Troodon;* 8, *Albertosaurus;* 13, *Dromiceiomimus.* Ankylosaurs: 2, *Panoplosaurus.* Pachycephalosaurs: 4, *Stegoceras.* Ceratopsians: 7, *Styracosaurus;* 9, *Centrosaurus;* 11, *Chasmosaurus.* Hadrosaurids: 12, *Corythosaurus;* 15, *Parasaurolophuyss.* Drawing by Tracy Ford.

ankylosaur (*Minmi;* Molnar 1980b, 1991), both of which appear to be aberrant. Small ornithischians (Rich and Rich 1989) are not known in sufficient completeness for a character analysis to determine relationships within hypsilophodont and basal iguanodont dinosaurs (Weishampel and Heinrich 1992: 162). In our opinion, other Australasian dinosaurian materials (cf. Coombs and Molnar 1981; Molnar et al. 1985; Wiffen and Molnar 1989; Rich and Rich 1993) are insufficiently complete to permit assignment to any family group with reasonable certitude. The relictual occurrence of large labyrinthodont amphibians (Rich and Rich 1993; Warren 1993) is consistent with an ecologic or zoogeographic isolation of Australia.

Figure 41.2. Late Cretaceous (Rio Colorado Formation) dinosaurs of South America. Snakes: 7, cf. *Dinilysia.* Crocodilians: 9, *Notosuchus.* Birds: 2, *Patagopteryx;* 5, *Neuquenornis.* Theropods: 3, *Alvarezsaurus;* 4, unidentified abelisaur. Sauropods: 1, 6, *Neuquensaurus;* 8, *Antarctosaurus.* Drawing by Tracy Ford.

Relatively completely preserved Late Cretaceous dinosaurian materials are essentially limited to southern South America (Fig. 41.2), so that the biogeographic effects of the fragmentation of Gondwana remain unknown (cf. Russell 1994, 1995, and references cited therein). Abelisaurid and titanosaurid materials (as well as materials of other tetrapods) from the Late Cretaceous of India, Madagascar, and Argentina suggest to one of us (J.F.B.) that the zoogeographic differentiation of Africa and South America did not proceed rapidly. Trackways and incomplete skeletal remains reveal the presence of theropods and sauropods of sizes equal to or exceeding those of *Tyrannosaurus* (Coria and Salgado 1995; Sereno et al. 1996) and *Apatosaurus* (Calvo 1991; Bonaparte and Coria 1993; also in Kenya [Harris and Russell 1985]). The paleogeographic implications of Antarctic records of a small ornithischian (Milner et al. 1992) and a nodosaurid (Gasparini et al. 1987; Coombs and Maryańska 1990: 477) are uncertain.

Comparable forms are known from the Late Cretaceous of Europe, which at the time was zoogeographically linked to the Southern Hemisphere (see below).

Faunal interchange occurred between the northern and southern continents during the Cretaceous. In Europe, the presence of Gondwana taxa (madtsoiid snakes, abelisaurids and diversified titanosaurids, and possibly baryonychids/spinosaurids; Astibia et al. 1990; Charig and Milner 1990: 139; Buffetaut and Le Loeuff 1991; Le Loeuff 1991, 1995; Le Loeuff and Buffetaut 1991; Russell 1994; Rage 1995) is suggestive of at least intermittent contacts between Europe and Africa through much of Cretaceous time. Indeed, a peculiar ornithomimosaur of uncertain affinities from Spain (Pérez-Moreno et al. 1994) may also reflect Gondwanan origins. However, dwarfed and/or somewhat aberrant hypsilophodonts or iguanodonts, hadrosaurids, and nodosaurs on islands in southern Europe during Late Cretaceous time also suggest mid-Cretaceous biogeographic links with the Northern Hemisphere (Weishampel et al. 1991, 1993; Pereda-Superbiola 1992). Small teeth and bones referred to the Deinonychosauria (Buffetaut et al. 1986; Telles Antunes and Sigogneau-Russell 1991) require more complete material in order to confirm the presence of the group within the Cretaceous of Europe (cf. Le Loeuff 1991: 99; Le Loeuff and Buffetaut 1991: 587); dromaeosaurs have been identified in the middle Cretaceous of Africa (Rauhut and Werner 1995).

In the Western Hemisphere, microvertebrate remains also suggest the existence of intermittent and incomplete connections between South and North America during Campanian and Maastrichtian time. As in the case of Africa and Europe, dispersal from the south to the north probably exceeded that in the opposite direction (Gayet et al. 1992, 1996, and references cited therein; see also Denton and O'Neill 1993). Bonaparte (1986; see also Gayet et al. 1992) suggests that hadrosaurids and ceratopsids entered South America during this interval along an inter-American route. One of us (D.A.R.) is currently persuaded that none of the known dinosaurian material compellingly suppports generic-level inter-American dispersal during Campano-Maastrichtian time. Dispersal routes linking the continents of the Western Hemisphere through Europe and possibly India may have been present (Russell 1994). Microvertebrates from the terminal Cretaceous of India have been cited as evidence of Asian immigration prior to the end of the Cretaceous (Jaeger et al. 1989; Prasad and Rage 1991), although some of these forms have been identified in Madagascar and may be Gondwanan in origin (Asher and Krause 1994; Gao 1994).

Present understanding of dinosaurian biogeography is greatly impeded by the incompleteness of the known dinosaurian record. Identifications based on incomplete materials should be used with caution in drawing biogeographic inferences. Proceeding from the coarsest resolution to the finest permitted by existing information:

1. Cosmopolitan Early Jurassic Pangaean assemblages slowly differentiated into distinct Late Cretaceous Northern and Southern Hemisphere assemblages, which were separated by a generally arid equatorial zone (Ziegler et al. 1987).

2. Central Asia was isolated during Early and Middle Jurassic time, but by Late Cretaceous time North America was biogeographically dominated by Asia. Earlier in Cretaceous time, European archipelagos reflected the biogeographic proximity of both northern and southern landmasses, similar to the manner in which the proximity of Asia and Australia is reflected in the East Indies today. By Late Cretaceous time, however, African elements dominated European assemblages.

3. During the Cretaceous, biogeographic subregions within Asia and North America included coastal plain–deltaic communities in western Asia and eastern Cordillera, semi-arid to arid continental environments in central Asia, latitudinally differentiated coastal environments in eastern Cordillera, and an Appalachia which was partly isolated by the Western Interior Seaway.

4. In the Southern Hemisphere, tenuous evidence suggests that Australia may have been somewhat isolated by mid-Cretaceous time, but the southern record is too incomplete to reveal the biogeographic effects of the Late Cretaceous fragmentation of the southern supercontinent.

5. The dominant direction of immigration across terminal Cretaceous links between the northern and southern continents apparently was from the south to the north.

References

Asher, R. J., and D. W. Krause. 1994. The first pre-Holocene (Cretaceous) record of Anura from Madagascar. *Journal of Vertebrate Paleontology* 14 (Supplement to no. 3): 15A.

Astibia, H.; E. Buffetaut; A. D. Buscalioni; H. Cappetta; C. Corral; R. Estes; F. Garcia-Carmilla; J. J. Jaeger; E. Jiminez-Fuentes; J. Le Loeuff; J. M. Mazin; X. Orue-Etxebarria; J. Pereda-Superbiola; J. E. Powell; J. C. Rage; J. Rodriguez-Lazaro; J. L. Sanz; and H. Tong. 1990. The fossil vertebrates from Lano (Basque Country, Spain): New evidence on the composition and affinities of the Late Cretaceous continental faunas of Europe. *Terra Nova* 2: 460–466.

Attridge, J.; A. W. Crompton; and F. A. Jenkins. 1985. The southern African Liassic prosauropod *Massospondylus* discovered in North America. *Journal of Vertebrate Paleontology* 5: 128–132.

Baghai, N. L. 1994. Palynology and paleobotany of the Aguja Formation (Campanian), Big Bend National Park, Texas. *Geological Society of America Abstracts with Program,* Rocky Mountain Section, pp. 2–3.

Bakhurina, N. N. 1993. Early Cretaceous pterosaurs from western Mongolia and the evolutionary history of the Dsungaripteroidea. *Journal of Vertebrate Paleontology* 13 (Supplement to no. 3): 24A.

Bakker, R. T.; D. Kralis; J. Siegwarth; and J. Filla. 1992. *Edmarka rex,* a new, gigantic theropod dinosaur from the middle Morrison Formation, Late Jurassic of the Como Bluff outcrop region. *Hunteria* 2 (9): 1–24.

Bartholomai, A., and R. E. Molnar. 1981. *Muttaburrasaurus,* a new iguanodontid (Ornithischia: Ornithopoda) dinosaur from the Lower Cretaceous of Australia. *Memoirs of the Queensland Museum* 20: 319–349.

Blows, W. T. 1987. The armoured dinosaur *Polocanthus foxi* from the Lower Cretaceous of the Isle of Wight. *Palaeontology* 30: 557–580.

Bonaparte, J. F. 1986. History of the terrestrial Cretaceous vertebrates of Gondwana. IV Congresso Argentino de Paleontolgía y Biostratigrafía, Mendoza, Argentina, 1986, 2: 63–95.

Bonaparte, J. F. 1990. New Late Cretaceous mammals from the Los Alamitos Formation, northern Patagonia. *National Geographic Research* 6: 63–93.

Bonaparte, J. F. 1991. Los vertebrados fosiles de la Formacion Rio Colorado, de la Ciudad de Neuquen y Cercanias, Cretacico Superior, Argentina. *Revista del Museo Argentino de Ciencias Naturales, Paleontologia* 4: 16–123.

Bonaparte, J. F. 1995. *Dinosaurios de America del Sur.* Buenos Aires: Museo Argentino de Ciencias Naturales "Bernardino Rivadavia."

Bonaparte, J. F., and R. A. Coria. 1993. Un nuevo y gigantesco sauropodo titanosaurio de la Formacion Rio Limay (Albiano-Cenomanio) de la Provincia del Neuquen, Argentina. *Ameghiniana* 30: 271–282.

Bonaparte, J. F., and Z. Kielan-Jaworowska. 1987. Late Cretaceous dinosaur and mammal faunas of Laurasia and Gondwana. Tyrrell Museum of Palaeontology Occasional Paper 3: 24–29.

Bonaparte, J. F., and J. E. Powell. 1980. A continental assemblage of tetrapods from the Upper Cretaceous of Argentina (Sauropoda-Coelurosauria-Carnosauria-Aves). *Mémoires de la Société Géologique* 139: 19–28.

Bonaparte, J. F.; F. E. Novas; and R. A. Coria. 1990. *Carnotaurus sastrei* Bonaparte, the horned, lightly built carnosaur from the Middle Cretaceous of Patagonia. Natural History Museum of Los Angeles County Contributions in Science 416.

Britt, B. B., and K. L. Stadtman. 1996. The Early Cretaceous Dalton Wells dinosaur fauna and the earliest North American titanosaurid sauropod. *Journal of Vertebrate Paleontology* 16 (Supplement to no. 3): 24A.

Buffetaut, E., and J. Le Loeuff. 1991. Late Cretaceous dinosaur faunas of Europe: Some correlation problems. *Cretaceous Research* 12: 159–176.

Buffetaut, E.; B. Marandat; and B. Sigé. 1986. Découverte de dents de deinonychosaures (Saurischia, Theropoda) dans le Crétacé supérieur du sud de la France. *Comptes Rendus de l'Académie des Sciences,* Paris, Série II, 303: 1393–1396.

Calvo, J. O. 1991. Huellas de dinosaurios en la Formacion Rio Limay (Albiano-Cenomaniano?), Picun Leufu, Provincia de Neuquen, Republica Argentina (Ornithischia-Saurischia-Sauropoda-Theropoda). *Ameghiniana* 28: 241–258.

Calvo, J. O., and J. F. Bonaparte. 1991. *Andesaurus delgadoi* gen. et sp. nov. (Saurischia-Sauropoda), dinosaurio Titanosauridae de la Formacion Rio Limay (Albiano-Cenomaniano), Neuquen, Argentina. *Ameghiniana* 28: 303–310.

Calvo, J. O., and L. Salgado. 1991. Posible registro de *Rebbachisaurus* Lavocat (Sauropoda) en el Cretacico medio de Patagonia. *Ameghiniana* 28: 404.

Charig, A. J., and A. C. Milner. 1990. The systematic position of *Baryonyx walkeri* in the light of Gauthier's reclassification of the Theropoda. In K. Carpenter and P. J. Currie (eds.), *Dinosaur Systematics: Approaches and Perspectives,* pp. 127–140. Cambridge: Cambridge University Press.

Chiappe, L. 1993. Enantiornithine (Aves) tarsometatarsi from the Cretaceous Lecho Formation of northwestern Argentina. *American Museum Novitates* 3083: 1–27.

Chure, D. J. 1995. The teeth of small theropods from the Morrison Formation (Upper Jurassic: Kimmeridgian, UT). *Journal of Vertebrate Paleontology* 15 (Supplement to no. 3): 23A.

Chure, D. J., and J. H. Madsen. 1993. A tyrannosaurid-like braincase from the Cleveland-Lloyd Dinosaur Quarry (CLDQ), Emery County, UT (Morrison Formation; Late Jurassic). *Journal of Vertebrate Paleontology* 13 (Supplement to no. 3): 30A.

Chure, D. J.; J. H. Madsen; and B. B. Britt. 1993. New data on theropod dinosaurs from the Late Jurassic Morrison Fm (MF). *Journal of Vertebrate Paleontology* 13 (Supplement to no. 3): 30A.

Coombs, W. P., and T. Maryańska. 1990. Ankylosauria. In D. B. Weishampel, P. Dodson, and H. Osmólska (eds.), *The Dinosauria,* pp. 456–483. Berkeley: University of California Press.

Coombs, W. P., and R. E. Molnar. 1981. Sauropoda (Reptilia, Saurischia) from the Cretaceous of Queensland. *Memoirs of the Queensland Museum* 20: 351–373.

Coria, R. A., and L. Salgado. 1995. A new giant carnivorous dinosaur from the Cretaceous of Patagonia. *Nature* 377: 224–226.

Currie, P. J., and X.-J. Zhao. 1994. A new carnosaur (Dinosauria, Theropoda) from the Jurassic of Xinjiang, People's Republic of China. *Canadian Journal of Earth Sciences* 30: 2037–2081.

Dashzeveg D.; M. J. Novacek; M. A. Norell; J. M. Clark; L. M. Chiappe; A. Davidson; M. C. McKenna; L. Dingus; C. Swisher; and P. Altangerel. 1995. Extraordinary preservation in a new vertebrate assemblage from the Late Cretaceous of Mongolia. *Nature* 374: 446–449.

Denton, R. K., and R. C. O'Neill. 1993. "Precocious" squamates from the Late Cretaceous of New Jersey, including the earliest record of a North American iguanian. *Journal of Vertebrate Paleontology* 13 (Supplement to no. 3): 32A–33A.

Dodson, P. 1990. Counting dinosaurs: How many kinds were there? *Proceedings of the National Academy of Sciences, U.S.A.* 87: 7608–7612.

Elzanowski, A., and P. Wellnhofer. 1993. Skull of *Archaeornithoides* from the Upper Cretaceous of Mongolia. *American Journal of Science* 293A: 235–252.

Evans, S. E., and A. R. Milner. 1993. Frogs and salamanders from the Upper Jurassic Morrison Formation (Quarry Nine, Como Bluff) of North America. *Journal of Vertebrate Paleontology* 13: 24–30.

Forster, C. A. 1996. The fragmentation of Gondwana: Using dinosaurs to test biogeographic hypotheses. Sixth North American Paleontological Convention Abstracts of Papers. Special Publication 8: 127. Washington, D.C.: Paleontological Society.

Forster, C. A., and P. C. Sereno. 1994. Phylogenetic analysis of hadrosaurid dinosaurs. *Journal of Vertebrate Paleontology* 14 (Supplement to no. 3): 25A.

Forster, C. A.; P. C. Sereno; T. W. Evans; and T. Rowe. 1993. A complete skull of *Chasmosaurus mariscalensis* (Dinosauria: Ceratopsidae) from the Aguja Formation (late Campanian) of west Texas. *Journal of Vertebrate Paleontology* 13: 161–170.

Galton, P. M., and J. A. Jensen. 1979. Remains of ornithopod dinosaurs from the Lower Cretaceous of North America. *Brigham Young University, Geology Series* 25: 1–10.

Gao, K. J. 1994. First discovery of Late Cretaceous cordylids (Squamata) from Madagascar. *Journal of Vertebrate Paleontology* 14 (Supplement to no. 3): 26A.

Gasparini, Z. B. de; E. Olivero; R. Scasso; and C. Rinaldi. 1987. Un ankylosaurio (Reptilia: Ornithischia) Campaniano en el continente Antartico. Anais do X Congresso Brasileiro de Paleontolgia, Rio de Janeiro, 1987, pp. 131–141.

Gayet, M.; J. C. Rage; T. Sempere; and P. Y. Gagnier. 1992. Modalités des échanges de vertébrés continentaux entre l'Amerique du Nord et l'Amerique du Sud au Crétacé supérieur et au Paléocène. *Bulletin de la Société géologique de France* 1963: 781–791.

Gayet, M.; J.-C. Rage; T. Sempere; and L. G. Marshall. 1996. Cretaceous and Paleocene Pan-American interchanges of continental vertebrates. Sixth North American Paleontological Convention Abstracts of Papers. Special Publication 8: 137. Washington, D.C.: Paleontological Society.

Grandstaff, B. S., and D. C. Parris. 1993. Distribution of taxa in an estuarine fauna from the Late Cretaceous of New Jersey (Ellisdale). *Journal of Vertebrate Paleontology* 13 (Supplement to no. 3): 38A.

Hammer, W. R. In press. Dinosaurs on ice: Jurassic dinosaurs from Antarctica. To be published in D. L. Wolberg and E. Stump (eds.), *Dinofest II*. Tempe: Arizona State University.

Hammer, W. R., and W. J. Hickerson. 1993. A new Jurassic dinosaur fauna from Antarctica. *Journal of Vertebrate Paleontology* 13 (Supplement to no. 3): 40A.

Harris, J. M., and D. A. Russell. 1985. Preliminary notes on the occurrence of dinosaurs in the Turkana Grits of Northern Kenya. Unpublished report submitted to Amoco Petroleum Company, Houston, Texas. 22 pp.

Holtz, T. R. Jr. 1993. Paleobiogeography of Late Mesozoic dinosaurs: Implications for paleoecology. *Journal of Vertebrate Paleontology* 13 (Supplement to no. 3): 42A.

Holtz, T. R. Jr. 1996. Late Mesozoic dinosaurian biogeography and diversity: Lineage based approaches. Sixth North American Paleontological Convention Abstracts of Papers. Special Publication 8: 177. Washington, D.C.: Paleontological Society.

Howse, S. C. B., and A. R. Milner. 1993. *Ornithodesma*: A maniraptoran theropod dinosaur from the Lower Cretaceous of the Isle of Wight, England. *Palaeontology* 36: 425–437.

Hu, S. J. 1993. A new Theropoda (*Dilophosaurus sinensis* sp. nov.) from Yunan, China. *Vertebrata PalAsiatica* 31: 65–69. (In Chinese with an English abstract.)

Hutchison, J. H. 1993. *Avisaurus*: A "dinosaur" grows wings. *Journal of Vertebrate Paleontology* 13 (Supplement to no. 3): 43A.

Jacobs, L. L. 1995. *Lone Star Dinosaurs*. College Station: Texas A & M University Press.

Jacobs, L. L. 1996. The pattern of terrestrial fauna change in the mid-Cretaceous of North America. Sixth North American Paleontological Convention Abstracts of Papers. Special Publication 8: 193. Washington, D.C.: Paleontological Society.

Jaeger, J. J.; V. Courtillot; and P. Tapponier. 1989. Paleontological view of the ages of the Deccan Traps, the Cretaceous/Tertiary boundary, and the India-Asia collision. *Geology* 17: 316–319.

Jerzykiewicz, T., and D. A. Russell. 1991. Late Mesozoic stratigraphy and vertebrates of the Gobi Basin. *Cretaceous Research* 12: 345–377.

Dinosaurian Faunas of the Later Mesozoic   657

Kellner, A. W. A. 1994. Comments on the paleobiogeography of Cretaceous archosaurs during the opening of the Atlantic Ocean. *Acta Geologica Leopoldensia* 17: 615–625.

Kellner, A. W. A., and D. de A. Campos. 1996. First Early Cretaceous theropod dinosaur from Brazil with comments on Spinosauridae. *Neues Jahrbuch für Geologie und Paläontologie Abhandlungen* 199: 151–166.

Kirkland, J. I. 1993. Polacanthid nodosaurs from the Upper Jurassic and Lower Cretaceous of the east-central Colorado Plateau. *Journal of Vertebrate Paleontology* 13 (Supplement to no. 3): 44A.

Kirkland, J. I. 1996. Biogeography of western North America's mid-Cretaceous dinosaur faunas: Losing European ties and the first great Asian–North American interchange. *Journal of Vertebrate Paleontology* 16 (Supplementa to no. 3): 45A.

Kirkland, J. I.; D. Burge; B. B. Britt; and W. Blows. 1993. The earliest Cretaceous (Barremian?) dinosaur fauna found to date on the Colorado Plateau. *Journal of Vertebrate Paleontology* 13 (Supplement to no. 3): 45A.

Kirkland, J. I., and J. M. Parrish. 1995. Theropod teeth from the Lower and Middle Cretaceous of Utah. *Journal of Vertebrate Paleontology* 15 (Supplement to no. 3): 39A.

Lamb, J. P.; L. M. Chiappe; and P. G. D. Ericson. 1993. A marine enantiornithine from the Cretaceous of Alabama. *Journal of Vertebrate Paleontology* 13 (Supplement to no. 3): 45A.

Lehman, T. M. 1987. Late Maastrichtian paleoenvironments and dinosaur biogeography in the Western Interior of North America. *Palaeogeography, Palaeoclimatology, Palaeoecology* 60: 189–217.

Lehman, T. M. In press. Late Campanian dinosaur biogeography in the Western Interior of North America. To be published in D. L. Wolberg and E. Stump (eds.), *Dinofest II*. Tempe: Arizona State University.

Le Loeuff, J. 1991. The Campano-Maastrichtian vertebrate faunas from southern Europe and their relationships with other faunas in the world: Palaeobiogeographical implications. *Cretaceous Research* 12: 93–114.

Le Loeuff, J. 1995. *Ampelosaurus atacis* (nov. gen., nov. sp.), un nouveau Titanosauridae (Dinosauria, Sauropoda) du Crétacé supérieur de la Haute Vallée de l'Aude (France). *Comptes Rendus de l'Académie des Sciences,* Paris, Série II, 321: 693–699.

Le Loeuff, J., and E. Buffetaut. 1991. *Tarascosaurus salluvicus* nov. gen., nov. sp., dinosaure théropode du Crétacé supérieur du sud de la France. *Géobios* 25: 585–594.

McIntosh, J. S. 1990. Sauropoda. In D. B. Weishampel, P. Dodson, and H. Osmólska (eds.), *The Dinosauria*, pp. 345–401. Berkeley: University of California Press.

Milner, A. C.; J. J. Hooker; and S. E. K. Sequeira. 1992. An ornithopod dinosaur from the Upper Cretaceous of the Antarctic Peninsula. *Journal of Vertebrate Paleontology* 12 (Supplement to no. 3): 44A.

Molnar, R. E. 1980a. Australian late Mesozoic terrestrial tetrapods: Some implications. *Mémoires de la Société géologique de France* 139: 131–143.

Molnar, R. E. 1980b. An ankylosaur (Ornithischia: Reptilia) from the Lower Cretaceous of Queensland. *Memoirs of the Queensland Museum* 20: 77–87.

Molnar, R. E. 1986. An enatiornithine bird from the Lower Cretaceous of Queensland, Australia. *Nature* 322: 736–738.

Molnar, R. E. 1991. A nearly complete articulated ankylosaur from Queensland, Australia. *Journal of Vertebrate Paleontology* 11 (Supplement to no. 3): 47A.

Molnar, R. E.; T. F. Flannery; and T. H. V. Rich. 1985. Aussie *Allosaurus* after all. *Journal of Paleontology* 59: 1511–1513.

Nessov, L. A. 1991a. Giant flying lizards of the family Azhdarchidae. Part 1: Morphology, systematics. *Vestnik Leningradskogo universiteta* 1991 (2): 14–23. (In Russian.).

Nessov, L. A. 1991b. Cretaceous vertebrates of the Asiatic part of the Soviet Union. Geological Association of Canada, Mineralogical Association of Canada, Joint Annual Meeting with Society of Economic Geologists, Toronto, Program with Abstracts 16: A89.

Nessov, L. A., and L. B. Golovneva. 1987. The evolution of ecosystems in the course of historical changes in faunas and floras. *Proceedings of the 29th Session of the All-Union Paleontological Society,* pp. 22–28. Leningrad: Nauka. (In Russian.)

Norman, D. B. 1995. Ornithopods from Mongolia: new observations. *Journal of Vertebrate Paleontology* 15 (Supplement to no. 3): 46A.

Norman, D. B., and D. B. Weishampel. 1990. Iguanodontidae and related ornithopods. In D. B. Weishampel, P. Dodson, and H. Osmólska (eds.), *The Dinosauria,* pp. 510–533. Berkeley: University of California Press.

Padian, K. 1989. Presence of the dinosaur *Scelidosaurus* indicates Jurassic age for the Kayenta Formation (Glen Canyon Group, northern Arizona). *Geology* 17: 438–441.

Parrish, J. M., and J. G. Eaton. 1991. Diversity and evolution of dinosaurs in the Cretaceous of the Kaipirowits Plateau, Utah. *Journal of Vertebrate Paleontology* 11 (Supplement to no. 3): 50A.

Peng, G. Z. 1992. Jurassic ornithopod *Agilisaurus louderbacki* (Ornithopoda: Fabrosauridae) from Zigong, Sichuan, China. *Vertebrata PalAsiatica* 30: 39–53. (In Chinese with an English abstract.)

Pereda-Superbiola, J. 1992. A revised census of European Late Cretaceous nodosaurids (Ornithischia: Ankylosauria): Last occurrence and possible extinction scenarios. *Terra Nova* 4: 641–648.

Pereda-Superbiola, J. 1994. *Polacanthus* (Ornithischia, Ankylosauria), a transatlantic armoured dinosaur from the Early Cretaceous of Europe and North America. *Palaeontographica* A 232: 133–159.

Pérez-Moreno, B.; J. L. Sanz; A. D. Buscalioni; J. J. Moratalla; F. Ortega; and D. Rasskin-Gutman. 1994. A unique multitoothed ornithomimosaur from the Lower Cretaceous of Spain. *Nature* 370: 363–367.

Perle A.; M. A. Norell; L. M. Chiappe; and J. M. Clark. 1993. Flightless bird from the Cretaceous of Mongolia. *Nature* 362: 623–626.

Prasad, G. V. R., and J. C. Rage. 1991. A discoglossid frog in the latest Cretaceous (Maastrichtian) of India: Further evidence for a terrestrial route between India and Laurasia in the latest Cretaceous. *Comptes Rendus de l'Académie des Sciences, Paris,* Série II, 313: 273–278.

Raath, M. A., and J. S. McIntosh. 1987. Sauropod dinosaurs from the Central Zambezi Valley, Zimbabwe, and the age of the Kadzi Formation. *South African Journal of Geology* 90: 107–119.

Rage, J.-C. 1995. Les Madtsoiidae (Reptilia: Serpentes) du Crétacé supérieur d'Europe: Témoins gondwaniens d'une dispersion transtéthysienne. *Comptes Rendus de l'Académie des Sciences,* Paris, Série II, 322: 603–608.

Rauhut, O. W. M., and C. Werner. 1995. First record of the family Dromaeosauridae (Dinosauria: Theropoda) in the Cretaceous of Gondwana (Wadi Milk Formation, northern Sudan). *Paläontologische Zeitschrift* 69: 475–489.

Rich, P. V., and T. H. V. Rich. 1993. Australia's polar dinosaurs. *Scientific American* 269: 50–55.

Rich, T. H. V., and P. V. Rich. 1989. Polar dinosaurs and biotas of the Early Cretaceous of southeastern Australia. *National Geographic Research* 5: 15–53.

Rowe, T.; R. L. Cifelli; T. M. Lehman; and A. Weil. 1992. The Campanian Terlingua local fauna, with a summary of other vertebrates from the Aguja Formation, Trans-Pecos Texas. *Journal of Vertebrate Paleontology* 12: 472–493.

Russell, D. A. 1988. A check list of North American marine Cretaceous vertebrates including fresh water fishes. Occasional Paper of the Tyrrell Museum of Palaeontology 4.

Russell, D. A. 1994. The role of central Asia in dinosaurian biogeography. *Canadian Journal of Earth Sciences* 30: 2002–2012.

Russell, D. A. 1995. China and the lost worlds of the dinosaurian era. *Historical Biology* 10: 3–12.

Russell, D. A. 1996. Isolated dinosaur bones from the Middle Cretaceous of the Tafilalt, Morocco. *Bulletin du Muséum national d'Histoire naturelle* Paris 18 (c): 171–224.

Dinosaurian Faunas of the Later Mesozoic   659

Russell, D. A., and Z. M. Dong. 1994. A nearly complete skeleton of a troodontid dinosaur from the Early Cretaceous of the Ordos Basin, Inner Mongolia, China. *Canadian Journal of Earth Sciences* 30: 2163–2173.

Russell, D. A., and Z. Zheng. 1994. A large mamenchisaurid from the Junggar Basin, Xinjiang, People's Republic of China. *Canadian Journal of Earth Sciences* 30: 2082–2095.

Salgado, L., and J. F. Bonaparte. 1991. Un nuevo sauropodo Dicraeosauridae, *Amargasaurus cazaui* gen. et sp. nov., de la Formacion La Amarga, Neocomiano de la Provincia de Neuquen, Argentina. *Ameghiniana* 28: 333–346.

Sampson, S.; C. A. Forster; D. W. Krause; P. Dodson; and F. Ravoavy. 1996. New dinosaur discoveries from the Late Cretaceous of Madagascar: Implications for Gondwanan biogeography. Sixth North American Paleontological Convention Abstracts of Papers. Special Publication 8: 336. Washington, D.C.: Paleontological Society.

Sanz, J. L.; L. M. Chiappe; and J. F. Bonaparte. 1993. The Spanish Lower Cretaceous bird *Concornis lacustris* reevaluated. *Journal of Vertebrate Paleontology* 13 (Supplement to no. 3): 56A.

Schwimmer, D. R. In press. Late Cretaceous dinosaurs in eastern U.S.A.: One big faunal province with western connections. To be published in D. L. Wolberg and E. Stump (eds.), *Dinofest II*. Tempe: Arizona State University.

Schwimmer, D. R.; G. D. Williams; J. L. Dobie; and W. G. Siesser. 1993. Late Cretaceous dinosaurs from the Blufftown Formation in western Georgia and eastern Alabama. *Journal of Paleontology* 67: 288–296.

Sereno, P. C. 1991. Ruling reptiles and wandering continents: A global look at dinosaur evolution. *GSA Today* 1: 141–145.

Sereno, P. C.; D. B. Dutheil; M. Iarochene; H. C. E. Larsson; G. H. Lyon; P. M. Magwene; C. A. Sidor; D. J. Varricchio; and J. A. Wilson. 1996. Predatory dinosaurs from the Sahara and Late Cretaceous faunal differentiation. *Science* 272: 986–991.

Sereno, P. C.; J. A. Wilson; H. C. E. Larsson; D. B. Dutheil; and H.-D. Sues. 1994. Early Cretaceous dinosaurs from the Sahara. *Science* 266: 267–271.

Shubin, N. H., and H. D. Sues. 1991. Biogeography of Early Mesozoic continental tetrapods: Patterns and implications. *Paleobiology* 17: 214–230.

Stromer, E. 1915. Wirbeltier-Reste der Baharije-Stufe (unterestes Cenoman). Part 3: Das Original des Theropoden *Spinosaurus aegyptiacus* nov. gen. nov spec. *Abhandlungen der Königlich Bayerischen Akademie der Wissenschaften Mathematisch-physikalische Klasse* 28: 1–32.

Sun A.; Li J.; Ye X.; Dong Z.; and Hou L. 1992. *The Chinese Fossil Reptiles and Their Kin*. Beijing: Science Press.

Taquet, P. 1976. *Géologie et Paléontologie du Gisement de Gadoufaoua (Aptien de Niger)*. Paris: Cahiers de Paléontologie Éditions du Centre National de la Recherche Scientifique.

Telles Antunes, M., and D. Sigogneau-Russell. 1991. Nouvelles données sur les dinosaures du Crétacé supérieur du Portugal. *Comptes Rendus de l'Académie des Sciences, Paris*, Série II, 313: 113–119.

Warren, A. 1993. Cretaceous temnospondyl. *Journal of Vertebrate Paleontology* 13 (Supplement to no. 3): 61A.

Weishampel, D. B., and P. R. Bjork. 1989. The first indisputable remains of *Iguanodon* from North America: *Iguanodon lakotaensis* n. sp. *Journal of Vertebrate Paleontology* 9: 56–66.

Weishampel, D. B., and R. E. Heinrich. 1992. Systematics of Hypsilophodontidae and basal Iguanodontia (Dinosauria: Ornithopoda). *Historical Biology* 6: 159–184.

Weishampel, D. B.; P. Dodson; and H. Osmólska (eds.). 1990. *The Dinosauria*. Berkeley: University of California Press.

Weishampel, D. B.; D. Grigorescu; and D. B. Norman. 1991. The dinosaurs of Transylvania. *National Geographic Research and Exploration* 7: 196–215.

Weishampel, D. B.; D. B. Norman; and D. Grigorescu. 1993. *Telmatosaurus transsylvanicus* from the Late Cretaceous of Romania: The most basal hadrosaurid dinosaur. *Palaeontology* 36: 361–385.

Wiffen, J., and R. E. Molnar. 1989. An Upper Cretaceous ornithopod from New Zealand. *Géobios* 22: 531–536.

Yaccobucci, M. 1990. Phylogeny and biogeography of ornithomimisauria. *Journal of Vertebrate Paleontology* 10 (Supplement to no. 3): 51A.

Ziegler, A. M.; A. L. Raymond; T. C. Gierlowski; M. A. Horrell; D. B. Rowley; and A. L. Lottes. 1987. Coal, climate and terrestrial productivity: The present and Early Cretaceous compared. In A. C. Scott (ed.), *Coal and Coal-Bearing Strata: Recent Advances*, pp. 25–49. Geological Society of London, Special Publication 32.

The Extinction of the Dinosaurs: A Dialogue between a Catastrophist and a Gradualist

Dale A. Russell
and
Peter Dodson

42

The concept of extinction as a significant process in the history of life emerged early in the nineteenth century with the realization that many organisms in the fossil record were no longer living (Rudwick 1985). Georges Cuvier, studying invertebrate and vertebrate fossils of the Paris Basin, observed abrupt changes between organic remains preserved in succeeding sedimentary series. He postulated the intervention of violent events or catastrophes that destroyed one biota and cleared the way for another. Conversely, Charles Lyell saw extensive sedimentary sequences containing slightly differing molluscan assemblages as demonstrating gradual change on a time scale of millions of years. Charles Darwin was deeply impressed with Lyell's uniformitarianism, and formulated his theory of evolution by natural selection in terms of gradual change. The gradualist paradigm was reinforced by numerous subsequent paleontological studies, such as the classic interpretations of the history of horses by Othniel Charles Marsh and William Diller Matthew.

Until recently, the extinction of the dinosaurs was the object of unconstrained speculation rather than systematic study. Many theories were advanced (see review in Dodson and Tatarinov 1990), most of which were from a gradualistic point of view. An early attempt to invoke an extraterrestrial cause for dinosaurian extinction (a nearby supernova; Russell and Tucker 1971) did not generate wide support, but did prepare the conceptual stage for the asteroid/comet impact hypothesis of Alvarez et al. (1980). The latter was based on the discovery of anomalously high concentrations of iridium at the Cretaceous-Tertiary boundary in marine strata, initially near Gubbio, Italy, and later at many localities around the world. The impact extinction hypothesis was immediately and vigorously challenged by Clemens (e.g., Clemens et al. 1981; Archibald and Clemens 1982; Clemens 1982), who was arguing from the fossil vertebrate record. Although it is now widely accepted that an important impact occurred at the end of the Cretaceous (Ward 1995), disagreement continues as to its biological effects (Archibald 1996). The long-enduring debate between gradualism and catastrophism in the history of life has become focused on the demise of the dinosaurs.

Two Scenarios

The dinosaur extinction debate thus evokes two contrasting scenarios. A gradualist scenario envisages a changing-world hypothesis, and draws heavily on the observation that terrestrial vertebrate communities, absent the dinosaurs, show substantial continuity across the Cretaceous-Tertiary boundary. Healthy, diverse dinosaur populations existed some millions of years before the end of the Cretaceous. In response to a variety of biological, physical, climatological, oceanographic, volcanologic, and tectonic factors, no one of which was necessarily decisive, dinosaurs went into decline. At first the decline was barely noticeable. Later, populations were markedly reduced in diversity compared to their earlier states, with reduced populations. They were now more susceptible to extinction by physical causes. The efficient cause of extinction thus may be irrelevant. What is of interest is why they had become so susceptible to extinction after 160 million years of success.

In contrast, the impact extinction scenario postulates that healthy populations of dinosaurs were exterminated by the effects of the collision of a large bolide with the earth. The impact probably produced a variety of severe stresses, the more important of which may have included an interval of planet-wide darkness and acid rain. When terrestrial communities reformed following the catastrophe, dinosaurs were no longer present.

The two scenarios contrast greatly in nearly every way. It would seem that definitive resolution is to be expected. Was the tempo of dinosaurian extinction restricted to the geologic instant of a bolide impact, or did it span several hundred thousand to several million years? Can the primary data of the dinosaurian record provide insight into the causes of their extinction, or is the extinction of dinosaurs better revealed from other sources of data? The present authors are proponents of opposing interpretations of the extinction of the dinosaurs (for example, compare the catastrophism of Russell [1982, 1984, and 1989] with the gradualism favored by Dodson and Tatarinov [1990]). Our disparate viewpoints nevertheless contain a broad consensus. It is our hope that others will find our points of agreement and disagreement to be as stimulating as we have.

Points of Agreement

1. For more than 160 million years, dinosaurs were the largest land-dwelling animals on all of the major land areas of the world. During this time they underwent a major adaptive radiation, and the functional beauty of their skeletal structures is an enduring marvel. The ultimate disappearance of the dinosaurs in no way negates their signal importance in the history of life on land, nor should they serve as cultural icons for failure.

2. In a phylogenetic sense, dinosaurs are not extinct, for birds are theropodan descendants (but see Feduccia 1996 for a dissenting view). For the purposes of this review, however, the term *dinosaur* connotes what cladists might term "non-avian dinosauromorph." We thus (unrepentantly) use a paraphyletic rather than a monophyletic (holophyletic) "Dinosauria." Whatever the scientific merits of the latter, the former is widely understood, and avoids such circumlocutions as "non-avian dinosaur."

3. The terminal-Cretaceous extinctions collectively rate among the five greatest extinctions of all time (Sepkoski 1992). The biotic turnover was planetary in scope, and varyingly affected terrestrial, aquatic and marine ecosystems.

4. Mass extinction is a complex phenomenon. It should not be presumed that all organisms on land and in the seas that became extinct during the Maastrichtian did so for the same reason, or at the same tempo. Demonstration of a catastrophic extinction for one group (e.g., planktonic foraminiferans) does not constitute evidence for catastrophic extinction of another (e.g., salamanders).

5. The dinosaurian record, as presently known, shows a peak in global diversity during early Maastrichtian time (Dodson 1990), suggesting that the causes of the decline of the dinosaurs occurred within the final 3 million years of the Cretaceous.

6. One can seldom make statistically significant statements on trends in dinosaurian diversity because of the limited nature of the skeletal record. Fewer than 1,000 articulated skeletons or partial skeletons of dinosaurs are available worldwide to document the final 10 million years of their existence.

7. Beyond North America, dinosaurs of Maastrichtian age are known from all of the continents of the world except Australia (although a New Zealand occurrence has been documented). Maastrichtian records of uncertain substage correlation include Alaska (Clemens and Nelms 1993), Antarctica (Campanian-Maastrichtian *fidé* Hooker et al. 1991), Argentina and Bolivia (Gayet et al. 1992), China (Mateer and Chen 1992), Mongolia (Jerzykiewicz and Russell 1991), New Zealand (Wiffen and Molnar 1988, 1989), and Siberia (Nessov and Starkov 1992). Late Maastrichtian records include Belgium and the Crimea (Russell 1982), eastern North America (Russell 1982), Egypt (Barthel and Herrmann-Degen 1981), France and adjacent Spain (Weishampel 1990; Buffetaut and Le Loeuff 1991; Feist 1991; Galbrun et al. 1993), India (Jaeger et al. 1989), Romania (Weishampel et al. 1991), and Siberia (Nessov and Starkov 1992). Many of these records refer to occurrences of fragmentary or disassociated bones; only the Mongolian (Nemegt) badlands have yielded relatively complete skeletons and an adequate sample of the assemblage to which the dinosaurs belonged. However, the terminal-Cretaceous–basal Paleocene (Maastrichtian-Danian) record in Mongolia is interrupted by a sedimentary hiatus. Taken together, these records demonstrate the worldwide survival of dinosaurs only into Maastrichtian time. In no case do they document a temporal series of two or more Maastrichtian assemblages, or a Maastrichtian-Danian succession.

8. By far the most complete record of biotic changes across the Cretaceous-Paleocene boundary in terrestrial environments is confined to the Western Interior of North America. Sedimentation was relatively continuous across the northern portion of this region, and well-sampled fossil assemblages are available in relative abundance.

9. Mesozoic chronofaunas characterized by the dominance of dinosaurs ended with the Cretaceous Period. Dinosaurs, which included all land

vertebrates exceeding 25 kilograms in weight, were not a trivial component of the Hell Creek assemblage. There is no evidence that dinosaurian herbivores were in the process of being replaced by large-bodied mammalian herbivores, such as the 500-kilogram *Coryphodon* of the early Cenozoic, during Hell Creek time.

10. Eustatic sea level changes near the end of the Cretaceous did not exceed those which occurred earlier during the dinosaurian era (cf. Haq et al. 1988), and thus were probably not a primary cause of dinosaurian extinctions.

11. It is currently tenable to postulate that the final disappearance of the dinosaurs coincided with a bolide strike (see Hildebrand 1993).

12. There is at present no compelling evidence that any dinosaur survived into Paleocene time (e.g., Rigby et al. 1987; Van Valen 1988), although neither of us would rule out the possibility. Indeed, one of the most amazing features of the end-Cretaceous extinctions is the absence of a credible record of any dinosaurian taxon from strata of early Tertiary age anywhere.

13. A signal aspect of the extinctions is the survival of many different varieties of organisms in terrestrial ecosystems. Survival patterns place important constraints on the magnitude of any physical stresses involved (e.g., Buffetaut 1990). For example, widespread freezing of soils is thereby precluded. Moreover, the ecology of organisms that survived and of those that became extinct may provide insight into the nature of the environmental stresses that led to the observed extinctions.

Points Supporting Alternative Models of Extinction: A Catastrophic Decline of Dinosaurs

Dale A. Russell

1. Within the Western Interior of North America (see above), the diversity of large dinosaurs in sediments of Hell Creek and equivalent strata of late Maastrichtian age may not have been as great as in the underlying Horseshoe Canyon Formation and equivalent strata of early Maastrichtian age. However, disarticulated teeth and bones indicate that the diversity of smaller, more derived dinosaurs was comparable (cf. late Maastrichtian occurrences of *Ornithomimus* and *Struthiomimus* [Russell 1972]; Dromaeosauridae [Carpenter 1982]; *Aublysodon*, *Richardoestesia*, and Troodontidae [Currie et al. 1990]; cf. *Chirostenotes* [D.A.R., personal observation]; *Stygimoloch* and *Stegoceras* [Goodwin 1989]).

2. Few articulated specimens of large dinosaurs have been collected from relatively limited exposures of the Scollard Formation, a Hell Creek equivalent which outcrops in the Red Deer Valley of Alberta. However, specimens are evidently as abundantly preserved there (4.7 per sq km) as in the highly productive badlands of Campanian age in Dinosaur Provincial Park, Alberta (3.9 per sq km; Béland and Russell 1978).

3. The broad expansion of Hell Creek–age deltas toward the east (Gill and Cobban 1973), warmer climates (Johnson and Hickey 1990), and change to dicot-dominated canopy forests (Wing et al. 1993; K. R. Johnson, personal communication 1993) constitute environmental changes of a magnitude sufficient to produce important changes between the older

Campano-Maastrichtian and younger Hell Creek dinosaurian communities. The latter assemblages were possibly representative of more densely forested environments (Russell 1989).

4. It has not been feasible to resolve increments of time of less than about 1 million years using relatively rare dinosaurian fossils (Sheehan et al. 1991). Large changes in leaf, pollen, and spore assemblages (Johnson 1992) are associated with trace-element anomalies linked to a bolide impact and time scales of days to tens to thousands of years. The bolide trace element signature is global in distribution (Hildebrand 1993).

5. Apparently in both marine and terrestrial environments, the Cretaceous ended with an abrupt collapse in green plant productivity associated with the bolide trace-element signature. Marine and terrestrial animals belonging to food chains based on organic detritus tended to dominate post-extinction assemblages. However, those dependent directly or indirectly on living plant tissues (e.g., dinosaurs on land, and planktonic foraminifera and mosasaurs in the sea) are postulated to have died on time scales consistent with starvation (Arthur et al. 1987; Sheehan and Fastovsky 1992; Olsson and Liu 1993).

A relatively parsimonious interpretation of the foregoing points is that dinosaur-dominated assemblages prospered in the Western Interior of North America until they were altered by regional topographic and climatic changes which began in middle Maastrichtian time. Several million years after they had achieved a new balance regionally, these assemblages were decimated by a catastrophic environmental deterioration resulting from the impact of a comet. Like the bolide trace-element signature, dinosaurian extermination was global in extent.

According to this interpretation, the late Maastrichtian dinosaur record is not special. It was preceded by an interval (Campanian through early Maastrichtian time) for which the record was more complete, generating the illusion of subsequent decline. This interpretation predicts that the extinction of the dinosaurs is unrelated to gradual changes in dinosaurian diversity, or to environmental stresses that would have precluded the survival of many Hell Creek microvertebrates. It also predicts that, should a relatively complete record be documented at other terrestrial sites around the world, no dinosaur-dominated assemblages will ever be found stratigraphically above the bolide signature.

Points Supporting Alternative Models of Extinction: A Gradual Decline of Dinosaurs

Peter Dodson

1. The record of non-dinosaurian terrestrial and freshwater aquatic vertebrates, including fishes, amphibians, turtles, lizards, champsosaurs, crocodiles, multituberculates, and placental mammals, shows substantial continuity across the Cretaceous-Tertiary boundary (Hutchison and Archibald 1986; Sloan et al. 1986; Sullivan 1987; Archibald and Bryant 1990; Archibald 1996; MacLeod et al. 1997). Plant communities also show continuity (McIver 1991), although significant disruptions have been noted (Johnson et al. 1989; Johnson 1992). These observations suggest

that terrestrial communities did not suffer a devastating catastrophe, but responded to changing environmental conditions as noted by various authors (e.g., Sloan et al. 1986; Johnson 1992).

2. Significant environmental changes occurred in the marine realm *during* the Maastrichtian (Barrera 1994; Ward 1995; MacLeod et al. 1997) up to 6 million years prior to the Cretaceous-Tertiary boundary. While some important Maastrichtian extinctions appear to have been abrupt or even catastrophic (particularly those of planktonic foraminifera), others, including those of reef-forming rudistid clams, inoceramid clams, and belemnites, took place up to 3 million years prior to the end of Maastrichtian time (Kauffman 1988; Ward 1990, 1995). Until recently (e.g., Ward et al. 1986), it was thought that ammonites also disappeared at least several hundred thousand years before the end of the Cretaceous. A number of species were then recorded at the Cretaceous-Tertiary boundary, and it can now be claimed that ammonites were victims of a catastrophe (e.g., Marshall 1995). However, it is clear that by the end of the Maastrichtian, ammonites were already greatly reduced in comparison with their diversity during Campanian or early Maastrichtian time.

The observation that dinosaurs became extinct during the Maastrichtian does not imply that they perished in a global terminal-Maastrichtian catastrophe. Many discussions of dinosaur extinction (e.g., Alvarez and Asaro 1990; Courtillot 1990; Glen 1990) have *assumed* that because geological, geochemical, geophysical, and astrophysical evidence has been documented for a terminal-Cretaceous catastrophe, the case for a catastrophic dinosaur extinction is also documented. Does dinosaur extinction best correspond to the planktonic foraminiferan model of catastrophic extinction or to the rudistid/inoceramid model of gradual disappearance?

3. The fossil record of dinosaurs, however incomplete (Dodson 1990; Dodson and Dawson 1991), can provide some insight into the nature of the extinction of dinosaurs. Articulated specimens provide the best taxonomic resolution, but do so at the expense of statistically significant sample sizes. Articulated specimens provide striking evidence of an apparently worldwide decline in dinosaur diversity during the late Maastrichtian. The number of recorded dinosaur genera during this interval is only about 30 percent of that during early Maastrichtian time. In the late Maastrichtian there are about eighteen well-characterized genera of dinosaurs, fourteen of which are from western North America.

Only in western North America can the presence of apparently healthy dinosaurian communities be demonstrated during the late Maastrichtian. Was this region an oasis in a changing world? Sheehan et al. (1991) and Sheehan and Fastovsky (1992) have examined the stratigraphic distribution of isolated dinosaur bones, which are diagnostic only to family level. It has been argued on the basis of this low-resolution component of the fossil record that dinosaur diversity was undiminished during the 2.25-million-year interval represented by the terminal-Cretaceous Hell Creek Formation of Montana. Hurlbert and Archibald (1995) have argued vigorously that the statistics used by Sheehan et al. (1991) are insufficient to support the claimed decline in diversity. The record of articulated dinosaurs suggests that the large dinosaur fauna of Hell Creek was dominated by a few common species (*Edmontosaurus, Triceratops, Tyrannosaurus*), unlike the more evenly diversified assemblages of the Judith River (Oldman) and Horseshoe Canyon formations of late Campanian and early Maastrichtian age from Alberta (Russell 1984; Dodson 1990; Weishampel 1990).

The worldwide terminal-Cretaceous decline in dinosaur diversity is mirrored in the stratigraphic sequence exposed along the Red Deer River of Alberta. This is the only region in the world where three successive dinosaur-bearing formations occur. A tabulation of articulated skeletons from the Judith River to the Horseshoe Canyon to the Scollard Formation suggests a diversity decline from thirty to eighteen to nine genera, respectively (Dodson 1990). Moreover, three important groups of previously successful dinosaurs had disappeared by the late Maastrichtian: the Centrosaurinae (short-frilled ceratopsids), the Lambeosaurinae (crested hadrosaurids), and the Nodosauridae (armored dinosaurs). Were these dinosaurs, like rudistids, inoceramids, and belemnites in the seas, bellwethers of increasing environmental stress?

4. In the Pyrenees, dinosaurs are reported to have disappeared from the fossil record between 1 million and 350,000 years before the end of the Cretaceous (Hansen 1990, 1991; Galbrun et al. 1993). Interestingly, reports of diversity trends in dinosaur egg assemblages show a parallel decline in diversity. In the Aix Basin in southeastern France, five egg types in the early Maastrichtian (Rognacian) are succeeded by two egg types in the middle Maastrichtian, but only one egg type survives into the late Maastrichtian (Vianey-Liaud et al. 1994). Erben et al. (1979) failed to note the taxonomic change in egg types through the Maastrichtian, mistakenly attributing all to *Hypselosaurus*. They thereby falsely concluded that eggshell thinning contributed to dinosaur extinction. In the Nanxiong Basin, Guandong Province, in southwestern China, twelve egg types are documented in the early Maastrichtian, but only a single type survives to the Cretaceous-Tertiary boundary, where it becomes extinct (Zhao 1994).

5. Gradual extinction resulted from a variety of environmental stresses, no one of which was in itself sufficient to cause the extinction of the dinosaurs. In western North America, such stresses included a general trend toward cooling temperatures (notwithstanding a short-term peak in temperature at the boundary itself), decreasing equability (Axelrod and Bailey 1968), the draining of the epicontinental seaway, orogeny, and volcanism (Courtillot 1990; Hansen 1990, 1991). Even biological factors such as temperature-dependent sex determination (Paladino et al. 1989) may have been a contributing factor, if ecological segregation of upland dinosaur breeding grounds (Horner 1984; Horner and Gorman 1988) from those of lowland turtles and crocodiles was maintained.

6. A logical sequel of the above points is the prediction that dinosaur-dominated assemblages of post-Cretaceous age may be found at low latitude. If the bolide impact scenario is correct, it is conceivable that dinosaurs survived at high latitude, far removed from the hypothesized Yucatan impact site.

The fossil record is consistent with a gradual worldwide disappearance of the dinosaurs. Yet most of the discussion of dinosaur extinction is centered on western North America, because here alone do vertebrate-bearing sediments of Paleocene age overlie latest Cretaceous dinosaur-bearing strata. It can be argued that if sections were better sampled, or if new boundary sections were found (perhaps in Argentina, China, Thailand, Antarctica, or elsewhere), a different picture of dinosaur extinction would emerge. On the basis of what is currently known, however, the overwhelming pattern of continuity of the terrestrial biota is too great to be compatible with fully apocalyptic accounts of a bolide impact scenario. Global wildfire (Wohlback et al. 1988) seems too far-fetched to be credible, and the survival of pH-sensitive aquatic vertebrates argues against acid rain as a major factor in extinction (D'Hondt et al. 1994). The possibility remains that a bolide impact constituted a coup de grâce for the last surviving dinosaur populations.

We have been pleasantly surprised to discover a broad area of common agreement. We concur that the latest Cretaceous dinosaurian record is far too incomplete to support either the catastrophic or the gradualistic model in a statistically meaningful manner. Indeed, Raup and Jablonski (1993) find that even the record of late Maastrichtian marine bivalve assemblages is too small to support statistical analyses. We differ in our assessment of which data are of greater significance. As we have described them, the two extinction models surely exist only in our imaginations. The truth lies in nature, which through the scientific method continually reveals ever-fascinating constellations of data which render the pursuit of scientific knowledge so enjoyable. We are confident that nature has much yet to teach us about the extinction of the dinosaurs.

References

Alvarez, L. W.; W. Alvarez; F. Asaro; and H. V. Michel. 1980. Extraterrestrial cause for the Cretaceous-Tertiary extinction. *Science* 208: 1095–1108.
Alvarez, W., and F. Asaro. 1990. An extraterrestrial impact. *Scientific American* 263 (4): 78–84.
Archibald, J. D. 1996. *Dinosaur Extinction and the End of an Era*. New York: Columbia University Press.
Archibald, J. D., and L. J. Bryant. 1990. Differential Cretaceous/Tertiary extinctions of nonmarine vertebrates: Evidence from northeastern Montana. Geological Society of America Special Paper 247: 549–562.
Archibald, J. D., and W. A. Clemens. 1982. Late Cretaceous extinctions. *American Scientist* 70: 377–385.
Arthur, M. A.; J. C. Zachos; and D. S. Jones. 1987. Primary productivity and the Cretaceous/Tertiary boundary event in the oceans. *Cretaceous Research* 8: 43–54.
Axelrod, D. I., and H. P. Bailey. 1968. Cretaceous dinosaur extinction. *Evolution* 22: 595–611.
Barrera, E. 1994. Global environmental changes preceding the Cretaceous-Tertiary boundary: Early-late Maastrichtian transition. *Geology* 22: 877–880.
Barthel, K. W., and W. Herrmann-Degen. 1981. Late Cretaceous and Early Tertiary stratigraphy in the Great Sand Sea and its SE margins (Farafra and Dakhla Oases), SW Desert, Egypt. *Mitteilungen der Bayerischen Staatssammlung für Paläontologie und Historische Geologie* 21: 141–182.
Béland, P., and D. A. Russell. 1978. Paleoecology of Dinosaur Provincial Park (Cretaceous), Alberta, interpreted from the distribution of articulated remains. *Canadian Journal of Earth Sciences* 15: 1012–1024.
Buffetaut, E. 1990. Vertebrate extinctions and survival across the Cretaceous-Tertiary boundary. *Tectonophysics* 171: 337–345.
Buffetaut, E., and J. Le Loeuff. 1991. Late Cretaceous dinosaur faunas of Europe: Some correlation problems. *Cretaceous Research* 12: 159–176.
Carpenter, K. 1982. Baby dinosaurs from the Late Cretaceous Lance and Hell Creek formations and a description of a new species of theropod. *University of Wyoming Contributions to Geology* 20: 123–134.
Clemens, W. A. 1982. Patterns of extinction and survival of the terrestrial biota during the Cretaceous/Tertiary transition. Geological Society of America Special Paper 190: 407–413.
Clemens, W. A.; J. D. Archibald; and L. J. Hickey. 1981. Out with a whimper not a bang. *Paleobiology* 7: 293–298.
Clemens, W. A., and L. G. Nelms. 1993. Paleoecological implications of Alaskan terrestrial vertebrate fauna in latest Cretaceous time at high paleolatitudes. *Geology* 21: 503–506.

Courtillot, V. E. 1990. A volcanic eruption. *Scientific American* 263 (4): 85–92.

Currie, P. J.; J. K. Rigby; and R. E. Sloan. 1990. Theropod teeth from the Judith River Formation of southern Alberta, Canada. In K. Carpenter and P. J. Currie (eds.), *Dinosaur Systematics: Perspectives and Approaches,* pp. 107–125. Cambridge: Cambridge University Press.

D'Hondt, S.; M. E. Q. Pilson; H. Sigurdsson; and S. Carey. 1994. Surface-water acidification and extinction at the Cretaceous-Tertiary boundary. *Geology* 22: 983–986.

Dodson, P. 1990. Counting dinosaurs: How many kinds were there? *Proceedings of the National Academy of Science* U.S.A. 87: 7608–7612.

Dodson, P., and S. D. Dawson. 1991. Making the fossil record of dinosaurs. *Modern Geology* 16: 3–15.

Dodson, P., and L. P. Tatarinov. 1990. Dinosaur extinction. In D. B. Weishampel, P. Dodson, and H. Osmólska (eds.), *The Dinosauria,* pp. 55–62. Berkeley: University of California Press.

Erben, H. K.; J. Hoefs; and K. H. Wedepohl. 1979. Paleobiological and isotopic studies of eggshells from a declining dinosaur species. *Paleobiology* 4: 380–414.

Feduccia, A. 1996. *The Origin and Evolution of Birds.* New Haven, Conn.: Yale University Press.

Feist, M. 1991. Charophytes at the Cretaceous-Tertiary boundary. Geology Society of America Abstracts with Programs 23 (5): A358.

Galbrun, B.; M. Feist; F. Columbo; R. Rocchia; and Y. Tambareau. 1993. Magnetostratigraphy and biostratigraphy of Cretaceous-Tertiary continental deposits, Ager Basin, Province of Lerida, Spain. *Palaeogeography, Palaeoclimatology, Palaeoecology* 102: 41–52.

Gayet, M.; L. G. Marshall; and T. Sempere. 1992. The Mesozoic and Paleocene vertebrates of Bolivia and their stratigraphic context: A review. *Revista Técnica de Yacimientos Petrolíferos Fiscales de Bolivia* 12 (3–4): 393–433.

Gill, J. R., and W. A. Cobban. 1973. Stratigraphy and geologic history of the Montana Group and equivalent rocks, Montana, Wyoming and North and South Dakota. U.S. Geological Survey Professional Paper 776: 1–37.

Glen, W. 1990. What killed the dinosaurs? *American Scientist* 78: 354–369.

Goodwin, M. B. 1989. New occurrences of pachycephalosaurid dinosaurs from the Hell Creek Formation, Garfield County, Montana. *Journal of Vertebrate Paleontology* 9 (Supplement to no. 3): 23A.

Hansen, H. J. 1990. Diachronous extinctions at the K/T boundary: A scenario. Geological Society of America Special Paper 247: 417–424.

Hansen, H. J. 1991. Diachronous disappearance of marine and terrestrial biota at the Cretaceous-Tertiary boundary. *Contributions from the Paleontological Museum University of Oslo* 364: 31–32.

Haq, B. U.; J. Hardenbol; and P. R. Vail. 1988. Mesozoic and Cenozoic chronostratigraphy and cycles of sea-level change. Society of Economic Paleontologists and Mineralogists Special Publication 42: 71–108.

Hildebrand, A. R. 1993. The Cretaceous/Tertiary boundary impact (or the dinosaurs didn't have a chance). *Journal of the Royal Society of Canada* 87: 77–118.

Hooker, J. J.; A. C. Milner; and S. E. K. Sequeira. 1991. An ornithopod dinosaur from the Late Cretaceous of west Antarctica. *Antarctic Research* 3: 331–332.

Horner, J. R. 1984. Three ecologically distinct vertebrate faunal communities from the Late Cretaceous Two Medicine Formation of Montana, with discussion of evolutionary pressures induced by interior seaway fluctuations. Montana Geological Society 1984 Field Conference, Northwestern Montana, pp. 299–303.

Horner, J. R., and J. Gorman. 1988. *Digging Dinosaurs.* New York: Workman.

Hurlbert, S. H., and J. D. Archibald. 1995. No statistical evidence for sudden (or gradual) extinction of dinosaurs. *Geology* 23: 881–884.

Hutchison, J. H., and J. D. Archibald. 1986. Diversity of turtles across the Cretaceous/Tertiary boundary in northeastern Montana. *Palaeogeography, Palaeoclimatology, Palaeoecology* 55: 1–22.

Jablonski, D. 1991. Extinctions: A paleontological perspective. *Science* 253: 754–757.

Jaeger, J. J.; V. Courtillot; and P. Tapponier. 1989. Paleontological view of the ages of the Deccan Traps, the Cretaceous/Tertiary boundary, and the India-Asia collision. *Geology* 17: 316–319.

Jerzykiewicz, T., and D. A. Russell. 1991. Late Mesozoic stratigraphy and vertebrates of the Gobi Basin. *Cretaceous Research* 12: 345–377.

Johnson, K. R. 1992. Leaf-fossil evidence for extensive floral extinction at the Cretaceous-Tertiary boundary, North Dakota, USA. *Cretaceous Research* 13: 91–117.

Johnson, K. R., and L. J. Hickey. 1990. Megafloral change across the Cretaceous/ Tertiary boundary in the northern Great Plains and Rocky Mountains, U.S.A. Geological Society of America Special Paper 247: 433–444.

Johnson, K. R.; D. L. Nichols; M. Attrep; and C. J. Orth. 1989. High-resolution leaf-fossil record spanning the Cretaceous/Tertiary boundary. *Nature* 340: 708–711.

Kauffman, E. G. 1988. The dynamics of marine stepwise mass extinction. *Revista Española de Paleontología extraordinario*: 54–71.

MacLeod, N.; P. F. Rawson; P. L. Forey; F. T. Banner; M. K. Boudagher-Fadel; P. R. Brown; J. A. Burnett; P. Chambers; S. Culver; S. E. Evans; C. Jeffery; M. A. Kaminski; A. R. Lord; A. C. Milner; A. R. Milner; N. Morris; E. Owen; B. R. Rosen; A. B. Smith; P. D. Taylor; E. Urquhart; and J. R. Young. 1997. The Cretaceous-Tertiary biotic transition. *Journal of the Geological Society, London* 154: 265–292.

Marshall, C. R. 1995. Distinguishing between sudden and gradual extinctions in the fossil record: Predicting the position of the Cretaceous-Tertiary iridium anomaly using the ammonite fossil record on Seymour Island, Antarctica. *Geology* 23: 731–734.

Mateer, N. J., and P. J. Chen. 1992. A review of the nonmarine Cretaceous-Tertiary transition in China. *Cretaceous Research* 13: 81–90.

McIver, E. E. 1991. Floristic change in the northern deciduous forests of western Canada during the Maastrichtian and Paleocene. Geological Society of America Abstracts with Program 23 (5): A358.

Nessov, L. A., and A. I. Starkov. 1992. Cretaceous vertebrates from the Gusino-ozerskaia Basin of Transbaikalia and their value for determining the age and depositional environment of the sediments. *Geologiia i geofizika*, no. 6: 10–19. (In Russian.)

Olsson, R. K., and C. J. Liu. 1993. Controversies on the placement of the Cretaceous-Paleocene boundary and the K/P mass extinction of planktonic formanifera. *Palaios* 8: 127–139.

Paladino, F. V.; P. Dodson; J. K. Hammond; and J. R. Spotila. 1989. Temperature-dependent sex determination in dinosaurs? Implications for population dynamics and extinction. Geological Society of America Special Publication 238: 63–70.

Raup, D. M., and D. Jablonski. 1993. Geography of end-Cretaceous marine bivalve extinctions. *Science* 260: 971–973.

Rigby, J. K. Jr.; K. R. Newman; J. Smit; S. Van der Kars; R. E. Sloan; and J. K. Rigby. 1987. Dinosaurs from the Paleocene part of the Hell Creek Formation, McCone County, Montana. *Palaios* 2: 296–302.

Rudwick, M. J. S. 1985. *The Meaning of Fossils*. Chicago: University of Chicago Press.

Russell, D. A. 1972. Ostrich dinosaurs from the Late Cretaceous of North America. *Canadian Journal of Earth Sciences* 9: 375–402.

Russell, D. A. 1982. A paleontological consensus on the extinction of the dinosaurs? Geological Society of America Special Paper 190: 401–405.

Russell, D. A. 1984. The gradual decline of the dinosaurs: Fact or fallacy? *Nature* 307: 360–361.

Russell, D. A. 1989. *An Odyssey in Time: the Dinosaurs of North America*. Toronto: University of Toronto Press.

Russell, D. A., and W. Tucker. 1971. Supernovae and the extinction of the dinosaurs. *Nature* 229: 553–554.

Sepkoski, J. J. 1992. Phylogenetic and ecologic patterns in the Phanerozoic history of marine biodiversity. In N. Eldredge (ed.), *Systematics, Ecology and the Biodiversity Crisis,* pp. 77–100. New York: Columbia University Press.

Sheehan, P. M., and D. E. Fastovsky. 1992. Major extinctions of land-dwelling vertebrates at the Cretaceous-Tertiary boundary, eastern Montana. *Geology* 20: 556–560.

Sheehan, P. M.; D. E. Fastovsky; R. G. Hoffmann; C. B. Berghaus; and D. L. Gabriel. 1991. Sudden extinction of the dinosaurs: Latest Cretaceous, upper Great Plains, U.S.A. *Science* 254: 835–839.

Sloan, R. E.; J. K. Rigby Jr.; L. M. Van Valen; and D. Gabriel. 1986. Gradual dinosaur extinction and simultaneous ungulate radiation in the Hell Creek Formation. *Science* 232: 629–633.

Sullivan, R. M. 1987. A reassessment of reptilian diversity across the Cretaceous-Tertiary boundary. *Natural History Museum of Los Angeles County Contributions to Science* 391: 1–26.

Van Valen, L. M. 1988. Paleocene dinosaurs or Cretaceous ungulates in South America. *Evolutionary Monographs* 10: 1–79.

Vianey-Liaud, M.; P. Mallan; O. Buscail; and C. Montgelard. 1994. Review of French dinosaur eggshells: Morphology, structure, mineral and organic composition. In K. Carpenter, K. F. Hirsch, and J. R. Horner (eds.), *Dinosaur Eggs and Babies,* pp. 151–183. Cambridge: Cambridge University Press.

Ward, P. D. 1990. The Cretaceous/Tertiary extinctions in the marine realm: A 1990 perspective. Geological Society of America Special Paper 247: 425–432.

Ward, P. D. 1995. After the fall: Lessons and directions from the K/T debate. *Palaios* 10: 530–538.

Ward, P. D.; J. Wiedmann; and J. F. Mount. 1986. Maastrichtian molluscan biostratigraphy and extinction patterns in a Cretaceous/Tertiary boundary section exposed at Zumaya, Spain. *Geology* 14: 899–903.

Weishampel, D. B. 1990. Dinosaurian distribution. In D. B. Weishampel, P. Dodson, and H. Osmólska (eds.), *The Dinosauria,* pp. 63–139. Berkeley: University of California Press.

Weishampel, D. B.; D. Grigorescu; and D. B. Norman. 1991. The dinosaurs of Transylvania. *National Geographic Research and Exploration* 7: 196–215.

Wiffen, J., and R. E. Molnar. 1988. First pterosaur from New Zealand. *Alcheringa* 12: 53–59.

Wiffen, J., and R. E. Molnar. 1989. An Upper Cretaceous ornithopod from New Zealand. *Geobios* 22: 531–536.

Wing, S. L.; L. J. Hickey; and C. C. Swisher. 1993. Implications of an exceptional fossil flora for Late Cretaceous vegetation. *Nature* 363: 342–344.

Wohlback, W. S.; I. Gilmour; E. Anders; C. J. Orth; and R. R. Brooks. 1988. Global fire at the Cretaceous-Tertiary boundary. *Nature* 334: 665–669.

Zhao, Z. 1994. Dinosaur eggs in China: On the structure and evolution of eggshells. In K. Carpenter, K. F. Hirsch, and J. R. Horner (eds.), *Dinosaur Eggs and Babies,* pp. 184–203. Cambridge: Cambridge University Press.

Dinosaurs and the Media

For a moment I wondered where I could have seen that ungainly shape, that arched back with triangular fringes along it, that strange bird-like head held close to the ground. Then it came back to me. It was the stegosaurus—the very creature which Maple White had preserved in his sketch-book, and which had been the first object which arrested the attention of Challenger! There he was—perhaps the very specimen which the American artist had encountered. The ground shook beneath his tremendous weight, and his gulpings of water resounded through the still night. For five minutes he was so close to my rock that by stretching out my hand I could have touched the hideous waving hackles upon his back. Then he lumbered away and was lost among the boulders.

—Sir Arthur Conan Doyle, *The Lost World*

Dinosaurs (of the non-avian variety, at least) may be gone, but they are by no means forgotten. Countless short stories, novels, and films have considered what it would be like to find living dinosaurs in some isolated corner of the world. Unfortunately, very little of the earth remains remote anymore, and the possibility of Mesozoic animals surviving in "lost worlds" in the wilds of South America or on forgotten islands in the Indian Ocean is close to nil. But that doesn't deter our hopes of seeing *Tyrannosaurus* or *Velociraptor* in the flesh: If naturally surviving dinosaurs are no longer an option, maybe we can resurrect the fearfully great reptiles by means of genetic engineering! Or so the current thinking about populating lost worlds with living dinosaurs goes. . . .

Part Six

Having paid our professional dues by covering the various aspects of dinosaur science, we end the book with a chapter that may seem out of place: a survey of how dinosaurs have been depicted in the popular media. The editors admit that they were (we hope the past tense is appropriate here) teenage geeks who loved the movies of Willis O'Brien, Ray Harryhausen, and Jim Danforth, the novels of Sir Arthur Conan Doyle and Edgar Rice Burroughs, and comic books such as "Tor" and "Turok," as much as we did the more sober scientific output of professional paleontologists (e.g., E. H. Colbert's little classic, *The Dinosaur Book*). The way dinosaurs were portrayed in the popular media played a major role in stimulating our interest in dinosaur science, and it seems only appropriate to pay tribute to that influence in our book's final chapter.

Dinosaurs in science fiction stories, movies, comic books, on trading cards and postage stamps—it's all here. Read this chapter just for the fun of it. There won't be a test on this material. We promise.

Dinosaurs and the Media

Donald F. Glut and M. K. Brett-Surman

43

Why Are Dinosaurs So Popular?

For more than a century (since the publication by O. C. Marsh in the 1880s of his skeletal restoration of *"Brontosaurus"*), dinosaurs have been the most famous of all the animals. No other creatures have so captured the imaginations of both children and adults. Adults are often amazed (and bewildered) when children know the names of dinosaurs before they know the names of the streets on which they live. Not infrequently, those children can also spell those names. Even the word *dinosaur* evokes visions of an age long ago when "monsters" were real. Certainly that is one of the keys to their popularity: Dinosaurs were "real monsters," yet they are harmless to us today. Equally significant, perhaps, dinosaurs are fun. There is no single aspect about them that does not appeal to someone. Today these animals, extinct for 65 million years, are among the greatest of educational tools, especially in schools where "science phobia" runs rampant. Where else can a student combine hard facts from such diverse sciences as geology, biology, history, physics, and ecology and not be bored? With adults, the popularity of dinosaurs can be explained quite simply: Dinosaurs represent everything we loved as children—adventure, power, time travel, science, mystery, lost worlds, and even a certain (and somehow pleasing) "inner chill."

The Science Fiction Dinosaur

Dinosaurs have often appeared in science fiction stories, one of the earliest important examples being the novel *The Lost World* (1912) by Sherlock Holmes creator Sir Arthur Conan Doyle (a lavishly illustrated, annotated version of which was recently published [Pilot and Rodin 1996]), with its lost plateau nestled away in the jungles of South America (Fig. 43.1). Edgar Rice Burroughs, the creator of Tarzan and other "pulp" magazine features, used dinosaurs in a number of his imaginative tales, most notably in the Pellucidar series (beginning with *At the Earth's Core* in 1914, in which the "lost world" scenario was transported to a hollow-earth environment); the Caspak series (starting off with *The Land That Time Forgot* in 1918); and *Tarzan the Terrible* (1921), set in a lost land in Africa. Regrettably, dinosaurs have most often been used in these stories in much the same way that Hollywood has used them: as vicious monsters that kill or must be killed. The most egregious use of dinosaurs as monsters was in the coarse pages of the mass-produced science fiction and fantasy "pulp" magazines of the 1930s and 1940s, often lurid "potboilers" churned out for a usually indiscriminating readership. Only in the last ten years have dinosaurs appeared in science fiction tales as animals instead of crazed killers. Once just a background foil, dinosaurs now appear as main characters (Sawyer 1992, 1993) in stories. They even have their own anthologies (Resnick and Greenberg 1993; Dann and Dozois 1995; see also Box 43.1 at the end of this chapter and Figs. 43.1, 43.2).

The Hollywood Dinosaur

Most people's first encounter with a dinosaur is in a motion picture. Unfortunately, Hollywood's portrayal of dinosaurs has rarely introduced them as they really were (*The Lost World* [1925] and *Jurassic Park* [1993] being notable exceptions); see below. One of the earliest dinosaur films was a silent animated cartoon entitled *Gertie the Dinosaur* (1912). In this short subject (which runs about ten minutes), an *Apatosaurus* was portrayed as the comical and quite lovable pet of Winsor McCay, the famous cartoonist who drew and shot the film. Gertie amazed audiences with such stunts as drinking dry a lake, leaving behind a dry Grand Canyon–sized hole.

Silent movies about prehistoric life, mostly featuring "cavemen" in Stone Age settings, were popular from 1913 to 1919, inspired in part by some early science fiction stories. Dinosaurs were first portrayed as villains in one of these pictures, *Brute Force* (1913), a "Stone Age" epic made by motion picture pioneer D. W. Griffith. The film was a sequel to *Man's Genesis*, which Griffith had made the previous year with cavemen but no dinosaurs. The dinosaur in *Brute Force*, portrayed by a life-sized mock-up of a *Ceratosaurus*, menaced cavemen in front of their caves, and was introduced on the screen by a title card reading "One of the perils of prehistoric apartment life."

The animal's appearance as a threat to early man reinforced the common misconception (persistent to this day among the unenlightened) that dino-

Figure 43.1. Dinosaur paperback covers courtesy of Ace Books, New York, except *The Lost World,* which is courtesy of Doubleday Books, New York.

Dinosaurs and the Media   *677*

Figure 43.2. Dinosaur pulp magazine covers. Upper left courtesy of *Weird Tales;* upper right courtesy of *Amazing Stories;* bottom row courtesy of Warren Magazines Inc.

saurs and "cavemen" coexisted, perhaps in some generalized "prehistoric world" (cf. Rudwick 1992). *Brute Force* established the movie formula according to which dinosaurs attacked and/or killed anyone or anything that moved. In later films, dinosaurs would also be depicted as monsters spending most of their time doing little more than walking into view, only to become locked "in mortal combat" with some other prehistoric animal.

The first feature-length motion picture to include dinosaurs was *The Lost World* (1925), based on the popular novel by Sir Arthur Conan Doyle (not to be confused with the more recent film based on the identically titled novel by Michael Crichton). *The Lost World* is a pivotal film of this genre for several reasons, including its attempt to portray dinosaurs as real animals (rather than monsters) going about their everyday, mundane lives. For example, rather than featuring just one individual of a species (the norm in most dinosaur films to this day), the film sometimes showed dinosaurs in small groups or entire herds.

Another important feature of this movie was the *way* that dinosaurs were brought to life. The dinosaur models used in *The Lost World* were animated one frame at a time by Willis O'Brien, who perfected this process (Archer 1993), which is known today as "stop motion" or "dimensional animation." O'Brien had already made a number of earlier dinosaur films, most notably a series of slapstick comedy shorts produced for Thomas Edison's Motion Picture Company a decade earlier, and also *The Ghost of Slumber Mountain* (1919), the first movie featuring "realistic" dinosaurs, designed under supervision of American Museum of Natural History paleontologist Barnum Brown.

Among the Mesozoic menagerie created for *The Lost World*, the most menacing dinosaur was the theropod *Allosaurus*, its appearance based directly on a watercolor painting done for the American Museum by the great paleontological artist Charles R. Knight (Czerkas and Glut 1982). At one point in the film, an *Allosaurus* fights a losing battle with an "*Agathaumas*" (a mostly imaginary reconstruction based on an old Knight painting, which in turn was based on scrappy fossil remains), after which this horned dinosaur is killed by the considerably larger and stockier theropod *Tyrannosaurus* (its only appearance in the film; Fig. 43.3).

The *Lost World* models, sculpted by Marcel Delgado, were all based on paintings and sculptures by Knight, who worked directly with scientists in creating his prehistoric images, basing them on what was considered to be paleontologically accurate at the time. In this context, given the information known in the 1920s, *The Lost World* remains one of the most (if not *the* most) accurate and influential dinosaur films ever made (Fig. 43.4).

In the early 1930s, the public was made "dinosaur-aware" by the advertising campaign utilizing dinosaurs launched by the Sinclair Refining Company. (Barnum Brown cleverly capitalized on this campaign by obtaining funds from Sinclair to finance his dinosaur-collecting expeditions.) During this period, dinosaurs were popular attractions at service stations, as World's Fair exhibits, in stamp albums, and at the United States' first "Dinosaur Park" (atop a hill overlooking Rapid City, South Dakota), and also in films.

Emerging from this period was one of the most influential dinosaur-related movies of all, *King Kong* (1933; Fig. 43.5). In some ways a successor to *The Lost World* (with a number of story parallels), it was also a film that would inspire many imitators over the succeeding years. Although *King Kong* was primarily about the discovery and capture of a giant "prehistoric" gorilla on the lost Skull Island, it was the island's Mesozoic reptiles (again, one of each species) that left the longest-lasting

Dinosaurs and the Media   679

Figure 43.3. *Tyrannosaurus* and *"Agathaumas"* from *The Lost World* (First National, 1925).

Glut and Brett-Surman

Figure 43.4. *Allosaurus* and
"Trachodon" from *The Lost World*
(First National, 1925).

Figure 43.5. King Kong confronts *Stegosaurus* in a publicity shot from a scene left out of the final cut (*King Kong,* 1933, photo courtesy of Turner Broadcasting Corporation).

impression on many people who saw it. Again, the team of Delgado and O'Brien, who had both greatly improved their techniques, created a Mesozoic menagerie based on the works of Knight. In a sequence that to this day remains one of the most dramatic of its kind ever committed to film, Kong, the King of Skull Island, battles to the death the "King of Dinosaurs," *Tyrannosaurus.* (Although O'Brien reportedly called this dinosaur an *Allosaurus,* the model was clearly based directly on a famous watercolor of *Tyrannosaurus* painted by Knight in the early 1900s, complete with the incorrectly depicted three-fingered hand and misplaced eye socket. The *King Kong* shooting script and the novelization by Delos W. Lovelace simply refer to this creature as a large carnivorous dinosaur.) However, although the Jurassic and Cretaceous animals in this film were extremely lifelike, and were portrayed in environments somewhat close to their original habitats, they were all—because of demands by producer Merian C. Cooper—depicted as at least twice their true size.

Dinosaurs continued to be popular in Hollywood movies into the 1940s. The original *One Million* B.C. (1940) was another "Stone Age" epic influenced in some ways by Griffith, who was originally slated to direct it. The film carried into the forties the popular myth that cavemen

Dinosaurs and the Media   683

and dinosaurs lived at the same time. Save for a *Triceratops* (a pig decorated with rubber ceratopsian enhancements) and a *Tyrannosaurus* (stuntman Paul Stader in a rubber dinosaur costume), most of the "dinosaurs" in this film were portrayed by live modern-day reptiles. That same year, Walt Disney's animation masterpiece *Fantasia*, although a cartoon, presented dinosaurs with reasonable accuracy, comparatively speaking, but there were some glaring errors. Most notably, dinosaurs from different Mesozoic periods and different places (as well as animals that were extinct before the first dinosaur appeared) were shown living at the same time and in the same place, again perpetuating the popular notion of some generalized "prehistoric world." Also, the *Tyrannosaurus*, like most movie representations of this dinosaur (including those in *King Kong* and *One Million* B.C.), was depicted as having three—not the correct two—claws on each hand.

More motion pictures featuring dinosaurs and other extinct creatures were filmed during the late 1940s and early 1950s, but the one with the biggest impact on the public was *The Beast from 20,000 Fathoms* (1953; Fig. 43.6). The movie was made to capitalize on an early 1950s reissue of *King Kong*, its major plot focus having been inspired by the last acts of both *The Lost World* and *King Kong* (i.e., a giant prehistoric creature running amouck in a modern-day city). It also played upon the public's growing Cold War fears, as the so-called "Beast" was awakened from suspended animation when an atomic-bomb test melted the iceberg imprisoning it since Mesozoic times. The film's stop-motion special effects were created by former Willis O'Brien protégé Ray Harryhausen (Fig. 43.7). In a conscious attempt not to duplicate the ending of *The Lost World*, Harryhausen "invented" his own impossibly huge dinosaur-like reptile, which was named "Rhedosaurus," a quadrupedal, spike-backed creature superficially resembling an enormous tuatara. Clearly, *The Beast from 20,000 Fathoms* firmly established the type of "dinosaur" movie whose main purpose was to show gigantic "prehistoric monsters" engaged in spectacular (or as spectacular as the special-effects budgets would allow) scenes of downtown metropolitan destruction. Indeed, Harryhausen's "Rhedosaurus" constituted the template for countless later films, including two British efforts, *The Giant Behemoth* (1959; British title *Behemoth, the Sea Monster*) and *Gorgo* (1960), both of which were helmed by Eugene Lourie, the director of *The Beast from 20,000 Fathoms*.

In 1954 Karel Zeman directed the first stop-action dinosaur movie to use a traveling matte during the special-effects process—a cinematic first (Robert Walters, personal communication). *Journey to the Beginning of Time* was a Czech film with an opening sequence filmed at the American Museum of Natural History in New York. After a visit to the museum, several children rent a rowboat in Central Park to spend some time on the "lake" pondering their visit. As they drift around a bend in the lake, they find themselves on a river. The farther they travel downstream, the further back in time they travel. This movie was fairly accurate for its day, and was one of the first to place the prehistoric "actors" in their correct geologic time intervals.

Most popular of all movie successors of the "Rhedosaurus" was *Godzilla, King of the Monsters* (1954; U.S. release 1956), known in its native Japan as *Gojira* (after the nickname of an employee of the Toho studio where the film was made). This movie took the basic concept—an impossibly huge dinosaur-like monster is revived by an atomic blast and then attacks humankind—a giant step further. Godzilla was no mere dinosaur. He was a monster mutated by H-bomb testing. Now possessing deadly incendiary breath in addition to size and strength, Godzilla became

Figure 43.6. The original model of the "Rhedosaurus" by Ray Harryhausen, from the movie *The Beast from 20,000 Fathoms.* Photograph courtesy of Ray Harryhausen and Warner Brothers Studio.

Figure 43.7. Special-effects legend Ray Harryhausen and his model of *Ceratosaurus* from *The Animal World*. Photograph courtesy of Ray Harryhausen.

a virtually unstoppable force of nature gone wild. Godzilla was, in fact, a metaphor for the Japanese people, who knew well the horrors inflicted by nuclear bombs. However, unlike the "Rhedosaurus" and the dinosaurs of the Lost World and Skull Island, Godzilla was not portrayed by animated models, but rather by human actor Harua Nakajima wearing a stiflingly hot rubber suit. *Godzilla* proved popular enough to spur a seemingly endless stream of sequels.

Dinosaurs have continued to be popular staples in motion pictures. In 1993, the film *Jurassic Park*, adapted from the best-selling novel by Michael Crichton, attempted to portray dinosaurs as marvelous animals (as well as terrifying monsters), bringing them "to life" through a combination of state-of-the-art special-effects techniques (including computer graphics imagery, or CGI). The film proved to be the top-grossing motion picture of all time, making more than a billion dollars at the box office, and further establishing the dinosaur as an ageless movie star with seemingly unending appeal. (For a selective listing of dinosaurs in the movies, see Box 43.2 at the end of this chapter and also Glut 1980; Bleiler 1990; Senn and Johnson 1992; and Warren 1982.)

The Electronic Dinosaur

With the advent of the home-video market, and with a VCR in almost every home, a number of documentary-style programs about dinosaurs have been made. Many of these are first broadcast on television, usually over the Public Broadcasting Service (PBS). Not all of them make it to the video market. Unfortunately, the scripts for these programs are usually not reviewed by professional paleontologists, and many factual errors and outdated concepts are presented as the latest facts. The best of these educational programs are, so far, *Dinosaurs* (hosted by David Susuki of the Canadian National Broadcasting Company as part of the series *Nature of Things*), *Lost Worlds, Vanished Lives* (hosted by David Attenborough), and *Dinosaurs and DNA* (hosted by Jeff Goldblum as part of the PBS series *Nova*). So many new productions are in progress that it is impossible to keep up with them. Caveat emptor!

Dinosaurs are now appearing on CD-ROM at a rapid rate. As of the summer of 1996, there were more than ten titles, with prices from $25 to $60. Unlike videos, these computer programs are updated and re-released, sometimes within the same year. It is impossible to recommend any one of them over another, simply because they all change in quality and content with each new release. Dinosaurs also appear on CD-ROM clip-art packages, but this market lags far behind the text-oriented programs. Outdated and illogical copyright laws have so far prevented most professional artists from being able to market their museum-quality restorations on CD-ROM.

The fastest-growing area for the "e-dinosaur" is on the World Wide Web (WWW). See Box 43.3 at the end of the chapter for a partial list of some of the first sites.

Dinosaurs as Marketing Devices: Dinosaurs Sell!

This axiom has been used (and misused) by Madison Avenue and elsewhere to sell virtually everything and anything, whether it has any relationship to dinosaurs or not. As already noted, the Sinclair Refining Company realized the value of exploiting dinosaurs to sell oil and gasoline as far back as the early 1930s. Unfortunately, because so many inferior products have been targeted at very young children, dinosaurs are often incorrectly thought of as for kids only, and/or as something cheap and outdated. In reality, adults do purchase (and often collect) dinosaur-related products. However, the travesty that "only children are interested in dinosaurs" has been perpetuated by poor marketing strategies, usually instigated by naive company executives.

Given the vast amount of dinosaur "product" that has been released up to the present day, it is unfortunate that the bulk of this product has been of inferior quality and geared to the youngest and least discriminating of dinosaur-item consumers. Ironically, museum gift shops, which, one might assume, should reflect the integrity of the sciences represented by the museum, are notorious for selling scientifically inaccurate or cheaply made products simply because they cost less and will sell more readily to shop visitors than the more expensive better-quality items. Accurate museum-quality models are often passed over by shop managers and buyers in favor of cheaply made, mass-produced dinosaur models whose blatant inaccuracies make the paleontologist cringe. A parent may be more prone to buy a child a rubber *Triceratops* depicted with sharp talons and a wide-open mouth filled with an array of sharp teeth than a more accurate—and probably more expensive—figure cast in bronze or pewter.

The first dinosaur models to be marketed to the public were sculpted during the mid–1800s by Benjamin Waterhouse Hawkins. Sold through the Ward's catalogue of scientific supplies, these plaster-cast figures were miniature replicas of the "life-sized" models that Hawkins made for the Crystal Palace grounds in Sydenham, London, based on the ideas of Sir Richard Owen (cf. McCarthy and Gilbert 1995; Torrens, chap. 14 of this volume). Although inaccurate by what is known today about dinosaurs and other Mesozoic reptiles, these figures were nicely sculpted and well made, and remain historically significant.

Among the most enduring and popular series of model dinosaurs to be sold in museum shops was that issued by the SRG company during the 1940s (and still available today in some shops). For many years, before miniature dinosaurs began to be mass-produced as toys, the metal-cast SRG figures were among the only small replicas of dinosaurs available to the general public. (Earlier, a series of plaster dinosaur sculptures made by paleontologist Charles Whitney Gilmore during the early twentieth century were widely distributed to different museums, but not to the public; these still occupy places in museum collections and displays.)

In more recent years, myriad manufacturing companies have issued various series of dinosaur figures for sale in museum shops, toy shops, and even (a relatively recent phenomenon) stores devoted entirely to dinosaur-related merchandise. Utilizing modern mass-production techniques, somewhat accurate miniature dinosaur figures are now being manufactured, often backed by the museums themselves. Among these is an ongoing series of plastic figures first put out by the Natural History Museum (London) in the 1970s. This was followed by the release of hard-rubber figures in the 1980s and 1990s by the Carnegie Museum of Natural History, and of the Wenzel/LoRusso models for the Boston Museum of Science. Of course, especially with the "dino-mania" originating with high-profile projects such as *Jurassic Park,* dinosaur figures and other merchandise continue to flood the retail outlets. Most of these products, as in past years, show little regard for attempts at scientific accuracy, failing to reflect current paleontological knowledge in favor of feeding a hungry market perceived by those in charge as simply "for kids only."

Fortunately, museum-quality reproductions are now available from professional sculptors such as Wayne Barlow, Donna Braginetz, Brian Franczak, John Gurche, Jim Gurney, Mark Hallett, Doug Henderson, Takeda Katashi, Eleanor Kish, Dan LoRusso, Tony McVey, Mike Milbourne, Bruce Mohn, Gregory Paul, John Sibbick, Michael Skrepnick, Paul Sorton, Jan Sovak, William Stout, Mike Trcic, Bob Walters, and Greg Wenzel, to name all too few.

Dinosaurs on Postage Stamps

Although stamp collecting has been a major hobby since the 1800s, dinosaur stamp collecting has blossomed as its own subfield only within the last twenty years. Dinosaur stamps, as a "topical," are now big enough to warrant their own advertising in stamp magazines. Several international stamp-collecting houses offer a "new issue service" just for dinosaur stamps. For a fee, they will automatically mail directly to your home all the new dinosaur stamps issued by most countries.

Dinosaurs first appeared on postage stamps in 1958, with a picture of *Lufengosaurus* (People's Republic of China; Fig. 43.8). Many countries have put dinosaurs on their stamps even though the dinosaurs depicted never lived there. Some countries have blatantly copied the works of such famous paleo-artists as Knight, Burian, Parker, Hallett, Paul, and Gurche. The range of portrayals has included skeletons, footprints, murals, coats of arms, silhouettes, life restorations, and cartoons, and of course a "Hollywood dinosaur." Some of the more interesting, and possibly valuable, stamps are those with mistakes—such as the use of the wrong name for a dinosaur.

Box 43.4 (at the end of this chapter) represents a collector's guide to all the dinosaur stamps up to 1992, the 150th anniversary of the word *dinosaur*. Stamps are listed by country, which is how stamp dealers also maintain their collections.

Dinosaur Trading Cards

For some mysterious reason, dinosaurs have never enjoyed the popularity on non-sports trading cards that would be expected for the world's most famous animals. A series of cards featuring dinosaurs and other extinct animals (mostly reproductions of paintings by artists such as Charles R. Knight) were released by Nu-Cards Sales in 1961 (Fig. 43.9), but they did not enjoy the popularity of other card series (e.g., those featuring movie monsters) being distributed at the same time. It was only with the "dinosaur fever" resulting from the 1993 release of *Jurassic Park* that many collector sets of dinosaur trading cards swept the market. Yet amazingly enough, few of these sets are based on museum-exhibit dinosaur skeletons! Many of them consist of poorly drawn or badly copied restorations taken (or "redrawn") from the work of other artists. Most sets appear as either "dinosaurs" or "prehistoric animals," with both types using any large extinct animal as a dinosaur. Some, but not all, of the more important trading-card sets for the collector are listed in Box 43.5 at the end of this chapter.

Dinosaurs in Comics

It seems fitting (and is probably not coincidental) that comic strips featuring dinosaurs made their first significant appearance during the early 1930s, when Sinclair's dinosaur-based advertising campaign was at its peak and *King Kong* was a virtually new movie. *Alley Oop,* a long-running newspaper series which began in 1934 (NEA Services), was created by cartoonist V. T. Hamlin, himself an amateur paleontologist. The title character was a strong caveman who rode a made-up dinosaur named Dinny (ironically, recent discoveries indicate that some sauropods may have had spines along the back like those sported by Dinny). Oop's stories started off as standard "Stone Age" comedy adventures, but later incorporated the science-fiction concept of time travel.

The 1950s were boom years for dinosaurs in comic books. Dinosaurs and other prehistoric animals, surviving in the lost land of Pal-ul-don (originally created in Burroughs's novel *Tarzan the Terrible*), frequently turned up in the Tarzan comic book series (Western Publishing Company). *Thun'da*, which premiered in 1952 from Magazine Enterprises (ME) and originally was written and drawn by Frank Frazetta, featured a modern man who became a Tarzan-type hero in a prehistoric African jungle. *Tor,* the creation of Joe Kubert, first appeared in 1953. This short-lived series starred a caveman with a conscience who struggled to survive in an anachronistic world of early humans and dinosaurs. A popular long-running series originating in 1954 was *Turok, Son of Stone* (Western Publishing Company), about two pre-Columbian Indians who stumbled upon a Grand Canyon–like "Lost Valley" populated by dinosaurs (usually referred to as "honkers"), cavemen, and other extinct denizens. The two Native Americans spent most of their time encountering these creatures, while at the same time searching for an exit out of Lost Valley.

During the sixties, dinosaurs remained popular in comic books. The series *The War That Time Forgot,* about men in uniform fighting prehistoric creatures on lost islands during World War II, began in *Star Spangled War Stories* (DC Comics) in 1960. *Kona* (Dell Publishing Company) was an unusual series pitting Kona, the Neanderthal monarch of Monster Isle, against dinosaurs and other menaces, usually involving bizarre storylines.

Comic books based on prehistoric time periods flourished during the 1970s, including Jack Kirby's *Devil Dinosaur* (starring a red tyrannosaur) and *Skull the Slayer* (both Marvel Comics Group), *Kong the Untamed* and *Warlord* (both DC), and *Tragg and the Sky Gods* (Western), the latter created by one of the present writers (Glut), and combining prehistoric adventure with science fiction.

The late 1980s and early 1990s saw another resurgence in dinosaur-related comics, among which was the independently published *Dinosaur Rex,* Tom Mason's satiric *Dinosaurs for Hire,* and Mark Schultz's very popular *Cadillacs and Dinosaurs* (Fig. 43.10), which became a franchise that included a Saturday-morning television cartoon show. In 1993, several series of comic books were spawned from the movie *Jurassic Park.* Also in the wake of that blockbuster film were brand-new comic-book titles, including Ricardo Delgado's "docudrama" mini-series *Age of Reptiles* (Dark Horse Comics), *Dinosaurs: A Celebration* (Marvel), and a revived *Turok, Dinosaur Hunter* (Valiant), with the hero reinvented as a "Rambo"-type character transported—along with dinosaurs—through time into our modern world.

Dinosaurs in Popular Books and Magazines

The first dinosaur dictionary was published by Donald F. Glut in 1972. It was the first compilation in a popular book of dinosaurian genera arranged "from A to Z," each entry having a block of capsulized information about the genus. This format has since been copied numerous times by other authors. The first "textbook" about dinosaurs was by W. E. Swinton in 1960.

The first popular book aimed at the dinosaur aficionado was *The Dinosaur Book,* written by paleontologist Edwin H. Colbert and published by the American Museum of Natural History in 1945. For many years, this was the only accessible popular book on the subject. Its publication firmly linked the name Colbert with the word *dinosaur,* although until that time the author had mainly specialized in fossil mammals. In succeeding years, many good dinosaur books have been published for general readers.

Up until 1990, the vast majority of dinosaur books were written by non-paleontologists. Unfortunately, this led to the publication of outdated

Figure 43.9. Dinosaur trading cards. *Top row, from left to right:* Dinosaurs Attack! (Topps, 1988); Dinosaurs (The Dino-Card Co., 1987); William Stout (artist) (Comic Images, 1993). *Middle row, from left to right:* DinoCardz (The DinoCardz Co., 1992); Infant Earth (Kitchen Sink Press, 1993); Dinosaur Nation (Kitchen Sink Press, 1993). *Bottom row, from left to right:* Dinosaurs (Nu-Cards, 1961); Jurassic Park, series I (Topps, 1993); Dinosaur, The Greatest Cards Unearthed (Mun-War Enterprises, 1993).

15 PARASAUROLOPHUS

24 SWIFT KILLERS!

PREHISTORIC SEA GIANT RAGES AGAINST CITY!

THEY COULDN'T BELIEVE THEIR EYES!

THEY COULDN'T ESCAPE THE TERROR!

AND NEITHER WILL YOU!

WARNER BROS.

The Beast From 20,000 Fathoms

CAST OF THOUSANDS!

JURASSIC PARK

THE DEADLIEST DINOSAUR

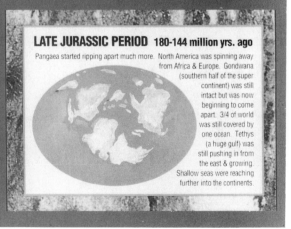

LATE JURASSIC PERIOD 180-144 million yrs. ago

Pangaea started ripping apart much more. North America was spinning away from Africa & Europe. Gondwana (southern half of the super continent) was still intact but was now beginning to come apart. 3/4 of world was still covered by one ocean. Tethys (a huge gulf) was still pushing in from the east & growing. Shallow seas were reaching further into the continents.

Figure 43.10. Dinosaurs on comic book covers. Upper left courtesy of DC Comics; upper right courtesy of Apple Comics; lower left courtesy of Kitchen Sink Press; lower right courtesy of Gold Key Comics.

information as accepted fact. A "bad" dinosaur book can be recognized by the following telltale signs: the author (1) uses outdated names such as "Anatosaurus" and "Brontosaurus"; (2) accepts the asteroid theory, or any extinction theory, as proven or solved; (3) cites only recent work from other popular books—not the technical literature; (4) bases the book on interviews with dinosaur paleontologists and accepts everything said literally; and (5) does not provide any counterevidence against the ideas presented.

The first modern dinosaur book to be up to date, with excellent illustrations, appeared in 1985: *The Illustrated Encyclopedia of Dinosaurs* by David B. Norman. It has served as a model for many subsequent dinosaur books. Some notable recent popular dinosaur books include Bakker 1986, Horner and Gorman 1988, Paul 1988, Horner and Lessem 1993, Jacobs 1993 and 1995, Lessem and Glut 1993, Gillette 1994, Lucas 1994, Colbert 1995, Lockley and Hunt 1995, and Fastovsky and Weishampel 1996.

Dinosaur magazines have also appeared with more frequency, but unfortunately they are aimed at the juvenile audience. There is a definite need for a magazine written for adults who are interested in dinosaurs! The one exception has been the excellent magazine *Kyoryugaky Saizensen* ("Dino-Frontline"), published from 1993 to 1996 by editor Masaaki Inoue for Gakken Mook in Tokyo, Japan. Regrettably there is no English version at this time.

Dinosaurs Today

The Dinosaur Society, founded in the early 1990s by Don Lessem, a popular science writer, is the first non-profit society devoted to dinosaurs. It also acts as a clearinghouse for the public and as a go-between for organizations and industries that need the technical expertise of a dinosaur paleontologist. The society also publishes a newsletter for the public and a report for fellow scientists and educators.

Universities today also have gotten into the dinosaur game by offering courses on dinosaurs. The first such offerings were at Stockton State College (New Jersey) and the University of California, Berkeley, in the late 1970s, followed quickly by George Washington University in 1980. These courses act not just to bolster the enrollment of geology departments (sometimes by the hundreds) but also as, sometimes regrettably, the only true college "biology" course to which non-science undergraduates are exposed. Museums have also expanded their adult education and outreach programs by establishing "parapaleontologist" programs. The Denver Museum of Natural History's program is the model for many other museums.

Dinosaurs are also on the Internet in a big way. Not only is there a discussion group just for dinosaurs, but there are numerous FTP and WWW sites loaded with files about dinosaurs, collections, pictures, and virtual tours of "exhibition halls" (see Box 43.3 for a list of sites). Many museums are also redoing their dinosaur halls to keep up with the flood of new information and scientifically accurate restorations now available to the public. In recent years the American Museum in New York, the Houston Museum, the Natural History Museum in London, the Academy of Natural Sciences in Philadelphia, the Denver Museum of Natural History, and the Field Museum in Chicago have renovated their dinosaur halls.

With more than seventy professional dinosaur paleontologists worldwide (from a low of about fifteen in the 1970s), there is now ample opportunity in many countries for the public to participate in the greatest detective story of all time—the story of the Mesozoic. This story begins in the field.

What to Do If You Find a Dinosaur Bone

Today it is possible, for the first time, for most of the public to actually go on a professional dinosaur expedition. Not only do museums and universities run programs for the public, but so do professional expeditioners. A list of expeditions available to the public is published annually in the newsletter of the Dinosaur Society [1-800-346-6366].

Whether you are by yourself or part of an expedition, the first thing to do is get permission from the landowner. All fossils are the property of the landowner until the owner says otherwise. There are six basic kinds of land in the United States: federal land, state land, local land, Native American land, corporate land, and private land. Each type exists under its own set of laws. You must know the laws before collecting fossils. Your state universities, natural history museums, government agencies, and rock clubs should all have handouts that cover this area. The Society of Vertebrate Paleontology may also be able to assist here.

When collecting, it is important to follow a "code of ethics." Many organizations and agencies have their own codes for their lands, but some simple rules and common sense apply: Always get permission in advance; do not leave a mess or big holes; document the locality information with pictures and on maps; tell a paleontologist about the site to prevent the loss of scientific information; do not collect any bones if you do not know what you are doing; and do not litter.

Collecting dinosaurs in the field is the most educational fun you can have—so go out and enjoy yourself!

References

Archer, S. 1993. *Willis O'Brien: Special Effects Genius.* Jefferson, N.C.: McFarland and Company.

Bakker, R. T. 1986. *The Dinosaur Heresies: New Theories Unlocking the Mystery of the Dinosaurs and Their Extinction.* New York: William Morrow.

Baldwin, S., and B. Halstead. 1991. *Dinosaur Stamps of the World.* Witham, Essex: Baldwin's Books.

Bleiler, E. F. 1990. *Science-Fiction: The Early Years.* Kent, Ohio: Kent State University Press.

Brett-Surman, M. K. 1991. Dinosaurs on stamps. *Biophilately* 40 (4): 10–19.

Colbert, E. H. 1945. *The Dinosaur Book: The Ruling Reptiles and Their Relatives.* Handbook no. 14. New York: American Museum of Natural History.

Colbert, E. H. 1995. *The Little Dinosaurs of Ghost Ranch.* New York: Columbia University Press.

Czerkas, S., and D. Glut. 1982. *Dinosaurs, Mammoths and Cavemen: The Art of Charles Knight.* New York: E. P. Dutton.

Czerkas, S. J., and E. C. Olson (eds.). 1987. *Dinosaurs Past and Present.* Seattle: Natural History Museum of Los Angeles County/University of Washington Press.

Dann, J., and G. Dozois. 1995. *Dinosaurs II.* New York: Ace Books.

Fastovsky, D. E., and D. B. Weishampel. 1996. *The Evolution and Extinction of the Dinosaurs.* Cambridge: Cambridge University Press.

Gillette, D. D. 1994. *Seismosaurus the Earth Shaker.* New York: Columbia University Press.

Glut, D., 1975. *The Dinosaur Dictionary.* Secaucus, N.J.: Citadel Press.

Glut, D. 1980. *The Dinosaur Scrapbook.* Secaucus, N.J.: Citadel Press.

Hasegawa, Y., and Y. Shiraki. 1994. *Dinosaurs Resurrected.* Tokyo: Mirai Bunkasha.

Horner, J. R., and J. Gorman. 1988. *Digging Dinosaurs.* New York: Workman Publishing.

Horner, J. R., and D. Lessem. 1993. *The Complete T. rex.* New York: Simon and Schuster.

Jacobs, L. 1993. *Quest for the African Dinosaurs: Ancient Roots of the Modern World.* New York: Villard Books.

Jacobs, L. 1995. *Lone Star Dinosaurs.* College Station: Texas A & M University Press.

Lamont, A. 1947. Paleontology in literature. *Quarry Manager's Journal* 30: 432–441, 542–551.

Lessem, D., and D. F. Glut. 1993. *The Dinosaur Society Dinosaur Encyclopedia.* New York: Random House.

Lockley, M., and A. P. Hunt. 1995. *Dinosaur Tracks and Other Fossil Footprints of the Western United States.* New York: Columbia University Press.

Lucas, S. G. 1994. *Dinosaurs: The Textbook.* Dubuque, Iowa: William C. Brown Publishers. 2nd ed., 1977.

McCarthy, S., and M. Gilbert. 1995. *The Crystal Palace Dinosaurs: The Story of the World's First Prehistoric Sculptures.* Anerly Hill, London: Crystal Palace Foundation.

Norman, D. B. 1985. *The Illustrated Encyclopedia of Dinosaurs.* London: Salamander Books.

Paul, G. S. 1988. *Predatory Dinosaurs of the World: A Complete Illustrated Guide.* New York: Simon and Schuster.

Pilot, R., and A. Rodin. 1996. *The Illustrated Lost World.* Indianapolis: Wessex Press.

Resnick, M., and M. H. Greenberg (eds.) 1993. *Dinosaur Fantastic.* New York: DAW Paperbacks.

Rudwick, M. J. S. 1992. *Scenes from Deep Time: Early Pictorial Representations of the Prehistoric World.* Chicago: University of Chicago Press.

Sarjeant, W. A. S. 1994. Geology in fiction. In W. A. S. Sarjeant (ed.), *Useful and Curious Geological Inquiries beyond the World,* pp. 318–337. 19th International History of Geology INHIGEO Symposium, Sydney, Australia.

Sawyer, R. J. 1992. *Far-Seer.* New York: Ace Paperbacks.

Sawyer, R. J. 1993. *Fossil Hunter.* New York: Ace Paperbacks.

Scully, V.; R. F. Zallinger; L. J. Hickey; and J. H. Ostrom. 1990. *The Great Dinosaur Mural at Yale: The Age of Reptiles.* New York: Harry N. Abrams.

Senn, B., and J. Johnson. 1992. *Fantastic Cinema Subject Guide.* Jefferson, N.C.: McFarland and Company.

Warren, B. 1982. *Keep Watching the Skies!* 2 vols. Jefferson, N.C.: McFarland and Company.

This compilation will be continually updated at http://www.dinosauria.com /jdp/misc/fiction.htm

Aldiss, B. 1958. "Poor Little Warrior." Reprinted in Silverberg et al. 1982.

Aldiss, B. 1967. *Crytozoic.* London: Sphere Books.

Aldiss, B. 1985. *The Malacian Tapestry.* New York: Berkley Books.

Allen, R. M. 1993. "Evolving Conspiracy." Reprinted in Resnick and Greenberg 1993.

Anderson, P. 1958. "Wildcat." Reprinted in Silverberg et al. 1982.

Andrews, A. 1993. "Day of the Dancing Dinosaur." *Science Fiction Age* 1 (3): 35–40.

Anthony, P. 1970. *Orn.* London: Corgi Books.

Arthur, R. 1940. "Tomb of Time." *Thrilling Wonder Stories* (November).

Ash, P. 1966. "Wings of a Bat." Reprinted in Silverberg et al. 1982.

Ashwell, P. 1996. "Bonehead." *Analog* (July).

Asimov, I. 1950. "Day of the Hunters." Reprinted in Silverberg et al. 1982; also in Dann and Dozois 1995.

Asimov, I. 1958. "A Statue for Father." Reprinted in Silverberg et al. 1982.

Astor, J. J. 1894. *A Journey in Other Worlds.* New York: Appleton Books.

Bakker, R. T. 1995. *Raptor Red.* New York: Bantam Books.

Barnes, A. K. 1937. "The Hothouse Planet." *Startling Stories* (September). Reprint, September 1949.

Barshofsky, P. 1930. "One Prehistoric Night." *Wonder Stories* (November).

Benford, G. 1992. "Rumbling Earth." *Aboriginal Science Fiction* 31 (Summer): 8–13.

Benford, G. 1992. "Shakers of the Earth." In Preiss and Silverberg 1992.

Bennett, R. A. 1916. "The Bowl of Baal." *All Around Magazine* (November 1916–February 1917).

Bierce, A. 1909. "For the Ahkoond."

In *The Collected Works of Ambrose Bierce,* vol. 1. New York: Neale Publishing Co.

Bishop, M. 1992. "Herding with the Hadrosaurs." In Preiss and Silverberg 1992; also in Dann and Dozois 1995.

Bradbury, R. 1983. *Dinosaur Tales.* New York: Bantam Books.

Bradbury, R. 1983. "Besides a Dinosaur, What Ya Wanna Be When You Grow Up?" Reprinted in Preiss and Silverberg 1992.

Branham, R. V. 1995. "Dinosaur Pliés." In Dann and Dozois 1995.

Bray, Lady E. O. 1921. *Old Time and the Boy; or, Prehistoric Wonderland.* London: Allenson Books.

Bridges, T. C. 1923. *Men of the Mist.* London: Collins Books.

Brown, P. E. 1908. "The Diplodocus." *New Broadway Magazine* (August).

Buckley, B. 1978. "The Runners." Reprinted in Dann and Dozois 1990.

Burroughs, E. R. 1918. *The Land That Time Forgot.* New York: Ace Books. This and subsequent novels are part of the Caspak series.

Burroughs, E. R. 1921. *Tarzan The Terrible.* Chicago: McClurg Books. Also in *Argosy All Story* [February 12–March 26]).

Burroughs, E. R. 1922. *At the Earth's Core.* New York: Ace Books. This and subsequent novels are part of the Pellucidar series.

Cabot, J. Y. 1942. "Blitzkrieg in the Past." *Amazing Stories* (July 1942).

Cadigan, P. 1993. "Dino Trend." In Resnick and Greenberg 1993.

Carroll, D. N. 1934. "When Reptiles Ruled." *Wonder Stories* (November).

Carter, L. 1979. *Journey to the Underground World.* New York: DAW Books.

Carter, L. 1980. *Zanthodon.* New York: DAW Books.

Carter, L. 1981. *Hurok of the Stone Age.* New York: DAW Books.

Casper, S. 1993. "Betrayal." In Resnick and Greenberg 1993.

Chesney, W. 1898. "The Crimson Beast." In *The Adventures of a Solicitor.* London: James Bowden Books.

Chilson, R. 1976. *The Shores of Kansas.* New York: Popular Library.

Ciencin, S. 1995. *Dinotopia: Windchaser.* New York: Random House, Bullseye Books.

Ciencin, S. 1995. *Dinotopia: Lost City.* New York: Random House, Bullseye Books.

Clarke, A. C. 1952. "Time's Arrow." Reprinted in Dann and Dozois 1990.

Crichton, M. 1990. *Jurassic Park.* New York: Alfred A. Knopf.

Crichton, M. 1995. *The Lost World.* New York: Alfred A. Knopf.

Dann, J., and G. Dozois. 1981. "A Change in the Weather." Reprinted in Dann and Dozois 1990.

Dann, J., and G. Dozois (eds.). 1990. *Dinosaurs!* New York: Ace Books.

Dann, J., and G. Dozois (eds.). 1995. *Dinosaurs II.* New York: Ace Books.

Davidson, A. 1989. "The Odd Old Bird." Reprinted in Dann and Dozois 1995.

de Camp, L. S. 1963. *A Gun for Dinosaur.* New York: Curtis Books.

de Camp, L. S. 1992. "The Big Splash." Reprinted in Dann and Dozois 1995.

de Camp, L. S. 1993. *Rivers of Time.* Riverdale, N.Y.: Baen Books.

de Camp, L. S., and C. C. de Camp. 1988. *The Stones of Nomuru.* Virginia Beach: Donning Publishing Co.

Dedman, S. 1986. "Mesozoic Error." *Aphelion,* no. 4 (Spring).

Dedman, S. 1993. "Vigil." *Fantasy and Science Fiction* (August).

Dedman, S. 1994. "Desired Dragons." *Alien Shores* (June).

Dedman, S. 1996. "Miniatures." *Eidolon* (March).

Dedman, S. 1997. "Sarcophagus." In *A Horror Story a Day.* Rockleigh, N.J.: Barnes and Noble.

Dehan, R. 1917. "The Great Beast of Kafue." In *Under the Hermes.* New York: Dodd-Mead.

Delaney, J. H. 1989. "Survival Course." *Analog* 109 (6): 92–110.

Delaplace, B. 1993. "Fellow Passengers." In Resnick and Greenberg 1993.

Dent, G. 1926. *The Emperor of IF.* London: Heinemann.

DiChario, N. A. 1993. "Whilst Slept the Sauropod." In Resnick and Greenberg 1993.

Doyle, Sir A. C. 1912. *The Lost World.* New York: Hodder and Stoughton.

Drake, D. 1982. *Time Safari.* New York: Tor Books.

Drake, D. 1984. *Birds of Prey.* New York: Tor Books.

Drake, D. 1993. *Tyrannosaur.* New York: Tor Books.

Efremov, I. A. 1946. *A Meeting over Tuscarora.* London: Hutchinson and Co.

Farber, S. N. 1988. "The Last Thunder Horse West of the Mississippi." Reprinted in Dann and Dozois 1990.

Farber, S. N. 1991. "The Sixty-five Million Year Sleep." *Amazing Stories* 66 (2): 53–56.

Farley, R. M. 1929. "Radio Flyers." *Argosy All Story Weekly* (May 11).

Fawcett, B. 1993. "After the Comet." In Resnick and Greenberg 1993.

Fawcett, E. D. 1894. *Swallowed by an Earthquake.* London: Edwin Arnold.

Feeley, G. 1993. "Thirteen Ways of Looking at a Dinosaur." In Resnick and Greenberg 1993.

Gauger, R. 1987. *Charon's Ark.* New York: Del Rey Books.

Gerrold, D. 1978. *Deathbeast.* New York: Popular Library.

Gerrold, D. 1993. "Rex." In Resnick and Greenberg 1993.

Glut, D. 1976. *Spawn.* Toronto: Laser Books.

Gottfried, F. D. 1980. "Hermes to the Ages." Reprinted in Silverberg et al. 1982.

Grimes, L. 1994. *Dinosaur Nexus.* New York: Avon Books.

Gurney, J. 1992. *Dinotopia.* Atlanta: Turner Publishing.

Gurney, J. 1995. *Dinotopia II: The World Beneath.* Atlanta: Turner Publishing.

Hamilton, E. 1929. "The Abysmal Invaders." *Weird Tales* (June).

Hansen, L. T. 1941. "Lords of the Underworld." *Amazing Stories* (April).

Harrison, H. 1970. "The Ever-Branching Tree." Reprinted in Silverberg et al. 1982.

Harrison, Harry. 1984–1988. The West of Eden Trilogy. New York: Bantam Books.

Harrison, H. 1992. "Dawn of the End-

less Night." In Preiss and Silverberg 1992.

Hering, H. A. 1899. "Silas P. Cornu's Diving DIVINING Rod." *Cassell's Family Magazine* (June).

Hernandez, L. 1993. "Pteri." In Resnick and Greenberg 1993.

Hulke, M. 1976. *Dr. Who and the Dinosaur Invasion.* London: Target Books.

Hyne, C. J. C. 1900. *The Lost Continent.* London: Hutchinson.

Jablonski, D. (ed.). 1981. *Behold the Mighty Dinosaur.* New York: Elsevier/Nelson Books.

Jacobson, M. 1991. *Gojira.* New York: Atlantic Monthly Press.

Jones, W. K. 1927. "The Beast of the Yungas." *Weird Tales* (September).

Kelly, J. P. 1990. "Mr. Boy." *Asimov's Science Fiction Magazine* (June).

Kelly, J. P. 1995. "Think Like a Dinosaur." *Asimov's Science Fiction Magazine* 19 (7): 10–32.

Kerr, K. 1993. "The Skull's Tale." In Resnick and Greenberg 1993.

Keyhoe, D. E. 1926. "Through The Vortex." *Weird Tales* (July).

Knight, H. A. 1984. *Carnosaur.* London: W. H. Allen and Co.

Koja, K., and B. N. Malzberg. 1993. "Rex Tremandae Majestatis." In Resnick and Greenberg 1993.

Lackey, M., and L. Dixon. 1993. "Last Rights." In Resnick and Greenberg 1993.

Landis, G. A. 1985. "Dinosaurs." Reprinted in Dann and Dozois 1990.

Landis, G. A. 1992. "Embracing the Alien." *Analog* 92 (13): 10–39.

Laumer, K. 1971. *Dinosaur Beach.* New York: DAW Books.

Leigh, S. 1992. *Dinosaur World.* New York: Avon Books.

Leigh, S. 1993. *Dinosaur Planet.* New York: Avon Books.

Leigh, S. 1994. *Dinosaur Warriors.* New York: Avon Books.

Leigh, S. 1995. *Dinosaur Conquest.* New York: Avon Books.

Leigh, S., and J. J. Miller. 1993. *Dinosaur Samurai.* New York: Avon Books.

Leigh, S., and J. J. Miller. 1995. *Dinosaur Empire.* New York: Avon Books.

Lindow, S. J. 1992. "Through Dinosaur Eyes." *Isaac Asimov's Science Fiction Magazine* 17 (1): 88–89.

Little, C. 1900. "Dick and Dr. Dan." *Happy Days* (March 17–May 5).

Longyear, B. 1989. *The Homecoming.* New York: Walker and Company.

Malzberg, B. 1992. "Major League Triceratops." In Preiss and Silverberg 1992.

Marsten, R. 1953. *Danger: Dinosaurs!* Philadelphia: J. C. Winston and Co.

Mash, R. 1983. *How to Keep Dinosaurs.* New York: Penguin Books.

McCaffrey, A. 1978. *Dinosaur Planet.* New York: Del Rey Books.

McCaffrey, A. 1984. *Dinosaur Planet Survivors.* New York: Del Rey Books.

McCoy, M. 1996. *Indiana Jones and the Dinosaur Eggs.* New York: Bantam Books.

McDowell, I. 1994. "Bernie." Reprinted in Dann and Dozois 1995.

Meacham, B. 1993. "On Tiptoe." In Resnick and Greenberg 1993.

Merritt, A. G. 1931. *The Face in the Abyss.* New York: Liveright.

Mill, J. 1854. *The Fossil Spirit: A Boy's Dream of Geology.* London: Darton Books.

Milne, R. D. 1882. "The Iguanodon's Egg." *The Argonaut* (April 1).

Milne, R. D. 1882. "The Hatching of the Iguanodon." *The Argonaut* (April 8).

Mimersheim, J. 1993. "The Pangaean Principle." In Resnick and Greenberg 1993.

Murray Chapman, C. H. 1924. *Dragons at Home.* London: Wells Gardner and Darton.

Obruchev, V. A. 1924. *Plutonia: An Adventure through Prehistory.* London: Lawrence and Wishart. English ed. copyright 1957.

O'Donnell, K. Jr. 1993. "'Saur Spot." In Resnick and Greenberg 1993.

Pelkie, J. W. 1945. "King of the Dinosaurs." *Fantastic Adventures* (October).

Petticolas, A. 1949. "Dinosaur Destroyer." *Amazing Stories* (January).

Phillips, A. 1929. "Death of the Moon." *Amazing Stories* (February).

Phillpotts, E. 1901. "A Story without an End. In *Fancy Free.* London: Methuen.

Pierce, H. 1989. *The Thirteenth Majestral.* New York: Tor Books.

Pierce, H. 1989. *Dinosaur Park.* New York: Tor Books.

Pope, G. 1894. *Romances of the Planets: N.1, Journey to Mars.* G. W. Dillingham.

Powell, F. 1906. *The Wolf Men.* London: Cassell.

Preiss, B., and R. Silverberg (eds.). 1992. *The Ultimate Dinosaur Book.* New York: Bantam Books.

Preuss, P. 1992. "Rhea's Time." In Preiss and Silverberg 1992.

Resnick, L. 1993. "Curren's Song." In Resnick and Greenberg 1993.

Resnick, M., and M. H. Greenberg (eds.). 1993. *Dinosaur Fantastic.* New York: DAW Paperback Books.

Rivkin, J. F. 1992. *Age of Dinosaurs (Tyrannosaurus Rex).* New York: Roc Paperbacks, Penguin Books.

Robertson, R. G. Y. 1991. "The Virgin and the Dinosaur." Reprinted in Dann and Dozois 1995.

Robertson, R. G. Y. 1995. "Ontogeny Recapitulates Phylogeny."In Dann and Dozois 1995.

Robertson, R. G. Y. 1996. "The Virgin and the Dinosaur." New York: Avon Books.

Robeson, K. 1933. *Land of Terror.* New York: Street and Smith Publications.

Robinson, F. M. 1993. "The Greatest Dying." In Resnick and Greenberg 1993.

Roof, K. M. 1930. "A Million Years After." *Weird Tales* (November).

Rousseau, V. 1920. "The Eye of the Balamok." *Argosy All-Story Weekly* (issue unknown).

Rusch, K. K. 1993. "Chameleon." In Resnick and Greenberg 1993.

Sagara, M. M. 1993. "Shadow of a Change." In Resnick and Greenberg 1993.

Savile, F. M. 1899. *Beyond the Great South Wall: The Secret of the Antarctic.* New York: New Amsterdam Book Co.

Sawyer, R. J. 1981. "If I'm Here, Imagine Where They Sent My Luggage." *The Village Voice* (January 14).

Sawyer, R. J. 1987. "Uphill Climb." *Amazing Stories* (March).

Sawyer, R. J. 1992. *Far-Seer* (Quintaglio #1). New York: Ace Paperbacks.

Sawyer, R. J. 1993. *Fossil Hunter* (Quintaglio #2). New York: Ace Paperbacks.

Sawyer, R. J. 1993. "Just Like Old Times." In Resnick and Greenberg 1993; also in Dann and Dozois 1995.

Sawyer, R. J. 1994. *Foreigner* (Quintaglio #3). New York: Ace Paperbacks.

Sawyer, R. J. 1994. *End of an Era.* New York: Ace Books.

Schow, D. J. 1987. *Sedalia.* Lincoln City, Ore.: Pulphouse Publishing.

Schultz, M. 1989. *Cadillacs and Dinosaurs.* Northampton, Mass.: Kitchen Sink Press.

Shadwell, T. 1991. *Dinosaur Trackers.* New York: Harper Paperbacks.

Sheckley, R. 1993. "Disquisitions on the Dinosaur." In Resnick and Greenberg 1993.

Sheffield, C. 1992. "The Feynman Saltation." In Preiss and Silverberg 1992.

Sherman, J. 1993. "Wise One's Tale." In Resnick and Greenberg 1993.

Silverberg, R. 1980. "Our Lady of the Sauropods." Reprinted in Silverberg et al. 1982.

Silverberg, R. 1987. *Project Pendulum.* New York: Bantam Paperbacks.

Silverberg, R. 1992. "The Way to Spook City." *Playboy* (August).

Silverberg, R. 1992. "Hunters in the Forest." In Preiss and Silverberg 1992.

Silverberg, R.; C. Waugh; and M. H. Greenberg (eds.). 1982. *The Science Fictional Dinosaur.* New York: Avon Books.

Simak, C. D. 1995. "Small Deer." In Dann and Dozois 1995.

Simpson, G. G. 1995. *The Dechronization of Sam Magruder.* New York: St. Martin's Press.

Smith, D. W. 1993. "Cutting Down Fred." In Resnick and Greenberg 1993.

Snyder, M. 1995. *Dinotopia: Hatchling.* New York: Random House, Bullseye Books.

Stables, W. G. 1906. *The City at the Pole.* London: James Nisbet and Co.

Steele, A. 1990. "Trembling Earth." Reprinted in Dann and Dozois 1995.

Stith, J. E. 1993. "One Giant Step." In Resnick and Greenberg 1993.

Sullivan, T. 1987. "Dinosaur on a Bicycle." Reprinted in Dann and Dozois 1990.

Taine, J. 1944. "The Greatest Adventure." *Famous Fantastic Mysteries* (June).

Tarr, J. 1993. "Revenants." In Resnick and Greenberg 1993.

Tem, S. R. 1987. "Dinosaur." Reprinted in Dann and Dozois 1990.

Tiptree, J. Jr. 1970. "The Night-Blooming Saurian." Reprinted in Dann and Dozois 1990.

Turtledove, H. 1985. "Hatching Season." Reprinted in Dann and Dozois 1990.

Turtledove, H. 1992. "The Green Buffalo." In Preiss and Silverberg 1992.

Utley, S. 1976. "Getting Away." Reprinted in Dann and Dozois 1990.

Vornholt, J. 1995. *Dinotopia: River Quest.* New York: Random House, Bullseye Books.

Waldrop, H. 1982. "Green Brother." Reprinted in Dann and Dozois 1990.

Wallis, B. 1930. "The Primeval Pit." *Weird Tales* (December).

Wells, R. 1969. *The Parasaurians.* New York: Berkley/Medallion Books.

Williams, R. M. 1943. "The Lost Warship." *Amazing Stories* (January).

Williams, W. J. 1991. *Dinosaurs.* Lincoln City, Ore.: Pulphouse Publishing.

Willis, C. 1991. "In the Late Cretaceous." Reprinted in Preiss and Silverberg 1992.

Wilson, F. P. 1989. *Dydeetown World.* New York: Simon and Schuster.

Winter, R. B. 1938. *Hal Hardy in the Lost Valley of the Giants.* Racine, Wis.: Whitman Publishing Co.

Wolverton, D. 1992. "Siren Song at Midnight." In Preiss and Silverberg 1992.

Wu, W. 1993. *Robots in Time.* New York: Avon Books.

Yep, L. 1986. *Monster Makers Inc.* New York: Signet Books.

Young, R. F. 1964. "When Time Was New." Reprinted in Silverberg et al. 1982.

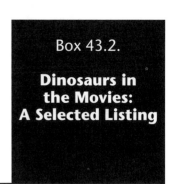

Box 43.2.

**Dinosaurs in
the Movies:
A Selected Listing**

Along the Moonbeam Trail (1920)
Animal World, The (1956)
At the Earth's Core (1976)
Baby, Secret of the Lost Legend (1983)
Beast of Hollow Mountain, The (1956)
Beast from 20,000 Fathoms, The (1953)
Bellow Durmiente, El (1953)
Birth of a Flivver (1916, 1-reeler for Thomas Edison pictures)
Brute Force (1913)
Carnosaur (1993)
Carnosaur II (1995)
Caveman (1981)
Curious Pets of Our Ancestors (1917, 1-reeler for Thomas Edison pictures)
Dinosaur and the Missing Link, The (1915, 1-reeler for Thomas Edison pictures)
Dinosaur Island (1993)
Dinosaur Valley Girls (1996)
Dinosaurs from the Deep (1995)
Dinosaurs, the Terrible Lizards (1970, an educational film)
Dinosaurus! (1960)
Doctor Mordrid (1993)
Evolution (1923)
Fantasia (1940)
Fig Leaves (1926)
Future Wars (1995)
Galaxy of Dinosaurs (1992)

Gertie the Dinosaur (1912, cartoon)
Ghost of Slumber Mountain, The (1919)
Giant Behemoth, The (1959)
Godzilla, King of the Monsters (1954)
Gorgo (1960)
In Prehistoric Days (reissue of *Brute Force*, 1913)
Isla de los Dinosaurios, La (1967)
Journey to the Beginning of Time (1954)
Jurassic Park (1993)
King Dinosaur (1955)
King Kong (1933)
King of the Kongo (1929, serial)
Land before Time, The (1988, cartoon)
Land That Time Forgot, The (1974)
Land Unknown, The (1957)
Last Dinosaur, The (1977)
Lost Continent (1951)
Lost Whirl, The (1927)
Lost World, The (1925)
Lost World, The (1960)
Lost World, The (1993)
Lost World, The (Jurassic Park II) (1997)
Morpheus Mike (1916, 1-reeler for Thomas Edison pictures)
Mystery Of Life, The (1931)
Nymphoid Barbarian in Dinosaur Hell (1991)
One Million B.C. (1940)
One Million Years B.C. (1966)

On Moonshine Mountain (1914)
Pathé Review (1923, a short)
People That Time Forgot, The (1977)
Planet of the Dinosaurs (1977)
Planeta Burg (1962)
Prehistoric Man, The (1908)
Prehistoric Poultry (1917, 1-reeler for Thomas Edison pictures)
Prehysteria (1994)
Prehysteria II (1995)
Reptilicus (1963)
Return to the Lost World (1993)
R.F.D., 10,000 B.C. (1917, 1-reeler for Thomas Edison pictures)
Robot Monster (1953)
The Savage (1926)
Secret of the Loch (1934)
Son of Kong, The (1933)
Super Mario Brothers (1993)
Tarzan's Desert Mystery (1943)
Three Ages, The (1923)
Two Lost Worlds (1950)
Unknown Island (1948)
Valley of Gwangi, The (1969)
Voyage to the Planet of Prehistoric Women (1968)
We're Back! (1993, cartoon)
When Dinosaurs Ruled the Earth (1970)
When Time Began (1976)
Women of the Prehistoric Planet (1968)

The following dinosaur, geology, and paleontology sites were some of the first to go online:

Natural History Resources
http://www.ucmp.berkeley.edu/subway/nathist.html

The Royal Tyrrell Museum, Drumheller, Alberta, Canada
http://tyrrell.magtech.ab.ca/

Earthnet Info Server (Illinois State Geological Survey)
http://denr1.igis.uiuc.edu/isgsroot/dinos/

The Museum of Paleontology, Berkeley (University of California)
http://www.ucmp.berkeley.edu/

The Paleontological Institute, Russian Academy of Sciences, Moscow
http://www.ucmp.berkeley.edu/pin/pin.html

The Natural History Museum, London
http://www.nhm.ac.uk/

The Society of Vertebrate Paleontology News Bulletin
http://eteweb.lscf.ucsb.edu/svp/

The Smithsonian Institution
http://nmnh.si.edu/departments/paleo.html

The Dinosaur Society
http://www.dinosociety.org/

Note: The http sites change every so often as hardware and software are upgraded. Please check the node addresses with any of the many Internet search engines. The Dinosaur Bulletin Board (newsgroup) is at listproc@usc.edu.

Box 43.4.

Dinosaurs on Stamps (1842–1992)

Compiled by
M. K. Brett-Surman

This compilation is offered as a checklist for the collector and lists all the non-avian dinosaurs on stamps (from 1842 to 1992) on which the actual bones of dinosaurs are pictured, or on which whole restorations of dinosaurs are used. It excludes footprints, cartoons, silhouettes, and unofficial issues such as the famous Sinclair Dinosaur Stamps. Scientifically invalid names (such as *"Brontosaurus,"* which appears on many stamps) are enclosed in quotation marks to denote that they are used as popular names. Avians, such as *Archaeopteryx*, are excluded.

The best books now in print on dinosaur stamps are *Dinosaur Stamps of the World* by Baldwin and Halstead (1991) and *Dinosaurs Resurrected* by Hasegawa and Shiraki (1994). The best philatelic periodical that covers dinosaurs on stamps is *Biophilately*. This list is revised from the December 1991 issue. (See also Andrew Scott, "Geology on Stamps: Dinomania," *Geology Today* [January–February 1994].)

Superscript numbers refer to endnotes. Countries inside quotation marks are not actual countries but usually states or "regions" within countries that issue stamps for profit. They are not recognized by international stamp organizations.

"Aden," 1968: *Tyrannosaurus,*[1] *"Brontosaurus"*
Afghanistan, 1988: *Styracosaurus, Pentaceratops, Stegosaurus, Ceratosaurus*
Antigua-Barbuda, 1992: *Allosaurus, Brachiosaurus, "Brontosaurus," Stegosaurus, Deinonychus, Tyrannosaurus, Triceratops, Protoceratops, Parasaurolophus*

Argentina, 1992: *Amargosaurus, Carnotaurus*
Belgium, 1966: *Iguanodon*
Benin, 1984: *"Anatosaurus,"*[2] *"Brontosaurus"*
1985: *Tyrannosaurus, Stegosaurus*
Brazil, 1991: "theropod," "sauropod"
British Antarctic Territories, 1991: hypsilophodont
Bulgaria, 1990: *"Brontosaurus," Stegosaurus, Protoceratops, Triceratops*
Cambodia (Kampuchea), 1986: *Brachiosaurus, Tarbosaurus*[3]
Canada, 1989: *Albertosaurus*
Central African Republic, 1988: *"Brontosaurus," Triceratops, Ankylosaurus, Stegosaurus, Tyrannosaurus, Corythosaurus, Allosaurus, Brachiosaurus*[4]
People's Republic of China, 1958: *Lufengosaurus*
People's Republic of Congo, 1970: *Kentrosaurus,*[5] *Brachiosaurus*
1975: *Ornithomimus, Tyrannosaurus, Stegosaurus*
Cuba, 1985: *"Brontosaurus," Iguanodon, Stegosaurus, Monoclonius, Corythosaurus, Tyrannosaurus, Triceratops, Euoplocephalus,*[6] *Styracosaurus, Saurolophus, "Anatosaurus"*
Dahomey, 1974: *Stegosaurus,*[7] *Tyrannosaurus*
"Dhufar," 1975: *Iguanodon, Tarbosaurus*[8]
Dominica, 1992: *Camptosaurus, Stegosaurus, Tyrannosaurus, Euoplocephalus, Torosaurus* (2 stamps), *Parasaurolophus, Corythosaurus* (2 stamps), *Edmontosaurus*
Equatorial Guinea, 1975: *Styracosaurus, Stegosaurus, Corythosaurus, Ankylosaurus, Triceratops, Diplodocus*
"Fujeria," 1968: *Triceratops, Plateosaurus, Stegosaurus, Allosaurus*

1972: *Triceratops, Stegosaurus, "Brontosaurus"*
Gambia, 1992: *Fabrosaurus, Allosaurus* (2 stamps), *Cetiosaurus, Camptosaurus, Dryosaurus, Kentrosaurus, Deinonychus, Spinosaurus* (2 stamps), *Saurolophus, Ornithomimus*
Germany, Berlin, 1977: *Iguanodon* (4 stamps)
Germany, East, 1990: *Dicraeosaurus, Kentrosaurus,*[9] *Dysalatosaurus, Brachiosaurus* (2 stamps)
Ghana, 1992: *Coelophysis* (2 stamps), *Anchisaurus, Heterodontosaurus, Elaphrosaurus, Iguanodon, Ouranosaurus, "Anatosaurus"*
Great Britain, 1991: *Iguanodon, Stegosaurus, Tyrannosaurus, Protoceratops, Triceratops*[10]
Guinea, 1987: *Iguanodon, Stegosaurus, Triceratops*[11]
Guinea-Bissau, 1989: *"Trachodon,"*[12] *Tyrannosaurus, Stegosaurus*
Hungary, 1986: *"Brontosaurus"*
1990: *Tarbosaurus, "Brontosaurus," Stegosaurus*
Korea, North, 1980: *Stegosaurus,*[13] *Tyrannosaurus*
1991: *"Brontosaurus," Stegosaurus, Allosaurus*
Kuwait, 1982: sauropod, sauropod[14]
Laos, 1988: *Tyrannosaurus, Ceratosaurus, Iguanodon, Euoplocephalus*(?), *"Trachodon"*[15]
Lesotho, 1992: *Procompsognathus, Plateosaurus, Massospondylus, Lesothosaurus* (2 stamps), *Ceratosaurus, Stegosaurus, Gasosaurus*
Malagasy Republic, 1989: *Tyrannosaurus, Stegosaurus, Triceratops, Saurolophus*
Maldive Islands, 1972: *Stegosaurus, Diplodocus, Triceratops, Tyrannosaurus* (2 stamps)[16]
1992: *Scelidosaurus, Allosaurus,*

Brachiosaurus, Apatosaurus, Mamenchisaurus, Stegosaurus, Deinonychus, Tenontosaurus, Iguanodon, Tyrannosaurus, "Stenonychosaurus," Euoplocephalus, Triceratops, Styracosaurus, "Monoclonius," hadrosaurs, Parasaurolophus, "Anatosaurus"

Mali, 1984: Iguanodon, Iguanodon, Triceratops

"Manama," 1971: Stegosaurus, Plateosaurus, Styracosaurus, Allosaurus, "Brontosaurus"

Mauritania, 1986: Iguanodon, Apatosaurus, Polacanthus(?)

Mongolia, 1967: Tarbosaurus, Talarurus, Protoceratops, Saurolophus
1977: Psittacosaurus
1990: Chasmosaurus, Stegosaurus, Probactrosaurus, Opisthocoelicauda, Iguanodon, Tarbosaurus, Mamenchisaurus, Allosaurus, "Ultrasaurus"[17]

Montserrat, 1992: Coelophysis, "Brontosaurus," Diplodocus, Tyrannosaurus

Morocco, 1988: Cetiosaurus
Nicaragua, 1987: Triceratops[18]
Niger, 1976: Ouranosaurus[19]
Niuafo'ou, 1989: Stegosaurus
"Oman," 1975: Megalosaurus, Triceratops

Poland, 1965: "Brontosaurus," Stegosaurus, Brachiosaurus, Styracosaurus, Corythosaurus, Tyrannosaurus
1980: Tarbosaurus

Saint Thomas and Prince Islands, 1982: Parasaurolophus, Stegosaurus, Triceratops, "Brontosaurus," Tyrannosaurus

San Marino, 1965: "Brontosaurus," Brachiosaurus, Tyrannosaurus, Stegosaurus, Iguanodon, Triceratops

Sierra Leone, 1992: Brachiosaurus, Kentrosaurus, Hypsilophodon, Iguanodon, "Trachodon"

Soviet Union, 1990: Saurolophus
Sweden, 1992: Plateosaurus
Tanzania, 1988: Plateosaurus, "Brontosaurus," Stegosaurus
1991: Stegosaurus, Triceratops, Edmontosaurus, Plateosaurus, Diplodocus, Iguanodon, Silvisaurus[20]
1992: Coelophysis, Anchisaurus, Heterodontosaurus, Camarasaurus, Lesothosaurus, Ceratosaurus, Cetiosaurus, Stegosaurus, Baryonyx (misspelled on stamp), Iguanodon, Spinosaurus, Pachycephalosaurus,

Saltasaurus, Allosaurus, Ornithomimus, Dryosaurus

Thailand, 1992: tyrannosaurid, sauropod

Uganda, 1992: Megalosaurus (2 stamps), Brachiosaurus (2 stamps), Kentrosaurus, Hypsilophodon

United States,[21] 1970: Stegosaurus, Camptosaurus, Allosaurus, Compsognathus, Apatosaurus

1989: Tyrannosaurus, Stegosaurus, "Brontosaurus"

Vietnam, 1979: "Brontosaurus," Iguanodon, Tyrannosaurus, Stegosaurus, Triceratops
1984: Diplodocus, Styracosaurus, Corythosaurus, Allosaurus, Brachiosaurus
1991: "Gorgosaurus," Ceratosaurus, Ankylosaurus (2 stamps)[22]

West Sahara, 1992: Megalosaurus, Stegosaurus, Psittacosaurus, Tarbosaurus, Pachycephalosaurus, Struthiomimus, Nemegtosaurus, Ankylosaurus, Protoceratops, Saurolophus

Yemen, 1971: Iguanodon
1990: Tyrannosaurus

Notes

1. The animal listed as "Dinosaurus" is probably Tyrannosaurus. The word "Dinosaurus" is not currently a valid name for any genus or species of dinosaur and is probably a junior synonym of Plateosaurus.

2. All species of "Anatosaurus" were assimilated into the name Edmontosaurus in 1979 with the exception of "Anatosaurus" copei. The generic name of this taxon was changed to Anatotitan in 1990. The name "Anatosaurus" is no longer used.

3. The pictures of Tarbosaurus and Brachiosaurus are taken from the works of Zdeněk Burian, a famous Czechoslovakian artist. The same picture of Tarbosaurus also appears on the 1975 issue from Dhufar.

4. Brachiosaurus is incorrectly restored. In this animal the forelimbs are longer than the hind limbs.

5. The proper name for Kentruosaurus is Kentrosaurus. An ornithomimid is in the background. The restoration is too small and generalized for a proper identification.

6. The 1985 set uses the Spanish

version of the names of the dinosaurs. In the 1987 set, Euoplocephalus is incorrectly restored without its tail club.

7. The Stegosaurus stamp is incorrectly labeled Crétacé (for the Cretaceous Period, 135 to 65 million years ago). It is actually from the Jurassic Period (about 200 to 135 million years ago).

8. The artwork for these two dinosaurs is taken from the works of Zdeněk Burian.

9. The proper name for Kentruosaurus is Kentrosaurus.

10. This set is labeled "Owen's Dinosaurs." When Richard Owen coined the word Dinosauria in 1842, he based it on three dinosaurs: Iguanodon, Hylaeosaurus, and Megalosaurus. The only stamp in this set that applies is Iguanodon. The other dinosaurs in this set were discovered after Owen died and have nothing to do with him. Most of the dinosaurs are improperly restored. In dinosaurs, the shoulder blade lies mostly parallel to the backbone, not perpendicular to it as in mammals, and as incorrectly seen here.

11. Triceratops is on a minisheet with many dinosaurs in the background. Several of these dinosaurs are taken from the artwork of both Burian and Zallinger.

12. The name "Trachodon" is no longer used in paleontology. The original material upon which the name was erected in 1856 has turned out to be from two different types of dinosaurs. This makes the name useless for scientific purposes, and it is now regarded as a nomen dubium (dubious name).

13. In the Stegosaurus stamp there appears to be one of the ankylosaurian dinosaurs in the background.

14. In this stamp the dinosaur appears to be one of the sauropods (the group to which Diplodocus belongs). It is too small and generalized to be properly identified, although Baldwin and Halstead (1991) call it Plateosaurus.

15. In this set, the names for Tyrannosaurus and "Trachodon" have been switched and appear on the wrong stamps. The Scolosaurus stamp is identified as Euoplocephalus in Baldwin and Halstead 1991, but the tail

club and head shape are closer to those of *Scolosaurus*. Most of the artwork in this set is copied from the famous Czechoslovakian artist Zdeněk Burian.

16. The inspiration for the artwork in this set appears to be the famous Rudolph Zallinger mural at Yale University.

17. *"Ultrasaurus"* is considered by most paleontologists to be a larger version of *Brachiosaurus*. This Mongolian set has some stamps which copy the art of John Gurche and Mark Hallett, two famous American artists. The original pieces of art can be seen in Czerkas and Olson 1987.

18. This stamp is part of a series taken from the famous Charles R. Knight murals; see Czerkas and Glut 1982 for more information about Knight's work.

19. The 60f denomination stamp pictures *Ouranosaurus*. Above the word *dinosaur* is the word *Archaeologie*. This points out one of the most popular misconceptions about dinosaurs. Archaeology is a subdivision of anthropology, and deals only with human beings, and thus merely the last 4 million years of time. Paleontology deals with *all* fossils and covers the last 3.5 billion years of time. Paleontologists dig up dinosaurs; archaeologists do not.

20. In this set, *Silvisaurus* is misspelled *Silviasaurus*, and it is incorrectly restored. There are no spikes along the tail. The restorations are highly inaccurate.

21. The first set, from 1970, is based on the legendary Zallinger Mural on display at the Peabody Museum of Natural History at Yale University. A popular book on the mural has been published (Scully et al. 1990) that includes a foldout of the entire masterpiece. The second set, from 1989, contains the infamous *"Brontosaurus"* stamp. The artist, John Gurche, is considered to be one of the best artists at dinosaur restorations. The original paintings for this stamp are only 1.5 times larger than the actual stamps!

22. Two stamps are labeled *"Ankylosaurus,"* one valued at 1,000d and one at 2,000d. The second stamp does *not* depict *Ankylosaurus* but is actually much more similar to the related dinosaur *Saichania*. Another stamp in the set, valued at 3,000d, is *Edaphosaurus*, which is often mistaken for a dinosaur. It is actually a member of the Synapsida, or mammal-like reptiles, and is more closely related to mammals than to the dinosaurs. The 100d stamp is labeled *Gorgosaurus*. This name was replaced by *Albertosaurus*, but *Gorgosaurus* as a valid taxon may be revived.

Dinosaurs and the Media 705

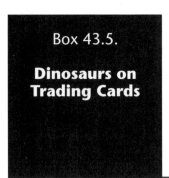

Box 43.5.

Dinosaurs on Trading Cards

Non-Sport Trading Cards

DINOSAURS (Nu-Cards, 1961); DINOSAURS (Golden Press, 1961); DINOSAURS (Milwaukee Public Museum, 1982–1990); BABY (Topps, 1984); DINOSAURS (The Dino-Card Co., 1987); DINOSAURS (Illuminations Inc., 1987); DINOSAURS (Ace, ca. 1987); DINOSAUR ACTION (Illuminations, 1987); DINOSAURS ATTACK! (Topps, 1988); DINOCARDZ (DinoCardz Co., 1992); WILLIAM STOUT (Comic Images, 1993, 1994, 1996); DINOSAUR, THE GREATEST CARDS UNEARTHED (Mun-War Enterprises, 1993); DINOSAURS (First Glance Productions, 1993); JURASSIC PARK (series I & II, Topps, 1993); INFANT EARTH (Kitchen Sink Press, 1993); DINOSAUR NATION (Kitchen Sink Press, 1993); ESCAPE OF THE DINOSAURS (Dynamic Marketing, 1993); DINOSAURS SWAP-IT CARDS—Series 1 (Orbis, 1993); DINOSAURS OF THE MESOZOIC (mostly by artist Brian Franczak, Redstone Marketing, 1994); DINOTOPIA (Collect-A-Card, 1995); CARNEGIE MUSEUM (Acme Studios, 1995); JURASSIC DINOSAURS (Dover Pub., 1995); CRETACEOUS DINOSAURS (Dover Pub., 1996); DINOSAURS (Wal-Mart, 1996); THE LOST WORLD (Topps, 1997); GODZILLA (Futami, year unknown).

Tea/Cigarette Cards

PREHISTORIC ANIMALS IN DIFFERENT AGES (Liebig, 1892); THE PREHISTORIC WORLD (Liebig, 1921); PREHISTORIC ANIMALS (Edwards Ringer & Bigg, 1924); PREHISTORIC ANIMALS (British American Tobacco Co. Ltd., 1931); DINOSAURS (Liebig, 1959); PREHISTORIC MONSTERS (W. Shipton Ltd., 1959); DINOSAURS (Cadet Sweets, 1961); PREHISTORIC ANIMALS (Cooper & Co. Stores Ltd., series I & II, 1962); PREHISTORIC ANIMALS (H. Chappel & Co., series I & II, 1962); DINOSAURS (Brooke Bond Tea Card set, 1963); PREHISTORIC ANIMALS (Milk Marketing Board, 1964); PREHISTORIC ANIMALS (Charter Tea & Coffee Co. Ltd., series I & II, 1965); PREHISTORIC ANIMALS (Sunblest Tea, series I & II, 1966); PREHISTORIC ANIMALS (Clover Dairies Ltd., 1966); PREHISTORIC ANIMALS (Gower & Burgons, 1967); PREHISTORIC ANIMALS (Quaker Oats, 1967); PREHISTORIC ANIMALS (Goodies Ltd., 1969); AGE OF THE DINOSAUR (Cadbury Schweppes, 1971); PREHISTORIC ANIMALS (T. Wall & Sons, 1971); PREHISTORIC ANIMALS (Brooke Bond & Co., 1972); PREHISTORIC ANIMALS (Rowntree & Co., 1978); AGE OF DINOSAURS (George Basset & Co., 1979); PREHISTORIC MONSTERS AND THE PRESENT (Kellogg Co. of Great Britain, 1985).

Appendix:
A Chronological History of Dinosaur Paleontology

This appendix is compiled from numerous sources, and summarizes some of the key personalities, events, and discoveries that have shaped our understanding of dinosaur paleontology. The appendix consists of two parts. Part 1 presents a chronological list of some of the more important historical developments in dinosaur paleontology. Part 2 synthesizes this information by organizing the history of dinosaur studies into a series of "ages," each characterized by certain defining features of the sciences at that time.

M. K. Brett-Surman

Part 1: A Chronology of Dinosaur Studies

300 B.C.E. (date approximate) Chang Qu writes about dinosaur ("dragon") bones in Wucheng (Sichuan), China.

1677 The first report of a "human thigh bone of one of the giants mentioned in the Bible" by Reverend Plot of England.

1763 R. Brookes publishes a figure referred to *Scrotum humanum* based on Reverend Plot's 1676 figure. This specimen, believed to be the distal end of a femur, is now referred to a megalosaurid.

1787 First dinosaur found in Gloucester County, New Jersey, by Matlack (Matelock?) and Caspar Wistar of Philadelphia. A description is read before the American Philosophical Society on October 5, 1787, but it will not be published for 75 years. It is reported at a meeting with Ben Franklin at Franklin's house. The specimen is believed to be at the Academy of Natural Sciences in Philadelphia.

1800 Pliny Moody (a student at Williams College) locates fossils on his farm in Connecticut. These dinosaur footprints were called "footprints of Noah's Raven" at the time by Harvard and Yale scientists, even though they were 1 foot long.

1803–1806 Lewis and Clark collect a Cretaceous fish skull later named *Saurocephalus* by Harlan. They also report a "fish rib" 3 feet long and 3 inches in circumference in an area where the Hell Creek Formation now outcrops. It is most likely a dinosaur, but the fossil cannot be found today. Benjamin Silliman and Amos Eaton of Yale University establish natural history as a profession through their popular lectures around New England. Their speaking fees help support their work.

1818 Solomon Elsworth collects dinosaur bones from the Connecticut Valley Triassic but mistakes them for human bones. They are now in the Yale University collections.

1822 James Parkinson publishes the name *Megalosaurus,* but without providing a description. This predates the presentation by Buckland before the Geological Society of London in 1824, where he announces the name *Megalosaurus* and provides a description.

1824 William Buckland announces *Megalosaurus,* based on jaws and teeth, before a meeting of the Geological Society of London on February 20. This is accepted as the premiere of the name even though Parkinson published the name (without description) in 1822.

1825 Gideon Mantell (a country doctor) names *Iguanodon* on the basis of a tooth found (at least by some accounts) by his wife while he was attending a patient. It is later named *I. anglicum* by Holl (1829). Cuvier says the tooth is that of a rhino, but Mantell publishes it as a "lizard," and Cuvier later admits that he was wrong. Mantell's wife subsequently leaves him. He gives up his practice, moves to London, and hunts fossils full-time. His house becomes so jammed with fossils that he later sells his entire collection to the British Museum (Natural History) for $24,000. Mantell's coat of arms for Maidstone (his residence) includes an *Iguanodon.*

1829 James Louis Macie Smithson dies in Genoa, Italy. He was the illegitimate son of the Duke of Northumberland (Hugh Smithson Percy). Smithson's will leaves everything to his nephew; however, in the case that his nephew dies without heirs, the will stipulates that the money will go "to the United States of America, to found at Washington, under the name of the Smithsonian Institution, an establishment for the increase and diffusion of knowledge among men." This will finally occur in 1846.

1830 Charles Lyell's *Principles Of Geology* makes geology a professional science instead of a gentleman's hobby. Lyell was taught by Reverend William Buckland (who described *Megalosaurus*). Lyell coins the word *palaeontology* ("discourse on ancient things") and recognizes this field as its own science (1829–1833).

1833 Mantell names *Hylaeosaurus.*

1836 Richard Owen is appointed first Hunterian Professor of Comparative Anatomy and Physiology of the Royal College of Surgeons in London.

1837 von Meyer names *Plateosaurus,* but it will not be recognized as a dinosaur until much later.

1841 Owen names *Cetiosaurus* but regards it as a marine reptile. It is not transferred to the Dinosauria until 1869.

1842 Sir Richard Owen coins the word *dinosaur* on the basis of only three partially known genera (*Megalosaurus, Iguanodon,* and *Hylaeosaurus*). Eight other fossil reptiles have already been named that will later be transferred to the Dinosauria.

1850s The geologist James Hall teaches Joseph T. Leidy and Ferdinand Vandeveer Hayden. They both boarded at his house in New York. Hayden is later sent to the Dakota Territories by Hall to collect fossils to compare with fossils from New York.

Hayden finds the first American dinosaur in 1854 in the Judith River deposits of the western U.S., but it remains undescribed until the time of Cope.

1852 Mantell describes the first known dinosaur skin, from the forelimb of *Pelorosaurus becklesii.*

1853 A dinner is held on December 31, 1853, inside Waterhouse Hawkins's life-sized model of *Iguanodon.* Richard Owen presides.

Leidy arranges with Spencer Baird of the United States National Museum (Smithsonian Institution) to have all government survey fossils sent to him at the Philadelphia Academy of Natural Sciences so that he can study them. He does so from 1847 to 1866, after which (in the 1870s) Cope and Marsh push him out of the field. Leidy gets Ferdinand Vandeveer Hayden appointed to head up the early government surveys out west. Hayden also explores the Indian lands for the War Department.

1854 The exhibit of Crystal Palace dinosaurs, based on Waterhouse Hawkins's sculptures, moves to Sydenham, England.

1855 Ferdinand Vandiveer Hayden discovers dinosaur remains in the Montana territories. They will be named in 1856 by Leidy as the first American dinosaurs.

The first skeleton of *Archaeopteryx* is found in Germany. The specimen will not be recognized as a bird until 1970.

1856 *Trachodon, Paleoscincus, Troodon,* and *Deinodon* are named.

Owen becomes superintendent of the Natural History Section of the British Museum.

1858 Edward Hitchcock publishes his monograph on the Triassic and Jurassic footprints of the Connecticut Valley.

Hadrosaurus is named. It is the first dinosaur skeleton more than 50 percent complete to be excavated in the United States.

Joseph Leidy's restoration of *Hadrosaurus* as a biped is the first dinosaur restored in this stance, and it causes a furor. This helps Cope to decide to become a paleontologist.

1859 Darwin publishes *The Origin of Species,* outlining his theory of evolution by natural selection.

1860 Cope (age 19) goes to the Smithsonian, where he publishes thirty-one papers in a single year!

While Cope and Marsh are still friends, Cope names a fish after Marsh: *Ptyonius marshii.*

The first specimen of a feather of *Archaeopteryx* is found.

1861 The second specimen of *Archaeopteryx* is found. It is virtually complete, and will become known as the "London specimen."

1863 Marsh studies in Europe. He meets Cope in Berlin.

1866 Marsh asks his Uncle, George Peabody, if he will endow a museum at Yale. Marsh writes to Silliman at Yale University, saying that his Uncle George Peabody will give $150,000 for a museum. He also lobbies for a chair of paleontology for himself. The Peabody Museum of Natural History at Yale University is founded.

Cope names his first dinosaur: *Laelaps.* The name is already taken by another animal (preoccupied), so Marsh renames it *Dryptosaurus.*

1867 Both Cope and Marsh publish papers noting the similarity of dinosaurs to birds. Huxley later picks up on this. They publish again on this subject in 1871.

1868 Benjamin Waterhouse Hawkins comes to the United States to build a "Paleozoic Museum." He erects a cast of *Hadrosaurus,* the first dinosaur in North America to be free-mounted, assisted by Cope and Leidy. The first real (not a cast) dinosaur skeleton will not be mounted until 1901.

1869 The American Museum of Natural History is created through the efforts of Albert S. Bickmore and Samuel Tilden.

Marsh exposes the Cardiff Giant hoax, and John Wellesley Powell conquers the Colorado River. Both are headline news features around the world, and cause the federal government to increase funding for surveys of the West.

U. S. Grant is a good friend of General Dodge, the Union Pacific engineer who is also a good friend of Hayden. Two months after Grant is elected president, the Hayden Survey becomes the core of the future United States Geological Survey, with an annual federal budget of $10,000.

Dinosaur bones are found at Garden Park, Colorado.

1874 The American Museum begins construction of its building at Central Park West in New York. President Grant lays the cornerstone.

G. M. Dawson collects dinosaur bones from Saskatchewan and Alberta. Cope later publishes these finds. These are the first-discovered Canadian dinosaurs.

1876 The Peabody Museum of Natural History opens its doors to the public.

1877 The American Museum opens in New York City.

The first bone from the Morrison area in Colorado is named *"Poicilopleuron"* (later changed to *Poikilopleuron*) by Leidy, on the basis of a partial caudal vertebra collected by the Hayden Survey.

Arthur Lakes (originally from England), a local teacher (and a painter), sends bones that he finds at the Dakota Hogback in Colorado to Marsh, but when he gets no reply, he writes to Cope. When Marsh finds out about the letter to Cope, he sends Benjamin Mudge to Colorado. Marsh outbids Cope. A few weeks later, Marsh is already publishing some results based on this scrappy material! *Atlantosaurus montanus* is based on two and a half caudal vertebrae.

Oramel Lucas (a graduate of Oberlin College and a schoolteacher at Canyon City) sends bones to both Marsh and Cope (at 10 cents per pound), but Cope hires Lucas immediately. Cope names this material *Camarasaurus supremus.*

Marsh hears about Canyon City and sends Mudge and Samuel W. Williston there to dig in the same beds (at $40 a month). Marsh then sends Williston to Como Bluff, Wyoming, to investigate stories about bones there too, based on letters from two men named Reed and Carlin (who use the pseudonyms "Harlow & Edwards"). Marsh hires Reed and Carlin to dig at Como for $90 a month. Como Bluff is near the famous Virginian Saloon at Medicine Bow. Reed is a section foreman for the Union Pacific Railroad, and later becomes curator at the University of Wyoming Museum.

On the basis of only samples from Reed and Carlin, Marsh names *Stegosaurus, Apatosaurus, Allosaurus,* and *Nanosaurus!*

Marsh believes that Como Bluff is 7 million years old, and that sauropods could rear up on their hind legs like kangaroos. Cope is getting better material at Garden Park, Colorado, than Marsh is getting at Morrison, so Mudge asks Marsh to send Williston to Garden Park and abandon Morrison. Williston later goes to Como to check out the stories based on letters by Reed and Carlin. Several hours after Williston arrives, he writes Marsh about a bonebed 7 *miles* long! Carlin and Frank Williston (S. W. Williston's brother) later sell out and go over to Cope.

The third, and most famous, specimen of *Archaeopteryx* is found. It will become known as the "Berlin specimen."

1878 Dinosaurs are discovered in the Fosse Sainte-Barbe coal mine near the town of Bernissart in Belgium. Thirty-nine articulated skeletons of

Iguanodon are discovered at a depth of 1056 feet. The mine will be closed for three years to allow the dinosaurs to be dug out—with the full cooperation of the management! They will be studied by Louis Dollo for most of the rest of his life. Dollo removes the spike from the dinosaur's nose, where Waterhouse Hawkins had mistakenly placed it, and correctly makes it a thumb spike. He also reconstructs *Iguanodon* as a biped like Leidy's *Hadrosaurus*.

Arthur Lakes arrives at Como Bluff, Wyoming, from Morrison, Colorado.

1879 Lakes and Reed, at Como Bluff, often feud because Lakes draws maps and pictures and does not dig all the time like Reed. *Stegosaurus* and *Camptosaurus* are found at the newly discovered Quarry 13 at Como Bluff, while *"Brontosaurus"* (today correctly called *Apatosaurus*) is found at Quarries 10 and 11.

1880s Marsh rides into a Sioux council meeting (with a few terrified graduate students). He meets Red Cloud, Crazy Horse, Sitting Bull, and Gall. He gives a feast for them, and tells them that he is looking for the "thunder horses" or "ghost horses" (fossils). After the field season, he comes back to their camp to show them his fossils, and to prove to them that he was not collecting gold. Red Cloud trusts him ever after. Later Marsh and his friends at the New York *Herald* blow the whistle on the corruption in the Bureau of Indian Affairs. Red Cloud comes to Yale, and even addresses Congress!

1881 Reed's brother is killed while swimming in the Little Medicine Bow River. Marsh sends $100 for burial costs. Reed begins to lose interest in digging dinosaurs.

G. M. Dawson and R. G. McConnell find dinosaur bones on the Red Deer River and near Lethbridge, both in Alberta, Canada.

Richard Owen opens the British Museum (Natural History) in South Kensington, London, after many years of lobbying the British government and Queen Victoria.

1882 A fully mounted skeleton of *Iguanodon* is completed at Bernissart. Marsh publishes his classification of the dinosaurs. It is the first classification, and forms the early basis of the modern classification.

Marsh is appointed the official vertebrate paleontologist of the U.S. Geological Survey. He is given about $15,000 per year in salary, plus money for thirty-five collectors, nine preparators, eight assistants, and freight charges for shipping fossils back from the field.

1883 Reed resigns at Como Bluff, and a man named Kenny takes over until 1885. Fred Brown then runs the operation until 1889, when all work at Como Bluff stops.

Felch finds the first nearly complete skeletons of *Allosaurus* and *Ceratosaurus* at Canyon City, Colorado.

Marsh spends both government money and his own money on the excavations at Garden Park.

1884 J. B. Tyrrell finds *"Laelaps"* (Cope) in western Canada, a dinosaur which is later named *Albertosaurus sarcophagus* by Osborn.

Upon his retirement, Richard Owen is knighted by Queen Victoria.

1886 Marsh's field crews at Canyon City, led by Felch, discover the first complete *Stegosaurus stenops*. It has an uncrushed skull and shows that the dorsal armor plates were held in a series of alternating plates in two rows. The skeleton will wind up at the Smithsonian Institution, displayed as it was found in the field (hence earning the nickname "the roadkill"). This will be the most complete specimen of *Stegosaurus* known until 1992.

1887 H. G. Seeley names the two orders of dinosaurs, the Ornithischia and Saurischia, on the basis of features in the pelvis.

1888 Dinosaur bones found only miles away from Pliny Moody's farm in Connecticut are labeled a "giant killed in Noah's Flood."

T. C. Weston reports large numbers of dinosaur bones along the Red Deer River in Alberta, and collects another *Albertosaurus* skull from the Edmonton Formation.

Marsh names the first of the horned dinosaurs, *"Bison" alticornus,* on the basis of a pair of horns from the Denver Formation in Colorado. This will later become part of *Triceratops*.

1889 Cope is ordered to turn all his fossils from the Hayden Survey over to the government. He thinks they will all wind up at Yale because Marsh is the official paleontologist of the U.S. Geological Survey. Cope calls the New York *Herald* and "blows the whistle" on Marsh, making many accusations. Not all of his charges are substantiated, and some are false, but Cope's actions set off a national scandal. Marsh replies in the same newspaper one week later.

T. C. Weston finds many skeletons along the Red Deer River.

John Bell Hatcher, who works for Marsh, begins his legendary collecting of *Triceratops* in the Lance Creek Formation of Wyoming.

1890 On January 12, 1890, Cope charges Marsh with collusion with John W. Powell and the U.S. Geological Survey. On January 19, Marsh, in response to Cope's charges in the New York *Herald*, fills an entire page of the newspaper with an account of Cope's professional blunders, and attributes all of Cope's accusations against him to envy.

Two young paleontologists, H. F. Osborn and W. B. Scott, go to Yale to see fossils. Marsh thinks they are spies for Cope, so he hides the good stuff and has a student show them only mediocre material. Marsh hides behind crates, giving hand signals to his technician to guide the visitors through the collections without letting them see any good fossils.

In October, Osborn joins the staff of the American Museum in New York. Initially he draws no salary and he pays the AMNH $1,500 to do field work!

1890s J. B. Hatcher's brother-in-law is Oscar A. Peterson! They both move from Princeton to the Carnegie Museum in Pittsburgh in 1899. Elmer Riggs is William Berryman Scott's student!

G. Jepsen is Sinclair's student.

C. H. Sternberg invents the Rice-goo-and-burlap bandage for fossils. Marsh uses plaster of paris as a bandage.

George Baur says that *Ceratosaurus* has a pathological fusion of the metatarsals. This was known to Marsh, but he deliberately didn't mention it, and instead describes the condition in *Ceratosaurus* as normal fusion of the bones, in order to justify the dinosaur's unique nature and make it closer to birds in morphology. Baur's statement is one of the first citations of paleopathology in the literature before the work of Moodie.

1891 Osborn starts work at the American Museum and founds its Department of Vertebrate Paleontology; he also has an appointment at Columbia University. Williston recommends his star student, Barnum Brown, for a job at the American Museum; Brown will become the greatest dinosaur field collector of the twentieth century. Cope similarly recommends Jacob Wortman. Walter Granger starts work for the museum as a taxidermist!

1893 Religious fundamentalists and an Arizona senator get funding cut off to the U.S. Geological Survey for doing "silly research on birds with teeth." Marsh has to resign from the survey and give up fossil material collected under its aegis to the United States National Museum (Smithsonian).

Osborn pioneers the modern museum concept by selling postcards and photos, using free-mounts instead of wall-mounts and bare glass cases. He hires Charles Knight (in 1897) and E. Christman to do accurate artwork of prehistoric animals for scientific papers and the popular galleries.

W. B. Scott of Princeton University hires the legendary collector J. B. Hatcher away from Yale University.

1896 The Carnegie Museum of Natural History is founded in Pittsburgh.

1897 Charles Knight paints *Laelaps* (*Dryptosaurus*) in the "fighting cock" pose, a painting that becomes famous as the first to depict dinosaurs as fully "warm-blooded."

Lawrence Lambe starts collecting dinosaurs along the Red Deer River. This is the first systematic collection of dinosaurs by a vertebrate paleontologist in Canada.

Brown, Granger, and Wortman of the American Museum of Natural History in New York explore the Medicine Bow area of Wyoming. They decide to explore the next day "by that cabin on the hill." As they approach it, they realize that it is made from bones! The surrounding area becomes the site of the famous Bone Cabin Quarry, and it yields 490 specimens of Morrison dinosaurs.

1898 Excavations begin at Bone Cabin Quarry.

1900 Hatcher and Utterback reopen the Marsh quarries at Garden Park, Colorado.

Brachiosaurus is found at Grand Junction, Colorado, by Elmer Riggs of the Field Museum in Chicago.

H. F. Osborn and museum director H. C. Bumpus of the American Museum establish the *American Museum Journal,* which will later become the magazine *Natural History.* This is the first museum journal for natural history aimed at public education. Osborn and Bumpus also establish guidebooks and leaflets for their exhibits.

1901 Beecher at Yale University's Peabody Museum of Natural History mounts *Edmontosaurus* (then called *"Claosaurus annectens"*) in a fully erect, bipedal stance, posed in full run. This again shows dinosaurs as very active animals, and is the first skeleton of real dinosaur bone to be mounted in the Western Hemisphere.

1902 Barnum Brown discovers the first *Tyrannosaurus rex* skeleton in the Hell Creek area of Montana. This specimen becomes the type of the species and is later traded to the Carnegie Museum in Pittsburgh, where it is now on exhibit.

Lawrence Lambe publishes the results of his Red Deer River collections.

J. P. Morgan commissions Charles Knight to paint murals at the American Museum depicting ancient life.

1903 Charles Whitney Gilmore joins the staff of the Smithsonian Institution.

Elmer Riggs of the Field Museum in Chicago officially establishes the name *Apatosaurus* as the correct name for *"Brontosaurus."*

1905 The American Museum mounts the first *"Brontosaurus"* skeleton in history. Although the name *"Brontosaurus"* was discarded in favor of the more correct name *Apatosaurus* two years earlier, the museum labels will not be corrected for decades.

The Carnegie Museum distributes casts of its famous *Diplodocus* skeleton to various museums all over the world. The first one mounted is for the British Museum in London. It causes a sensation all over the Western world, and results in fifty scientific papers. *Diplodocus* becomes the most famous dinosaur until "dethroned" by *Tyrannosaurus.*

1906 H. F. Osborn is offered the job of secretary of the Smithsonian Institution, but he rejects it. Osborn's friend and benefactor J. P. Morgan establishes a fund for vertebrate paleontology at the American Museum. Morgan also contributes to the purchase of the Cope Collection.

1907 Eberhard Fraas of the Stuttgart Museum collects at Tendaguru in Tanzania (then known as Tanganyika).

The skeleton of an *Allosaurus* feeding on the carcass of an *Apatosaurus* is mounted at the American Museum.

1908 Barnum Brown collects a third (and fairly complete) *T. rex* skeleton in the Hell Creek area. This specimen will be put on exhibit in New York, and will become the most famous mount of all the dinosaurs.

H. F. Osborn becomes president of the American Museum.

C. H. Sternberg finds the famous "duckbill mummy" (now referred to *Edmontosaurus*). The specimen is on display at the American Museum.

1909 A rancher in Alberta tells Barnum Brown about the many dinosaurs on his ranch on the Red Deer River.

On August 17, Earl Douglass, who works for the Carnegie Museum, finds articulated *Apatosaurus* dorsals at Split Mountain, north of Jensen, Utah. This very same *Apatosaurus* is now on display in the Carnegie Museum. It takes Douglass and his helpers six years to get the skeleton out of the rock and mount it at the Carnegie. It takes seven years to excavate the west half of the quarry, and six more years to do the east half. Douglass works the quarry for Carnegie from 1909 to 1922. He then spends two years working it for the University of Utah and the Smithsonian Institution. The Smithsonian's mounted *Diplodocus* is collected during this time. These excavations touch only the top and sides of the famous hogback ridge where the quarry face of Dinosaur National Monument is now situated. The present quarry and visitor's center covers only the middle-lower-central part. Famous workers in the "golden age" at the quarry are Earl Douglass, J. Leroy Kaye, Golden York, Jacob Kay, and George Goodrich (the bearded gentleman seen in an oft-reproduced photograph of the site). Beginning in the 1950s, legendary preparators Tobe Wilkins and Jim Adams worked at the monument with paleontologist Ted White, followed by Russell King in the 1970s, and Dan Chure from the 1980s to the present time.

1910 From 1910 until 1917, Barnum Brown collects along the Red Deer River, with very friendly "competition" from the Sternbergs. They all get excellent skeletons and help each other.

The American Museum mounts the first skeleton of *Tyrannosaurus rex.* Photographs of this mount will appear in countless dinosaur books. The skeleton will be remounted, to reflect modern thinking about the dinosaur's stance, in the 1990s.

1912 The work at Tendaguru comes to an end.

1915 On October 4, President Woodrow Wilson declares the Earl Douglass Quarry (80 acres) as Dinosaur National Monument. This is also intended to keep it free from homesteaders and mining operations.

1916 The British ship *Mt. Temple* is sunk by a German U-boat. Part of the cargo was two specimens of *Corythosaurus* that had been collected by Charles Sternberg.

1917 Yale's old Peabody Museum of Natural History is demolished to make way for a new building.

1919 The most complete sauropod ever found, a juvenile *Camarasaurus*, is collected for the Carnegie Museum at Dinosaur National Monument.

1921 The Museum of Paleontology at the University of California, Berkeley, is founded. It will become a major center for research and will produce some of the most famous vertebrate paleontologists.

1922 The American Museum starts its Central Asiatic Expeditions in Mongolia and the Gobi Desert. The expedition is led by Roy Chapman Andrews.

1923 The American Museum's Gobi Expedition finds *Protoceratops* and the first dinosaur nests at the Flaming Cliffs. They also find the first skull of *Velociraptor*.

1924 Osborn names *Velociraptor* and many other dinosaurs found in Mongolia by the American Museum expeditions.

1925 This is the last year the American Museum will find dinosaurs in the Gobi Desert (until operations resume some sixty years later).

The present building of the Peabody Museum of Natural History opens to the public.

1927 C. C. Young (Yang Zhong-jian) becomes China's first professional vertebrate paleontologist.

The Cleveland-Lloyd dinosaur quarry in Utah is found. It is one of the largest Jurassic predator traps known, and contains the remains of numerous individuals of *Allosaurus*.

W. D. Matthew founds the Department of Paleontology at the University of California, Berkeley. It is the only separate university department in the world devoted to paleontology.

1929 The first center for vertebrate paleontology in China is founded by C. C. Young.

Carl Wiman describes the first Chinese sauropod, *Euhelopus* (originally *Helopus*).

The first dinosaur footprint is excavated in Shanxi Province, China.

1932 The Howe Quarry (containing the skeletons of several Morrison Formation dinosaurs) is discovered on Barker Howe's Ranch near Shell, Wyoming, during a visit by Barnum Brown.

1934 The American Museum of Natural History in New York begins excavations at the Howe Quarry.

1935 H. F. Osborn dies.

1938 R. T. Bird of the American Museum learns of Cretaceous dinosaur footprint sites in the bed of the Paluxy River near Glen Rose, Texas.

1940 The Society of Vertebrate Paleontology is founded by A. S. Romer. There are approximately forty members. The first meeting is held at Harvard University.

1941 Barnum Brown retires from the American Museum in New York, but continues to work. He is the greatest collector of dinosaurs of all time, and was employed by the museum for sixty-six years.

1942 Rudolf Zallinger begins the *Age of Reptiles* mural at the Yale Peabody Museum.

Dilophosaurus and what will later be named *Nanotyrannus* are found.

1943 Raymond Cowles proposes that dinosaurs became extinct as a result of overheating. Cowles was a student of Charles Bogert, and helped Bogert and Colbert with an influential early paper on dinosaur physiology.

Allied bombers destroy the type of *"Dysalatosaurus" lettow-verbecki* in the Humboldt Museum in Berlin.

1944 G. G. Simpson publishes the first textbook to integrate the new "Synthetic Theory" of evolution and vertebrate paleontology.

Allied bombers destroy the types of *Aegyptosaurus, Carcharodonto-saurus, Bahariasaurus,* and *Spinosaurus.* They were housed in the Bavarian State Museum in Munich, Germany.

1947 George Whitaker and E. H. Colbert discover *Coelophysis* at the famous Ghost Ranch Quarry in New Mexico. This is the largest mass accumulation of well-preserved theropod skeletons anywhere in the world.

Rudolf Zallinger completes the *Age of Reptiles* mural at the Yale Peabody Museum.

1948 The two most famous skeletons of *Coelophysis* are found at Ghost Ranch by George Whitaker and Carl Sorenson of the American Museum. These skeletons have juveniles inside them, indicating that this species was cannibalistic.

Tarbosaurus is found by J. Eaglon of the Soviet Union.

1949 Glen Jepsen, Ernst Mayr, and G. G. Simpson publish the second textbook for paleontologists to integrate vertebrate paleontology and evolutionary theory.

Rudolf Zallinger wins the Pulitzer Prize for his 110-foot dinosaur mural at the Yale Peabody Museum.

1951 The first Chinese-only dinosaur expedition is inaugurated. Its members find new taxa and vast fossil deposits in the Shantung area.

The fourth specimen of *Archaeopteyx* is found.

1953 The Jurassic hall of dinosaurs is reopened at the American Museum.

China starts the world's first journal devoted exclusively to vertebrate paleontology, *Vertebrata PalAsiatica.* The Institute of Vertebrate Paleontology and Paleoanthropology is founded by C. C. Young (Yang Zhong-jian).

1956 The fifth, or "Maxberg," specimen of *Archaeopteryx* is found.

M. W. De Laubenfels publishes a paper in the *Journal of Paleontology* hypothesizing that the dinosaurs became extinct as the result of an asteroid impact.

1957 Enlow and Brown find Haversian Systems in dinosaur bone. This will later be used as evidence that dinosaurs were "warm-blooded."

1958 The first dinosaur postage stamp is issued by China. It features *Lufengosaurus.*

1959 Paleontologist Osvaldo Reig and a goat-herder named Victorino Herrera discover the second oldest dinosaur, *Herrerasaurus,* in Argentina.

1961 Geologists from Shell Oil discover hadrosaur bones in the Colville River area of the North Slope of Alaska.

1962 Petrified Forest National Park is created. It holds many famous Triassic fossils.

Dong Zhiming joins the Institute of Vertebrate Paleontology and Paleoanthropology in Beijing, China.

1965 Philip Taquet begins work in Niger and discovers new dinosaurs such as *Ouranosaurus*.

1967 The Cleveland-Lloyd dinosaur quarry in Utah is declared a National Natural Landmark.

1969 John H. Ostrom (a former student of Colbert) says that dinosaurs may have been warm-blooded, and therefore are not good indicators of Mesozoic climate. Ostrom publishes a description of *Deinonychus*.

1972 Jim Jensen of Brigham Young University finds the Dry Mesa Quarry in Utah. It produces some of the best sauropod material in the world, and later becomes famous when *"Supersaurus"* and *"Ultrasauros"* are named.

1974 R. T. Bakker and P. M. Galton reunite the Saurischia and Ornithischia by resurrecting the Dinosauria as an official taxon.

1975 *Scientific American* publishes an article by Robert T. Bakker called "Dinosaur Renaissance," summarizing Bakker's ideas about dinosaur endothermy. This sparks a new era of dinosaur paleontology.

1976 J. O. Farlow and two engineers publish an experimental study speculating that the bony plates of *Stegosaurus* functioned as thermoregulatory devices.

1977 Colbert's Ghost Ranch *Coelophysis* quarry is designated a National Landmark.

1979 Dinosaur Provincial Park in Alberta (the area where Brown and the Sternbergs collected many beautiful Late Cretaceous dinosaur skeletons) is designated a UNESCO World Heritage Site.

The Dashanpu Quarries are discovered in Zigong, China. These constitute the richest Middle Jurassic dinosaur site in the world.

1980 The American Association for the Advancement of Science publishes a symposium volume entitled *A Cold Look at the Warm-Blooded Dinosaurs,* edited by R. D. K. Thomas and E. C. Olson, that critically evaluates the idea of dinosaur endothermy.

Nobel Laureate Luis Alvarez and others publish a paper hypothesizing that an asteroid impact caused the extinctions at the end of the Cretaceous.

1982 *Stegosaurus stenops* is named the state fossil of Colorado.

1984 The first dinosaur-only art show takes place in Boston.

1985 The Tyrrell Museum of Palaeontology in Drumheller, Alberta, opens on September 25, with Philip Currie as head of dinosaur research.

1986 A Sino-Canadian agreement is reached that calls for a five-year plan to excavate dinosaurs in both countries, and for exploration and training of paleontologists.

The first symposium devoted exclusively to dinosaur systematics is held at the Tyrrell Museum of Palaeontology in Alberta. The results are published in a symposium volume in 1990, edited by Kenneth Carpenter and Philip Currie.

The first symposium dealing with the study of dinosaur trace fossils is held at the New Mexico Museum of Natural History in Albuquerque; papers from this symposium will be published in 1989, edited by David D. Gillette and Martin G. Lockley.

Paul Sereno and Jacques Gauthier publish landmark papers on the cladistic classification of the Ornithischia and the Saurischia, respectively.

John R. Horner becomes the first paleontologist to receive the coveted McArthur Foundation Award, for his work on dinosaur nesting behavior.

1987 "Dinosaurs Past and Present" is the first international art show to tour the world, showcasing the best dinosaur art from the world's leading paleontological artists.

The Zigong Dinosaur Museum opens. It is the first museum in Asia devoted just to dinosaurs.

The sixth specimen of *Archaeopteryx* is found.

1988 Paul Sereno discovers a complete skull and skeleton of *Herrerasaurus*.

1989 The Department of Paleontology at the University of California, Berkeley, merges with the Biology Department to become the Department of Integrative Biology. The Paleontology Department had been the only separate department devoted to paleontology in the whole history of academe.

Robert Gaston finds a quarry of dinosaur fossils in Utah. From this pit will come *Utahraptor* and a new genus of nodosaurid.

Kevin Pope and Charles Duller of NASA discover the Chicxulub Sinkholes in Mexico. Later, Adriana Ocampo recognizes that the site has the classic characteristics of an impact site.

1990 The Black Hills Institute finds the largest known skull and skeleton to date of *T. rex*. They nickname it "Sue" after its discoverer, Susan Hendrickson.

The first book on dinosaurs for professional paleontologists, *The Dinosauria*, edited by David Weishampel, Peter Dodson, and Halszka Osmólska, is published.

1991 *Eoraptor* is discovered by Ricardo Martinez in Argentina.

Utahraptor is found by Jim Kirkland in Utah.

Hadrosaurus foulkii is named the official state fossil of New Jersey.

The Denver Museum of Natural History starts its "parapaleontologist" program to train amateurs in the skills of finding and collecting fossils.

1992 The seventh specimen of *Archaeopteryx* is found.

Bryan Small of the Denver Museum of Natural History finds the most complete specimen of *Stegosaurus stenops* yet discovered at Canyon City. This specimen verifies Gilmore's interpretation, based on the Smithsonian's "roadkill" specimen, that the plates were held in two alternating rows.

An embryonic *Camptosaurus* is discovered at Dinosaur National Monument.

1993 The American Museum of Natural History finds more than 13 troodontid skeletons, 147 mammals, and 175 mammals at Ukhaa-Tolgod, Gobi Desert. Also found is an *Oviraptor* nest with an adult in the brooding position, but this find is not announced to the public until 1995. There are also more than 100 uncollected dinosaur specimens. This is one of the greatest Cretaceous finds in history.

Mrs. Lin Spearpoint of the Isle Of Wight, England, finds the most complete skeleton of *Polacanthus* ever found.

The movie *Jurassic Park* is released. It is the first movie in decades that pictures dinosaurs as animals, not "monsters." Using a new level of special effects, it becomes the top moneymaking movie of all time.

Paul Sereno names *Eoraptor* as the "first" dinosaur, and Jim Kirkland names *Utahraptor.*

The Smithsonian Institution sends out its first Dinosaur Expedition since 1938, to Shell, Wyoming. It is led by Drs. M. K. Brett-Surman and Nicholas Hotton III.

1994 The first *T. rex* skeleton in Saskatchewan is found by Robert Gebhardt.

Wyoming chooses *Triceratops* as its state fossil.

The American Museum Expedition in Mongolia discovers that the eggs long attributed to *Protoceratops* actually belong to *Oviraptor.*

The first textbook about dinosaur eggs and babies appears.

William E. Swinton, author (in 1970) of the first dinosaur textbook, dies.

The *Hadrosaurus foulkii* quarry in Haddonfield, New Jersey, is designated a National Historical Landmark.

1995 The largest theropod dinosaur yet found, *Giganotosaurus* from Argentina, is described.

Mesozoic bird taxonomy is revised, and many new genera are described.

Many new Cretaceous dinosaurs are announced from Utah and Africa from stages that were previously sparsely represented.

The trial of the owners of the Black Hills Institute for Geological Research is concluded. The individuals had been prosecuted for alleged improprieties in the collection of the *Tyrannosaurus* specimen known as Sue. Many find the whole affair unpalatable.

1996 Paul Sereno reveals the skull of an African theropod, *Carcharodontosaurus,* that rivals that of *Tyrannosaurus* in size.

The Cleveland-Lloyd Quarry in Utah is ransacked, and valuable fossils are stolen.

The earliest fossils of the family Tyrannosauridae are found in Thailand by French paleontologists.

Comparative sizes of *Giganotosaurus* (top) and *Tyrannosaurus* (bottom). Drawing © 1997 by Greg Paul.

Appendix   719

New kinds of dinosaurs and birds are found in Madagascar.

Reports of a feathered dinosaur related to *Compsognathus* cause a sensation at the meeting of the Society of Vertebrate Paleontology at the American Museum.

1997 A team of paleontologists and ornithologists visit China to examine the allegedly feathered theropod *Sinosauropteryx*. They confirm that the dinosaur is a compsognathid. However, they find no compelling evidence that the structures running along the dinosaur's back and the top and bottom edges of its tail are feathers. Controversy over the specimen continues.

Part 2: The Ages of Dinosaur Paleontology

To summarize the preceding chronology in a more concise fashion, dinosaur paleontology can be divided, somewhat arbitrarily, into four distinct "ages." These ages are based on key events that started the field in new directions, whether through the direct progress of paleontologists or by happenstance.

I. *The Heroic Period* (1820–1899) Individual effort marks this period, characterized by the first "scientific publications" on a massive scale. Paleontology becomes its own science distinct from "natural history."

II. *The Classical Period* (1899–1929) (from the death of Marsh to the stock market crash) Museum efforts dominate vertebrate paleontology. Individuals can no longer afford to fund entire expeditions, and museums also need to hire specialists for specimen preparation and the construction of exhibits. Only institutions as large as museums can afford full dinosaur collecting and research programs. Museums go mainly for exhibit specimens as one of their primary goals in the field, followed by the collection of research specimens. Henry Fairfield Osborn leads the trend toward establishing specialties in collections management, field work, preparation, traveling exhibits, and adult education.

III. *The Modern Age* (1933–1969) (from the end of the "Great Depression" to Ostrom's landmark paper on the inappropriate use of dinosaurs as Mesozoic climate indicators) Macroevolution, Neo-Darwinism, plate tectonics, functional morphology, and radiometric dating become the leading scientific unifying concepts for paleontologists. University dominance of paleontology characterizes this age, as museum budgets drop as a result of the Great Depression. There is a transitional period to the next age, 1969–1975, when professionals rethink dinosaur physiology, but the public is not yet generally aware of the changes.

IV. *The Renaissance* (1975–) This age begins with Bakker's *Scientific American* article. Cladistics, paleoecology, computer-based multimorphometric programs, eclectic/multidisciplinary theorizing, CT scanning, and a great increase in field work (with the highest rate of new genera being found) define this age. The public now embraces dinosaurs as "endothermic," but paleontologists and other scientists continue the debate. Professional dinosaur paleontologists also start to actively write for the public.

Glossary

Acetabulum: the hip socket, where the thighbone (femur) articulates with the pelvis.

Adenosine triphosphate (*ATP*): the biochemical fuel for cellular metabolism.

Age: a subdivision of a geologic epoch.

Aggradational deposits: sedimentary layers that accumulate during the filling in of a depositional basin.

Agnathans: jawless fishes.

Agonistic display: behaviors used to threaten or drive off rivals.

Allopatric speciation: the formation of new species after the geographic isolation of different populations of what was formerly a single species. Over time the separated populations become genetically different enough that interbreeding is no longer possible. The species of finches on the Galapagos Islands provide a good example.

Amnion: one of the internal membranes that characterize the amniote egg.

Amniote egg: the shelled egg of advanced land vertebrates, characterized by a series of membranes (amnion, chorion, etc.) that surround, protect, and nourish the developing embryo.

Amniotes: tetrapods that reproduce by means of the amniote egg or its derivatives; they include reptiles, birds, and mammals.

Analogous: used to describe structures in different organisms that serve the same function but are not derived from the same ancestral structure. The wings of birds, bats, and pterosaurs are good examples.

Anamniotes: vertebrates that do not reproduce by means of the amniote egg; fishes and amphibians.

Anapsids: amniotes with a solid, unperforated skull in the temple region behind the eye, such as turtles.

Angiosperms: a group of seed plants in which the seed is surrounded by a fruit; the flowering plants.

Anterior: toward the front end.

Antibodies: proteins produced by the body to attack foreign invaders (e.g., germs).

Antorbital fenestra: an opening in the skull in front of the orbit and behind the external nares.

Antorbital fossa: a depression that surrounds the antorbital fenestra.

Apomorphy: a derived character.

Appendicular skeleton: bones of the limbs and limb girdles.

Archetype: Richard Owen's concept of a metaphysical, generalized body plan or blueprint, of which the observed body forms of living animals are a physical manifestation.

Archosauriforms: a subgroup of archosauromorphs that includes archosaurs and some earlier, closely related groups.

Archosauromorphs: a group of diapsids that includes rhynchosaurs, protorosaurs, trilophosaurs, and archosauriforms.

Archosaurs: derived archosauriforms, including dinosaurs, birds, pterosaurs, crocodilians, and their close relatives.

Articular: a bone toward the rear of the mandible, by which the dinosaurian lower jaw articulates with the quadrate bone of the cranium.

Astragalus: a large, proximally and medially positioned anklebone that articulates with the tibia.

Atlas: the first neck vertebra.

Autotrophs: organisms that manufacture their own food.

Axial skeleton: bones of the spine, trunk, and tail.

Binomial system: the practice of using two names (generic and specific) in the formal scientific name of a species.

Biochron: a short interval of geologic time defined on the basis of its fossil content.

Biomass: the amount of living material represented by the combined individual masses of all the animals of a population.

Brachial enlargement: an enlargement of the spinal cord to accommodate nerves that run to and from the forelimb.

Bradymetabolic: used to describe organisms with slow metabolic rates.

Braincase: a group of small, tightly sutured bones that surround and protect the brain.

Branch: on a cladogram, a line connecting a taxon to the node joining it to another taxon. The branch represents the divergence of a taxon from its closest relatives.

Bryophytes: mosses and their kin.

Calcaneum: a proximally and laterally positioned ankle bone that articulates with the fibula; forms the heel of mammals.

Calcareous layer: the hard outer portion of an eggshell.

Caliche: a lime deposit formed during the evaporation of pore water from soils.

Cancelli: spaces within endochondral bone, lined with endosteum and filled with marrow.

Capitulum: the ventral projection by which a rib articulates with a vertebra.

Carpals: bones of the wrist.

Caudal: toward the tail.

Caudals: tail vertebrae.

Centrum: the spool-shaped ventral portion of a vertebra.

Cervicals: neck vertebrae.

Chevrons: V-shaped bones beneath the caudal vertebrae.

Choanae: openings of the nasal tract in the roof of the mouth.

Chondrichthyes: fishes with a cartilaginous skeleton; sharks, skates, and rays.

Chondroblasts: cartilage-forming cells in the perichondrium.

Chondroclasts: cartilage-resorbing cells.

Cingulum: a shelf near the base of a tooth crown, just above the tooth root.

Clade: a genetically related group of organisms; also known as a monophyletic group.

Cladistics: phylogenetic systematics.

Clavicles: collarbones, which attach the shoulder girdle to the sternum.

Cleidoic egg: an egg enclosed in membranes and a shell; the amniote egg.

Condyle: a rounded, knobby joint.

Conifers: a group of gymnosperms that includes pines, larches, spruces, firs, and their relatives.

Convergences: characters shared by groups that do not reflect membership of the group in the same clade, acquired independently by members of the different groups.

Coprolites: fossilized feces.

Coracoid: a bone of the pectoral girdle, located ventral to the scapula.

Coronoid process: a projection of bone on the upper surface of the lower jaw, behind the tooth row, to which jaw-closing muscles attach.

Cranial: toward the head.

Cranium: the skeleton of the head without the lower jaws.

Crocodylotarsians: the archosaurian group that includes crocodilians and their close relatives.

Crurotarsians: in one classification scheme of archosaurs, a lineage including crocodylotarsians and ornithosuchids.

Cycadophytes: a group of gymnosperms; cycads and cycadeoids.

Definition: the meaning of a taxon name. It is defined by its member taxa or by a statement of ancestry.

Deltopectoral crest: a bony flange on the upper arm bone (humerus) that served for muscle attachment.

Dentary: the tooth-bearing bone of the mandible.

Dentine: a hard tissue forming the cores of teeth.

Diagenesis: chemical changes affecting sediments (and any contained bones) after burial.

Diagnosis: the way in which a taxon is recognized. Fossil vertebrate taxa are diagnosed on the basis of skeletal features.

Diaphragm: a sheet of muscle separating the chest cavity from the abdomen in synapsids.

Diaphysis: the shaft of a long bone.

Diapophyses: attachment sites of ribs to the vertebrae.

Diapsids: amniotes with two temporal openings on each side of the skull.

Diastema: a gap in the tooth row.

Diffuse idiopathic skeletal hyperostosis (DISH): formation of bone deposits in association with ligaments that connect vertebrae.

Digit: a toe or finger.

Digitigrade: used to describe animals that walk only on their toes, with the main hand and foot bones off the ground.

Diploid: used to describe cells or organisms that have two of each kind of chromosome in the cell nucleus.

Dispersalist biogeography: a school of thought that interprets the geographic distribution of organisms in terms of movements of the organisms themselves across the earth's surface.

Distal: away from the central portion of an animal.

Dorsal: toward the top (literally the back) of an animal.

Dorsals: vertebrae of the back.

Durham's Law: Even in the best of circumstances, only about 10 percent of the actual original biota is preserved as fossils.

Eburnation: a grooving of bone articular surfaces associated with severe osteoarthritis.

Ectotherm: an organism that derives most of its body heat from sources outside itself.

Edentulous: toothless.

Enamel: a hard tissue forming the outer covering of teeth.

Encephalization quotient: a comparison of the reconstructed brain size of a dinosaur with that expected for a crocodilian of the same body mass.

Endochondral bone: bone formed by the endosteum.

Endosacral enlargement: an enlargement of the opening in the sacrum for the spinal cord. This enlargement houses nerves that pass from the spinal cord to the hind limbs, and also a structure known as the glycogen body, the function of which is not understood.

Endosteum: a bone-forming tissue located in a bone's interior.

Endotherm: an organism that obtains most of its body heat from its own metabolism.

Eon: a very long interval of geologic time; the Phanerozoic Eon, for example, includes the Paleozoic, Mesozoic, and Cenozoic eras.

Epaxial tendons: ossified tendons located above the vertebral centra, running across the neural spines.

Epiphyses: the terminal ends of bones.

Epipophyses: bony processes at the rear end of vertebrae.

Epitopes: complex folded regions of molecules that are examined by antibodies to determine whether the molecules are from the organism itself or are foreign intruders (such as germs).

Epoch: a subdivision of a geologic period.

Era: a long interval of geologic time. The Mesozoic Era, the time of the dinosaurs, is subdivided into the Triassic, Jurassic, and Cretaceous periods. The Mesozoic Era is itself a subdivision of an even longer interval of time, the Phanerozoic Eon.

Euryapsids: Reptiles with a single temporal opening placed high on each side of the skull; probably a derived group of diapsids.

Evaporites: sedimentary rocks formed by the evaporation of seawater and the crystallization of minerals previously dissolved in the water.

Evolutionary systematics (evolutionary taxonomy, gradistics): an eclectic approach to classification based on the Linnean system and the overall morphological similarities among organisms; attempts to recognize ancestor-descendant relationships.

Exostosis: a bone growth resulting from partial periosteal detachment.

Femur: the thighbone.

Fenestra: an opening (or window) into a bone.

Fibro-lamellar bone: a rather open (filled with blood vessels) hard tissue characteristic of fast-growing bone.

Fibula: the smaller, outer bone of the lower hind leg.

Foramen magnum: a large opening in the back of the back of the skull, through which the spinal cord exits the skull.

Formation: a formally defined, mappable sedimentary rock unit.

Frontal: a bone of the skull roof, located just behind the nasal.

Furcula: the wishbone, a structure that connects the pectoral girdle with the sternum. Found in birds and some dinosaurs, it takes the place of, and may be derived from, the clavicles.

Gametophyte: a haploid plant that produces eggs and sperm.

Gastralia: belly ribs that help support the viscera.

Gastroliths: "stomach stones" found within the gut regions of herbivorous dinosaurs, presumably used to process food.

Genome: the sequence of nitrogenous bases in the DNA of an organism; each species has its own characteristic genome.

Ghost lineage: a hypothetical extension of the geologic range of a taxon, earlier than when the taxon is first seen in the geologic record, predicted on the basis of the earliest geologic occurrence of the taxon's sister taxon.

Gigantothermy: use of large body size, circulatory adjustments, and body insulation to maintain a high, constant body temperature with low rates of metabolism.

Glenoid: the socket in the shoulder girdle to which the humerus attaches.

Gnathostomes: jawed vertebrates.

Gymnosperms: a paraphyletic group of seed plants, including conifers, seed ferns, cycadophytes, and their relatives, in which the seed is not surrounded by a fruit.

Haploid: used to describe cells or organisms that have a single chromosome of each type in the cell nucleus.

Heterotrophs: organisms that cannot manufacture their own food, but instead must feed, directly or indirectly, on other organisms.

Homeotherm (*Homoiotherm*): an organism that maintains a fairly constant body temperature.

Homologous: a word used to describe anatomical structures in different organisms derived from the same structure in their common ancestor.

Homoplasy: A shared similarity between two taxa that is explained by convergence, character reversal, or chance.

Humerus: the upper arm bone.

Hyoids: throat bones located at the base of the tongue.

Hypantrum: a small, anterior projection at the base of the neural spine that articulates with the hyposphene of the preceding vertebra.

Hypaxial tendons: ossified tendons that run across the chevrons between caudal vertebrae.

Hyposphene: a small, posterior projection at the base of the neural spine that articulates with the hypantrum of the following vertebra.

Ichnocoenosis: a footprint assemblage.

Ichnofabric: the manner in which trace fossils affect the texture of a sedimentary deposit.

Ichnofacies: sedimentary deposits of a particular kind that repeatedly have the same distinctive track assemblages.

Ichnology: the study of footprints and other trace fossils.

Ichnotaxonomy: the naming and classification of trace fossils.

Ilium: the upper bone of the pelvis; attaches the pelvic girdle to the sacrum.

Ischium: the more posterior of the lower bones of the pelvis.

Jugal: the cheek bone, located posterior to the maxilla and below the orbit.

Labyrinthodonts: primitive amphibians whose teeth have complex infoldings of enamel.

Lacrimal (*lachrimal*): a bone positioned between the antorbital fenestra and the orbit.

Lamellar-zonal bone: a layered, rather dense hard tissue characteristic of slowly growing bone; often shows growth rings.

Lateral: away from the midline of an animal.

Lateral temporal fenestra: an opening in the side of the skull, behind the orbit.

Lepidosauromorphs: lizard-like diapsids.

Lepidosaurs: a subgroup of lepidosauromorphs that includes lizards, snakes, and the tuatara.

Lissamphibians: the modern amphibian groups; frogs, toads, salamanders, and caecilians.

Lumbar vertebrae: vertebrae of the lower back in mammals.

Mandible: the lower jaw.

Mantle: a thick region of the earth's interior, located beneath the planet's crust but external to the core.

Manus: collective term for the bones of the forefoot (or hand).

Maxilla: the posterior tooth-bearing bone of the upper jaw.

Maxillary fenestra: an opening in the skull in front of the antorbital fenestra of theropod dinosaurs.

Medial: toward the midline of an animal.

Megatracksites: single surfaces or thin packages of sedimentary beds that are rich in fossilized footprints over a large geographic area.

Metacarpals: bones of the forefoot or hand (excluding the fingers); bones of the "palm" of the hand.

Metaphysis: the region of a bone located between the diaphysis and the epiphysis.

Metatarsals: bones of the foot (excluding the toes).

Monophyletic groups: taxa composed of a single taxon and all of its descendants.

Multicameral lungs: lungs subdivided into many small chambers.

Nares: openings in the skull for the nostrils.

Nasal: a bone on the top of the skull, to the rear of the premaxilla.

Nested hierarchy: the arrangement of taxa into a series of larger and more inclusive groups.

Neural arch: the dorsal portion of a vertebra, located above the centrum, and surrounding the spinal cord.

Neural canal: an opening in a vertebra, located above the centrum, through which the spinal cord passes.

Neural spine: process projecting dorsally from a vertebral neural arch.

Node: the point where two or more lines in a cladogram meet; in cladistics, a node constitutes a taxon that contains all of the descendant taxa that ultimately meet at that node.

Node-based definitions: taxon definitions that take the form "the most recent common ancestor of taxon X and taxon Y, and all descendants of that common ancestor."

Nomenclature: the official naming of taxa.

Notochord: a rod of stiff tissue running along the back of chordate animals at some point in their lives. The notochord is replaced by the vertebral column in vertebrates.

Obturator foramen: an opening in the pubis, located near the acetabulum.

Obturator process: a bony projection from the ischium.

Occipital condyle: a rounded, knobby joint by which the skull articulates with the vertebral column.

Occiput: the area in the back part of the skull where the neck attaches.

Olecranon: a process on the ulna for muscle attachment.

Olfactory turbinates: thin bones lined with sensory (olfactory) epithelia, located in the nasal passages.

Opisthocoelous: used to describe vertebral centra with convex anterior faces and concave posterior faces.

Opisthopubic: used to describe a pubis that is directed rearward.

Orbit: the opening in the skull for the eye.

Ornithodirans: the archosaurian lineage that includes dinosaurs and birds.

Ornithosuchians: in one classification scheme of archosaurs, a group that includes ornithosuchids and ornithodirans.

Ossified tendons: bony tissues that connect across vertebrae to strengthen the backbone.

Osteichthyes: bony fishes.

Osteoarthritis: a painful condition characterized by the formation of osteophytes at joints, eburnation, and increases in bone density.

Osteoblasts: bone-forming cells.

Osteochondroma: an exostosis with a cap of cartilage.

Osteoclasts: bone-resorbing cells.

Osteocytes: cells involved in the maintenance of bony tissue.

Osteoderms: bones that form in the skin (e.g., stegosaur plates and ankylosaur scutes).

Osteophytes: overgrowths of bone that form at articular surfaces.

Pace: the distance between two successive footprints of the opposite feet (right to left or left to right).

Palate: the bony roof of the mouth.

Palatine: one of the bones of the palate, positioned toward the front of the skull and lateral to the vomer.

Palichnostratigraphy: subdivision of the time intervals represented by sedimentary rocks on the basis of fossilized footprint assemblages.

Palpebral: a small bone in the eyelid.

Pangaea (Pangea): the supercontinent, composed of the modern, presently separated continents of the world, that existed during the late Paleozoic and early Mesozoic eras.

Parabronchial lung: the complex lung of birds, in which air flows in the same direction across the lung whether the bird inhales or exhales. The operation of the lung involves the use of extensive air sacs external to the lung itself.

Paraphyletic groups: taxa consisting of a single ancestor and some, but not all, of its descendants.

Parasagittal: parallel to an animal's midline.

Parataxonomy: a classification that is parallel to the Linnean taxonomic system. Parataxonomy does not reflect the actual taxonomic relationships of the organisms themselves, but rather classifies objects made by the organisms, such as footprints or eggs.

Parietal: a bone of the rear portion of the skull roof, located behind the frontal.

Parsimony: the scientific principle that the simplest explanation is the best for any phenomenon. In systematics, parsimony is the principle that the evolutionary tree that requires the smallest number of evolutionary changes is the most likely approximation to the true historical pattern of phylogeny.

Pectoral girdle: the complex of bones by which the forelimb attaches to the body; includes the scapula, coracoid, sternum, and clavicles.

Pelvic girdle (pelvis): the complex of bones by which the hind limb attaches to the body; includes the ilium, ischium, and pubis.

Pelycosaurs: a paraphyletic group of basal synapsids.

Perichondrium: a coating of tissue that lines the periphery of the growing cartilage precursor of a bone and deposits the cartilage.

Period: one of the major intervals of geologic time. The Triassic, Jurassic, and Cretaceous periods were the time intervals during which dinosaurs dominated terrestrial faunas. Periods are subdivided into epochs, and periods are grouped together into eras.

Periosteal bone: bone formed by the periosteum.

Periosteum: bone-forming tissue at the periphery of a growing bone.

Pes: collective term for the bones of the hind foot.

Phalanx (plural *phalanges*): bones of the fingers or toes.

Phylogenetic systematics (cladistics): an approach to classification based strictly on the interrelationships among clades of organisms.

Phylogeny: an evolutionary tree depicting ancestor-descendant relationships.

Plantigrade: a word used to describe animals (including humans) that walk flat-footed, with the metatarsals against the ground.

Glossary   727

Plate tectonics: the unifying theory of the earth sciences. The surface of the earth is shaped by the interaction of large tectonic plates composed of the crust and the outer mantle.

Pleurocoels: openings along the lateral surfaces of vertebrae into chambers inside the centrum and/or neural arch.

Pleurokinesis: the development of a hinge between the maxilla and the remainder of the skull, such that the maxillae swing outward when the jaws are closed.

Poikilotherms: animals whose body temperatures fluctuate in response to changing environmental temperatures.

Polymerase chain reaction: a technique that allows large-scale replication of selected gene segments.

Polyphyletic groups: groups with multiple ancestors; considered invalid by taxonomists of whatever persuasion.

Postacetabular process: the portion of the ilium posterior to the acetabulum.

Postcrania: collective term for all the bones of the skeleton other than the skull.

Posterior: toward the rear end.

Postzygapophyses: bony projections at the rear ends of vertebral neural arches that articulate with prezygapophyses of the following vertebrae.

Preacetabular process: the portion of the ilium anterior to the acetabulum.

Predentary: a bone at the front end of the lower jaw in ornithischians.

Premaxilla: the anterior tooth-bearing bone of the upper jaw.

Prepubis: an anteriorly directed process of the pubis.

Presacrals: vertebrae of the neck and body trunk.

Prezygapophyses: bony projections at the front ends of vertebral neural arches that articulate with postzygapophyses of the preceding vertebrae.

Primary osteons: structures formed by layers of bone that are deposited inward from the walls of tunnels surrounding blood vessels in newly formed bone.

Primitive characters: characters found in all members of a group under study, and possibly in taxa outside that group.

Procoelous: used to describe vertebral centra with concave anterior faces and convex posterior faces.

Promaxillary fenestra: an opening in the skull in front of the maxillary fenestra and the antorbital fenestra in theropod dinosaurs.

Proximal: toward the central portion of an animal.

Pseudo-acromion process: a spur of bone on the shoulder blade (scapula) for muscle attachment.

Pteridophytes: a paraphyletic group of vascular plants that includes ferns, horsetails, and club mosses.

Pteridosperms: fern-like plants that reproduce by seeds.

Pterygoid: a large bone in the rear part of the roof of the mouth, located behind the vomer and lateral to the braincase.

Pubis: the more anterior of the lower bones of the pelvis.

Quadrate: a large bone at the rear of the skull, to which the dinosaurian lower jaw articulates.

Radius: the smaller, more anteriorly placed bone of the forearm.

Red beds: reddish-colored sedimentary rocks.

Regional heterothermy: having a different temperature in different parts of the body.

Respiratory turbinates: thin, complex structures of bone or cartilage, lined with respiratory epithelia and located in the nasal airway.

Reversal: a transformation of a character in an advanced lineage back to the ancestral state.

Rhamphotheca: a horny covering over the anterior tips of the upper and lower jaws.

Rostral: in ceratopsians, a bone located in front of the premaxilla in the upper jaw.

Sacrals: pelvic vertebrae.

Sacrum: a structure formed from the fusion of the sacral vertebrae.

Scapula: shoulder blade.

Sea-floor spreading: the creation of new oceanic crust at mid-ocean ridges, crust that then moves laterally away from the ridge.

Sexual display: behaviors used to attract a mate.

Shared derived characters (synapomorphies): characters shared by two or more descendant taxa that depart from the primitive configuration of the characters.

Shell membrane: the inner, organic layer of an eggshell.

Shell units: abutting and interlocking components of the hard, calcareous layer of an eggshell.

Sister taxon (sister group): a taxon that shares a splitting event with another taxon is the sister taxon of the latter. Two sister taxa share a common node on a cladogram.

Spondyloarthropathy: a form of arthritis in which bones at joints fuse together.

Spondylosis deformans: a condition characterized by the growth of bony spurs from the margins of vertebral centra.

Squamosal: a bone on the posterior surface of the skull.

Standard metabolic rate (SMR): the minimal rate of metabolism necessary for life, based on a resting animal that is not digesting food, under specified temperature conditions.

Stem-based definitions: taxon definitions that take the form "taxon X and all organisms sharing a more recent common ancestor with taxon X than with taxon Y."

Sternum: the breastbone, formed by the fusion of a series of bones along the ventral edge of the trunk.

Stride: the distance between two successive prints of the same foot (right to right or left to left).

Subnarial foramen: a small opening in the skull below the nose.

Supraorbitals: small bones along the upper rim of the eye opening (orbit) of the skull.

Supratemporal fenestra: an opening in the top of the skull, behind the orbit.

Sutures: immovable joints between bones.

Symphysis: a joint between two bones, connected by fibrous tissues, that allows limited movement between the bones.

Synapomorphies: shared derived characters.

Synapsids: amniotes with a single, laterally placed opening low on either side of the skull, behind the orbit; includes mammals.

Syndesmophytes: overgrowths of bone that bridge vertebral centra.

Synonymy: different names assigned to the same taxon.

Synsacrum: a structure formed by fusion of several sacral vertebrae to form a single unit.

Systematics: the scientific study of the diversity of organisms within and among clades.

Tachymetabolic: used to describe animals with rapid metabolic rates.

Tarsals: ankle bones.

Taxon: a named group of organisms.

Taxonomy: the scientific practice and study of labeling and ordering like groups of organisms.

Tetrapods: the four-footed, land-living vertebrates (including secondarily aquatic forms).

Thagomizer: collective term for the tail spikes of stegosaurs.

Therapsids: a group of non-mammalian synapsids, known in non- cladistic parlance as pre-mammals, or mammal-like reptiles.

Thoracic vertebrae: vertebrae of the chest region in mammals.

Tibia: the larger, more medial bone of the lower hind leg.

Tillites: sedimentary rocks formed from the consolidation of glacial deposits.

Trabeculae: bony struts forming the framework of endochondral bone.

Tuberculum: the dorsal projection by which a rib articulates with a vertebra.

Type specimen: the actual individual specimen first used to name a new taxon.

Ulna: one of two bones of the forearm; it is larger and more posteriorly located than the radius, and forms the elbow joint.

Undertracks: footprints formed by the transmission of the trackmaker's weight into buried sediment layers, deforming those layers. The animal makes a "true" print on the surface across which it actually walks, but also a series of undertracks on underlying sediment surfaces.

Unguals: the terminal bones of fingers or toes; they support, and lie underneath, horny claws or nails.

Ureotelic: used to describe animals (synapsids) whose nitrogenous wastes are released in the form of urea.

Uricotelic: used to describe animals (reptiles) whose nitrogenous wastes are released in the form of uric acid.

Vascular plants: plants with a skeleton of conducting tissues that distribute water and food products throughout the body of the plant.

Ventral: toward the bottom (literally belly) of an animal.

Vertebra: one of the bones of the backbone.

Vicariance biogeography: a school of thought that interprets the history and geographic distribution of organisms in relation to the history and movements of continents and islands.

Viscera: the digestive tract and other internal organs; "guts."

Vomer: one of the bones of the palate, located at the anterior end of the skull.

Zygote: a fertilized egg.

Contributors

R. McNeill Alexander is a leading authority on the application of engineering principles to the interpretation of animals as living machines. He has published numerous technical articles and books, including *Dynamics of Dinosaurs and Other Extinct Giants* (1989).

Reese E. Barrick has pioneered the application of isotopic ratios in fossil bones to the interpretation of dinosaur physiology.

Michael J. Benton is one of the most prolific workers in the field of vertebrate paleontology. He has published extensively on the evolution of Early Mesozoic vertebrates and patterns of vertebrate diversity over geologic time.

José F. Bonaparte is the leading authority on Mesozoic vertebrates of South America. He has published numerous technical articles and monographs on dinosaurs, therapsids, and other Mesozoic animals.

M. K. Brett-Surman is Assistant Professorial Lecturer in Geology at the George Washington University and museum specialist at the National Museum of Natural History of the Smithsonian Institution. He has named three duckbill dinosaurs—*Anatotitan, Gilmoreosaurus,* and *Secernosaurus.* As a consultant, he has worked with paleoartists, publishers, and television crews. He is also museum specialist for dinosaurs at the Smithsonian Institution.

Kenneth Carpenter is an authority on dinosaurs and Mesozoic marine reptiles. He has worked on dinosaur exhibits at several museums, including the Philadelphia Academy of Natural Sciences, the Museum of the Rockies, and the Denver Museum of Natural History. He has edited important collections of papers dealing with dinosaurs, including *Dinosaur Systematics: Approaches and Perspectives* with Philip J. Currie (1990).

Ralph E. Chapman specializes in the use of computers to analyze shapes of animal skeletons, the better to understand their function and evolution. He is particularly interested in the bony head domes of pachycephalosaurs.

Karen Chin is a recognized leader in the study of an oddly neglected topic: the fossilized droppings of prehistoric animals. She has analyzed the chemistry and components of these trace fossils, the better to understand the diets of dinosaurs, and their ecological interactions with other members of Mesozoic biological communities.

Edwin H. Colbert is the grand old man of dinosaur paleontology. He has led fossil-collecting expeditions to all parts of the globe, and has published many articles, monographs, and books about dinosaurs and other extinct vertebrates. His most recent book is *The Little Dinosaurs of Ghost Ranch* (1995).

731

Philip J. Currie is one of the world's leading authorities on carnivorous dinosaurs. He has collected dinosaur fossils in western Canada and China, and has published numerous articles describing these finds.

Peter Dodson is interested in ways of reconstructing the composition of dinosaur communities, and in the evolution of horned dinosaurs. He is the author of *The Horned Dinosaurs: A Natural History* (1996).

James O. Farlow is broadly interested in many aspects of dinosaur biology. He has published numerous articles on such theoretical topics as dinosaur physiology, feeding interactions in dinosaur communities, and body size distributions in dinosaur faunas. He is also interested in the problem of correlating dinosaur footprints with the kinds of animals that made them. His most recent book (with Ralph E. Molnar) is *The Great Hunters: Meat-Eating Dinosaurs and Their World* (1995).

Catherine A. Forster specializes in the application of quantitative analytical techniques to the interpretation of dinosaur skeletal structure. She is most interested in ornithopods and horned dinosaurs, and is a leading authority on the systematics of *Triceratops*.

Peter M. Galton has published more articles about dinosaurs than any other living paleontologist. He is most interested in the evolution of plant-eating dinosaurs, particularly prosauropods and stegosaurs.

Nicholas Geist works on the evolution of endothermy in the archosaur-bird lineage. He is currently investigating avian respiratory physiology, particularly the structure, function, and evolution of respiratory turbinates in modern birds.

David D. Gillette has studied the skeletons and tracks of dinosaurs and other fossil vertebrates at many localities in the western U.S. He supervised the collection of the skeleton of *Seismosaurus*, one of the biggest sauropods, and described this work in *Seismosaurus: The Earth Shaker* (1994). He also put together (with Martin G. Lockley) the most important collection of articles dealing with dinosaur footprints, *Dinosaur Tracks and Traces* (1989).

Donald F. Glut is a science fiction writer with numerous books and scripts to his credit (including the novelization of *The Empire Strikes Back*). He is the author of *The New Dinosaur Dictionary* (1982), and his "Iridium Band" has released three albums of dinosaur-oriented rock and roll music: *Dinosaur Tracks, More Dinosaur Tracks,* and *Dinosaur Tracks Again*.

Douglas Henderson is one of the best-known dinosaur artists. His work has appeared in numerous museum exhibits and books, including *Dawn of the Dinosaurs: The Triassic in Petrified Forest* (1988).

Willem Hillenius is interested in the origins of endothermy. He combines analyses of comparative morphology and physiology of modern animals with the study of fossils in order to understand the processes of evolution of stamina and homeothermy in mammals, birds, and their ancestors.

Karl F. Hirsch (deceased), an amateur with no formal training in paleontology, was one of the world's experts on the study of dinosaur eggs. He assisted Kenneth Carpenter and John R. Horner in editing *Dinosaur Eggs and Babies* (1994).

Thomas R. Holtz, Jr., specializes in the evolution and functional morphology of meat-eating dinosaurs (especially the Tyrannisauridae), and is also interested in the geographic distribution of dinosaur faunas of the later Mesozoic Era. He is currently teaching at the University of Maryland.

Terry Jones's interests include vertebrate functional morphology, physiology, and paleontology. He is currently studying the evolution of amniote lung morphology and activity metabolism.

John R. Lavas has written a history of dinosaur hunting in central Asia, *Dragons from the Dunes: The Search for Dinosaurs in the Gobi Desert* (1993).

Andrew Leitch uses computerized tomography to explore the internal structures of dinosaur skulls and other bones.

Martin G. Lockley is the world's most prolific authority on dinosaur footprints. He has done field work at tracksites all over the world. His most recent book (with Adrian P. Hunt) is *Dinosaur Tracks and Other Fossil Footprints of the Western United States* (1995).

Teresa Maryańska has participated in important dinosaur-collecting expeditions in Mongolia. She has written numerous technical papers and monographs on ankylosaurs, pachycephalosaurs, ceratopsians, and hadrosaurs.

John S. McIntosh, a retired physicist, is the world's leading authority on the anatomy, systematics, and evolution of sauropod dinosaurs. He assisted John H. Ostrom in writing *Marsh's Dinosaurs: The Collections from Como Bluff* (1966).

Ralph E. Molnar is broadly interested in many aspects of dinosaur paleontology: functional morphology, evolution, and biogeography. He has published numerous articles on dinosaur faunas of Australia.

Michael Morales specializes in the stratigraphy and vertebrate faunas of the late Paleozoic and early Mesozoic, and is especially interested in labyrinthodont amphibians and Permo-Triassic vertebrate footprints of the American Southwest.

Frank V. Paladino is a comparative physiologist who has worked on temperature regulation in mammals, birds, and reptiles. He has collaborated with James R. Spotila in studies of the natural history and thermal biology of sea turtles, particularly the leatherback turtle, and has used satellites to track migration patterns of these reptiles.

J. Michael Parrish is an authority on the functional morphology and evolution of non-dinosaurian archosaurs.

R. E. H. Reid is one of the leading authorities on the interpretation of the internal microscopic structure of dinosaur bone. He has published several articles detailing the implications of his findings for interpretations of dinosaur growth patterns and physiology.

Bruce M. Rothschild is an M.D. interested in arthritis and other bone ailments of his patients, both human and alive, and nonhuman and extinct. He has published several articles and books about diseases and injuries recorded in fossil bone, most notably (with Larry D. Martin) *Paleopathology: Disease in the Fossil Record* (1993).

John Ruben is a physiologist who specializes in thermoregulation and exercise metabolism in terrestrial vertebrates. With Willem Hillenius he has discovered features of the inner bones of the snout that provide insight into the rates of metabolism of dinosaurs and other extinct vertebrates.

Dale A. Russell has collected dinosaurs from sites around the world, and has published numerous articles describing his work. He has also written papers speculating about dinosaur paleoecology and the evolution of intelligence. He is the author of *An Odyssey in Time: the Dinosaurs of North America* (1989).

Scott Sampson is interested in the interaction between behaviors related to courtship and reproduction and the evolution of vertebrate species. His research considers this general topic in the specific context of the evolution of horned dinosaurs.

William A. S. Sarjeant has written extensively on the subjects of marine micropaleontology, vertebrate trace fossils, and the history of geology. He is also a recognized authority on the life and times of Sherlock Holmes, and is a fine musician to boot.

Mary Higby Schweitzer's work has focused on the search for organic molecules in fossilized bone, in the hope of disovering materials that will aid in interpreting the evolutionary relationships of dinosaurs.

Paul C. Sereno has examined dinosaurs in the field and lab all over the world, and has led significant collecting expeditions to South America and Africa. He is interested in reconstructing the pattern of dinosaurian evolution, and in relating it to the changing geography of the Mesozoic world.

William J. Showers uses stable isotope geochemistry to investigate a variety of oceanographic and geologic research problems. He has studied the life cycles and shell geochemistry of tiny marine organisms known as foraminiferans, and participated in the development of the silver phosphate technique for analyzing the Oxygen–18 content of phosphates.

James R. Spotila is a leading authority on the application of biophysical models to the temperature regulation of terrestrial vertebrates. He has published several articles speculating about the thermal biology of dinosaurs.

Michael K. Stoskopf is the author/editor of the leading clinical textbook in fish medicine. He is internationally known for his research on the health and welfare of wild animals. His current research efforts focus on environmental health risk assessment and physiological adjustments to environmental parameters.

Hans-Dieter Sues has published numerous articles dealing with the evolution of dinosaurs, therapsids, and other extinct vertebrates. He is especially interested in vertebrate faunas of the early Mesozoic Era.

Bruce H. Tiffney is a paleobotanist interested in the evolutionary interactions between plants and terrestrial vertebrates.

Hugh Torrens specializes in the history of geology, and has made significant contributions to our understanding of the circumstances under which the dinosaurs came to be recognized as a distinctive group of animals.

Jacques VanHeerden is interested in early Mesozoic dinosaur faunas from southern Africa.

Darla K. Zelenitsky studies fossil eggs and eggshells from southern Alberta. She described the first-known pathological hadrosaur eggshell, as well as the oldest-known bird eggshell from North America. She is presently studying embryonic remains of hadrosaurs.

Index

Blood pressure, 450. *See also* Hemodynamics
Body size. *See also* Gigantism
coprolites and, 377, 379
engineering perspective on athleticism and, 416–20
balance and center of gravity, 420–22
size limits and, 414–16
weapons and, 422–23
food quality and herbivores, 363
homeothermy and, 454–55
Jurassic plant community and, 364–65
metabolic rate and, 460–61
of sauropods, 285–87
sexual dimorphism and, 390
thermoregulation and, 471, 487
Bogert, Charles, 716
Bonaparte, José F., 48, 269
Bone. *See also* Fossils; Paleopathology
growth of, 404–11
hemodynamics and, 462–67
oxygen isotopes in
diagenesis and ratios of, 479–82
fossil species, 482–85
intrabone and interbone variability, 476–77
physiology and, 485–88
ratios of in modern animals, 477–79
Bone-associated proteins, 142
Bonebeds, and evidence of behavior, 388–89
Bone Cabin site (Wyoming), 26, 266, 713
Bony fishes, 610
Books, representations of dinosaurs in popular, 691
Bookstein, Fred, 121
Boreogomphodon, 636
Borissyak, A. A., 35
Borogovia, 530, 535
Borsuk-Bialynicka, Magdalena, 39
Boston Museum of Science, 688
Bothriospondylus, 266, 274
Bottorff, William, 114
Bouvier, Marianne, 462
Bowfin, 611
Brachiosaurids, 274, 279
Brachiosaurus
body size of, 281, 286, 414, 415, 421, 464
discovery of skeletons at Tendaguru (Tanzania), 20, 43, 44, 266–67
evolution of, 274
feeding and biology of, 285
flesh restoration of, 277
forelimbs of, 271
growth of, 410, 466
height of and heart, 450
osteology of, 78
phylogenetic analysis of, 128–32
skull of, 269
tail of, 282
Brachyceratops, 327
Brachychampsa, 650
Brachychirotherium, 252

Brachyopids, 612
Bradymetabolism, 453, 485
Braginetz, Donna, 688
Braincase, 84
Brain size. *See* Encephalization Quotient
Branca, Wilhelm, 19, 20
Branches, and phylogenetic analyses, 128
Brazil, aboriginal peoples and dinosaur footprints in, 3. *See also* Caturrita Formation; Santa Maria Formation
Brett-Surman, M. K., 719
British Association for the Advancement of Science (BAAS), 177–79
British Museum of Natural History, 711
Broderip, William John, 182
Brontopodus, 271, 537
"*Brontosaurus.*" *See Apatosaurus*
Brookes, Richard, 5, 707
Brown, Barnum, 26–27, 29, 31, 297, 679, 712, 714–15, 716
Brown, Fred, 711
Brute Force (film, 1913), 676, 679
Buckland, William, 7–8, 9, 11, 13–14, 24, 95, 708
Buettneria, 634
Buffrénil, Vivian de, 302
Bumpus, H. C., 713
Burian, Zdeněk, 690, 704, 705
Burlap-and-plaster method, for fossil preservation, 70–71
Burroughs, Edgar Rice, 674, 676

Cadillacs and Dinosaurs (comic book), 691
Caecilians, 612
Caenagnathidae, 230
Caenagnathus, 224, 230
Calcaneum, 89
Calcified cartilage, and bone growth, 405–408
Calcite cement, 481, 482
Caliches, 594
Callovian Age, continental positions during, 588
Calvo, Jorge, 48
Camarasaurids, 274
Camarasaurus
body size of, 286
early discovery of fossils of, 266, 710, 715
flesh restoration of head of, 270
homeothermy and, 486
osteology of, 274
quadrupedality of, 544
sexual dimorphism in, 440
skeletal reconstruction of, 278
skull of, 269
tail vertebrae of, 271
Campanian Age, continental positions during, 589
Camptosaurs, 334–35
Camptosaurus, 334, 335, 340, 431, 433, 436, 437, 531, 711, 718

Canada, and history of dinosaur study, 28–30, 41, 710, 713, 718. *See also* Alberta
Canadian-Chinese Dinosaur Project, 41, 718
Cannibalism, and theropods, 226, 377
Capitosaurs, 611
Capitulum, 86
Carbonate, in bone, 481, 482
Carcharodontosaurus, 46, 415, 716, 719
Cardiff Giant hoax, 710
Cardiodon, 264
Cardiovascular system, and modeling of dinosaur physiology, 496–98
Carlson, William, 114
Carmelopodus, 570
Carnegie, Andrew, 27–28, 150
Carnegie Museum of Natural History (Pittsburgh), 688, 713, 714
Carnegie Quarry (Utah), 27–28
Carnian Age
continental positions during, 587
evolution of first dinosaurs and, 208, 213
tetrapod assemblages of, 630–36, 640
Carnian-Norian extinction event, 211–12
Carnivores. *See also* Diet; Hunting
coprolites of, 377
jaws and teeth of, 255
Carnosauria
cladogram of, 104
growth rings on bones of, 465
theropods and, 229, 230
Carnotaurus, 43, 219, 220, 222
Carpals, 87
Carpenter, Kenneth, 717
Cartilage, and growth, 403–404, 405–408
Cartilaginous fishes, 610
Case, Ted, 458
Casts, of dinosaur footprints, 523–24
Catastrophism, as theory of extinction, 580, 662–69
Caturrita Formation (Brazil), 630
Caumont, A. de, 10–11
Cavernous bones, and hemodynamics, 466–67
CD-ROM, 114, 687
Census studies, and footprints, 559–63, 573
Centrosaurinae, 327, 385, 388–89, 432, 668
Centrosaurus
bonebed accumulations of, 327
bone fractures and, 432, 435
cladogram of horned dinosaurs and, 100
horns and behavior of, 385, 386
manus and pes of, 529
museum exhibit of, 163
skeletal reconstruction of, 325
species assemblages in Late Cretaceous and, 650
Cerapoda, 99, 339

Ceratopsians
  behavior of, 385
  bonebeds and, 388–89
  classification of armored dinosaurs
    and, 293–94
  fractures of bone, 432, 434, 435
  as herbivores, 367
  horns and neck frills of, 321–28, 422
  sexual dimorphism in, 389, 390
  teeth of, 83, 85
  vertebral fusion in, 441, 442
Ceratopsipes, 544–45
Ceratosaurus, *218*, 220, *222*, 229,
  384, *411*, 484, 711, 712
Cervical bars, and vertebral fusion,
  441
Cetiosaurus, 9, 11, 186, 264–65, 267,
  273, 279, 410, 708
Chalimia, 636
Champsosaurids, 616
Champsosaurus, *615*, 616, *651*
Chanaresuchus, *632*
Chang Qu, 707
Chapman, Ralph E., 121
Charig, A. J., 272
Chasmosaurinae, 100, 327, 328, 385
Chasmosaurus
  archosaurs and, 195
  cladogram of, 100
  as example of chasmosaurine, 328
  forelimbs of, *88*
  hind limbs of, *90*
  pelvic girdle of, *89*
  tetrapod assemblage and, *651*
Chatterjea, *635*
Chatterjee, Sankar, 45–46
Cheeks, of stegosaurs, 297
Cheirolepidiaceae, 359
Chemical hardeners, and fossil preser-
  vation, 70, 75
Chialingosaurus, 296
Chigutisaurs, 612
Chilantaisaurus, 237
China, and history of dinosaur study,
  4, 35, 41–42, 715, 716, 718, 720.
  See also Asia; Dashanpu Quarries
Chindesaurus, *635*
Chinese language, alphabetization of
  names, x
Chinle Formation (New Mexico), 169–
  72, 212
Chirostenotes, *530*, *650*
Chondrichthyes, 610
Chondroclasts, 404
Chondrosarcoma, 443
Chordata, 608
Christman, E., 713
Chromatography, 145
Chungkingosaurus, 296
Chure, Dan, 714
Cladistic system, of phylogenetic sys-
  tematics, 99–105, 128–32
Cladogram
  of Amniota, *105*
  of Carnosauria and Allosauroidea,
    104
  hypothetical example of, *101*

of Marginocephalia, *318*
of ornithopods, *339*
of Prosauropoda, 257–59
of sauropodomorphs, *102*
of *Triceratops* and *Chasmosaurus*,
  100
Clark, William, 6, 707
Classification. *See* Systematics; Taxonomy
Cleveland-Lloyd Dinosaur Quarry
  (Utah), 65, 228, 715, 717, 719
Clevosaurus, 637
Climate. *See also* Paleoclimatology
  gradualist theory of extinction and,
    668
  metabolic rates and homeothermy,
    486
Club mosses, 355
Cluster analysis, 126–27
Coal, and paleoclimatology, 594
Coelacanths, 611
Coelophysis
  behavior of, 226, 228, 377
  bonebed assemblage of, 212, 228,
    229
  early discovery of fossils of, 32, 716
  manus and pes of, *530*
  paleoillustration and, 169–72
  sexual dimorphism of, 390
  skeletal reconstruction of, *168*, *218*
  theropod assemblages and, 631
  Triassic-Jurassic boundary and in-
    crease in fossils of, 201
Coelurosauria, and theropods, 229,
  230
Colbert, Edwin H., 32, 47, 716
Cold-blooded animals. *See* Ectothermy
Collagen, 142
Collection Forum (journal), 76
Collection management, for fossil
  bones, 76
Collinson, Peter, 5
Colonialism, and study of dinosaurs
  in Third World, 23, 44
Color patterns, as mating signals, 385,
  387
Colubrifer, *630*
Comics, representations of dinosaurs
  in, 690–91, *694*
Commonwealth of Independent States
  (CIS), study of dinosaurs in, 42
Community dynamics, and footprints,
  562–63
Como Bluff (Wyoming), 710, 711
Comparative anatomy, and museum
  exhibits of dinosaur skeletons, 155–
  57
Competition, and evolution of dino-
  saurs, 637–38
Compsognathus, 220, 377, *378*, *530*,
  720
Computed axial tomography (CT),
  114, 514–15
Computer-Aided Design (CAD), 68,
  113
Computer-assisted tomography (CAT),
  397, 428

Computer databases, 116–17
Conifers, 357, 359, 361, 379
Conservation, of fossil bones, 75–76
Continental drift
  biogeography and, 582–86, 587–90
  climate and, 593–94
  evolution and, 601
  southern continents and, 44
Continental rotation, *586*
Convergences, in evolution, 103, 387
Conybeare, William Daniel, 7
Coombs, Walter P., 283, 307, 457–58
Cooper, Merian C., 683
Cope, Edward Drinker, 25–26, 191,
  265, 267, 270, 349, 528, 709, 710,
  712, 713
Cope's Rule, of evolution, 340
Coprolites, 377, 379, 380
Copulation, and bone fractures, 433.
  See also Reproduction
Coracoid, 87
Coria, Rudolpho, 48
Correlation tools, and footprints,
  570–71
Coryphodon, 665
Corythosaurus
  forelimbs of, *88*
  hind limbs of, *90*
  museum exhibit of, *159*, *160*
  nasal chambers of, 342
  paleoecology and, 225
  pelvic girdle of, *89*
  species assemblage of Late Creta-
    ceous and, *651*
Courtship behavior, and mating signals, 390–91. *See also* Head-butting
Cowles, Raymond, 716
Cranial ornamentation, of theropods,
  220–21
Craterosaurus, 296
Crests, of hadrosaurs, 342
Cretaceous Period
  continental positions during, *588*,
    589
  formation and fragmentation of Gon-
    dwana, 649–54
  formation of Laurasia, 648–49
  geologic time system and, 107, 108,
    109
  paleobotany of, 364–68
Cretaceous-Tertiary (K-T) boundary,
  mass extinction, 580
Crichton, Michael, 686
Crocodiles
  body mass and metabolic rate, *461*
  eggs of, 399
  growth of, 410
  heart and arteries of, *452*
  locomotion of, 194, 201, 545
  molecular paleontology and rela-
    tionship of to dinosaurs, 139
Crocodylomorpha, 200–201, 616
Crocodylotarsi, 197, 198–99
Croizat, Leon, 591–92
Crompton, A. W., 272
Crurotarsi, 197
Cryolophosaurus, 50, 221, 645

Cryptodirans, 613
Crystal Palace Company, 186–87, 709
Culture. *See also* Popular culture
classes in gradistic system of taxonomy and, 98
early interpretations of dinosaur remains and, 3–4, 35
Cupressaceae, 357
Curation, of fossil bones, 75–76
Curator (journal), 76
Currie, Philip, 32, 41, 717
Cutler, Alan, 119
Cutler, William E., 20
Cuvier, Georges, 6, 7–8, 12, 13, 16, 175, 176, 538, 662, 708
Cycadeoidea, 357
Cycadophytes, 357–58, 359, 361
Cynodonts, 624, 629
Cynognathus, 629
Czekanowskiales, 358

Dakota megatracksite, 569, 573
Danforth, Jim, 674
Dark Horse Comics, 691
Darwin, Charles, 13, 47–48, 79, 662, 709
Dashanpu Quarries (China), 717
Dashzeveg Demberlyin, 39, 40
Daspletosaurus, 82, 86
Data management, new technologies in, 115–18
Datousaurus, 269
Davitashvili, L. S., 302
Davy, Humphrey, 176, 177
Dawson, George, 28–29, 710, 711
DC Comics, 691
Deinodon, 25, 709
Deinonychosauria, 231
Deinonychus
bone fractures in, 432
first discovery of, 32
lung and axial skeleton of, *516*
manus and pes of, 224, *225, 230, 529*
pack behavior and, 228
predator/prey interaction and, 374, *375*
restoration of, *218*
tail of, 91
vertebrae of, 91, *219*
De Laubenfels, M. W., 716
Delgado, Marcel, 679, 683
Delgado, Ricardo, 691
Dell Publishing Company, 691
Deltadromeus, 46
Delta values, for oxygen isotopes, 475
Deltopectoral crest, 310
Demberlyin, Dashzeveg, 39, 40
Dental pathology, 435
Denver Museum of Natural History, 695, 718
De Pauw, Louis, 22
Depéret, Charles, 46
Depositional environments, and footprints, 566
Dermal armor, of stegosaurs, 300–301. *See also* Armored dinosaurs

Dermochelys, 450, 451
Desiccation, of newly excavated fossils, 70
Desmatosuchus, 200, *634*
Desmond, Adrian, 177
Devil Dinosaur (comic book), 691
Devil's Coulee (Alberta), 225
Diademodon, 629
Diagenesis, 427–28, 479–82
Diaphragm, 497–98
Diapsids, 192–93, 614
Dibothrosuchus, 637
Diceratops, 433
Dicquemare, Abbé, 6
Dicraeosaurus, 266–67, 279, *283*
Dicroidium, 212, 638
Dicynodonts, 623–24
Diet. *See also* Carnivores; Feeding; Herbivores
of ankylosaurs, 315
coprolites and, 377, 379, 380
evidence of, 371–72, 379, 381
footprints and, 371
fossil assemblages and predator/prey interactions, 372–74
of prosauropods, 254–56
of sauropods, 284
stomach contents and, 376–77
Diffuse idiopathic skeletal hyperostosis (DISH), 438–40, 441, 443
Digital (computer) reconstruction techniques, and bone pathology, 428
Digitigrade stance, 90, 195
Digits, osteology of, 87, 89–90
Dilophosaurus, 221, *222,* 229, 436, 637, 716
Dinilysia, 653
Dinodontosaurus, 629, *633*
Dinosauria, definition of, 205–207
Dinosaur National Monument (Utah), 266, 573, 715, 718
Dinosaur Provincial Park (Alberta), 225, 717
Dinosaur Rex (comic book), 691
Dinosaurs
genera of from essentially complete skeletal material, *646–47*
geographical and chronological framework
basic concepts of biogeography, 579–80, 581–605
catastropic and gradualist explanations of extinction, 662–69
dinosaurian faunas of Late Mesozoic, 644–55
tetrapods of Early Mesozoic, 627–40
groups of
ankylosaurs, 307–15
introduction to, 173–74
marginocephalians, 317–28
origins and early evolution, 204–13
ornithopods, 330–44
politics and invention of dinosaurs, 175–88
prosauropods, 242–60

sauropods, 264–87
segnosaurs, 234–40
stegosaurs, 291–302
theropods, 216–31
history of study of
in Asia, 34–42
contributions of nineteenth-century European researchers, 12–23
earliest discoveries, 3–11
introduction to, 1–2
in North America during nineteenth century, 24–33
southern continents and, 43–51
in media and popular culture, 673–96
origins of term, ix, 175, 243, 704, 708
paleobiology of
diet and coprolites, 371–81
ectothermy and endothermy, 449–71, 505–17
eggs and, 394–401
engineering perspective on body size, 414–24
footprints and ecological/environmental information, 554–74
growth and, 403–12
modeling of physiology, 491–502
oxygen isotopes in bone, 474–88
plants as food and habitat, 352–68
social behavior of, 383–91
paleontology and procedures for study of
footprints and, 519–48
geologic time, 107–10
hunting for fossil bones, 64–76
introduction to, 61–63
molecular paleontology, 136–47
museum exhibits, 150–64
osteology, 78–91
paleoillustration, 165–72
taxonomy and systematics, 92–105
technology and, 112–32
paleopathology, 426–43
Dinosaurs: A Celebration (comic book), 691
Dinosaurs and DNA (television), 687
Dinosaurs for Hire (comic book), 691
Dinosaur Society, 695, 696
Dinosaur State Park (Connecticut), *520*
Dinosaur Valley State Park (Texas), *537,* 565
Dinoturbation index, 567
Diplodocids, 274, 278–79
Diplodocus
body size, 414, 421–22
diffuse idiopathic skeletal hyperostosis, 439, 440
discovery of fossils of, 27–28, 266, 714
evolution of, 274, 278–79
feeding of, 285
head of, 270
museum exhibits of, 150, 266
posture of, 282–83, 544

Index   741

Genasauria, 99
Genera, definition of, 94
General circulation models (GCMs), 598
Geneva Lens Measure, 397
Genus, definition of, 93, 94
Geochemistry, and paleoclimatology, 594
Geographic distribution. *See also* Biogeography
  of ornithopods, 338–39
  paleophytogeography and, 361–62
  of stegosaurs, 295–96
Geographic Information Systems (GIS), 113
Geographic isolation, 600
Geological Society of London, 175–76, 178, 181, 186
Geologic maps, and location of fossil bones, 64, 65, 72
Geologic time. *See also* Stratigraphy; Temporal distribution
  ghost lineages, 108, 110
  Mesozoic time scale, *109*
  system of, 107–108
Geology. *See* Continental drift; Fossilization; Geologic time; Sedimentary rocks
George Washington University, 695
Germany, and history of dinosaur study, 11, 17–20
Gerrothorax, 612
Gertie the Dinosaur (film, 1912), 676
Ghost lineages, in geologic time, 108, 110
Ghost prints, 558
Ghost Ranch (New Mexico), 212, 228, 716, 717
Ghost of Slumber Mountain, The (film, 1919), 679
Giant Behemoth, The (film, 1959), 684
Gigandipus, 526–27
Giganotosaurus, 48, 415, 719
Gigantism
  physiology and evolution of, 498–501
  sauropods and, 286–87
Gigantosaurus, 95
Gigantothermy, 487, 499, 500
Gilbert, Davies, 14
Gillette, David, 33, 113, 718
Gilmore, Charles, 28, 544, 713
Ginkgo, 356, 358, 359
Global Positioning Systems (GPS), 113
Glues, and repair of fossil bones, 75
Glyptal, 75
Gnathostomes, 608
Gnetophytes, 358
Gobi Desert (Mongolia), and history of dinosaur study, 35–40, 715, 719
Godzilla, King of the Monsters (film, 1954), 684, 686
Goldman, Leon, 443
Gomani, Elizabeth M., 47
Gomphodont cynodonts, 636

Gondwana. *See also* Continental drift
  formation and fragmentation of during Cretaceous, 649–51
  prosauropod genera and, 243
Goodrich, George, 714
Gopher system, and Internet, 117
Gorgo (film, 1960), 684
Gorgosaurus, 217, *221*, 705
Gout, 438
Goyocephale, 319
Gracilisuchus, 629, *632*
Gradistic system, of taxonomy
  definition and diagnosis of different groups, 103–105
  evolutionary systematics and, 97–99
Gradualism, and theories of extinction, 580, 662–69
Gradzinski, R., *39*
Grallator, 570
Granger, Walter, 26, 36, 41, 155, 713
Grant, Robert, 177–78
Grant, Ulysses S., 710
Gravitholus, 121
Gravity, center of and body size, 420–22
Gray, S. W., 120
Greece, interpretation of fossils in classical, 4
Greek language, and taxonomy, 93
Gresslyosaurus, 243
Grid system, for mapping
  of excavation, 67
  of footprints, 522
Griffith, D. W., 676
Ground-penetrating radar, 72, *73*
Growth, of dinosaurs
  bonebeds and patterns of, 388–89
  early development and further, 403–405
  endochondral bone and epiphyses, 405–408
  exostoses and paleopathology, 429–30
  patterns in end of, 466
  periosteal bone, 408–11
Growth rings, in bone, 408–10, *412*, 464–66, 470
Gualosuchus, *632*
Gurche, John, 688, 690, 705
Gymnosperms, 356–59, 361, 364

Habitat, flora and definition of, 352. *See also* Ecology; Environment
Hadrosauridae, subfamilies of, 96–97, 335–36
Hadrosaurinae, 335, 336
Hadrosaurs
  behavior of, 384–85
  fractures of bone, 433
  geographic distribution of, 338
  as herbivores, 367
  migration of, 229
  morphometric analysis of, 122, 123, 124
  phalanges of, 430
  taxonomy of, 335
  teeth of, 85

vertebral fusion in, 442
Hadrosaurus, 25, 151, 153, *154*, 528
Hagfishes, 610
Hall, James, 708
Hallett, Mark, 168, 688, 690, 705
Hamlin, V. T., 690
Hammer, William, 50
Haplocanthosaurus, 266, *272*, 273
Haramyoids, 624–25
Hard-shelled eggs, 395, 399
Harpymimus, 221
Harryhausen, Ray, 674, 684, *686*
Hatcher, John Bell, 266, 282, 712, 713
Haughton, S. H., 47, 95
Hauterivian-Barremian Ages, continental positions during, *588*
Haversian Systems, 716
Hawkins, B. Waterhouse, 151, 153, 162–63, 188, 688, 708, 709, 711
Hay, O. P., 282
Hayden, Ferdinand V., 25, 708, 709
Head-butting, and pachycephalosaurs, 320–21, 384, 388, 423. *See also* Behavior; Fighting
Heart. *See* Cardiovascular system
Hedin, Sven, 41
Hell Creek assemblage (Montana), 665–66, 667, 707
Helopus. See *Euhelopus*
Hemangioma, 443
Hemiprotosuchus, 636
Hemodynamics, of dinosaurs. *See also* Ectothermy; Endothermy
  cavernous bone and, 466–67
  evidence of from bone, 462–66
  height and heart, 450–52
  intermediate position between ectothermy and endothermy, 468–71
  interpretations of, 449–50
  model of for dinosaurs, 459–62
  warm blood versus cold blood, 453–59
Hemoglobin, 142, 147
Henderson, Doug, 688
Hendrickson, Susan, 718
Hennig, Edwin, 99, 296. *See also* Cladistic system
Herbivores. *See also* Diet; Feeding
  angiosperms and, 360
  evolution of plants and, 353, 362–63
  gymnosperms and, 358–59
  jaws and teeth of, *255*
  ornithopods as, 330
  segnosaurians as, 239
Herding behavior. *See also* Pack behavior
  footprints and, 565
  of ornithopods, 343
Herrerasaurus, 48, *207*, 208–209, 209–10, 216, 217, *221*, 630, 716
Hesperosuchus, *634*
Heterodontosaurids, 333
Heterodontosaurus, *332*, 339, 340, *531*
High-performance liquid chromatography (HPLC), 145–46

Index   743

Laboratory
  footprints and, 524, 526
  new technologies in, 114–15
  preparation and preservation of fossil bones, 74–75
Labyrinthodonts, 611–12
Lagerpeton, 48, 197, 617–18, 629, *633*
Lagosuchids, 197
Lagosuchus, 617–18, *633*
Lakes, Arthur, 26, 267, 710, 711
Lambe, Lawrence M., 29, 713
Lambeosaurinae, 335, 336, 384–85, 389, 668
Lambeosaurus, 225, 435
Lamellar-zonal bone, 463
Lampreys, 609
Landmarks
  as homologous features in osteology, 82
  morphometric studies and, 121, 122, 123, 124
Land That Time Forgot, The (Burroughs, 1918), 676
Laos, dinosaur finds in, 42
Lapparent, Albert-Félix de, 23, 46
Latex, and footprint casts, 524
Latin, and taxonomy, 93, 96
Laurasia, formation of, 648–49
Lavocat, R., 46
Lawyers, 350–51
Leakey, Louis S. B., 20
Leaves, and evidence of past climate, 595–96
Lefeld, J., *39*
Legs, and locomotion of theropods, 223–24. *See also* Forelimbs; Hind limbs
Leidy, Joseph, 25, 95, 151, 153, 331, 528, 708, 709
Leonardi, Giuseppe, 33
Leonardo da Vinci, 4
Lepidosauria, 192–93
Lepidosauromorphs, 192–93, 614
Lepidotes, 610
Lepkowski, M., 39
Leptoceratops, 321, 323, *324*, 441
Leptocycas, 357
Leptosuchus, 634
Lesothosaurus, 332, 337, 339, *531*
Lessem, Don, 695
Leurospondylus, 650
Lewis, Meriwether, 6, 707
Lewisuchus, 633
Lexovisaurus, 300, 302
Lhuyd, Edward, 4–5
Life restorations, of dinosaurs, 163
Ligaments, ossification of, 438–40
Linné, Carol von (Carolus Linnaeus), 5, 93. *See also* Gradistic system
Lissamphibia, 612
Lister, Martin, 4
Lizards
  heart and arteries of, *452*
  of Mesozoic Era, 614, 622
Lobe-finned fishes, 611
Lockley, Martin, 33, 718

Locomotion. *See also* Bipedality; Quadrupedality; Speed
  of archosaurs, 194
  of ceratopsids, 327
  of crocodilians, 201
  footprints and, 543–48
  of ornithopods, 342–43
  of prosauropods, 250–53, 259–60
  of stegosaurs, 302
  of theropods, 223–24
Longfellow, Henry Wadsworth, 526
Longisquama, 618, *619*
Lookijnova, M., *37*
Los Colorados Formation (Argentina), 636
Lost World, The (Doyle, 1912), 676
Lost World, The (film, 1925), 679
Lost World (television series), 687
Lotosaurus, 200
Lourie, Eugene, 684
Lovelace, Delos W., 683
Lucas, Oramel, 710
Lucas, William, 179
Lufengosaurus, 637, 690
Lull, Richard M., 6, 26, 27, 163, 297, 349
Lungfishes, 611
Lungs, of dinosaurs, 498. *See also* Respiration
Luperosuchus, 629, *633*
Lycopsids, 355
Lydekker, Richard, 45, 266
Lyell, Charles, 8, 186, 662, 708
Lystrosaurus, 628, 629, *630*

Maastrichtian Age
  continental positions during, *589*
  marine organisms and extinction, 667
  as peak of dinosaur diversity, 664
MacClade (computer program), 130, 131, *132*
Macrodontophion, 11
Madagascar, dinosaur fossils from, 46
Madsen, James, 32
Magazine Enterprises (ME), 690
Magazines, images of dinosaurs in, 691
Magnetic Resonance Imaging (MRI), 429
Magnetometry, 74
Maiasaura
  behavior of, 391
  coprolites of, 379
  discovery of, 32
  eggs of, 344, 401
  as herbivore, 367
  herding behavior and, 343
  museum exhibit of, *162*
Majungatholus, 604–605
Malawi, dinosaur research in, 47
Malawisaurus, 47, 95
Malecki, J., *39*
Maleev, E. A., 37, *38*
Maleevus, 312
Mamenchisaurus, 269, *273*, 274, *275*
Mammals

body mass and physiology of, *461*, *497*
  bone growth in, 407, 408
  endothermy and evolution of, 469–70
  of Mesozoic Era, 624–25
  modeling of dinosaur physiology and, 494
  radiation of in early Tertiary, 368
  respiratory turbinates of, 510–11
  taxonomy and, 105
  vertebral column of, 86
Manakin, Colonel, 35
Mandible, 85
Manicouagan impact crater (Quebec), 212–13, 640
Maniraptora, 230
Mantell, Mary Ann, 8, 14
Mantell, Gideon
  contributions of to study of dinosaurs, 14–15, 24, 265, 708
  discovery of *Iguanodon*, 8–9, 11, 95, 296, 708
  early fossil finds in England, 9–10
  politics and invention of "dinosaurs," 176, 177, 178, 180, 181–82, 184, 185–89
Manus
  footprints and, 529, 530, 531
  osteology of, 87
Mapping
  of footprints, *521*, 522, 559, 560
  of fossil excavation sites, 67–68
Marasuchus, 48, 197, 198, 629, *632*
Marginocephalians
  ceratopsians and, 321–28
  evolution of, 317–18
  pachycephalosaurs and, 319–21
Marine organisms
  Maastrichtian extinctions and, 667
  polar regions and evolution of, 602
  post-extinction assemblages and, 666
  reptiles of Mesozoic Era as, 620–22
  vicariance biogeography and, 590
Marketing, and images of dinosaurs, 687–88
Markgraf, Richard, 23
Marsh, O. C.
  "*Brontosaurus*" controversy and, 267, 268, 282, 675
  contributions to study of dinosaurs, 265–66, 709, 710, 711
  discovery and description of *Stegosaurus*, 291–92, 293, 299
  early fossil discoveries by, 25–26, 324–25, 331
  extinction theories and, 662
  footprints of sauropods and, 283, 528
  rivalry with Cope, 265, 267, 710–11, 712–13
  on sauropod classification, 279
Marsupials, 624, 625
Martinez, Ricardo, 718
Martinez, Rubén, 48
Marvel Comics Group, 691

Novas, Fernando, 48, 49
Novojilov, N., 37
Nowinski, Aleksandr, 39
Nu-Cards, 690
Nuclear magnetic resonance imaging, 146–47

O'Brien, Willis, 674, 679, 683
Obruchev, Vladimir, 35
Ocampo, Adriana, 718
Occipital condyle, 84
Ohmdenosaurus, 272
Olduvai Gorge, 47
Oligokyphus, 637
Olivieria, *631*
Olsen, George, 26, 29, 36
Olson, E. C., 717
Omeisaurus, 41, 269, 274, *286*
Omosaurus, 292, 296
One Million B.C. (film, 1940), 683–84
Ophthalmosaurus, *620*
Opisthocoelicaudia, 274
Orlov, Yuri, 37
Ornamented domes, of pachycephalosaurs, 319–21
Ornategulum, *610*
Ornatotholus, 319
Ornithischians
  behavior of, 384
  bipedality and classification of, 331
  early evolution of dinosaurs and, 205
  flora of Cretaceous and, 366–68
  flora of Jurassic and, 364
  flora of Late Triassic and, 363
  footprints of, 535
  ossified tendons of, 91
  pelvic girdle of, 88
  Saurischia and, 209–10
  sexual dimorphism, 390
  teeth of, 85
Ornithodira, 197, 617–18
Ornithomimids, 221, 222, 223–24, 226
Ornithomimus, *218*, *220*, *512*, *513*, *514*, 515
Ornithopods
  behavior of, 384
  bone growth in, 406–407
  characteristics of, 330
  cladogram of, *339*
  classification of, 331–36
  evolution of, 339–41
  fabrosaurs and, 337–38
  footprints of, 534–35, 548, 565
  functional morphology of, 341–43
  future of study of, 344
  geographic distribution of, 338–39
  growth curves for, 564
  history of discovery, 331
  skin and eggs of, 343–44
Ornithosuchidae, 197, 199–200, 616–17
Orodromeus, 344, 484
Osborn, Henry Fairfield, 26, 35–36, 266, 268, 282, 293, 600, 712, 713, 714, 715, 721

Osmólska, Halszka, 39, 718
Ossified tendons. *See also* Ligaments
  of ornithischians, 91
  of ornithopods, 343
Osteoarthritis, 436
Osteoblasts, 404
Osteocalcin, 142
Osteoclasts, 405
Osteoderms, 90–91
Osteology, of dinosaurs
  anatomical terminology, 80–81
  sections of skeleton, 81–91
  as source of information on extinct species, 78–79
Osteomalacia, 434, 435
Osteomyelitis, 436
Osteophytes, 437
Ostrom, John, 31, 32, 349, 555, 717
Otozoum, 533, 570
Ouranosaurus, 46, 223, 335, *336*, 384, 603
Outgroups, and phylogenetic analysis, 131
Outline methods, in morphometric studies, 121
Oviraptor, 36, 40, *218*, 221, *223*, 324, *400*, 401, 718, 719
Oviraptorids, 221–22
Owen, Richard, 349
  comparative anatomy and, 79
  contributions to study of dinosaurs, 12, 15, 24, 292, 708, 709, 711
  Darwin and, 48
  discovery and description of *Cetiosaurus*, 9, 11, 264
  footprints and, 538
  museum exhibits and, 162
  origins of term *dinosaur*, ix, 175, 243, 704, 708
  paleopathology and, 349
  politics and invention of "dinosaurs," 175–88
  taxonomy of dinosaurs and, 96, 205
Oxidation, of fossils, 70
Oxygen consumption, per unit mass, 456
Oxygen isotopes, in bone
  diagenesis and ratios of, 479–82
  endothermy and, 457
  fossil species and, 482–85
  intrabone and interbone variability in, 476–77
  paleoclimatology and, 594–95
  physiology and, 485–88
  ratios of in modern animals, 477–79
  thermoregulation, metabolic rates, and, 474–76

Pachycephalosaurs, 120, 319–21, 331, 384, 388, 422–23
Pachycephalosaurus, 319, 320–21
Pachygenelus, 637
Pachyrhinosaurus, 327, 432, 435, 441
Pack behavior, of theropods, 228
Paddlefish, 611
Paleobiology. *See also* Paleobotany
  of ankylosaurs, 313–15

behavior and
  biomechanical considerations, 388
  bonebeds and growth patterns, 388–89
  evidence for, 385, 387
  living animals as models for, 387–88
  mating signals, 390–91
  new interpretations of, 383–85
  sexual dimorphism, 389–90
diet and
  coprolites, 377, 379, 380
  evidence of, 371–72, 379, 381
  fossil assemblages and predator/prey interactions, 372–74
  stomach contents, 376–77
  tooth marks on bone, 374, 376
eggs and
  classification of, 397–400
  definition and characteristics of, 394–95
  identification of, 401
  study of, 395–97
endothermy and respiratory turbinates, 508–17
engineering perspective on body size
  athleticism and, 416–20
  balance and center of gravity, 420–22
  size limits and, 414–16
  weapons and, 422–23
growth and
  early development and further, 403–405
  endochondral bone and epiphyses, 405–408
  periosteal bone, 408–11
hemodynamics and
  cavernous bone and, 466–67
  evidence of from bone, 462–66
  height and heart, 450–52
  intermediate position between ectothermy and endothermy, 468–71
  model of, 459–62
  theories on, 449–50
  warm blood versus cold blood, 453–59
introduction to concepts of, 347–51
modeling of physiology
  cardiovascular constraints, 496–98
  gigantism and evolution, 498–501
  living animals as templates for, 491–93, 494–96
  tetrapods and, 493–94
oxygen isotopes in bone
  diagenesis and ratios of, 479–82
  fossil species and, 482–85
  intrabone and interbone variability in, 476–77
  physiology and, 485–88
  ratios in modern animals, 477–79
  thermoregulation, metabolic rates, and, 474–76
  of sauropods, 281–87
  of segnosaurians, 239–40
  of stegosaurs, 302

Index   747

Acknowledgment

The editors and Indiana University Press extend their thanks to the many paleoartists and illustrators who contributed to this book and whose work adds so much to the presentation of scientific information about dinosaurs.

Donna Braginetz
Kenneth Carpenter
Susan Durning
Larry Felder
Tracy Ford
Brian Franczak
James Gurney
Mark Hallett
Douglas Henderson
Scott Hocknull
Berislav Kržič

Bruce J. Mohn
Bill Parsons
Gregory S. Paul
Michael W. Skrepnick
Rick Spears
Robert F. Walters
Bill Watterson
Gregory C. Wenzel
James E. Whitcraft
Jeremy White

Design considerations did not permit credit to be given for chapter opening and title page art. We give that credit here. Title pages: Robert F. Walters, Jim Whitcraft (*T. rex*). Chapter openings: Robert F. Walters, chapters 1, 2, 3, 4, 5, 6, 8, 15, 16, 17, 20, 21, 23, 24, 35, 39, 40, 41; James Whitcraft, chapters 7, 9, 11, 26, 29, 36, 37, 38, 42, 43; Gregory S. Paul, chapters 10, 18, 32; Gregory Wenzel, chapters 22, 25, 28; Kenneth Carpenter, chapters 12, 30; Douglas Henderson, chapter 13; Bruce J. Mohn, chapter 19; Jeremy White, chapter 27; Rick Spears, chapter 31; Berislav Kržič, chapter 33. The drawings are used with the permission of the artists and are copyrighted by them. The drawings that open chapters 14 and 34 are in the public domain.

Editor: **Jane Lyle**
Book and Jacket Designer: **Sharon L. Sklar**
Sponsoring Editor: **Robert Sloan**
Typeface: **Sabon/Stone Sans**
Compositor: **Greg Delisle**
Printer: **Maple Vail Book Manufacturing Co.**
Component Printer: **Phoenix Color**